T0201636

Convex Optimization

Convex Optimization

Stephen Boyd

Department of Electrical Engineering
Stanford University

Lieven Vandenberghe

Electrical Engineering Department
University of California, Los Angeles

CAMBRIDGE
UNIVERSITY PRESS

CAMBRIDGE
UNIVERSITY PRESS

University Printing House, Cambridge CB2 8BS, United Kingdom

One Liberty Plaza, 20th Floor, New York, NY 10006, USA

477 Williamstown Road, Port Melbourne, VIC 3207, Australia

314-321, 3rd Floor, Plot 3, Splendor Forum, Jasola District Centre,
New Delhi – 110025, India

103 Penang Road, #05–06/07, Visioncrest Commercial, Singapore 238467

Cambridge University Press is part of the University of Cambridge.

It furthers the University's mission by disseminating knowledge in the pursuit of
education, learning and research at the highest international levels of excellence.

www.cambridge.org
Information on this title: www.cambridge.org/9780521833783

© Cambridge University Press 2004

This publication is in copyright. Subject to statutory exception
and to the provisions of relevant collective licensing agreements,
no reproduction of any part may take place without
the written permission of Cambridge University Press.

First published 2004
28th printing 2022

Printed in the United Kingdom by TJ Books Limited, Padstow Cornwall

A catalogue record for this publication is available from the British Library

Library of Congress Cataloguing-in-Publication data

Boyd, Stephen P.
Convex Optimization / Stephen Boyd & Lieven Vandenberghe
    p. cm.
Includes bibliographical references and index.
ISBN 0 521 83378 7
1. Mathematical optimization. 2. Convex functions. I. Vandenberghe, Lieven. II. Title.

QA402.5.B69 2004
519.6–dc22   2003063284

ISBN 978-0-521-83378-3 Hardback

Cambridge University Press has no responsiblity for the persistency or accuracy of URLs
for external or third-party internet websites referred to in this publication, and does not
guarantee that any content on such websites is, or will remain, accurate or appropriate.

For

Anna, Nicholas, and Nora

Daniël and Margriet

Contents

Preface

This book is about *convex optimization*, a special class of mathematical optimization problems, which includes least-squares and linear programming problems. It is well known that least-squares and linear programming problems have a fairly complete theory, arise in a variety of applications, and can be solved numerically very efficiently. The basic point of this book is that the same can be said for the larger class of convex optimization problems.

While the mathematics of convex optimization has been studied for about a century, several related recent developments have stimulated new interest in the topic. The first is the recognition that interior-point methods, developed in the 1980s to solve linear programming problems, can be used to solve convex optimization problems as well. These new methods allow us to solve certain new classes of convex optimization problems, such as semidefinite programs and second-order cone programs, almost as easily as linear programs.

The second development is the discovery that convex optimization problems (beyond least-squares and linear programs) are more prevalent in practice than was previously thought. Since 1990 many applications have been discovered in areas such as automatic control systems, estimation and signal processing, communications and networks, electronic circuit design, data analysis and modeling, statistics, and finance. Convex optimization has also found wide application in combinatorial optimization and global optimization, where it is used to find bounds on the optimal value, as well as approximate solutions. We believe that many other applications of convex optimization are still waiting to be discovered.

There are great advantages to recognizing or formulating a problem as a convex optimization problem. The most basic advantage is that the problem can then be solved, very reliably and efficiently, using interior-point methods or other special methods for convex optimization. These solution methods are reliable enough to be embedded in a computer-aided design or analysis tool, or even a real-time reactive or automatic control system. There are also theoretical or conceptual advantages of formulating a problem as a convex optimization problem. The associated dual problem, for example, often has an interesting interpretation in terms of the original problem, and sometimes leads to an efficient or distributed method for solving it.

We think that convex optimization is an important enough topic that everyone who uses computational mathematics should know at least a little bit about it. In our opinion, convex optimization is a natural next topic after advanced linear algebra (topics like least-squares, singular values), and linear programming.

Goal of this book

For many general purpose optimization methods, the typical approach is to just try out the method on the problem to be solved. The full benefits of convex optimization, in contrast, only come when the problem is known ahead of time to be convex. Of course, many optimization problems are not convex, and it can be difficult to recognize the ones that are, or to reformulate a problem so that it is convex.

> *Our main goal is to help the reader develop a working knowledge of convex optimization, i.e., to develop the skills and background needed to recognize, formulate, and solve convex optimization problems.*

Developing a working knowledge of convex optimization can be mathematically demanding, especially for the reader interested primarily in applications. In our experience (mostly with graduate students in electrical engineering and computer science), the investment often pays off well, and sometimes very well.

There are several books on linear programming, and general nonlinear programming, that focus on problem formulation, modeling, and applications. Several other books cover the theory of convex optimization, or interior-point methods and their complexity analysis. This book is meant to be something in between, a book on general convex optimization that focuses on problem formulation and modeling.

We should also mention what this book is *not*. It is not a text primarily about convex analysis, or the mathematics of convex optimization; several existing texts cover these topics well. Nor is the book a survey of algorithms for convex optimization. Instead we have chosen just a few good algorithms, and describe only simple, stylized versions of them (which, however, do work well in practice). We make no attempt to cover the most recent state of the art in interior-point (or other) methods for solving convex problems. Our coverage of numerical implementation issues is also highly simplified, but we feel that it is adequate for the potential user to develop working implementations, and we do cover, in some detail, techniques for exploiting structure to improve the efficiency of the methods. We also do not cover, in more than a simplified way, the complexity theory of the algorithms we describe. We do, however, give an introduction to the important ideas of self-concordance and complexity analysis for interior-point methods.

Audience

This book is meant for the researcher, scientist, or engineer who uses mathematical optimization, or more generally, computational mathematics. This includes, naturally, those working directly in optimization and operations research, and also many others who use optimization, in fields like computer science, economics, finance, statistics, data mining, and many fields of science and engineering. Our primary focus is on the latter group, the potential *users* of convex optimization, and not the (less numerous) experts in the field of convex optimization.

The only background required of the reader is a good knowledge of advanced calculus and linear algebra. If the reader has seen basic mathematical analysis (*e.g.*, norms, convergence, elementary topology), and basic probability theory, he or she should be able to follow every argument and discussion in the book. We hope that readers who have not seen analysis and probability, however, can still get all of the essential ideas and important points. Prior exposure to numerical computing or optimization is not needed, since we develop all of the needed material from these areas in the text or appendices.

Using this book in courses

We hope that this book will be useful as the primary or alternate textbook for several types of courses. Since 1995 we have been using drafts of this book for graduate courses on linear, nonlinear, and convex optimization (with engineering applications) at Stanford and UCLA. We are able to cover most of the material, though not in detail, in a one quarter graduate course. A one semester course allows for a more leisurely pace, more applications, more detailed treatment of theory, and perhaps a short student project. A two quarter sequence allows an expanded treatment of the more basic topics such as linear and quadratic programming (which are very useful for the applications oriented student), or a more substantial student project.

This book can also be used as a reference or alternate text for a more traditional course on linear and nonlinear optimization, or a course on control systems (or other applications area), that includes some coverage of convex optimization. As the secondary text in a more theoretically oriented course on convex optimization, it can be used as a source of simple practical examples.

Acknowledgments

We have been developing the material for this book for almost a decade. Over the years we have benefited from feedback and suggestions from many people, including our own graduate students, students in our courses, and our colleagues at Stanford, UCLA, and elsewhere. Unfortunately, space limitations and shoddy record keeping do not allow us to name everyone who has contributed. However, we wish to particularly thank A. Aggarwal, V. Balakrishnan, A. Bernard, B. Bray, R. Cottle, A. d'Aspremont, J. Dahl, J. Dattorro, D. Donoho, J. Doyle, L. El Ghaoui, P. Glynn, M. Grant, A. Hansson, T. Hastie, A. Lewis, M. Lobo, Z.-Q. Luo, M. Mesbahi, W. Naylor, P. Parrilo, I. Pressman, R. Tibshirani, B. Van Roy, L. Xiao, and Y. Ye. J. Jalden and A. d'Aspremont contributed the time-frequency analysis example in §6.5.4, and the consumer preference bounding example in §6.5.5, respectively. P. Parrilo suggested exercises 4.4 and 4.56. Newer printings benefited greatly from Igal Sason's meticulous reading of the book.

We want to single out two others for special acknowledgment. Arkadi Nemirovski incited our original interest in convex optimization, and encouraged us to write this book. We also want to thank Kishan Baheti for playing a critical role in the development of this book. In 1994 he encouraged us to apply for a National Science Foundation combined research and curriculum development grant, on convex optimization with engineering applications, and this book is a direct (if delayed) consequence.

Stephen Boyd   *Stanford, California*
Lieven Vandenberghe   *Los Angeles, California*

Chapter 1

Introduction

In this introduction we give an overview of mathematical optimization, focusing on the special role of convex optimization. The concepts introduced informally here will be covered in later chapters, with more care and technical detail.

1.1 Mathematical optimization

A *mathematical optimization problem*, or just *optimization problem*, has the form

$$
\begin{array}{ll}
\text{minimize} & f_0(x) \\
\text{subject to} & f_i(x) \le b_i, \quad i = 1, \dots, m.
\end{array}
\tag{1.1}
$$

Here the vector $x = (x_1, \dots, x_n)$ is the *optimization variable* of the problem, the function $f_0 : \mathbf{R}^n \to \mathbf{R}$ is the *objective* function, the functions $f_i : \mathbf{R}^n \to \mathbf{R}$, $i = 1, \dots, m$, are the (inequality) *constraint functions*, and the constants $b_1, \dots, b_m$ are the limits, or bounds, for the constraints. A vector $x^\star$ is called *optimal*, or a *solution* of the problem (1.1), if it has the smallest objective value among all vectors that satisfy the constraints: for any z with $f_1(z) \le b_1, \dots, f_m(z) \le b_m$, we have $f_0(z) \ge f_0(x^\star)$.

We generally consider families or classes of optimization problems, characterized by particular forms of the objective and constraint functions. As an important example, the optimization problem (1.1) is called a *linear program* if the objective and constraint functions $f_0, \dots, f_m$ are linear, *i.e.*, satisfy

$$
f_i(\alpha x + \beta y) = \alpha f_i(x) + \beta f_i(y)
\tag{1.2}
$$

for all x, $y \in \mathbf{R}^n$ and all α, $\beta \in \mathbf{R}$. If the optimization problem is not linear, it is called a *nonlinear program*.

This book is about a class of optimization problems called *convex optimization problems*. A convex optimization problem is one in which the objective and constraint functions are convex, which means they satisfy the inequality

$$
f_i(\alpha x + \beta y) \le \alpha f_i(x) + \beta f_i(y)
\tag{1.3}
$$

for all x, $y \in \mathbf{R}^n$ and all α, $\beta \in \mathbf{R}$ with $\alpha + \beta = 1$, $\alpha \geq 0$, $\beta \geq 0$. Comparing (1.3) and (1.2), we see that convexity is more general than linearity: inequality replaces the more restrictive equality, and the inequality must hold only for certain values of α and β. Since any linear program is therefore a convex optimization problem, we can consider convex optimization to be a generalization of linear programming.

1.1.1   Applications

The optimization problem (1.1) is an abstraction of the problem of making the best possible choice of a vector in $\mathbf{R}^n$ from a set of candidate choices. The variable x represents the choice made; the constraints $f_i(x) \leq b_i$ represent firm requirements or specifications that limit the possible choices, and the objective value $f_0(x)$ represents the cost of choosing x. (We can also think of $-f_0(x)$ as representing the value, or utility, of choosing x.) A solution of the optimization problem (1.1) corresponds to a choice that has minimum cost (or maximum utility), among all choices that meet the firm requirements.

In *portfolio optimization*, for example, we seek the best way to invest some capital in a set of n assets. The variable x_i represents the investment in the ith asset, so the vector $x \in \mathbf{R}^n$ describes the overall portfolio allocation across the set of assets. The constraints might represent a limit on the budget (*i.e.*, a limit on the total amount to be invested), the requirement that investments are nonnegative (assuming short positions are not allowed), and a minimum acceptable value of expected return for the whole portfolio. The objective or cost function might be a measure of the overall risk or variance of the portfolio return. In this case, the optimization problem (1.1) corresponds to choosing a portfolio allocation that minimizes risk, among all possible allocations that meet the firm requirements.

Another example is *device sizing* in electronic design, which is the task of choosing the width and length of each device in an electronic circuit. Here the variables represent the widths and lengths of the devices. The constraints represent a variety of engineering requirements, such as limits on the device sizes imposed by the manufacturing process, timing requirements that ensure that the circuit can operate reliably at a specified speed, and a limit on the total area of the circuit. A common objective in a device sizing problem is the total power consumed by the circuit. The optimization problem (1.1) is to find the device sizes that satisfy the design requirements (on manufacturability, timing, and area) and are most power efficient.

In *data fitting*, the task is to find a model, from a family of potential models, that best fits some observed data and prior information. Here the variables are the parameters in the model, and the constraints can represent prior information or required limits on the parameters (such as nonnegativity). The objective function might be a measure of misfit or prediction error between the observed data and the values predicted by the model, or a statistical measure of the unlikeliness or implausibility of the parameter values. The optimization problem (1.1) is to find the model parameter values that are consistent with the prior information, and give the smallest misfit or prediction error with the observed data (or, in a statistical framework, are most likely).

An amazing variety of practical problems involving decision making (or system design, analysis, and operation) can be cast in the form of a mathematical optimization problem, or some variation such as a multicriterion optimization problem. Indeed, mathematical optimization has become an important tool in many areas. It is widely used in engineering, in electronic design automation, automatic control systems, and optimal design problems arising in civil, chemical, mechanical, and aerospace engineering. Optimization is used for problems arising in network design and operation, finance, supply chain management, scheduling, and many other areas. The list of applications is still steadily expanding.

For most of these applications, mathematical optimization is used as an aid to a human decision maker, system designer, or system operator, who supervises the process, checks the results, and modifies the problem (or the solution approach) when necessary. This human decision maker also carries out any actions suggested by the optimization problem, *e.g.*, buying or selling assets to achieve the optimal portfolio.

A relatively recent phenomenon opens the possibility of many other applications for mathematical optimization. With the proliferation of computers embedded in products, we have seen a rapid growth in *embedded optimization*. In these embedded applications, optimization is used to automatically make real-time choices, and even carry out the associated actions, with no (or little) human intervention or oversight. In some application areas, this blending of traditional automatic control systems and embedded optimization is well under way; in others, it is just starting. Embedded real-time optimization raises some new challenges: in particular, it requires solution methods that are extremely reliable, and solve problems in a predictable amount of time (and memory).

1.1.2   Solving optimization problems

A *solution method* for a class of optimization problems is an algorithm that computes a solution of the problem (to some given accuracy), given a particular problem from the class, *i.e.*, an *instance* of the problem. Since the late 1940s, a large effort has gone into developing algorithms for solving various classes of optimization problems, analyzing their properties, and developing good software implementations. The effectiveness of these algorithms, *i.e.*, our ability to solve the optimization problem (1.1), varies considerably, and depends on factors such as the particular forms of the objective and constraint functions, how many variables and constraints there are, and special structure, such as *sparsity*. (A problem is *sparse* if each constraint function depends on only a small number of the variables).

Even when the objective and constraint functions are smooth (for example, polynomials) the general optimization problem (1.1) is surprisingly difficult to solve. Approaches to the general problem therefore involve some kind of compromise, such as very long computation time, or the possibility of not finding the solution. Some of these methods are discussed in §1.4.

There are, however, some important exceptions to the general rule that most optimization problems are difficult to solve. For a few problem classes we have effective algorithms that can reliably solve even large problems, with hundreds or thousands of variables and constraints. Two important and well known examples, described in §1.2 below (and in detail in chapter 4), are least-squares problems and linear programs. It is less well known that convex optimization is another exception to the rule: Like least-squares or linear programming, there are very effective algorithms that can reliably and efficiently solve even large convex problems.

1.2   Least-squares and linear programming

In this section we describe two very widely known and used special subclasses of convex optimization: least-squares and linear programming. (A complete technical treatment of these problems will be given in chapter 4.)

1.2.1   Least-squares problems

A *least-squares* problem is an optimization problem with no constraints (*i.e.*, $m = 0$) and an objective which is a sum of squares of terms of the form $a_i^T x - b_i$:

$$\text{minimize} \quad f_0(x) = \|Ax - b\|_2^2 = \sum_{i=1}^k (a_i^T x - b_i)^2. \tag{1.4}$$

Here $A \in \mathbf{R}^{k \times n}$ (with $k \geq n$), a_i^T are the rows of A, and the vector $x \in \mathbf{R}^n$ is the optimization variable.

Solving least-squares problems

The solution of a least-squares problem (1.4) can be reduced to solving a set of linear equations,

$$(A^T A)x = A^T b,$$

so we have the analytical solution $x = (A^T A)^{-1} A^T b$. For least-squares problems we have good algorithms (and software implementations) for solving the problem to high accuracy, with very high reliability. The least-squares problem can be solved in a time approximately proportional to $n^2 k$, with a known constant. A current desktop computer can solve a least-squares problem with hundreds of variables, and thousands of terms, in a few seconds; more powerful computers, of course, can solve larger problems, or the same size problems, faster. (Moreover, these solution times will decrease exponentially in the future, according to Moore's law.) Algorithms and software for solving least-squares problems are reliable enough for embedded optimization.

   In many cases we can solve even larger least-squares problems, by exploiting some special structure in the coefficient matrix A. Suppose, for example, that the matrix A is *sparse*, which means that it has far fewer than kn nonzero entries. By exploiting sparsity, we can usually solve the least-squares problem much faster than order $n^2 k$. A current desktop computer can solve a sparse least-squares problem with tens of thousands of variables, and hundreds of thousands of terms, in around a minute (although this depends on the particular sparsity pattern).

For extremely large problems (say, with millions of variables), or for problems with exacting real-time computing requirements, solving a least-squares problem can be a challenge. But in the vast majority of cases, we can say that existing methods are very effective, and extremely reliable. Indeed, we can say that solving least-squares problems (that are not on the boundary of what is currently achievable) is a (mature) *technology*, that can be reliably used by many people who do not know, and do not need to know, the details.

Using least-squares

The least-squares problem is the basis for regression analysis, optimal control, and many parameter estimation and data fitting methods. It has a number of statistical interpretations, *e.g.*, as maximum likelihood estimation of a vector x, given linear measurements corrupted by Gaussian measurement errors.

Recognizing an optimization problem as a least-squares problem is straightforward; we only need to verify that the objective is a quadratic function (and then test whether the associated quadratic form is positive semidefinite). While the basic least-squares problem has a simple fixed form, several standard techniques are used to increase its flexibility in applications.

In *weighted least-squares*, the weighted least-squares cost

$$\sum_{i=1}^{k} w_i (a_i^T x - b_i)^2,$$

where $w_1, \ldots, w_k$ are positive, is minimized. (This problem is readily cast and solved as a standard least-squares problem.) Here the weights w_i are chosen to reflect differing levels of concern about the sizes of the terms $a_i^T x - b_i$, or simply to influence the solution. In a statistical setting, weighted least-squares arises in estimation of a vector x, given linear measurements corrupted by errors with unequal variances.

Another technique in least-squares is *regularization*, in which extra terms are added to the cost function. In the simplest case, a positive multiple of the sum of squares of the variables is added to the cost function:

$$\sum_{i=1}^{k} (a_i^T x - b_i)^2 + \rho \sum_{i=1}^{n} x_i^2,$$

where $\rho > 0$. (This problem too can be formulated as a standard least-squares problem.) The extra terms penalize large values of x, and result in a sensible solution in cases when minimizing the first sum only does not. The parameter ρ is chosen by the user to give the right trade-off between making the original objective function $\sum_{i=1}^{k} (a_i^T x - b_i)^2$ small, while keeping $\sum_{i=1}^{n} x_i^2$ not too big. Regularization comes up in statistical estimation when the vector x to be estimated is given a prior distribution.

Weighted least-squares and regularization are covered in chapter 6; their statistical interpretations are given in chapter 7.

1.2.2   Linear programming

Another important class of optimization problems is *linear programming*, in which
the objective and all constraint functions are linear:

$$\begin{array}{ll} \text{minimize} & c^T x \\ \text{subject to} & a_i^T x \le b_i, \quad i = 1, \dots, m. \end{array} \tag{1.5}$$

Here the vectors $c, a_1, \dots, a_m \in \mathbf{R}^n$ and scalars $b_1, \dots, b_m \in \mathbf{R}$ are problem parameters that specify the objective and constraint functions.

Solving linear programs

There is no simple analytical formula for the solution of a linear program (as there
is for a least-squares problem), but there are a variety of very effective methods for
solving them, including Dantzig's simplex method, and the more recent interior-
point methods described later in this book. While we cannot give the exact number
of arithmetic operations required to solve a linear program (as we can for least-
squares), we can establish rigorous bounds on the number of operations required
to solve a linear program, to a given accuracy, using an interior-point method. The
complexity in practice is order $n^2 m$ (assuming $m \ge n$) but with a constant that is
less well characterized than for least-squares. These algorithms are quite reliable,
although perhaps not quite as reliable as methods for least-squares. We can easily
solve problems with hundreds of variables and thousands of constraints on a small
desktop computer, in a matter of seconds. If the problem is sparse, or has some
other exploitable structure, we can often solve problems with tens or hundreds of
thousands of variables and constraints.

As with least-squares problems, it is still a challenge to solve extremely large
linear programs, or to solve linear programs with exacting real-time computing requirements. But, like least-squares, we can say that solving (most) linear programs
is a mature technology. Linear programming solvers can be (and are) embedded in
many tools and applications.

Using linear programming

Some applications lead directly to linear programs in the form (1.5), or one of
several other standard forms. In many other cases the original optimization problem does not have a standard linear program form, but can be transformed to an
equivalent linear program (and then, of course, solved) using techniques covered in
detail in chapter 4.

As a simple example, consider the *Chebyshev approximation problem*:

$$\text{minimize} \quad \max_{i=1,\dots,k} |a_i^T x - b_i|. \tag{1.6}$$

Here $x \in \mathbf{R}^n$ is the variable, and $a_1, \dots, a_k \in \mathbf{R}^n$, $b_1, \dots, b_k \in \mathbf{R}$ are parameters
that specify the problem instance. Note the resemblance to the least-squares prob-
lem (1.4). For both problems, the objective is a measure of the size of the terms
$a_i^T x - b_i$. In least-squares, we use the sum of squares of the terms as objective,
whereas in Chebyshev approximation, we use the maximum of the absolute values.

One other important distinction is that the objective function in the Chebyshev approximation problem (1.6) is not differentiable; the objective in the least-squares problem (1.4) is quadratic, and therefore differentiable.

The Chebyshev approximation problem (1.6) can be solved by solving the linear program

$$
\begin{aligned}
\text{minimize} \quad & t \\
\text{subject to} \quad & a_i^T x - t \le b_i, \quad i = 1, \dots, k \\
& -a_i^T x - t \le -b_i, \quad i = 1, \dots, k,
\end{aligned}
\tag{1.7}
$$

with variables $x \in \mathbf{R}^n$ and $t \in \mathbf{R}$. (The details will be given in chapter 6.) Since linear programs are readily solved, the Chebyshev approximation problem is therefore readily solved.

Anyone with a working knowledge of linear programming would recognize the Chebyshev approximation problem (1.6) as one that can be reduced to a linear program. For those without this background, though, it might not be obvious that the Chebyshev approximation problem (1.6), with its nondifferentiable objective, can be formulated and solved as a linear program.

While recognizing problems that can be reduced to linear programs is more involved than recognizing a least-squares problem, it is a skill that is readily acquired, since only a few standard tricks are used. The task can even be partially automated; some software systems for specifying and solving optimization problems can automatically recognize (some) problems that can be reformulated as linear programs.

1.3   Convex optimization

A convex optimization problem is one of the form

$$
\begin{aligned}
\text{minimize} \quad & f_0(x) \\
\text{subject to} \quad & f_i(x) \le b_i, \quad i = 1, \dots, m,
\end{aligned}
\tag{1.8}
$$

where the functions $f_0, \dots, f_m : \mathbf{R}^n \to \mathbf{R}$ are convex, i.e., satisfy

$$
f_i(\alpha x + \beta y) \le \alpha f_i(x) + \beta f_i(y)
$$

for all x, $y \in \mathbf{R}^n$ and all α, $\beta \in \mathbf{R}$ with $\alpha + \beta = 1$, $\alpha \ge 0$, $\beta \ge 0$. The least-squares problem (1.4) and linear programming problem (1.5) are both special cases of the general convex optimization problem (1.8).

1.3.1   Solving convex optimization problems

There is in general no analytical formula for the solution of convex optimization problems, but (as with linear programming problems) there are very effective methods for solving them. Interior-point methods work very well in practice, and in some cases can be proved to solve the problem to a specified accuracy with a number of operations that does not exceed a polynomial of the problem dimensions. (This is covered in chapter 11.)

We will see that interior-point methods can solve the problem (1.8) in a number of steps or iterations that is almost always in the range between 10 and 100. Ignoring any structure in the problem (such as sparsity), each step requires on the order of

$$\max\{n^3, n^2 m, F\}$$

operations, where F is the cost of evaluating the first and second derivatives of the objective and constraint functions $f_0, \ldots, f_m$.

Like methods for solving linear programs, these interior-point methods are quite reliable. We can easily solve problems with hundreds of variables and thousands of constraints on a current desktop computer, in at most a few tens of seconds. By exploiting problem structure (such as sparsity), we can solve far larger problems, with many thousands of variables and constraints.

We cannot yet claim that solving general convex optimization problems is a mature technology, like solving least-squares or linear programming problems. Research on interior-point methods for general nonlinear convex optimization is still a very active research area, and no consensus has emerged yet as to what the best method or methods are. But it is reasonable to expect that solving general convex optimization problems will become a technology within a few years. And for some subclasses of convex optimization problems, for example second-order cone programming or geometric programming (studied in detail in chapter 4), it is fair to say that interior-point methods are approaching a technology.

1.3.2   Using convex optimization

Using convex optimization is, at least conceptually, very much like using least-squares or linear programming. If we can formulate a problem as a convex optimization problem, then we can solve it efficiently, just as we can solve a least-squares problem efficiently. With only a bit of exaggeration, we can say that, if you formulate a practical problem as a convex optimization problem, then you have solved the original problem.

There are also some important differences. Recognizing a least-squares problem is straightforward, but recognizing a convex function can be difficult. In addition, there are many more tricks for transforming convex problems than for transforming linear programs. Recognizing convex optimization problems, or those that can be transformed to convex optimization problems, can therefore be challenging. The main goal of this book is to give the reader the background needed to do this. Once the skill of recognizing or formulating convex optimization problems is developed, you will find that surprisingly many problems can be solved via convex optimization.

The challenge, and art, in using convex optimization is in recognizing and formulating the problem. Once this formulation is done, solving the problem is, like least-squares or linear programming, (almost) technology.

1.4   Nonlinear optimization

Nonlinear optimization (or nonlinear programming) is the term used to describe an optimization problem when the objective or constraint functions are not linear, but not known to be convex. Sadly, there are no effective methods for solving the general nonlinear programming problem (1.1). Even simple looking problems with as few as ten variables can be extremely challenging, while problems with a few hundreds of variables can be intractable. Methods for the general nonlinear programming problem therefore take several different approaches, each of which involves some compromise.

1.4.1   Local optimization

In *local optimization*, the compromise is to give up seeking the optimal x, which minimizes the objective over all feasible points. Instead we seek a point that is only locally optimal, which means that it minimizes the objective function among feasible points that are near it, but is not guaranteed to have a lower objective value than all other feasible points. A large fraction of the research on general nonlinear programming has focused on methods for local optimization, which as a consequence are well developed.

Local optimization methods can be fast, can handle large-scale problems, and are widely applicable, since they only require differentiability of the objective and constraint functions. As a result, local optimization methods are widely used in applications where there is value in finding a good point, if not the very best. In an engineering design application, for example, local optimization can be used to improve the performance of a design originally obtained by manual, or other, design methods.

There are several disadvantages of local optimization methods, beyond (possibly) not finding the true, globally optimal solution. The methods require an initial guess for the optimization variable. This initial guess or starting point is critical, and can greatly affect the objective value of the local solution obtained. Little information is provided about how far from (globally) optimal the local solution is. Local optimization methods are often sensitive to algorithm parameter values, which may need to be adjusted for a particular problem, or family of problems.

Using a local optimization method is trickier than solving a least-squares problem, linear program, or convex optimization problem. It involves experimenting with the choice of algorithm, adjusting algorithm parameters, and finding a good enough initial guess (when one instance is to be solved) or a method for producing a good enough initial guess (when a family of problems is to be solved). Roughly speaking, local optimization methods are more art than technology. Local optimization is a well developed art, and often very effective, but it is nevertheless an art. In contrast, there is little art involved in solving a least-squares problem or a linear program (except, of course, those on the boundary of what is currently possible).

An interesting comparison can be made between local optimization methods for nonlinear programming, and convex optimization. Since differentiability of the ob-

jective and constraint functions is the only requirement for most local optimization
methods, formulating a practical problem as a nonlinear optimization problem is
relatively straightforward. The art in local optimization is in solving the problem
(in the weakened sense of finding a locally optimal point), once it is formulated.
In convex optimization these are reversed: The art and challenge is in problem
formulation; once a problem is formulated as a convex optimization problem, it is
relatively straightforward to solve it.

1.4.2   Global optimization

In *global optimization*, the true global solution of the optimization problem (1.1)
is found; the compromise is efficiency. The worst-case complexity of global optimization methods grows exponentially with the problem sizes n and m; the hope
is that in practice, for the particular problem instances encountered, the method is
far faster. While this favorable situation does occur, it is not typical. Even small
problems, with a few tens of variables, can take a very long time (*e.g.*, hours or
days) to solve.

Global optimization is used for problems with a small number of variables, where
computing time is not critical, and the value of finding the true global solution is
very high. One example from engineering design is *worst-case analysis* or *verification* of a high value or safety-critical system. Here the variables represent uncertain
parameters, that can vary during manufacturing, or with the environment or operating condition. The objective function is a utility function, *i.e.*, one for which
smaller values are worse than larger values, and the constraints represent prior
knowledge about the possible parameter values. The optimization problem (1.1) is
the problem of finding the *worst-case* values of the parameters. If the worst-case
value is acceptable, we can certify the system as safe or reliable (with respect to
the parameter variations).

A local optimization method can rapidly find a set of parameter values that
is bad, but not guaranteed to be the absolute worst possible. If a local optimization method finds parameter values that yield unacceptable performance, it has
succeeded in determining that the system is not reliable. But a local optimization
method cannot certify the system as reliable; it can only fail to find bad parameter
values. A global optimization method, in contrast, will find the absolute worst values of the parameters, and if the associated performance is acceptable, can certify
the system as safe. The cost is computation time, which can be very large, even
for a relatively small number of parameters. But it may be worth it in cases where
the value of certifying the performance is high, or the cost of being wrong about
the reliability or safety is high.

1.4.3   Role of convex optimization in nonconvex problems

In this book we focus primarily on convex optimization problems, and applications
that can be reduced to convex optimization problems. But convex optimization
also plays an important role in problems that are *not* convex.

Initialization for local optimization

One obvious use is to combine convex optimization with a local optimization method. Starting with a nonconvex problem, we first find an approximate, but convex, formulation of the problem. By solving this approximate problem, which can be done easily and without an initial guess, we obtain the exact solution to the approximate convex problem. This point is then used as the starting point for a local optimization method, applied to the original nonconvex problem.

Convex heuristics for nonconvex optimization

Convex optimization is the basis for several heuristics for solving nonconvex problems. One interesting example we will see is the problem of finding a *sparse* vector x (*i.e.*, one with few nonzero entries) that satisfies some constraints. While this is a difficult combinatorial problem, there are some simple heuristics, based on convex optimization, that often find fairly sparse solutions. (These are described in chapter 6.)

Another broad example is given by *randomized algorithms*, in which an approximate solution to a nonconvex problem is found by drawing some number of candidates from a probability distribution, and taking the best one found as the approximate solution. Now suppose the family of distributions from which we will draw the candidates is parametrized, *e.g.*, by its mean and covariance. We can then pose the question, which of these distributions gives us the smallest expected value of the objective? It turns out that this problem is sometimes a convex problem, and therefore efficiently solved. (See, *e.g.*, exercise 11.23.)

Bounds for global optimization

Many methods for global optimization require a cheaply computable lower bound on the optimal value of the nonconvex problem. Two standard methods for doing this are based on convex optimization. In *relaxation*, each nonconvex constraint is replaced with a looser, but convex, constraint. In *Lagrangian relaxation*, the Lagrangian dual problem (described in chapter 5) is solved. This problem is convex, and provides a lower bound on the optimal value of the nonconvex problem.

1.5   Outline

The book is divided into three main parts, titled *Theory*, *Applications*, and *Algorithms*.

1.5.1   Part I: Theory

In part I, *Theory*, we cover basic definitions, concepts, and results from convex analysis and convex optimization. We make no attempt to be encyclopedic, and skew our selection of topics toward those that we think are useful in recognizing and formulating convex optimization problems. This is classical material, almost all of which can be found in other texts on convex analysis and optimization. We make no attempt to give the most general form of the results; for that the reader can refer to any of the standard texts on convex analysis.

Chapters 2 and 3 cover convex sets and convex functions, respectively. We give some common examples of convex sets and functions, as well as a number of convex calculus rules, *i.e.*, operations on sets and functions that preserve convexity. Combining the basic examples with the convex calculus rules allows us to form (or perhaps more importantly, recognize) some fairly complicated convex sets and functions.

In chapter 4, *Convex optimization problems*, we give a careful treatment of optimization problems, and describe a number of transformations that can be used to reformulate problems. We also introduce some common subclasses of convex optimization, such as linear programming and geometric programming, and the more recently developed second-order cone programming and semidefinite programming.

Chapter 5 covers Lagrangian duality, which plays a central role in convex optimization. Here we give the classical Karush-Kuhn-Tucker conditions for optimality, and a local and global sensitivity analysis for convex optimization problems.

1.5.2   Part II: Applications

In part II, *Applications*, we describe a variety of applications of convex optimization, in areas like probability and statistics, computational geometry, and data fitting. We have described these applications in a way that is accessible, we hope, to a broad audience. To keep each application short, we consider only simple cases, sometimes adding comments about possible extensions. We are sure that our treatment of some of the applications will cause experts to cringe, and we apologize to them in advance. But our goal is to convey the flavor of the application, quickly and to a broad audience, and not to give an elegant, theoretically sound, or complete treatment. Our own backgrounds are in electrical engineering, in areas like control systems, signal processing, and circuit analysis and design. Although we include these topics in the courses we teach (using this book as the main text), only a few of these applications are broadly enough accessible to be included here.

The aim of part II is to show the reader, by example, how convex optimization can be applied in practice.

1.5.3   Part III: Algorithms

In part III, *Algorithms*, we describe numerical methods for solving convex optimization problems, focusing on Newton's algorithm and interior-point methods. Part III is organized as three chapters, which cover unconstrained optimization, equality constrained optimization, and inequality constrained optimization, respectively. These chapters follow a natural hierarchy, in which solving a problem is reduced to solving a sequence of simpler problems. Quadratic optimization problems (including, *e.g.*, least-squares) form the base of the hierarchy; they can be solved exactly by solving a set of linear equations. Newton's method, developed in chapters 9 and 10, is the next level in the hierarchy. In Newton's method, solving an unconstrained or equality constrained problem is reduced to solving a sequence of quadratic problems. In chapter 11, we describe interior-point methods, which form the top level of the hierarchy. These methods solve an inequality constrained problem by solving a sequence of unconstrained, or equality constrained, problems.

Overall we cover just a handful of algorithms, and omit entire classes of good methods, such as quasi-Newton, conjugate-gradient, bundle, and cutting-plane algorithms. For the methods we do describe, we give simplified variants, and not the latest, most sophisticated versions. Our choice of algorithms was guided by several criteria. We chose algorithms that are simple (to describe and implement), but also reliable and robust, and effective and fast enough for most problems.

Many users of convex optimization end up using (but not developing) standard software, such as a linear or semidefinite programming solver. For these users, the material in part III is meant to convey the basic flavor of the methods, and give some ideas of their basic attributes. For those few who will end up developing new algorithms, we think that part III serves as a good introduction.

1.5.4 Appendices

There are three appendices. The first lists some basic facts from mathematics that we use, and serves the secondary purpose of setting out our notation. The second appendix covers a fairly particular topic, optimization problems with quadratic objective and one quadratic constraint. These are nonconvex problems that nevertheless can be effectively solved, and we use the results in several of the applications described in part II.

The final appendix gives a brief introduction to numerical linear algebra, concentrating on methods that can exploit problem structure, such as sparsity, to gain efficiency. We do not cover a number of important topics, including roundoff analysis, or give any details of the methods used to carry out the required factorizations. These topics are covered by a number of excellent texts.

1.5.5 Comments on examples

In many places in the text (but particularly in parts II and III, which cover applications and algorithms, respectively) we illustrate ideas using specific examples. In some cases, the examples are chosen (or designed) specifically to illustrate our point; in other cases, the examples are chosen to be 'typical'. This means that the examples were chosen as samples from some obvious or simple probability distribution. The dangers of drawing conclusions about algorithm performance from a few tens or hundreds of randomly generated examples are well known, so we will not repeat them here. These examples are meant only to give a rough idea of algorithm performance, or a rough idea of how the computational effort varies with problem dimensions, and not as accurate predictors of algorithm performance. In particular, your results may vary from ours.

1.5.6   Comments on exercises

Each chapter concludes with a set of exercises. Some involve working out the details of an argument or claim made in the text. Others focus on determining, or establishing, convexity of some given sets, functions, or problems; or more generally, convex optimization problem formulation. Some chapters include numerical exercises, which require some (but not much) programming in an appropriate high level language. The difficulty level of the exercises is mixed, and varies without warning from quite straightforward to rather tricky.

1.6   Notation

Our notation is more or less standard, with a few exceptions. In this section we describe our basic notation; a more complete list appears on page 697.

We use $\mathbf{R}$ to denote the set of real numbers, $\mathbf{R}_+$ to denote the set of nonnegative real numbers, and $\mathbf{R}_{++}$ to denote the set of positive real numbers. The set of real n-vectors is denoted $\mathbf{R}^n$, and the set of real $m \times n$ matrices is denoted $\mathbf{R}^{m \times n}$. We delimit vectors and matrices with square brackets, with the components separated by space. We use parentheses to construct column vectors from comma separated lists. For example, if a, b, $c \in \mathbf{R}$, we have

$$(a, b, c) = \begin{bmatrix} a \\ b \\ c \end{bmatrix} = [\, a \quad b \quad c \,]^T,$$

which is an element of $\mathbf{R}^3$. The symbol $\mathbf{1}$ denotes a vector all of whose components are one (with dimension determined from context). The notation x_i can refer to the ith component of the vector x, or to the ith element of a set or sequence of vectors $x_1, x_2, \ldots$. The context, or the text, makes it clear which is meant.

We use $\mathbf{S}^k$ to denote the set of symmetric $k \times k$ matrices, $\mathbf{S}^k_+$ to denote the set of symmetric positive semidefinite $k \times k$ matrices, and $\mathbf{S}^k_{++}$ to denote the set of symmetric positive definite $k \times k$ matrices. The curled inequality symbol $\succeq$ (and its strict form $\succ$) is used to denote generalized inequality: between vectors, it represents componentwise inequality; between symmetric matrices, it represents matrix inequality. With a subscript, the symbol $\preceq_K$ (or $\prec_K$) denotes generalized inequality with respect to the cone K (explained in §2.4.1).

Our notation for describing functions deviates a bit from standard notation, but we hope it will cause no confusion. We use the notation $f : \mathbf{R}^p \to \mathbf{R}^q$ to mean that f is an $\mathbf{R}^q$-valued function on some *subset* of $\mathbf{R}^p$, specifically, its *domain*, which we denote $\mathbf{dom}\, f$. We can think of our use of the notation $f : \mathbf{R}^p \to \mathbf{R}^q$ as a declaration of the function *type*, as in a computer language: $f : \mathbf{R}^p \to \mathbf{R}^q$ means that the function f takes as argument a real p-vector, and returns a real q-vector. The set $\mathbf{dom}\, f$, the domain of the function f, specifies the subset of $\mathbf{R}^p$ of points x for which $f(x)$ is defined. As an example, we describe the logarithm function as $\log : \mathbf{R} \to \mathbf{R}$, with $\mathbf{dom}\, \log = \mathbf{R}_{++}$. The notation $\log : \mathbf{R} \to \mathbf{R}$ means that the logarithm function accepts and returns a real number; $\mathbf{dom}\log = \mathbf{R}_{++}$ means that the logarithm is defined only for positive numbers.

We use $\mathbf{R}^n$ as a generic finite-dimensional vector space. We will encounter several other finite-dimensional vector spaces, *e.g.*, the space of polynomials of a variable with a given maximum degree, or the space $\mathbf{S}^k$ of symmetric $k \times k$ matrices. By identifying a basis for a vector space, we can always identify it with $\mathbf{R}^n$ (where n is its dimension), and therefore the generic results, stated for the vector space $\mathbf{R}^n$, can be applied. We usually leave it to the reader to translate general results or statements to other vector spaces. For example, any linear function $f : \mathbf{R}^n \to \mathbf{R}$ can be represented in the form $f(x) = c^T x$, where $c \in \mathbf{R}^n$. The corresponding statement for the vector space $\mathbf{S}^k$ can be found by choosing a basis and translating. This results in the statement: any linear function $f : \mathbf{S}^k \to \mathbf{R}$ can be represented in the form $f(X) = \mathbf{tr}(CX)$, where $C \in \mathbf{S}^k$.

Bibliography

Least-squares is a very old subject; see, for example, the treatise written (in Latin) by Gauss in the 1820s, and recently translated by Stewart [Gau95]. More recent work includes the books by Lawson and Hanson [LH95] and Björck [Bjö96]. References on linear programming can be found in chapter 4.

There are many good texts on local methods for nonlinear programming, including Gill, Murray, and Wright [GMW81], Nocedal and Wright [NW99], Luenberger [Lue84], and Bertsekas [Ber99].

Global optimization is covered in the books by Horst and Pardalos [HP94], Pinter [Pin95], and Tuy [Tuy98]. Using convex optimization to find bounds for nonconvex problems is an active research topic, and addressed in the books above on global optimization, the book by Ben-Tal and Nemirovski [BTN01, §4.3], and the survey by Nesterov, Wolkowicz, and Ye [NWY00]. Some notable papers on this subject are Goemans and Williamson [GW95], Nesterov [Nes00, Nes98], Ye [Ye99], and Parrilo [Par03]. Randomized methods are discussed in Motwani and Raghavan [MR95].

Convex analysis, the mathematics of convex sets, functions, and optimization problems, is a well developed subfield of mathematics. Basic references include the books by Rockafellar [Roc70], Hiriart-Urruty and Lemaréchal [HUL93, HUL01], Borwein and Lewis [BL00], and Bertsekas, Nedić, and Ozdaglar [Ber03]. More references on convex analysis can be found in chapters 2–5.

Nesterov and Nemirovski [NN94] were the first to point out that interior-point methods can solve many convex optimization problems; see also the references in chapter 11. The book by Ben-Tal and Nemirovski [BTN01] covers modern convex optimization, interior-point methods, and applications.

Solution methods for convex optimization that we do not cover in this book include subgradient methods [Sho85], bundle methods [HUL93], cutting-plane methods [Kel60, EM75, GLY96], and the ellipsoid method [Sho91, BGT81].

The idea that convex optimization problems are tractable is not new. It has long been recognized that the theory of convex optimization is far more straightforward (and complete) than the theory of general nonlinear optimization. In this context Rockafellar stated, in his 1993 SIAM Review survey paper [Roc93],

> In fact the great watershed in optimization isn't between linearity and nonlinearity, but convexity and nonconvexity.

The first formal argument that convex optimization problems are easier to solve than general nonlinear optimization problems was made by Nemirovski and Yudin, in their 1983 book *Problem Complexity and Method Efficiency in Optimization* [NY83]. They showed that the information-based complexity of convex optimization problems is far lower than that of general nonlinear optimization problems. A more recent book on this topic is Vavasis [Vav91].

The low (theoretical) complexity of interior-point methods is integral to modern research in this area. Much of the research focuses on proving that an interior-point (or other) method can solve some class of convex optimization problems with a number of operations that grows no faster than a polynomial of the problem dimensions and $\log(1/\epsilon)$, where $\epsilon > 0$ is the required accuracy. (We will see some simple results like these in chapter 11.) The first comprehensive work on this topic is the book by Nesterov and Nemirovski [NN94]. Other books include Ben-Tal and Nemirovski [BTN01, lecture 5] and Renegar [Ren01]. The polynomial-time complexity of interior-point methods for various convex optimization problems is in marked contrast to the situation for a number of nonconvex optimization problems, for which all known algorithms require, in the worst case, a number of operations that is exponential in the problem dimensions.

Convex optimization has been used in many applications areas, too numerous to cite here. Convex analysis is central in economics and finance, where it is the basis of many results. For example the separating hyperplane theorem, together with a no-arbitrage assumption, is used to deduce the existence of prices and risk-neutral probabilities (see, *e.g.*, Luenberger [Lue95, Lue98] and Ross [Ros99]). Convex optimization, especially our ability to solve semidefinite programs, has recently received particular attention in automatic control theory. Applications of convex optimization in control theory can be found in the books by Boyd and Barratt [BB91], Boyd, El Ghaoui, Feron, and Balakrishnan [BEFB94], Dahleh and Diaz-Bobillo [DDB95], El Ghaoui and Niculescu [EN00], and Dullerud and Paganini [DP00]. A good example of embedded (convex) optimization is model predictive control, an automatic control technique that requires the solution of a (convex) quadratic program at each step. Model predictive control is now widely used in the chemical process control industry; see Morari and Zafirou [MZ89]. Another applications area where convex optimization (and especially, geometric programming) has a long history is electronic circuit design. Research papers on this topic include Fishburn and Dunlop [FD85], Sapatnekar, Rao, Vaidya, and Kang [SRVK93], and Hershenson, Boyd, and Lee [HBL01]. Luo [Luo03] gives a survey of applications in signal processing and communications. More references on applications of convex optimization can be found in chapters 4 and 6–8.

High quality implementations of recent interior-point methods for convex optimization problems are available in the LOQO [Van97] and MOSEK [MOS02] software packages, and the codes listed in chapter 11. Software systems for specifying optimization problems include AMPL [FGK99] and GAMS [BKMR98]. Both provide some support for recognizing problems that can be transformed to linear programs.

Part I

Theory

Chapter 2

Convex sets

2.1   Affine and convex sets

2.1.1   Lines and line segments

Suppose $x_1 \neq x_2$ are two points in $\mathbf{R}^n$. Points of the form

$$y = \theta x_1 + (1 - \theta)x_2,$$

where $\theta \in \mathbf{R}$, form the *line* passing through x_1 and x_2. The parameter value $\theta = 0$ corresponds to $y = x_2$, and the parameter value $\theta = 1$ corresponds to $y = x_1$. Values of the parameter θ between 0 and 1 correspond to the (closed) *line segment* between x_1 and x_2.

Expressing y in the form

$$y = x_2 + \theta(x_1 - x_2)$$

gives another interpretation: y is the sum of the *base point* x_2 (corresponding to $\theta = 0$) and the *direction* $x_1 - x_2$ (which points from x_2 to x_1) scaled by the parameter θ. Thus, θ gives the fraction of the way from x_2 to x_1 where y lies. As θ increases from 0 to 1, the point y moves from x_2 to x_1; for $\theta > 1$, the point y lies on the line beyond x_1. This is illustrated in figure 2.1.

2.1.2   Affine sets

A set $C \subseteq \mathbf{R}^n$ is *affine* if the line through any two distinct points in C lies in C, *i.e.*, if for any x_1, $x_2 \in C$ and $\theta \in \mathbf{R}$, we have $\theta x_1 + (1 - \theta)x_2 \in C$. In other words, C contains the linear combination of any two points in C, provided the coefficients in the linear combination sum to one.

This idea can be generalized to more than two points. We refer to a point of the form $\theta_1 x_1 + \cdots + \theta_k x_k$, where $\theta_1 + \cdots + \theta_k = 1$, as an *affine combination* of the points x_1, ..., x_k. Using induction from the definition of affine set (*i.e.*, that it contains every affine combination of two points in it), it can be shown that

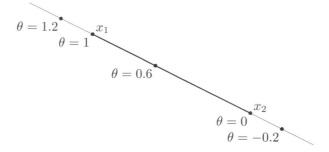

Figure 2.1 The line passing through x_1 and x_2 is described parametrically by $\theta x_1 + (1-\theta)x_2$, where θ varies over $\mathbf{R}$. The line segment between x_1 and x_2, which corresponds to θ between 0 and 1, is shown darker.

an affine set contains every affine combination of its points: If C is an affine set, $x_1, \ldots, x_k \in C$, and $\theta_1 + \cdots + \theta_k = 1$, then the point $\theta_1 x_1 + \cdots + \theta_k x_k$ also belongs to C.

If C is an affine set and $x_0 \in C$, then the set

$$V = C - x_0 = \{x - x_0 \mid x \in C\}$$

is a subspace, $i.e.$, closed under sums and scalar multiplication. To see this, suppose v_1, $v_2 \in V$ and α, $\beta \in \mathbf{R}$. Then we have $v_1 + x_0 \in C$ and $v_2 + x_0 \in C$, and so

$$\alpha v_1 + \beta v_2 + x_0 = \alpha(v_1 + x_0) + \beta(v_2 + x_0) + (1 - \alpha - \beta)x_0 \in C,$$

since C is affine, and $\alpha + \beta + (1 - \alpha - \beta) = 1$. We conclude that $\alpha v_1 + \beta v_2 \in V$, since $\alpha v_1 + \beta v_2 + x_0 \in C$.

Thus, the affine set C can be expressed as

$$C = V + x_0 = \{v + x_0 \mid v \in V\},$$

$i.e.$, as a subspace plus an offset. The subspace V associated with the affine set C does not depend on the choice of x_0, so x_0 can be chosen as any point in C. We define the *dimension* of an affine set C as the dimension of the subspace $V = C - x_0$, where x_0 is any element of C.

Example 2.1 *Solution set of linear equations.* The solution set of a system of linear equations, $C = \{x \mid Ax = b\}$, where $A \in \mathbf{R}^{m \times n}$ and $b \in \mathbf{R}^m$, is an affine set. To show this, suppose x_1, $x_2 \in C$, $i.e.$, $Ax_1 = b$, $Ax_2 = b$. Then for any θ, we have

$$
\begin{aligned}
A(\theta x_1 + (1-\theta)x_2) &= \theta A x_1 + (1-\theta)A x_2 \\
&= \theta b + (1-\theta)b \\
&= b,
\end{aligned}
$$

which shows that the affine combination $\theta x_1 + (1-\theta)x_2$ is also in C. The subspace associated with the affine set C is the nullspace of A.

We also have a converse: every affine set can be expressed as the solution set of a system of linear equations.

The set of all affine combinations of points in some set $C \subseteq \mathbf{R}^n$ is called the *affine hull* of C, and denoted $\mathbf{aff}\, C$:

$$\mathbf{aff}\, C = \{\theta_1 x_1 + \cdots + \theta_k x_k \mid x_1, \ldots, x_k \in C, \ \theta_1 + \cdots + \theta_k = 1\}.$$

The affine hull is the smallest affine set that contains C, in the following sense: if S is any affine set with $C \subseteq S$, then $\mathbf{aff}\, C \subseteq S$.

2.1.3   Affine dimension and relative interior

We define the *affine dimension* of a set C as the dimension of its affine hull. Affine dimension is useful in the context of convex analysis and optimization, but is not always consistent with other definitions of dimension. As an example consider the unit circle in $\mathbf{R}^2$, *i.e.*, $\{x \in \mathbf{R}^2 \mid x_1^2 + x_2^2 = 1\}$. Its affine hull is all of $\mathbf{R}^2$, so its affine dimension is two. By most definitions of dimension, however, the unit circle in $\mathbf{R}^2$ has dimension one.

If the affine dimension of a set $C \subseteq \mathbf{R}^n$ is less than n, then the set lies in the affine set $\mathbf{aff}\, C \neq \mathbf{R}^n$. We define the *relative interior* of the set C, denoted $\mathbf{relint}\, C$, as its interior relative to $\mathbf{aff}\, C$:

$$\mathbf{relint}\, C = \{x \in C \mid B(x, r) \cap \mathbf{aff}\, C \subseteq C \text{ for some } r > 0\},$$

where $B(x, r) = \{y \mid \|y - x\| \leq r\}$, the ball of radius r and center x in the norm $\|\cdot\|$. (Here $\|\cdot\|$ is any norm; all norms define the same relative interior.) We can then define the *relative boundary* of a set C as $\mathbf{cl}\, C \setminus \mathbf{relint}\, C$, where $\mathbf{cl}\, C$ is the closure of C.

Example 2.2   Consider a square in the (x_1, x_2)-plane in $\mathbf{R}^3$, defined as

$$C = \{x \in \mathbf{R}^3 \mid -1 \leq x_1 \leq 1, \ -1 \leq x_2 \leq 1, \ x_3 = 0\}.$$

Its affine hull is the (x_1, x_2)-plane, *i.e.*, $\mathbf{aff}\, C = \{x \in \mathbf{R}^3 \mid x_3 = 0\}$. The interior of C is empty, but the relative interior is

$$\mathbf{relint}\, C = \{x \in \mathbf{R}^3 \mid -1 < x_1 < 1, \ -1 < x_2 < 1, \ x_3 = 0\}.$$

Its boundary (in $\mathbf{R}^3$) is itself; its relative boundary is the wire-frame outline,

$$\{x \in \mathbf{R}^3 \mid \max\{|x_1|, |x_2|\} = 1, \ x_3 = 0\}.$$

2.1.4   Convex sets

A set C is *convex* if the line segment between any two points in C lies in C, *i.e.*, if for any $x_1, \ x_2 \in C$ and any θ with $0 \leq \theta \leq 1$, we have

$$\theta x_1 + (1 - \theta)x_2 \in C.$$

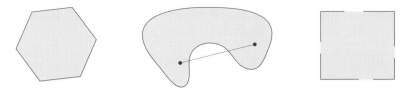

Figure 2.2 Some simple convex and nonconvex sets. *Left.* The hexagon, which includes its boundary (shown darker), is convex. *Middle.* The kidney shaped set is not convex, since the line segment between the two points in the set shown as dots is not contained in the set. *Right.* The square contains some boundary points but not others, and is not convex.

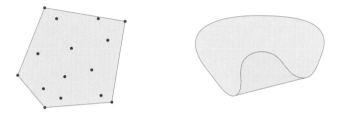

Figure 2.3 The convex hulls of two sets in $\mathbf{R}^2$. *Left.* The convex hull of a set of fifteen points (shown as dots) is the pentagon (shown shaded). *Right.* The convex hull of the kidney shaped set in figure 2.2 is the shaded set.

Roughly speaking, a set is convex if every point in the set can be seen by every other point, along an unobstructed straight path between them, where unobstructed means lying in the set. Every affine set is also convex, since it contains the entire line between any two distinct points in it, and therefore also the line segment between the points. Figure 2.2 illustrates some simple convex and nonconvex sets in $\mathbf{R}^2$.

We call a point of the form $\theta_1 x_1 + \cdots + \theta_k x_k$, where $\theta_1 + \cdots + \theta_k = 1$ and $\theta_i \geq 0$, $i = 1, \ldots, k$, a *convex combination* of the points $x_1, \ldots, x_k$. As with affine sets, it can be shown that a set is convex if and only if it contains every convex combination of its points. A convex combination of points can be thought of as a *mixture* or *weighted average* of the points, with θ_i the fraction of x_i in the mixture.

The *convex hull* of a set C, denoted $\mathbf{conv}\, C$, is the set of all convex combinations of points in C:

$$\mathbf{conv}\, C = \{\theta_1 x_1 + \cdots + \theta_k x_k \mid x_i \in C,\ \theta_i \geq 0,\ i = 1, \ldots, k,\ \theta_1 + \cdots + \theta_k = 1\}.$$

As the name suggests, the convex hull $\mathbf{conv}\, C$ is always convex. It is the smallest convex set that contains C: If B is any convex set that contains C, then $\mathbf{conv}\, C \subseteq B$. Figure 2.3 illustrates the definition of convex hull.

The idea of a convex combination can be generalized to include infinite sums, integrals, and, in the most general form, probability distributions. Suppose $\theta_1, \theta_2, \ldots$

satisfy

$$\theta_i \geq 0, \quad i = 1, 2, \ldots, \qquad \sum_{i=1}^{\infty} \theta_i = 1,$$

and $x_1, x_2, \ldots \in C$, where $C \subseteq \mathbf{R}^n$ is convex. Then

$$\sum_{i=1}^{\infty} \theta_i x_i \in C,$$

if the series converges. More generally, suppose $p : \mathbf{R}^n \to \mathbf{R}$ satisfies $p(x) \geq 0$ for all $x \in C$ and $\int_C p(x)\, dx = 1$, where $C \subseteq \mathbf{R}^n$ is convex. Then

$$\int_C p(x)x\, dx \in C,$$

if the integral exists.

In the most general form, suppose $C \subseteq \mathbf{R}^n$ is convex and x is a random vector with $x \in C$ with probability one. Then $\mathbf{E}\, x \in C$. Indeed, this form includes all the others as special cases. For example, suppose the random variable x only takes on the two values x_1 and x_2, with $\mathbf{prob}(x = x_1) = \theta$ and $\mathbf{prob}(x = x_2) = 1 - \theta$, where $0 \leq \theta \leq 1$. Then $\mathbf{E}\, x = \theta x_1 + (1 - \theta)x_2$, and we are back to a simple convex combination of two points.

2.1.5 Cones

A set C is called a *cone*, or *nonnegative homogeneous*, if for every $x \in C$ and $\theta \geq 0$ we have $\theta x \in C$. A set C is a *convex cone* if it is convex and a cone, which means that for any x_1, $x_2 \in C$ and θ_1, $\theta_2 \geq 0$, we have

$$\theta_1 x_1 + \theta_2 x_2 \in C.$$

Points of this form can be described geometrically as forming the two-dimensional pie slice with apex 0 and edges passing through x_1 and x_2. (See figure 2.4.)

A point of the form $\theta_1 x_1 + \cdots + \theta_k x_k$ with $\theta_1, \ldots, \theta_k \geq 0$ is called a *conic combination* (or a *nonnegative linear combination*) of $x_1, \ldots, x_k$. If x_i are in a convex cone C, then every conic combination of x_i is in C. Conversely, a set C is a convex cone if and only if it contains all conic combinations of its elements. Like convex (or affine) combinations, the idea of conic combination can be generalized to infinite sums and integrals.

The *conic hull* of a set C is the set of all conic combinations of points in C, i.e.,

$$\{\theta_1 x_1 + \cdots + \theta_k x_k \mid x_i \in C,\ \theta_i \geq 0,\ i = 1, \ldots, k\},$$

which is also the smallest convex cone that contains C (see figure 2.5).

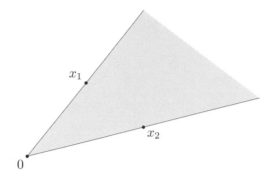

Figure 2.4 The pie slice shows all points of the form $\theta_1 x_1 + \theta_2 x_2$, where θ_1, $\theta_2 \geq 0$. The apex of the slice (which corresponds to $\theta_1 = \theta_2 = 0$) is at 0; its edges (which correspond to $\theta_1 = 0$ or $\theta_2 = 0$) pass through the points x_1 and x_2.

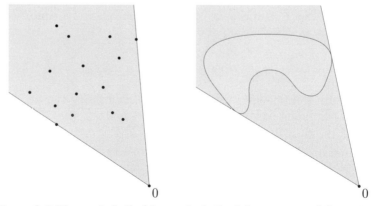

Figure 2.5 The conic hulls (shown shaded) of the two sets of figure 2.3.

2.2 Some important examples

In this section we describe some important examples of convex sets which we will encounter throughout the rest of the book. We start with some simple examples.

- The empty set $\emptyset$, any single point (*i.e.*, singleton) $\{x_0\}$, and the whole space $\mathbf{R}^n$ are affine (hence, convex) subsets of $\mathbf{R}^n$.

- Any line is affine. If it passes through zero, it is a subspace, hence also a convex cone.

- A line segment is convex, but not affine (unless it reduces to a point).

- A *ray*, which has the form $\{x_0 + \theta v \mid \theta \geq 0\}$, where $v \neq 0$, is convex, but not affine. It is a convex cone if its base x_0 is 0.

- Any subspace is affine, and a convex cone (hence convex).

2.2.1 Hyperplanes and halfspaces

A *hyperplane* is a set of the form

$$\{x \mid a^T x = b\},$$

where $a \in \mathbf{R}^n$, $a \neq 0$, and $b \in \mathbf{R}$. Analytically it is the solution set of a nontrivial linear equation among the components of x (and hence an affine set). Geometrically, the hyperplane $\{x \mid a^T x = b\}$ can be interpreted as the set of points with a constant inner product to a given vector a, or as a hyperplane with *normal vector* a; the constant $b \in \mathbf{R}$ determines the offset of the hyperplane from the origin. This geometric interpretation can be understood by expressing the hyperplane in the form

$$\{x \mid a^T(x - x_0) = 0\},$$

where x_0 is any point in the hyperplane (*i.e.*, any point that satisfies $a^T x_0 = b$). This representation can in turn be expressed as

$$\{x \mid a^T(x - x_0) = 0\} = x_0 + a^\perp,$$

where $a^\perp$ denotes the orthogonal complement of a, *i.e.*, the set of all vectors orthogonal to it:

$$a^\perp = \{v \mid a^T v = 0\}.$$

This shows that the hyperplane consists of an offset x_0, plus all vectors orthogonal to the (normal) vector a. These geometric interpretations are illustrated in figure 2.6.

A hyperplane divides $\mathbf{R}^n$ into two *halfspaces*. A (closed) halfspace is a set of the form

$$\{x \mid a^T x \leq b\}, \tag{2.1}$$

where $a \neq 0$, *i.e.*, the solution set of one (nontrivial) linear inequality. Halfspaces are convex, but not affine. This is illustrated in figure 2.7.

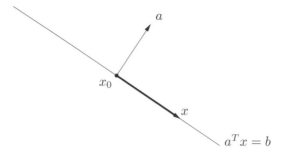

Figure 2.6 Hyperplane in $\mathbf{R}^2$, with normal vector a and a point x_0 in the hyperplane. For any point x in the hyperplane, $x - x_0$ (shown as the darker arrow) is orthogonal to a.

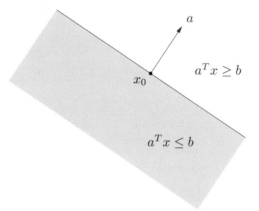

Figure 2.7 A hyperplane defined by $a^T x = b$ in $\mathbf{R}^2$ determines two halfspaces. The halfspace determined by $a^T x \geq b$ (not shaded) is the halfspace extending in the direction a. The halfspace determined by $a^T x \leq b$ (which is shown shaded) extends in the direction $-a$. The vector a is the outward normal of this halfspace.

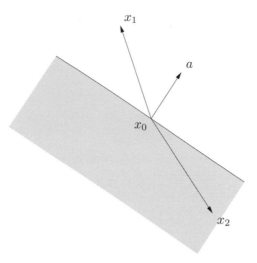

Figure 2.8 The shaded set is the halfspace determined by $a^T(x - x_0) \leq 0$. The vector $x_1 - x_0$ makes an acute angle with a, so x_1 is not in the halfspace. The vector $x_2 - x_0$ makes an obtuse angle with a, and so is in the halfspace.

The halfspace (2.1) can also be expressed as

$$\{x \mid a^T(x - x_0) \leq 0\}, \tag{2.2}$$

where x_0 is any point on the associated hyperplane, *i.e.*, satisfies $a^T x_0 = b$. The representation (2.2) suggests a simple geometric interpretation: the halfspace consists of x_0 plus any vector that makes an obtuse (or right) angle with the (outward normal) vector a. This is illustrated in figure 2.8.

The boundary of the halfspace (2.1) is the hyperplane $\{x \mid a^T x = b\}$. The set $\{x \mid a^T x < b\}$, which is the interior of the halfspace $\{x \mid a^T x \leq b\}$, is called an *open halfspace*.

2.2.2   Euclidean balls and ellipsoids

A *(Euclidean) ball* (or just ball) in $\mathbf{R}^n$ has the form

$$B(x_c, r) = \{x \mid \|x - x_c\|_2 \leq r\} = \{x \mid (x - x_c)^T(x - x_c) \leq r^2\},$$

where $r > 0$, and $\| \cdot \|_2$ denotes the Euclidean norm, *i.e.*, $\|u\|_2 = (u^T u)^{1/2}$. The vector x_c is the *center* of the ball and the scalar r is its *radius*; $B(x_c, r)$ consists of all points within a distance r of the center x_c. Another common representation for the Euclidean ball is

$$B(x_c, r) = \{x_c + ru \mid \|u\|_2 \leq 1\}.$$

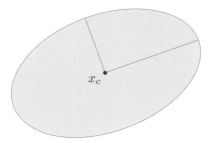

Figure 2.9 An ellipsoid in $\mathbf{R}^2$, shown shaded. The center x_c is shown as a
dot, and the two semi-axes are shown as line segments.

A Euclidean ball is a convex set: if $\|x_1 - x_c\|_2 \le r$, $\|x_2 - x_c\|_2 \le r$, and
$0 \le \theta \le 1$, then

$$
\begin{aligned}
\|\theta x_1 + (1-\theta)x_2 - x_c\|_2 &= \|\theta(x_1 - x_c) + (1-\theta)(x_2 - x_c)\|_2 \\
&\le \theta\|x_1 - x_c\|_2 + (1-\theta)\|x_2 - x_c\|_2 \\
&\le r.
\end{aligned}
$$

(Here we use the homogeneity property and triangle inequality for $\|\cdot\|_2$; see §A.1.2.)

A related family of convex sets is the *ellipsoids*, which have the form

$$
\mathcal{E} = \{x \mid (x - x_c)^T P^{-1}(x - x_c) \le 1\}, \tag{2.3}
$$

where $P = P^T \succ 0$, i.e., P is symmetric and positive definite. The vector $x_c \in \mathbf{R}^n$
is the *center* of the ellipsoid. The matrix P determines how far the ellipsoid extends
in every direction from x_c; the lengths of the semi-axes of $\mathcal{E}$ are given by $\sqrt{\lambda_i}$, where
λ_i are the eigenvalues of P. A ball is an ellipsoid with $P = r^2 I$. Figure 2.9 shows
an ellipsoid in $\mathbf{R}^2$.

Another common representation of an ellipsoid is

$$
\mathcal{E} = \{x_c + Au \mid \|u\|_2 \le 1\}, \tag{2.4}
$$

where A is square and nonsingular. In this representation we can assume without
loss of generality that A is symmetric and positive definite. By taking $A = P^{1/2}$,
this representation gives the ellipsoid defined in (2.3). When the matrix A in (2.4)
is symmetric positive semidefinite but singular, the set in (2.4) is called a *degenerate
ellipsoid*; its affine dimension is equal to the rank of A. Degenerate ellipsoids are
also convex.

2.2.3   Norm balls and norm cones

Suppose $\|\cdot\|$ is any norm on $\mathbf{R}^n$ (see §A.1.2). From the general properties of norms it
can be shown that a *norm ball* of radius r and center x_c, given by $\{x \mid \|x - x_c\| \le r\}$,
is convex. The *norm cone* associated with the norm $\|\cdot\|$ is the set

$$
C = \{(x, t) \mid \|x\| \le t\} \subseteq \mathbf{R}^{n+1}.
$$

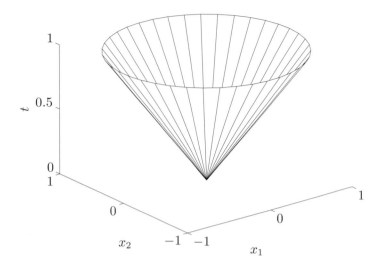

Figure 2.10 Boundary of second-order cone in $\mathbf{R}^3$, $\{(x_1, x_2, t) \mid (x_1^2 + x_2^2)^{1/2} \leq t\}$.

It is (as the name suggests) a convex cone.

Example 2.3 *The second-order cone is the norm cone for the Euclidean norm, i.e.,*

$$
\begin{aligned}
C &= \{(x,t) \in \mathbf{R}^{n+1} \mid \|x\|_2 \leq t\} \\
&= \left\{ \begin{bmatrix} x \\ t \end{bmatrix} \middle| \begin{bmatrix} x \\ t \end{bmatrix}^T \begin{bmatrix} I & 0 \\ 0 & -1 \end{bmatrix} \begin{bmatrix} x \\ t \end{bmatrix} \leq 0, \ t \geq 0 \right\}.
\end{aligned}
$$

The second-order cone is also known by several other names. It is called the *quadratic cone*, since it is defined by a quadratic inequality. It is also called the *Lorentz cone* or *ice-cream cone*. Figure 2.10 shows the second-order cone in $\mathbf{R}^3$.

2.2.4   Polyhedra

A *polyhedron* is defined as the solution set of a finite number of linear equalities and inequalities:

$$
\mathcal{P} = \{x \mid a_j^T x \leq b_j, \ j = 1, \ldots, m, \ c_j^T x = d_j, \ j = 1, \ldots, p\}. \tag{2.5}
$$

A polyhedron is thus the intersection of a finite number of halfspaces and hyperplanes. Affine sets (*e.g.*, subspaces, hyperplanes, lines), rays, line segments, and halfspaces are all polyhedra. It is easily shown that polyhedra are convex sets. A bounded polyhedron is sometimes called a *polytope*, but some authors use the opposite convention (*i.e.*, polytope for any set of the form (2.5), and polyhedron

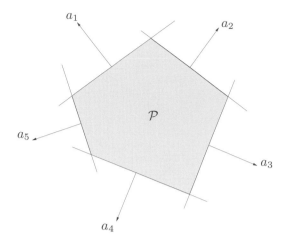

Figure 2.11 The polyhedron $\mathcal{P}$ (shown shaded) is the intersection of five halfspaces, with outward normal vectors $a_1, \ldots, a_5$.

when it is bounded). Figure 2.11 shows an example of a polyhedron defined as the intersection of five halfspaces.

It will be convenient to use the compact notation

$$\mathcal{P} = \{x \mid Ax \preceq b, \ Cx = d\} \tag{2.6}$$

for (2.5), where

$$A = \begin{bmatrix} a_1^T \\ \vdots \\ a_m^T \end{bmatrix}, \qquad C = \begin{bmatrix} c_1^T \\ \vdots \\ c_p^T \end{bmatrix},$$

and the symbol $\preceq$ denotes *vector inequality* or *componentwise inequality* in $\mathbf{R}^m$: $u \preceq v$ means $u_i \leq v_i$ for $i = 1, \ldots, m$.

Example 2.4 The *nonnegative orthant* is the set of points with nonnegative components, *i.e.*,

$$\mathbf{R}_+^n = \{x \in \mathbf{R}^n \mid x_i \geq 0, \ i = 1, \ldots, n\} = \{x \in \mathbf{R}^n \mid x \succeq 0\}.$$

(Here $\mathbf{R}_+$ denotes the set of nonnegative numbers: $\mathbf{R}_+ = \{x \in \mathbf{R} \mid x \geq 0\}$.) The nonnegative orthant is a polyhedron and a cone (and therefore called a *polyhedral cone*).

Simplexes

Simplexes are another important family of polyhedra. Suppose the $k + 1$ points $v_0, \ldots, v_k \in \mathbf{R}^n$ are *affinely independent*, which means $v_1 - v_0, \ldots, v_k - v_0$ are linearly independent. The simplex determined by them is given by

$$C = \mathbf{conv}\{v_0, \ldots, v_k\} = \{\theta_0 v_0 + \cdots + \theta_k v_k \mid \theta \succeq 0, \ \mathbf{1}^T \theta = 1\}, \tag{2.7}$$

where $\mathbf{1}$ denotes the vector with all entries one. The affine dimension of this simplex is k, so it is sometimes referred to as a k-dimensional simplex in $\mathbf{R}^n$.

Example 2.5 *Some common simplexes.* A 1-dimensional simplex is a line segment; a 2-dimensional simplex is a triangle (including its interior); and a 3-dimensional simplex is a tetrahedron.

The *unit simplex* is the n-dimensional simplex determined by the zero vector and the unit vectors, *i.e.*, $0, e_1, \ldots, e_n \in \mathbf{R}^n$. It can be expressed as the set of vectors that satisfy

$$x \succeq 0, \qquad \mathbf{1}^T x \leq 1.$$

The *probability simplex* is the $(n-1)$-dimensional simplex determined by the unit vectors $e_1, \ldots, e_n \in \mathbf{R}^n$. It is the set of vectors that satisfy

$$x \succeq 0, \qquad \mathbf{1}^T x = 1.$$

Vectors in the probability simplex correspond to probability distributions on a set with n elements, with x_i interpreted as the probability of the ith element.

To describe the simplex (2.7) as a polyhedron, *i.e.*, in the form (2.6), we proceed as follows. By definition, $x \in C$ if and only if $x = \theta_0 v_0 + \theta_1 v_1 + \cdots + \theta_k v_k$ for some $\theta \succeq 0$ with $\mathbf{1}^T \theta = 1$. Equivalently, if we define $y = (\theta_1, \ldots, \theta_k)$ and

$$B = \left[\begin{array}{ccc} v_1 - v_0 & \cdots & v_k - v_0 \end{array}\right] \in \mathbf{R}^{n \times k},$$

we can say that $x \in C$ if and only if

$$x = v_0 + By \tag{2.8}$$

for some $y \succeq 0$ with $\mathbf{1}^T y \leq 1$. Now we note that affine independence of the points $v_0, \ldots, v_k$ implies that the matrix B has rank k. Therefore there exists a nonsingular matrix $A = (A_1, A_2) \in \mathbf{R}^{n \times n}$ such that

$$AB = \left[\begin{array}{c} A_1 \\ A_2 \end{array}\right] B = \left[\begin{array}{c} I \\ 0 \end{array}\right].$$

Multiplying (2.8) on the left with A, we obtain

$$A_1 x = A_1 v_0 + y, \qquad A_2 x = A_2 v_0.$$

From this we see that $x \in C$ if and only if $A_2 x = A_2 v_0$, and the vector $y = A_1 x - A_1 v_0$ satisfies $y \succeq 0$ and $\mathbf{1}^T y \leq 1$. In other words we have $x \in C$ if and only if

$$A_2 x = A_2 v_0, \qquad A_1 x \succeq A_1 v_0, \qquad \mathbf{1}^T A_1 x \leq 1 + \mathbf{1}^T A_1 v_0,$$

which is a set of linear equalities and inequalities in x, and so describes a polyhedron.

Convex hull description of polyhedra

The convex hull of the finite set $\{v_1, \ldots, v_k\}$ is

$$\mathbf{conv}\{v_1, \ldots, v_k\} = \{\theta_1 v_1 + \cdots + \theta_k v_k \mid \theta \succeq 0, \ \mathbf{1}^T \theta = 1\}.$$

This set is a polyhedron, and bounded, but (except in special cases, *e.g.*, a simplex) it is not simple to express it in the form (2.5), *i.e.*, by a set of linear equalities and inequalities.

A generalization of this convex hull description is

$$\{\theta_1 v_1 + \cdots + \theta_k v_k \mid \theta_1 + \cdots + \theta_m = 1, \ \theta_i \geq 0, \ i = 1, \ldots, k\}, \qquad (2.9)$$

where $m \leq k$. Here we consider nonnegative linear combinations of v_i, but only the first m coefficients are required to sum to one. Alternatively, we can interpret (2.9) as the convex hull of the points $v_1, \ldots, v_m$, plus the conic hull of the points $v_{m+1}, \ldots, v_k$. The set (2.9) defines a polyhedron, and conversely, every polyhedron can be represented in this form (although we will not show this).

The question of how a polyhedron is represented is subtle, and has very important practical consequences. As a simple example consider the unit ball in the ℓ_∞-norm in $\mathbf{R}^n$,

$$C = \{x \mid |x_i| \leq 1, \ i = 1, \ldots, n\}.$$

The set C can be described in the form (2.5) with $2n$ linear inequalities $\pm e_i^T x \leq 1$, where e_i is the ith unit vector. To describe it in the convex hull form (2.9) requires at least 2^n points:

$$C = \mathbf{conv}\{v_1, \ldots, v_{2^n}\},$$

where $v_1, \ldots, v_{2^n}$ are the 2^n vectors all of whose components are 1 or -1. Thus the size of the two descriptions differs greatly, for large n.

2.2.5 The positive semidefinite cone

We use the notation $\mathbf{S}^n$ to denote the set of symmetric $n \times n$ matrices,

$$\mathbf{S}^n = \{X \in \mathbf{R}^{n \times n} \mid X = X^T\},$$

which is a vector space with dimension $n(n+1)/2$. We use the notation $\mathbf{S}^n_+$ to denote the set of symmetric positive semidefinite matrices:

$$\mathbf{S}^n_+ = \{X \in \mathbf{S}^n \mid X \succeq 0\},$$

and the notation $\mathbf{S}^n_{++}$ to denote the set of symmetric positive definite matrices:

$$\mathbf{S}^n_{++} = \{X \in \mathbf{S}^n \mid X \succ 0\}.$$

(This notation is meant to be analogous to $\mathbf{R}_+$, which denotes the nonnegative reals, and $\mathbf{R}_{++}$, which denotes the positive reals.)

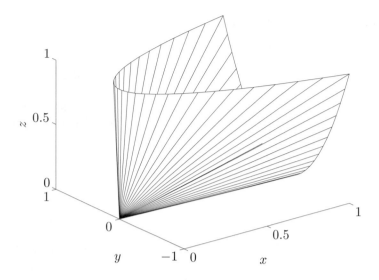

Figure 2.12 Boundary of positive semidefinite cone in $\mathbf{S}^2$.

The set $\mathbf{S}^n_+$ is a convex cone: if $\theta_1, \theta_2 \geq 0$ and A, $B \in \mathbf{S}^n_+$, then $\theta_1 A + \theta_2 B \in \mathbf{S}^n_+$. This can be seen directly from the definition of positive semidefiniteness: for any $x \in \mathbf{R}^n$, we have

$$x^T(\theta_1 A + \theta_2 B)x = \theta_1 x^T A x + \theta_2 x^T B x \geq 0,$$

if $A \succeq 0$, $B \succeq 0$ and θ_1, $\theta_2 \geq 0$.

Example 2.6 *Positive semidefinite cone in* $\mathbf{S}^2$. We have

$$X = \begin{bmatrix} x & y \\ y & z \end{bmatrix} \in \mathbf{S}^2_+ \quad \Longleftrightarrow \quad x \geq 0, \quad z \geq 0, \quad xz \geq y^2.$$

The boundary of this cone is shown in figure 2.12, plotted in $\mathbf{R}^3$ as (x, y, z).

2.3   Operations that preserve convexity

In this section we describe some operations that preserve convexity of sets, or allow us to construct convex sets from others. These operations, together with the simple examples described in §2.2, form a calculus of convex sets that is useful for determining or establishing convexity of sets.

2.3.1   Intersection

Convexity is preserved under intersection: if S_1 and S_2 are convex, then $S_1 \cap S_2$ is convex. This property extends to the intersection of an infinite number of sets: if S_α is convex for every $\alpha \in \mathcal{A}$, then $\bigcap_{\alpha \in \mathcal{A}} S_\alpha$ is convex. (Subspaces, affine sets, and convex cones are also closed under arbitrary intersections.) As a simple example, a polyhedron is the intersection of halfspaces and hyperplanes (which are convex), and therefore is convex.

Example 2.7 The positive semidefinite cone $\mathbf{S}_+^n$ can be expressed as

$$\bigcap_{z \neq 0} \{X \in \mathbf{S}^n \mid z^T X z \geq 0\}.$$

For each $z \neq 0$, $z^T X z$ is a (not identically zero) linear function of X, so the sets

$$\{X \in \mathbf{S}^n \mid z^T X z \geq 0\}$$

are, in fact, halfspaces in $\mathbf{S}^n$. Thus the positive semidefinite cone is the intersection of an infinite number of halfspaces, and so is convex.

Example 2.8 We consider the set

$$S = \{x \in \mathbf{R}^m \mid |p(t)| \leq 1 \text{ for } |t| \leq \pi/3\}, \tag{2.10}$$

where $p(t) = \sum_{k=1}^m x_k \cos kt$. The set S can be expressed as the intersection of an infinite number of *slabs*: $S = \bigcap_{|t| \leq \pi/3} S_t$, where

$$S_t = \{x \mid -1 \leq (\cos t, \ldots, \cos mt)^T x \leq 1\},$$

and so is convex. The definition and the set are illustrated in figures 2.13 and 2.14, for $m = 2$.

In the examples above we establish convexity of a set by expressing it as a (possibly infinite) intersection of halfspaces. We will see in §2.5.1 that a converse holds: *every* closed convex set S is a (usually infinite) intersection of halfspaces. In fact, a closed convex set S is the intersection of all halfspaces that contain it:

$$S = \bigcap \{\mathcal{H} \mid \mathcal{H} \text{ halfspace, } S \subseteq \mathcal{H}\}.$$

2.3.2   Affine functions

Recall that a function $f : \mathbf{R}^n \to \mathbf{R}^m$ is *affine* if it is a sum of a linear function and a constant, *i.e.*, if it has the form $f(x) = Ax + b$, where $A \in \mathbf{R}^{m \times n}$ and $b \in \mathbf{R}^m$. Suppose $S \subseteq \mathbf{R}^n$ is convex and $f : \mathbf{R}^n \to \mathbf{R}^m$ is an affine function. Then the image of S under f,

$$f(S) = \{f(x) \mid x \in S\},$$

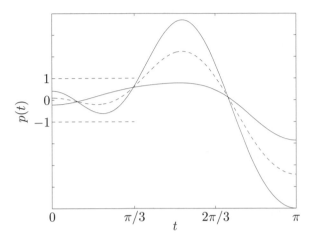

Figure 2.13 Three trigonometric polynomials associated with points in the set S defined in (2.10), for $m = 2$. The trigonometric polynomial plotted with dashed line type is the average of the other two.

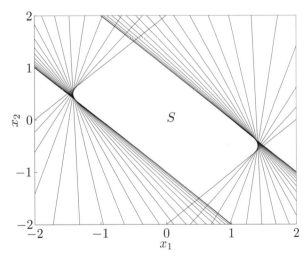

Figure 2.14 The set S defined in (2.10), for $m = 2$, is shown as the white area in the middle of the plot. The set is the intersection of an infinite number of slabs (20 of which are shown), hence convex.

is convex. Similarly, if $f : \mathbf{R}^k \to \mathbf{R}^n$ is an affine function, the *inverse image* of S under f,

$$f^{-1}(S) = \{x \mid f(x) \in S\},$$

is convex.

Two simple examples are *scaling* and *translation*. If $S \subseteq \mathbf{R}^n$ is convex, $\alpha \in \mathbf{R}$, and $a \in \mathbf{R}^n$, then the sets αS and $S + a$ are convex, where

$$\alpha S = \{\alpha x \mid x \in S\}, \qquad S + a = \{x + a \mid x \in S\}.$$

The *projection* of a convex set onto some of its coordinates is convex: if $S \subseteq \mathbf{R}^m \times \mathbf{R}^n$ is convex, then

$$T = \{x_1 \in \mathbf{R}^m \mid (x_1, x_2) \in S \text{ for some } x_2 \in \mathbf{R}^n\}$$

is convex.

The *sum* of two sets is defined as

$$S_1 + S_2 = \{x + y \mid x \in S_1, \ y \in S_2\}.$$

If S_1 and S_2 are convex, then $S_1 + S_2$ is convex. To see this, if S_1 and S_2 are convex, then so is the direct or Cartesian product

$$S_1 \times S_2 = \{(x_1, x_2) \mid x_1 \in S_1, \ x_2 \in S_2\}.$$

The image of this set under the linear function $f(x_1, x_2) = x_1 + x_2$ is the sum $S_1 + S_2$.

We can also consider the *partial sum* of S_1, $S_2 \in \mathbf{R}^n \times \mathbf{R}^m$, defined as

$$S = \{(x, y_1 + y_2) \mid (x, y_1) \in S_1, \ (x, y_2) \in S_2\},$$

where $x \in \mathbf{R}^n$ and $y_i \in \mathbf{R}^m$. For $m = 0$, the partial sum gives the intersection of S_1 and S_2; for $n = 0$, it is set addition. Partial sums of convex sets are convex (see exercise 2.16).

Example 2.9 *Polyhedron.* The polyhedron $\{x \mid Ax \preceq b, \ Cx = d\}$ can be expressed as the inverse image of the Cartesian product of the nonnegative orthant and the origin under the affine function $f(x) = (b - Ax, d - Cx)$:

$$\{x \mid Ax \preceq b, \ Cx = d\} = \{x \mid f(x) \in \mathbf{R}_+^m \times \{0\}\}.$$

Example 2.10 *Solution set of linear matrix inequality.* The condition

$$A(x) = x_1 A_1 + \cdots + x_n A_n \preceq B, \tag{2.11}$$

where B, $A_i \in \mathbf{S}^m$, is called a *linear matrix inequality* (LMI) in x (Note the similarity to an ordinary linear inequality,

$$a^T x = x_1 a_1 + \cdots + x_n a_n \leq b,$$

with b, $a_i \in \mathbf{R}$.)

The solution set of a linear matrix inequality, $\{x \mid A(x) \preceq B\}$, is convex. Indeed, it is the inverse image of the positive semidefinite cone under the affine function $f : \mathbf{R}^n \to \mathbf{S}^m$ given by $f(x) = B - A(x)$.

Example 2.11 *Hyperbolic cone.* The set

$$\{x \mid x^T P x \le (c^T x)^2, \ c^T x \ge 0\}$$

where $P \in \mathbf{S}_+^n$ and $c \in \mathbf{R}^n$, is convex, since it is the inverse image of the second-order cone,

$$\{(z,t) \mid z^T z \le t^2, \ t \ge 0\},$$

under the affine function $f(x) = (P^{1/2} x, c^T x)$.

Example 2.12 *Ellipsoid.* The ellipsoid

$$\mathcal{E} = \{x \mid (x - x_c)^T P^{-1}(x - x_c) \le 1\},$$

where $P \in \mathbf{S}_{++}^n$, is the image of the unit Euclidean ball $\{u \mid \|u\|_2 \le 1\}$ under the affine mapping $f(u) = P^{1/2} u + x_c$. (It is also the inverse image of the unit ball under the affine mapping $g(x) = P^{-1/2}(x - x_c)$.)

2.3.3   Linear-fractional and perspective functions

In this section we explore a class of functions, called *linear-fractional*, that is more general than affine but still preserves convexity.

The perspective function

We define the *perspective function* $P : \mathbf{R}^{n+1} \to \mathbf{R}^n$, with domain $\mathbf{dom}\, P = \mathbf{R}^n \times \mathbf{R}_{++}$, as $P(z,t) = z/t$. (Here $\mathbf{R}_{++}$ denotes the set of positive numbers: $\mathbf{R}_{++} = \{x \in \mathbf{R} \mid x > 0\}$.) The perspective function scales or normalizes vectors so the last component is one, and then drops the last component.

Remark 2.1 We can interpret the perspective function as the action of a *pin-hole camera*. A pin-hole camera (in $\mathbf{R}^3$) consists of an opaque horizontal plane $x_3 = 0$, with a single pin-hole at the origin, through which light can pass, and a horizontal image plane $x_3 = -1$. An object at x, above the camera (*i.e.*, with $x_3 > 0$), forms an image at the point $-(x_1/x_3, x_2/x_3, 1)$ on the image plane. Dropping the last component of the image point (since it is always -1), the image of a point at x appears at $y = -(x_1/x_3, x_2/x_3) = -P(x)$ on the image plane. This is illustrated in figure 2.15.

If $C \subseteq \mathbf{dom}\, P$ is convex, then its image

$$P(C) = \{P(x) \mid x \in C\}$$

is convex. This result is certainly intuitive: a convex object, viewed through a pin-hole camera, yields a convex image. To establish this fact we show that line segments are mapped to line segments under the perspective function. (This too

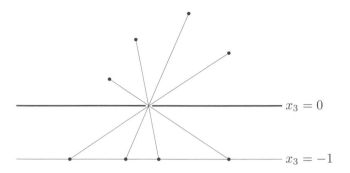

Figure 2.15 Pin-hole camera interpretation of perspective function. The dark horizontal line represents the plane $x_3 = 0$ in $\mathbf{R}^3$, which is opaque, except for a pin-hole at the origin. Objects or light sources above the plane appear on the image plane $x_3 = -1$, which is shown as the lighter horizontal line. The mapping of the position of a source to the position of its image is related to the perspective function.

makes sense: a line segment, viewed through a pin-hole camera, yields a line segment image.) Suppose that $x = (\tilde{x}, x_{n+1})$, $y = (\tilde{y}, y_{n+1}) \in \mathbf{R}^{n+1}$ with $x_{n+1} > 0$, $y_{n+1} > 0$. Then for $0 \le \theta \le 1$,

$$P(\theta x + (1 - \theta)y) = \frac{\theta \tilde{x} + (1 - \theta)\tilde{y}}{\theta x_{n+1} + (1 - \theta)y_{n+1}} = \mu P(x) + (1 - \mu)P(y),$$

where

$$\mu = \frac{\theta x_{n+1}}{\theta x_{n+1} + (1 - \theta)y_{n+1}} \in [0, 1].$$

This correspondence between θ and μ is monotonic: as θ varies between 0 and 1 (which sweeps out the line segment $[x, y]$), μ varies between 0 and 1 (which sweeps out the line segment $[P(x), P(y)]$). This shows that $P([x, y]) = [P(x), P(y)]$.

Now suppose C is convex with $C \subseteq \mathbf{dom}\, P$ (*i.e.*, $x_{n+1} > 0$ for all $x \in C$), and x, $y \in C$. To establish convexity of $P(C)$ we need to show that the line segment $[P(x), P(y)]$ is in $P(C)$. But this line segment is the image of the line segment $[x, y]$ under P, and so lies in $P(C)$.

The inverse image of a convex set under the perspective function is also convex: if $C \subseteq \mathbf{R}^n$ is convex, then

$$P^{-1}(C) = \{(x, t) \in \mathbf{R}^{n+1} \mid x/t \in C, \ t > 0\}$$

is convex. To show this, suppose $(x, t) \in P^{-1}(C)$, $(y, s) \in P^{-1}(C)$, and $0 \le \theta \le 1$. We need to show that

$$\theta(x, t) + (1 - \theta)(y, s) \in P^{-1}(C),$$

i.e., that

$$\frac{\theta x + (1 - \theta)y}{\theta t + (1 - \theta)s} \in C$$

$(\theta t + (1 - \theta)s > 0$ is obvious). This follows from

$$\frac{\theta x + (1 - \theta)y}{\theta t + (1 - \theta)s} = \mu(x/t) + (1 - \mu)(y/s),$$

where

$$\mu = \frac{\theta t}{\theta t + (1 - \theta)s} \in [0, 1].$$

Linear-fractional functions

A *linear-fractional function* is formed by composing the perspective function with an affine function. Suppose $g : \mathbf{R}^n \to \mathbf{R}^{m+1}$ is affine, *i.e.*,

$$g(x) = \begin{bmatrix} A \\ c^T \end{bmatrix} x + \begin{bmatrix} b \\ d \end{bmatrix}, \tag{2.12}$$

where $A \in \mathbf{R}^{m \times n}$, $b \in \mathbf{R}^m$, $c \in \mathbf{R}^n$, and $d \in \mathbf{R}$. The function $f : \mathbf{R}^n \to \mathbf{R}^m$ given by $f = P \circ g$, *i.e.*,

$$f(x) = (Ax + b)/(c^T x + d), \qquad \mathbf{dom}\, f = \{x \mid c^T x + d > 0\}, \tag{2.13}$$

is called a *linear-fractional* (or *projective*) function. If $c = 0$ and $d > 0$, the domain of f is $\mathbf{R}^n$, and f is an affine function. So we can think of affine and linear functions as special cases of linear-fractional functions.

Remark 2.2 *Projective interpretation.* It is often convenient to represent a linear-fractional function as a matrix

$$Q = \begin{bmatrix} A & b \\ c^T & d \end{bmatrix} \in \mathbf{R}^{(m+1) \times (n+1)} \tag{2.14}$$

that acts on (multiplies) points of form $(x, 1)$, which yields $(Ax + b, c^T x + d)$. This result is then scaled or normalized so that its last component is one, which yields $(f(x), 1)$.

This representation can be interpreted geometrically by associating $\mathbf{R}^n$ with a set of rays in $\mathbf{R}^{n+1}$ as follows. With each point z in $\mathbf{R}^n$ we associate the (open) ray $\mathcal{P}(z) = \{t(z, 1) \mid t > 0\}$ in $\mathbf{R}^{n+1}$. The last component of this ray takes on positive values. Conversely any ray in $\mathbf{R}^{n+1}$, with base at the origin and last component which takes on positive values, can be written as $\mathcal{P}(v) = \{t(v, 1) \mid t \geq 0\}$ for some $v \in \mathbf{R}^n$. This (projective) correspondence $\mathcal{P}$ between $\mathbf{R}^n$ and the halfspace of rays with positive last component is one-to-one and onto.

The linear-fractional function (2.13) can be expressed as

$$f(x) = \mathcal{P}^{-1}(Q\mathcal{P}(x)).$$

Thus, we start with $x \in \mathbf{dom}\, f$, *i.e.*, $c^T x + d > 0$. We then form the ray $\mathcal{P}(x)$ in $\mathbf{R}^{n+1}$. The linear transformation with matrix Q acts on this ray to produce another ray $Q\mathcal{P}(x)$. Since $x \in \mathbf{dom}\, f$, the last component of this ray assumes positive values. Finally we take the inverse projective transformation to recover $f(x)$.

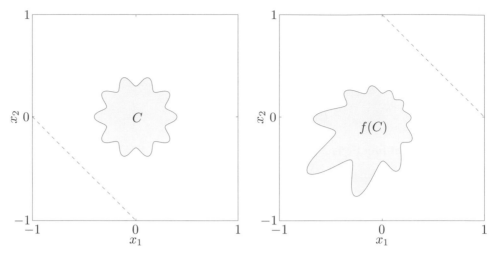

Figure 2.16 *Left.* A set $C \subseteq \mathbf{R}^2$. The dashed line shows the boundary of the domain of the linear-fractional function $f(x) = x/(x_1 + x_2 + 1)$ with **dom** $f = \{(x_1, x_2) \mid x_1 + x_2 + 1 > 0\}$. *Right.* Image of C under f. The dashed line shows the boundary of the domain of f^{-1}.

Like the perspective function, linear-fractional functions preserve convexity. If C is convex and lies in the domain of f (*i.e.*, $c^T x + d > 0$ for $x \in C$), then its image $f(C)$ is convex. This follows immediately from results above: the image of C under the affine mapping (2.12) is convex, and the image of the resulting set under the perspective function P, which yields $f(C)$, is convex. Similarly, if $C \subseteq \mathbf{R}^m$ is convex, then the inverse image $f^{-1}(C)$ is convex.

Example 2.13 *Conditional probabilities.* Suppose u and v are random variables that take on values in $\{1, \dots, n\}$ and $\{1, \dots, m\}$, respectively, and let p_{ij} denote **prob**$(u = i, v = j)$. Then the conditional probability $f_{ij} = $ **prob**$(u = i | v = j)$ is given by

$$f_{ij} = \frac{p_{ij}}{\sum_{k=1}^{n} p_{kj}}.$$

Thus f is obtained by a linear-fractional mapping from p.

It follows that if C is a convex set of joint probabilities for (u, v), then the associated set of conditional probabilities of u given v is also convex.

Figure 2.16 shows a set $C \subseteq \mathbf{R}^2$, and its image under the linear-fractional function

$$f(x) = \frac{1}{x_1 + x_2 + 1} x, \qquad \mathbf{dom}\, f = \{(x_1, x_2) \mid x_1 + x_2 + 1 > 0\}.$$

2.4   Generalized inequalities

2.4.1   Proper cones and generalized inequalities

A cone $K \subseteq \mathbf{R}^n$ is called a *proper cone* if it satisfies the following:

- K is convex.

- K is closed.

- K is *solid*, which means it has nonempty interior.

- K is *pointed*, which means that it contains no line (or equivalently, $x \in K$, $-x \in K \implies x = 0$).

A proper cone K can be used to define a *generalized inequality*, which is a partial ordering on $\mathbf{R}^n$ that has many of the properties of the standard ordering on $\mathbf{R}$. We associate with the proper cone K the partial ordering on $\mathbf{R}^n$ defined by

$$x \preceq_K y \iff y - x \in K.$$

We also write $x \succeq_K y$ for $y \preceq_K x$. Similarly, we define an associated strict partial ordering by

$$x \prec_K y \iff y - x \in \mathbf{int}\, K,$$

and write $x \succ_K y$ for $y \prec_K x$. (To distinguish the generalized inequality $\preceq_K$ from the strict generalized inequality, we sometimes refer to $\preceq_K$ as the nonstrict generalized inequality.)

When $K = \mathbf{R}_+$, the partial ordering $\preceq_K$ is the usual ordering $\leq$ on $\mathbf{R}$, and the strict partial ordering $\prec_K$ is the same as the usual strict ordering $<$ on $\mathbf{R}$. So generalized inequalities include as a special case ordinary (nonstrict and strict) inequality in $\mathbf{R}$.

Example 2.14 *Nonnegative orthant and componentwise inequality.* The nonnegative orthant $K = \mathbf{R}_+^n$ is a proper cone. The associated generalized inequality $\preceq_K$ corresponds to componentwise inequality between vectors: $x \preceq_K y$ means that $x_i \leq y_i$, $i = 1, \ldots, n$. The associated strict inequality corresponds to componentwise strict inequality: $x \prec_K y$ means that $x_i < y_i$, $i = 1, \ldots, n$.

The nonstrict and strict partial orderings associated with the nonnegative orthant arise so frequently that we drop the subscript $\mathbf{R}_+^n$; it is understood when the symbol $\preceq$ or $\prec$ appears between vectors.

Example 2.15 *Positive semidefinite cone and matrix inequality.* The positive semidefinite cone $\mathbf{S}_+^n$ is a proper cone in $\mathbf{S}^n$. The associated generalized inequality $\preceq_K$ is the usual matrix inequality: $X \preceq_K Y$ means $Y - X$ is positive semidefinite. The interior of $\mathbf{S}_+^n$ (in $\mathbf{S}^n$) consists of the positive definite matrices, so the strict generalized inequality also agrees with the usual strict inequality between symmetric matrices: $X \prec_K Y$ means $Y - X$ is positive definite.

Here, too, the partial ordering arises so frequently that we drop the subscript: for symmetric matrices we write simply $X \preceq Y$ or $X \prec Y$. It is understood that the generalized inequalities are with respect to the positive semidefinite cone.

Example 2.16 *Cone of polynomials nonnegative on* $[0, 1]$. Let K be defined as

$$K = \{c \in \mathbf{R}^n \mid c_1 + c_2 t + \cdots + c_n t^{n-1} \geq 0 \text{ for } t \in [0, 1]\}, \qquad (2.15)$$

i.e., K is the cone of (coefficients of) polynomials of degree $n-1$ that are nonnegative on the interval $[0, 1]$. It can be shown that K is a proper cone; its interior is the set of coefficients of polynomials that are positive on the interval $[0, 1]$.

Two vectors c, $d \in \mathbf{R}^n$ satisfy $c \preceq_K d$ if and only if

$$c_1 + c_2 t + \cdots + c_n t^{n-1} \leq d_1 + d_2 t + \cdots + d_n t^{n-1}$$

for all $t \in [0, 1]$.

Properties of generalized inequalities

A generalized inequality $\preceq_K$ satisfies many properties, such as

- $\preceq_K$ *is preserved under addition*: if $x \preceq_K y$ and $u \preceq_K v$, then $x + u \preceq_K y + v$.

- $\preceq_K$ *is transitive*: if $x \preceq_K y$ and $y \preceq_K z$ then $x \preceq_K z$.

- $\preceq_K$ *is preserved under nonnegative scaling*: if $x \preceq_K y$ and $\alpha \geq 0$ then $\alpha x \preceq_K \alpha y$.

- $\preceq_K$ *is reflexive*: $x \preceq_K x$.

- $\preceq_K$ *is antisymmetric*: if $x \preceq_K y$ and $y \preceq_K x$, then $x = y$.

- $\preceq_K$ *is preserved under limits*: if $x_i \preceq_K y_i$ for $i = 1, 2, \ldots$, $x_i \to x$ and $y_i \to y$ as $i \to \infty$, then $x \preceq_K y$.

The corresponding strict generalized inequality $\prec_K$ satisfies, for example,

- if $x \prec_K y$ then $x \preceq_K y$.

- if $x \prec_K y$ and $u \preceq_K v$ then $x + u \prec_K y + v$.

- if $x \prec_K y$ and $\alpha > 0$ then $\alpha x \prec_K \alpha y$.

- $x \not\prec_K x$.

- if $x \prec_K y$, then for u and v small enough, $x + u \prec_K y + v$.

These properties are inherited from the definitions of $\preceq_K$ and $\prec_K$, and the properties of proper cones; see exercise 2.30.

2.4.2   Minimum and minimal elements

The notation of generalized inequality (*i.e.*, $\preceq_K$, $\prec_K$) is meant to suggest the analogy to ordinary inequality on $\mathbf{R}$ (*i.e.*, $\leq$, $<$). While many properties of ordinary inequality do hold for generalized inequalities, some important ones do not. The most obvious difference is that $\leq$ on $\mathbf{R}$ is a *linear ordering*: any two points are *comparable*, meaning either $x \leq y$ or $y \leq x$. This property does not hold for other generalized inequalities. One implication is that concepts like minimum and maximum are more complicated in the context of generalized inequalities. We briefly discuss this in this section.

We say that $x \in S$ is the *minimum* element of S (with respect to the generalized inequality $\preceq_K$) if for every $y \in S$ we have $x \preceq_K y$. We define the *maximum* element of a set S, with respect to a generalized inequality, in a similar way. If a set has a minimum (maximum) element, then it is unique. A related concept is *minimal element*. We say that $x \in S$ is a *minimal* element of S (with respect to the generalized inequality $\preceq_K$) if $y \in S$, $y \preceq_K x$ only if $y = x$. We define *maximal* element in a similar way. A set can have many different minimal (maximal) elements.

We can describe minimum and minimal elements using simple set notation. A point $x \in S$ is the minimum element of S if and only if

$$S \subseteq x + K.$$

Here $x + K$ denotes all the points that are comparable to x and greater than or equal to x (according to $\preceq_K$). A point $x \in S$ is a minimal element if and only if

$$(x - K) \cap S = \{x\}.$$

Here $x - K$ denotes all the points that are comparable to x and less than or equal to x (according to $\preceq_K$); the only point in common with S is x.

For $K = \mathbf{R}_+$, which induces the usual ordering on $\mathbf{R}$, the concepts of minimal and minimum are the same, and agree with the usual definition of the minimum element of a set.

Example 2.17 Consider the cone $\mathbf{R}_+^2$, which induces componentwise inequality in $\mathbf{R}^2$. Here we can give some simple geometric descriptions of minimal and minimum elements. The inequality $x \preceq y$ means y is above and to the right of x. To say that $x \in S$ is the minimum element of a set S means that all other points of S lie above and to the right. To say that x is a minimal element of a set S means that no other point of S lies to the left and below x. This is illustrated in figure 2.17.

Example 2.18 *Minimum and minimal elements of a set of symmetric matrices.* We associate with each $A \in \mathbf{S}_{++}^n$ an ellipsoid centered at the origin, given by

$$\mathcal{E}_A = \{x \mid x^T A^{-1} x \leq 1\}.$$

We have $A \preceq B$ if and only if $\mathcal{E}_A \subseteq \mathcal{E}_B$.

Let $v_1, \ldots, v_k \in \mathbf{R}^n$ be given and define

$$S = \{P \in \mathbf{S}_{++}^n \mid v_i^T P^{-1} v_i \leq 1, \; i = 1, \ldots, k\},$$

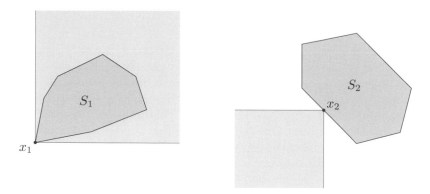

Figure 2.17 *Left.* The set S_1 has a minimum element x_1 with respect to componentwise inequality in $\mathbf{R}^2$. The set $x_1 + K$ is shaded lightly; x_1 is the minimum element of S_1 since $S_1 \subseteq x_1 + K$. *Right.* The point x_2 is a minimal point of S_2. The set $x_2 - K$ is shown lightly shaded. The point x_2 is minimal because $x_2 - K$ and S_2 intersect only at x_2.

which corresponds to the set of ellipsoids that contain the points $v_1, \ldots, v_k$. The set S does not have a minimum element: for any ellipsoid that contains the points $v_1, \ldots, v_k$ we can find another one that contains the points, and is not comparable to it. An ellipsoid is minimal if it contains the points, but no smaller ellipsoid does. Figure 2.18 shows an example in $\mathbf{R}^2$ with $k = 2$.

2.5   Separating and supporting hyperplanes

2.5.1   Separating hyperplane theorem

In this section we describe an idea that will be important later: the use of hyperplanes or affine functions to separate convex sets that do not intersect. The basic result is the *separating hyperplane theorem*: Suppose C and D are two convex sets that do not intersect, *i.e.*, $C \cap D = \emptyset$. Then there exist $a \neq 0$ and b such that $a^T x \leq b$ for all $x \in C$ and $a^T x \geq b$ for all $x \in D$. In other words, the affine function $a^T x - b$ is nonpositive on C and nonnegative on D. The hyperplane $\{x \mid a^T x = b\}$ is called a *separating hyperplane* for the sets C and D, or is said to *separate* the sets C and D. This is illustrated in figure 2.19.

Proof of separating hyperplane theorem

Here we consider a special case, and leave the extension of the proof to the general case as an exercise (exercise 2.22). We assume that the (Euclidean) *distance* between C and D, defined as

$$\mathbf{dist}(C, D) = \inf\{\|u - v\|_2 \mid u \in C, \ v \in D\},$$

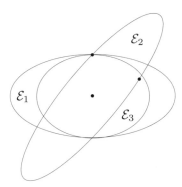

Figure 2.18 Three ellipsoids in $\mathbf{R}^2$, centered at the origin (shown as the lower dot), that contain the points shown as the upper dots. The ellipsoid $\mathcal{E}_1$ is not minimal, since there exist ellipsoids that contain the points, and are smaller (*e.g.*, $\mathcal{E}_3$). $\mathcal{E}_3$ is not minimal for the same reason. The ellipsoid $\mathcal{E}_2$ is minimal, since no other ellipsoid (centered at the origin) contains the points and is contained in $\mathcal{E}_2$.

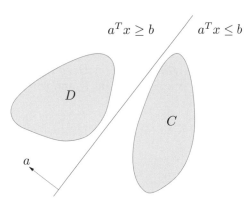

Figure 2.19 The hyperplane $\{x \mid a^T x = b\}$ separates the disjoint convex sets C and D. The affine function $a^T x - b$ is nonpositive on C and nonnegative on D.

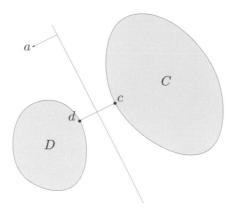

Figure 2.20 Construction of a separating hyperplane between two convex sets. The points $c \in C$ and $d \in D$ are the pair of points in the two sets that are closest to each other. The separating hyperplane is orthogonal to, and bisects, the line segment between c and d.

is positive, and that there exist points $c \in C$ and $d \in D$ that achieve the minimum distance, $i.e.$, $\|c - d\|_2 = \mathbf{dist}(C, D)$. (These conditions are satisfied, for example, when C and D are closed and one set is bounded.)

Define

$$a = d - c, \qquad b = \frac{\|d\|_2^2 - \|c\|_2^2}{2}.$$

We will show that the affine function

$$f(x) = a^T x - b = (d - c)^T (x - (1/2)(d + c))$$

is nonpositive on C and nonnegative on D, $i.e.$, that the hyperplane $\{x \mid a^T x = b\}$ separates C and D. This hyperplane is perpendicular to the line segment between c and d, and passes through its midpoint, as shown in figure 2.20.

We first show that f is nonnegative on D. The proof that f is nonpositive on C is similar (or follows by swapping C and D and considering $-f$). Suppose there were a point $u \in D$ for which

$$f(u) = (d - c)^T (u - (1/2)(d + c)) < 0. \tag{2.16}$$

We can express $f(u)$ as

$$f(u) = (d - c)^T (u - d + (1/2)(d - c)) = (d - c)^T (u - d) + (1/2)\|d - c\|_2^2.$$

We see that (2.16) implies $(d - c)^T (u - d) < 0$. Now we observe that

$$\frac{d}{dt} \|d + t(u - d) - c\|_2^2 \bigg|_{t=0} = 2(d - c)^T (u - d) < 0,$$

so for some small $t > 0$, with $t \le 1$, we have

$$\|d + t(u - d) - c\|_2 < \|d - c\|_2,$$

i.e., the point $d + t(u - d)$ is closer to c than d is. Since D is convex and contains d and u, we have $d + t(u - d) \in D$. But this is impossible, since d is assumed to be the point in D that is closest to C.

Example 2.19 *Separation of an affine and a convex set.* Suppose C is convex and D is affine, *i.e.*, $D = \{Fu + g \mid u \in \mathbf{R}^m\}$, where $F \in \mathbf{R}^{n \times m}$. Suppose C and D are disjoint, so by the separating hyperplane theorem there are $a \neq 0$ and b such that $a^T x \leq b$ for all $x \in C$ and $a^T x \geq b$ for all $x \in D$.

Now $a^T x \geq b$ for all $x \in D$ means $a^T F u \geq b - a^T g$ for all $u \in \mathbf{R}^m$. But a linear function is bounded below on $\mathbf{R}^m$ only when it is zero, so we conclude $a^T F = 0$ (and hence, $b \leq a^T g$).

Thus we conclude that there exists $a \neq 0$ such that $F^T a = 0$ and $a^T x \leq a^T g$ for all $x \in C$.

Strict separation

The separating hyperplane we constructed above satisfies the stronger condition that $a^T x < b$ for all $x \in C$ and $a^T x > b$ for all $x \in D$. This is called *strict separation* of the sets C and D. Simple examples show that in general, disjoint convex sets need not be strictly separable by a hyperplane (even when the sets are closed; see exercise 2.23). In many special cases, however, strict separation can be established.

Example 2.20 *Strict separation of a point and a closed convex set.* Let C be a closed convex set and $x_0 \notin C$. Then there exists a hyperplane that strictly separates x_0 from C.

To see this, note that the two sets C and $B(x_0, \epsilon)$ do not intersect for some $\epsilon > 0$. By the separating hyperplane theorem, there exist $a \neq 0$ and b such that $a^T x \leq b$ for $x \in C$ and $a^T x \geq b$ for $x \in B(x_0, \epsilon)$.

Using $B(x_0, \epsilon) = \{x_0 + u \mid \|u\|_2 \leq \epsilon\}$, the second condition can be expressed as

$$a^T(x_0 + u) \geq b \quad \text{for all} \quad \|u\|_2 \leq \epsilon.$$

The u that minimizes the lefthand side is $u = -\epsilon a / \|a\|_2$; using this value we have

$$a^T x_0 - \epsilon \|a\|_2 \geq b.$$

Therefore the affine function

$$f(x) = a^T x - b - \epsilon \|a\|_2 / 2$$

is negative on C and positive at x_0.

As an immediate consequence we can establish a fact that we already mentioned above: a closed convex set is the intersection of all halfspaces that contain it. Indeed, let C be closed and convex, and let S be the intersection of all halfspaces containing C. Obviously $x \in C \Rightarrow x \in S$. To show the converse, suppose there exists $x \in S$, $x \notin C$. By the strict separation result there exists a hyperplane that strictly separates x from C, *i.e.*, there is a halfspace containing C but not x. In other words, $x \notin S$.

Converse separating hyperplane theorems

The converse of the separating hyperplane theorem (*i.e.*, existence of a separating hyperplane implies that C and D do not intersect) is not true, unless one imposes additional constraints on C or D, even beyond convexity. As a simple counterexample, consider $C = D = \{0\} \subseteq \mathbf{R}$. Here the hyperplane $x = 0$ separates C and D.

By adding conditions on C and D various converse separation theorems can be derived. As a very simple example, suppose C and D are convex sets, with C open, and there exists an affine function f that is nonpositive on C and nonnegative on D. Then C and D are disjoint. (To see this we first note that f must be negative on C; for if f were zero at a point of C then f would take on positive values near the point, which is a contradiction. But then C and D must be disjoint since f is negative on C and nonnegative on D.) Putting this converse together with the separating hyperplane theorem, we have the following result: any two convex sets C and D, at least one of which is open, are disjoint if and only if there exists a separating hyperplane.

Example 2.21 *Theorem of alternatives for strict linear inequalities.* We derive the necessary and sufficient conditions for solvability of a system of strict linear inequalities

$$Ax \prec b. \tag{2.17}$$

These inequalities are infeasible if and only if the (convex) sets

$$C = \{b - Ax \mid x \in \mathbf{R}^n\}, \qquad D = \mathbf{R}^m_{++} = \{y \in \mathbf{R}^m \mid y \succ 0\}$$

do not intersect. The set D is open; C is an affine set. Hence by the result above, C and D are disjoint if and only if there exists a separating hyperplane, *i.e.*, a nonzero $\lambda \in \mathbf{R}^m$ and $\mu \in \mathbf{R}$ such that $\lambda^T y \le \mu$ on C and $\lambda^T y \ge \mu$ on D.

Each of these conditions can be simplified. The first means $\lambda^T(b - Ax) \le \mu$ for all x. This implies (as in example 2.19) that $A^T\lambda = 0$ and $\lambda^T b \le \mu$. The second inequality means $\lambda^T y \ge \mu$ for all $y \succ 0$. This implies $\mu \le 0$ and $\lambda \succeq 0$, $\lambda \ne 0$.

Putting it all together, we find that the set of strict inequalities (2.17) is infeasible if and only if there exists $\lambda \in \mathbf{R}^m$ such that

$$\lambda \ne 0, \qquad \lambda \succeq 0, \qquad A^T\lambda = 0, \qquad \lambda^T b \le 0. \tag{2.18}$$

This is also a system of linear inequalities and linear equations in the variable $\lambda \in \mathbf{R}^m$. We say that (2.17) and (2.18) form a pair of *alternatives*: for any data A and b, exactly one of them is solvable.

2.5.2 Supporting hyperplanes

Suppose $C \subseteq \mathbf{R}^n$, and x_0 is a point in its boundary $\mathbf{bd}\, C$, *i.e.*,

$$x_0 \in \mathbf{bd}\, C = \mathbf{cl}\, C \setminus \mathbf{int}\, C.$$

If $a \ne 0$ satisfies $a^T x \le a^T x_0$ for all $x \in C$, then the hyperplane $\{x \mid a^T x = a^T x_0\}$ is called a *supporting hyperplane* to C at the point x_0. This is equivalent to saying

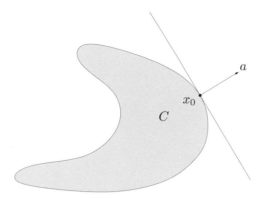

Figure 2.21 The hyperplane $\{x \mid a^T x = a^T x_0\}$ supports C at x_0.

that the point x_0 and the set C are separated by the hyperplane $\{x \mid a^T x = a^T x_0\}$. The geometric interpretation is that the hyperplane $\{x \mid a^T x = a^T x_0\}$ is tangent to C at x_0, and the halfspace $\{x \mid a^T x \le a^T x_0\}$ contains C. This is illustrated in figure 2.21.

A basic result, called the *supporting hyperplane theorem*, states that for any nonempty convex set C, and any $x_0 \in \mathbf{bd}\, C$, there exists a supporting hyperplane to C at x_0. The supporting hyperplane theorem is readily proved from the separating hyperplane theorem. We distinguish two cases. If the interior of C is nonempty, the result follows immediately by applying the separating hyperplane theorem to the sets $\{x_0\}$ and $\mathbf{int}\, C$. If the interior of C is empty, then C must lie in an affine set of dimension less than n, and any hyperplane containing that affine set contains C and x_0, and is a (trivial) supporting hyperplane.

There is also a partial converse of the supporting hyperplane theorem: If a set is closed, has nonempty interior, and has a supporting hyperplane at every point in its boundary, then it is convex. (See exercise 2.27.)

2.6   Dual cones and generalized inequalities

2.6.1   Dual cones

Let K be a cone. The set

$$K^* = \{y \mid x^T y \ge 0 \text{ for all } x \in K\} \tag{2.19}$$

is called the *dual cone* of K. As the name suggests, K^* is a cone, and is always convex, even when the original cone K is not (see exercise 2.31).

Geometrically, $y \in K^*$ if and only if $-y$ is the normal of a hyperplane that supports K at the origin. This is illustrated in figure 2.22.

Example 2.22 *Subspace.* The dual cone of a subspace $V \subseteq \mathbf{R}^n$ (which is a cone) is its orthogonal complement $V^\perp = \{y \mid y^T v = 0 \text{ for all } v \in V\}$.

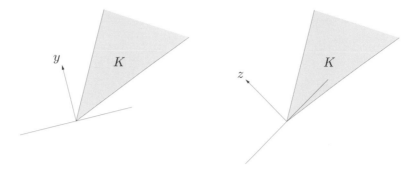

Figure 2.22 *Left.* The halfspace with inward normal y contains the cone K, so $y \in K^*$. *Right.* The halfspace with inward normal z does not contain K, so $z \notin K^*$.

Example 2.23 *Nonnegative orthant.* The cone $\mathbf{R}_+^n$ is its own dual:

$$y^T x \geq 0 \text{ for all } x \succeq 0 \iff y \succeq 0.$$

We call such a cone *self-dual*.

Example 2.24 *Positive semidefinite cone.* On the set of symmetric $n \times n$ matrices $\mathbf{S}^n$, we use the standard inner product $\mathbf{tr}(XY) = \sum_{i,j=1}^n X_{ij} Y_{ij}$ (see §A.1.1). The positive semidefinite cone $\mathbf{S}_+^n$ is self-dual, *i.e.*, for X, $Y \in \mathbf{S}^n$,

$$\mathbf{tr}(XY) \geq 0 \text{ for all } X \succeq 0 \iff Y \succeq 0.$$

We will establish this fact.

Suppose $Y \notin \mathbf{S}_+^n$. Then there exists $q \in \mathbf{R}^n$ with

$$q^T Y q = \mathbf{tr}(qq^T Y) < 0.$$

Hence the positive semidefinite matrix $X = qq^T$ satisfies $\mathbf{tr}(XY) < 0$; it follows that $Y \notin (\mathbf{S}_+^n)^*$.

Now suppose X, $Y \in \mathbf{S}_+^n$. We can express X in terms of its eigenvalue decomposition as $X = \sum_{i=1}^n \lambda_i q_i q_i^T$, where (the eigenvalues) $\lambda_i \geq 0$, $i = 1, \dots, n$. Then we have

$$\mathbf{tr}(YX) = \mathbf{tr}\left(Y \sum_{i=1}^n \lambda_i q_i q_i^T\right) = \sum_{i=1}^n \lambda_i q_i^T Y q_i \geq 0.$$

This shows that $Y \in (\mathbf{S}_+^n)^*$.

Example 2.25 *Dual of a norm cone.* Let $\|\cdot\|$ be a norm on $\mathbf{R}^n$. The dual of the associated cone $K = \{(x, t) \in \mathbf{R}^{n+1} \mid \|x\| \leq t\}$ is the cone defined by the dual norm, *i.e.*,

$$K^* = \{(u, v) \in \mathbf{R}^{n+1} \mid \|u\|_* \leq v\},$$

where the dual norm is given by $\|u\|_* = \sup\{u^T x \mid \|x\| \le 1\}$ (see (A.1.6)).

To prove the result we have to show that

$$x^T u + tv \ge 0 \text{ whenever } \|x\| \le t \iff \|u\|_* \le v. \qquad (2.20)$$

Let us start by showing that the righthand condition on (u, v) implies the lefthand condition. Suppose $\|u\|_* \le v$, and $\|x\| \le t$ for some $t > 0$. (If $t = 0$, x must be zero, so obviously $u^T x + vt \ge 0$.) Applying the definition of the dual norm, and the fact that $\|-x/t\| \le 1$, we have

$$u^T(-x/t) \le \|u\|_* \le v,$$

and therefore $u^T x + vt \ge 0$.

Next we show that the lefthand condition in (2.20) implies the righthand condition in (2.20). Suppose $\|u\|_* > v$, $i.e.$, that the righthand condition does not hold. Then by the definition of the dual norm, there exists an x with $\|x\| \le 1$ and $x^T u > v$. Taking $t = 1$, we have

$$u^T(-x) + v < 0,$$

which contradicts the lefthand condition in (2.20).

Dual cones satisfy several properties, such as:

- K^* is closed and convex.

- $K_1 \subseteq K_2$ implies $K_2^* \subseteq K_1^*$.

- If K has nonempty interior, then K^* is pointed.

- If the closure of K is pointed then K^* has nonempty interior.

- K^{**} is the closure of the convex hull of K. (Hence if K is convex and closed, $K^{**} = K$.)

(See exercise 2.31.) These properties show that if K is a proper cone, then so is its dual K^*, and moreover, that $K^{**} = K$.

2.6.2   Dual generalized inequalities

Now suppose that the convex cone K is proper, so it induces a generalized inequality $\preceq_K$. Then its dual cone K^* is also proper, and therefore induces a generalized inequality. We refer to the generalized inequality $\preceq_{K^*}$ as the *dual* of the generalized inequality $\preceq_K$.

Some important properties relating a generalized inequality and its dual are:

- $x \preceq_K y$ if and only if $\lambda^T x \le \lambda^T y$ for all $\lambda \succeq_{K^*} 0$.

- $x \prec_K y$ if and only if $\lambda^T x < \lambda^T y$ for all $\lambda \succeq_{K^*} 0$, $\lambda \ne 0$.

Since $K = K^{**}$, the dual generalized inequality associated with $\preceq_{K^*}$ is $\preceq_K$, so these properties hold if the generalized inequality and its dual are swapped. As a specific example, we have $\lambda \preceq_{K^*} \mu$ if and only if $\lambda^T x \le \mu^T x$ for all $x \succeq_K 0$.

Example 2.26 *Theorem of alternatives for linear strict generalized inequalities.* Suppose $K \subseteq \mathbf{R}^m$ is a proper cone. Consider the strict generalized inequality

$$Ax \prec_K b, \tag{2.21}$$

where $x \in \mathbf{R}^n$.

We will derive a theorem of alternatives for this inequality. Suppose it is infeasible, *i.e.*, the affine set $\{b - Ax \mid x \in \mathbf{R}^n\}$ does not intersect the open convex set $\mathbf{int}\, K$. Then there is a separating hyperplane, *i.e.*, a nonzero $\lambda \in \mathbf{R}^m$ and $\mu \in \mathbf{R}$ such that $\lambda^T (b - Ax) \leq \mu$ for all x, and $\lambda^T y \geq \mu$ for all $y \in \mathbf{int}\, K$. The first condition implies $A^T \lambda = 0$ and $\lambda^T b \leq \mu$. The second condition implies $\lambda^T y \geq \mu$ for all $y \in K$, which can only happen if $\lambda \in K^*$ and $\mu \leq 0$.

Putting it all together we find that if (2.21) is infeasible, then there exists λ such that

$$\lambda \neq 0, \qquad \lambda \succeq_{K^*} 0, \qquad A^T \lambda = 0, \qquad \lambda^T b \leq 0. \tag{2.22}$$

Now we show the converse: if (2.22) holds, then the inequality system (2.21) cannot be feasible. Suppose that both inequality systems hold. Then we have $\lambda^T (b - Ax) > 0$, since $\lambda \neq 0$, $\lambda \succeq_{K^*} 0$, and $b - Ax \succ_K 0$. But using $A^T \lambda = 0$ we find that $\lambda^T (b - Ax) = \lambda^T b \leq 0$, which is a contradiction.

Thus, the inequality systems (2.21) and (2.22) are alternatives: for any data A, b, exactly one of them is feasible. (This generalizes the alternatives (2.17), (2.18) for the special case $K = \mathbf{R}_+^m$.)

2.6.3   Minimum and minimal elements via dual inequalities

We can use dual generalized inequalities to characterize minimum and minimal elements of a (possibly nonconvex) set $S \subseteq \mathbf{R}^m$ with respect to the generalized inequality induced by a proper cone K.

Dual characterization of minimum element

We first consider a characterization of the *minimum* element: x is the minimum element of S, with respect to the generalized inequality $\preceq_K$, if and only if for all $\lambda \succ_{K^*} 0$, x is the unique minimizer of $\lambda^T z$ over $z \in S$. Geometrically, this means that for any $\lambda \succ_{K^*} 0$, the hyperplane

$$\{z \mid \lambda^T (z - x) = 0\}$$

is a strict supporting hyperplane to S at x. (By strict supporting hyperplane, we mean that the hyperplane intersects S only at the point x.) Note that convexity of the set S is *not* required. This is illustrated in figure 2.23.

To show this result, suppose x is the minimum element of S, *i.e.*, $x \preceq_K z$ for all $z \in S$, and let $\lambda \succ_{K^*} 0$. Let $z \in S$, $z \neq x$. Since x is the minimum element of S, we have $z - x \succeq_K 0$. From $\lambda \succ_{K^*} 0$ and $z - x \succeq_K 0$, $z - x \neq 0$, we conclude $\lambda^T (z - x) > 0$. Since z is an arbitrary element of S, not equal to x, this shows that x is the unique minimizer of $\lambda^T z$ over $z \in S$. Conversely, suppose that for all $\lambda \succ_{K^*} 0$, x is the unique minimizer of $\lambda^T z$ over $z \in S$, but x is not the minimum

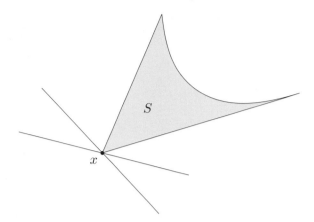

Figure 2.23 Dual characterization of minimum element. The point x is the minimum element of the set S with respect to $\mathbf{R}^2_+$. This is equivalent to: for every $\lambda \succ 0$, the hyperplane $\{z \mid \lambda^T(z - x) = 0\}$ strictly supports S at x, *i.e.*, contains S on one side, and touches it only at x.

element of S. Then there exists $z \in S$ with $z \not\succeq_K x$. Since $z - x \not\succeq_K 0$, there exists $\tilde{\lambda} \succeq_{K^*} 0$ with $\tilde{\lambda}^T(z - x) < 0$. Hence $\lambda^T(z - x) < 0$ for $\lambda \succ_{K^*} 0$ in the neighborhood of $\tilde{\lambda}$. This contradicts the assumption that x is the unique minimizer of $\lambda^T z$ over S.

Dual characterization of minimal elements

We now turn to a similar characterization of *minimal elements*. Here there is a gap between the necessary and sufficient conditions. If $\lambda \succ_{K^*} 0$ and x minimizes $\lambda^T z$ over $z \in S$, then x is minimal. This is illustrated in figure 2.24.

To show this, suppose that $\lambda \succ_{K^*} 0$, and x minimizes $\lambda^T z$ over S, but x is not minimal, *i.e.*, there exists a $z \in S$, $z \neq x$, and $z \preceq_K x$. Then $\lambda^T(x - z) > 0$, which contradicts our assumption that x is the minimizer of $\lambda^T z$ over S.

The converse is in general false: a point x can be minimal in S, but not a minimizer of $\lambda^T z$ over $z \in S$, for any λ, as shown in figure 2.25. This figure suggests that convexity plays an important role in the converse, which is correct. Provided the set S is convex, we can say that for any minimal element x there exists a nonzero $\lambda \succeq_{K^*} 0$ such that x minimizes $\lambda^T z$ over $z \in S$.

To show this, suppose x is minimal, which means that $((x - K) \setminus \{x\}) \cap S = \emptyset$. Applying the separating hyperplane theorem to the convex sets $(x - K) \setminus \{x\}$ and S, we conclude that there is a $\lambda \neq 0$ and μ such that $\lambda^T(x - y) \leq \mu$ for all $y \in K$, and $\lambda^T z \geq \mu$ for all $z \in S$. From the first inequality we conclude $\lambda \succeq_{K^*} 0$. Since $x \in S$ and $x \in x - K$, we have $\lambda^T x = \mu$, so the second inequality implies that μ is the minimum value of $\lambda^T z$ over S. Therefore, x is a minimizer of $\lambda^T z$ over S, where $\lambda \neq 0$, $\lambda \succeq_{K^*} 0$.

This converse theorem cannot be strengthened to $\lambda \succ_{K^*} 0$. Examples show that a point x can be a minimal point of a convex set S, but not a minimizer of

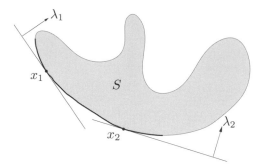

Figure 2.24 A set $S \subseteq \mathbf{R}^2$. Its set of minimal points, with respect to $\mathbf{R}^2_+$, is shown as the darker section of its (lower, left) boundary. The minimizer of $\lambda_1^T z$ over S is x_1, and is minimal since $\lambda_1 \succ 0$. The minimizer of $\lambda_2^T z$ over S is x_2, which is another minimal point of S, since $\lambda_2 \succ 0$.

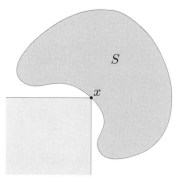

Figure 2.25 The point x is a minimal element of $S \subseteq \mathbf{R}^2$ with respect to $\mathbf{R}^2_+$. However there exists no λ for which x minimizes $\lambda^T z$ over $z \in S$.

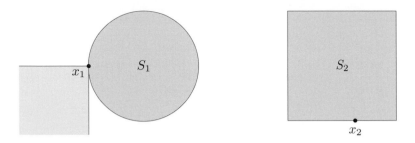

Figure 2.26 *Left.* The point $x_1 \in S_1$ is minimal, but is not a minimizer of $\lambda^T z$ over S_1 for any $\lambda \succ 0$. (It does, however, minimize $\lambda^T z$ over $z \in S_1$ for $\lambda = (1,0)$.) *Right.* The point $x_2 \in S_2$ is *not* minimal, but it does minimize $\lambda^T z$ over $z \in S_2$ for $\lambda = (0,1) \succeq 0$.

$\lambda^T z$ over $z \in S$ for any $\lambda \succ_{K^*} 0$. (See figure 2.26, left.) Nor is it true that any minimizer of $\lambda^T z$ over $z \in S$, with $\lambda \succeq_{K^*} 0$, is minimal (see figure 2.26, right.)

Example 2.27 *Pareto optimal production frontier.* We consider a product which requires n resources (such as labor, electricity, natural gas, water) to manufacture. The product can be manufactured or produced in many ways. With each production method, we associate a *resource vector* $x \in \mathbf{R}^n$, where x_i denotes the amount of resource i consumed by the method to manufacture the product. We assume that $x_i \geq 0$ (*i.e.*, resources are consumed by the production methods) and that the resources are valuable (so using less of any resource is preferred).

The *production set* $P \subseteq \mathbf{R}^n$ is defined as the set of all resource vectors x that correspond to some production method.

Production methods with resource vectors that are minimal elements of P, with respect to componentwise inequality, are called *Pareto optimal* or *efficient*. The set of minimal elements of P is called the *efficient production frontier*.

We can give a simple interpretation of Pareto optimality. We say that one production method, with resource vector x, is *better* than another, with resource vector y, if $x_i \leq y_i$ for all i, and for some i, $x_i < y_i$. In other words, one production method is better than another if it uses no more of each resource than another method, and for at least one resource, actually uses less. This corresponds to $x \preceq y$, $x \neq y$. Then we can say: A production method is Pareto optimal or efficient if there is no better production method.

We can find Pareto optimal production methods (*i.e.*, minimal resource vectors) by minimizing

$$\lambda^T x = \lambda_1 x_1 + \cdots + \lambda_n x_n$$

over the set P of production vectors, using any λ that satisfies $\lambda \succ 0$.

Here the vector λ has a simple interpretation: λ_i is the *price* of resource i. By minimizing $\lambda^T x$ over P we are finding the overall cheapest production method (for the resource prices λ_i). As long as the prices are positive, the resulting production method is guaranteed to be efficient.

These ideas are illustrated in figure 2.27.

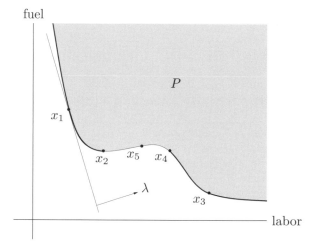

Figure 2.27 The production set P, for a product that requires labor and fuel to produce, is shown shaded. The two dark curves show the efficient production frontier. The points x_1, x_2 and x_3 are efficient. The points x_4 and x_5 are not (since in particular, x_2 corresponds to a production method that uses no more fuel, and less labor). The point x_1 is also the minimum cost production method for the price vector λ (which is positive). The point x_2 is efficient, but cannot be found by minimizing the total cost $\lambda^T x$ for any price vector $\lambda \succeq 0$.

Bibliography

Minkowski is generally credited with the first systematic study of convex sets, and the introduction of fundamental concepts such as supporting hyperplanes and the supporting hyperplane theorem, the Minkowski distance function (exercise 3.34), extreme points of a convex set, and many others.

Some well known early surveys are Bonnesen and Fenchel [BF48], Eggleston [Egg58], Klee [Kle63], and Valentine [Val64]. More recent books devoted to the geometry of convex sets include Lay [Lay82] and Webster [Web94]. Klee [Kle71], Fenchel [Fen83], Tikhomorov [Tik90], and Berger [Ber90] give very readable overviews of the history of convexity and its applications throughout mathematics.

Linear inequalities and polyhedral sets are studied extensively in connection with the linear programming problem, for which we give references at the end of chapter 4. Some landmark publications in the history of linear inequalities and linear programming are Motzkin [Mot33], von Neumann and Morgenstern [vNM53], Kantorovich [Kan60], Koopmans [Koo51], and Dantzig [Dan63]. Dantzig [Dan63, Chapter 2] includes an historical survey of linear inequalities, up to around 1963.

Generalized inequalities were introduced in nonlinear optimization during the 1960s (see Luenberger [Lue69, §8.2] and Isii [Isi64]), and are used extensively in cone programming (see the references in chapter 4). Bellman and Fan [BF63] is an early paper on sets of generalized linear inequalities (with respect to the positive semidefinite cone).

For extensions and a proof of the separating hyperplane theorem we refer the reader to Rockafellar [Roc70, part III], and Hiriart-Urruty and Lemaréchal [HUL93, volume 1, §III4]. Dantzig [Dan63, page 21] attributes the term *theorem of the alternative* to von Neumann and Morgenstern [vNM53, page 138]. For more references on theorems of alternatives, see chapter 5.

The terminology of example 2.27 (including Pareto optimality, efficient production, and the price interpretation of λ) is discussed in detail by Luenberger [Lue95].

Convex geometry plays a prominent role in the classical theory of moments (Krein and Nudelman [KN77], Karlin and Studden [KS66]). A famous example is the duality between the cone of nonnegative polynomials and the cone of power moments; see exercise 2.37.

Exercises

Definition of convexity

2.1 Let $C \subseteq \mathbf{R}^n$ be a convex set, with $x_1, \dots, x_k \in C$, and let $\theta_1, \dots, \theta_k \in \mathbf{R}$ satisfy $\theta_i \geq 0$, $\theta_1 + \cdots + \theta_k = 1$. Show that $\theta_1 x_1 + \cdots + \theta_k x_k \in C$. (The definition of convexity is that this holds for $k = 2$; you must show it for arbitrary k.) *Hint.* Use induction on k.

2.2 Show that a set is convex if and only if its intersection with any line is convex. Show that a set is affine if and only if its intersection with any line is affine.

2.3 *Midpoint convexity.* A set C is *midpoint convex* if whenever two points a, b are in C, the average or midpoint $(a + b)/2$ is in C. Obviously a convex set is midpoint convex. It can be proved that under mild conditions midpoint convexity implies convexity. As a simple case, prove that if C is closed and midpoint convex, then C is convex.

2.4 Show that the convex hull of a set S is the intersection of all convex sets that contain S. (The same method can be used to show that the conic, or affine, or linear hull of a set S is the intersection of all conic sets, or affine sets, or subspaces that contain S.)

Examples

2.5 What is the distance between two parallel hyperplanes $\{x \in \mathbf{R}^n \mid a^T x = b_1\}$ and $\{x \in \mathbf{R}^n \mid a^T x = b_2\}$?

2.6 *When does one halfspace contain another?* Give conditions under which

$$\{x \mid a^T x \leq b\} \subseteq \{x \mid \tilde{a}^T x \leq \tilde{b}\}$$

(where $a \neq 0$, $\tilde{a} \neq 0$). Also find the conditions under which the two halfspaces are equal.

2.7 *Voronoi description of halfspace.* Let a and b be distinct points in $\mathbf{R}^n$. Show that the set of all points that are closer (in Euclidean norm) to a than b, i.e., $\{x \mid \|x - a\|_2 \leq \|x - b\|_2\}$, is a halfspace. Describe it explicitly as an inequality of the form $c^T x \leq d$. Draw a picture.

2.8 Which of the following sets S are polyhedra? If possible, express S in the form $S = \{x \mid Ax \preceq b, \ Fx = g\}$.

   (a) $S = \{y_1 a_1 + y_2 a_2 \mid -1 \leq y_1 \leq 1, \ -1 \leq y_2 \leq 1\}$, where $a_1, a_2 \in \mathbf{R}^n$.

   (b) $S = \{x \in \mathbf{R}^n \mid x \succeq 0, \ \mathbf{1}^T x = 1, \ \sum_{i=1}^n x_i a_i = b_1, \ \sum_{i=1}^n x_i a_i^2 = b_2\}$, where $a_1, \dots, a_n \in \mathbf{R}$ and $b_1, b_2 \in \mathbf{R}$.

   (c) $S = \{x \in \mathbf{R}^n \mid x \succeq 0, \ x^T y \leq 1 \text{ for all } y \text{ with } \|y\|_2 = 1\}$.

   (d) $S = \{x \in \mathbf{R}^n \mid x \succeq 0, \ x^T y \leq 1 \text{ for all } y \text{ with } \sum_{i=1}^n |y_i| = 1\}$.

2.9 *Voronoi sets and polyhedral decomposition.* Let $x_0, \dots, x_K \in \mathbf{R}^n$. Consider the set of points that are closer (in Euclidean norm) to x_0 than the other x_i, i.e.,

$$V = \{x \in \mathbf{R}^n \mid \|x - x_0\|_2 \leq \|x - x_i\|_2, \ i = 1, \dots, K\}.$$

V is called the *Voronoi region* around x_0 with respect to $x_1, \dots, x_K$.

   (a) Show that V is a polyhedron. Express V in the form $V = \{x \mid Ax \preceq b\}$.

   (b) Conversely, given a polyhedron P with nonempty interior, show how to find $x_0, \dots, x_K$ so that the polyhedron is the Voronoi region of x_0 with respect to $x_1, \dots, x_K$.

   (c) We can also consider the sets

$$V_k = \{x \in \mathbf{R}^n \mid \|x - x_k\|_2 \leq \|x - x_i\|_2, \ i \neq k\}.$$

   The set V_k consists of points in $\mathbf{R}^n$ for which the closest point in the set $\{x_0, \dots, x_K\}$ is x_k.

The sets $V_0, \ldots, V_K$ give a polyhedral decomposition of $\mathbf{R}^n$. More precisely, the sets V_k are polyhedra, $\bigcup_{k=0}^{K} V_k = \mathbf{R}^n$, and $\mathbf{int}\, V_i \cap \mathbf{int}\, V_j = \emptyset$ for $i \neq j$, i.e., V_i and V_j intersect at most along a boundary.

Suppose that $P_1, \ldots, P_m$ are polyhedra such that $\bigcup_{i=1}^{m} P_i = \mathbf{R}^n$, and $\mathbf{int}\, P_i \cap \mathbf{int}\, P_j = \emptyset$ for $i \neq j$. Can this polyhedral decomposition of $\mathbf{R}^n$ be described as the Voronoi regions generated by an appropriate set of points?

2.10 *Solution set of a quadratic inequality.* Let $C \subseteq \mathbf{R}^n$ be the solution set of a quadratic inequality,
$$C = \{x \in \mathbf{R}^n \mid x^T A x + b^T x + c \leq 0\},$$
with $A \in \mathbf{S}^n$, $b \in \mathbf{R}^n$, and $c \in \mathbf{R}$.

  (a) Show that C is convex if $A \succeq 0$.

  (b) Show that the intersection of C and the hyperplane defined by $g^T x + h = 0$ (where $g \neq 0$) is convex if $A + \lambda g g^T \succeq 0$ for some $\lambda \in \mathbf{R}$.

Are the converses of these statements true?

2.11 *Hyperbolic sets.* Show that the *hyperbolic set* $\{x \in \mathbf{R}_+^2 \mid x_1 x_2 \geq 1\}$ is convex. As a generalization, show that $\{x \in \mathbf{R}_+^n \mid \prod_{i=1}^{n} x_i \geq 1\}$ is convex. *Hint.* If $a, b \geq 0$ and $0 \leq \theta \leq 1$, then $a^\theta b^{1-\theta} \leq \theta a + (1 - \theta) b$; see §3.1.9.

2.12 Which of the following sets are convex?

  (a) A *slab*, i.e., a set of the form $\{x \in \mathbf{R}^n \mid \alpha \leq a^T x \leq \beta\}$.

  (b) A *rectangle*, i.e., a set of the form $\{x \in \mathbf{R}^n \mid \alpha_i \leq x_i \leq \beta_i, \ i = 1, \ldots, n\}$. A rectangle is sometimes called a *hyperrectangle* when $n > 2$.

  (c) A *wedge*, i.e., $\{x \in \mathbf{R}^n \mid a_1^T x \leq b_1, \ a_2^T x \leq b_2\}$.

  (d) The set of points closer to a given point than a given set, i.e.,
  $$\{x \mid \|x - x_0\|_2 \leq \|x - y\|_2 \text{ for all } y \in S\}$$
  where $S \subseteq \mathbf{R}^n$.

  (e) The set of points closer to one set than another, i.e.,
  $$\{x \mid \mathbf{dist}(x, S) \leq \mathbf{dist}(x, T)\},$$
  where $S, T \subseteq \mathbf{R}^n$, and
  $$\mathbf{dist}(x, S) = \inf\{\|x - z\|_2 \mid z \in S\}.$$

  (f) [HUL93, volume 1, page 93] The set $\{x \mid x + S_2 \subseteq S_1\}$, where $S_1, S_2 \subseteq \mathbf{R}^n$ with S_1 convex.

  (g) The set of points whose distance to a does not exceed a fixed fraction θ of the distance to b, i.e., the set $\{x \mid \|x - a\|_2 \leq \theta \|x - b\|_2\}$. You can assume $a \neq b$ and $0 \leq \theta \leq 1$.

2.13 *Conic hull of outer products.* Consider the set of rank-k *outer products*, defined as $\{XX^T \mid X \in \mathbf{R}^{n \times k}, \ \mathbf{rank}\, X = k\}$. Describe its conic hull in simple terms.

2.14 *Expanded and restricted sets.* Let $S \subseteq \mathbf{R}^n$, and let $\| \cdot \|$ be a norm on $\mathbf{R}^n$.

  (a) For $a \geq 0$ we define S_a as $\{x \mid \mathbf{dist}(x, S) \leq a\}$, where $\mathbf{dist}(x, S) = \inf_{y \in S} \|x - y\|$. We refer to S_a as S *expanded* or *extended* by a. Show that if S is convex, then S_a is convex.

  (b) For $a \geq 0$ we define $S_{-a} = \{x \mid B(x, a) \subseteq S\}$, where $B(x, a)$ is the ball (in the norm $\| \cdot \|$), centered at x, with radius a. We refer to S_{-a} as S *shrunk* or *restricted* by a, since S_{-a} consists of all points that are at least a distance a from $\mathbf{R}^n \backslash S$. Show that if S is convex, then S_{-a} is convex.

2.15 *Some sets of probability distributions.* Let x be a real-valued random variable with $\mathbf{prob}(x = a_i) = p_i$, $i = 1, \ldots, n$, where $a_1 < a_2 < \cdots < a_n$. Of course $p \in \mathbf{R}^n$ lies in the standard probability simplex $P = \{p \mid \mathbf{1}^T p = 1, \ p \succeq 0\}$. Which of the following conditions are convex in p? (That is, for which of the following conditions is the set of $p \in P$ that satisfy the condition convex?)

(a) $\alpha \leq \mathbf{E} f(x) \leq \beta$, where $\mathbf{E} f(x)$ is the expected value of $f(x)$, i.e., $\mathbf{E} f(x) = \sum_{i=1}^{n} p_i f(a_i)$. (The function $f : \mathbf{R} \to \mathbf{R}$ is given.)

(b) $\mathbf{prob}(x > \alpha) \leq \beta$.

(c) $\mathbf{E} |x^3| \leq \alpha \mathbf{E} |x|$.

(d) $\mathbf{E} x^2 \leq \alpha$.

(e) $\mathbf{E} x^2 \geq \alpha$.

(f) $\mathbf{var}(x) \leq \alpha$, where $\mathbf{var}(x) = \mathbf{E}(x - \mathbf{E} x)^2$ is the variance of x.

(g) $\mathbf{var}(x) \geq \alpha$.

(h) $\mathbf{quartile}(x) \geq \alpha$, where $\mathbf{quartile}(x) = \inf\{\beta \mid \mathbf{prob}(x \leq \beta) \geq 0.25\}$.

(i) $\mathbf{quartile}(x) \leq \alpha$.

Operations that preserve convexity

2.16 Show that if S_1 and S_2 are convex sets in $\mathbf{R}^{m \times n}$, then so is their partial sum

$$S = \{(x, y_1 + y_2) \mid x \in \mathbf{R}^m, \ y_1, \ y_2 \in \mathbf{R}^n, (x, y_1) \in S_1, \ (x, y_2) \in S_2\}.$$

2.17 *Image of polyhedral sets under perspective function.* In this problem we study the image of hyperplanes, halfspaces, and polyhedra under the perspective function $P(x, t) = x/t$, with $\mathbf{dom}\, P = \mathbf{R}^n \times \mathbf{R}_{++}$. For each of the following sets C, give a simple description of

$$P(C) = \{v/t \mid (v, t) \in C, \ t > 0\}.$$

(a) The polyhedron $C = \mathbf{conv}\{(v_1, t_1), \ldots, (v_K, t_K)\}$ where $v_i \in \mathbf{R}^n$ and $t_i > 0$.

(b) The hyperplane $C = \{(v, t) \mid f^T v + gt = h\}$ (with f and g not both zero).

(c) The halfspace $C = \{(v, t) \mid f^T v + gt \leq h\}$ (with f and g not both zero).

(d) The polyhedron $C = \{(v, t) \mid Fv + gt \preceq h\}$.

2.18 *Invertible linear-fractional functions.* Let $f : \mathbf{R}^n \to \mathbf{R}^n$ be the linear-fractional function

$$f(x) = (Ax + b)/(c^T x + d), \qquad \mathbf{dom}\, f = \{x \mid c^T x + d > 0\}.$$

Suppose the matrix

$$Q = \begin{bmatrix} A & b \\ c^T & d \end{bmatrix}$$

is nonsingular. Show that f is invertible and that f^{-1} is a linear-fractional mapping. Give an explicit expression for f^{-1} and its domain in terms of A, b, c, and d. *Hint.* It may be easier to express f^{-1} in terms of Q.

2.19 *Linear-fractional functions and convex sets.* Let $f : \mathbf{R}^m \to \mathbf{R}^n$ be the linear-fractional function

$$f(x) = (Ax + b)/(c^T x + d), \qquad \mathbf{dom}\, f = \{x \mid c^T x + d > 0\}.$$

In this problem we study the inverse image of a convex set C under f, i.e.,

$$f^{-1}(C) = \{x \in \mathbf{dom}\, f \mid f(x) \in C\}.$$

For each of the following sets $C \subseteq \mathbf{R}^n$, give a simple description of $f^{-1}(C)$.

(a) The halfspace $C = \{y \mid g^T y \leq h\}$ (with $g \neq 0$).

(b) The polyhedron $C = \{y \mid Gy \preceq h\}$.

(c) The ellipsoid $\{y \mid y^T P^{-1} y \leq 1\}$ (where $P \in \mathbf{S}_{++}^n$).

(d) The solution set of a linear matrix inequality, $C = \{y \mid y_1 A_1 + \cdots + y_n A_n \preceq B\}$, where $A_1, \ldots, A_n, B \in \mathbf{S}^p$.

Separation theorems and supporting hyperplanes

2.20 *Strictly positive solution of linear equations.* Suppose $A \in \mathbf{R}^{m \times n}$, $b \in \mathbf{R}^m$, with $b \in \mathcal{R}(A)$. Show that there exists an x satisfying

$$x \succ 0, \qquad Ax = b$$

if and only if there exists no λ with

$$A^T \lambda \succeq 0, \qquad A^T \lambda \neq 0, \qquad b^T \lambda \leq 0.$$

Hint. First prove the following fact from linear algebra: $c^T x = d$ for all x satisfying $Ax = b$ if and only if there is a vector λ such that $c = A^T \lambda$, $d = b^T \lambda$.

2.21 *The set of separating hyperplanes.* Suppose that C and D are disjoint subsets of $\mathbf{R}^n$. Consider the set of $(a, b) \in \mathbf{R}^{n+1}$ for which $a^T x \leq b$ for all $x \in C$, and $a^T x \geq b$ for all $x \in D$. Show that this set is a convex cone (which is the singleton $\{0\}$ if there is no hyperplane that separates C and D).

2.22 Finish the proof of the separating hyperplane theorem in §2.5.1: Show that a separating hyperplane exists for two disjoint convex sets C and D. You can use the result proved in §2.5.1, *i.e.*, that a separating hyperplane exists when there exist points in the two sets whose distance is equal to the distance between the two sets.

Hint. If C and D are disjoint convex sets, then the set $\{x - y \mid x \in C, \ y \in D\}$ is convex and does not contain the origin.

2.23 Give an example of two closed convex sets that are disjoint but cannot be strictly separated.

2.24 *Supporting hyperplanes.*

(a) Express the closed convex set $\{x \in \mathbf{R}_+^2 \mid x_1 x_2 \geq 1\}$ as an intersection of halfspaces.

(b) Let $C = \{x \in \mathbf{R}^n \mid \|x\|_\infty \leq 1\}$, the ℓ_∞-norm unit ball in $\mathbf{R}^n$, and let $\hat{x}$ be a point in the boundary of C. Identify the supporting hyperplanes of C at $\hat{x}$ explicitly.

2.25 *Inner and outer polyhedral approximations.* Let $C \subseteq \mathbf{R}^n$ be a closed convex set, and suppose that $x_1, \ldots, x_K$ are on the boundary of C. Suppose that for each i, $a_i^T (x - x_i) = 0$ defines a supporting hyperplane for C at x_i, *i.e.*, $C \subseteq \{x \mid a_i^T (x - x_i) \leq 0\}$. Consider the two polyhedra

$$P_{\text{inner}} = \mathbf{conv}\{x_1, \ldots, x_K\}, \qquad P_{\text{outer}} = \{x \mid a_i^T (x - x_i) \leq 0, \ i = 1, \ldots, K\}.$$

Show that $P_{\text{inner}} \subseteq C \subseteq P_{\text{outer}}$. Draw a picture illustrating this.

2.26 *Support function.* The support function of a set $C \subseteq \mathbf{R}^n$ is defined as

$$S_C(y) = \sup\{y^T x \mid x \in C\}.$$

(We allow $S_C(y)$ to take on the value $+\infty$.) Suppose that C and D are closed convex sets in $\mathbf{R}^n$. Show that $C = D$ if and only if their support functions are equal.

2.27 *Converse supporting hyperplane theorem.* Suppose the set C is closed, has nonempty interior, and has a supporting hyperplane at every point in its boundary. Show that C is convex.

Convex cones and generalized inequalities

2.28 *Positive semidefinite cone for n = 1, 2, 3.* Give an explicit description of the positive semidefinite cone $\mathbf{S}^n_+$, in terms of the matrix coefficients and ordinary inequalities, for $n = 1, 2, 3$. To describe a general element of $\mathbf{S}^n$, for $n = 1, 2, 3$, use the notation

$$
x_1, \qquad
\begin{bmatrix} x_1 & x_2 \\ x_2 & x_3 \end{bmatrix}, \qquad
\begin{bmatrix} x_1 & x_2 & x_3 \\ x_2 & x_4 & x_5 \\ x_3 & x_5 & x_6 \end{bmatrix}.
$$

2.29 *Cones in $\mathbf{R}^2$.* Suppose $K \subseteq \mathbf{R}^2$ is a closed convex cone.

    (a) Give a simple description of K in terms of the polar coordinates of its elements $(x = r(\cos\phi, \sin\phi)$ with $r \geq 0)$.

    (b) Give a simple description of K^*, and draw a plot illustrating the relation between K and K^*.

    (c) When is K pointed?

    (d) When is K proper (hence, defines a generalized inequality)? Draw a plot illustrating what $x \preceq_K y$ means when K is proper.

2.30 *Properties of generalized inequalities.* Prove the properties of (nonstrict and strict) generalized inequalities listed in §2.4.1.

2.31 *Properties of dual cones.* Let K^* be the dual cone of a convex cone K, as defined in (2.19). Prove the following.

    (a) K^* is indeed a convex cone.

    (b) $K_1 \subseteq K_2$ implies $K_2^* \subseteq K_1^*$.

    (c) K^* is closed.

    (d) The interior of K^* is given by $\mathbf{int}\, K^* = \{y \mid y^T x > 0 \text{ for all } x \in \mathbf{cl}\, K\}$.

    (e) If K has nonempty interior then K^* is pointed.

    (f) K^{**} is the closure of K. (Hence if K is closed, $K^{**} = K$.)

    (g) If the closure of K is pointed then K^* has nonempty interior.

2.32 Find the dual cone of $\{Ax \mid x \succeq 0\}$, where $A \in \mathbf{R}^{m \times n}$.

2.33 *The monotone nonnegative cone.* We define the *monotone nonnegative cone* as

$$
K_{\mathrm{m}+} = \{x \in \mathbf{R}^n \mid x_1 \geq x_2 \geq \cdots \geq x_n \geq 0\}.
$$

i.e., all nonnegative vectors with components sorted in nonincreasing order.

    (a) Show that $K_{\mathrm{m}+}$ is a proper cone.

    (b) Find the dual cone $K_{\mathrm{m}+}^*$. *Hint.* Use the identity

$$
\sum_{i=1}^n x_i y_i = (x_1 - x_2)y_1 + (x_2 - x_3)(y_1 + y_2) + (x_3 - x_4)(y_1 + y_2 + y_3) + \cdots
$$
$$
+ (x_{n-1} - x_n)(y_1 + \cdots + y_{n-1}) + x_n(y_1 + \cdots + y_n).
$$

2.34 *The lexicographic cone and ordering.* The *lexicographic cone* is defined as

$$
K_{\mathrm{lex}} = \{0\} \cup \{x \in \mathbf{R}^n \mid x_1 = \cdots = x_k = 0, \ x_{k+1} > 0, \text{ for some } k, \ 0 \leq k < n\},
$$

i.e., all vectors whose first nonzero coefficient (if any) is positive.

    (a) Verify that K_{lex} is a cone, but *not* a proper cone.

(b) We define the *lexicographic ordering* on $\mathbf{R}^n$ as follows: $x \preceq_{\text{lex}} y$ if and only if $y - x \in K_{\text{lex}}$. (Since K_{lex} is not a proper cone, the lexicographic ordering is not a generalized inequality.) Show that the lexicographic ordering is a *linear ordering*: for any x, $y \in \mathbf{R}^n$, either $x \preceq_{\text{lex}} y$ or $y \preceq_{\text{lex}} x$. Therefore any set of vectors can be sorted with respect to the lexicographic cone, which yields the familiar sorting used in dictionaries.

(c) Find K_{lex}^*.

2.35 *Copositive matrices.* A matrix $X \in \mathbf{S}^n$ is called *copositive* if $z^T X z \geq 0$ for all $z \succeq 0$. Verify that the set of copositive matrices is a proper cone. Find its dual cone.

2.36 *Euclidean distance matrices.* Let $x_1, \ldots, x_n \in \mathbf{R}^k$. The matrix $D \in \mathbf{S}^n$ defined by $D_{ij} = \|x_i - x_j\|_2^2$ is called a *Euclidean distance matrix*. It satisfies some obvious properties such as $D_{ij} = D_{ji}$, $D_{ii} = 0$, $D_{ij} \geq 0$, and (from the triangle inequality) $D_{ik}^{1/2} \leq D_{ij}^{1/2} + D_{jk}^{1/2}$. We now pose the question: When is a matrix $D \in \mathbf{S}^n$ a Euclidean distance matrix (for some points in $\mathbf{R}^k$, for some k)? A famous result answers this question: $D \in \mathbf{S}^n$ is a Euclidean distance matrix if and only if $D_{ii} = 0$ and $x^T D x \leq 0$ for all x with $\mathbf{1}^T x = 0$. (See §8.3.3.)

Show that the set of Euclidean distance matrices is a convex cone.

2.37 *Nonnegative polynomials and Hankel LMIs.* Let K_{pol} be the set of (coefficients of) nonnegative polynomials of degree $2k$ on $\mathbf{R}$:

$$K_{\text{pol}} = \{x \in \mathbf{R}^{2k+1} \mid x_1 + x_2 t + x_3 t^2 + \cdots + x_{2k+1} t^{2k} \geq 0 \text{ for all } t \in \mathbf{R}\}.$$

(a) Show that K_{pol} is a proper cone.

(b) A basic result states that a polynomial of degree $2k$ is nonnegative on $\mathbf{R}$ if and only if it can be expressed as the sum of squares of two polynomials of degree k or less. In other words, $x \in K_{\text{pol}}$ if and only if the polynomial

$$p(t) = x_1 + x_2 t + x_3 t^2 + \cdots + x_{2k+1} t^{2k}$$

can be expressed as

$$p(t) = r(t)^2 + s(t)^2,$$

where r and s are polynomials of degree k.

Use this result to show that

$$K_{\text{pol}} = \left\{ x \in \mathbf{R}^{2k+1} \;\middle|\; x_i = \sum_{m+n=i+1} Y_{mn} \text{ for some } Y \in \mathbf{S}_+^{k+1} \right\}.$$

In other words, $p(t) = x_1 + x_2 t + x_3 t^2 + \cdots + x_{2k+1} t^{2k}$ is nonnegative if and only if there exists a matrix $Y \in \mathbf{S}_+^{k+1}$ such that

$$
\begin{aligned}
x_1 &= Y_{11} \\
x_2 &= Y_{12} + Y_{21} \\
x_3 &= Y_{13} + Y_{22} + Y_{31} \\
&\vdots \\
x_{2k+1} &= Y_{k+1,k+1}.
\end{aligned}
$$

(c) Show that $K_{\text{pol}}^* = K_{\text{han}}$ where

$$K_{\text{han}} = \{z \in \mathbf{R}^{2k+1} \mid H(z) \succeq 0\}$$

and

$$
H(z) = \begin{bmatrix}
z_1 & z_2 & z_3 & \cdots & z_k & z_{k+1} \\
z_2 & z_3 & z_4 & \cdots & z_{k+1} & z_{k+2} \\
z_3 & z_4 & z_5 & \cdots & z_{k+2} & z_{k+4} \\
\vdots & \vdots & \vdots & \ddots & \vdots & \vdots \\
z_k & z_{k+1} & z_{k+2} & \cdots & z_{2k-1} & z_{2k} \\
z_{k+1} & z_{k+2} & z_{k+3} & \cdots & z_{2k} & z_{2k+1}
\end{bmatrix}.
$$

(This is the *Hankel matrix* with coefficients $z_1, \ldots, z_{2k+1}$.)

(d) Let K_{mom} be the conic hull of the set of all vectors of the form $(1, t, t^2, \ldots, t^{2k})$, where $t \in \mathbf{R}$. Show that $y \in K_{\mathrm{mom}}$ if and only if $y_1 \geq 0$ and

$$ y = y_1 (1, \mathbf{E}\, u, \mathbf{E}\, u^2, \ldots, \mathbf{E}\, u^{2k}) $$

for some random variable u. In other words, the elements of K_{mom} are nonnegative multiples of the moment vectors of all possible distributions on $\mathbf{R}$. Show that $K_{\mathrm{pol}} = K_{\mathrm{mom}}^*$.

(e) Combining the results of (c) and (d), conclude that $K_{\mathrm{han}} = \mathbf{cl}\, K_{\mathrm{mom}}$.

As an example illustrating the relation between K_{mom} and K_{han}, take $k = 2$ and $z = (1, 0, 0, 0, 1)$. Show that $z \in K_{\mathrm{han}}$, $z \notin K_{\mathrm{mom}}$. Find an explicit sequence of points in K_{mom} which converge to z.

2.38 [Roc70, pages 15, 61] *Convex cones constructed from sets.*

(a) The *barrier cone* of a set C is defined as the set of all vectors y such that $y^T x$ is bounded above over $x \in C$. In other words, a nonzero vector y is in the barrier cone if and only if it is the normal vector of a halfspace $\{x \mid y^T x \leq \alpha\}$ that contains C. Verify that the barrier cone is a convex cone (with no assumptions on C).

(b) The *recession cone* (also called *asymptotic cone*) of a set C is defined as the set of all vectors y such that for each $x \in C$, $x - ty \in C$ for all $t \geq 0$. Show that the recession cone of a convex set is a convex cone. Show that if C is nonempty, closed, and convex, then the recession cone of C is the dual of the barrier cone.

(c) The *normal cone* of a set C at a boundary point x_0 is the set of all vectors y such that $y^T (x - x_0) \leq 0$ for all $x \in C$ (*i.e.*, the set of vectors that define a supporting hyperplane to C at x_0). Show that the normal cone is a convex cone (with no assumptions on C). Give a simple description of the normal cone of a polyhedron $\{x \mid Ax \preceq b\}$ at a point in its boundary.

2.39 *Separation of cones.* Let K and $\tilde{K}$ be two convex cones whose interiors are nonempty and disjoint. Show that there is a nonzero y such that $y \in K^*$, $-y \in \tilde{K}^*$.

Chapter 3

Convex functions

3.1 Basic properties and examples

3.1.1 Definition

A function $f : \mathbf{R}^n \rightarrow \mathbf{R}$ is *convex* if $\mathbf{dom}\, f$ is a convex set and if for all x, $y \in \mathbf{dom}\, f$, and θ with $0 \le \theta \le 1$, we have

$$f(\theta x + (1 - \theta)y) \le \theta f(x) + (1 - \theta)f(y). \tag{3.1}$$

Geometrically, this inequality means that the line segment between $(x, f(x))$ and $(y, f(y))$, which is the *chord* from x to y, lies above the graph of f (figure 3.1). A function f is *strictly convex* if strict inequality holds in (3.1) whenever $x \ne y$ and $0 < \theta < 1$. We say f is *concave* if $-f$ is convex, and *strictly concave* if $-f$ is strictly convex.

For an affine function we always have equality in (3.1), so all affine (and therefore also linear) functions are both convex and concave. Conversely, any function that is convex and concave is affine.

A function is convex if and only if it is convex when restricted to any line that intersects its domain. In other words f is convex if and only if for all $x \in \mathbf{dom}\, f$ and

Figure 3.1 Graph of a convex function. The chord (*i.e.*, line segment) between any two points on the graph lies above the graph.

all v, the function $g(t) = f(x + tv)$ is convex (on its domain, $\{t \mid x + tv \in \mathbf{dom}\, f\}$). This property is very useful, since it allows us to check whether a function is convex by restricting it to a line.

The *analysis* of convex functions is a well developed field, which we will not pursue in any depth. One simple result, for example, is that a convex function is continuous on the relative interior of its domain; it can have discontinuities only on its relative boundary.

3.1.2 Extended-value extensions

It is often convenient to extend a convex function to all of $\mathbf{R}^n$ by defining its value to be ∞ outside its domain. If f is convex we define its *extended-value extension* $\tilde{f} : \mathbf{R}^n \to \mathbf{R} \cup \{\infty\}$ by

$$\tilde{f}(x) = \begin{cases} f(x) & x \in \mathbf{dom}\, f \\ \infty & x \notin \mathbf{dom}\, f. \end{cases}$$

The extension $\tilde{f}$ is defined on all $\mathbf{R}^n$, and takes values in $\mathbf{R} \cup \{\infty\}$. We can recover the domain of the original function f from the extension $\tilde{f}$ as $\mathbf{dom}\, f = \{x \mid \tilde{f}(x) < \infty\}$.

The extension can simplify notation, since we do not need to explicitly describe the domain, or add the qualifier 'for all $x \in \mathbf{dom}\, f$' every time we refer to $f(x)$. Consider, for example, the basic defining inequality (3.1). In terms of the extension $\tilde{f}$, we can express it as: for $0 < \theta < 1$,

$$\tilde{f}(\theta x + (1 - \theta)y) \le \theta \tilde{f}(x) + (1 - \theta)\tilde{f}(y)$$

for *any* x and y. (For $\theta = 0$ or $\theta = 1$ the inequality always holds.) Of course here we must interpret the inequality using extended arithmetic and ordering. For x and y both in $\mathbf{dom}\, f$, this inequality coincides with (3.1); if either is outside $\mathbf{dom}\, f$, then the righthand side is ∞, and the inequality therefore holds. As another example of this notational device, suppose f_1 and f_2 are two convex functions on $\mathbf{R}^n$. The pointwise sum $f = f_1 + f_2$ is the function with domain $\mathbf{dom}\, f = \mathbf{dom}\, f_1 \cap \mathbf{dom}\, f_2$, with $f(x) = f_1(x) + f_2(x)$ for any $x \in \mathbf{dom}\, f$. Using extended-value extensions we can simply say that for any x, $\tilde{f}(x) = \tilde{f}_1(x) + \tilde{f}_2(x)$. In this equation the domain of f has been automatically defined as $\mathbf{dom}\, f = \mathbf{dom}\, f_1 \cap \mathbf{dom}\, f_2$, since $\tilde{f}(x) = \infty$ whenever $x \notin \mathbf{dom}\, f_1$ or $x \notin \mathbf{dom}\, f_2$. In this example we are relying on extended arithmetic to automatically define the domain.

In this book we will use the same symbol to denote a convex function and its extension, whenever there is no harm from the ambiguity. This is the same as assuming that all convex functions are implicitly extended, *i.e.*, are defined as ∞ outside their domains.

Example 3.1 *Indicator function of a convex set.* Let $C \subseteq \mathbf{R}^n$ be a convex set, and consider the (convex) function I_C with domain C and $I_C(x) = 0$ for all $x \in C$. In other words, the function is identically zero on the set C. Its extended-value extension

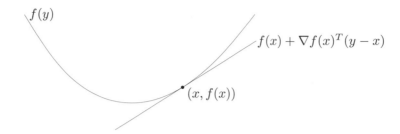

Figure 3.2 If f is convex and differentiable, then $f(x)+\nabla f(x)^T(y-x) \leq f(y)$ for all $x,\ y \in \mathbf{dom}\, f$.

is given by

$$
\tilde{I}_C(x) = \left\{ \begin{array}{ll} 0 & x \in C \\ \infty & x \notin C. \end{array} \right.
$$

The convex function $\tilde{I}_C$ is called the *indicator function* of the set C.

We can play several notational tricks with the indicator function $\tilde{I}_C$. For example the problem of minimizing a function f (defined on all of $\mathbf{R}^n$, say) on the set C is the same as minimizing the function $f + \tilde{I}_C$ over all of $\mathbf{R}^n$. Indeed, the function $f + \tilde{I}_C$ is (by our convention) f restricted to the set C.

In a similar way we can extend a concave function by defining it to be $-\infty$ outside its domain.

3.1.3   First-order conditions

Suppose f is differentiable (*i.e.*, its gradient ∇f exists at each point in $\mathbf{dom}\, f$, which is open). Then f is convex if and only if $\mathbf{dom}\, f$ is convex and

$$
f(y) \geq f(x) + \nabla f(x)^T(y-x) \tag{3.2}
$$

holds for all $x,\ y \in \mathbf{dom}\, f$. This inequality is illustrated in figure 3.2.

The affine function of y given by $f(x)+\nabla f(x)^T(y-x)$ is, of course, the first-order Taylor approximation of f near x. The inequality (3.2) states that for a convex function, the first-order Taylor approximation is in fact a *global underestimator* of the function. Conversely, if the first-order Taylor approximation of a function is always a global underestimator of the function, then the function is convex.

The inequality (3.2) shows that from *local information* about a convex function (*i.e.*, its value and derivative at a point) we can derive *global information* (*i.e.*, a global underestimator of it). This is perhaps the most important property of convex functions, and explains some of the remarkable properties of convex functions and convex optimization problems. As one simple example, the inequality (3.2) shows that if $\nabla f(x) = 0$, then for all $y \in \mathbf{dom}\, f$, $f(y) \geq f(x)$, *i.e.*, x is a global minimizer of the function f.

Strict convexity can also be characterized by a first-order condition: f is strictly convex if and only if $\mathbf{dom}\, f$ is convex and for $x,\ y \in \mathbf{dom}\, f$, $x \neq y$, we have

$$f(y) > f(x) + \nabla f(x)^T (y - x). \tag{3.3}$$

For concave functions we have the corresponding characterization: f is concave if and only if $\mathbf{dom}\, f$ is convex and

$$f(y) \leq f(x) + \nabla f(x)^T (y - x)$$

for all $x,\ y \in \mathbf{dom}\, f$.

Proof of first-order convexity condition

To prove (3.2), we first consider the case $n = 1$: We show that a differentiable function $f : \mathbf{R} \to \mathbf{R}$ is convex if and only if

$$f(y) \geq f(x) + f'(x)(y - x) \tag{3.4}$$

for all x and y in $\mathbf{dom}\, f$.

Assume first that f is convex and $x,\ y \in \mathbf{dom}\, f$. Since $\mathbf{dom}\, f$ is convex (*i.e.*, an interval), we conclude that for all $0 < t \leq 1$, $x + t(y - x) \in \mathbf{dom}\, f$, and by convexity of f,

$$f(x + t(y - x)) \leq (1 - t)f(x) + tf(y).$$

If we divide both sides by t, we obtain

$$f(y) \geq f(x) + \frac{f(x + t(y - x)) - f(x)}{t},$$

and taking the limit as $t \to 0$ yields (3.4).

To show sufficiency, assume the function satisfies (3.4) for all x and y in $\mathbf{dom}\, f$ (which is an interval). Choose any $x \neq y$, and $0 \leq \theta \leq 1$, and let $z = \theta x + (1 - \theta)y$. Applying (3.4) twice yields

$$f(x) \geq f(z) + f'(z)(x - z), \qquad f(y) \geq f(z) + f'(z)(y - z).$$

Multiplying the first inequality by θ, the second by $1 - \theta$, and adding them yields

$$\theta f(x) + (1 - \theta)f(y) \geq f(z),$$

which proves that f is convex.

Now we can prove the general case, with $f : \mathbf{R}^n \to \mathbf{R}$. Let $x,\ y \in \mathbf{R}^n$ and consider f restricted to the line passing through them, *i.e.*, the function defined by $g(t) = f(ty + (1 - t)x)$, so $g'(t) = \nabla f(ty + (1 - t)x)^T (y - x)$.

First assume f is convex, which implies g is convex, so by the argument above we have $g(1) \geq g(0) + g'(0)$, which means

$$f(y) \geq f(x) + \nabla f(x)^T (y - x).$$

Now assume that this inequality holds for any x and y, so if $ty + (1 - t)x \in \mathbf{dom}\, f$ and $\tilde{t}y + (1 - \tilde{t})x \in \mathbf{dom}\, f$, we have

$$f(ty + (1 - t)x) \geq f(\tilde{t}y + (1 - \tilde{t})x) + \nabla f(\tilde{t}y + (1 - \tilde{t})x)^T (y - x)(t - \tilde{t}),$$

i.e., $g(t) \geq g(\tilde{t}) + g'(\tilde{t})(t - \tilde{t})$. We have seen that this implies that g is convex.

3.1.4 Second-order conditions

We now assume that f is twice differentiable, that is, its *Hessian* or second derivative $\nabla^2 f$ exists at each point in $\mathbf{dom}\, f$, which is open. Then f is convex if and only if $\mathbf{dom}\, f$ is convex and its Hessian is positive semidefinite: for all $x \in \mathbf{dom}\, f$,

$$\nabla^2 f(x) \succeq 0.$$

For a function on $\mathbf{R}$, this reduces to the simple condition $f''(x) \geq 0$ (and $\mathbf{dom}\, f$ convex, *i.e.*, an interval), which means that the derivative is nondecreasing. The condition $\nabla^2 f(x) \succeq 0$ can be interpreted geometrically as the requirement that the graph of the function have positive (upward) curvature at x. We leave the proof of the second-order condition as an exercise (exercise 3.8).

Similarly, f is concave if and only if $\mathbf{dom}\, f$ is convex and $\nabla^2 f(x) \preceq 0$ for all $x \in \mathbf{dom}\, f$. Strict convexity can be partially characterized by second-order conditions. If $\nabla^2 f(x) \succ 0$ for all $x \in \mathbf{dom}\, f$, then f is strictly convex. The converse, however, is not true: for example, the function $f : \mathbf{R} \rightarrow \mathbf{R}$ given by $f(x) = x^4$ is strictly convex but has zero second derivative at $x = 0$.

Example 3.2 *Quadratic functions.* Consider the quadratic function $f : \mathbf{R}^n \rightarrow \mathbf{R}$, with $\mathbf{dom}\, f = \mathbf{R}^n$, given by

$$f(x) = (1/2)x^T P x + q^T x + r,$$

with $P \in \mathbf{S}^n$, $q \in \mathbf{R}^n$, and $r \in \mathbf{R}$. Since $\nabla^2 f(x) = P$ for all x, f is convex if and only if $P \succeq 0$ (and concave if and only if $P \preceq 0$).

For quadratic functions, strict convexity is easily characterized: f is strictly convex if and only if $P \succ 0$ (and strictly concave if and only if $P \prec 0$).

Remark 3.1 The separate requirement that $\mathbf{dom}\, f$ be convex cannot be dropped from the first- or second-order characterizations of convexity and concavity. For example, the function $f(x) = 1/x^2$, with $\mathbf{dom}\, f = \{x \in \mathbf{R} \mid x \neq 0\}$, satisfies $f''(x) > 0$ for all $x \in \mathbf{dom}\, f$, but is not a convex function.

3.1.5 Examples

We have already mentioned that all linear and affine functions are convex (and concave), and have described the convex and concave quadratic functions. In this section we give a few more examples of convex and concave functions. We start with some functions on $\mathbf{R}$, with variable x.

- *Exponential.* e^{ax} is convex on $\mathbf{R}$, for any $a \in \mathbf{R}$.

- *Powers.* x^a is convex on $\mathbf{R}_{++}$ when $a \geq 1$ or $a \leq 0$, and concave for $0 \leq a \leq 1$.

- *Powers of absolute value.* $|x|^p$, for $p \geq 1$, is convex on $\mathbf{R}$.

- *Logarithm.* $\log x$ is concave on $\mathbf{R}_{++}$.

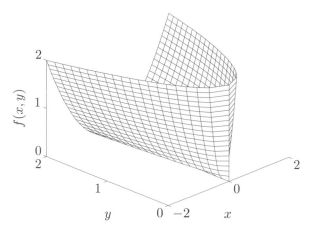

Figure 3.3 Graph of $f(x,y) = x^2/y$.

- *Negative entropy.* $x \log x$ (either on $\mathbf{R}_{++}$, or on $\mathbf{R}_+$, defined as 0 for $x = 0$) is convex.

Convexity or concavity of these examples can be shown by verifying the basic inequality (3.1), or by checking that the second derivative is nonnegative or nonpositive. For example, with $f(x) = x \log x$ we have

$$f'(x) = \log x + 1, \qquad f''(x) = 1/x,$$

so that $f''(x) > 0$ for $x > 0$. This shows that the negative entropy function is (strictly) convex.

We now give a few interesting examples of functions on $\mathbf{R}^n$.

- *Norms.* Every norm on $\mathbf{R}^n$ is convex.

- *Max function.* $f(x) = \max\{x_1, \ldots, x_n\}$ is convex on $\mathbf{R}^n$.

- *Quadratic-over-linear function.* The function $f(x, y) = x^2/y$, with

$$\mathbf{dom}\, f = \mathbf{R} \times \mathbf{R}_{++} = \{(x, y) \in \mathbf{R}^2 \mid y > 0\},$$

  is convex (figure 3.3).

- *Log-sum-exp.* The function $f(x) = \log(e^{x_1} + \cdots + e^{x_n})$ is convex on $\mathbf{R}^n$. This function can be interpreted as a differentiable (in fact, analytic) approximation of the max function, since

$$\max\{x_1, \ldots, x_n\} \le f(x) \le \max\{x_1, \ldots, x_n\} + \log n$$

  for all x. (The second inequality is tight when all components of x are equal.) Figure 3.4 shows f for $n = 2$.

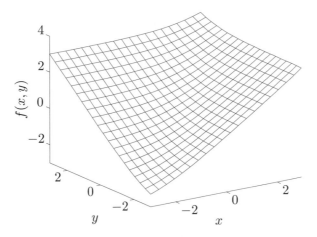

Figure 3.4 Graph of $f(x, y) = \log(e^x + e^y)$.

- *Geometric mean.* The geometric mean $f(x) = \left(\prod_{i=1}^{n} x_i\right)^{1/n}$ is concave on $\mathbf{dom}\, f = \mathbf{R}_{++}^n$.

- *Log-determinant.* The function $f(X) = \log \det X$ is concave on $\mathbf{dom}\, f = \mathbf{S}_{++}^n$.

Convexity (or concavity) of these examples can be verified in several ways, such as directly verifying the inequality (3.1), verifying that the Hessian is positive semidefinite, or restricting the function to an arbitrary line and verifying convexity of the resulting function of one variable.

Norms.   If $f : \mathbf{R}^n \to \mathbf{R}$ is a norm, and $0 \le \theta \le 1$, then

$$f(\theta x + (1 - \theta)y) \le f(\theta x) + f((1 - \theta)y) = \theta f(x) + (1 - \theta)f(y).$$

The inequality follows from the triangle inequality, and the equality follows from homogeneity of a norm.

Max function.   The function $f(x) = \max_i x_i$ satisfies, for $0 \le \theta \le 1$,

$$
\begin{aligned}
f(\theta x + (1 - \theta)y) &= \max_i(\theta x_i + (1 - \theta)y_i) \\
&\le \theta \max_i x_i + (1 - \theta) \max_i y_i \\
&= \theta f(x) + (1 - \theta)f(y).
\end{aligned}
$$

Quadratic-over-linear function.   To show that the quadratic-over-linear function $f(x, y) = x^2/y$ is convex, we note that (for $y > 0$),

$$\nabla^2 f(x, y) = \frac{2}{y^3} \begin{bmatrix} y^2 & -xy \\ -xy & x^2 \end{bmatrix} = \frac{2}{y^3} \begin{bmatrix} y \\ -x \end{bmatrix} \begin{bmatrix} y \\ -x \end{bmatrix}^T \succeq 0.$$

Log-sum-exp.   The Hessian of the log-sum-exp function is

$$\nabla^2 f(x) = \frac{1}{(\mathbf{1}^T z)^2} \left((\mathbf{1}^T z) \, \mathbf{diag}(z) - zz^T \right),$$

where $z = (e^{x_1}, \ldots, e^{x_n})$. To verify that $\nabla^2 f(x) \succeq 0$ we must show that for all v, $v^T \nabla^2 f(x) v \geq 0$, i.e.,

$$v^T \nabla^2 f(x) v = \frac{1}{(\mathbf{1}^T z)^2} \left(\left(\sum_{i=1}^n z_i \right) \left(\sum_{i=1}^n v_i^2 z_i \right) - \left(\sum_{i=1}^n v_i z_i \right)^2 \right) \geq 0.$$

But this follows from the Cauchy-Schwarz inequality $(a^T a)(b^T b) \geq (a^T b)^2$ applied to the vectors with components $a_i = v_i \sqrt{z_i}$, $b_i = \sqrt{z_i}$.

Geometric mean.   In a similar way we can show that the geometric mean $f(x) = \left(\prod_{i=1}^n x_i \right)^{1/n}$ is concave on $\mathbf{dom}\, f = \mathbf{R}_{++}^n$. Its Hessian $\nabla^2 f(x)$ is given by

$$\frac{\partial^2 f(x)}{\partial x_k^2} = -(n-1) \frac{\left(\prod_{i=1}^n x_i \right)^{1/n}}{n^2 x_k^2}, \qquad \frac{\partial^2 f(x)}{\partial x_k \partial x_l} = \frac{\left(\prod_{i=1}^n x_i \right)^{1/n}}{n^2 x_k x_l} \quad \text{for } k \neq l,$$

and can be expressed as

$$\nabla^2 f(x) = -\frac{\prod_{i=1}^n x_i^{1/n}}{n^2} \left(n \, \mathbf{diag}(1/x_1^2, \ldots, 1/x_n^2) - qq^T \right)$$

where $q_i = 1/x_i$. We must show that $\nabla^2 f(x) \preceq 0$, i.e., that

$$v^T \nabla^2 f(x) v = -\frac{\prod_{i=1}^n x_i^{1/n}}{n^2} \left(n \sum_{i=1}^n v_i^2/x_i^2 - \left(\sum_{i=1}^n v_i/x_i \right)^2 \right) \leq 0$$

for all v. Again this follows from the Cauchy-Schwarz inequality $(a^T a)(b^T b) \geq (a^T b)^2$, applied to the vectors $a = \mathbf{1}$ and $b_i = v_i/x_i$.

Log-determinant.   For the function $f(X) = \log \det X$, we can verify concavity by considering an arbitrary line, given by $X = Z + tV$, where Z, $V \in \mathbf{S}^n$. We define $g(t) = f(Z + tV)$, and restrict g to the interval of values of t for which $Z + tV \succ 0$. Without loss of generality, we can assume that $t = 0$ is inside this interval, i.e., $Z \succ 0$. We have

$$
\begin{aligned}
g(t) &= \log \det(Z + tV) \\
&= \log \det(Z^{1/2}(I + tZ^{-1/2}VZ^{-1/2})Z^{1/2}) \\
&= \sum_{i=1}^n \log(1 + t\lambda_i) + \log \det Z
\end{aligned}
$$

where $\lambda_1, \ldots, \lambda_n$ are the eigenvalues of $Z^{-1/2}VZ^{-1/2}$. Therefore we have

$$g'(t) = \sum_{i=1}^n \frac{\lambda_i}{1 + t\lambda_i}, \qquad g''(t) = -\sum_{i=1}^n \frac{\lambda_i^2}{(1 + t\lambda_i)^2}.$$

Since $g''(t) \leq 0$, we conclude that f is concave.

3.1.6   Sublevel sets

The α-*sublevel set* of a function $f : \mathbf{R}^n \to \mathbf{R}$ is defined as

$$C_\alpha = \{x \in \mathbf{dom}\, f \mid f(x) \leq \alpha\}.$$

Sublevel sets of a convex function are convex, for any value of α. The proof is immediate from the definition of convexity: if x, $y \in C_\alpha$, then $f(x) \leq \alpha$ and $f(y) \leq \alpha$, and so $f(\theta x + (1-\theta)y) \leq \alpha$ for $0 \leq \theta \leq 1$, and hence $\theta x + (1-\theta)y \in C_\alpha$.

The converse is not true: a function can have all its sublevel sets convex, but not be a convex function. For example, $f(x) = -e^x$ is not convex on $\mathbf{R}$ (indeed, it is strictly concave) but all its sublevel sets are convex.

If f is concave, then its α-*superlevel set*, given by $\{x \in \mathbf{dom}\, f \mid f(x) \geq \alpha\}$, is a convex set. The sublevel set property is often a good way to establish convexity of a set, by expressing it as a sublevel set of a convex function, or as the superlevel set of a concave function.

Example 3.3 The geometric and arithmetic means of $x \in \mathbf{R}_+^n$ are, respectively,

$$G(x) = \left(\prod_{i=1}^n x_i \right)^{1/n}, \qquad A(x) = \frac{1}{n} \sum_{i=1}^n x_i,$$

(where we take $0^{1/n} = 0$ in our definition of G). The arithmetic-geometric mean inequality states that $G(x) \leq A(x)$.

Suppose $0 \leq \alpha \leq 1$, and consider the set

$$\{x \in \mathbf{R}_+^n \mid G(x) \geq \alpha A(x)\},$$

i.e., the set of vectors with geometric mean at least as large as a factor α times the arithmetic mean. This set is convex, since it is the 0-superlevel set of the function $G(x) - \alpha A(x)$, which is concave. In fact, the set is positively homogeneous, so it is a convex cone.

3.1.7   Epigraph

The graph of a function $f : \mathbf{R}^n \to \mathbf{R}$ is defined as

$$\{(x, f(x)) \mid x \in \mathbf{dom}\, f\},$$

which is a subset of $\mathbf{R}^{n+1}$. The *epigraph* of a function $f : \mathbf{R}^n \to \mathbf{R}$ is defined as

$$\mathbf{epi}\, f = \{(x, t) \mid x \in \mathbf{dom}\, f, \ f(x) \leq t\},$$

which is a subset of $\mathbf{R}^{n+1}$. ('Epi' means 'above' so epigraph means 'above the graph'.) The definition is illustrated in figure 3.5.

The link between convex sets and convex functions is via the epigraph: A function is convex if and only if its epigraph is a convex set. A function is concave if and only if its *hypograph*, defined as

$$\mathbf{hypo}\, f = \{(x, t) \mid t \leq f(x)\},$$

is a convex set.

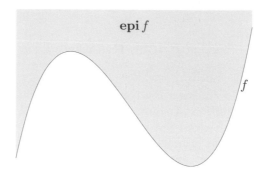

Figure 3.5 Epigraph of a function f, shown shaded. The lower boundary, shown darker, is the graph of f.

Example 3.4 *Matrix fractional function.* The function $f : \mathbf{R}^n \times \mathbf{S}^n \to \mathbf{R}$, defined as

$$f(x, Y) = x^T Y^{-1} x$$

is convex on $\mathbf{dom}\, f = \mathbf{R}^n \times \mathbf{S}^n_{++}$. (This generalizes the quadratic-over-linear function $f(x, y) = x^2/y$, with $\mathbf{dom}\, f = \mathbf{R} \times \mathbf{R}_{++}$.)

One easy way to establish convexity of f is via its epigraph:

$$
\begin{aligned}
\mathbf{epi}\, f &= \{(x, Y, t) \mid Y \succ 0,\ x^T Y^{-1} x \le t\} \\
&= \left\{ (x, Y, t) \ \middle| \ \begin{bmatrix} Y & x \\ x^T & t \end{bmatrix} \succeq 0,\ Y \succ 0 \right\},
\end{aligned}
$$

using the Schur complement condition for positive semidefiniteness of a block matrix (see §A.5.5). The last condition is a linear matrix inequality in (x, Y, t), and therefore **epi** f is convex.

For the special case $n = 1$, the matrix fractional function reduces to the quadratic-over-linear function x^2/y, and the associated LMI representation is

$$\begin{bmatrix} y & x \\ x & t \end{bmatrix} \succeq 0, \qquad y > 0$$

(the graph of which is shown in figure 3.3).

Many results for convex functions can be proved (or interpreted) geometrically using epigraphs, and applying results for convex sets. As an example, consider the first-order condition for convexity:

$$f(y) \ge f(x) + \nabla f(x)^T (y - x),$$

where f is convex and $x,\ y \in \mathbf{dom}\, f$. We can interpret this basic inequality geometrically in terms of **epi** f. If $(y, t) \in \mathbf{epi}\, f$, then

$$t \ge f(y) \ge f(x) + \nabla f(x)^T (y - x).$$

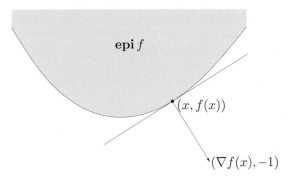

Figure 3.6 For a differentiable convex function f, the vector $(\nabla f(x), -1)$ defines a supporting hyperplane to the epigraph of f at x.

We can express this as:

$$(y, t) \in \mathbf{epi}\, f \implies \left[\begin{array}{c} \nabla f(x) \\ -1 \end{array}\right]^T \left(\left[\begin{array}{c} y \\ t \end{array}\right] - \left[\begin{array}{c} x \\ f(x) \end{array}\right]\right) \leq 0.$$

This means that the hyperplane defined by $(\nabla f(x), -1)$ supports $\mathbf{epi}\, f$ at the boundary point $(x, f(x))$; see figure 3.6.

3.1.8   Jensen's inequality and extensions

The basic inequality (3.1), *i.e.*,

$$f(\theta x + (1 - \theta)y) \leq \theta f(x) + (1 - \theta)f(y),$$

is sometimes called *Jensen's inequality*. It is easily extended to convex combinations of more than two points: If f is convex, $x_1, \ldots, x_k \in \mathbf{dom}\, f$, and $\theta_1, \ldots, \theta_k \geq 0$ with $\theta_1 + \cdots + \theta_k = 1$, then

$$f(\theta_1 x_1 + \cdots + \theta_k x_k) \leq \theta_1 f(x_1) + \cdots + \theta_k f(x_k).$$

As in the case of convex sets, the inequality extends to infinite sums, integrals, and expected values. For example, if $p(x) \geq 0$ on $S \subseteq \mathbf{dom}\, f$, $\int_S p(x)\, dx = 1$, then

$$f\left(\int_S p(x)x\, dx\right) \leq \int_S f(x)p(x)\, dx,$$

provided the integrals exist. In the most general case we can take any probability measure with support in $\mathbf{dom}\, f$. If x is a random variable such that $x \in \mathbf{dom}\, f$ with probability one, and f is convex, then we have

$$f(\mathbf{E}\, x) \leq \mathbf{E}\, f(x), \tag{3.5}$$

provided the expectations exist. We can recover the basic inequality (3.1) from this general form, by taking the random variable x to have support $\{x_1, x_2\}$, with

$\mathbf{prob}(x = x_1) = \theta$, $\mathbf{prob}(x = x_2) = 1 - \theta$. Thus the inequality (3.5) characterizes convexity: If f is not convex, there is a random variable x, with $x \in \mathbf{dom}\, f$ with probability one, such that $f(\mathbf{E}\, x) > \mathbf{E}\, f(x)$.

All of these inequalities are now called *Jensen's inequality*, even though the inequality studied by Jensen was the very simple one

$$f\left(\frac{x+y}{2}\right) \leq \frac{f(x) + f(y)}{2}.$$

Remark 3.2   We can interpret (3.5) as follows. Suppose $x \in \mathbf{dom}\, f \subseteq \mathbf{R}^n$ and z is any zero mean random vector in $\mathbf{R}^n$. Then we have

$$\mathbf{E}\, f(x + z) \geq f(x).$$

Thus, randomization or dithering (*i.e.*, adding a zero mean random vector to the argument) cannot decrease the value of a convex function on average.

3.1.9   Inequalities

Many famous inequalities can be derived by applying Jensen's inequality to some appropriate convex function. (Indeed, convexity and Jensen's inequality can be made the foundation of a theory of inequalities.) As a simple example, consider the arithmetic-geometric mean inequality:

$$\sqrt{ab} \leq (a+b)/2 \tag{3.6}$$

for $a, b \geq 0$. The function $-\log x$ is convex; Jensen's inequality with $\theta = 1/2$ yields

$$-\log\left(\frac{a+b}{2}\right) \leq \frac{-\log a - \log b}{2}.$$

Taking the exponential of both sides yields (3.6).

As a less trivial example we prove Hölder's inequality: for $p > 1$, $1/p + 1/q = 1$, and $x,\, y \in \mathbf{R}^n$,

$$\sum_{i=1}^{n} x_i y_i \leq \left(\sum_{i=1}^{n} |x_i|^p\right)^{1/p} \left(\sum_{i=1}^{n} |y_i|^q\right)^{1/q}.$$

By convexity of $-\log x$, and Jensen's inequality with general θ, we obtain the more general arithmetic-geometric mean inequality

$$a^\theta b^{1-\theta} \leq \theta a + (1-\theta) b,$$

valid for $a,\, b \geq 0$ and $0 \leq \theta \leq 1$. Applying this with

$$a = \frac{|x_i|^p}{\sum_{j=1}^{n} |x_j|^p}, \qquad b = \frac{|y_i|^q}{\sum_{j=1}^{n} |y_j|^q}, \qquad \theta = 1/p,$$

yields

$$\left(\frac{|x_i|^p}{\sum_{j=1}^{n} |x_j|^p}\right)^{1/p} \left(\frac{|y_i|^q}{\sum_{j=1}^{n} |y_j|^q}\right)^{1/q} \leq \frac{|x_i|^p}{p \sum_{j=1}^{n} |x_j|^p} + \frac{|y_i|^q}{q \sum_{j=1}^{n} |y_j|^q}.$$

Summing over i then yields Hölder's inequality.

3.2   Operations that preserve convexity

In this section we describe some operations that preserve convexity or concavity of functions, or allow us to construct new convex and concave functions. We start with some simple operations such as addition, scaling, and pointwise supremum, and then describe some more sophisticated operations (some of which include the simple operations as special cases).

3.2.1   Nonnegative weighted sums

Evidently if f is a convex function and $\alpha \geq 0$, then the function αf is convex. If f_1 and f_2 are both convex functions, then so is their sum $f_1 + f_2$. Combining nonnegative scaling and addition, we see that the set of convex functions is itself a convex cone: a nonnegative weighted sum of convex functions,

$$f = w_1 f_1 + \cdots + w_m f_m,$$

is convex. Similarly, a nonnegative weighted sum of concave functions is concave. A nonnegative, nonzero weighted sum of strictly convex (concave) functions is strictly convex (concave).

These properties extend to infinite sums and integrals. For example if $f(x, y)$ is convex in x for each $y \in \mathcal{A}$, and $w(y) \geq 0$ for each $y \in \mathcal{A}$, then the function g defined as

$$g(x) = \int_{\mathcal{A}} w(y) f(x, y) \, dy$$

is convex in x (provided the integral exists).

The fact that convexity is preserved under nonnegative scaling and addition is easily verified directly, or can be seen in terms of the associated epigraphs. For example, if $w \geq 0$ and f is convex, we have

$$\mathbf{epi}(wf) = \begin{bmatrix} I & 0 \\ 0 & w \end{bmatrix} \mathbf{epi}\, f,$$

which is convex because the image of a convex set under a linear mapping is convex.

3.2.2   Composition with an affine mapping

Suppose $f : \mathbf{R}^n \to \mathbf{R}$, $A \in \mathbf{R}^{n \times m}$, and $b \in \mathbf{R}^n$. Define $g : \mathbf{R}^m \to \mathbf{R}$ by

$$g(x) = f(Ax + b),$$

with $\mathbf{dom}\, g = \{x \mid Ax + b \in \mathbf{dom}\, f\}$. Then if f is convex, so is g; if f is concave, so is g.

3.2.3   Pointwise maximum and supremum

If f_1 and f_2 are convex functions then their *pointwise maximum* f, defined by

$$f(x) = \max\{f_1(x), f_2(x)\},$$

with $\mathbf{dom}\, f = \mathbf{dom}\, f_1 \cap \mathbf{dom}\, f_2$, is also convex. This property is easily verified: if $0 \le \theta \le 1$ and $x,\ y \in \mathbf{dom}\, f$, then

$$
\begin{aligned}
f(\theta x + (1-\theta)y) &= \max\{f_1(\theta x + (1-\theta)y), f_2(\theta x + (1-\theta)y)\} \\
&\le \max\{\theta f_1(x) + (1-\theta)f_1(y), \theta f_2(x) + (1-\theta)f_2(y)\} \\
&\le \theta \max\{f_1(x), f_2(x)\} + (1-\theta)\max\{f_1(y), f_2(y)\} \\
&= \theta f(x) + (1-\theta)f(y),
\end{aligned}
$$

which establishes convexity of f. It is easily shown that if $f_1, \ldots, f_m$ are convex, then their pointwise maximum

$$f(x) = \max\{f_1(x), \ldots, f_m(x)\}$$

is also convex.

Example 3.5 *Piecewise-linear functions.* The function

$$f(x) = \max\{a_1^T x + b_1, \ldots, a_L^T x + b_L\}$$

defines a piecewise-linear (or really, affine) function (with L or fewer regions). It is convex since it is the pointwise maximum of affine functions.

The converse can also be shown: any piecewise-linear convex function with L or fewer regions can be expressed in this form. (See exercise 3.29.)

Example 3.6 *Sum of r largest components.* For $x \in \mathbf{R}^n$ we denote by $x_{[i]}$ the ith largest component of x, *i.e.,*

$$x_{[1]} \ge x_{[2]} \ge \cdots \ge x_{[n]}$$

are the components of x sorted in nonincreasing order. Then the function

$$f(x) = \sum_{i=1}^{r} x_{[i]},$$

i.e., the sum of the r largest elements of x, is a convex function. This can be seen by writing it as

$$f(x) = \sum_{i=1}^{r} x_{[i]} = \max\{x_{i_1} + \cdots + x_{i_r} \mid 1 \le i_1 < i_2 < \cdots < i_r \le n\},$$

i.e., the maximum of all possible sums of r different components of x. Since it is the pointwise maximum of $n!/(r!(n-r)!)$ linear functions, it is convex.

As an extension it can be shown that the function $\sum_{i=1}^{r} w_i x_{[i]}$ is convex, provided $w_1 \ge w_2 \ge \cdots \ge w_r \ge 0$. (See exercise 3.19.)

The pointwise maximum property extends to the pointwise supremum over an infinite set of convex functions. If for each $y \in \mathcal{A}$, $f(x, y)$ is convex in x, then the function g, defined as

$$g(x) = \sup_{y \in \mathcal{A}} f(x, y) \tag{3.7}$$

is convex in x. Here the domain of g is

$$\mathbf{dom}\, g = \{x \mid (x, y) \in \mathbf{dom}\, f \text{ for all } y \in \mathcal{A}, \ \sup_{y \in \mathcal{A}} f(x, y) < \infty\}.$$

Similarly, the pointwise infimum of a set of concave functions is a concave function.

In terms of epigraphs, the pointwise supremum of functions corresponds to the intersection of epigraphs: with f, g, and $\mathcal{A}$ as defined in (3.7), we have

$$\mathbf{epi}\, g = \bigcap_{y \in \mathcal{A}} \mathbf{epi}\, f(\cdot, y).$$

Thus, the result follows from the fact that the intersection of a family of convex sets is convex.

Example 3.7 *Support function of a set.* Let $C \subseteq \mathbf{R}^n$, with $C \neq \emptyset$. The *support function* S_C associated with the set C is defined as

$$S_C(x) = \sup\{x^T y \mid y \in C\}$$

(and, naturally, $\mathbf{dom}\, S_C = \{x \mid \sup_{y \in C} x^T y < \infty\}$).

For each $y \in C$, $x^T y$ is a linear function of x, so S_C is the pointwise supremum of a family of linear functions, hence convex.

Example 3.8 *Distance to farthest point of a set.* Let $C \subseteq \mathbf{R}^n$. The distance (in any norm) to the farthest point of C,

$$f(x) = \sup_{y \in C} \|x - y\|,$$

is convex. To see this, note that for any y, the function $\|x - y\|$ is convex in x. Since f is the pointwise supremum of a family of convex functions (indexed by $y \in C$), it is a convex function of x.

Example 3.9 *Least-squares cost as a function of weights.* Let $a_1, \ldots, a_n \in \mathbf{R}^m$. In a weighted least-squares problem we minimize the objective function $\sum_{i=1}^n w_i (a_i^T x - b_i)^2$ over $x \in \mathbf{R}^m$. We refer to w_i as *weights*, and allow negative w_i (which opens the possibility that the objective function is unbounded below).

We define the (optimal) *weighted least-squares cost* as

$$g(w) = \inf_x \sum_{i=1}^n w_i (a_i^T x - b_i)^2,$$

with domain

$$\mathbf{dom}\, g = \left\{ w \ \middle| \ \inf_x \sum_{i=1}^n w_i (a_i^T x - b_i)^2 > -\infty \right\}.$$

Since g is the infimum of a family of linear functions of w (indexed by $x \in \mathbf{R}^m$), it is a concave function of w.

We can derive an explicit expression for g, at least on part of its domain. Let $W = \mathbf{diag}(w)$, the diagonal matrix with elements $w_1, \ldots, w_n$, and let $A \in \mathbf{R}^{n \times m}$ have rows a_i^T, so we have

$$g(w) = \inf_x (Ax - b)^T W (Ax - b) = \inf_x (x^T A^T W A x - 2b^T W A x + b^T W b).$$

From this we see that if $A^T W A \not\succeq 0$, the quadratic function is unbounded below in x, so $g(w) = -\infty$, i.e., $w \notin \mathbf{dom}\, g$. We can give a simple expression for g when $A^T W A \succ 0$ (which defines a strict linear matrix inequality), by analytically minimizing the quadratic function:

$$
\begin{aligned}
g(w) &= b^T W b - b^T W A (A^T W A)^{-1} A^T W b \\
&= \sum_{i=1}^{n} w_i b_i^2 - \sum_{i=1}^{n} w_i^2 b_i^2 a_i^T \left(\sum_{j=1}^{n} w_j a_j a_j^T \right)^{-1} a_i.
\end{aligned}
$$

Concavity of g from this expression is not immediately obvious (but does follow, for example, from convexity of the matrix fractional function; see example 3.4).

Example 3.10 *Maximum eigenvalue of a symmetric matrix.* The function $f(X) = \lambda_{\max}(X)$, with $\mathbf{dom}\, f = \mathbf{S}^m$, is convex. To see this, we express f as

$$f(X) = \sup\{y^T X y \mid \|y\|_2 = 1\},$$

i.e., as the pointwise supremum of a family of linear functions of X (i.e., $y^T X y$) indexed by $y \in \mathbf{R}^m$.

Example 3.11 *Norm of a matrix.* Consider $f(X) = \|X\|_2$ with $\mathbf{dom}\, f = \mathbf{R}^{p \times q}$, where $\| \cdot \|_2$ denotes the spectral norm or maximum singular value. Convexity of f follows from

$$f(X) = \sup\{u^T X v \mid \|u\|_2 = 1,\ \|v\|_2 = 1\},$$

which shows it is the pointwise supremum of a family of linear functions of X.

As a generalization suppose $\| \cdot \|_a$ and $\| \cdot \|_b$ are norms on $\mathbf{R}^p$ and $\mathbf{R}^q$, respectively. The induced norm of a matrix $X \in \mathbf{R}^{p \times q}$ is defined as

$$\|X\|_{a,b} = \sup_{v \neq 0} \frac{\|Xv\|_a}{\|v\|_b}.$$

(This reduces to the spectral norm when both norms are Euclidean.) The induced norm can be expressed as

$$
\begin{aligned}
\|X\|_{a,b} &= \sup\{\|Xv\|_a \mid \|v\|_b = 1\} \\
&= \sup\{u^T X v \mid \|u\|_{a*} = 1,\ \|v\|_b = 1\},
\end{aligned}
$$

where $\| \cdot \|_{a*}$ is the dual norm of $\| \cdot \|_a$, and we use the fact that

$$\|z\|_a = \sup\{u^T z \mid \|u\|_{a*} = 1\}.$$

Since we have expressed $\|X\|_{a,b}$ as a supremum of linear functions of X, it is a convex function.

Representation as pointwise supremum of affine functions

The examples above illustrate a good method for establishing convexity of a function: by expressing it as the pointwise supremum of a family of affine functions. Except for a technical condition, a converse holds: almost every convex function can be expressed as the pointwise supremum of a family of affine functions. For example, if $f : \mathbf{R}^n \to \mathbf{R}$ is convex, with $\mathbf{dom}\, f = \mathbf{R}^n$, then we have

$$f(x) = \sup\{g(x) \mid g \text{ affine}, \ g(z) \le f(z) \text{ for all } z\}.$$

In other words, f is the pointwise supremum of the set of all affine global underestimators of it. We give the proof of this result below, and leave the case where $\mathbf{dom}\, f \ne \mathbf{R}^n$ as an exercise (exercise 3.28).

Suppose f is convex with $\mathbf{dom}\, f = \mathbf{R}^n$. The inequality

$$f(x) \ge \sup\{g(x) \mid g \text{ affine}, \ g(z) \le f(z) \text{ for all } z\}$$

is clear, since if g is any affine underestimator of f, we have $g(x) \le f(x)$. To establish equality, we will show that for each $x \in \mathbf{R}^n$, there is an affine function g, which is a global underestimator of f, and satisfies $g(x) = f(x)$.

The epigraph of f is, of course, a convex set. Hence we can find a supporting hyperplane to it at $(x, f(x))$, i.e., $a \in \mathbf{R}^n$ and $b \in \mathbf{R}$ with $(a, b) \ne 0$ and

$$\begin{bmatrix} a \\ b \end{bmatrix}^T \begin{bmatrix} x - z \\ f(x) - t \end{bmatrix} \le 0$$

for all $(z, t) \in \mathbf{epi}\, f$. This means that

$$a^T(x - z) + b(f(x) - f(z) - s) \le 0 \tag{3.8}$$

for all $z \in \mathbf{dom}\, f = \mathbf{R}^n$ and all $s \ge 0$ (since $(z, t) \in \mathbf{epi}\, f$ means $t = f(z) + s$ for some $s \ge 0$). For the inequality (3.8) to hold for all $s \ge 0$, we must have $b \ge 0$. If $b = 0$, then the inequality (3.8) reduces to $a^T(x - z) \le 0$ for all $z \in \mathbf{R}^n$, which implies $a = 0$ and contradicts $(a, b) \ne 0$. We conclude that $b > 0$, i.e., that the supporting hyperplane is not vertical.

Using the fact that $b > 0$ we rewrite (3.8) for $s = 0$ as

$$g(z) = f(x) + (a/b)^T(x - z) \le f(z)$$

for all z. The function g is an affine underestimator of f, and satisfies $g(x) = f(x)$.

3.2.4   Composition

In this section we examine conditions on $h : \mathbf{R}^k \to \mathbf{R}$ and $g : \mathbf{R}^n \to \mathbf{R}^k$ that guarantee convexity or concavity of their composition $f = h \circ g : \mathbf{R}^n \to \mathbf{R}$, defined by

$$f(x) = h(g(x)), \qquad \mathbf{dom}\, f = \{x \in \mathbf{dom}\, g \mid g(x) \in \mathbf{dom}\, h\}.$$

Scalar composition

We first consider the case $k = 1$, so $h : \mathbf{R} \to \mathbf{R}$ and $g : \mathbf{R}^n \to \mathbf{R}$. We can restrict ourselves to the case $n = 1$ (since convexity is determined by the behavior of a function on arbitrary lines that intersect its domain).

To discover the composition rules, we start by assuming that h and g are twice differentiable, with $\mathbf{dom}\, g = \mathbf{dom}\, h = \mathbf{R}$. In this case, convexity of f reduces to $f'' \geq 0$ (meaning, $f''(x) \geq 0$ for all $x \in \mathbf{R}$).

The second derivative of the composition function $f = h \circ g$ is given by

$$f''(x) = h''(g(x))g'(x)^2 + h'(g(x))g''(x). \tag{3.9}$$

Now suppose, for example, that g is convex (so $g'' \geq 0$) and h is convex and nondecreasing (so $h'' \geq 0$ and $h' \geq 0$). It follows from (3.9) that $f'' \geq 0$, *i.e.*, f is convex. In a similar way, the expression (3.9) gives the results:

> f is convex if h is convex and nondecreasing, and g is convex,
>
> f is convex if h is convex and nonincreasing, and g is concave,
>
> f is concave if h is concave and nondecreasing, and g is concave, $\qquad$ (3.10)
>
> f is concave if h is concave and nonincreasing, and g is convex.

These statements are valid when the functions g and h are twice differentiable and have domains that are all of $\mathbf{R}$. It turns out that very similar composition rules hold in the general case $n > 1$, without assuming differentiability of h and g, or that $\mathbf{dom}\, g = \mathbf{R}^n$ and $\mathbf{dom}\, h = \mathbf{R}$:

> f is convex if h is convex, $\tilde{h}$ is nondecreasing, and g is convex,
>
> f is convex if h is convex, $\tilde{h}$ is nonincreasing, and g is concave,
>
> f is concave if h is concave, $\tilde{h}$ is nondecreasing, and g is concave, $\qquad$ (3.11)
>
> f is concave if h is concave, $\tilde{h}$ is nonincreasing, and g is convex.

Here $\tilde{h}$ denotes the extended-value extension of the function h, which assigns the value ∞ $(-\infty)$ to points not in $\mathbf{dom}\, h$ for h convex (concave). The only difference between these results, and the results in (3.10), is that we require that the *extended-value extension* function $\tilde{h}$ be nonincreasing or nondecreasing, on all of $\mathbf{R}$.

To understand what this means, suppose h is convex, so $\tilde{h}$ takes on the value ∞ outside $\mathbf{dom}\, h$. To say that $\tilde{h}$ is nondecreasing means that for *any* x, $y \in \mathbf{R}$, with $x < y$, we have $\tilde{h}(x) \leq \tilde{h}(y)$. In particular, this means that if $y \in \mathbf{dom}\, h$, then $x \in \mathbf{dom}\, h$. In other words, the domain of h extends infinitely in the negative direction; it is either $\mathbf{R}$, or an interval of the form $(-\infty, a)$ or $(-\infty, a]$. In a similar way, to say that h is convex and $\tilde{h}$ is nonincreasing means that h is nonincreasing and $\mathbf{dom}\, h$ extends infinitely in the positive direction. This is illustrated in figure 3.7.

Example 3.12 Some simple examples will illustrate the conditions on h that appear in the composition theorems.

- The function $h(x) = \log x$, with $\mathbf{dom}\, h = \mathbf{R}_{++}$, is concave and satisfies $\tilde{h}$ nondecreasing.

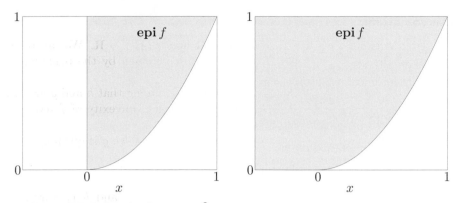

Figure 3.7 *Left.* The function x^2, with domain $\mathbf{R}_+$, is convex and nonde-creasing on its domain, but its extended-value extension is *not* nondecreasing. *Right.* The function $\max\{x,0\}^2$, with domain $\mathbf{R}$, is convex, and its extended-value extension is nondecreasing.

- The function $h(x) = x^{1/2}$, with $\mathbf{dom}\,h = \mathbf{R}_+$, is concave and satisfies the condition $\tilde{h}$ nondecreasing.
- The function $h(x) = x^{3/2}$, with $\mathbf{dom}\,h = \mathbf{R}_+$, is convex but *does not* satisfy the condition $\tilde{h}$ nondecreasing. For example, we have $\tilde{h}(-1) = \infty$, but $\tilde{h}(1) = 1$.
- The function $h(x) = x^{3/2}$ for $x \geq 0$, and $h(x) = 0$ for $x < 0$, with $\mathbf{dom}\,h = \mathbf{R}$, is convex and *does* satisfy the condition $\tilde{h}$ nondecreasing.

The composition results (3.11) can be proved directly, without assuming dif-ferentiability, or using the formula (3.9). As an example, we will prove the fol-lowing composition theorem: if g is convex, h is convex, and $\tilde{h}$ is nondecreasing, then $f = h \circ g$ is convex. Assume that x, $y \in \mathbf{dom}\,f$, and $0 \leq \theta \leq 1$. Since x, $y \in \mathbf{dom}\,f$, we have that x, $y \in \mathbf{dom}\,g$ and $g(x)$, $g(y) \in \mathbf{dom}\,h$. Since $\mathbf{dom}\,g$ is convex, we conclude that $\theta x + (1-\theta)y \in \mathbf{dom}\,g$, and from convexity of g, we have

$$g(\theta x + (1-\theta)y) \leq \theta g(x) + (1-\theta)g(y). \tag{3.12}$$

Since $g(x)$, $g(y) \in \mathbf{dom}\,h$, we conclude that $\theta g(x) + (1-\theta)g(y) \in \mathbf{dom}\,h$, *i.e.*, the righthand side of (3.12) is in $\mathbf{dom}\,h$. Now we use the assumption that $\tilde{h}$ is nondecreasing, which means that its domain extends infinitely in the negative direction. Since the righthand side of (3.12) is in $\mathbf{dom}\,h$, we conclude that the lefthand side, *i.e.*, $g(\theta x+(1-\theta)y) \in \mathbf{dom}\,h$. This means that $\theta x+(1-\theta)y \in \mathbf{dom}\,f$. At this point, we have shown that $\mathbf{dom}\,f$ is convex.

Now using the fact that $\tilde{h}$ is nondecreasing and the inequality (3.12), we get

$$h(g(\theta x + (1-\theta)y)) \leq h(\theta g(x) + (1-\theta)g(y)). \tag{3.13}$$

From convexity of h, we have

$$h(\theta g(x) + (1-\theta)g(y)) \leq \theta h(g(x)) + (1-\theta)h(g(y)). \tag{3.14}$$

Putting (3.13) and (3.14) together, we have

$$h(g(\theta x + (1 - \theta)y)) \le \theta h(g(x)) + (1 - \theta)h(g(y)).$$

which proves the composition theorem.

Example 3.13 *Simple composition results.*

- If g is convex then $\exp g(x)$ is convex.
- If g is concave and positive, then $\log g(x)$ is concave.
- If g is concave and positive, then $1/g(x)$ is convex.
- If g is convex and nonnegative and $p \ge 1$, then $g(x)^p$ is convex.
- If g is convex then $-\log(-g(x))$ is convex on $\{x \mid g(x) < 0\}$.

Remark 3.3 The requirement that monotonicity hold for the extended-value extension $\tilde{h}$, and not just the function h, cannot be removed. For example, consider the function $g(x) = x^2$, with $\mathbf{dom}\, g = \mathbf{R}$, and $h(x) = 0$, with $\mathbf{dom}\, h = [1, 2]$. Here g is convex, and h is convex and nondecreasing. But the function $f = h \circ g$, given by

$$f(x) = 0, \qquad \mathbf{dom}\, f = [-\sqrt{2}, -1] \cup [1, \sqrt{2}],$$

is not convex, since its domain is not convex. Here, of course, the function $\tilde{h}$ is *not* nondecreasing.

Vector composition

We now turn to the more complicated case when $k \ge 1$. Suppose

$$f(x) = h(g(x)) = h(g_1(x), \dots, g_k(x)),$$

with $h : \mathbf{R}^k \to \mathbf{R}$, $g_i : \mathbf{R}^n \to \mathbf{R}$. Again without loss of generality we can assume $n = 1$. As in the case $k = 1$, we start by assuming the functions are twice differentiable, with $\mathbf{dom}\, g = \mathbf{R}$ and $\mathbf{dom}\, h = \mathbf{R}^k$, in order to discover the composition rules. We have

$$f''(x) = g'(x)^T \nabla^2 h(g(x)) g'(x) + \nabla h(g(x))^T g''(x), \qquad (3.15)$$

which is the vector analog of (3.9). Again the issue is to determine conditions under which $f''(x) \ge 0$ for all x (or $f''(x) \le 0$ for all x for concavity). From (3.15) we can derive many rules, for example:

f is convex if h is convex, h is nondecreasing in each argument, and g_i are convex,

f is convex if h is convex, h is nonincreasing in each argument, and g_i are concave,

f is concave if h is concave, h is nondecreasing in each argument, and g_i are concave.

As in the scalar case, similar composition results hold in general, with $n > 1$, no assumption of differentiability of h or g, and general domains. For the general results, the monotonicity condition on h must hold for the extended-value extension $\tilde{h}$.

To understand the meaning of the condition that the extended-value extension $\tilde{h}$ be monotonic, we consider the case where $h : \mathbf{R}^k \to \mathbf{R}$ is convex, and $\tilde{h}$ nondecreasing, i.e., whenever $u \preceq v$, we have $\tilde{h}(u) \leq \tilde{h}(v)$. This implies that if $v \in \mathbf{dom}\, h$, then so is u: the domain of h must extend infinitely in the $-\mathbf{R}_+^k$ directions. We can express this compactly as $\mathbf{dom}\, h - \mathbf{R}_+^k = \mathbf{dom}\, h$.

Example 3.14 *Vector composition examples.*

- Let $h(z) = z_{[1]} + \cdots + z_{[r]}$, the sum of the r largest components of $z \in \mathbf{R}^k$. Then h is convex and nondecreasing in each argument. Suppose $g_1, \ldots, g_k$ are convex functions on $\mathbf{R}^n$. Then the composition function $f = h \circ g$, i.e., the pointwise sum of the r largest g_i's, is convex.

- The function $h(z) = \log(\sum_{i=1}^k e^{z_i})$ is convex and nondecreasing in each argument, so $\log(\sum_{i=1}^k e^{g_i})$ is convex whenever g_i are.

- For $0 < p \leq 1$, the function $h(z) = (\sum_{i=1}^k z_i^p)^{1/p}$ on $\mathbf{R}_+^k$ is concave, and its extension (which has the value $-\infty$ for $z \not\succeq 0$) is nondecreasing in each component. So if g_i are concave and nonnegative, we conclude that $f(x) = (\sum_{i=1}^k g_i(x)^p)^{1/p}$ is concave.

- Suppose $p \geq 1$, and $g_1, \ldots, g_k$ are convex and nonnegative. Then the function $(\sum_{i=1}^k g_i(x)^p)^{1/p}$ is convex.

  To show this, we consider the function $h : \mathbf{R}^k \to \mathbf{R}$ defined as

  $$h(z) = \left(\sum_{i=1}^k \max\{z_i, 0\}^p \right)^{1/p},$$

  with $\mathbf{dom}\, h = \mathbf{R}^k$, so $h = \tilde{h}$. This function is convex, and nondecreasing, so we conclude $h(g(x))$ is a convex function of x. For $z \succeq 0$, we have $h(z) = (\sum_{i=1}^k z_i^p)^{1/p}$, so our conclusion is that $(\sum_{i=1}^k g_i(x)^p)^{1/p}$ is convex.

- The geometric mean $h(z) = (\prod_{i=1}^k z_i)^{1/k}$ on $\mathbf{R}_+^k$ is concave and its extension is nondecreasing in each argument. It follows that if $g_1, \ldots, g_k$ are nonnegative concave functions, then so is their geometric mean, $(\prod_{i=1}^k g_i)^{1/k}$.

3.2.5   Minimization

We have seen that the maximum or supremum of an arbitrary family of convex functions is convex. It turns out that some special forms of minimization also yield convex functions. If f is convex in (x, y), and C is a convex nonempty set, then the function

$$g(x) = \inf_{y \in C} f(x, y) \tag{3.16}$$

is convex in x, provided $g(x) > -\infty$ for some x (which implies $g(x) > -\infty$ for all x). The domain of g is the projection of $\textbf{dom}\, f$ on its x-coordinates, i.e.,

$$\textbf{dom}\, g = \{x \mid (x, y) \in \textbf{dom}\, f \text{ for some } y \in C\}.$$

We prove this by verifying Jensen's inequality for x_1, $x_2 \in \textbf{dom}\, g$. Let $\epsilon > 0$. Then there are y_1, $y_2 \in C$ such that $f(x_i, y_i) \le g(x_i) + \epsilon$ for $i = 1$, 2. Now let $\theta \in [0, 1]$. We have

$$
\begin{aligned}
g(\theta x_1 + (1 - \theta)x_2) &= \inf_{y \in C} f(\theta x_1 + (1 - \theta)x_2, y) \\
&\le f(\theta x_1 + (1 - \theta)x_2, \theta y_1 + (1 - \theta)y_2) \\
&\le \theta f(x_1, y_1) + (1 - \theta)f(x_2, y_2) \\
&\le \theta g(x_1) + (1 - \theta)g(x_2) + \epsilon.
\end{aligned}
$$

Since this holds for any $\epsilon > 0$, we have

$$g(\theta x_1 + (1 - \theta)x_2) \le \theta g(x_1) + (1 - \theta)g(x_2).$$

The result can also be seen in terms of epigraphs. With f, g, and C defined as in (3.16), and assuming the infimum over $y \in C$ is attained for each x, we have

$$\textbf{epi}\, g = \{(x, t) \mid (x, y, t) \in \textbf{epi}\, f \text{ for some } y \in C\}.$$

Thus $\textbf{epi}\, g$ is convex, since it is the projection of a convex set on some of its components.

Example 3.15 *Schur complement.* Suppose the quadratic function

$$f(x, y) = x^T A x + 2x^T B y + y^T C y,$$

(where A and C are symmetric) is convex in (x, y), which means

$$\begin{bmatrix} A & B \\ B^T & C \end{bmatrix} \succeq 0.$$

We can express $g(x) = \inf_y f(x, y)$ as

$$g(x) = x^T (A - BC^\dagger B^T)x,$$

where $C^\dagger$ is the pseudo-inverse of C (see §A.5.4). By the minimization rule, g is convex, so we conclude that $A - BC^\dagger B^T \succeq 0$.

If C is invertible, i.e., $C \succ 0$, then the matrix $A - BC^{-1}B^T$ is called the *Schur complement* of C in the matrix

$$\begin{bmatrix} A & B \\ B^T & C \end{bmatrix}$$

(see §A.5.5).

Example 3.16 *Distance to a set.* The distance of a point x to a set $S \subseteq \mathbf{R}^n$, in the norm $\|\cdot\|$, is defined as

$$\textbf{dist}(x, S) = \inf_{y \in S} \|x - y\|.$$

The function $\|x - y\|$ is convex in (x, y), so if the set S is convex, the distance function $\textbf{dist}(x, S)$ is a convex function of x.

Example 3.17 Suppose h is convex. Then the function g defined as

$$g(x) = \inf\{h(y) \mid Ay = x\}$$

is convex. To see this, we define f by

$$f(x,y) = \begin{cases} h(y) & \text{if } Ay = x \\ \infty & \text{otherwise,} \end{cases}$$

which is convex in (x,y). Then g is the minimum of f over y, and hence is convex. (It is not hard to show directly that g is convex.)

3.2.6   Perspective of a function

If $f : \mathbf{R}^n \to \mathbf{R}$, then the *perspective* of f is the function $g : \mathbf{R}^{n+1} \to \mathbf{R}$ defined by

$$g(x,t) = tf(x/t),$$

with domain

$$\mathbf{dom}\, g = \{(x,t) \mid x/t \in \mathbf{dom}\, f, \ t > 0\}.$$

The perspective operation preserves convexity: If f is a convex function, then so is its perspective function g. Similarly, if f is concave, then so is g.

This can be proved several ways, for example, direct verification of the defining inequality (see exercise 3.33). We give a short proof here using epigraphs and the perspective mapping on $\mathbf{R}^{n+1}$ described in §2.3.3 (which will also explain the name 'perspective'). For $t > 0$ we have

$$
\begin{aligned}
(x,t,s) \in \mathbf{epi}\, g \quad &\Longleftrightarrow \quad tf(x/t) \leq s \\
&\Longleftrightarrow \quad f(x/t) \leq s/t \\
&\Longleftrightarrow \quad (x/t, s/t) \in \mathbf{epi}\, f.
\end{aligned}
$$

Therefore $\mathbf{epi}\, g$ is the inverse image of $\mathbf{epi}\, f$ under the perspective mapping that takes (u,v,w) to $(u,w)/v$. It follows (see §2.3.3) that $\mathbf{epi}\, g$ is convex, so the function g is convex.

Example 3.18 *Euclidean norm squared.* The perspective of the convex function $f(x) = x^T x$ on $\mathbf{R}^n$ is

$$g(x,t) = t(x/t)^T (x/t) = \frac{x^T x}{t},$$

which is convex in (x,t) for $t > 0$.

We can deduce convexity of g using several other methods. First, we can express g as the sum of the quadratic-over-linear functions x_i^2/t, which were shown to be convex in §3.1.5. We can also express g as a special case of the matrix fractional function $x^T (tI)^{-1} x$ (see example 3.4).

Example 3.19 *Negative logarithm.* Consider the convex function $f(x) = -\log x$ on $\mathbf{R}_{++}$. Its perspective is

$$g(x,t) = -t\log(x/t) = t\log(t/x) = t\log t - t\log x,$$

and is convex on $\mathbf{R}_{++}^2$. The function g is called the *relative entropy* of t and x. For $x = 1$, g reduces to the negative entropy function.

From convexity of g we can establish convexity or concavity of several interesting related functions. First, the relative entropy of two vectors $u,\ v \in \mathbf{R}_{++}^n$, defined as

$$\sum_{i=1}^{n} u_i \log(u_i/v_i),$$

is convex in (u,v), since it is a sum of relative entropies of $u_i,\ v_i$.

A closely related function is the *Kullback-Leibler divergence* between $u,\ v \in \mathbf{R}_{++}^n$, given by

$$D_{\mathrm{kl}}(u,v) = \sum_{i=1}^{n} \left(u_i \log(u_i/v_i) - u_i + v_i \right), \qquad (3.17)$$

which is convex, since it is the relative entropy plus a linear function of (u,v). The Kullback-Leibler divergence satisfies $D_{\mathrm{kl}}(u,v) \geq 0$, and $D_{\mathrm{kl}}(u,v) = 0$ if and only if $u = v$, and so can be used as a measure of deviation between two positive vectors; see exercise 3.13. (Note that the relative entropy and the Kullback-Leibler divergence are the same when u and v are probability vectors, *i.e.*, satisfy $\mathbf{1}^T u = \mathbf{1}^T v = 1$.)

If we take $v_i = \mathbf{1}^T u$ in the relative entropy function, we obtain the concave (and homogeneous) function of $u \in \mathbf{R}_{++}^n$ given by

$$\sum_{i=1}^{n} u_i \log(\mathbf{1}^T u/u_i) = (\mathbf{1}^T u)\sum_{i=1}^{n} z_i \log(1/z_i),$$

where $z = u/(\mathbf{1}^T u)$, which is called the *normalized entropy* function. The vector $z = u/\mathbf{1}^T u$ is a normalized vector or probability distribution, since its components sum to one; the normalized entropy of u is $\mathbf{1}^T u$ times the entropy of this normalized distribution.

Example 3.20 Suppose $f : \mathbf{R}^m \to \mathbf{R}$ is convex, and $A \in \mathbf{R}^{m \times n}$, $b \in \mathbf{R}^m$, $c \in \mathbf{R}^n$, and $d \in \mathbf{R}$. We define

$$g(x) = (c^T x + d)f\left((Ax + b)/(c^T x + d)\right),$$

with

$$\mathbf{dom}\,g = \{x \mid c^T x + d > 0,\ (Ax + b)/(c^T x + d) \in \mathbf{dom}\,f\}.$$

Then g is convex.

3.3   The conjugate function

In this section we introduce an operation that will play an important role in later chapters.

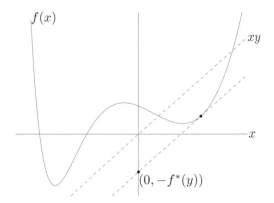

Figure 3.8 A function $f : \mathbf{R} \to \mathbf{R}$, and a value $y \in \mathbf{R}$. The conjugate function $f^*(y)$ is the maximum gap between the linear function yx and $f(x)$, as shown by the dashed line in the figure. If f is differentiable, this occurs at a point x where $f'(x) = y$.

3.3.1   Definition and examples

Let $f : \mathbf{R}^n \to \mathbf{R}$. The function $f^* : \mathbf{R}^n \to \mathbf{R}$, defined as

$$f^*(y) = \sup_{x \in \mathbf{dom}\, f} \left(y^T x - f(x) \right), \tag{3.18}$$

is called the *conjugate* of the function f. The domain of the conjugate function consists of $y \in \mathbf{R}^n$ for which the supremum is finite, *i.e.*, for which the difference $y^T x - f(x)$ is bounded above on $\mathbf{dom}\, f$. This definition is illustrated in figure 3.8.

We see immediately that f^* is a convex function, since it is the pointwise supremum of a family of convex (indeed, affine) functions of y. This is true whether or not f is convex. (Note that when f is convex, the subscript $x \in \mathbf{dom}\, f$ is not necessary since, by convention, $y^T x - f(x) = -\infty$ for $x \notin \mathbf{dom}\, f$.)

We start with some simple examples, and then describe some rules for conjugating functions. This allows us to derive an analytical expression for the conjugate of many common convex functions.

Example 3.21   We derive the conjugates of some convex functions on $\mathbf{R}$.

- *Affine function.* $f(x) = ax + b$. As a function of x, $yx - ax - b$ is bounded if and only if $y = a$, in which case it is constant. Therefore the domain of the conjugate function f^* is the singleton $\{a\}$, and $f^*(a) = -b$.

- *Negative logarithm.* $f(x) = -\log x$, with $\mathbf{dom}\, f = \mathbf{R}_{++}$. The function $xy + \log x$ is unbounded above if $y \geq 0$ and reaches its maximum at $x = -1/y$ otherwise. Therefore, $\mathbf{dom}\, f^* = \{y \mid y < 0\} = -\mathbf{R}_{++}$ and $f^*(y) = -\log(-y) - 1$ for $y < 0$.

- *Exponential.* $f(x) = e^x$. $xy - e^x$ is unbounded if $y < 0$. For $y > 0$, $xy - e^x$ reaches its maximum at $x = \log y$, so we have $f^*(y) = y \log y - y$. For $y = 0$,

$f^*(y) = \sup_x -e^x = 0$. In summary, $\mathbf{dom}\, f^* = \mathbf{R}_+$ and $f^*(y) = y\log y - y$ (with the interpretation $0\log 0 = 0$).

- *Negative entropy.* $f(x) = x\log x$, with $\mathbf{dom}\, f = \mathbf{R}_+$ (and $f(0) = 0$). The function $xy - x\log x$ is bounded above on $\mathbf{R}_+$ for all y, hence $\mathbf{dom}\, f^* = \mathbf{R}$. It attains its maximum at $x = e^{y-1}$, and substituting we find $f^*(y) = e^{y-1}$.

- *Inverse.* $f(x) = 1/x$ on $\mathbf{R}_{++}$. For $y > 0$, $yx - 1/x$ is unbounded above. For $y = 0$ this function has supremum 0; for $y < 0$ the supremum is attained at $x = (-y)^{-1/2}$. Therefore we have $f^*(y) = -2(-y)^{1/2}$, with $\mathbf{dom}\, f^* = -\mathbf{R}_+$.

Example 3.22 *Strictly convex quadratic function.* Consider $f(x) = \frac{1}{2}x^T Q x$, with $Q \in \mathbf{S}^n_{++}$. The function $y^T x - \frac{1}{2}x^T Q x$ is bounded above as a function of x for all y. It attains its maximum at $x = Q^{-1}y$, so

$$f^*(y) = \frac{1}{2}y^T Q^{-1}y.$$

Example 3.23 *Log-determinant.* We consider $f(X) = \log\det X^{-1}$ on $\mathbf{S}^n_{++}$. The conjugate function is defined as

$$f^*(Y) = \sup_{X\succ 0} \left(\mathbf{tr}(YX) + \log\det X\right),$$

since $\mathbf{tr}(YX)$ is the standard inner product on $\mathbf{S}^n$. We first show that $\mathbf{tr}(YX) + \log\det X$ is unbounded above unless $Y \prec 0$. If $Y \not\prec 0$, then Y has an eigenvector v, with $\|v\|_2 = 1$, and eigenvalue $\lambda \geq 0$. Taking $X = I + tvv^T$ we find that

$$\mathbf{tr}(YX) + \log\det X = \mathbf{tr}\,Y + t\lambda + \log\det(I + tvv^T) = \mathbf{tr}\,Y + t\lambda + \log(1+t),$$

which is unbounded above as $t \to \infty$.

Now consider the case $Y \prec 0$. We can find the maximizing X by setting the gradient with respect to X equal to zero:

$$\nabla_X \left(\mathbf{tr}(YX) + \log\det X\right) = Y + X^{-1} = 0$$

(see §A.4.1), which yields $X = -Y^{-1}$ (which is, indeed, positive definite). Therefore we have

$$f^*(Y) = \log\det(-Y)^{-1} - n,$$

with $\mathbf{dom}\, f^* = -\mathbf{S}^n_{++}$.

Example 3.24 *Indicator function.* Let I_S be the indicator function of a (not necessarily convex) set $S \subseteq \mathbf{R}^n$, i.e., $I_S(x) = 0$ on $\mathbf{dom}\, I_S = S$. Its conjugate is

$$I_S^*(y) = \sup_{x\in S} y^T x,$$

which is the support function of the set S.

Example 3.25 *Log-sum-exp function.* To derive the conjugate of the log-sum-exp function $f(x) = \log(\sum_{i=1}^{n} e^{x_i})$, we first determine the values of y for which the maximum over x of $y^T x - f(x)$ is attained. By setting the gradient with respect to x equal to zero, we obtain the condition

$$y_i = \frac{e^{x_i}}{\sum_{j=1}^{n} e^{x_j}}, \quad i = 1, \ldots, n.$$

These equations are solvable for x if and only if $y \succ 0$ and $\mathbf{1}^T y = 1$. By substituting the expression for y_i into $y^T x - f(x)$ we obtain $f^*(y) = \sum_{i=1}^{n} y_i \log y_i$. This expression for f^* is still correct if some components of y are zero, as long as $y \succeq 0$ and $\mathbf{1}^T y = 1$, and we interpret $0 \log 0$ as 0.

In fact the domain of f^* is exactly given by $\mathbf{1}^T y = 1$, $y \succeq 0$. To show this, suppose that a component of y is negative, say, $y_k < 0$. Then we can show that $y^T x - f(x)$ is unbounded above by choosing $x_k = -t$, and $x_i = 0$, $i \neq k$, and letting t go to infinity. If $y \succeq 0$ but $\mathbf{1}^T y \neq 1$, we choose $x = t\mathbf{1}$, so that

$$y^T x - f(x) = t\mathbf{1}^T y - t - \log n.$$

If $\mathbf{1}^T y > 1$, this grows unboundedly as $t \to \infty$; if $\mathbf{1}^T y < 1$, it grows unboundedly as $t \to -\infty$.

In summary,

$$f^*(y) = \begin{cases} \sum_{i=1}^{n} y_i \log y_i & \text{if } y \succeq 0 \text{ and } \mathbf{1}^T y = 1 \\ \infty & \text{otherwise.} \end{cases}$$

In other words, the conjugate of the log-sum-exp function is the negative entropy function, restricted to the probability simplex.

Example 3.26 *Norm.* Let $\|\cdot\|$ be a norm on $\mathbf{R}^n$, with dual norm $\|\cdot\|_*$. We will show that the conjugate of $f(x) = \|x\|$ is

$$f^*(y) = \begin{cases} 0 & \|y\|_* \leq 1 \\ \infty & \text{otherwise,} \end{cases}$$

i.e., the conjugate of a norm is the indicator function of the dual norm unit ball.

If $\|y\|_* > 1$, then by definition of the dual norm, there is a $z \in \mathbf{R}^n$ with $\|z\| \leq 1$ and $y^T z > 1$. Taking $x = tz$ and letting $t \to \infty$, we have

$$y^T x - \|x\| = t(y^T z - \|z\|) \to \infty,$$

which shows that $f^*(y) = \infty$. Conversely, if $\|y\|_* \leq 1$, then we have $y^T x \leq \|x\|\|y\|_*$ for all x, which implies for all x, $y^T x - \|x\| \leq 0$. Therefore $x = 0$ is the value that maximizes $y^T x - \|x\|$, with maximum value 0.

Example 3.27 *Norm squared.* Now consider the function $f(x) = (1/2)\|x\|^2$, where $\|\cdot\|$ is a norm, with dual norm $\|\cdot\|_*$. We will show that its conjugate is $f^*(y) = (1/2)\|y\|_*^2$. From $y^T x \leq \|y\|_* \|x\|$, we conclude

$$y^T x - (1/2)\|x\|^2 \leq \|y\|_* \|x\| - (1/2)\|x\|^2$$

for all x. The righthand side is a quadratic function of $\|x\|$, which has maximum value $(1/2)\|y\|_*^2$. Therefore for all x, we have

$$y^T x - (1/2)\|x\|^2 \leq (1/2)\|y\|_*^2,$$

which shows that $f^*(y) \leq (1/2)\|y\|_*^2$.

To show the other inequality, let x be any vector with $y^T x = \|y\|_*\|x\|$, scaled so that $\|x\| = \|y\|_*$. Then we have, for this x,

$$y^T x - (1/2)\|x\|^2 = (1/2)\|y\|_*^2,$$

which shows that $f^*(y) \geq (1/2)\|y\|_*^2$.

Example 3.28 *Revenue and profit functions.* We consider a business or enterprise that consumes n resources and produces a product that can be sold. We let $r = (r_1, \ldots, r_n)$ denote the vector of resource quantities consumed, and $S(r)$ denote the sales revenue derived from the product produced (as a function of the resources consumed). Now let p_i denote the price (per unit) of resource i, so the total amount paid for resources by the enterprise is $p^T r$. The profit derived by the firm is then $S(r) - p^T r$. Let us fix the prices of the resources, and ask what is the maximum profit that can be made, by wisely choosing the quantities of resources consumed. This maximum profit is given by

$$M(p) = \sup_r \left(S(r) - p^T r \right).$$

The function $M(p)$ gives the maximum profit attainable, as a function of the resource prices. In terms of conjugate functions, we can express M as

$$M(p) = (-S)^*(-p).$$

Thus the maximum profit (as a function of resource prices) is closely related to the conjugate of gross sales (as a function of resources consumed).

3.3.2 Basic properties

Fenchel's inequality

From the definition of conjugate function, we immediately obtain the inequality

$$f(x) + f^*(y) \geq x^T y$$

for all x, y. This is called *Fenchel's inequality* (or *Young's inequality* when f is differentiable).

For example with $f(x) = (1/2)x^T Q x$, where $Q \in \mathbf{S}_{++}^n$, we obtain the inequality

$$x^T y \leq (1/2)x^T Q x + (1/2)y^T Q^{-1} y.$$

Conjugate of the conjugate

The examples above, and the name 'conjugate', suggest that the conjugate of the conjugate of a convex function is the original function. This is the case provided a technical condition holds: if f is convex, and f is closed (*i.e.*, **epi** f is a closed set; see §A.3.3), then $f^{**} = f$. For example, if **dom** $f = \mathbf{R}^n$, then we have $f^{**} = f$, *i.e.*, the conjugate of the conjugate of f is f again (see exercise 3.39).

Differentiable functions

The conjugate of a differentiable function f is also called the *Legendre transform* of f. (To distinguish the general definition from the differentiable case, the term *Fenchel conjugate* is sometimes used instead of conjugate.)

Suppose f is convex and differentiable, with $\mathbf{dom}\, f = \mathbf{R}^n$. Any maximizer x^* of $y^T x - f(x)$ satisfies $y = \nabla f(x^*)$, and conversely, if x^* satisfies $y = \nabla f(x^*)$, then x^* maximizes $y^T x - f(x)$. Therefore, if $y = \nabla f(x^*)$, we have

$$f^*(y) = x^{*T} \nabla f(x^*) - f(x^*).$$

This allows us to determine $f^*(y)$ for any y for which we can solve the gradient equation $y = \nabla f(z)$ for z.

We can express this another way. Let $z \in \mathbf{R}^n$ be arbitrary and define $y = \nabla f(z)$. Then we have

$$f^*(y) = z^T \nabla f(z) - f(z).$$

Scaling and composition with affine transformation

For $a > 0$ and $b \in \mathbf{R}$, the conjugate of $g(x) = af(x) + b$ is $g^*(y) = af^*(y/a) - b$.

Suppose $A \in \mathbf{R}^{n \times n}$ is nonsingular and $b \in \mathbf{R}^n$. Then the conjugate of $g(x) = f(Ax + b)$ is

$$g^*(y) = f^*(A^{-T} y) - b^T A^{-T} y,$$

with $\mathbf{dom}\, g^* = A^T \mathbf{dom}\, f^*$.

Sums of independent functions

If $f(u, v) = f_1(u) + f_2(v)$, where f_1 and f_2 are convex functions with conjugates f_1^* and f_2^*, respectively, then

$$f^*(w, z) = f_1^*(w) + f_2^*(z).$$

In other words, the conjugate of the sum of *independent* convex functions is the sum of the conjugates. ('Independent' means they are functions of different variables.)

3.4   Quasiconvex functions

3.4.1   Definition and examples

A function $f : \mathbf{R}^n \to \mathbf{R}$ is called *quasiconvex* (or *unimodal*) if its domain and all its sublevel sets

$$S_\alpha = \{x \in \mathbf{dom}\, f \mid f(x) \le \alpha\},$$

for $\alpha \in \mathbf{R}$, are convex. A function is *quasiconcave* if $-f$ is quasiconvex, *i.e.*, every superlevel set $\{x \mid f(x) \ge \alpha\}$ is convex. A function that is both quasiconvex and quasiconcave is called *quasilinear*. If a function f is quasilinear, then its domain, and every level set $\{x \mid f(x) = \alpha\}$ is convex.

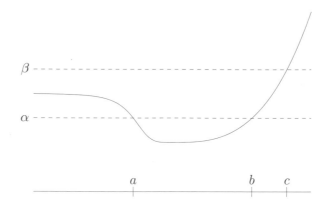

Figure 3.9 A quasiconvex function on $\mathbf{R}$. For each α, the α-sublevel set S_α is convex, *i.e.*, an interval. The sublevel set S_α is the interval $[a, b]$. The sublevel set S_β is the interval $(-\infty, c]$.

For a function on $\mathbf{R}$, quasiconvexity requires that each sublevel set be an interval (including, possibly, an infinite interval). An example of a quasiconvex function on $\mathbf{R}$ is shown in figure 3.9.

Convex functions have convex sublevel sets, and so are quasiconvex. But simple examples, such as the one shown in figure 3.9, show that the converse is not true.

Example 3.29 Some examples on $\mathbf{R}$:

- *Logarithm.* $\log x$ on $\mathbf{R}_{++}$ is quasiconvex (and quasiconcave, hence quasilinear).
- *Ceiling function.* $\mathrm{ceil}(x) = \inf\{z \in \mathbf{Z} \mid z \geq x\}$ is quasiconvex (and quasiconcave).

These examples show that quasiconvex functions can be concave, or discontinuous. We now give some examples on $\mathbf{R}^n$.

Example 3.30 *Length of a vector.* We define the *length* of $x \in \mathbf{R}^n$ as the largest index of a nonzero component, *i.e.*,

$$f(x) = \max\{i \mid x_i \neq 0\}.$$

(We define the length of the zero vector to be zero.) This function is quasiconvex on $\mathbf{R}^n$, since its sublevel sets are subspaces:

$$f(x) \leq \alpha \iff x_i = 0 \text{ for } i = \lfloor \alpha \rfloor + 1, \ldots, n.$$

Example 3.31 Consider $f : \mathbf{R}^2 \to \mathbf{R}$, with $\mathbf{dom}\, f = \mathbf{R}_+^2$ and $f(x_1, x_2) = x_1 x_2$. This function is neither convex nor concave since its Hessian

$$\nabla^2 f(x) = \begin{bmatrix} 0 & 1 \\ 1 & 0 \end{bmatrix}$$

is indefinite; it has one positive and one negative eigenvalue. The function f is quasiconcave, however, since the superlevel sets

$$\{x \in \mathbf{R}_+^2 \mid x_1 x_2 \geq \alpha\}$$

are convex sets for all α. (Note, however, that f is *not* quasiconcave on $\mathbf{R}^2$.)

Example 3.32 *Linear-fractional function.* The function

$$f(x) = \frac{a^T x + b}{c^T x + d},$$

with $\mathbf{dom}\, f = \{x \mid c^T x + d > 0\}$, is quasiconvex, and quasiconcave, *i.e.*, quasilinear. Its α-sublevel set is

$$
\begin{aligned}
S_\alpha &= \{x \mid c^T x + d > 0, \ (a^T x + b)/(c^T x + d) \leq \alpha\} \\
&= \{x \mid c^T x + d > 0, \ a^T x + b \leq \alpha(c^T x + d)\},
\end{aligned}
$$

which is convex, since it is the intersection of an open halfspace and a closed halfspace. (The same method can be used to show its superlevel sets are convex.)

Example 3.33 *Distance ratio function.* Suppose $a, b \in \mathbf{R}^n$, and define

$$f(x) = \frac{\|x - a\|_2}{\|x - b\|_2},$$

i.e., the ratio of the Euclidean distance to a to the distance to b. Then f is quasiconvex on the halfspace $\{x \mid \|x - a\|_2 \leq \|x - b\|_2\}$. To see this, we consider the α-sublevel set of f, with $\alpha \leq 1$ since $f(x) \leq 1$ on the halfspace $\{x \mid \|x - a\|_2 \leq \|x - b\|_2\}$. This sublevel set is the set of points satisfying

$$\|x - a\|_2 \leq \alpha \|x - b\|_2.$$

Squaring both sides, and rearranging terms, we see that this is equivalent to

$$(1 - \alpha^2)x^T x - 2(a - \alpha^2 b)^T x + a^T a - \alpha^2 b^T b \leq 0.$$

This describes a convex set (in fact a Euclidean ball) if $\alpha \leq 1$.

Example 3.34 *Internal rate of return.* Let $x = (x_0, x_1, \ldots, x_n)$ denote a cash flow sequence over n periods, where $x_i > 0$ means a payment to us in period i, and $x_i < 0$ means a payment by us in period i. We define the *present value* of a cash flow, with interest rate $r \geq 0$, to be

$$\mathrm{PV}(x, r) = \sum_{i=0}^{n} (1 + r)^{-i} x_i.$$

(The factor $(1 + r)^{-i}$ is a *discount factor* for a payment by or to us in period i.)

Now we consider cash flows for which $x_0 < 0$ and $x_0 + x_1 + \cdots + x_n > 0$. This means that we start with an investment of $|x_0|$ in period 0, and that the total of the remaining cash flow, $x_1 + \cdots + x_n$, (not taking any discount factors into account) exceeds our initial investment.

For such a cash flow, $\mathrm{PV}(x, 0) > 0$ and $\mathrm{PV}(x, r) \to x_0 < 0$ as $r \to \infty$, so it follows that for at least one $r \geq 0$, we have $\mathrm{PV}(x, r) = 0$. We define the *internal rate of return* of the cash flow as the smallest interest rate $r \geq 0$ for which the present value is zero:

$$\mathrm{IRR}(x) = \inf\{r \geq 0 \mid \mathrm{PV}(x, r) = 0\}.$$

Internal rate of return is a quasiconcave function of x (restricted to $x_0 < 0$, $x_1 + \cdots + x_n > 0$). To see this, we note that

$$\mathrm{IRR}(x) \geq R \iff \mathrm{PV}(x, r) > 0 \text{ for } 0 \leq r < R.$$

The lefthand side defines the R-superlevel set of IRR. The righthand side is the intersection of the sets $\{x \mid \mathrm{PV}(x, r) > 0\}$, indexed by r, over the range $0 \leq r < R$. For each r, $\mathrm{PV}(x, r) > 0$ defines an open halfspace, so the righthand side defines a convex set.

3.4.2   Basic properties

The examples above show that quasiconvexity is a considerable generalization of convexity. Still, many of the properties of convex functions hold, or have analogs, for quasiconvex functions. For example, there is a variation on Jensen's inequality that characterizes quasiconvexity: A function f is quasiconvex if and only if **dom** f is convex and for any x, $y \in \mathbf{dom}\, f$ and $0 \leq \theta \leq 1$,

$$f(\theta x + (1 - \theta)y) \leq \max\{f(x), f(y)\}, \tag{3.19}$$

i.e., the value of the function on a segment does not exceed the maximum of its values at the endpoints. The inequality (3.19) is sometimes called Jensen's inequality for quasiconvex functions, and is illustrated in figure 3.10.

Example 3.35   *Cardinality of a nonnegative vector.* The *cardinality* or *size* of a vector $x \in \mathbf{R}^n$ is the number of nonzero components, and denoted $\mathbf{card}(x)$. The function **card** is quasiconcave on $\mathbf{R}^n_+$ (but not $\mathbf{R}^n$). This follows immediately from the modified Jensen inequality

$$\mathbf{card}(x + y) \geq \min\{\mathbf{card}(x), \mathbf{card}(y)\},$$

which holds for x, $y \succeq 0$.

Example 3.36   *Rank of positive semidefinite matrix.* The function $\mathbf{rank}\, X$ is quasiconcave on $\mathbf{S}^n_+$. This follows from the modified Jensen inequality (3.19),

$$\mathbf{rank}(X + Y) \geq \min\{\mathbf{rank}\, X, \mathbf{rank}\, Y\}$$

which holds for X, $Y \in \mathbf{S}^n_+$. (This can be considered an extension of the previous example, since $\mathbf{rank}(\mathbf{diag}(x)) = \mathbf{card}(x)$ for $x \succeq 0$.)

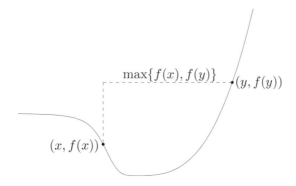

Figure 3.10 A quasiconvex function on $\mathbf{R}$. The value of f between x and y is no more than $\max\{f(x), f(y)\}$.

Like convexity, quasiconvexity is characterized by the behavior of a function f on lines: f is quasiconvex if and only if its restriction to any line intersecting its domain is quasiconvex. In particular, quasiconvexity of a function can be verified by restricting it to an arbitrary line, and then checking quasiconvexity of the resulting function on $\mathbf{R}$.

Quasiconvex functions on $\mathbf{R}$

We can give a simple characterization of quasiconvex functions on $\mathbf{R}$. We consider continuous functions, since stating the conditions in the general case is cumbersome. A continuous function $f : \mathbf{R} \to \mathbf{R}$ is quasiconvex if and only if at least one of the following conditions holds:

- f is nondecreasing

- f is nonincreasing

- there is a point $c \in \mathbf{dom}\, f$ such that for $t \le c$ (and $t \in \mathbf{dom}\, f$), f is nonincreasing, and for $t \ge c$ (and $t \in \mathbf{dom}\, f$), f is nondecreasing.

The point c can be chosen as any point which is a global minimizer of f. Figure 3.11 illustrates this.

3.4.3   Differentiable quasiconvex functions

First-order conditions

Suppose $f : \mathbf{R}^n \to \mathbf{R}$ is differentiable. Then f is quasiconvex if and only if $\mathbf{dom}\, f$ is convex and for all $x, y \in \mathbf{dom}\, f$

$$f(y) \le f(x) \Longrightarrow \nabla f(x)^T (y - x) \le 0. \qquad (3.20)$$

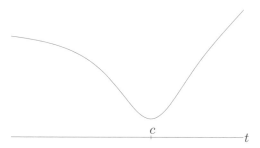

Figure 3.11 A quasiconvex function on **R**. The function is nonincreasing for $t \leq c$ and nondecreasing for $t \geq c$.

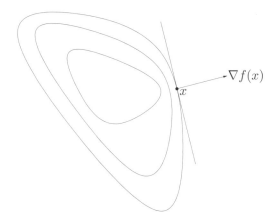

Figure 3.12 Three level curves of a quasiconvex function f are shown. The vector $\nabla f(x)$ defines a supporting hyperplane to the sublevel set $\{z \mid f(z) \leq f(x)\}$ at x.

This is the analog of inequality (3.2), for quasiconvex functions. We leave the proof as an exercise (exercise 3.43).

The condition (3.20) has a simple geometric interpretation when $\nabla f(x) \neq 0$. It states that $\nabla f(x)$ defines a supporting hyperplane to the sublevel set $\{y \mid f(y) \leq f(x)\}$, at the point x, as illustrated in figure 3.12.

While the first-order condition for convexity (3.2), and the first-order condition for quasiconvexity (3.20) are similar, there are some important differences. For example, if f is convex and $\nabla f(x) = 0$, then x is a global minimizer of f. But this statement is *false* for quasiconvex functions: it is possible that $\nabla f(x) = 0$, but x is not a global minimizer of f.

Second-order conditions

Now suppose f is twice differentiable. If f is quasiconvex, then for all $x \in \mathbf{dom}\, f$, and all $y \in \mathbf{R}^n$, we have

$$y^T \nabla f(x) = 0 \implies y^T \nabla^2 f(x) y \geq 0. \tag{3.21}$$

For a quasiconvex function on $\mathbf{R}$, this reduces to the simple condition

$$f'(x) = 0 \implies f''(x) \geq 0,$$

i.e., at any point with zero slope, the second derivative is nonnegative. For a quasiconvex function on $\mathbf{R}^n$, the interpretation of the condition (3.21) is a bit more complicated. As in the case $n = 1$, we conclude that whenever $\nabla f(x) = 0$, we must have $\nabla^2 f(x) \succeq 0$. When $\nabla f(x) \neq 0$, the condition (3.21) means that $\nabla^2 f(x)$ is positive semidefinite on the $(n-1)$-dimensional subspace $\nabla f(x)^\perp$. This implies that $\nabla^2 f(x)$ can have at most one negative eigenvalue.

As a (partial) converse, if f satisfies

$$y^T \nabla f(x) = 0 \implies y^T \nabla^2 f(x) y > 0 \tag{3.22}$$

for all $x \in \mathbf{dom}\, f$ and all $y \in \mathbf{R}^n$, $y \neq 0$, then f is quasiconvex. This condition is the same as requiring $\nabla^2 f(x)$ to be positive definite for any point with $\nabla f(x) = 0$, and for all other points, requiring $\nabla^2 f(x)$ to be positive definite on the $(n-1)$-dimensional subspace $\nabla f(x)^\perp$.

Proof of second-order conditions for quasiconvexity

By restricting the function to an arbitrary line, it suffices to consider the case in which $f : \mathbf{R} \to \mathbf{R}$.

We first show that if $f : \mathbf{R} \to \mathbf{R}$ is quasiconvex on an interval (a, b), then it must satisfy (3.21), *i.e.*, if $f'(c) = 0$ with $c \in (a, b)$, then we must have $f''(c) \geq 0$. If $f'(c) = 0$ with $c \in (a, b)$, $f''(c) < 0$, then for small positive ϵ we have $f(c - \epsilon) < f(c)$ and $f(c + \epsilon) < f(c)$. It follows that the sublevel set $\{x \mid f(x) \leq f(c) - \epsilon\}$ is disconnected for small positive ϵ, and therefore not convex, which contradicts our assumption that f is quasiconvex.

Now we show that if the condition (3.22) holds, then f is quasiconvex. Assume that (3.22) holds, *i.e.*, for each $c \in (a, b)$ with $f'(c) = 0$, we have $f''(c) > 0$. This means that whenever the function f' crosses the value 0, it is strictly increasing. Therefore it can cross the value 0 at most once. If f' does not cross the value 0 at all, then f is either nonincreasing or nondecreasing on (a, b), and therefore quasiconvex. Otherwise it must cross the value 0 exactly once, say at $c \in (a, b)$. Since $f''(c) > 0$, it follows that $f'(t) \leq 0$ for $a < t \leq c$, and $f'(t) \geq 0$ for $c \leq t < b$. This shows that f is quasiconvex.

3.4.4   Operations that preserve quasiconvexity

Nonnegative weighted maximum

A nonnegative weighted maximum of quasiconvex functions, *i.e.*,

$$f = \max\{w_1 f_1, \dots, w_m f_m\},$$

with $w_i \geq 0$ and f_i quasiconvex, is quasiconvex. The property extends to the general pointwise supremum

$$f(x) = \sup_{y \in C}(w(y)g(x, y))$$

where $w(y) \geq 0$ and $g(x, y)$ is quasiconvex in x for each y. This fact can be easily verified: $f(x) \leq \alpha$ if and only if

$$w(y)g(x, y) \leq \alpha \text{ for all } y \in C,$$

i.e., the α-sublevel set of f is the intersection of the α-sublevel sets of the functions $w(y)g(x, y)$ in the variable x.

Example 3.37 *Generalized eigenvalue.* The *maximum generalized eigenvalue* of a pair of symmetric matrices (X, Y), with $Y \succ 0$, is defined as

$$\lambda_{\max}(X, Y) = \sup_{u \neq 0} \frac{u^T X u}{u^T Y u} = \sup\{\lambda \mid \det(\lambda Y - X) = 0\}.$$

(See §A.5.3). This function is quasiconvex on $\mathbf{dom}\, f = \mathbf{S}^n \times \mathbf{S}^n_{++}$.

To see this we consider the expression

$$\lambda_{\max}(X, Y) = \sup_{u \neq 0} \frac{u^T X u}{u^T Y u}.$$

For each $u \neq 0$, the function $u^T X u / u^T Y u$ is linear-fractional in (X, Y), hence a quasiconvex function of (X, Y). We conclude that $\lambda_{\max}$ is quasiconvex, since it is the supremum of a family of quasiconvex functions.

Composition

If $g : \mathbf{R}^n \to \mathbf{R}$ is quasiconvex and $h : \mathbf{R} \to \mathbf{R}$ is nondecreasing, then $f = h \circ g$ is quasiconvex.

The composition of a quasiconvex function with an affine or linear-fractional transformation yields a quasiconvex function. If f is quasiconvex, then $g(x) = f(Ax + b)$ is quasiconvex, and $\tilde{g}(x) = f((Ax + b)/(c^T x + d))$ is quasiconvex on the set

$$\{x \mid c^T x + d > 0, \ (Ax + b)/(c^T x + d) \in \mathbf{dom}\, f\}.$$

Minimization

If $f(x, y)$ is quasiconvex jointly in x and y and C is a convex set, then the function

$$g(x) = \inf_{y \in C} f(x, y)$$

is quasiconvex.

To show this, we need to show that $\{x \mid g(x) \leq \alpha\}$ is convex, where $\alpha \in \mathbf{R}$ is arbitrary. From the definition of g, $g(x) \leq \alpha$ if and only if for any $\epsilon > 0$ there exists a $y \in C$ with $f(x, y) \le \alpha + \epsilon$. Now let x_1 and x_2 be two points in the α-sublevel set of g. Then for any $\epsilon > 0$, there exists $y_1, y_2 \in C$ with

$$f(x_1, y_1) \le \alpha + \epsilon, \qquad f(x_2, y_2) \le \alpha + \epsilon,$$

and since f is quasiconvex in x and y, we also have

$$f(\theta x_1 + (1 - \theta)x_2, \theta y_1 + (1 - \theta)y_2) \le \alpha + \epsilon,$$

for $0 \le \theta \le 1$. Hence $g(\theta x_1 + (1 - \theta)x_2) \le \alpha$, which proves that $\{x \mid g(x) \le \alpha\}$ is convex.

3.4.5   Representation via family of convex functions

In the sequel, it will be convenient to represent the sublevel sets of a quasiconvex function f (which are convex) via inequalities of convex functions. We seek a family of convex functions $\phi_t : \mathbf{R}^n \to \mathbf{R}$, indexed by $t \in \mathbf{R}$, with

$$f(x) \le t \iff \phi_t(x) \le 0, \tag{3.23}$$

i.e., the t-sublevel set of the quasiconvex function f is the 0-sublevel set of the convex function ϕ_t. Evidently ϕ_t must satisfy the property that for all $x \in \mathbf{R}^n$, $\phi_t(x) \le 0 \implies \phi_s(x) \le 0$ for $s \ge t$. This is satisfied if for each x, $\phi_t(x)$ is a nonincreasing function of t, i.e., $\phi_s(x) \le \phi_t(x)$ whenever $s \ge t$.

To see that such a representation always exists, we can take

$$\phi_t(x) = \begin{cases} 0 & f(x) \le t \\ \infty & \text{otherwise,} \end{cases}$$

i.e., ϕ_t is the indicator function of the t-sublevel of f. Obviously this representation is not unique; for example if the sublevel sets of f are closed, we can take

$$\phi_t(x) = \mathbf{dist}\,(x, \{z \mid f(z) \le t\})\,.$$

We are usually interested in a family ϕ_t with nice properties, such as differentiability.

Example 3.38   *Convex over concave function.* Suppose p is a convex function, q is a concave function, with $p(x) \ge 0$ and $q(x) > 0$ on a convex set C. Then the function f defined by $f(x) = p(x)/q(x)$, on C, is quasiconvex.

Here we have

$$f(x) \le t \iff p(x) - tq(x) \le 0,$$

so we can take $\phi_t(x) = p(x) - tq(x)$ for $t \ge 0$. For each t, ϕ_t is convex and for each x, $\phi_t(x)$ is decreasing in t.

3.5   Log-concave and log-convex functions

3.5.1   Definition

A function $f : \mathbf{R}^n \to \mathbf{R}$ is *logarithmically concave* or *log-concave* if $f(x) > 0$ for all $x \in \mathbf{dom}\, f$ and $\log f$ is concave. It is said to be *logarithmically convex* or *log-convex* if $\log f$ is convex. Thus f is log-convex if and only if $1/f$ is log-concave. It is convenient to allow f to take on the value zero, in which case we take $\log f(x) = -\infty$. In this case we say f is log-concave if the extended-value function $\log f$ is concave.

We can express log-concavity directly, without logarithms: a function $f : \mathbf{R}^n \to \mathbf{R}$, with convex domain and $f(x) > 0$ for all $x \in \mathbf{dom}\, f$, is log-concave if and only if for all x, $y \in \mathbf{dom}\, f$ and $0 \le \theta \le 1$, we have

$$f(\theta x + (1 - \theta)y) \ge f(x)^\theta f(y)^{1-\theta}.$$

In particular, the value of a log-concave function at the average of two points is at least the *geometric mean* of the values at the two points.

From the composition rules we know that e^h is convex if h is convex, so a log-convex function is convex. Similarly, a nonnegative concave function is log-concave. It is also clear that a log-convex function is quasiconvex and a log-concave function is quasiconcave, since the logarithm is monotone increasing.

Example 3.39   *Some simple examples of log-concave and log-convex functions.*

- *Affine function.* $f(x) = a^T x + b$ is log-concave on $\{x \mid a^T x + b > 0\}$.
- *Powers.* $f(x) = x^a$, on $\mathbf{R}_{++}$, is log-convex for $a \le 0$, and log-concave for $a \ge 0$.
- *Exponentials.* $f(x) = e^{ax}$ is log-convex and log-concave.
- The cumulative distribution function of a Gaussian density,

$$\Phi(x) = \frac{1}{\sqrt{2\pi}} \int_{-\infty}^{x} e^{-u^2/2} \, du,$$

  is log-concave (see exercise 3.54).
- *Gamma function.* The Gamma function,

$$\Gamma(x) = \int_{0}^{\infty} u^{x-1} e^{-u} \, du,$$

  is log-convex for $x \ge 1$ (see exercise 3.52).
- *Determinant.* $\det X$ is log concave on $\mathbf{S}_{++}^n$.
- *Determinant over trace.* $\det X / \mathbf{tr}\, X$ is log concave on $\mathbf{S}_{++}^n$ (see exercise 3.49).

Example 3.40   *Log-concave density functions.*   Many common probability density functions are log-concave. Two examples are the multivariate normal distribution,

$$f(x) = \frac{1}{\sqrt{(2\pi)^n \det \Sigma}} e^{-\frac{1}{2}(x-\bar{x})^T \Sigma^{-1}(x-\bar{x})}$$

(where $\bar{x} \in \mathbf{R}^n$ and $\Sigma \in \mathbf{S}_{++}^n$), and the exponential distribution on $\mathbf{R}_+^n$,

$$ f(x) = \left(\prod_{i=1}^n \lambda_i \right) e^{-\lambda^T x} $$

(where $\lambda \succ 0$). Another example is the uniform distribution over a convex set C,

$$ f(x) = \begin{cases} 1/\alpha & x \in C \\ 0 & x \notin C \end{cases} $$

where $\alpha = \mathbf{vol}(C)$ is the volume (Lebesgue measure) of C. In this case $\log f$ takes on the value $-\infty$ outside C, and $-\log \alpha$ on C, hence is concave.

As a more exotic example consider the Wishart distribution, defined as follows. Let $x_1, \ldots, x_p \in \mathbf{R}^n$ be independent Gaussian random vectors with zero mean and covariance $\Sigma \in \mathbf{S}^n$, with $p > n$. The random matrix $X = \sum_{i=1}^p x_i x_i^T$ has the Wishart density

$$ f(X) = a \, (\det X)^{(p-n-1)/2} \, e^{-\frac{1}{2}\, \mathbf{tr}(\Sigma^{-1} X)}, $$

with $\mathbf{dom}\, f = \mathbf{S}_{++}^n$, and a is a positive constant. The Wishart density is log-concave, since

$$ \log f(X) = \log a + \frac{p-n-1}{2} \log \det X - \frac{1}{2} \mathbf{tr}(\Sigma^{-1} X), $$

which is a concave function of X.

3.5.2   Properties

Twice differentiable log-convex/concave functions

Suppose f is twice differentiable, with $\mathbf{dom}\, f$ convex, so

$$ \nabla^2 \log f(x) = \frac{1}{f(x)} \nabla^2 f(x) - \frac{1}{f(x)^2} \nabla f(x) \nabla f(x)^T. $$

We conclude that f is log-convex if and only if for all $x \in \mathbf{dom}\, f$,

$$ f(x) \nabla^2 f(x) \succeq \nabla f(x) \nabla f(x)^T, $$

and log-concave if and only if for all $x \in \mathbf{dom}\, f$,

$$ f(x) \nabla^2 f(x) \preceq \nabla f(x) \nabla f(x)^T. $$

Multiplication, addition, and integration

Log-convexity and log-concavity are closed under multiplication and positive scaling. For example, if f and g are log-concave, then so is the pointwise product $h(x) = f(x)g(x)$, since $\log h(x) = \log f(x) + \log g(x)$, and $\log f(x)$ and $\log g(x)$ are concave functions of x.

Simple examples show that the sum of log-concave functions is not, in general, log-concave. Log-convexity, however, is preserved under sums. Let f and g be log-convex functions, i.e., $F = \log f$ and $G = \log g$ are convex. From the composition rules for convex functions, it follows that

$$ \log\left(\exp F + \exp G\right) = \log(f + g) $$

is convex. Therefore the sum of two log-convex functions is log-convex.

More generally, if $f(x,y)$ is log-convex in x for each $y \in C$ then

$$g(x) = \int_C f(x,y) \, dy$$

is log-convex.

Example 3.41   *Laplace transform of a nonnegative function and the moment and cumulant generating functions.* Suppose $p : \mathbf{R}^n \to \mathbf{R}$ satisfies $p(x) \geq 0$ for all x. The Laplace transform of p,

$$P(z) = \int p(x) e^{-z^T x} \, dx,$$

is log-convex on $\mathbf{R}^n$. (Here $\operatorname{\mathbf{dom}} P$ is, naturally, $\{z \mid P(z) < \infty\}$.)

Now suppose p is a density, *i.e.*, satisfies $\int p(x) \, dx = 1$. The function $M(z) = P(-z)$ is called the *moment generating function* of the density. It gets its name from the fact that the moments of the density can be found from the derivatives of the moment generating function, evaluated at $z = 0$, *e.g.*,

$$\nabla M(0) = \mathbf{E}\, v, \qquad \nabla^2 M(0) = \mathbf{E}\, vv^T,$$

where v is a random variable with density p.

The function $\log M(z)$, which is convex, is called the *cumulant generating function* for p, since its derivatives give the cumulants of the density. For example, the first and second derivatives of the cumulant generating function, evaluated at zero, are the mean and covariance of the associated random variable:

$$\nabla \log M(0) = \mathbf{E}\, v, \qquad \nabla^2 \log M(0) = \mathbf{E}(v - \mathbf{E}\, v)(v - \mathbf{E}\, v)^T.$$

Integration of log-concave functions

In some special cases log-concavity is preserved by integration. If $f : \mathbf{R}^n \times \mathbf{R}^m \to \mathbf{R}$ is log-concave, then

$$g(x) = \int f(x,y) \, dy$$

is a log-concave function of x (on $\mathbf{R}^n$). (The integration here is over $\mathbf{R}^m$.) A proof of this result is not simple; see the references.

This result has many important consequences, some of which we describe in the rest of this section. It implies, for example, that marginal distributions of log-concave probability densities are log-concave. It also implies that log-concavity is closed under convolution, *i.e.*, if f and g are log-concave on $\mathbf{R}^n$, then so is the convolution

$$(f * g)(x) = \int f(x - y)g(y) \, dy.$$

(To see this, note that $g(y)$ and $f(x-y)$ are log-concave in (x,y), hence the product $f(x - y)g(y)$ is; then the integration result applies.)

Suppose $C \subseteq \mathbf{R}^n$ is a convex set and w is a random vector in $\mathbf{R}^n$ with log-concave probability density p. Then the function

$$f(x) = \mathbf{prob}(x + w \in C)$$

is log-concave in x. To see this, express f as

$$f(x) = \int g(x + w)p(w) \, dw,$$

where g is defined as

$$g(u) = \begin{cases} 1 & u \in C \\ 0 & u \notin C, \end{cases}$$

(which is log-concave) and apply the integration result.

Example 3.42 The *cumulative distribution function* of a probability density function $f : \mathbf{R}^n \to \mathbf{R}$ is defined as

$$F(x) = \mathbf{prob}(w \preceq x) = \int_{-\infty}^{x_n} \cdots \int_{-\infty}^{x_1} f(z) \, dz_1 \cdots dz_n,$$

where w is a random variable with density f. If f is log-concave, then F is log-concave. We have already encountered a special case: the cumulative distribution function of a Gaussian random variable,

$$f(x) = \frac{1}{\sqrt{2\pi}} \int_{-\infty}^{x} e^{-t^2/2} \, dt,$$

is log-concave. (See example 3.39 and exercise 3.54.)

Example 3.43 *Yield function.* Let $x \in \mathbf{R}^n$ denote the nominal or target value of a set of parameters of a product that is manufactured. Variation in the manufacturing process causes the parameters of the product, when manufactured, to have the value $x + w$, where $w \in \mathbf{R}^n$ is a random vector that represents manufacturing variation, and is usually assumed to have zero mean. The *yield* of the manufacturing process, as a function of the nominal parameter values, is given by

$$Y(x) = \mathbf{prob}(x + w \in S),$$

where $S \subseteq \mathbf{R}^n$ denotes the set of acceptable parameter values for the product, *i.e.*, the product *specifications*.

If the density of the manufacturing error w is log-concave (for example, Gaussian) and the set S of product specifications is convex, then the yield function Y is log-concave. This implies that the *α-yield region*, defined as the set of nominal parameters for which the yield exceeds α, is convex. For example, the 95% yield region

$$\{x \mid Y(x) \geq 0.95\} = \{x \mid \log Y(x) \geq \log 0.95\}$$

is convex, since it is a superlevel set of the concave function $\log Y$.

Example 3.44 *Volume of polyhedron.* Let $A \in \mathbf{R}^{m \times n}$. Define

$$P_u = \{x \in \mathbf{R}^n \mid Ax \preceq u\}.$$

Then its volume $\mathbf{vol}\, P_u$ is a log-concave function of u.

To prove this, note that the function

$$\Psi(x,u) = \begin{cases} 1 & Ax \preceq u \\ 0 & \text{otherwise,} \end{cases}$$

is log-concave. By the integration result, we conclude that

$$\int \Psi(x,u)\, dx = \mathbf{vol}\, P_u$$

is log-concave.

3.6   Convexity with respect to generalized inequalities

We now consider generalizations of the notions of monotonicity and convexity, using generalized inequalities instead of the usual ordering on $\mathbf{R}$.

3.6.1   Monotonicity with respect to a generalized inequality

Suppose $K \subseteq \mathbf{R}^n$ is a proper cone with associated generalized inequality $\preceq_K$. A function $f : \mathbf{R}^n \to \mathbf{R}$ is called *K-nondecreasing* if

$$x \preceq_K y \Longrightarrow f(x) \le f(y),$$

and *K-increasing* if

$$x \preceq_K y, \ x \ne y \Longrightarrow f(x) < f(y).$$

We define *K-nonincreasing* and *K-decreasing* functions in a similar way.

Example 3.45 *Monotone vector functions.* A function $f : \mathbf{R}^n \to \mathbf{R}$ is nondecreasing with respect to $\mathbf{R}^n_+$ if and only if

$$x_1 \le y_1, \ldots, x_n \le y_n \implies f(x) \le f(y)$$

for all x, y. This is the same as saying that f, when restricted to any component x_i (*i.e.*, x_i is considered the variable while x_j for $j \ne i$ are fixed), is nondecreasing.

Example 3.46 *Matrix monotone functions.* A function $f : \mathbf{S}^n \to \mathbf{R}$ is called *matrix monotone* (increasing, decreasing) if it is monotone with respect to the positive semidefinite cone. Some examples of matrix monotone functions of the variable $X \in \mathbf{S}^n$:

- **tr**(WX), where $W \in \mathbf{S}^n$, is matrix nondecreasing if $W \succeq 0$, and matrix increasing if $W \succ 0$ (it is matrix nonincreasing if $W \preceq 0$, and matrix decreasing if $W \prec 0$).
- **tr**(X^{-1}) is matrix decreasing on $\mathbf{S}^n_{++}$.
- $\det X$ is matrix increasing on $\mathbf{S}^n_{++}$, and matrix nondecreasing on $\mathbf{S}^n_+$.

Gradient conditions for monotonicity

Recall that a differentiable function $f : \mathbf{R} \to \mathbf{R}$, with convex (*i.e.*, interval) domain, is nondecreasing if and only if $f'(x) \geq 0$ for all $x \in \mathbf{dom}\, f$, and increasing if $f'(x) > 0$ for all $x \in \mathbf{dom}\, f$ (but the converse is not true). These conditions are readily extended to the case of monotonicity with respect to a generalized inequality. A differentiable function f, with convex domain, is K-nondecreasing if and only if

$$\nabla f(x) \succeq_{K^*} 0 \tag{3.24}$$

for all $x \in \mathbf{dom}\, f$. Note the difference with the simple scalar case: the gradient must be nonnegative in the *dual* inequality. For the strict case, we have the following: If

$$\nabla f(x) \succ_{K^*} 0 \tag{3.25}$$

for all $x \in \mathbf{dom}\, f$, then f is K-increasing. As in the scalar case, the converse is not true.

Let us prove these first-order conditions for monotonicity. First, assume that f satisfies (3.24) for all x, but is not K-nondecreasing, *i.e.*, there exist x, y with $x \preceq_K y$ and $f(y) < f(x)$. By differentiability of f there exists a $t \in [0,1]$ with

$$\frac{d}{dt} f(x + t(y - x)) = \nabla f(x + t(y - x))^T (y - x) < 0.$$

Since $y - x \in K$ this means

$$\nabla f(x + t(y - x)) \notin K^*,$$

which contradicts our assumption that (3.24) is satisfied everywhere. In a similar way it can be shown that (3.25) implies f is K-increasing.

It is also straightforward to see that it is necessary that (3.24) hold everywhere. Assume (3.24) does not hold for $x = z$. By the definition of dual cone this means there exists a $v \in K$ with

$$\nabla f(z)^T v < 0.$$

Now consider $h(t) = f(z + tv)$ as a function of t. We have $h'(0) = \nabla f(z)^T v < 0$, and therefore there exists $t > 0$ with $h(t) = f(z + tv) < h(0) = f(z)$, which means f is not K-nondecreasing.

3.6.2 Convexity with respect to a generalized inequality

Suppose $K \subseteq \mathbf{R}^m$ is a proper cone with associated generalized inequality $\preceq_K$. We say $f : \mathbf{R}^n \to \mathbf{R}^m$ is K-*convex* if for all x, y, and $0 \leq \theta \leq 1$,

$$f(\theta x + (1 - \theta)y) \preceq_K \theta f(x) + (1 - \theta)f(y).$$

The function is *strictly K-convex* if

$$f(\theta x + (1-\theta)y) \prec_K \theta f(x) + (1-\theta)f(y)$$

for all $x \neq y$ and $0 < \theta < 1$. These definitions reduce to ordinary convexity and strict convexity when $m = 1$ (and $K = \mathbf{R}_+$).

Example 3.47 *Convexity with respect to componentwise inequality.* A function $f : \mathbf{R}^n \to \mathbf{R}^m$ is convex with respect to componentwise inequality (*i.e.*, the generalized inequality induced by $\mathbf{R}^m_+$) if and only if for all x, y and $0 \leq \theta \leq 1$,

$$f(\theta x + (1-\theta)y) \preceq \theta f(x) + (1-\theta)f(y),$$

i.e., each component f_i is a convex function. The function f is strictly convex with respect to componentwise inequality if and only if each component f_i is strictly convex.

Example 3.48 *Matrix convexity.* Suppose f is a symmetric matrix valued function, *i.e.*, $f : \mathbf{R}^n \to \mathbf{S}^m$. The function f is convex with respect to matrix inequality if

$$f(\theta x + (1-\theta)y) \preceq \theta f(x) + (1-\theta)f(y)$$

for any x and y, and for $\theta \in [0,1]$. This is sometimes called *matrix convexity*. An equivalent definition is that the scalar function $z^T f(x)z$ is convex for all vectors z. (This is often a good way to prove matrix convexity). A matrix function is strictly matrix convex if

$$f(\theta x + (1-\theta)y) \prec \theta f(x) + (1-\theta)f(y)$$

when $x \neq y$ and $0 < \theta < 1$, or, equivalently, if $z^T f z$ is strictly convex for every $z \neq 0$.

Some examples:

- The function $f(X) = XX^T$ where $X \in \mathbf{R}^{n \times m}$ is matrix convex, since for fixed z the function $z^T XX^T z = \|X^T z\|_2^2$ is a convex quadratic function of (the components of) X. For the same reason, $f(X) = X^2$ is matrix convex on $\mathbf{S}^n$.

- The function X^p is matrix convex on $\mathbf{S}^n_{++}$ for $1 \leq p \leq 2$ or $-1 \leq p \leq 0$, and matrix concave for $0 \leq p \leq 1$.

- The function $f(X) = e^X$ is *not* matrix convex on $\mathbf{S}^n$, for $n \geq 2$.

Many of the results for convex functions have extensions to K-convex functions. As a simple example, a function is K-convex if and only if its restriction to any line in its domain is K-convex. In the rest of this section we list a few results for K-convexity that we will use later; more results are explored in the exercises.

Dual characterization of K-convexity

A function f is K-convex if and only if for every $w \succeq_{K^*} 0$, the (real-valued) function $w^T f$ is convex (in the ordinary sense); f is strictly K-convex if and only if for every nonzero $w \succeq_{K^*} 0$ the function $w^T f$ is strictly convex. (These follow directly from the definitions and properties of dual inequality.)

Differentiable K-convex functions

A differentiable function f is K-convex if and only if its domain is convex, and for all x, $y \in \mathbf{dom}\, f$,

$$f(y) \succeq_K f(x) + Df(x)(y - x).$$

(Here $Df(x) \in \mathbf{R}^{m \times n}$ is the derivative or Jacobian matrix of f at x; see §A.4.1.) The function f is strictly K-convex if and only if for all x, $y \in \mathbf{dom}\, f$ with $x \neq y$,

$$f(y) \succ_K f(x) + Df(x)(y - x).$$

Composition theorem

Many of the results on composition can be generalized to K-convexity. For example, if $g : \mathbf{R}^n \to \mathbf{R}^p$ is K-convex, $h : \mathbf{R}^p \to \mathbf{R}$ is convex, and $\tilde{h}$ (the extended-value extension of h) is K-nondecreasing, then $h \circ g$ is convex. This generalizes the fact that a nondecreasing convex function of a convex function is convex. The condition that $\tilde{h}$ be K-nondecreasing implies that $\mathbf{dom}\, h - K = \mathbf{dom}\, h$.

Example 3.49 The quadratic matrix function $g : \mathbf{R}^{m \times n} \to \mathbf{S}^n$ defined by

$$g(X) = X^T A X + B^T X + X^T B + C,$$

where $A \in \mathbf{S}^m$, $B \in \mathbf{R}^{m \times n}$, and $C \in \mathbf{S}^n$, is convex when $A \succeq 0$.

The function $h : \mathbf{S}^n \to \mathbf{R}$ defined by $h(Y) = -\log\det(-Y)$ is convex and increasing on $\mathbf{dom}\, h = -\mathbf{S}^n_{++}$.

By the composition theorem, we conclude that

$$f(X) = -\log\det(-(X^T A X + B^T X + X^T B + C))$$

is convex on

$$\mathbf{dom}\, f = \{X \in \mathbf{R}^{m \times n} \mid X^T A X + B^T X + X^T B + C \prec 0\}.$$

This generalizes the fact that

$$-\log(-(ax^2 + bx + c))$$

is convex on

$$\{x \in \mathbf{R} \mid ax^2 + bx + c < 0\},$$

provided $a \geq 0$.

Bibliography

The standard reference on convex analysis is Rockafellar [Roc70]. Other books on convex functions are Stoer and Witzgall [SW70], Roberts and Varberg [RV73], Van Tiel [vT84], Hiriart-Urruty and Lemaréchal [HUL93], Ekeland and Témam [ET99], Borwein and Lewis [BL00], Florenzano and Le Van [FL01], Barvinok [Bar02], and Bertsekas, Nedić, and Ozdaglar [Ber03]. Most nonlinear programming texts also include chapters on convex functions (see, for example, Mangasarian [Man94], Bazaraa, Sherali, and Shetty [BSS93], Bertsekas [Ber99], Polyak [Pol87], and Peressini, Sullivan, and Uhl [PSU88]).

Jensen's inequality appears in [Jen06]. A general study of inequalities, in which Jensen's inequality plays a central role, is presented by Hardy, Littlewood, and Pólya [HLP52], and Beckenbach and Bellman [BB65].

The term *perspective function* is from Hiriart-Urruty and Lemaréchal [HUL93, volume 1, page 100]. For the definitions in example 3.19 (relative entropy and Kullback-Leibler divergence), and the related exercise 3.13, see Cover and Thomas [CT91].

Some important early references on quasiconvex functions (as well as other extensions of convexity) are Nikaidô [Nik54], Mangasarian [Man94, chapter 9], Arrow and Enthoven [AE61], Ponstein [Pon67], and Luenberger [Lue68]. For a more comprehensive reference list, we refer to Bazaraa, Sherali, and Shetty [BSS93, page 126].

Prékopa [Pré80] gives a survey of log-concave functions. Log-convexity of the Laplace transform is mentioned in Barndorff-Nielsen [BN78, §7]. For a proof of the integration result of log-concave functions, see Prékopa [Pré71, Pré73].

Generalized inequalities are used extensively in the recent literature on cone programming, starting with Nesterov and Nemirovski [NN94, page 156]; see also Ben-Tal and Nemirovski [BTN01] and the references at the end of chapter 4. Convexity with respect to generalized inequalities also appears in the work of Luenberger [Lue69, §8.2] and Isii [Isi64]. Matrix monotonicity and matrix convexity are attributed to Löwner [Löw34], and are discussed in detail by Davis [Dav63], Roberts and Varberg [RV73, page 216] and Marshall and Olkin [MO79, §16E]. For the result on convexity and concavity of the function X^p in example 3.48, see Bondar [Bon94, theorem 16.1]. For a simple example that demonstrates that e^X is not matrix convex, see Marshall and Olkin [MO79, page 474].

Exercises

Definition of convexity

3.1 Suppose $f : \mathbf{R} \to \mathbf{R}$ is convex, and a, $b \in \mathbf{dom}\, f$ with $a < b$.

(a) Show that

$$f(x) \le \frac{b - x}{b - a} f(a) + \frac{x - a}{b - a} f(b)$$

for all $x \in [a, b]$.

(b) Show that

$$\frac{f(x) - f(a)}{x - a} \le \frac{f(b) - f(a)}{b - a} \le \frac{f(b) - f(x)}{b - x}$$

for all $x \in (a, b)$. Draw a sketch that illustrates this inequality.

(c) Suppose f is differentiable. Use the result in (b) to show that

$$f'(a) \le \frac{f(b) - f(a)}{b - a} \le f'(b).$$

Note that these inequalities also follow from (3.2):

$$f(b) \ge f(a) + f'(a)(b - a), \qquad f(a) \ge f(b) + f'(b)(a - b).$$

(d) Suppose f is twice differentiable. Use the result in (c) to show that $f''(a) \ge 0$ and $f''(b) \ge 0$.

3.2 *Level sets of convex, concave, quasiconvex, and quasiconcave functions.* Some level sets of a function f are shown below. The curve labeled 1 shows $\{x \mid f(x) = 1\}$, etc.

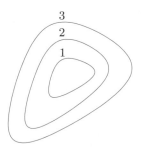

Could f be convex (concave, quasiconvex, quasiconcave)? Explain your answer. Repeat for the level curves shown below.

3.3 *Inverse of an increasing convex function.* Suppose $f : \mathbf{R} \to \mathbf{R}$ is increasing and convex on its domain (a, b). Let g denote its inverse, *i.e.*, the function with domain $(f(a), f(b))$ and $g(f(x)) = x$ for $a < x < b$. What can you say about convexity or concavity of g?

3.4 [RV73, page 15] Show that a continuous function $f : \mathbf{R}^n \to \mathbf{R}$ is convex if and only if for every line segment, its average value on the segment is less than or equal to the average of its values at the endpoints of the segment: For every $x, \ y \in \mathbf{R}^n$,

$$\int_0^1 f(x + \lambda(y - x)) \, d\lambda \le \frac{f(x) + f(y)}{2}.$$

3.5 [RV73, page 22] *Running average of a convex function.* Suppose $f : \mathbf{R} \to \mathbf{R}$ is convex, with $\mathbf{R}_+ \subseteq \mathbf{dom} \, f$. Show that its *running average* F, defined as

$$F(x) = \frac{1}{x} \int_0^x f(t) \, dt, \qquad \mathbf{dom} \, F = \mathbf{R}_{++},$$

is convex. *Hint.* For each s, $f(sx)$ is convex in x, so $\int_0^1 f(sx) \, ds$ is convex.

3.6 *Functions and epigraphs.* When is the epigraph of a function a halfspace? When is the epigraph of a function a convex cone? When is the epigraph of a function a polyhedron?

3.7 Suppose $f : \mathbf{R}^n \to \mathbf{R}$ is convex with $\mathbf{dom} \, f = \mathbf{R}^n$, and bounded above on $\mathbf{R}^n$. Show that f is constant.

3.8 *Second-order condition for convexity.* Prove that a twice differentiable function f is convex if and only if its domain is convex and $\nabla^2 f(x) \succeq 0$ for all $x \in \mathbf{dom} \, f$. *Hint.* First consider the case $f : \mathbf{R} \to \mathbf{R}$. You can use the first-order condition for convexity (which was proved on page 70).

3.9 *Second-order conditions for convexity on an affine set.* Let $F \in \mathbf{R}^{n \times m}$, $\hat{x} \in \mathbf{R}^n$. The *restriction* of $f : \mathbf{R}^n \to \mathbf{R}$ to the affine set $\{Fz + \hat{x} \mid z \in \mathbf{R}^m\}$ is defined as the function $\tilde{f} : \mathbf{R}^m \to \mathbf{R}$ with

$$\tilde{f}(z) = f(Fz + \hat{x}), \qquad \mathbf{dom} \, \tilde{f} = \{z \mid Fz + \hat{x} \in \mathbf{dom} \, f\}.$$

Suppose f is twice differentiable with a convex domain.

(a) Show that $\tilde{f}$ is convex if and only if for all $z \in \mathbf{dom} \, \tilde{f}$

$$F^T \nabla^2 f(Fz + \hat{x}) F \succeq 0.$$

(b) Suppose $A \in \mathbf{R}^{p \times n}$ is a matrix whose nullspace is equal to the range of F, *i.e.*, $AF = 0$ and $\mathbf{rank} \, A = n - \mathbf{rank} \, F$. Show that $\tilde{f}$ is convex if and only if for all $z \in \mathbf{dom} \, \tilde{f}$ there exists a $\lambda \in \mathbf{R}$ such that

$$\nabla^2 f(Fz + \hat{x}) + \lambda A^T A \succeq 0.$$

Hint. Use the following result: If $B \in \mathbf{S}^n$ and $A \in \mathbf{R}^{p \times n}$, then $x^T B x \ge 0$ for all $x \in \mathcal{N}(A)$ if and only if there exists a λ such that $B + \lambda A^T A \succeq 0$.

3.10 *An extension of Jensen's inequality.* One interpretation of Jensen's inequality is that randomization or dithering hurts, *i.e.*, raises the average value of a convex function: For f convex and v a zero mean random variable, we have $\mathbf{E} f(x_0 + v) \ge f(x_0)$. This leads to the following conjecture. If f_0 is convex, then the larger the variance of v, the larger $\mathbf{E} f(x_0 + v)$.

(a) Give a counterexample that shows that this conjecture is false. Find zero mean random variables v and w, with $\mathbf{var}(v) > \mathbf{var}(w)$, a convex function f, and a point x_0, such that $\mathbf{E} f(x_0 + v) < \mathbf{E} f(x_0 + w)$.

(b) The conjecture is true when v and w are scaled versions of each other. Show that $\mathbf{E} f(x_0 + tv)$ is monotone increasing in $t \geq 0$, when f is convex and v is zero mean.

3.11 *Monotone mappings.* A function $\psi : \mathbf{R}^n \to \mathbf{R}^n$ is called *monotone* if for all x, $y \in \mathbf{dom}\,\psi$,

$$(\psi(x) - \psi(y))^T (x - y) \geq 0.$$

(Note that 'monotone' as defined here is not the same as the definition given in §3.6.1. Both definitions are widely used.) Suppose $f : \mathbf{R}^n \to \mathbf{R}$ is a differentiable convex function. Show that its gradient ∇f is monotone. Is the converse true, *i.e.*, is every monotone mapping the gradient of a convex function?

3.12 Suppose $f : \mathbf{R}^n \to \mathbf{R}$ is convex, $g : \mathbf{R}^n \to \mathbf{R}$ is concave, $\mathbf{dom}\,f = \mathbf{dom}\,g = \mathbf{R}^n$, and for all x, $g(x) \leq f(x)$. Show that there exists an affine function h such that for all x, $g(x) \leq h(x) \leq f(x)$. In other words, if a concave function g is an underestimator of a convex function f, then we can fit an affine function between f and g.

3.13 *Kullback-Leibler divergence and the information inequality.* Let D_{kl} be the Kullback-Leibler divergence, as defined in (3.17). Prove the *information inequality*: $D_{\mathrm{kl}}(u, v) \geq 0$ for all u, $v \in \mathbf{R}^n_{++}$. Also show that $D_{\mathrm{kl}}(u, v) = 0$ if and only if $u = v$.

Hint. The Kullback-Leibler divergence can be expressed as

$$D_{\mathrm{kl}}(u, v) = f(u) - f(v) - \nabla f(v)^T (u - v),$$

where $f(v) = \sum_{i=1}^{n} v_i \log v_i$ is the negative entropy of v.

3.14 *Convex-concave functions and saddle-points.* We say the function $f : \mathbf{R}^n \times \mathbf{R}^m \to \mathbf{R}$ is *convex-concave* if $f(x, z)$ is a concave function of z, for each fixed x, and a convex function of x, for each fixed z. We also require its domain to have the product form $\mathbf{dom}\,f = A \times B$, where $A \subseteq \mathbf{R}^n$ and $B \subseteq \mathbf{R}^m$ are convex.

(a) Give a second-order condition for a twice differentiable function $f : \mathbf{R}^n \times \mathbf{R}^m \to \mathbf{R}$ to be convex-concave, in terms of its Hessian $\nabla^2 f(x, z)$.

(b) Suppose that $f : \mathbf{R}^n \times \mathbf{R}^m \to \mathbf{R}$ is convex-concave and differentiable, with $\nabla f(\tilde{x}, \tilde{z}) = 0$. Show that the *saddle-point property* holds: for all x, z, we have

$$f(\tilde{x}, z) \leq f(\tilde{x}, \tilde{z}) \leq f(x, \tilde{z}).$$

Show that this implies that f satisfies the *strong max-min property*:

$$\sup_z \inf_x f(x, z) = \inf_x \sup_z f(x, z)$$

(and their common value is $f(\tilde{x}, \tilde{z})$).

(c) Now suppose that $f : \mathbf{R}^n \times \mathbf{R}^m \to \mathbf{R}$ is differentiable, but not necessarily convex-concave, and the saddle-point property holds at $\tilde{x}, \tilde{z}$:

$$f(\tilde{x}, z) \leq f(\tilde{x}, \tilde{z}) \leq f(x, \tilde{z})$$

for all x, z. Show that $\nabla f(\tilde{x}, \tilde{z}) = 0$.

Examples

3.15 *A family of concave utility functions.* For $0 < \alpha \leq 1$ let

$$u_\alpha(x) = \frac{x^\alpha - 1}{\alpha},$$

with $\mathbf{dom}\,u_\alpha = \mathbf{R}_+$. We also define $u_0(x) = \log x$ (with $\mathbf{dom}\,u_0 = \mathbf{R}_{++}$).

(a) Show that for $x > 0$, $u_0(x) = \lim_{\alpha \to 0} u_\alpha(x)$.

(b) Show that u_α are concave, monotone increasing, and all satisfy $u_\alpha(1) = 0$.

These functions are often used in economics to model the benefit or utility of some quantity of goods or money. Concavity of u_α means that the marginal utility (*i.e.*, the increase in utility obtained for a fixed increase in the goods) decreases as the amount of goods increases. In other words, concavity models the effect of *satiation*.

3.16 For each of the following functions determine whether it is convex, concave, quasiconvex, or quasiconcave.

(a) $f(x) = e^x - 1$ on $\mathbf{R}$.

(b) $f(x_1, x_2) = x_1 x_2$ on $\mathbf{R}_{++}^2$.

(c) $f(x_1, x_2) = 1/(x_1 x_2)$ on $\mathbf{R}_{++}^2$.

(d) $f(x_1, x_2) = x_1/x_2$ on $\mathbf{R}_{++}^2$.

(e) $f(x_1, x_2) = x_1^2/x_2$ on $\mathbf{R} \times \mathbf{R}_{++}$.

(f) $f(x_1, x_2) = x_1^\alpha x_2^{1-\alpha}$, where $0 \le \alpha \le 1$, on $\mathbf{R}_{++}^2$.

3.17 Suppose $p < 1$, $p \ne 0$. Show that the function

$$f(x) = \left(\sum_{i=1}^n x_i^p \right)^{1/p}$$

with $\mathbf{dom}\, f = \mathbf{R}_{++}^n$ is concave. This includes as special cases $f(x) = (\sum_{i=1}^n x_i^{1/2})^2$ and the *harmonic mean* $f(x) = (\sum_{i=1}^n 1/x_i)^{-1}$. *Hint.* Adapt the proofs for the log-sum-exp function and the geometric mean in §3.1.5.

3.18 Adapt the proof of concavity of the log-determinant function in §3.1.5 to show the following.

(a) $f(X) = \mathbf{tr}\left(X^{-1}\right)$ is convex on $\mathbf{dom}\, f = \mathbf{S}_{++}^n$.

(b) $f(X) = (\det X)^{1/n}$ is concave on $\mathbf{dom}\, f = \mathbf{S}_{++}^n$.

3.19 *Nonnegative weighted sums and integrals.*

(a) Show that $f(x) = \sum_{i=1}^r \alpha_i x_{[i]}$ is a convex function of x, where $\alpha_1 \ge \alpha_2 \ge \cdots \ge \alpha_r \ge 0$, and $x_{[i]}$ denotes the ith largest component of x. (You can use the fact that $f(x) = \sum_{i=1}^k x_{[i]}$ is convex on $\mathbf{R}^n$.)

(b) Let $T(x, \omega)$ denote the trigonometric polynomial

$$T(x, \omega) = x_1 + x_2 \cos \omega + x_3 \cos 2\omega + \cdots + x_n \cos(n-1)\omega.$$

Show that the function

$$f(x) = -\int_0^{2\pi} \log T(x, \omega)\, d\omega$$

is convex on $\{x \in \mathbf{R}^n \mid T(x, \omega) > 0,\ 0 \le \omega \le 2\pi\}$.

3.20 *Composition with an affine function.* Show that the following functions $f : \mathbf{R}^n \to \mathbf{R}$ are convex.

(a) $f(x) = \|Ax - b\|$, where $A \in \mathbf{R}^{m \times n}$, $b \in \mathbf{R}^m$, and $\|\cdot\|$ is a norm on $\mathbf{R}^m$.

(b) $f(x) = -\left(\det(A_0 + x_1 A_1 + \cdots + x_n A_n)\right)^{1/m}$, on $\{x \mid A_0 + x_1 A_1 + \cdots + x_n A_n \succ 0\}$, where $A_i \in \mathbf{S}^m$.

(c) $f(X) = \mathbf{tr}\left(A_0 + x_1 A_1 + \cdots + x_n A_n\right)^{-1}$, on $\{x \mid A_0 + x_1 A_1 + \cdots + x_n A_n \succ 0\}$, where $A_i \in \mathbf{S}^m$. (Use the fact that $\mathbf{tr}(X^{-1})$ is convex on $\mathbf{S}_{++}^m$; see exercise 3.18.)

3.21 *Pointwise maximum and supremum.* Show that the following functions $f : \mathbf{R}^n \to \mathbf{R}$ are convex.

(a) $f(x) = \max_{i=1,\dots,k} \|A^{(i)} x - b^{(i)}\|$, where $A^{(i)} \in \mathbf{R}^{m \times n}$, $b^{(i)} \in \mathbf{R}^m$ and $\| \cdot \|$ is a norm on $\mathbf{R}^m$.

(b) $f(x) = \sum_{i=1}^r |x|_{[i]}$ on $\mathbf{R}^n$, where $|x|$ denotes the vector with $|x|_i = |x_i|$ (*i.e.*, $|x|$ is the absolute value of x, componentwise), and $|x|_{[i]}$ is the ith largest component of $|x|$. In other words, $|x|_{[1]}, |x|_{[2]}, \dots, |x|_{[n]}$ are the absolute values of the components of x, sorted in nonincreasing order.

3.22 *Composition rules.* Show that the following functions are convex.

(a) $f(x) = -\log(-\log(\sum_{i=1}^m e^{a_i^T x + b_i}))$ on $\mathbf{dom}\, f = \{x \mid \sum_{i=1}^m e^{a_i^T x + b_i} < 1\}$. You can use the fact that $\log(\sum_{i=1}^n e^{y_i})$ is convex.

(b) $f(x, u, v) = -\sqrt{uv - x^T x}$ on $\mathbf{dom}\, f = \{(x, u, v) \mid uv > x^T x, \ u, \ v > 0\}$. Use the fact that $x^T x / u$ is convex in (x, u) for $u > 0$, and that $-\sqrt{x_1 x_2}$ is convex on $\mathbf{R}_{++}^2$.

(c) $f(x, u, v) = -\log(uv - x^T x)$ on $\mathbf{dom}\, f = \{(x, u, v) \mid uv > x^T x, \ u, \ v > 0\}$.

(d) $f(x, t) = -(t^p - \|x\|_p^p)^{1/p}$ where $p > 1$ and $\mathbf{dom}\, f = \{(x, t) \mid t \geq \|x\|_p\}$. You can use the fact that $\|x\|_p^p / u^{p-1}$ is convex in (x, u) for $u > 0$ (see exercise 3.23), and that $-x^{1/p} y^{1 - 1/p}$ is convex on $\mathbf{R}_+^2$ (see exercise 3.16).

(e) $f(x, t) = -\log(t^p - \|x\|_p^p)$ where $p > 1$ and $\mathbf{dom}\, f = \{(x, t) \mid t > \|x\|_p\}$. You can use the fact that $\|x\|_p^p / u^{p-1}$ is convex in (x, u) for $u > 0$ (see exercise 3.23).

3.23 *Perspective of a function.*

(a) Show that for $p > 1$,

$$f(x, t) = \frac{|x_1|^p + \cdots + |x_n|^p}{t^{p-1}} = \frac{\|x\|_p^p}{t^{p-1}}$$

is convex on $\{(x, t) \mid t > 0\}$.

(b) Show that

$$f(x) = \frac{\|Ax + b\|_2^2}{c^T x + d}$$

is convex on $\{x \mid c^T x + d > 0\}$, where $A \in \mathbf{R}^{m \times n}$, $b \in \mathbf{R}^m$, $c \in \mathbf{R}^n$ and $d \in \mathbf{R}$.

3.24 *Some functions on the probability simplex.* Let x be a real-valued random variable which takes values in $\{a_1, \dots, a_n\}$ where $a_1 < a_2 < \cdots < a_n$, with $\mathbf{prob}(x = a_i) = p_i$, $i = 1, \dots, n$. For each of the following functions of p (on the probability simplex $\{p \in \mathbf{R}_+^n \mid \mathbf{1}^T p = 1\}$), determine if the function is convex, concave, quasiconvex, or quasiconcave.

(a) $\mathbf{E}\, x$.

(b) $\mathbf{prob}(x \geq \alpha)$.

(c) $\mathbf{prob}(\alpha \leq x \leq \beta)$.

(d) $\sum_{i=1}^n p_i \log p_i$, the negative entropy of the distribution.

(e) $\mathbf{var}\, x = \mathbf{E}(x - \mathbf{E}\, x)^2$.

(f) $\mathbf{quartile}(x) = \inf\{\beta \mid \mathbf{prob}(x \leq \beta) \geq 0.25\}$.

(g) The cardinality of the smallest set $\mathcal{A} \subseteq \{a_1, \dots, a_n\}$ with probability $\geq 90\%$. (By cardinality we mean the number of elements in $\mathcal{A}$.)

(h) The minimum width interval that contains 90% of the probability, *i.e.*,

$$\inf \{\beta - \alpha \mid \mathbf{prob}(\alpha \leq x \leq \beta) \geq 0.9\}.$$

3.25 *Maximum probability distance between distributions.* Let p, $q \in \mathbf{R}^n$ represent two probability distributions on $\{1, \ldots, n\}$ (so p, $q \succeq 0$, $\mathbf{1}^T p = \mathbf{1}^T q = 1$). We define the *maximum probability distance* $d_{\mathrm{mp}}(p, q)$ between p and q as the maximum difference in probability assigned by p and q, over all events:

$$d_{\mathrm{mp}}(p, q) = \max\{|\, \mathbf{prob}(p, C) - \mathbf{prob}(q, C)| \mid C \subseteq \{1, \ldots, n\}\}.$$

Here $\mathbf{prob}(p, C)$ is the probability of C, under the distribution p, *i.e.*, $\mathbf{prob}(p, C) = \sum_{i \in C} p_i$.

Find a simple expression for d_{mp}, involving $\|p - q\|_1 = \sum_{i=1}^n |p_i - q_i|$, and show that d_{mp} is a convex function on $\mathbf{R}^n \times \mathbf{R}^n$. (Its domain is $\{(p, q) \mid p, q \succeq 0, \mathbf{1}^T p = \mathbf{1}^T q = 1\}$, but it has a natural extension to all of $\mathbf{R}^n \times \mathbf{R}^n$.)

3.26 *More functions of eigenvalues.* Let $\lambda_1(X) \geq \lambda_2(X) \geq \cdots \geq \lambda_n(X)$ denote the eigenvalues of a matrix $X \in \mathbf{S}^n$. We have already seen several functions of the eigenvalues that are convex or concave functions of X.

- The maximum eigenvalue $\lambda_1(X)$ is convex (example 3.10). The minimum eigenvalue $\lambda_n(X)$ is concave.
- The sum of the eigenvalues (or trace), $\mathbf{tr}\, X = \lambda_1(X) + \cdots + \lambda_n(X)$, is linear.
- The sum of the inverses of the eigenvalues (or trace of the inverse), $\mathbf{tr}(X^{-1}) = \sum_{i=1}^n 1/\lambda_i(X)$, is convex on $\mathbf{S}^n_{++}$ (exercise 3.18).
- The geometric mean of the eigenvalues, $(\det X)^{1/n} = (\prod_{i=1}^n \lambda_i(X))^{1/n}$, and the logarithm of the product of the eigenvalues, $\log \det X = \sum_{i=1}^n \log \lambda_i(X)$, are concave on $X \in \mathbf{S}^n_{++}$ (exercise 3.18 and page 74).

In this problem we explore some more functions of eigenvalues, by exploiting variational characterizations.

(a) *Sum of k largest eigenvalues.* Show that $\sum_{i=1}^k \lambda_i(X)$ is convex on $\mathbf{S}^n$. *Hint.* [HJ85, page 191] Use the variational characterization

$$\sum_{i=1}^k \lambda_i(X) = \sup\{\mathbf{tr}(V^T X V) \mid V \in \mathbf{R}^{n \times k}, V^T V = I\}.$$

(b) *Geometric mean of k smallest eigenvalues.* Show that $(\prod_{i=n-k+1}^n \lambda_i(X))^{1/k}$ is concave on $\mathbf{S}^n_{++}$. *Hint.* [MO79, page 513] For $X \succ 0$, we have

$$\left(\prod_{i=n-k+1}^n \lambda_i(X) \right)^{1/k} = \frac{1}{k} \inf\{\mathbf{tr}(V^T X V) \mid V \in \mathbf{R}^{n \times k}, \det V^T V = 1\}.$$

(c) *Log of product of k smallest eigenvalues.* Show that $\sum_{i=n-k+1}^n \log \lambda_i(X)$ is concave on $\mathbf{S}^n_{++}$. *Hint.* [MO79, page 513] For $X \succ 0$,

$$\prod_{i=n-k+1}^n \lambda_i(X) = \inf\left\{ \prod_{i=1}^k (V^T X V)_{ii} \;\middle|\; V \in \mathbf{R}^{n \times k}, V^T V = I \right\}.$$

3.27 *Diagonal elements of Cholesky factor.* Each $X \in \mathbf{S}^n_{++}$ has a unique *Cholesky factorization* $X = LL^T$, where L is lower triangular, with $L_{ii} > 0$. Show that L_{ii} is a concave function of X (with domain $\mathbf{S}^n_{++}$).

Hint. L_{ii} can be expressed as $L_{ii} = (w - z^T Y^{-1} z)^{1/2}$, where

$$\begin{bmatrix} Y & z \\ z^T & w \end{bmatrix}$$

is the leading $i \times i$ submatrix of X.

Operations that preserve convexity

3.28 *Expressing a convex function as the pointwise supremum of a family of affine functions.* In this problem we extend the result proved on page 83 to the case where $\mathbf{dom}\, f \neq \mathbf{R}^n$. Let $f : \mathbf{R}^n \to \mathbf{R}$ be a convex function. Define $\tilde{f} : \mathbf{R}^n \to \mathbf{R}$ as the pointwise supremum of all affine functions that are global underestimators of f:

$$\tilde{f}(x) = \sup\{g(x) \mid g \text{ affine}, \ g(z) \le f(z) \text{ for all } z\}.$$

(a) Show that $f(x) = \tilde{f}(x)$ for $x \in \mathbf{int}\,\mathbf{dom}\, f$.

(b) Show that $f = \tilde{f}$ if f is closed (*i.e.*, $\mathbf{epi}\, f$ is a closed set; see §A.3.3).

3.29 *Representation of piecewise-linear convex functions.* A function $f : \mathbf{R}^n \to \mathbf{R}$, with $\mathbf{dom}\, f = \mathbf{R}^n$, is called *piecewise-linear* if there exists a partition of $\mathbf{R}^n$ as

$$\mathbf{R}^n = X_1 \cup X_2 \cup \cdots \cup X_L,$$

where $\mathbf{int}\, X_i \neq \emptyset$ and $\mathbf{int}\, X_i \cap \mathbf{int}\, X_j = \emptyset$ for $i \neq j$, and a family of affine functions $a_1^T x + b_1, \ldots, a_L^T x + b_L$ such that $f(x) = a_i^T x + b_i$ for $x \in X_i$.

Show that this means that $f(x) = \max\{a_1^T x + b_1, \ldots, a_L^T x + b_L\}$.

3.30 *Convex hull or envelope of a function.* The *convex hull* or *convex envelope* of a function $f : \mathbf{R}^n \to \mathbf{R}$ is defined as

$$g(x) = \inf\{t \mid (x,t) \in \mathbf{conv}\,\mathbf{epi}\, f\}.$$

Geometrically, the epigraph of g is the convex hull of the epigraph of f.

Show that g is the largest convex underestimator of f. In other words, show that if h is convex and satisfies $h(x) \le f(x)$ for all x, then $h(x) \le g(x)$ for all x.

3.31 [Roc70, page 35] *Largest homogeneous underestimator.* Let f be a convex function. Define the function g as

$$g(x) = \inf_{\alpha > 0} \frac{f(\alpha x)}{\alpha}.$$

(a) Show that g is homogeneous ($g(tx) = tg(x)$ for all $t \ge 0$).

(b) Show that g is the largest homogeneous underestimator of f: If h is homogeneous and $h(x) \le f(x)$ for all x, then we have $h(x) \le g(x)$ for all x.

(c) Show that g is convex.

3.32 *Products and ratios of convex functions.* In general the product or ratio of two convex functions is not convex. However, there are some results that apply to functions on $\mathbf{R}$. Prove the following.

(a) If f and g are convex, both nondecreasing (or nonincreasing), and positive functions on an interval, then fg is convex.

(b) If f, g are concave, positive, with one nondecreasing and the other nonincreasing, then fg is concave.

(c) If f is convex, nondecreasing, and positive, and g is concave, nonincreasing, and positive, then f/g is convex.

3.33 *Direct proof of perspective theorem.* Give a direct proof that the perspective function g, as defined in §3.2.6, of a convex function f is convex: Show that $\mathbf{dom}\, g$ is a convex set, and that for (x,t), $(y,s) \in \mathbf{dom}\, g$, and $0 \le \theta \le 1$, we have

$$g(\theta x + (1-\theta)y, \theta t + (1-\theta)s) \le \theta g(x,t) + (1-\theta)g(y,s).$$

3.34 *The Minkowski function.* The *Minkowski function* of a convex set C is defined as

$$M_C(x) = \inf\{t > 0 \mid t^{-1}x \in C\}.$$

(a) Draw a picture giving a geometric interpretation of how to find $M_C(x)$.

(b) Show that M_C is homogeneous, i.e., $M_C(\alpha x) = \alpha M_C(x)$ for $\alpha \geq 0$.

(c) What is $\mathbf{dom}\, M_C$?

(d) Show that M_C is a convex function.

(e) Suppose C is also closed, bounded, symmetric (if $x \in C$ then $-x \in C$), and has nonempty interior. Show that M_C is a norm. What is the corresponding unit ball?

3.35 *Support function calculus.* Recall that the support function of a set $C \subseteq \mathbf{R}^n$ is defined as $S_C(y) = \sup\{y^T x \mid x \in C\}$. On page 81 we showed that S_C is a convex function.

(a) Show that $S_B = S_{\mathbf{conv}\, B}$.

(b) Show that $S_{A+B} = S_A + S_B$.

(c) Show that $S_{A \cup B} = \max\{S_A, S_B\}$.

(d) Let B be closed and convex. Show that $A \subseteq B$ if and only if $S_A(y) \leq S_B(y)$ for all y.

Conjugate functions

3.36 Derive the conjugates of the following functions.

(a) *Max function.* $f(x) = \max_{i=1,\ldots,n} x_i$ on $\mathbf{R}^n$.

(b) *Sum of largest elements.* $f(x) = \sum_{i=1}^{r} x_{[i]}$ on $\mathbf{R}^n$.

(c) *Piecewise-linear function on* $\mathbf{R}$. $f(x) = \max_{i=1,\ldots,m}(a_i x + b_i)$ on $\mathbf{R}$. You can assume that the a_i are sorted in increasing order, i.e., $a_1 \leq \cdots \leq a_m$, and that none of the functions $a_i x + b_i$ is redundant, i.e., for each k there is at least one x with $f(x) = a_k x + b_k$.

(d) *Power function.* $f(x) = x^p$ on $\mathbf{R}_{++}$, where $p > 1$. Repeat for $p < 0$.

(e) *Negative geometric mean.* $f(x) = -(\prod x_i)^{1/n}$ on $\mathbf{R}_{++}^n$.

(f) *Negative generalized logarithm for second-order cone.* $f(x,t) = -\log(t^2 - x^T x)$ on $\{(x,t) \in \mathbf{R}^n \times \mathbf{R} \mid \|x\|_2 < t\}$.

3.37 Show that the conjugate of $f(X) = \mathbf{tr}(X^{-1})$ with $\mathbf{dom}\, f = \mathbf{S}_{++}^n$ is given by

$$f^*(Y) = -2\,\mathbf{tr}(-Y)^{1/2}, \qquad \mathbf{dom}\, f^* = -\mathbf{S}_+^n.$$

Hint. The gradient of f is $\nabla f(X) = -X^{-2}$.

3.38 *Young's inequality.* Let $f : \mathbf{R} \to \mathbf{R}$ be an increasing function, with $f(0) = 0$, and let g be its inverse. Define F and G as

$$F(x) = \int_0^x f(a)\, da, \qquad G(y) = \int_0^y g(a)\, da.$$

Show that F and G are conjugates. Give a simple graphical interpretation of Young's inequality,

$$xy \leq F(x) + G(y).$$

3.39 *Properties of conjugate functions.*

(a) *Conjugate of convex plus affine function.* Define $g(x) = f(x) + c^T x + d$, where f is convex. Express g^* in terms of f^* (and c, d).

(b) *Conjugate of perspective.* Express the conjugate of the perspective of a convex function f in terms of f^*.

(c) *Conjugate and minimization.* Let $f(x,z)$ be convex in (x,z) and define $g(x) = \inf_z f(x,z)$. Express the conjugate g^* in terms of f^*.

As an application, express the conjugate of $g(x) = \inf_z\{h(z) \mid Az+b=x\}$, where h is convex, in terms of h^*, A, and b.

(d) *Conjugate of conjugate.* Show that the conjugate of the conjugate of a closed convex function is itself: $f = f^{**}$ if f is closed and convex. (A function is closed if its epigraph is closed; see §A.3.3.) *Hint.* Show that f^{**} is the pointwise supremum of all affine global underestimators of f. Then apply the result of exercise 3.28.

3.40 *Gradient and Hessian of conjugate function.* Suppose $f : \mathbf{R}^n \to \mathbf{R}$ is convex and twice continuously differentiable. Suppose $\bar{y}$ and $\bar{x}$ are related by $\bar{y} = \nabla f(\bar{x})$, and that $\nabla^2 f(\bar{x}) \succ 0$.

(a) Show that $\nabla f^*(\bar{y}) = \bar{x}$.

(b) Show that $\nabla^2 f^*(\bar{y}) = \nabla^2 f(\bar{x})^{-1}$.

3.41 *Conjugate of negative normalized entropy.* Show that the conjugate of the negative normalized entropy

$$f(x) = \sum_{i=1}^n x_i \log(x_i/\mathbf{1}^T x),$$

with $\mathbf{dom}\, f = \mathbf{R}^n_{++}$, is given by

$$f^*(y) = \begin{cases} 0 & \sum_{i=1}^n e^{y_i} \le 1 \\ +\infty & \text{otherwise.} \end{cases}$$

Quasiconvex functions

3.42 *Approximation width.* Let $f_0, \ldots, f_n : \mathbf{R} \to \mathbf{R}$ be given continuous functions. We consider the problem of approximating f_0 as a linear combination of $f_1, \ldots, f_n$. For $x \in \mathbf{R}^n$, we say that $f = x_1 f_1 + \cdots + x_n f_n$ approximates f_0 with tolerance $\epsilon > 0$ over the interval $[0, T]$ if $|f(t) - f_0(t)| \le \epsilon$ for $0 \le t \le T$. Now we choose a fixed tolerance $\epsilon > 0$ and define the *approximation width* as the largest T such that f approximates f_0 over the interval $[0, T]$:

$$W(x) = \sup\{T \mid |x_1 f_1(t) + \cdots + x_n f_n(t) - f_0(t)| \le \epsilon \text{ for } 0 \le t \le T\}.$$

Show that W is quasiconcave.

3.43 *First-order condition for quasiconvexity.* Prove the first-order condition for quasiconvexity given in §3.4.3: A differentiable function $f : \mathbf{R}^n \to \mathbf{R}$, with $\mathbf{dom}\, f$ convex, is quasiconvex if and only if for all $x, y \in \mathbf{dom}\, f$,

$$f(y) \le f(x) \implies \nabla f(x)^T (y - x) \le 0.$$

Hint. It suffices to prove the result for a function on $\mathbf{R}$; the general result follows by restriction to an arbitrary line.

3.44 *Second-order conditions for quasiconvexity.* In this problem we derive alternate representations of the second-order conditions for quasiconvexity given in §3.4.3. Prove the following.

(a) A point $x \in \mathbf{dom}\, f$ satisfies (3.21) if and only if there exists a σ such that

$$\nabla^2 f(x) + \sigma \nabla f(x) \nabla f(x)^T \succeq 0. \qquad (3.26)$$

It satisfies (3.22) for all $y \ne 0$ if and only if there exists a σ such

$$\nabla^2 f(x) + \sigma \nabla f(x) \nabla f(x)^T \succ 0. \qquad (3.27)$$

Hint. We can assume without loss of generality that $\nabla^2 f(x)$ is diagonal.

(b) A point $x \in \mathbf{dom}\, f$ satisfies (3.21) if and only if either $\nabla f(x) = 0$ and $\nabla^2 f(x) \succeq 0$, or $\nabla f(x) \neq 0$ and the matrix

$$H(x) = \begin{bmatrix} \nabla^2 f(x) & \nabla f(x) \\ \nabla f(x)^T & 0 \end{bmatrix}$$

has exactly one negative eigenvalue. It satisfies (3.22) for all $y \neq 0$ if and only if $H(x)$ has exactly one nonpositive eigenvalue.

Hint. You can use the result of part (a). The following result, which follows from the eigenvalue interlacing theorem in linear algebra, may also be useful: If $B \in \mathbf{S}^n$ and $a \in \mathbf{R}^n$, then

$$\lambda_n \left(\begin{bmatrix} B & a \\ a^T & 0 \end{bmatrix} \right) \geq \lambda_n(B).$$

3.45 Use the first and second-order conditions for quasiconvexity given in §3.4.3 to verify quasiconvexity of the function $f(x) = -x_1 x_2$, with $\mathbf{dom}\, f = \mathbf{R}^2_{++}$.

3.46 *Quasilinear functions with domain $\mathbf{R}^n$.* A function on $\mathbf{R}$ that is quasilinear (*i.e.*, quasiconvex and quasiconcave) is monotone, *i.e.*, either nondecreasing or nonincreasing. In this problem we consider a generalization of this result to functions on $\mathbf{R}^n$.

Suppose the function $f : \mathbf{R}^n \to \mathbf{R}$ is quasilinear and continuous with $\mathbf{dom}\, f = \mathbf{R}^n$. Show that it can be expressed as $f(x) = g(a^T x)$, where $g : \mathbf{R} \to \mathbf{R}$ is monotone and $a \in \mathbf{R}^n$. In other words, a quasilinear function with domain $\mathbf{R}^n$ must be a monotone function of a linear function. (The converse is also true.)

Log-concave and log-convex functions

3.47 Suppose $f : \mathbf{R}^n \to \mathbf{R}$ is differentiable, $\mathbf{dom}\, f$ is convex, and $f(x) > 0$ for all $x \in \mathbf{dom}\, f$. Show that f is log-concave if and only if for all $x, y \in \mathbf{dom}\, f$,

$$\frac{f(y)}{f(x)} \leq \exp \left(\frac{\nabla f(x)^T (y - x)}{f(x)} \right).$$

3.48 Show that if $f : \mathbf{R}^n \to \mathbf{R}$ is log-concave and $a \geq 0$, then the function $g = f - a$ is log-concave, where $\mathbf{dom}\, g = \{x \in \mathbf{dom}\, f \mid f(x) > a\}$.

3.49 Show that the following functions are log-concave.

(a) *Logistic function:* $f(x) = e^x/(1 + e^x)$ with $\mathbf{dom}\, f = \mathbf{R}$.

(b) *Harmonic mean:*

$$f(x) = \frac{1}{1/x_1 + \cdots + 1/x_n}, \qquad \mathbf{dom}\, f = \mathbf{R}^n_{++}.$$

(c) *Product over sum:*

$$f(x) = \frac{\prod_{i=1}^n x_i}{\sum_{i=1}^n x_i}, \qquad \mathbf{dom}\, f = \mathbf{R}^n_{++}.$$

(d) *Determinant over trace:*

$$f(X) = \frac{\det X}{\mathbf{tr}\, X}, \qquad \mathbf{dom}\, f = \mathbf{S}^n_{++}.$$

3.50 *Coefficients of a polynomial as a function of the roots.* Show that the coefficients of a polynomial with real negative roots are log-concave functions of the roots. In other words, the functions $a_i : \mathbf{R}^n \to \mathbf{R}$, defined by the identity

$$s^n + a_1(\lambda)s^{n-1} + \cdots + a_{n-1}(\lambda)s + a_n(\lambda) = (s - \lambda_1)(s - \lambda_2)\cdots(s - \lambda_n),$$

are log-concave on $-\mathbf{R}_{++}^n$.

Hint. The function

$$S_k(x) = \sum_{1 \le i_1 < i_2 < \cdots < i_k \le n} x_{i_1} x_{i_2} \cdots x_{i_k},$$

with $\mathbf{dom}\, S_k \in \mathbf{R}_+^n$ and $1 \le k \le n$, is called the kth elementary symmetric function on $\mathbf{R}^n$. It can be shown that $S_k^{1/k}$ is concave (see [ML57]).

3.51 [BL00, page 41] Let p be a polynomial on $\mathbf{R}$, with all its roots real. Show that it is log-concave on any interval on which it is positive.

3.52 [MO79, §3.E.2] *Log-convexity of moment functions.* Suppose $f : \mathbf{R} \to \mathbf{R}$ is nonnegative with $\mathbf{R}_+ \subseteq \mathbf{dom}\, f$. For $x \ge 0$ define

$$\phi(x) = \int_0^\infty u^x f(u)\, du.$$

Show that ϕ is a log-convex function. (If x is a positive integer, and f is a probability density function, then $\phi(x)$ is the xth moment of the distribution.)

Use this to show that the Gamma function,

$$\Gamma(x) = \int_0^\infty u^{x-1} e^{-u}\, du,$$

is log-convex for $x \ge 1$.

3.53 Suppose x and y are independent random vectors in $\mathbf{R}^n$, with log-concave probability density functions f and g, respectively. Show that the probability density function of the sum $z = x + y$ is log-concave.

3.54 *Log-concavity of Gaussian cumulative distribution function.* The cumulative distribution function of a Gaussian random variable,

$$f(x) = \frac{1}{\sqrt{2\pi}} \int_{-\infty}^x e^{-t^2/2}\, dt,$$

is log-concave. This follows from the general result that the convolution of two log-concave functions is log-concave. In this problem we guide you through a simple self-contained proof that f is log-concave. Recall that f is log-concave if and only if $f''(x)f(x) \le f'(x)^2$ for all x.

(a) Verify that $f''(x)f(x) \le f'(x)^2$ for $x \ge 0$. That leaves us the hard part, which is to show the inequality for $x < 0$.

(b) Verify that for any t and x we have $t^2/2 \ge -x^2/2 + xt$.

(c) Using part (b) show that $e^{-t^2/2} \le e^{x^2/2 - xt}$. Conclude that, for $x < 0$,

$$\int_{-\infty}^x e^{-t^2/2}\, dt \le e^{x^2/2} \int_{-\infty}^x e^{-xt}\, dt.$$

(d) Use part (c) to verify that $f''(x)f(x) \le f'(x)^2$ for $x \le 0$.

3.55 *Log-concavity of the cumulative distribution function of a log-concave probability density.* In this problem we extend the result of exercise 3.54. Let $g(t) = \exp(-h(t))$ be a differentiable log-concave probability density function, and let

$$f(x) = \int_{-\infty}^{x} g(t)\, dt = \int_{-\infty}^{x} e^{-h(t)}\, dt$$

be its cumulative distribution. We will show that f is log-concave, *i.e.*, it satisfies $f''(x)f(x) \le (f'(x))^2$ for all x.

   (a) Express the derivatives of f in terms of the function h. Verify that $f''(x)f(x) \le (f'(x))^2$ if $h'(x) \ge 0$.

   (b) Assume that $h'(x) < 0$. Use the inequality

$$h(t) \ge h(x) + h'(x)(t - x)$$

   (which follows from convexity of h), to show that

$$\int_{-\infty}^{x} e^{-h(t)}\, dt \le \frac{e^{-h(x)}}{-h'(x)}.$$

   Use this inequality to verify that $f''(x)f(x) \le (f'(x))^2$ if $h'(x) < 0$.

3.56 *More log-concave densities.* Show that the following densities are log-concave.

   (a) [MO79, page 493] The *gamma density*, defined by

$$f(x) = \frac{\alpha^{\lambda}}{\Gamma(\lambda)} x^{\lambda - 1} e^{-\alpha x},$$

   with $\mathbf{dom}\, f = \mathbf{R}_+$. The parameters λ and α satisfy $\lambda \ge 1$, $\alpha > 0$.

   (b) [MO79, page 306] The *Dirichlet density*

$$f(x) = \frac{\Gamma(\mathbf{1}^T \lambda)}{\Gamma(\lambda_1) \cdots \Gamma(\lambda_{n+1})} x_1^{\lambda_1 - 1} \cdots x_n^{\lambda_n - 1} \left(1 - \sum_{i=1}^{n} x_i \right)^{\lambda_{n+1} - 1}$$

   with $\mathbf{dom}\, f = \{x \in \mathbf{R}_{++}^n \mid \mathbf{1}^T x < 1\}$. The parameter λ satisfies $\lambda \succeq \mathbf{1}$.

Convexity with respect to a generalized inequality

3.57 Show that the function $f(X) = X^{-1}$ is matrix convex on $\mathbf{S}_{++}^n$.

3.58 *Schur complement.* Suppose $X \in \mathbf{S}^n$ partitioned as

$$X = \begin{bmatrix} A & B \\ B^T & C \end{bmatrix},$$

where $A \in \mathbf{S}^k$. The *Schur complement* of X (with respect to A) is $S = C - B^T A^{-1} B$ (see §A.5.5). Show that the Schur complement, viewed as a function from $\mathbf{S}^n$ into $\mathbf{S}^{n-k}$, is matrix concave on $\mathbf{S}_{++}^n$.

3.59 *Second-order conditions for K-convexity.* Let $K \subseteq \mathbf{R}^m$ be a proper convex cone, with associated generalized inequality $\preceq_K$. Show that a twice differentiable function $f : \mathbf{R}^n \to \mathbf{R}^m$, with convex domain, is K-convex if and only if for all $x \in \mathbf{dom}\, f$ and all $y \in \mathbf{R}^n$,

$$\sum_{i,j=1}^{n} \frac{\partial^2 f(x)}{\partial x_i \partial x_j} y_i y_j \succeq_K 0,$$

i.e., the second derivative is a K-nonnegative bilinear form. (Here $\partial^2 f/\partial x_i \partial x_j \in \mathbf{R}^m$, with components $\partial^2 f_k/\partial x_i \partial x_j$, for $k = 1, \dots, m$; see §A.4.1.)

3.60 *Sublevel sets and epigraph of K-convex functions.* Let $K \subseteq \mathbf{R}^m$ be a proper convex cone with associated generalized inequality $\preceq_K$, and let $f : \mathbf{R}^n \to \mathbf{R}^m$. For $\alpha \in \mathbf{R}^m$, the α-sublevel set of f (with respect to $\preceq_K$) is defined as

$$C_\alpha = \{x \in \mathbf{R}^n \mid f(x) \preceq_K \alpha\}.$$

The epigraph of f, with respect to $\preceq_K$, is defined as the set

$$\mathbf{epi}_K f = \{(x, t) \in \mathbf{R}^{n+m} \mid f(x) \preceq_K t\}.$$

Show the following:

(a) If f is K-convex, then its sublevel sets C_α are convex for all α.

(b) f is K-convex if and only if $\mathbf{epi}_K f$ is a convex set.

Chapter 4

Convex optimization problems

4.1 Optimization problems

4.1.1 Basic terminology

We use the notation

$$
\begin{array}{ll}
\text{minimize} & f_0(x) \\
\text{subject to} & f_i(x) \le 0, \quad i = 1, \ldots, m \\
& h_i(x) = 0, \quad i = 1, \ldots, p
\end{array}
\tag{4.1}
$$

to describe the problem of finding an x that minimizes $f_0(x)$ among all x that satisfy the conditions $f_i(x) \le 0$, $i = 1, \ldots, m$, and $h_i(x) = 0$, $i = 1, \ldots, p$. We call $x \in \mathbf{R}^n$ the *optimization variable* and the function $f_0 : \mathbf{R}^n \to \mathbf{R}$ the *objective function* or *cost function*. The inequalities $f_i(x) \le 0$ are called *inequality constraints*, and the corresponding functions $f_i : \mathbf{R}^n \to \mathbf{R}$ are called the *inequality constraint functions*. The equations $h_i(x) = 0$ are called the *equality constraints*, and the functions $h_i : \mathbf{R}^n \to \mathbf{R}$ are the *equality constraint functions*. If there are no constraints (*i.e.*, $m = p = 0$) we say the problem (4.1) is *unconstrained*.

The set of points for which the objective and all constraint functions are defined,

$$
\mathcal{D} = \bigcap_{i=0}^{m} \operatorname{\mathbf{dom}} f_i \ \cap\ \bigcap_{i=1}^{p} \operatorname{\mathbf{dom}} h_i,
$$

is called the *domain* of the optimization problem (4.1). A point $x \in \mathcal{D}$ is *feasible* if it satisfies the constraints $f_i(x) \le 0$, $i = 1, \ldots, m$, and $h_i(x) = 0$, $i = 1, \ldots, p$. The problem (4.1) is said to be feasible if there exists at least one feasible point, and *infeasible* otherwise. The set of all feasible points is called the *feasible set* or the *constraint set*.

The *optimal value* $p^\star$ of the problem (4.1) is defined as

$$
p^\star = \inf \left\{ f_0(x) \mid f_i(x) \le 0, \ i = 1, \ldots, m, \ h_i(x) = 0, \ i = 1, \ldots, p \right\}.
$$

We allow $p^\star$ to take on the extended values $\pm\infty$. If the problem is infeasible, we have $p^\star = \infty$ (following the standard convention that the infimum of the empty set is ∞). If there are feasible points x_k with $f_0(x_k) \to -\infty$ as $k \to \infty$, then $p^\star = -\infty$, and we say the problem (4.1) is *unbounded below*.

Optimal and locally optimal points

We say $x^\star$ is an *optimal point*, or solves the problem (4.1), if $x^\star$ is feasible and $f_0(x^\star) = p^\star$. The set of all optimal points is the *optimal set*, denoted

$$X_{\mathrm{opt}} = \{x \mid f_i(x) \le 0, \ i = 1, \ldots, m, \ h_i(x) = 0, \ i = 1, \ldots, p, \ f_0(x) = p^\star\}.$$

If there exists an optimal point for the problem (4.1), we say the optimal value is *attained* or *achieved*, and the problem is *solvable*. If X_{opt} is empty, we say the optimal value is not attained or not achieved. (This always occurs when the problem is unbounded below.) A feasible point x with $f_0(x) \le p^\star + \epsilon$ (where $\epsilon > 0$) is called ϵ-*suboptimal*, and the set of all ϵ-suboptimal points is called the ϵ-*suboptimal set* for the problem (4.1).

We say a feasible point x is *locally optimal* if there is an $R > 0$ such that

$$\begin{aligned} f_0(x) = \inf\{f_0(z) \mid f_i(z) \le 0, \ i = 1, \ldots, m, \\ h_i(z) = 0, \ i = 1, \ldots, p, \ \|z - x\|_2 \le R\}, \end{aligned}$$

or, in other words, x solves the optimization problem

$$\begin{aligned} \text{minimize} \quad & f_0(z) \\ \text{subject to} \quad & f_i(z) \le 0, \quad i = 1, \ldots, m \\ & h_i(z) = 0, \quad i = 1, \ldots, p \\ & \|z - x\|_2 \le R \end{aligned}$$

with variable z. Roughly speaking, this means x minimizes f_0 over nearby points in the feasible set. The term 'globally optimal' is sometimes used for 'optimal' to distinguish between 'locally optimal' and 'optimal'. Throughout this book, however, optimal will mean globally optimal.

If x is feasible and $f_i(x) = 0$, we say the ith inequality constraint $f_i(x) \le 0$ is *active* at x. If $f_i(x) < 0$, we say the constraint $f_i(x) \le 0$ is *inactive*. (The equality constraints are active at all feasible points.) We say that a constraint is *redundant* if deleting it does not change the feasible set.

Example 4.1 We illustrate these definitions with a few simple unconstrained optimization problems with variable $x \in \mathbf{R}$, and $\mathbf{dom}\, f_0 = \mathbf{R}_{++}$.

- $f_0(x) = 1/x$: $p^\star = 0$, but the optimal value is not achieved.
- $f_0(x) = -\log x$: $p^\star = -\infty$, so this problem is unbounded below.
- $f_0(x) = x \log x$: $p^\star = -1/e$, achieved at the (unique) optimal point $x^\star = 1/e$.

Feasibility problems

If the objective function is identically zero, the optimal value is either zero (if the feasible set is nonempty) or ∞ (if the feasible set is empty). We call this the *feasibility problem,* and will sometimes write it as

$$
\begin{array}{ll}
\text{find} & x \\
\text{subject to} & f_i(x) \leq 0, \quad i = 1, \ldots, m \\
& h_i(x) = 0, \quad i = 1, \ldots, p.
\end{array}
$$

The feasibility problem is thus to determine whether the constraints are consistent, and if so, find a point that satisfies them.

4.1.2   Expressing problems in standard form

We refer to (4.1) as an optimization problem in *standard form.* In the standard form problem we adopt the convention that the righthand side of the inequality and equality constraints are zero. This can always be arranged by subtracting any nonzero righthand side: we represent the equality constraint $g_i(x) = \tilde{g}_i(x)$, for example, as $h_i(x) = 0$, where $h_i(x) = g_i(x) - \tilde{g}_i(x)$. In a similar way we express inequalities of the form $f_i(x) \geq 0$ as $-f_i(x) \leq 0$.

Example 4.2 *Box constraints.* Consider the optimization problem

$$
\begin{array}{ll}
\text{minimize} & f_0(x) \\
\text{subject to} & l_i \leq x_i \leq u_i, \quad i = 1, \ldots, n,
\end{array}
$$

where $x \in \mathbf{R}^n$ is the variable. The constraints are called *variable bounds* (since they give lower and upper bounds for each x_i) or *box constraints* (since the feasible set is a box).

We can express this problem in standard form as

$$
\begin{array}{ll}
\text{minimize} & f_0(x) \\
\text{subject to} & l_i - x_i \leq 0, \quad i = 1, \ldots, n \\
& x_i - u_i \leq 0, \quad i = 1, \ldots, n.
\end{array}
$$

There are $2n$ inequality constraint functions:

$$
f_i(x) = l_i - x_i, \quad i = 1, \ldots, n,
$$

and

$$
f_i(x) = x_{i-n} - u_{i-n}, \quad i = n + 1, \ldots, 2n.
$$

Maximization problems

We concentrate on the minimization problem by convention. We can solve the *maximization* problem

$$
\begin{array}{ll}
\text{maximize} & f_0(x) \\
\text{subject to} & f_i(x) \leq 0, \quad i = 1, \ldots, m \\
& h_i(x) = 0, \quad i = 1, \ldots, p
\end{array}
\tag{4.2}
$$

by minimizing the function $-f_0$ subject to the constraints. By this correspondence we can define all the terms above for the maximization problem (4.2). For example the optimal value of (4.2) is defined as

$$p^\star = \sup\{f_0(x) \mid f_i(x) \leq 0, \; i = 1, \ldots, m, \; h_i(x) = 0, \; i = 1, \ldots, p\},$$

and a feasible point x is ϵ-suboptimal if $f_0(x) \geq p^\star - \epsilon$. When the maximization problem is considered, the objective is sometimes called the *utility* or *satisfaction level* instead of the cost.

4.1.3   Equivalent problems

In this book we will use the notion of equivalence of optimization problems in an informal way. We call two problems *equivalent* if from a solution of one, a solution of the other is readily found, and vice versa. (It is possible, but complicated, to give a formal definition of equivalence.)

As a simple example, consider the problem

$$
\begin{array}{ll}
\text{minimize} & \tilde{f}(x) = \alpha_0 f_0(x) \\
\text{subject to} & \tilde{f}_i(x) = \alpha_i f_i(x) \leq 0, \quad i = 1, \ldots, m \\
& \tilde{h}_i(x) = \beta_i h_i(x) = 0, \quad i = 1, \ldots, p,
\end{array}
\tag{4.3}
$$

where $\alpha_i > 0$, $i = 0, \ldots, m$, and $\beta_i \neq 0$, $i = 1, \ldots, p$. This problem is obtained from the standard form problem (4.1) by scaling the objective and inequality constraint functions by positive constants, and scaling the equality constraint functions by nonzero constants. As a result, the feasible sets of the problem (4.3) and the original problem (4.1) are identical. A point x is optimal for the original problem (4.1) if and only if it is optimal for the scaled problem (4.3), so we say the two problems are equivalent. The two problems (4.1) and (4.3) are not, however, the same (unless α_i and β_i are all equal to one), since the objective and constraint functions differ.

We now describe some general transformations that yield equivalent problems.

Change of variables

Suppose $\phi : \mathbf{R}^n \to \mathbf{R}^n$ is one-to-one, with image covering the problem domain $\mathcal{D}$, i.e., $\phi(\mathbf{dom}\,\phi) \supseteq \mathcal{D}$. We define functions $\tilde{f}_i$ and $\tilde{h}_i$ as

$$\tilde{f}_i(z) = f_i(\phi(z)), \quad i = 0, \ldots, m, \qquad \tilde{h}_i(z) = h_i(\phi(z)), \quad i = 1, \ldots, p.$$

Now consider the problem

$$
\begin{array}{ll}
\text{minimize} & \tilde{f}_0(z) \\
\text{subject to} & \tilde{f}_i(z) \leq 0, \quad i = 1, \ldots, m \\
& \tilde{h}_i(z) = 0, \quad i = 1, \ldots, p,
\end{array}
\tag{4.4}
$$

with variable z. We say that the standard form problem (4.1) and the problem (4.4) are related by the *change of variable* or *substitution of variable* $x = \phi(z)$.

The two problems are clearly equivalent: if x solves the problem (4.1), then $z = \phi^{-1}(x)$ solves the problem (4.4); if z solves the problem (4.4), then $x = \phi(z)$ solves the problem (4.1).

Transformation of objective and constraint functions

Suppose that $\psi_0 : \mathbf{R} \to \mathbf{R}$ is monotone increasing, $\psi_1, \ldots, \psi_m : \mathbf{R} \to \mathbf{R}$ satisfy $\psi_i(u) \le 0$ if and only if $u \le 0$, and $\psi_{m+1}, \ldots, \psi_{m+p} : \mathbf{R} \to \mathbf{R}$ satisfy $\psi_i(u) = 0$ if and only if $u = 0$. We define functions $\tilde{f}_i$ and $\tilde{h}_i$ as the compositions

$$\tilde{f}_i(x) = \psi_i(f_i(x)), \quad i = 0, \ldots, m, \qquad \tilde{h}_i(x) = \psi_{m+i}(h_i(x)), \quad i = 1, \ldots, p.$$

Evidently the associated problem

$$
\begin{aligned}
\text{minimize} \quad & \tilde{f}_0(x) \\
\text{subject to} \quad & \tilde{f}_i(x) \le 0, \quad i = 1, \ldots, m \\
& \tilde{h}_i(x) = 0, \quad i = 1, \ldots, p
\end{aligned}
$$

and the standard form problem (4.1) are equivalent; indeed, the feasible sets are identical, and the optimal points are identical. (The example (4.3) above, in which the objective and constraint functions are scaled by appropriate constants, is the special case when all ψ_i are linear.)

Example 4.3 *Least-norm and least-norm-squared problems.* As a simple example consider the unconstrained Euclidean norm minimization problem

$$\text{minimize} \quad \|Ax - b\|_2, \tag{4.5}$$

with variable $x \in \mathbf{R}^n$. Since the norm is always nonnegative, we can just as well solve the problem

$$\text{minimize} \quad \|Ax - b\|_2^2 = (Ax - b)^T (Ax - b), \tag{4.6}$$

in which we minimize the square of the Euclidean norm. The problems (4.5) and (4.6) are clearly equivalent; the optimal points are the same. The two problems are not the same, however. For example, the objective in (4.5) is not differentiable at any x with $Ax - b = 0$, whereas the objective in (4.6) is differentiable for all x (in fact, quadratic).

Slack variables

One simple transformation is based on the observation that $f_i(x) \le 0$ if and only if there is an $s_i \ge 0$ that satisfies $f_i(x) + s_i = 0$. Using this transformation we obtain the problem

$$
\begin{aligned}
\text{minimize} \quad & f_0(x) \\
\text{subject to} \quad & s_i \ge 0, \quad i = 1, \ldots, m \\
& f_i(x) + s_i = 0, \quad i = 1, \ldots, m \\
& h_i(x) = 0, \quad i = 1, \ldots, p,
\end{aligned}
\tag{4.7}
$$

where the variables are $x \in \mathbf{R}^n$ and $s \in \mathbf{R}^m$. This problem has $n + m$ variables, m inequality constraints (the nonnegativity constraints on s_i), and $m + p$ equality constraints. The new variable s_i is called the *slack variable* associated with the original inequality constraint $f_i(x) \le 0$. Introducing slack variables replaces each inequality constraint with an equality constraint, and a nonnegativity constraint.

The problem (4.7) is equivalent to the original standard form problem (4.1). Indeed, if (x, s) is feasible for the problem (4.7), then x is feasible for the original problem, since $s_i = -f_i(x) \geq 0$. Conversely, if x is feasible for the original problem, then (x, s) is feasible for the problem (4.7), where we take $s_i = -f_i(x)$. Similarly, x is optimal for the original problem (4.1) if and only if (x, s) is optimal for the problem (4.7), where $s_i = -f_i(x)$.

Eliminating equality constraints

If we can explicitly parametrize all solutions of the equality constraints

$$h_i(x) = 0, \quad i = 1, \ldots, p, \tag{4.8}$$

using some parameter $z \in \mathbf{R}^k$, then we can *eliminate* the equality constraints from the problem, as follows. Suppose the function $\phi : \mathbf{R}^k \rightarrow \mathbf{R}^n$ is such that x satisfies (4.8) if and only if there is some $z \in \mathbf{R}^k$ such that $x = \phi(z)$. The optimization problem

$$
\begin{array}{ll}
\text{minimize} & \tilde{f}_0(z) = f_0(\phi(z)) \\
\text{subject to} & \tilde{f}_i(z) = f_i(\phi(z)) \leq 0, \quad i = 1, \ldots, m
\end{array}
$$

is then equivalent to the original problem (4.1). This transformed problem has variable $z \in \mathbf{R}^k$, m inequality constraints, and no equality constraints. If z is optimal for the transformed problem, then $x = \phi(z)$ is optimal for the original problem. Conversely, if x is optimal for the original problem, then (since x is feasible) there is at least one z such that $x = \phi(z)$. Any such z is optimal for the transformed problem.

Eliminating linear equality constraints

The process of eliminating variables can be described more explicitly, and easily carried out numerically, when the equality constraints are all linear, *i.e.*, have the form $Ax = b$. If $Ax = b$ is inconsistent, *i.e.*, $b \notin \mathcal{R}(A)$, then the original problem is infeasible. Assuming this is not the case, let x_0 denote any solution of the equality constraints. Let $F \in \mathbf{R}^{n \times k}$ be any matrix with $\mathcal{R}(F) = \mathcal{N}(A)$, so the general solution of the linear equations $Ax = b$ is given by $Fz + x_0$, where $z \in \mathbf{R}^k$. (We can choose F to be full rank, in which case we have $k = n - \mathbf{rank}\, A$.)

Substituting $x = Fz + x_0$ into the original problem yields the problem

$$
\begin{array}{ll}
\text{minimize} & f_0(Fz + x_0) \\
\text{subject to} & f_i(Fz + x_0) \leq 0, \quad i = 1, \ldots, m,
\end{array}
$$

with variable z, which is equivalent to the original problem, has no equality constraints, and $\mathbf{rank}\, A$ fewer variables.

Introducing equality constraints

We can also *introduce* equality constraints and new variables into a problem. Instead of describing the general case, which is complicated and not very illuminating, we give a typical example that will be useful later. Consider the problem

$$
\begin{array}{ll}
\text{minimize} & f_0(A_0 x + b_0) \\
\text{subject to} & f_i(A_i x + b_i) \leq 0, \quad i = 1, \ldots, m \\
& h_i(x) = 0, \quad i = 1, \ldots, p,
\end{array}
$$

where $x \in \mathbf{R}^n$, $A_i \in \mathbf{R}^{k_i \times n}$, and $f_i : \mathbf{R}^{k_i} \to \mathbf{R}$. In this problem the objective and constraint functions are given as compositions of the functions f_i with affine transformations defined by $A_i x + b_i$.

We introduce new variables $y_i \in \mathbf{R}^{k_i}$, as well as new equality constraints $y_i = A_i x + b_i$, for $i = 0, \ldots, m$, and form the equivalent problem

$$
\begin{array}{ll}
\text{minimize} & f_0(y_0) \\
\text{subject to} & f_i(y_i) \leq 0, \quad i = 1, \ldots, m \\
& y_i = A_i x + b_i, \quad i = 0, \ldots, m \\
& h_i(x) = 0, \quad i = 1, \ldots, p.
\end{array}
$$

This problem has $k_0 + \cdots + k_m$ new variables,

$$
y_0 \in \mathbf{R}^{k_0}, \quad \ldots, \quad y_m \in \mathbf{R}^{k_m},
$$

and $k_0 + \cdots + k_m$ new equality constraints,

$$
y_0 = A_0 x + b_0, \quad \ldots, \quad y_m = A_m x + b_m.
$$

The objective and inequality constraints in this problem are *independent*, i.e., involve different optimization variables.

Optimizing over some variables

We always have

$$
\inf_{x,y} f(x, y) = \inf_x \tilde{f}(x)
$$

where $\tilde{f}(x) = \inf_y f(x, y)$. In other words, we can always minimize a function by first minimizing over some of the variables, and then minimizing over the remaining ones. This simple and general principle can be used to transform problems into equivalent forms. The general case is cumbersome to describe and not illuminating, so we describe instead an example.

Suppose the variable $x \in \mathbf{R}^n$ is partitioned as $x = (x_1, x_2)$, with $x_1 \in \mathbf{R}^{n_1}$, $x_2 \in \mathbf{R}^{n_2}$, and $n_1 + n_2 = n$. We consider the problem

$$
\begin{array}{ll}
\text{minimize} & f_0(x_1, x_2) \\
\text{subject to} & f_i(x_1) \leq 0, \quad i = 1, \ldots, m_1 \\
& \tilde{f}_i(x_2) \leq 0, \quad i = 1, \ldots, m_2,
\end{array}
\tag{4.9}
$$

in which the constraints are independent, in the sense that each constraint function depends on x_1 or x_2. We first minimize over x_2. Define the function $\tilde{f}_0$ of x_1 by

$$
\tilde{f}_0(x_1) = \inf\{f_0(x_1, z) \mid \tilde{f}_i(z) \leq 0, \ i = 1, \ldots, m_2\}.
$$

The problem (4.9) is then equivalent to

$$
\begin{array}{ll}
\text{minimize} & \tilde{f}_0(x_1) \\
\text{subject to} & f_i(x_1) \leq 0, \quad i = 1, \ldots, m_1.
\end{array}
\tag{4.10}
$$

Example 4.4 *Minimizing a quadratic function with constraints on some variables.*
Consider a problem with strictly convex quadratic objective, with some of the variables unconstrained:

$$
\begin{aligned}
\text{minimize} \quad & x_1^T P_{11} x_1 + 2x_1^T P_{12} x_2 + x_2^T P_{22} x_2 \\
\text{subject to} \quad & f_i(x_1) \leq 0, \quad i = 1, \ldots, m,
\end{aligned}
$$

where P_{11} and P_{22} are symmetric. Here we can analytically minimize over x_2:

$$
\inf_{x_2} \left(x_1^T P_{11} x_1 + 2x_1^T P_{12} x_2 + x_2^T P_{22} x_2 \right) = x_1^T \left(P_{11} - P_{12} P_{22}^{-1} P_{12}^T \right) x_1
$$

(see §A.5.5). Therefore the original problem is equivalent to

$$
\begin{aligned}
\text{minimize} \quad & x_1^T \left(P_{11} - P_{12} P_{22}^{-1} P_{12}^T \right) x_1 \\
\text{subject to} \quad & f_i(x_1) \leq 0, \quad i = 1, \ldots, m.
\end{aligned}
$$

Epigraph problem form

The *epigraph form* of the standard problem (4.1) is the problem

$$
\begin{aligned}
\text{minimize} \quad & t \\
\text{subject to} \quad & f_0(x) - t \leq 0 \\
& f_i(x) \leq 0, \quad i = 1, \ldots, m \\
& h_i(x) = 0, \quad i = 1, \ldots, p,
\end{aligned}
\tag{4.11}
$$

with variables $x \in \mathbf{R}^n$ and $t \in \mathbf{R}$. We can easily see that it is equivalent to the original problem: (x, t) is optimal for (4.11) if and only if x is optimal for (4.1) and $t = f_0(x)$. Note that the objective function of the epigraph form problem is a *linear* function of the variables x, t.

The epigraph form problem (4.11) can be interpreted geometrically as an optimization problem in the 'graph space' (x, t): we minimize t over the epigraph of f_0, subject to the constraints on x. This is illustrated in figure 4.1.

Implicit and explicit constraints

By a simple trick already mentioned in §3.1.2, we can include any of the constraints *implicitly* in the objective function, by redefining its domain. As an extreme example, the standard form problem can be expressed as the *unconstrained* problem

$$
\text{minimize} \quad F(x),
\tag{4.12}
$$

where we define the function F as f_0, but with domain restricted to the feasible set:

$$
\mathbf{dom}\, F = \{ x \in \mathbf{dom}\, f_0 \mid f_i(x) \leq 0, \ i = 1, \ldots, m, \ h_i(x) = 0, \ i = 1, \ldots, p \},
$$

and $F(x) = f_0(x)$ for $x \in \mathbf{dom}\, F$. (Equivalently, we can define $F(x)$ to have value ∞ for x not feasible.) The problems (4.1) and (4.12) are clearly equivalent: they have the same feasible set, optimal points, and optimal value.

Of course this transformation is nothing more than a notational trick. Making the constraints implicit has not made the problem any easier to analyze or solve,

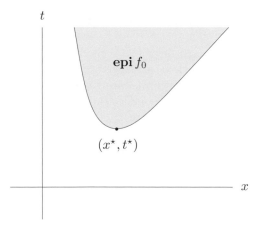

Figure 4.1 Geometric interpretation of epigraph form problem, for a problem with no constraints. The problem is to find the point in the epigraph (shown shaded) that minimizes t, *i.e.*, the 'lowest' point in the epigraph. The optimal point is $(x^\star, t^\star)$.

even though the problem (4.12) is, at least nominally, unconstrained. In some ways the transformation makes the problem more difficult. Suppose, for example, that the objective f_0 in the original problem is differentiable, so in particular its domain is open. The restricted objective function F is probably not differentiable, since its domain is likely not to be open.

Conversely, we will encounter problems with implicit constraints, which we can then make explicit. As a simple example, consider the unconstrained problem

$$\text{minimize} \quad f(x) \tag{4.13}$$

where the function f is given by

$$f(x) = \begin{cases} x^T x & Ax = b \\ \infty & \text{otherwise.} \end{cases}$$

Thus, the objective function is equal to the quadratic form $x^T x$ on the affine set defined by $Ax = b$, and ∞ off the affine set. Since we can clearly restrict our attention to points that satisfy $Ax = b$, we say that the problem (4.13) has an *implicit equality constraint* $Ax = b$ hidden in the objective. We can make the implicit equality constraint explicit, by forming the equivalent problem

$$\begin{aligned} \text{minimize} \quad & x^T x \\ \text{subject to} \quad & Ax = b. \end{aligned} \tag{4.14}$$

While the problems (4.13) and (4.14) are clearly equivalent, they are not the same. The problem (4.13) is unconstrained, but its objective function is not differentiable. The problem (4.14), however, has an equality constraint, but its objective and constraint functions are differentiable.

4.1.4   Parameter and oracle problem descriptions

For a problem in the standard form (4.1), there is still the question of how the objective and constraint functions are specified. In many cases these functions have some analytical or closed form, *i.e.*, are given by a formula or expression that involves the variable x as well as some parameters. Suppose, for example, the objective is quadratic, so it has the form $f_0(x) = (1/2)x^T P x + q^T x + r$. To specify the objective function we give the coefficients (also called *problem parameters* or *problem data*) $P \in \mathbf{S}^n$, $q \in \mathbf{R}^n$, and $r \in \mathbf{R}$. We call this a *parameter problem description*, since the specific problem to be solved (*i.e.*, the problem instance) is specified by giving the values of the parameters that appear in the expressions for the objective and constraint functions.

In other cases the objective and constraint functions are described by *oracle* models (which are also called *black box* or *subroutine* models). In an oracle model, we do not know f explicitly, but can evaluate $f(x)$ (and usually also some derivatives) at any $x \in \mathbf{dom}\, f$. This is referred to as *querying the oracle*, and is usually associated with some cost, such as time. We are also given some prior information about the function, such as convexity and a bound on its values. As a concrete example of an oracle model, consider an unconstrained problem, in which we are to minimize the function f. The function value $f(x)$ and its gradient $\nabla f(x)$ are evaluated in a subroutine. We can call the subroutine at any $x \in \mathbf{dom}\, f$, but do not have access to its source code. Calling the subroutine with argument x yields (when the subroutine returns) $f(x)$ and $\nabla f(x)$. Note that in the oracle model, we never really know the function; we only know the function value (and some derivatives) at the points where we have queried the oracle. (We also know some given prior information about the function, such as differentiability and convexity.)

In practice the distinction between a parameter and oracle problem description is not so sharp. If we are given a parameter problem description, we can construct an oracle for it, which simply evaluates the required functions and derivatives when queried. Most of the algorithms we study in part III work with an oracle model, but can be made more efficient when they are restricted to solve a specific parametrized family of problems.

4.2   Convex optimization

4.2.1   Convex optimization problems in standard form

A *convex optimization problem* is one of the form

$$
\begin{array}{ll}
\text{minimize} & f_0(x) \\
\text{subject to} & f_i(x) \le 0, \quad i = 1, \ldots, m \\
& a_i^T x = b_i, \quad i = 1, \ldots, p,
\end{array}
\qquad (4.15)
$$

where $f_0, \ldots, f_m$ are convex functions. Comparing (4.15) with the general standard form problem (4.1), the convex problem has three additional requirements:

- the objective function must be convex,

- the inequality constraint functions must be convex,

- the equality constraint functions $h_i(x) = a_i^T x - b_i$ must be affine.

We immediately note an important property: The feasible set of a convex optimization problem is convex, since it is the intersection of the domain of the problem

$$\mathcal{D} = \bigcap_{i=0}^{m} \mathbf{dom}\, f_i,$$

which is a convex set, with m (convex) sublevel sets $\{x \mid f_i(x) \leq 0\}$ and p hyperplanes $\{x \mid a_i^T x = b_i\}$. (We can assume without loss of generality that $a_i \neq 0$: if $a_i = 0$ and $b_i = 0$ for some i, then the ith equality constraint can be deleted; if $a_i = 0$ and $b_i \neq 0$, the ith equality constraint is inconsistent, and the problem is infeasible.) Thus, in a convex optimization problem, we minimize a convex objective function over a convex set.

If f_0 is quasiconvex instead of convex, we say the problem (4.15) is a (standard form) *quasiconvex optimization problem*. Since the sublevel sets of a convex or quasiconvex function are convex, we conclude that for a convex or quasiconvex optimization problem the ϵ-suboptimal sets are convex. In particular, the optimal set is convex. If the objective is strictly convex, then the optimal set contains at most one point.

Concave maximization problems

With a slight abuse of notation, we will also refer to

$$\begin{array}{ll}
\text{maximize} & f_0(x) \\
\text{subject to} & f_i(x) \leq 0, \quad i = 1,\ldots,m \\
& a_i^T x = b_i, \quad i = 1,\ldots,p,
\end{array} \qquad (4.16)$$

as a convex optimization problem if the objective function f_0 is concave, and the inequality constraint functions $f_1,\ldots,f_m$ are convex. This *concave maximization problem* is readily solved by minimizing the convex objective function $-f_0$. All of the results, conclusions, and algorithms that we describe for the minimization problem are easily transposed to the maximization case. In a similar way the maximization problem (4.16) is called *quasiconvex* if f_0 is quasiconcave.

Abstract form convex optimization problem

It is important to note a subtlety in our definition of convex optimization problem. Consider the example with $x \in \mathbf{R}^2$,

$$\begin{array}{ll}
\text{minimize} & f_0(x) = x_1^2 + x_2^2 \\
\text{subject to} & f_1(x) = x_1/(1 + x_2^2) \leq 0 \\
& h_1(x) = (x_1 + x_2)^2 = 0,
\end{array} \qquad (4.17)$$

which is in the standard form (4.1). This problem is *not* a convex optimization problem in standard form since the equality constraint function h_1 is not affine, and the inequality constraint function f_1 is not convex. Nevertheless the feasible set, which is $\{x \mid x_1 \le 0, \ x_1 + x_2 = 0\}$, is convex. So although in this problem we are minimizing a convex function f_0 over a convex set, it is not a convex optimization problem by our definition.

Of course, the problem is readily reformulated as

$$
\begin{array}{ll}
\text{minimize} & f_0(x) = x_1^2 + x_2^2 \\
\text{subject to} & \tilde{f}_1(x) = x_1 \le 0 \\
& \tilde{h}_1(x) = x_1 + x_2 = 0,
\end{array}
\tag{4.18}
$$

which is in standard convex optimization form, since f_0 and $\tilde{f}_1$ are convex, and $\tilde{h}_1$ is affine.

Some authors use the term *abstract convex optimization problem* to describe the (abstract) problem of minimizing a convex function over a convex set. Using this terminology, the problem (4.17) is an abstract convex optimization problem. *We will not use this terminology in this book.* For us, a convex optimization problem is not just one of minimizing a convex function over a convex set; it is also required that the feasible set be described specifically by a set of inequalities involving convex functions, and a set of linear equality constraints. The problem (4.17) is *not* a convex optimization problem, but the problem (4.18) *is* a convex optimization problem. (The two problems are, however, equivalent.)

Our adoption of the stricter definition of convex optimization problem does not matter much in practice. To solve the abstract problem of minimizing a convex function over a convex set, we need to find a description of the set in terms of convex inequalities and linear equality constraints. As the example above suggests, this is usually straightforward.

4.2.2   Local and global optima

A fundamental property of convex optimization problems is that any locally optimal point is also (globally) optimal. To see this, suppose that x is locally optimal for a convex optimization problem, *i.e.*, x is feasible and

$$
f_0(x) = \inf\{f_0(z) \mid z \text{ feasible}, \ \|z - x\|_2 \le R\},
\tag{4.19}
$$

for some $R > 0$. Now suppose that x is *not* globally optimal, *i.e.*, there is a feasible y such that $f_0(y) < f_0(x)$. Evidently $\|y - x\|_2 > R$, since otherwise $f_0(x) \le f_0(y)$. Consider the point z given by

$$
z = (1 - \theta)x + \theta y, \qquad \theta = \frac{R}{2\|y - x\|_2}.
$$

Then we have $\|z - x\|_2 = R/2 < R$, and by convexity of the feasible set, z is feasible. By convexity of f_0 we have

$$
f_0(z) \le (1 - \theta)f_0(x) + \theta f_0(y) < f_0(x),
$$

which contradicts (4.19). Hence there exists no feasible y with $f_0(y) < f_0(x)$, *i.e.*, x is globally optimal.

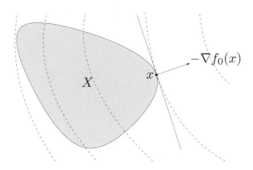

Figure 4.2 Geometric interpretation of the optimality condition (4.21). The feasible set X is shown shaded. Some level curves of f_0 are shown as dashed lines. The point x is optimal: $-\nabla f_0(x)$ defines a supporting hyperplane (shown as a solid line) to X at x.

It is not true that locally optimal points of quasiconvex optimization problems are globally optimal; see §4.2.5.

4.2.3   An optimality criterion for differentiable f_0

Suppose that the objective f_0 in a convex optimization problem is differentiable, so that for all $x, y \in \mathbf{dom}\, f_0$,

$$f_0(y) \geq f_0(x) + \nabla f_0(x)^T (y - x) \tag{4.20}$$

(see §3.1.3). Let X denote the feasible set, *i.e.*,

$$X = \{x \mid f_i(x) \leq 0, \ i = 1, \ldots, m, \ h_i(x) = 0, \ i = 1, \ldots, p\}.$$

Then x is optimal if and only if $x \in X$ and

$$\nabla f_0(x)^T (y - x) \geq 0 \text{ for all } y \in X. \tag{4.21}$$

This optimality criterion can be understood geometrically: If $\nabla f_0(x) \neq 0$, it means that $-\nabla f_0(x)$ defines a supporting hyperplane to the feasible set at x (see figure 4.2).

Proof of optimality condition

First suppose $x \in X$ and satisfies (4.21). Then if $y \in X$ we have, by (4.20), $f_0(y) \geq f_0(x)$. This shows x is an optimal point for (4.1).

Conversely, suppose x is optimal, but the condition (4.21) does not hold, *i.e.*, for some $y \in X$ we have
$$\nabla f_0(x)^T (y - x) < 0.$$

Consider the point $z(t) = ty + (1-t)x$, where $t \in [0,1]$ is a parameter. Since $z(t)$ is on the line segment between x and y, and the feasible set is convex, $z(t)$ is feasible. We claim that for small positive t we have $f_0(z(t)) < f_0(x)$, which will prove that x is not optimal. To show this, note that

$$\left. \frac{d}{dt} f_0(z(t)) \right|_{t=0} = \nabla f_0(x)^T (y-x) < 0,$$

so for small positive t, we have $f_0(z(t)) < f_0(x)$.

We will pursue the topic of optimality conditions in much more depth in chapter 5, but here we examine a few simple examples.

Unconstrained problems

For an unconstrained problem (*i.e.*, $m = p = 0$), the condition (4.21) reduces to the well known necessary and sufficient condition

$$\nabla f_0(x) = 0 \qquad (4.22)$$

for x to be optimal. While we have already seen this optimality condition, it is useful to see how it follows from (4.21). Suppose x is optimal, which means here that $x \in \mathbf{dom}\, f_0$, and for all feasible y we have $\nabla f_0(x)^T (y-x) \geq 0$. Since f_0 is differentiable, its domain is (by definition) open, so all y sufficiently close to x are feasible. Let us take $y = x - t\nabla f_0(x)$, where $t \in \mathbf{R}$ is a parameter. For t small and positive, y is feasible, and so

$$\nabla f_0(x)^T (y-x) = -t\|\nabla f_0(x)\|_2^2 \geq 0,$$

from which we conclude $\nabla f_0(x) = 0$.

There are several possible situations, depending on the number of solutions of (4.22). If there are no solutions of (4.22), then there are no optimal points; the optimal value of the problem is not attained. Here we can distinguish between two cases: the problem is unbounded below, or the optimal value is finite, but not attained. On the other hand we can have multiple solutions of the equation (4.22), in which case each such solution is a minimizer of f_0.

Example 4.5 *Unconstrained quadratic optimization.* Consider the problem of minimizing the quadratic function

$$f_0(x) = (1/2)x^T P x + q^T x + r,$$

where $P \in \mathbf{S}_+^n$ (which makes f_0 convex). The necessary and sufficient condition for x to be a minimizer of f_0 is

$$\nabla f_0(x) = Px + q = 0.$$

Several cases can occur, depending on whether this (linear) equation has no solutions, one solution, or many solutions.

- If $q \notin \mathcal{R}(P)$, then there is no solution. In this case f_0 is unbounded below.
- If $P \succ 0$ (which is the condition for f_0 to be strictly convex), then there is a unique minimizer, $x^\star = -P^{-1}q$.

- If P is singular, but $q \in \mathcal{R}(P)$, then the set of optimal points is the (affine) set $X_{\mathrm{opt}} = -P^\dagger q + \mathcal{N}(P)$, where $P^\dagger$ denotes the pseudo-inverse of P (see §A.5.4).

Example 4.6 *Analytic centering.* Consider the (unconstrained) problem of minimizing the (convex) function $f_0 : \mathbf{R}^n \to \mathbf{R}$, defined as

$$ f_0(x) = -\sum_{i=1}^{m} \log(b_i - a_i^T x), \qquad \mathbf{dom}\, f_0 = \{x \mid Ax \prec b\}, $$

where $a_1^T, \ldots, a_m^T$ are the rows of A. The function f_0 is differentiable, so the necessary and sufficient conditions for x to be optimal are

$$ Ax \prec b, \qquad \nabla f_0(x) = \sum_{i=1}^{m} \frac{1}{b_i - a_i^T x} a_i = 0. \tag{4.23} $$

(The condition $Ax \prec b$ is just $x \in \mathbf{dom}\, f_0$.) If $Ax \prec b$ is infeasible, then the domain of f_0 is empty. Assuming $Ax \prec b$ is feasible, there are still several possible cases (see exercise 4.2):

- There are no solutions of (4.23), and hence no optimal points for the problem. This occurs if and only if f_0 is unbounded below.

- There are many solutions of (4.23). In this case it can be shown that the solutions form an affine set.

- There is a unique solution of (4.23), *i.e.*, a unique minimizer of f_0. This occurs if and only if the open polyhedron $\{x \mid Ax \prec b\}$ is nonempty and bounded.

Problems with equality constraints only

Consider the case where there are equality constraints but no inequality constraints, *i.e.*,

$$ \begin{array}{ll} \text{minimize} & f_0(x) \\ \text{subject to} & Ax = b. \end{array} $$

Here the feasible set is affine. We assume that it is nonempty; otherwise the problem is infeasible. The optimality condition for a feasible x is that

$$ \nabla f_0(x)^T (y - x) \geq 0 $$

must hold for all y satisfying $Ay = b$. Since x is feasible, every feasible y has the form $y = x + v$ for some $v \in \mathcal{N}(A)$. The optimality condition can therefore be expressed as:

$$ \nabla f_0(x)^T v \geq 0 \text{ for all } v \in \mathcal{N}(A). $$

If a linear function is nonnegative on a subspace, then it must be zero on the subspace, so it follows that $\nabla f_0(x)^T v = 0$ for all $v \in \mathcal{N}(A)$. In other words,

$$ \nabla f_0(x) \perp \mathcal{N}(A). $$

Using the fact that $\mathcal{N}(A)^\perp = \mathcal{R}(A^T)$, this optimality condition can be expressed as $\nabla f_0(x) \in \mathcal{R}(A^T)$, *i.e.*, there exists a $\nu \in \mathbf{R}^p$ such that

$$\nabla f_0(x) + A^T \nu = 0.$$

Together with the requirement $Ax = b$ (*i.e.*, that x is feasible), this is the classical Lagrange multiplier optimality condition, which we will study in greater detail in chapter 5.

Minimization over the nonnegative orthant

As another example we consider the problem

$$\begin{array}{ll} \text{minimize} & f_0(x) \\ \text{subject to} & x \succeq 0, \end{array}$$

where the only inequality constraints are nonnegativity constraints on the variables.

The optimality condition (4.21) is then

$$x \succeq 0, \qquad \nabla f_0(x)^T (y - x) \geq 0 \text{ for all } y \succeq 0.$$

The term $\nabla f_0(x)^T y$, which is a linear function of y, is unbounded below on $y \succeq 0$, unless we have $\nabla f_0(x) \succeq 0$. The condition then reduces to $-\nabla f_0(x)^T x \geq 0$. But $x \succeq 0$ and $\nabla f_0(x) \succeq 0$, so we must have $\nabla f_0(x)^T x = 0$, *i.e.*,

$$\sum_{i=1}^n (\nabla f_0(x))_i x_i = 0.$$

Now each of the terms in this sum is the product of two nonnegative numbers, so we conclude that each term must be zero, *i.e.*, $(\nabla f_0(x))_i x_i = 0$ for $i = 1, \ldots, n$.

The optimality condition can therefore be expressed as

$$x \succeq 0, \qquad \nabla f_0(x) \succeq 0, \qquad x_i (\nabla f_0(x))_i = 0, \quad i = 1, \ldots, n.$$

The last condition is called *complementarity*, since it means that the sparsity patterns (*i.e.*, the set of indices corresponding to nonzero components) of the vectors x and $\nabla f_0(x)$ are complementary (*i.e.*, have empty intersection). We will encounter complementarity conditions again in chapter 5.

4.2.4   Equivalent convex problems

It is useful to see which of the transformations described in §4.1.3 preserve convexity.

Eliminating equality constraints

For a convex problem the equality constraints must be linear, *i.e.*, of the form $Ax = b$. In this case they can be eliminated by finding a particular solution x_0 of

$Ax = b$, and a matrix F whose range is the nullspace of A, which results in the problem

$$\begin{array}{ll} \text{minimize} & f_0(Fz + x_0) \\ \text{subject to} & f_i(Fz + x_0) \leq 0, \quad i = 1, \ldots, m, \end{array}$$

with variable z. Since the composition of a convex function with an affine function is convex, eliminating equality constraints preserves convexity of a problem. Moreover, the process of eliminating equality constraints (and reconstructing the solution of the original problem from the solution of the transformed problem) involves standard linear algebra operations.

At least in principle, this means we can restrict our attention to convex optimization problems which have no equality constraints. In many cases, however, it is better to retain the equality constraints, since eliminating them can make the problem harder to understand and analyze, or ruin the efficiency of an algorithm that solves it. This is true, for example, when the variable x has very large dimension, and eliminating the equality constraints would destroy sparsity or some other useful structure of the problem.

Introducing equality constraints

We can introduce new variables and equality constraints into a convex optimization problem, provided the equality constraints are linear, and the resulting problem will also be convex. For example, if an objective or constraint function has the form $f_i(A_i x + b_i)$, where $A_i \in \mathbf{R}^{k_i \times n}$, we can introduce a new variable $y_i \in \mathbf{R}^{k_i}$, replace $f_i(A_i x + b_i)$ with $f_i(y_i)$, and add the linear equality constraint $y_i = A_i x + b_i$.

Slack variables

By introducing slack variables we have the new constraints $f_i(x) + s_i = 0$. Since equality constraint functions must be affine in a convex problem, we must have f_i affine. In other words: introducing slack variables for *linear inequalities* preserves convexity of a problem.

Epigraph problem form

The epigraph form of the convex optimization problem (4.15) is

$$\begin{array}{ll} \text{minimize} & t \\ \text{subject to} & f_0(x) - t \leq 0 \\ & f_i(x) \leq 0, \quad i = 1, \ldots, m \\ & a_i^T x = b_i, \quad i = 1, \ldots, p. \end{array}$$

The objective is linear (hence convex) and the new constraint function $f_0(x) - t$ is also convex in (x, t), so the epigraph form problem is convex as well.

It is sometimes said that a linear objective is *universal* for convex optimization, since any convex optimization problem is readily transformed to one with linear objective. The epigraph form of a convex problem has several practical uses. By assuming the objective of a convex optimization problem is linear, we can simplify theoretical analysis. It can also simplify algorithm development, since an algorithm that solves convex optimization problems with linear objective can, using the transformation above, solve any convex optimization problem (provided it can handle the constraint $f_0(x) - t \le 0$).

Minimizing over some variables

Minimizing a convex function over some variables preserves convexity. Therefore, if f_0 in (4.9) is jointly convex in x_1 and x_2, and f_i, $i = 1, \ldots, m_1$, and $\tilde{f}_i$, $i = 1, \ldots, m_2$, are convex, then the equivalent problem (4.10) is convex.

4.2.5   Quasiconvex optimization

Recall that a quasiconvex optimization problem has the standard form

$$
\begin{array}{ll}
\text{minimize} & f_0(x) \\
\text{subject to} & f_i(x) \le 0, \quad i = 1, \ldots, m \\
& Ax = b,
\end{array}
\tag{4.24}
$$

where the inequality constraint functions $f_1, \ldots, f_m$ are convex, and the objective f_0 is quasiconvex (instead of convex, as in a convex optimization problem). (Quasiconvex constraint functions can be replaced with equivalent convex constraint functions, *i.e.*, constraint functions that are convex and have the same 0-sublevel set, as in §3.4.5.)

In this section we point out some basic differences between convex and quasiconvex optimization problems, and also show how solving a quasiconvex optimization problem can be reduced to solving a sequence of convex optimization problems.

Locally optimal solutions and optimality conditions

The most important difference between convex and quasiconvex optimization is that a quasiconvex optimization problem can have locally optimal solutions that are not (globally) optimal. This phenomenon can be seen even in the simple case of unconstrained minimization of a quasiconvex function on $\mathbf{R}$, such as the one shown in figure 4.3.

Nevertheless, a variation of the optimality condition (4.21) given in §4.2.3 does hold for quasiconvex optimization problems with differentiable objective function. Let X denote the feasible set for the quasiconvex optimization problem (4.24). It follows from the first-order condition for quasiconvexity (3.20) that x is optimal if

$$
x \in X, \qquad \nabla f_0(x)^T (y - x) > 0 \text{ for all } y \in X \setminus \{x\}.
\tag{4.25}
$$

There are two important differences between this criterion and the analogous one (4.21) for convex optimization:

- The condition (4.25) is only *sufficient* for optimality; simple examples show that it need not hold for an optimal point. In contrast, the condition (4.21) is necessary and sufficient for x to solve the convex problem.

- The condition (4.25) requires the gradient of f_0 to be nonzero, whereas the condition (4.21) does not. Indeed, when $\nabla f_0(x) = 0$ in the convex case, the condition (4.21) is satisfied, and x is optimal.

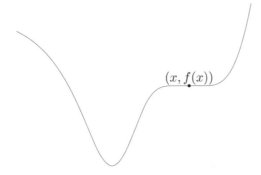

$(x, f(x))$

Figure 4.3 A quasiconvex function f on $\mathbf{R}$, with a locally optimal point x that is not globally optimal. This example shows that the simple optimality condition $f'(x) = 0$, valid for convex functions, does not hold for quasiconvex functions.

Quasiconvex optimization via convex feasibility problems

One general approach to quasiconvex optimization relies on the representation of the sublevel sets of a quasiconvex function via a family of convex inequalities, as described in §3.4.5. Let $\phi_t : \mathbf{R}^n \to \mathbf{R}$, $t \in \mathbf{R}$, be a family of convex functions that satisfy

$$f_0(x) \le t \iff \phi_t(x) \le 0,$$

and also, for each x, $\phi_t(x)$ is a nonincreasing function of t, i.e., $\phi_s(x) \le \phi_t(x)$ whenever $s \ge t$.

Let $p^\star$ denote the optimal value of the quasiconvex optimization problem (4.24). If the feasibility problem

$$
\begin{array}{ll}
\text{find} & x \\
\text{subject to} & \phi_t(x) \le 0 \\
& f_i(x) \le 0, \quad i = 1, \dots, m \\
& Ax = b,
\end{array}
\tag{4.26}
$$

is feasible, then we have $p^\star \le t$. Conversely, if the problem (4.26) is infeasible, then we can conclude $p^\star \ge t$. The problem (4.26) is a convex feasibility problem, since the inequality constraint functions are all convex, and the equality constraints are linear. Thus, we can check whether the optimal value $p^\star$ of a quasiconvex optimization problem is less than or more than a given value t by solving the convex feasibility problem (4.26). If the convex feasibility problem is feasible then we have $p^\star \le t$, and any feasible point x is feasible for the quasiconvex problem and satisfies $f_0(x) \le t$. If the convex feasibility problem is infeasible, then we know that $p^\star \ge t$.

This observation can be used as the basis of a simple algorithm for solving the quasiconvex optimization problem (4.24) using *bisection*, solving a convex feasibility problem at each step. We assume that the problem is feasible, and start with an interval $[l, u]$ known to contain the optimal value $p^\star$. We then solve the convex feasibility problem at its midpoint $t = (l + u)/2$, to determine whether the optimal value is in the lower or upper half of the interval, and update the interval accordingly. This produces a new interval, which also contains the optimal value, but has half the width of the initial interval. This is repeated until the width of the interval is small enough:

Algorithm 4.1 *Bisection method for quasiconvex optimization.*

given $l \leq p^\star$, $u \geq p^\star$, tolerance $\epsilon > 0$.

repeat
    1. $t := (l + u)/2$.
    2. Solve the convex feasibility problem (4.26).
    3. **if** (4.26) is feasible, $u := t$;   **else** $l := t$.
until $u - l \leq \epsilon$.

The interval $[l, u]$ is guaranteed to contain $p^\star$, *i.e.*, we have $l \leq p^\star \leq u$ at each step. In each iteration the interval is divided in two, *i.e.*, bisected, so the length of the interval after k iterations is $2^{-k}(u - l)$, where $u - l$ is the length of the initial interval. It follows that exactly $\lceil \log_2((u - l)/\epsilon) \rceil$ iterations are required before the algorithm terminates. Each step involves solving the convex feasibility problem (4.26).

4.3   Linear optimization problems

When the objective and constraint functions are all affine, the problem is called a *linear program* (LP). A general linear program has the form

$$
\begin{array}{ll}
\text{minimize} & c^T x + d \\
\text{subject to} & Gx \preceq h \\
& Ax = b,
\end{array}
\tag{4.27}
$$

where $G \in \mathbf{R}^{m \times n}$ and $A \in \mathbf{R}^{p \times n}$. Linear programs are, of course, convex optimization problems.

It is common to omit the constant d in the objective function, since it does not affect the optimal (or feasible) set. Since we can maximize an affine objective $c^T x + d$, by minimizing $-c^T x - d$ (which is still convex), we also refer to a maximization problem with affine objective and constraint functions as an LP.

The geometric interpretation of an LP is illustrated in figure 4.4. The feasible set of the LP (4.27) is a polyhedron $\mathcal{P}$; the problem is to minimize the affine function $c^T x + d$ (or, equivalently, the linear function $c^T x$) over $\mathcal{P}$.

Standard and inequality form linear programs

Two special cases of the LP (4.27) are so widely encountered that they have been given separate names. In a *standard form LP* the only inequalities are componen-

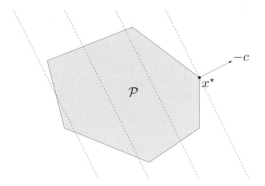

Figure 4.4 Geometric interpretation of an LP. The feasible set $\mathcal{P}$, which is a polyhedron, is shaded. The objective $c^T x$ is linear, so its level curves are hyperplanes orthogonal to c (shown as dashed lines). The point $x^\star$ is optimal; it is the point in $\mathcal{P}$ as far as possible in the direction $-c$.

twise nonnegativity constraints $x \succeq 0$:

$$
\begin{array}{ll}
\text{minimize} & c^T x \\
\text{subject to} & Ax = b \\
& x \succeq 0.
\end{array}
\tag{4.28}
$$

If the LP has no equality constraints, it is called an *inequality form LP*, usually written as

$$
\begin{array}{ll}
\text{minimize} & c^T x \\
\text{subject to} & Ax \preceq b.
\end{array}
\tag{4.29}
$$

Converting LPs to standard form

It is sometimes useful to transform a general LP (4.27) to one in standard form (4.28) (for example in order to use an algorithm for standard form LPs). The first step is to introduce slack variables s_i for the inequalities, which results in

$$
\begin{array}{ll}
\text{minimize} & c^T x + d \\
\text{subject to} & Gx + s = h \\
& Ax = b \\
& s \succeq 0.
\end{array}
$$

The second step is to express the variable x as the difference of two nonnegative variables x^+ and x^-, *i.e.*, $x = x^+ - x^-$, $x^+, x^- \succeq 0$. This yields the problem

$$
\begin{array}{ll}
\text{minimize} & c^T x^+ - c^T x^- + d \\
\text{subject to} & Gx^+ - Gx^- + s = h \\
& Ax^+ - Ax^- = b \\
& x^+ \succeq 0, \quad x^- \succeq 0, \quad s \succeq 0,
\end{array}
$$

which is an LP in standard form, with variables x^+, x^-, and s. (For equivalence of this problem and the original one (4.27), see exercise 4.10.)

These techniques for manipulating problems (along with many others we will see in the examples and exercises) can be used to formulate many problems as linear programs. With some abuse of terminology, it is common to refer to a problem that can be formulated as an LP as an LP, even if it does not have the form (4.27).

4.3.1   Examples

LPs arise in a vast number of fields and applications; here we give a few typical examples.

Diet problem

A healthy diet contains m different nutrients in quantities at least equal to $b_1, \ldots, b_m$. We can compose such a diet by choosing nonnegative quantities $x_1, \ldots, x_n$ of n different foods. One unit quantity of food j contains an amount a_{ij} of nutrient i, and has a cost of c_j. We want to determine the cheapest diet that satisfies the nutritional requirements. This problem can be formulated as the LP

$$
\begin{array}{ll}
\text{minimize} & c^T x \\
\text{subject to} & Ax \succeq b \\
& x \succeq 0.
\end{array}
$$

Several variations on this problem can also be formulated as LPs. For example, we can insist on an exact amount of a nutrient in the diet (which gives a linear equality constraint), or we can impose an upper bound on the amount of a nutrient, in addition to the lower bound as above.

Chebyshev center of a polyhedron

We consider the problem of finding the largest Euclidean ball that lies in a polyhedron described by linear inequalities,

$$
\mathcal{P} = \{ x \in \mathbf{R}^n \mid a_i^T x \le b_i, \ i = 1, \ldots, m \}.
$$

(The center of the optimal ball is called the *Chebyshev center* of the polyhedron; it is the point deepest inside the polyhedron, *i.e.*, farthest from the boundary; see §8.5.1.) We represent the ball as

$$
\mathcal{B} = \{ x_c + u \mid \|u\|_2 \le r \}.
$$

The variables in the problem are the center $x_c \in \mathbf{R}^n$ and the radius r; we wish to maximize r subject to the constraint $\mathcal{B} \subseteq \mathcal{P}$.

We start by considering the simpler constraint that $\mathcal{B}$ lies in one halfspace $a_i^T x \le b_i$, *i.e.*,

$$
\|u\|_2 \le r \implies a_i^T (x_c + u) \le b_i. \tag{4.30}
$$

Since

$$
\sup\{ a_i^T u \mid \|u\|_2 \le r \} = r \|a_i\|_2
$$

we can write (4.30) as

$$a_i^T x_c + r\|a_i\|_2 \le b_i, \tag{4.31}$$

which is a linear inequality in x_c and r. In other words, the constraint that the ball lies in the halfspace determined by the inequality $a_i^T x \le b_i$ can be written as a linear inequality.

Therefore $\mathcal{B} \subseteq \mathcal{P}$ if and only if (4.31) holds for all $i = 1, \ldots, m$. Hence the Chebyshev center can be determined by solving the LP

$$
\begin{array}{ll}
\text{maximize} & r \\
\text{subject to} & a_i^T x_c + r\|a_i\|_2 \le b_i, \quad i = 1, \ldots, m,
\end{array}
$$

with variables r and x_c. (For more on the Chebyshev center, see §8.5.1.)

Dynamic activity planning

We consider the problem of choosing, or planning, the activity levels of n activities, or sectors of an economy, over N time periods. We let $x_j(t) \ge 0$, $t = 1, \ldots, N$, denote the activity level of sector j, in period t. The activities both consume and produce products or goods in proportion to their activity levels. The amount of good i produced per unit of activity j is given by a_{ij}. Similarly, the amount of good i consumed per unit of activity j is b_{ij}. The total amount of goods produced in period t is given by $Ax(t) \in \mathbf{R}^m$, and the amount of goods consumed is $Bx(t) \in \mathbf{R}^m$. (Although we refer to these products as 'goods', they can also include unwanted products such as pollutants.)

The goods consumed in a period cannot exceed those produced in the previous period: we must have $Bx(t+1) \preceq Ax(t)$ for $t = 1, \ldots, N$. A vector $g_0 \in \mathbf{R}^m$ of initial goods is given, which constrains the first period activity levels: $Bx(1) \preceq g_0$. The (vectors of) excess goods not consumed by the activities are given by

$$
\begin{array}{rcl}
s(0) & = & g_0 - Bx(1) \\
s(t) & = & Ax(t) - Bx(t+1), \quad t = 1, \ldots, N-1 \\
s(N) & = & Ax(N).
\end{array}
$$

The objective is to maximize a discounted total value of excess goods:

$$c^T s(0) + \gamma c^T s(1) + \cdots + \gamma^N c^T s(N),$$

where $c \in \mathbf{R}^m$ gives the values of the goods, and $\gamma > 0$ is a discount factor. (The value c_i is negative if the ith product is unwanted, e.g., a pollutant; $|c_i|$ is then the cost of disposal per unit.)

Putting it all together we arrive at the LP

$$
\begin{array}{ll}
\text{maximize} & c^T s(0) + \gamma c^T s(1) + \cdots + \gamma^N c^T s(N) \\
\text{subject to} & x(t) \succeq 0, \quad t = 1, \ldots, N \\
& s(t) \succeq 0, \quad t = 0, \ldots, N \\
& s(0) = g_0 - Bx(1) \\
& s(t) = Ax(t) - Bx(t+1), \quad t = 1, \ldots, N-1 \\
& s(N) = Ax(N),
\end{array}
$$

with variables $x(1), \ldots, x(N)$, $s(0), \ldots, s(N)$. This problem is a standard form LP; the variables $s(t)$ are the slack variables associated with the constraints $Bx(t+1) \preceq Ax(t)$.

Chebyshev inequalities

We consider a probability distribution for a discrete random variable x on a set $\{u_1, \ldots, u_n\} \subseteq \mathbf{R}$ with n elements. We describe the distribution of x by a vector $p \in \mathbf{R}^n$, where

$$p_i = \mathbf{prob}(x = u_i),$$

so p satisfies $p \succeq 0$ and $\mathbf{1}^T p = 1$. Conversely, if p satisfies $p \succeq 0$ and $\mathbf{1}^T p = 1$, then it defines a probability distribution for x. We assume that u_i are known and fixed, but the distribution p is not known.

If f is any function of x, then

$$\mathbf{E}\, f = \sum_{i=1}^{n} p_i f(u_i)$$

is a linear function of p. If S is any subset of $\mathbf{R}$, then

$$\mathbf{prob}(x \in S) = \sum_{u_i \in S} p_i$$

is a linear function of p.

Although we do not know p, we are given prior knowledge of the following form: We know upper and lower bounds on expected values of some functions of x, and probabilities of some subsets of $\mathbf{R}$. This prior knowledge can be expressed as linear inequality constraints on p,

$$\alpha_i \leq a_i^T p \leq \beta_i, \quad i = 1, \ldots, m.$$

The problem is to give lower and upper bounds on $\mathbf{E}\, f_0(x) = a_0^T p$, where f_0 is some function of x.

To find a lower bound we solve the LP

$$\begin{array}{ll} \text{minimize} & a_0^T p \\ \text{subject to} & p \succeq 0, \quad \mathbf{1}^T p = 1 \\ & \alpha_i \leq a_i^T p \leq \beta_i, \quad i = 1, \ldots, m, \end{array}$$

with variable p. The optimal value of this LP gives the lowest possible value of $\mathbf{E}\, f_0(X)$ for any distribution that is consistent with the prior information. Moreover, the bound is sharp: the optimal solution gives a distribution that is consistent with the prior information and achieves the lower bound. In a similar way, we can find the best upper bound by maximizing $a_0^T p$ subject to the same constraints. (We will consider Chebyshev inequalities in more detail in §7.4.1.)

Piecewise-linear minimization

Consider the (unconstrained) problem of minimizing the piecewise-linear, convex function

$$f(x) = \max_{i=1,\ldots,m} (a_i^T x + b_i).$$

This problem can be transformed to an equivalent LP by first forming the epigraph problem,

$$\begin{array}{ll} \text{minimize} & t \\ \text{subject to} & \max_{i=1,\ldots,m} (a_i^T x + b_i) \leq t, \end{array}$$

and then expressing the inequality as a set of m separate inequalities:

$$\begin{array}{ll} \text{minimize} & t \\ \text{subject to} & a_i^T x + b_i \le t, \quad i = 1, \ldots, m. \end{array}$$

This is an LP (in inequality form), with variables x and t.

4.3.2   Linear-fractional programming

The problem of minimizing a ratio of affine functions over a polyhedron is called a *linear-fractional program*:

$$\begin{array}{ll} \text{minimize} & f_0(x) \\ \text{subject to} & Gx \preceq h \\ & Ax = b \end{array} \tag{4.32}$$

where the objective function is given by

$$f_0(x) = \frac{c^T x + d}{e^T x + f}, \qquad \mathbf{dom}\, f_0 = \{x \mid e^T x + f > 0\}.$$

The objective function is quasiconvex (in fact, quasilinear) so linear-fractional programs are quasiconvex optimization problems.

Transforming to a linear program

If the feasible set

$$\{x \mid Gx \preceq h, \ Ax = b, \ e^T x + f > 0\}$$

is nonempty, the linear-fractional program (4.32) can be transformed to an equivalent linear program

$$\begin{array}{ll} \text{minimize} & c^T y + dz \\ \text{subject to} & Gy - hz \preceq 0 \\ & Ay - bz = 0 \\ & e^T y + fz = 1 \\ & z \ge 0 \end{array} \tag{4.33}$$

with variables y, z.

To show the equivalence, we first note that if x is feasible in (4.32) then the pair

$$y = \frac{x}{e^T x + f}, \qquad z = \frac{1}{e^T x + f}$$

is feasible in (4.33), with the same objective value $c^T y + dz = f_0(x)$. It follows that the optimal value of (4.32) is greater than or equal to the optimal value of (4.33).

Conversely, if (y, z) is feasible in (4.33), with $z \ne 0$, then $x = y/z$ is feasible in (4.32), with the same objective value $f_0(x) = c^T y + dz$. If (y, z) is feasible in (4.33) with $z = 0$, and x_0 is feasible for (4.32), then $x = x_0 + ty$ is feasible in (4.32) for all $t \ge 0$. Moreover, $\lim_{t \to \infty} f_0(x_0 + ty) = c^T y + dz$, so we can find feasible points in (4.32) with objective values arbitrarily close to the objective value of (y, z). We conclude that the optimal value of (4.32) is less than or equal to the optimal value of (4.33).

Generalized linear-fractional programming

A generalization of the linear-fractional program (4.32) is the *generalized linear-fractional program* in which

$$f_0(x) = \max_{i=1,\ldots,r} \frac{c_i^T x + d_i}{e_i^T x + f_i}, \qquad \mathbf{dom}\, f_0 = \{x \mid e_i^T x + f_i > 0, \ i = 1, \ldots, r\}.$$

The objective function is the pointwise maximum of r quasiconvex functions, and therefore quasiconvex, so this problem is quasiconvex. When $r = 1$ it reduces to the standard linear-fractional program.

Example 4.7 *Von Neumann growth problem.*   We consider an economy with n sectors, and activity levels $x_i > 0$ in the current period, and activity levels $x_i^+ > 0$ in the next period. (In this problem we only consider one period.) There are m goods which are consumed, and also produced, by the activity: An activity level x consumes goods $Bx \in \mathbf{R}^m$, and produces goods Ax. The goods consumed in the next period cannot exceed the goods produced in the current period, *i.e.*, $Bx^+ \preceq Ax$. The *growth rate* in sector i, over the period, is given by x_i^+/x_i.

Von Neumann's growth problem is to find an activity level vector x that maximizes the minimum growth rate across all sectors of the economy. This problem can be expressed as a generalized linear-fractional problem

$$\begin{array}{ll} \text{maximize} & \min_{i=1,\ldots,n} x_i^+/x_i \\ \text{subject to} & x^+ \succeq 0 \\ & Bx^+ \preceq Ax \end{array}$$

with domain $\{(x, x^+) \mid x \succ 0\}$. Note that this problem is homogeneous in x and x^+, so we can replace the implicit constraint $x \succ 0$ by the explicit constraint $x \succeq \mathbf{1}$.

4.4   Quadratic optimization problems

The convex optimization problem (4.15) is called a *quadratic program* (QP) if the objective function is (convex) quadratic, and the constraint functions are affine. A quadratic program can be expressed in the form

$$\begin{array}{ll} \text{minimize} & (1/2)x^T P x + q^T x + r \\ \text{subject to} & Gx \preceq h \\ & Ax = b, \end{array} \tag{4.34}$$

where $P \in \mathbf{S}_+^n$, $G \in \mathbf{R}^{m \times n}$, and $A \in \mathbf{R}^{p \times n}$. In a quadratic program, we minimize a convex quadratic function over a polyhedron, as illustrated in figure 4.5.

If the objective in (4.15) as well as the inequality constraint functions are (convex) quadratic, as in

$$\begin{array}{ll} \text{minimize} & (1/2)x^T P_0 x + q_0^T x + r_0 \\ \text{subject to} & (1/2)x^T P_i x + q_i^T x + r_i \leq 0, \quad i = 1, \ldots, m \\ & Ax = b, \end{array} \tag{4.35}$$

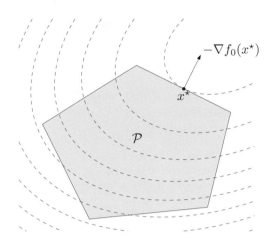

Figure 4.5 Geometric illustration of QP. The feasible set $\mathcal{P}$, which is a polyhedron, is shown shaded. The contour lines of the objective function, which is convex quadratic, are shown as dashed curves. The point $x^\star$ is optimal.

where $P_i \in \mathbf{S}_+^n$, $i = 0, 1 \ldots, m$, the problem is called a *quadratically constrained quadratic program* (QCQP). In a QCQP, we minimize a convex quadratic function over a feasible region that is the intersection of ellipsoids (when $P_i \succ 0$).

Quadratic programs include linear programs as a special case, by taking $P = 0$ in (4.34). Quadratically constrained quadratic programs include quadratic programs (and therefore also linear programs) as a special case, by taking $P_i = 0$ in (4.35), for $i = 1, \ldots, m$.

4.4.1   Examples

Least-squares and regression

The problem of minimizing the convex quadratic function

$$\|Ax - b\|_2^2 = x^T A^T A x - 2b^T A x + b^T b$$

is an (unconstrained) QP. It arises in many fields and has many names, *e.g.*, *regression analysis* or *least-squares approximation*. This problem is simple enough to have the well known analytical solution $x = A^\dagger b$, where $A^\dagger$ is the pseudo-inverse of A (see §A.5.4).

When linear inequality constraints are added, the problem is called *constrained regression* or *constrained least-squares*, and there is no longer a simple analytical solution. As an example we can consider regression with lower and upper bounds on the variables, *i.e.*,

$$\begin{array}{ll} \text{minimize} & \|Ax - b\|_2^2 \\ \text{subject to} & l_i \leq x_i \leq u_i, \quad i = 1, \ldots, n, \end{array}$$

which is a QP. (We will study least-squares and regression problems in far more depth in chapters 6 and 7.)

Distance between polyhedra

The (Euclidean) distance between the polyhedra $\mathcal{P}_1 = \{x \mid A_1 x \preceq b_1\}$ and $\mathcal{P}_2 = \{x \mid A_2 x \preceq b_2\}$ in $\mathbf{R}^n$ is defined as

$$\mathbf{dist}(\mathcal{P}_1, \mathcal{P}_2) = \inf\{\|x_1 - x_2\|_2 \mid x_1 \in \mathcal{P}_1, \ x_2 \in \mathcal{P}_2\}.$$

If the polyhedra intersect, the distance is zero.

To find the distance between $\mathcal{P}_1$ and $\mathcal{P}_2$, we can solve the QP

$$\begin{array}{ll} \text{minimize} & \|x_1 - x_2\|_2^2 \\ \text{subject to} & A_1 x_1 \preceq b_1, \quad A_2 x_2 \preceq b_2, \end{array}$$

with variables $x_1, \ x_2 \in \mathbf{R}^n$. This problem is infeasible if and only if one of the polyhedra is empty. The optimal value is zero if and only if the polyhedra intersect, in which case the optimal x_1 and x_2 are equal (and is a point in the intersection $\mathcal{P}_1 \cap \mathcal{P}_2$). Otherwise the optimal x_1 and x_2 are the points in $\mathcal{P}_1$ and $\mathcal{P}_2$, respectively, that are closest to each other. (We will study geometric problems involving distance in more detail in chapter 8.)

Bounding variance

We consider again the Chebyshev inequalities example (page 150), where the variable is an unknown probability distribution given by $p \in \mathbf{R}^n$, about which we have some prior information. The variance of a random variable $f(x)$ is given by

$$\mathbf{E} \, f^2 - (\mathbf{E} \, f)^2 = \sum_{i=1}^{n} f_i^2 p_i - \left(\sum_{i=1}^{n} f_i p_i \right)^2,$$

(where $f_i = f(u_i)$), which is a concave quadratic function of p.

It follows that we can maximize the variance of $f(x)$, subject to the given prior information, by solving the QP

$$\begin{array}{ll} \text{maximize} & \sum_{i=1}^{n} f_i^2 p_i - \left(\sum_{i=1}^{n} f_i p_i \right)^2 \\ \text{subject to} & p \succeq 0, \quad \mathbf{1}^T p = 1 \\ & \alpha_i \leq a_i^T p \leq \beta_i, \quad i = 1, \ldots, m. \end{array}$$

The optimal value gives the maximum possible variance of $f(x)$, over all distributions that are consistent with the prior information; the optimal p gives a distribution that achieves this maximum variance.

Linear program with random cost

We consider an LP,

$$\begin{array}{ll} \text{minimize} & c^T x \\ \text{subject to} & Gx \preceq h \\ & Ax = b, \end{array}$$

with variable $x \in \mathbf{R}^n$. We suppose that the cost function (vector) $c \in \mathbf{R}^n$ is *random*, with mean value $\bar{c}$ and covariance $\mathbf{E}(c - \bar{c})(c - \bar{c})^T = \Sigma$. (We assume for simplicity that the other problem parameters are deterministic.) For a given $x \in \mathbf{R}^n$, the cost $c^T x$ is a (scalar) random variable with mean $\mathbf{E}\, c^T x = \bar{c}^T x$ and variance

$$\mathbf{var}(c^T x) = \mathbf{E}(c^T x - \mathbf{E}\, c^T x)^2 = x^T \Sigma x.$$

In general there is a trade-off between small expected cost and small cost variance. One way to take variance into account is to minimize a linear combination of the expected value and the variance of the cost, *i.e.*,

$$\mathbf{E}\, c^T x + \gamma \, \mathbf{var}(c^T x),$$

which is called the *risk-sensitive cost*. The parameter $\gamma \geq 0$ is called the *risk-aversion parameter*, since it sets the relative values of cost variance and expected value. (For $\gamma > 0$, we are willing to trade off an increase in expected cost for a sufficiently large decrease in cost variance).

To minimize the risk-sensitive cost we solve the QP

$$\begin{array}{ll} \text{minimize} & \bar{c}^T x + \gamma x^T \Sigma x \\ \text{subject to} & Gx \preceq h \\ & Ax = b. \end{array}$$

Markowitz portfolio optimization

We consider a classical portfolio problem with n assets or stocks held over a period of time. We let x_i denote the amount of asset i held throughout the period, with x_i in dollars, at the price at the beginning of the period. A normal long position in asset i corresponds to $x_i > 0$; a short position in asset i (*i.e.*, the obligation to buy the asset at the end of the period) corresponds to $x_i < 0$. We let p_i denote the relative price change of asset i over the period, *i.e.*, its change in price over the period divided by its price at the beginning of the period. The overall return on the portfolio is $r = p^T x$ (given in dollars). The optimization variable is the portfolio vector $x \in \mathbf{R}^n$.

A wide variety of constraints on the portfolio can be considered. The simplest set of constraints is that $x_i \geq 0$ (*i.e.*, no short positions) and $\mathbf{1}^T x = B$ (*i.e.*, the total budget to be invested is B, which is often taken to be one).

We take a stochastic model for price changes: $p \in \mathbf{R}^n$ is a random vector, with known mean $\bar{p}$ and covariance Σ. Therefore with portfolio $x \in \mathbf{R}^n$, the return r is a (scalar) random variable with mean $\bar{p}^T x$ and variance $x^T \Sigma x$. The choice of portfolio x involves a trade-off between the mean of the return, and its variance.

The classical portfolio optimization problem, introduced by Markowitz, is the QP

$$\begin{array}{ll} \text{minimize} & x^T \Sigma x \\ \text{subject to} & \bar{p}^T x \geq r_{\min} \\ & \mathbf{1}^T x = 1, \quad x \succeq 0, \end{array}$$

where x, the portfolio, is the variable. Here we find the portfolio that minimizes the return variance (which is associated with the *risk* of the portfolio) subject to achieving a minimum acceptable mean return r_{min}, and satisfying the portfolio budget and no-shorting constraints.

Many extensions are possible. One standard extension, for example, is to allow short positions, *i.e.*, $x_i < 0$. To do this we introduce variables x_{long} and x_{short}, with

$$x_{long} \succeq 0, \qquad x_{short} \succeq 0, \qquad x = x_{long} - x_{short}, \qquad \mathbf{1}^T x_{short} \leq \eta \mathbf{1}^T x_{long}.$$

The last constraint limits the total short position at the beginning of the period to some fraction η of the total long position at the beginning of the period.

As another extension we can include linear transaction costs in the portfolio optimization problem. Starting from a given initial portfolio x_{init} we buy and sell assets to achieve the portfolio x, which we then hold over the period as described above. We are charged a transaction fee for buying and selling assets, which is proportional to the amount bought or sold. To handle this, we introduce variables u_{buy} and u_{sell}, which determine the amount of each asset we buy and sell before the holding period. We have the constraints

$$x = x_{init} + u_{buy} - u_{sell}, \qquad u_{buy} \succeq 0, \qquad u_{sell} \succeq 0.$$

We replace the simple budget constraint $\mathbf{1}^T x = 1$ with the condition that the initial buying and selling, including transaction fees, involves zero net cash:

$$(1 - f_{sell})\mathbf{1}^T u_{sell} = (1 + f_{buy})\mathbf{1}^T u_{buy}$$

Here the lefthand side is the total proceeds from selling assets, less the selling transaction fee, and the righthand side is the total cost, including transaction fee, of buying assets. The constants $f_{buy} \geq 0$ and $f_{sell} \geq 0$ are the transaction fee rates for buying and selling (assumed the same across assets, for simplicity).

The problem of minimizing return variance, subject to a minimum mean return, and the budget and trading constraints, is a QP with variables x, u_{buy}, u_{sell}.

4.4.2   Second-order cone programming

A problem that is closely related to quadratic programming is the *second-order cone program* (SOCP):

$$
\begin{array}{ll}
\text{minimize} & f^T x \\
\text{subject to} & \|A_i x + b_i\|_2 \leq c_i^T x + d_i, \quad i = 1, \dots, m \\
& Fx = g,
\end{array}
\tag{4.36}
$$

where $x \in \mathbf{R}^n$ is the optimization variable, $A_i \in \mathbf{R}^{n_i \times n}$, and $F \in \mathbf{R}^{p \times n}$. We call a constraint of the form

$$\|Ax + b\|_2 \leq c^T x + d,$$

where $A \in \mathbf{R}^{k \times n}$, a *second-order cone constraint*, since it is the same as requiring the affine function $(Ax + b, c^T x + d)$ to lie in the second-order cone in $\mathbf{R}^{k+1}$.

When $c_i = 0$, $i = 1, \dots, m$, the SOCP (4.36) is equivalent to a QCQP (which is obtained by squaring each of the constraints). Similarly, if $A_i = 0$, $i = 1, \dots, m$, then the SOCP (4.36) reduces to a (general) LP. Second-order cone programs are, however, more general than QCQPs (and of course, LPs).

Robust linear programming

We consider a linear program in inequality form,

$$\begin{array}{ll} \text{minimize} & c^T x \\ \text{subject to} & a_i^T x \le b_i, \quad i = 1, \dots, m, \end{array}$$

in which there is some uncertainty or variation in the parameters c, a_i, b_i. To simplify the exposition we assume that c and b_i are fixed, and that a_i are known to lie in given ellipsoids:

$$a_i \in \mathcal{E}_i = \{\bar{a}_i + P_i u \mid \|u\|_2 \le 1\},$$

where $P_i \in \mathbf{R}^{n \times n}$. (If P_i is singular we obtain 'flat' ellipsoids, of dimension **rank** P_i; $P_i = 0$ means that a_i is known perfectly.)

We will require that the constraints be satisfied for all possible values of the parameters a_i, which leads us to the *robust linear program*

$$\begin{array}{ll} \text{minimize} & c^T x \\ \text{subject to} & a_i^T x \le b_i \text{ for all } a_i \in \mathcal{E}_i, \quad i = 1, \dots, m. \end{array} \tag{4.37}$$

The robust linear constraint, $a_i^T x \le b_i$ for all $a_i \in \mathcal{E}_i$, can be expressed as

$$\sup\{a_i^T x \mid a_i \in \mathcal{E}_i\} \le b_i,$$

the lefthand side of which can be expressed as

$$\begin{aligned} \sup\{a_i^T x \mid a_i \in \mathcal{E}_i\} &= \bar{a}_i^T x + \sup\{u^T P_i^T x \mid \|u\|_2 \le 1\} \\ &= \bar{a}_i^T x + \|P_i^T x\|_2. \end{aligned}$$

Thus, the robust linear constraint can be expressed as

$$\bar{a}_i^T x + \|P_i^T x\|_2 \le b_i,$$

which is evidently a second-order cone constraint. Hence the robust LP (4.37) can be expressed as the SOCP

$$\begin{array}{ll} \text{minimize} & c^T x \\ \text{subject to} & \bar{a}_i^T x + \|P_i^T x\|_2 \le b_i, \quad i = 1, \dots, m. \end{array}$$

Note that the additional norm terms act as *regularization terms*; they prevent x from being large in directions with considerable uncertainty in the parameters a_i.

Linear programming with random constraints

The robust LP described above can also be considered in a statistical framework. Here we suppose that the parameters a_i are independent Gaussian random vectors, with mean $\bar{a}_i$ and covariance Σ_i. We require that each constraint $a_i^T x \le b_i$ should hold with a probability (or confidence) exceeding η, where $\eta \ge 0.5$, *i.e.*,

$$\mathbf{prob}(a_i^T x \le b_i) \ge \eta. \tag{4.38}$$

We will show that this probability constraint can be expressed as a second-order cone constraint.

Letting $u = a_i^T x$, with σ^2 denoting its variance, this constraint can be written as

$$\mathbf{prob}\left(\frac{u - \overline{u}}{\sigma} \le \frac{b_i - \overline{u}}{\sigma}\right) \ge \eta.$$

Since $(u - \overline{u})/\sigma$ is a zero mean unit variance Gaussian variable, the probability above is simply $\Phi((b_i - \overline{u})/\sigma)$, where

$$\Phi(z) = \frac{1}{\sqrt{2\pi}} \int_{-\infty}^{z} e^{-t^2/2} \, dt$$

is the cumulative distribution function of a zero mean unit variance Gaussian random variable. Thus the probability constraint (4.38) can be expressed as

$$\frac{b_i - \overline{u}}{\sigma} \ge \Phi^{-1}(\eta),$$

or, equivalently,

$$\overline{u} + \Phi^{-1}(\eta)\sigma \le b_i.$$

From $\overline{u} = \overline{a}_i^T x$ and $\sigma = (x^T \Sigma_i x)^{1/2}$ we obtain

$$\overline{a}_i^T x + \Phi^{-1}(\eta) \|\Sigma_i^{1/2} x\|_2 \le b_i.$$

By our assumption that $\eta \ge 1/2$, we have $\Phi^{-1}(\eta) \ge 0$, so this constraint is a second-order cone constraint.

In summary, the problem

$$
\begin{array}{ll}
\text{minimize} & c^T x \\
\text{subject to} & \mathbf{prob}(a_i^T x \le b_i) \ge \eta, \quad i = 1, \ldots, m
\end{array}
$$

can be expressed as the SOCP

$$
\begin{array}{ll}
\text{minimize} & c^T x \\
\text{subject to} & \overline{a}_i^T x + \Phi^{-1}(\eta) \|\Sigma_i^{1/2} x\|_2 \le b_i, \quad i = 1, \ldots, m.
\end{array}
$$

(We will consider robust convex optimization problems in more depth in chapter 6. See also exercises 4.13, 4.28, and 4.59.)

Example 4.8 *Portfolio optimization with loss risk constraints.* We consider again the classical Markowitz portfolio problem described above (page 155). We assume here that the price change vector $p \in \mathbf{R}^n$ is a Gaussian random variable, with mean $\overline{p}$ and covariance Σ. Therefore the return r is a Gaussian random variable with mean $\overline{r} = \overline{p}^T x$ and variance $\sigma_r^2 = x^T \Sigma x$.

Consider a *loss risk constraint* of the form

$$\mathbf{prob}(r \le \alpha) \le \beta, \tag{4.39}$$

where α is a given unwanted return level (*e.g.*, a large loss) and β is a given maximum probability.

As in the stochastic interpretation of the robust LP given above, we can express this constraint using the cumulative distribution function Φ of a unit Gaussian random variable. The inequality (4.39) is equivalent to

$$\bar{p}^T x + \Phi^{-1}(\beta) \, \|\Sigma^{1/2} x\|_2 \geq \alpha.$$

Provided $\beta \leq 1/2$ (*i.e.*, $\Phi^{-1}(\beta) \leq 0$), this loss risk constraint is a second-order cone constraint. (If $\beta > 1/2$, the loss risk constraint becomes nonconvex in x.)

The problem of maximizing the expected return subject to a bound on the loss risk (with $\beta \leq 1/2$), can therefore be cast as an SOCP with one second-order cone constraint:

$$
\begin{array}{ll}
\text{maximize} & \bar{p}^T x \\
\text{subject to} & \bar{p}^T x + \Phi^{-1}(\beta) \, \|\Sigma^{1/2} x\|_2 \geq \alpha \\
& x \succeq 0, \quad \mathbf{1}^T x = 1.
\end{array}
$$

There are many extensions of this problem. For example, we can impose several loss risk constraints, *i.e.*,

$$\mathbf{prob}(r \leq \alpha_i) \leq \beta_i, \quad i = 1, \ldots, k,$$

(where $\beta_i \leq 1/2$), which expresses the risks (β_i) we are willing to accept for various levels of loss (α_i).

Minimal surface

Consider a differentiable function $f : \mathbf{R}^2 \to \mathbf{R}$ with $\mathbf{dom}\, f = C$. The surface area of its graph is given by

$$A = \int_C \sqrt{1 + \|\nabla f(x)\|_2^2}\, dx = \int_C \|(\nabla f(x), 1)\|_2\, dx,$$

which is a convex functional of f. The *minimal surface problem* is to find the function f that minimizes A subject to some constraints, for example, some given values of f on the boundary of C.

We will approximate this problem by discretizing the function f. Let $C = [0,1] \times [0,1]$, and let f_{ij} denote the value of f at the point $(i/K, j/K)$, for $i, j = 0, \ldots, K$. An approximate expression for the gradient of f at the point $x = (i/K, j/K)$ can be found using forward differences:

$$\nabla f(x) \approx K \begin{bmatrix} f_{i+1,j} - f_{i,j} \\ f_{i,j+1} - f_{i,j} \end{bmatrix}.$$

Substituting this into the expression for the area of the graph, and approximating the integral as a sum, we obtain an approximation for the area of the graph:

$$A \approx A_{\text{disc}} = \frac{1}{K^2} \sum_{i,j=0}^{K-1} \left\| \begin{bmatrix} K(f_{i+1,j} - f_{i,j}) \\ K(f_{i,j+1} - f_{i,j}) \\ 1 \end{bmatrix} \right\|_2$$

The discretized area approximation A_{disc} is a convex function of f_{ij}.

We can consider a wide variety of constraints on f_{ij}, such as equality or inequality constraints on any of its entries (for example, on the boundary values), or on its moments. As an example, we consider the problem of finding the minimal area surface with fixed boundary values on the left and right edges of the square:

$$
\begin{array}{ll}
\text{minimize} & A_{\text{disc}} \\
\text{subject to} & f_{0j} = l_j, \quad j = 0, \dots, K \\
& f_{Kj} = r_j, \quad j = 0, \dots, K
\end{array}
\tag{4.40}
$$

where f_{ij}, $i, j = 0, \dots, K$, are the variables, and l_j, r_j are the given boundary values on the left and right sides of the square.

We can transform the problem (4.40) into an SOCP by introducing new variables t_{ij}, i, $j = 0, \dots, K-1$:

$$
\begin{array}{ll}
\text{minimize} & (1/K^2) \sum_{i,j=0}^{K-1} t_{ij} \\
\text{subject to} & \left\| \begin{bmatrix} K(f_{i+1,j} - f_{i,j}) \\ K(f_{i,j+1} - f_{i,j}) \\ 1 \end{bmatrix} \right\|_2 \leq t_{ij}, \quad i, \ j = 0, \dots, K-1 \\
& f_{0j} = l_j, \quad j = 0, \dots, K \\
& f_{Kj} = r_j, \quad j = 0, \dots, K.
\end{array}
$$

4.5 Geometric programming

In this section we describe a family of optimization problems that are *not* convex in their natural form. These problems can, however, be transformed to convex optimization problems, by a change of variables and a transformation of the objective and constraint functions.

4.5.1 Monomials and posynomials

A function $f : \mathbf{R}^n \to \mathbf{R}$ with $\mathbf{dom}\, f = \mathbf{R}_{++}^n$, defined as

$$
f(x) = c x_1^{a_1} x_2^{a_2} \cdots x_n^{a_n},
\tag{4.41}
$$

where $c > 0$ and $a_i \in \mathbf{R}$, is called a *monomial function*, or simply, a *monomial*. The exponents a_i of a monomial can be any real numbers, including fractional or negative, but the coefficient c can only be positive. (The term 'monomial' conflicts with the standard definition from algebra, in which the exponents must be non-negative integers, but this should not cause any confusion.) A sum of monomials, *i.e.*, a function of the form

$$
f(x) = \sum_{k=1}^{K} c_k x_1^{a_{1k}} x_2^{a_{2k}} \cdots x_n^{a_{nk}},
\tag{4.42}
$$

where $c_k > 0$, is called a *posynomial function* (with K terms), or simply, a *posynomial*.

Posynomials are closed under addition, multiplication, and nonnegative scaling. Monomials are closed under multiplication and division. If a posynomial is multiplied by a monomial, the result is a posynomial; similarly, a posynomial can be divided by a monomial, with the result a posynomial.

4.5.2   Geometric programming

An optimization problem of the form

$$
\begin{array}{ll}
\text{minimize} & f_0(x) \\
\text{subject to} & f_i(x) \le 1, \quad i = 1, \dots, m \\
& h_i(x) = 1, \quad i = 1, \dots, p
\end{array}
\tag{4.43}
$$

where $f_0, \dots, f_m$ are posynomials and $h_1, \dots, h_p$ are monomials, is called a *geometric program* (GP). The domain of this problem is $\mathcal{D} = \mathbf{R}^n_{++}$; the constraint $x \succ 0$ is implicit.

Extensions of geometric programming

Several extensions are readily handled. If f is a posynomial and h is a monomial, then the constraint $f(x) \le h(x)$ can be handled by expressing it as $f(x)/h(x) \le 1$ (since f/h is posynomial). This includes as a special case a constraint of the form $f(x) \le a$, where f is posynomial and $a > 0$. In a similar way if h_1 and h_2 are both nonzero monomial functions, then we can handle the equality constraint $h_1(x) = h_2(x)$ by expressing it as $h_1(x)/h_2(x) = 1$ (since h_1/h_2 is monomial). We can maximize a nonzero monomial objective function, by minimizing its inverse (which is also a monomial).

For example, consider the problem

$$
\begin{array}{ll}
\text{maximize} & x/y \\
\text{subject to} & 2 \le x \le 3 \\
& x^2 + 3y/z \le \sqrt{y} \\
& x/y = z^2,
\end{array}
$$

with variables x, y, $z \in \mathbf{R}$ (and the implicit constraint x, y, $z > 0$). Using the simple transformations described above, we obtain the equivalent standard form GP

$$
\begin{array}{ll}
\text{minimize} & x^{-1}y \\
\text{subject to} & 2x^{-1} \le 1, \quad (1/3)x \le 1 \\
& x^2 y^{-1/2} + 3y^{1/2}z^{-1} \le 1 \\
& xy^{-1}z^{-2} = 1.
\end{array}
$$

We will refer to a problem like this one, that is easily transformed to an equivalent GP in the standard form (4.43), also as a GP. (In the same way that we refer to a problem easily transformed to an LP as an LP.)

4.5.3   Geometric program in convex form

Geometric programs are not (in general) convex optimization problems, but they can be transformed to convex problems by a change of variables and a transformation of the objective and constraint functions.

We will use the variables defined as $y_i = \log x_i$, so $x_i = e^{y_i}$. If f is the monomial function of x given in (4.41), *i.e.*,

$$f(x) = c x_1^{a_1} x_2^{a_2} \cdots x_n^{a_n},$$

then

$$
\begin{aligned}
f(x) &= f(e^{y_1}, \ldots, e^{y_n}) \\
&= c(e^{y_1})^{a_1} \cdots (e^{y_n})^{a_n} \\
&= e^{a^T y + b},
\end{aligned}
$$

where $b = \log c$. The change of variables $y_i = \log x_i$ turns a monomial function into the exponential of an affine function.

Similarly, if f is the posynomial given by (4.42), *i.e.*,

$$f(x) = \sum_{k=1}^{K} c_k x_1^{a_{1k}} x_2^{a_{2k}} \cdots x_n^{a_{nk}},$$

then

$$f(x) = \sum_{k=1}^{K} e^{a_k^T y + b_k},$$

where $a_k = (a_{1k}, \ldots, a_{nk})$ and $b_k = \log c_k$. After the change of variables, a posynomial becomes a sum of exponentials of affine functions.

The geometric program (4.43) can be expressed in terms of the new variable y as

$$
\begin{array}{ll}
\text{minimize} & \sum_{k=1}^{K_0} e^{a_{0k}^T y + b_{0k}} \\
\text{subject to} & \sum_{k=1}^{K_i} e^{a_{ik}^T y + b_{ik}} \le 1, \quad i = 1, \ldots, m \\
& e^{g_i^T y + h_i} = 1, \quad i = 1, \ldots, p,
\end{array}
$$

where $a_{ik} \in \mathbf{R}^n$, $i = 0, \ldots, m$, contain the exponents of the posynomial inequality constraints, and $g_i \in \mathbf{R}^n$, $i = 1, \ldots, p$, contain the exponents of the monomial equality constraints of the original geometric program.

Now we transform the objective and constraint functions, by taking the logarithm. This results in the problem

$$
\begin{array}{lll}
\text{minimize} & \tilde{f}_0(y) = \log \left(\sum_{k=1}^{K_0} e^{a_{0k}^T y + b_{0k}} \right) & \\
\text{subject to} & \tilde{f}_i(y) = \log \left(\sum_{k=1}^{K_i} e^{a_{ik}^T y + b_{ik}} \right) \le 0, \quad i = 1, \ldots, m & \text{(4.44)} \\
& \tilde{h}_i(y) = g_i^T y + h_i = 0, \quad i = 1, \ldots, p.
\end{array}
$$

Since the functions $\tilde{f}_i$ are convex, and $\tilde{h}_i$ are affine, this problem is a convex optimization problem. We refer to it as a *geometric program in convex form*. To distinguish it from the original geometric program, we refer to (4.43) as a *geometric program in posynomial form*.

Note that the transformation between the posynomial form geometric program (4.43) and the convex form geometric program (4.44) does not involve any computation; the problem data for the two problems are the same. It simply changes the form of the objective and constraint functions.

If the posynomial objective and constraint functions all have only one term, *i.e.*, are monomials, then the convex form geometric program (4.44) reduces to a (general) linear program. We can therefore consider geometric programming to be a generalization, or extension, of linear programming.

4.5.4   Examples

Frobenius norm diagonal scaling

Consider a matrix $M \in \mathbf{R}^{n \times n}$, and the associated linear function that maps u into $y = Mu$. Suppose we scale the coordinates, *i.e.*, change variables to $\tilde{u} = Du$, $\tilde{y} = Dy$, where D is diagonal, with $D_{ii} > 0$. In the new coordinates the linear function is given by $\tilde{y} = DMD^{-1}\tilde{u}$.

Now suppose we want to choose the scaling in such a way that the resulting matrix, DMD^{-1}, is small. We will use the Frobenius norm (squared) to measure the size of the matrix:

$$
\begin{aligned}
\|DMD^{-1}\|_F^2 &= \mathbf{tr}\left(\left(DMD^{-1}\right)^T \left(DMD^{-1}\right)\right) \\
&= \sum_{i,j=1}^{n} \left(DMD^{-1}\right)_{ij}^2 \\
&= \sum_{i,j=1}^{n} M_{ij}^2 d_i^2 / d_j^2,
\end{aligned}
$$

where $D = \mathbf{diag}(d)$. Since this is a posynomial in d, the problem of choosing the scaling d to minimize the Frobenius norm is an unconstrained geometric program,

$$
\text{minimize} \quad \sum_{i,j=1}^{n} M_{ij}^2 d_i^2 / d_j^2,
$$

with variable d. The only exponents in this geometric program are 0, 2, and -2.

Design of a cantilever beam

We consider the design of a cantilever beam, which consists of N segments, numbered from right to left as $1, \ldots, N$, as shown in figure 4.6. Each segment has unit length and a uniform rectangular cross-section with width w_i and height h_i. A vertical load (force) F is applied at the right end of the beam. This load causes the beam to deflect (downward), and induces stress in each segment of the beam. We assume that the deflections are small, and that the material is linearly elastic, with Young's modulus E.

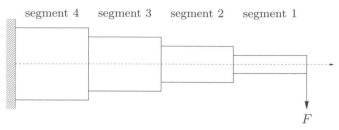

Figure 4.6 Segmented cantilever beam with 4 segments. Each segment has unit length and a rectangular profile. A vertical force F is applied at the right end of the beam.

The design variables in the problem are the widths w_i and heights h_i of the N segments. We seek to minimize the total volume of the beam (which is proportional to its weight),

$$w_1 h_1 + \cdots + w_N h_N,$$

subject to some design constraints. We impose upper and lower bounds on width and height of the segments,

$$w_{\min} \le w_i \le w_{\max}, \quad h_{\min} \le h_i \le h_{\max}, \quad i = 1, \ldots, N,$$

as well as the aspect ratios,

$$S_{\min} \le h_i/w_i \le S_{\max}.$$

In addition, we have a limit on the maximum allowable stress in the material, and on the vertical deflection at the end of the beam.

We first consider the maximum stress constraint. The maximum stress in segment i, which we denote σ_i, is given by $\sigma_i = 6iF/(w_i h_i^2)$. We impose the constraints

$$\frac{6iF}{w_i h_i^2} \le \sigma_{\max}, \quad i = 1, \ldots, N,$$

to ensure that the stress does not exceed the maximum allowable value $\sigma_{\max}$ anywhere in the beam.

The last constraint is a limit on the vertical deflection at the end of the beam, which we will denote y_1:

$$y_1 \le y_{\max}.$$

The deflection y_1 can be found by a recursion that involves the deflection and slope of the beam segments:

$$v_i = 12(i - 1/2)\frac{F}{E w_i h_i^3} + v_{i+1}, \qquad y_i = 6(i - 1/3)\frac{F}{E w_i h_i^3} + v_{i+1} + y_{i+1}, \quad (4.45)$$

for $i = N, N - 1, \ldots, 1$, with starting values $v_{N+1} = y_{N+1} = 0$. In this recursion, y_i is the deflection at the right end of segment i, and v_i is the slope at that point. We can use the recursion (4.45) to show that these deflection and slope quantities are in fact posynomial functions of the variables w and h. We first note that v_{N+1} and y_{N+1} are zero, and therefore posynomials. Now assume that v_{i+1} and y_{i+1} are posynomial functions of w and h. The lefthand equation in (4.45) shows that v_i is the sum of a monomial and a posynomial (*i.e.*, v_{i+1}), and therefore is a posynomial. From the righthand equation in (4.45), we see that the deflection y_i is the sum of a monomial and two posynomials (v_{i+1} and y_{i+1}), and so is a posynomial. In particular, the deflection at the end of the beam, y_1, is a posynomial.

The problem is then

$$
\begin{array}{ll}
\text{minimize} & \sum_{i=1}^{N} w_i h_i \\
\text{subject to} & w_{\min} \le w_i \le w_{\max}, \quad i = 1, \ldots, N \\
& h_{\min} \le h_i \le h_{\max}, \quad i = 1, \ldots, N \\
& S_{\min} \le h_i/w_i \le S_{\max}, \quad i = 1, \ldots, N \\
& 6iF/(w_i h_i^2) \le \sigma_{\max}, \quad i = 1, \ldots, N \\
& y_1 \le y_{\max},
\end{array}
\tag{4.46}
$$

with variables w and h. This is a GP, since the objective is a posynomial, and the constraints can all be expressed as posynomial inequalities. (In fact, the constraints can be all be expressed as monomial inequalities, with the exception of the deflection limit, which is a complicated posynomial inequality.)

When the number of segments N is large, the number of monomial terms appearing in the posynomial y_1 grows approximately as N^2. Another formulation of this problem, explored in exercise 4.31, is obtained by introducing $v_1, \ldots, v_N$ and $y_1, \ldots, y_N$ as variables, and including a modified version of the recursion as a set of constraints. This formulation avoids this growth in the number of monomial terms.

Minimizing spectral radius via Perron-Frobenius theory

Suppose the matrix $A \in \mathbf{R}^{n \times n}$ is elementwise nonnegative, *i.e.*, $A_{ij} \ge 0$ for $i, j = 1, \ldots, n$, and irreducible, which means that the matrix $(I + A)^{n-1}$ is elementwise positive. The Perron-Frobenius theorem states that A has a positive real eigenvalue λ_{pf} equal to its spectral radius, *i.e.*, the largest magnitude of its eigenvalues. The Perron-Frobenius eigenvalue λ_{pf} determines the asymptotic rate of growth or decay of A^k, as $k \to \infty$; in fact, the matrix $((1/\lambda_{\text{pf}})A)^k$ converges. Roughly speaking, this means that as $k \to \infty$, A^k grows like λ_{pf}^k, if $\lambda_{\text{pf}} > 1$, or decays like λ_{pf}^k, if $\lambda_{\text{pf}} < 1$.

A basic result in the theory of nonnegative matrices states that the Perron-Frobenius eigenvalue is given by

$$
\lambda_{\text{pf}} = \inf\{\lambda \mid Av \preceq \lambda v \text{ for some } v \succ 0\}
$$

(and moreover, that the infimum is achieved). The inequality $Av \preceq \lambda v$ can be expressed as

$$
\sum_{j=1}^{n} A_{ij} v_j/(\lambda v_i) \le 1, \quad i = 1, \ldots, n,
\tag{4.47}
$$

which is a set of posynomial inequalities in the variables A_{ij}, v_i, and λ. Thus, the condition that $\lambda_{\text{pf}} \le \lambda$ can be expressed as a set of posynomial inequalities in A, v, and λ. This allows us to solve some optimization problems involving the Perron-Frobenius eigenvalue using geometric programming.

Suppose that the entries of the matrix A are posynomial functions of some underlying variable $x \in \mathbf{R}^k$. In this case the inequalities (4.47) are posynomial inequalities in the variables $x \in \mathbf{R}^k$, $v \in \mathbf{R}^n$, and $\lambda \in \mathbf{R}$. We consider the problem of choosing x to minimize the Perron-Frobenius eigenvalue (or spectral radius) of A, possibly subject to posynomial inequalities on x,

$$
\begin{array}{ll}
\text{minimize} & \lambda_{\mathrm{pf}}(A(x)) \\
\text{subject to} & f_i(x) \le 1, \quad i = 1, \dots, p,
\end{array}
$$

where f_i are posynomials. Using the characterization above, we can express this problem as the GP

$$
\begin{array}{ll}
\text{minimize} & \lambda \\
\text{subject to} & \sum_{j=1}^{n} A_{ij} v_j / (\lambda v_i) \le 1, \quad i = 1, \dots, n \\
& f_i(x) \le 1, \quad i = 1, \dots, p,
\end{array}
$$

where the variables are x, v, and λ.

As a specific example, we consider a simple model for the population dynamics for a bacterium, with time or period denoted by $t = 0, 1, 2, \dots$, in hours. The vector $p(t) \in \mathbf{R}_+^4$ characterizes the population age distribution at period t: $p_1(t)$ is the total population between 0 and 1 hours old; $p_2(t)$ is the total population between 1 and 2 hours old; and so on. We (arbitrarily) assume that no bacteria live more than 4 hours. The population propagates in time as $p(t+1) = Ap(t)$, where

$$
A = \begin{bmatrix}
b_1 & b_2 & b_3 & b_4 \\
s_1 & 0 & 0 & 0 \\
0 & s_2 & 0 & 0 \\
0 & 0 & s_3 & 0
\end{bmatrix}.
$$

Here b_i is the birth rate among bacteria in age group i, and s_i is the survival rate from age group i into age group $i+1$. We assume that $b_i > 0$ and $0 < s_i < 1$, which implies that the matrix A is irreducible.

The Perron-Frobenius eigenvalue of A determines the asymptotic growth or decay rate of the population. If $\lambda_{\mathrm{pf}} < 1$, the population converges to zero like λ_{pf}^t, and so has a half-life of $-1 / \log_2 \lambda_{\mathrm{pf}}$ hours. If $\lambda_{\mathrm{pf}} > 1$ the population grows geometrically like λ_{pf}^t, with a doubling time of $1 / \log_2 \lambda_{\mathrm{pf}}$ hours. Minimizing the spectral radius of A corresponds to finding the fastest decay rate, or slowest growth rate, for the population.

As our underlying variables, on which the matrix A depends, we take c_1 and c_2, the concentrations of two chemicals in the environment that affect the birth and survival rates of the bacteria. We model the birth and survival rates as monomial functions of the two concentrations:

$$
\begin{aligned}
b_i &= b_i^{\mathrm{nom}} (c_1 / c_1^{\mathrm{nom}})^{\alpha_i} (c_2 / c_2^{\mathrm{nom}})^{\beta_i}, \quad i = 1, \dots, 4 \\
s_i &= s_i^{\mathrm{nom}} (c_1 / c_1^{\mathrm{nom}})^{\gamma_i} (c_2 / c_2^{\mathrm{nom}})^{\delta_i}, \quad i = 1, \dots, 3.
\end{aligned}
$$

Here, b_i^{nom} is nominal birth rate, s_i^{nom} is nominal survival rate, and c_i^{nom} is nominal concentration of chemical i. The constants α_i, β_i, γ_i, and δ_i give the effect on the birth and survival rates due to changes in the concentrations of the chemicals away from the nominal values. For example $\alpha_2 = -0.3$ and $\gamma_1 = 0.5$ means that an increase in concentration of chemical 1, over the nominal concentration, causes a decrease in the birth rate of bacteria that are between 1 and 2 hours old, and an increase in the survival rate of bacteria from 0 to 1 hours old.

We assume that the concentrations c_1 and c_2 can be independently increased or decreased (say, within a factor of 2), by administering drugs, and pose the problem of finding the drug mix that maximizes the population decay rate (*i.e.*, minimizes $\lambda_{\mathrm{pf}}(A)$). Using the approach described above, this problem can be posed as the GP

$$
\begin{aligned}
\text{minimize} \quad & \lambda \\
\text{subject to} \quad & b_1 v_1 + b_2 v_2 + b_3 v_3 + b_4 v_4 \leq \lambda v_1 \\
& s_1 v_1 \leq \lambda v_2 \\
& s_2 v_2 \leq \lambda v_3 \\
& s_3 v_3 \leq \lambda v_4 \\
& 1/2 \leq c_i/c_i^{\mathrm{nom}} \leq 2, \quad i = 1, 2 \\
& b_i = b_i^{\mathrm{nom}} (c_1/c_1^{\mathrm{nom}})^{\alpha_i} (c_2/c_2^{\mathrm{nom}})^{\beta_i}, \quad i = 1, \dots, 4 \\
& s_i = s_i^{\mathrm{nom}} (c_1/c_1^{\mathrm{nom}})^{\gamma_i} (c_2/c_2^{\mathrm{nom}})^{\delta_i}, \quad i = 1, \dots, 3,
\end{aligned}
$$

with variables b_i, s_i, c_i, v_i, and λ.

4.6   Generalized inequality constraints

One very useful generalization of the standard form convex optimization problem (4.15) is obtained by allowing the inequality constraint functions to be vector valued, and using generalized inequalities in the constraints:

$$
\begin{aligned}
\text{minimize} \quad & f_0(x) \\
\text{subject to} \quad & f_i(x) \preceq_{K_i} 0, \quad i = 1, \dots, m \\
& Ax = b,
\end{aligned}
\tag{4.48}
$$

where $f_0 : \mathbf{R}^n \to \mathbf{R}$, $K_i \subseteq \mathbf{R}^{k_i}$ are proper cones, and $f_i : \mathbf{R}^n \to \mathbf{R}^{k_i}$ are K_i-convex. We refer to this problem as a (standard form) *convex optimization problem with generalized inequality constraints*. Problem (4.15) is a special case with $K_i = \mathbf{R}_+$, $i = 1, \dots, m$.

Many of the results for ordinary convex optimization problems hold for problems with generalized inequalities. Some examples are:

- The feasible set, any sublevel set, and the optimal set are convex.

- Any point that is locally optimal for the problem (4.48) is globally optimal.

- The optimality condition for differentiable f_0, given in §4.2.3, holds without any change.

We will also see (in chapter 11) that convex optimization problems with generalized inequality constraints can often be solved as easily as ordinary convex optimization problems.

4.6.1   Conic form problems

Among the simplest convex optimization problems with generalized inequalities are the *conic form problems* (or *cone programs*), which have a linear objective and one inequality constraint function, which is affine (and therefore K-convex):

$$
\begin{array}{ll}
\text{minimize} & c^T x \\
\text{subject to} & Fx + g \preceq_K 0 \\
& Ax = b.
\end{array}
\tag{4.49}
$$

When K is the nonnegative orthant, the conic form problem reduces to a linear program. We can view conic form problems as a generalization of linear programs in which componentwise inequality is replaced with a generalized linear inequality.

Continuing the analogy to linear programming, we refer to the conic form problem

$$
\begin{array}{ll}
\text{minimize} & c^T x \\
\text{subject to} & x \succeq_K 0 \\
& Ax = b
\end{array}
$$

as a *conic form problem in standard form*. Similarly, the problem

$$
\begin{array}{ll}
\text{minimize} & c^T x \\
\text{subject to} & Fx + g \preceq_K 0
\end{array}
$$

is called a *conic form problem in inequality form*.

4.6.2   Semidefinite programming

When K is $\mathbf{S}^k_+$, the cone of positive semidefinite $k \times k$ matrices, the associated conic form problem is called a *semidefinite program* (SDP), and has the form

$$
\begin{array}{ll}
\text{minimize} & c^T x \\
\text{subject to} & x_1 F_1 + \cdots + x_n F_n + G \preceq 0 \\
& Ax = b,
\end{array}
\tag{4.50}
$$

where $G, F_1, \ldots, F_n \in \mathbf{S}^k$, and $A \in \mathbf{R}^{p \times n}$. The inequality here is a linear matrix inequality (see example 2.10).

If the matrices $G, F_1, \ldots, F_n$ are all diagonal, then the LMI in (4.50) is equivalent to a set of n linear inequalities, and the SDP (4.50) reduces to a linear program.

Standard and inequality form semidefinite programs

Following the analogy to LP, a *standard form SDP* has linear equality constraints, and a (matrix) nonnegativity constraint on the variable $X \in \mathbf{S}^n$:

$$
\begin{array}{ll}
\text{minimize} & \mathbf{tr}(CX) \\
\text{subject to} & \mathbf{tr}(A_i X) = b_i, \quad i = 1, \ldots, p \\
& X \succeq 0,
\end{array}
\tag{4.51}
$$

where C, $A_1, \ldots, A_p \in \mathbf{S}^n$. (Recall that $\mathbf{tr}(CX) = \sum_{i,j=1}^n C_{ij}X_{ij}$ is the form of a general real-valued linear function on $\mathbf{S}^n$.) This form should be compared to the standard form linear program (4.28). In LP and SDP standard forms, we minimize a linear function of the variable, subject to p linear equality constraints on the variable, and a nonnegativity constraint on the variable.

An *inequality form SDP*, analogous to an inequality form LP (4.29), has no equality constraints, and one LMI:

$$\begin{array}{ll} \text{minimize} & c^T x \\ \text{subject to} & x_1 A_1 + \cdots + x_n A_n \preceq B, \end{array}$$

with variable $x \in \mathbf{R}^n$, and parameters B, $A_1, \ldots, A_n \in \mathbf{S}^k$, $c \in \mathbf{R}^n$.

Multiple LMIs and linear inequalities

It is common to refer to a problem with linear objective, linear equality and inequality constraints, and several LMI constraints, *i.e.*,

$$\begin{array}{ll} \text{minimize} & c^T x \\ \text{subject to} & F^{(i)}(x) = x_1 F_1^{(i)} + \cdots + x_n F_n^{(i)} + G^{(i)} \preceq 0, \quad i = 1, \ldots, K \\ & Gx \preceq h, \qquad Ax = b, \end{array}$$

as an SDP as well. Such problems are readily transformed to an SDP, by forming a large block diagonal LMI from the individual LMIs and linear inequalities:

$$\begin{array}{ll} \text{minimize} & c^T x \\ \text{subject to} & \mathbf{diag}(Gx - h, F^{(1)}(x), \ldots, F^{(K)}(x)) \preceq 0 \\ & Ax = b. \end{array}$$

4.6.3   Examples

Second-order cone programming

The SOCP (4.36) can be expressed as a conic form problem

$$\begin{array}{ll} \text{minimize} & c^T x \\ \text{subject to} & -(A_i x + b_i, c_i^T x + d_i) \preceq_{K_i} 0, \quad i = 1, \ldots, m \\ & Fx = g, \end{array}$$

in which

$$K_i = \{(y, t) \in \mathbf{R}^{n_i+1} \mid \|y\|_2 \le t\},$$

i.e., the second-order cone in $\mathbf{R}^{n_i+1}$. This explains the name *second-order cone program* for the optimization problem (4.36).

Matrix norm minimization

Let $A(x) = A_0 + x_1 A_1 + \cdots + x_n A_n$, where $A_i \in \mathbf{R}^{p \times q}$. We consider the unconstrained problem

$$\text{minimize} \quad \|A(x)\|_2,$$

where $\| \cdot \|_2$ denotes the spectral norm (maximum singular value), and $x \in \mathbf{R}^n$ is the variable. This is a convex problem since $\|A(x)\|_2$ is a convex function of x.

Using the fact that $\|A\|_2 \leq s$ if and only if $A^T A \preceq s^2 I$ (and $s \geq 0$), we can express the problem in the form

$$
\begin{array}{ll}
\text{minimize} & s \\
\text{subject to} & A(x)^T A(x) \preceq sI,
\end{array}
$$

with variables x and s. Since the function $A(x)^T A(x) - sI$ is matrix convex in (x, s), this is a convex optimization problem with a single $q \times q$ matrix inequality constraint.

We can also formulate the problem using a single linear matrix inequality of size $(p + q) \times (p + q)$, using the fact that

$$
A^T A \preceq t^2 I \text{ (and } t \geq 0) \iff \begin{bmatrix} tI & A \\ A^T & tI \end{bmatrix} \succeq 0.
$$

(see §A.5.5). This results in the SDP

$$
\begin{array}{ll}
\text{minimize} & t \\
\text{subject to} & \begin{bmatrix} tI & A(x) \\ A(x)^T & tI \end{bmatrix} \succeq 0
\end{array}
$$

in the variables x and t.

Moment problems

Let t be a random variable in $\mathbf{R}$. The expected values $\mathbf{E}\, t^k$ (assuming they exist) are called the (power) *moments* of the distribution of t. The following classical results give a characterization of a moment sequence.

If there is a probability distribution on $\mathbf{R}$ such that $x_k = \mathbf{E}\, t^k$, $k = 0, \ldots, 2n$, then $x_0 = 1$ and

$$
H(x_0, \ldots, x_{2n}) = \begin{bmatrix}
x_0 & x_1 & x_2 & \cdots & x_{n-1} & x_n \\
x_1 & x_2 & x_3 & \cdots & x_n & x_{n+1} \\
x_2 & x_3 & x_4 & \cdots & x_{n+1} & x_{n+2} \\
\vdots & \vdots & \vdots & & \vdots & \vdots \\
x_{n-1} & x_n & x_{n+1} & \cdots & x_{2n-2} & x_{2n-1} \\
x_n & x_{n+1} & x_{n+2} & \cdots & x_{2n-1} & x_{2n}
\end{bmatrix} \succeq 0. \qquad (4.52)
$$

(The matrix H is called the *Hankel matrix* associated with $x_0, \ldots, x_{2n}$.) This is easy to see: Let $x_i = \mathbf{E}\, t^i$, $i = 0, \ldots, 2n$ be the moments of some distribution, and let $y = (y_0, y_1, \ldots y_n) \in \mathbf{R}^{n+1}$. Then we have

$$
y^T H(x_0, \ldots, x_{2n}) y = \sum_{i,j=0}^{n} y_i y_j \, \mathbf{E}\, t^{i+j} = \mathbf{E}(y_0 + y_1 t^1 + \cdots + y_n t^n)^2 \geq 0.
$$

The following partial converse is less obvious: If $x_0 = 1$ and $H(x) \succ 0$, then there exists a probability distribution on $\mathbf{R}$ such that $x_i = \mathbf{E}\, t^i$, $i = 0, \ldots, 2n$. (For a proof, see exercise 2.37.) Now suppose that $x_0 = 1$, and $H(x) \succeq 0$ (but possibly $H(x) \not\succ 0$), *i.e.*, the linear matrix inequality (4.52) holds, but possibly not strictly. In this case, there is a sequence of distributions on $\mathbf{R}$, whose moments converge to x. In summary: the condition that $x_0, \ldots, x_{2n}$ be the moments of some distribution on $\mathbf{R}$ (or the limit of the moments of a sequence of distributions) can be expressed as the linear matrix inequality (4.52) in the variable x, together with the linear equality $x_0 = 1$. Using this fact, we can cast some interesting problems involving moments as SDPs.

Suppose t is a random variable on $\mathbf{R}$. We do not know its distribution, but we do know some bounds on the moments, *i.e.*,

$$\underline{\mu}_k \leq \mathbf{E}\, t^k \leq \overline{\mu}_k, \quad k = 1, \ldots, 2n$$

(which includes, as a special case, knowing exact values of some of the moments). Let $p(t) = c_0 + c_1 t + \cdots + c_{2n} t^{2n}$ be a given polynomial in t. The expected value of $p(t)$ is linear in the moments $\mathbf{E}\, t^i$:

$$\mathbf{E}\, p(t) = \sum_{i=0}^{2n} c_i \,\mathbf{E}\, t^i = \sum_{i=0}^{2n} c_i x_i.$$

We can compute upper and lower bounds for $\mathbf{E}\, p(t)$,

$$
\begin{array}{ll}
\text{minimize (maximize)} & \mathbf{E}\, p(t) \\
\text{subject to} & \underline{\mu}_k \leq \mathbf{E}\, t^k \leq \overline{\mu}_k, \quad k = 1, \ldots, 2n,
\end{array}
$$

over all probability distributions that satisfy the given moment bounds, by solving the SDP

$$
\begin{array}{ll}
\text{minimize (maximize)} & c_1 x_1 + \cdots + c_{2n} x_{2n} \\
\text{subject to} & \underline{\mu}_k \leq x_k \leq \overline{\mu}_k, \quad k = 1, \ldots, 2n \\
& H(1, x_1, \ldots, x_{2n}) \succeq 0
\end{array}
$$

with variables $x_1, \ldots, x_{2n}$. This gives bounds on $\mathbf{E}\, p(t)$, over all probability distributions that satisfy the known moment constraints. The bounds are sharp in the sense that there exists a sequence of distributions, whose moments satisfy the given moment bounds, for which $\mathbf{E}\, p(t)$ converges to the upper and lower bounds found by these SDPs.

Bounding portfolio risk with incomplete covariance information

We consider once again the setup for the classical Markowitz portfolio problem (see page 155). We have a portfolio of n assets or stocks, with x_i denoting the amount of asset i that is held over some investment period, and p_i denoting the relative price change of asset i over the period. The change in total value of the portfolio is $p^T x$. The price change vector p is modeled as a random vector, with mean and covariance

$$\overline{p} = \mathbf{E}\, p, \qquad \Sigma = \mathbf{E}(p - \overline{p})(p - \overline{p})^T.$$

The change in value of the portfolio is therefore a random variable with mean $\overline{p}^T x$ and standard deviation $\sigma = (x^T \Sigma x)^{1/2}$. The risk of a large loss, *i.e.*, a change in portfolio value that is substantially below its expected value, is directly related to the standard deviation σ, and increases with it. For this reason the standard deviation σ (or the variance σ^2) is used as a measure of the risk associated with the portfolio.

In the classical portfolio optimization problem, the portfolio x is the optimization variable, and we minimize the risk subject to a minimum mean return and other constraints. The price change statistics $\bar{p}$ and Σ are known problem parameters. In the risk bounding problem considered here, we turn the problem around: we assume the portfolio x is known, but only partial information is available about the covariance matrix Σ. We might have, for example, an upper and lower bound on each entry:

$$L_{ij} \leq \Sigma_{ij} \leq U_{ij}, \quad i,\, j = 1, \ldots, n,$$

where L and U are given. We now pose the question: what is the maximum risk for our portfolio, over all covariance matrices consistent with the given bounds? We define the *worst-case variance* of the portfolio as

$$\sigma_{\mathrm{wc}}^2 = \sup\{x^T \Sigma x \mid L_{ij} \leq \Sigma_{ij} \leq U_{ij},\ i,j = 1, \ldots, n,\ \Sigma \succeq 0\}.$$

We have added the condition $\Sigma \succeq 0$, which the covariance matrix must, of course, satisfy.

We can find σ_{wc} by solving the SDP

$$
\begin{array}{ll}
\text{maximize} & x^T \Sigma x \\
\text{subject to} & L_{ij} \leq \Sigma_{ij} \leq U_{ij}, \quad i,\, j = 1, \ldots, n \\
& \Sigma \succeq 0
\end{array}
$$

with variable $\Sigma \in \mathbf{S}^n$ (and problem parameters x, L, and U). The optimal Σ is the worst covariance matrix consistent with our given bounds on the entries, where 'worst' means largest risk with the (given) portfolio x. We can easily construct a distribution for p that is consistent with the given bounds, and achieves the worst-case variance, from an optimal Σ for the SDP. For example, we can take $p = \bar{p} + \Sigma^{1/2} v$, where v is any random vector with $\mathbf{E}\, v = 0$ and $\mathbf{E}\, vv^T = I$.

Evidently we can use the same method to determine σ_{wc} for any prior information about Σ that is convex. We list here some examples.

- *Known variance of certain portfolios.* We might have equality constraints such as
$$u_k^T \Sigma u_k = \sigma_k^2,$$
where u_k and σ_k are given. This corresponds to prior knowledge that certain known portfolios (given by u_k) have known (or very accurately estimated) variance.

- *Including effects of estimation error.* If the covariance Σ is estimated from empirical data, the estimation method will give an estimate $\hat{\Sigma}$, and some information about the reliability of the estimate, such as a confidence ellipsoid. This can be expressed as
$$C(\Sigma - \hat{\Sigma}) \leq \alpha,$$
where C is a positive definite quadratic form on $\mathbf{S}^n$, and the constant α determines the confidence level.

- *Factor models.* The covariance might have the form

$$\Sigma = F\Sigma_{\text{factor}}F^T + D,$$

where $F \in \mathbf{R}^{n \times k}$, $\Sigma_{\text{factor}} \in \mathbf{S}^k$, and D is diagonal. This corresponds to a model of the price changes of the form

$$p = Fz + d,$$

where z is a random variable (the underlying *factors* that affect the price changes) and d_i are independent (additional volatility of each asset price). We assume that the factors are known. Since Σ is linearly related to Σ_{factor} and D, we can impose any convex constraint on them (representing prior information) and still compute σ_{wc} using convex optimization.

- *Information about correlation coefficients.* In the simplest case, the diagonal entries of Σ (*i.e.*, the volatilities of each asset price) are known, and bounds on correlation coefficients between price changes are known:

$$l_{ij} \leq \rho_{ij} = \frac{\Sigma_{ij}}{\Sigma_{ii}^{1/2}\Sigma_{jj}^{1/2}} \leq u_{ij}, \quad i,\, j = 1, \ldots, n.$$

Since Σ_{ii} are known, but Σ_{ij} for $i \neq j$ are not, these are linear inequalities.

Fastest mixing Markov chain on a graph

We consider an undirected graph, with nodes $1, \ldots, n$, and a set of edges

$$\mathcal{E} \subseteq \{1, \ldots, n\} \times \{1, \ldots, n\}.$$

Here $(i,j) \in \mathcal{E}$ means that nodes i and j are connected by an edge. Since the graph is undirected, $\mathcal{E}$ is symmetric: $(i,j) \in \mathcal{E}$ if and only if $(j,i) \in \mathcal{E}$. We allow the possibility of self-loops, *i.e.*, we can have $(i,i) \in \mathcal{E}$.

We define a Markov chain, with state $X(t) \in \{1, \ldots, n\}$, for $t \in \mathbf{Z}_+$ (the set of nonnegative integers), as follows. With each edge $(i,j) \in \mathcal{E}$ we associate a probability P_{ij}, which is the probability that X makes a transition between nodes i and j. State transitions can only occur across edges; we have $P_{ij} = 0$ for $(i,j) \notin \mathcal{E}$. The probabilities associated with the edges must be nonnegative, and for each node, the sum of the probabilities of links connected to the node (including a self-loop, if there is one) must equal one.

The Markov chain has transition probability matrix

$$P_{ij} = \mathbf{prob}(X(t+1) = i \mid X(t) = j), \quad i, j = 1, \ldots, n.$$

This matrix must satisfy

$$P_{ij} \geq 0, \quad i,\, j = 1, \ldots, n, \qquad \mathbf{1}^T P = \mathbf{1}^T, \qquad P = P^T, \qquad (4.53)$$

and also

$$P_{ij} = 0 \quad \text{for } (i,j) \notin \mathcal{E}. \qquad (4.54)$$

Since P is symmetric and $\mathbf{1}^T P = \mathbf{1}^T$, we conclude $P\mathbf{1} = \mathbf{1}$, so the uniform distribution $(1/n)\mathbf{1}$ is an equilibrium distribution for the Markov chain. Convergence of the distribution of $X(t)$ to $(1/n)\mathbf{1}$ is determined by the second largest (in magnitude) eigenvalue of P, $i.e.$, by $r = \max\{\lambda_2, -\lambda_n\}$, where

$$1 = \lambda_1 \geq \lambda_2 \geq \cdots \geq \lambda_n$$

are the eigenvalues of P. We refer to r as the $mixing\ rate$ of the Markov chain. If $r = 1$, then the distribution of $X(t)$ need not converge to $(1/n)\mathbf{1}$ (which means the Markov chain does not mix). When $r < 1$, the distribution of $X(t)$ approaches $(1/n)\mathbf{1}$ asymptotically as r^t, as $t \to \infty$. Thus, the smaller r is, the faster the Markov chain mixes.

The $fastest\ mixing\ Markov\ chain\ problem$ is to find P, subject to the constraints (4.53) and (4.54), that minimizes r. (The problem data is the graph, $i.e.$, $\mathcal{E}$.) We will show that this problem can be formulated as an SDP.

Since the eigenvalue $\lambda_1 = 1$ is associated with the eigenvector $\mathbf{1}$, we can express the mixing rate as the norm of the matrix P, restricted to the subspace $\mathbf{1}^\perp$: $r = \|QPQ\|_2$, where $Q = I - (1/n)\mathbf{1}\mathbf{1}^T$ is the matrix representing orthogonal projection on $\mathbf{1}^\perp$. Using the property $P\mathbf{1} = \mathbf{1}$, we have

$$
\begin{aligned}
r &= \|QPQ\|_2 \\
&= \|(I - (1/n)\mathbf{1}\mathbf{1}^T)P(I - (1/n)\mathbf{1}\mathbf{1}^T)\|_2 \\
&= \|P - (1/n)\mathbf{1}\mathbf{1}^T\|_2.
\end{aligned}
$$

This shows that the mixing rate r is a convex function of P, so the fastest mixing Markov chain problem can be cast as the convex optimization problem

$$
\begin{array}{ll}
\text{minimize} & \|P - (1/n)\mathbf{1}\mathbf{1}^T\|_2 \\
\text{subject to} & P\mathbf{1} = \mathbf{1} \\
& P_{ij} \geq 0, \quad i, j = 1, \dots, n \\
& P_{ij} = 0 \text{ for } (i,j) \notin \mathcal{E},
\end{array}
$$

with variable $P \in \mathbf{S}^n$. We can express the problem as an SDP by introducing a scalar variable t to bound the norm of $P - (1/n)\mathbf{1}\mathbf{1}^T$:

$$
\begin{array}{ll}
\text{minimize} & t \\
\text{subject to} & -tI \preceq P - (1/n)\mathbf{1}\mathbf{1}^T \preceq tI \\
& P\mathbf{1} = \mathbf{1} \\
& P_{ij} \geq 0, \quad i, j = 1, \dots, n \\
& P_{ij} = 0 \text{ for } (i,j) \notin \mathcal{E}.
\end{array}
\qquad (4.55)
$$

4.7   Vector optimization

4.7.1   General and convex vector optimization problems

In §4.6 we extended the standard form problem (4.1) to include vector-valued constraint functions. In this section we investigate the meaning of a vector-valued *objective function.* We denote a general *vector optimization problem* as

$$\begin{array}{lll} \text{minimize (with respect to } K) & f_0(x) & \\ \text{subject to} & f_i(x) \le 0, & i = 1, \ldots, m \\ & h_i(x) = 0, & i = 1, \ldots, p. \end{array} \qquad (4.56)$$

Here $x \in \mathbf{R}^n$ is the optimization variable, $K \subseteq \mathbf{R}^q$ is a proper cone, $f_0 : \mathbf{R}^n \to \mathbf{R}^q$ is the objective function, $f_i : \mathbf{R}^n \to \mathbf{R}$ are the inequality constraint functions, and $h_i : \mathbf{R}^n \to \mathbf{R}$ are the equality constraint functions. The only difference between this problem and the standard optimization problem (4.1) is that here, the objective function takes values in $\mathbf{R}^q$, and the problem specification includes a proper cone K, which is used to compare objective values. In the context of vector optimization, the standard optimization problem (4.1) is sometimes called a *scalar optimization problem.*

We say the vector optimization problem (4.56) is a *convex vector optimization problem* if the objective function f_0 is K-convex, the inequality constraint functions $f_1, \ldots, f_m$ are convex, and the equality constraint functions $h_1, \ldots, h_p$ are affine. (As in the scalar case, we usually express the equality constraints as $Ax = b$, where $A \in \mathbf{R}^{p \times n}$.)

What meaning can we give to the vector optimization problem (4.56)? Suppose x and y are two feasible points (*i.e.*, they satisfy the constraints). Their associated objective values, $f_0(x)$ and $f_0(y)$, are to be compared using the generalized inequality $\preceq_K$. We interpret $f_0(x) \preceq_K f_0(y)$ as meaning that x is 'better than or equal' in value to y (as judged by the objective f_0, with respect to K). The confusing aspect of vector optimization is that the two objective values $f_0(x)$ and $f_0(y)$ need not be comparable; we can have neither $f_0(x) \preceq_K f_0(y)$ nor $f_0(y) \preceq_K f_0(x)$, *i.e.*, neither is better than the other. This cannot happen in a scalar objective optimization problem.

4.7.2   Optimal points and values

We first consider a special case, in which the meaning of the vector optimization problem is clear. Consider the set of objective values of feasible points,

$$\mathcal{O} = \{f_0(x) \mid \exists x \in \mathcal{D}, \ f_i(x) \le 0, \ i = 1, \ldots, m, \ h_i(x) = 0, \ i = 1, \ldots, p\} \subseteq \mathbf{R}^q,$$

which is called the set of *achievable objective values*. If this set has a minimum element (see §2.4.2), *i.e.*, there is a feasible x such that $f_0(x) \preceq_K f_0(y)$ for all feasible y, then we say x is *optimal* for the problem (4.56), and refer to $f_0(x)$ as the *optimal value* of the problem. (When a vector optimization problem has an optimal value, it is unique.) If $x^\star$ is an optimal point, then $f_0(x^\star)$, the objective at $x^\star$, can be compared to the objective at every other feasible point, and is better than or equal to it. Roughly speaking, $x^\star$ is unambiguously a best choice for x, among feasible points.

A point $x^\star$ is optimal if and only if it is feasible and

$$\mathcal{O} \subseteq f_0(x^\star) + K \qquad (4.57)$$

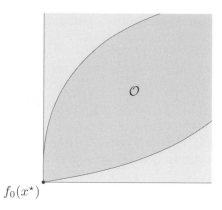

$f_0(x^\star)$

Figure 4.7 The set $\mathcal{O}$ of achievable values for a vector optimization with objective values in $\mathbf{R}^2$, with cone $K = \mathbf{R}_+^2$, is shown shaded. In this case, the point labeled $f_0(x^\star)$ is the optimal value of the problem, and $x^\star$ is an optimal point. The objective value $f_0(x^\star)$ can be compared to every other achievable value $f_0(y)$, and is better than or equal to $f_0(y)$. (Here, 'better than or equal to' means 'is below and to the left of'.) The lightly shaded region is $f_0(x^\star)+K$, which is the set of all $z \in \mathbf{R}^2$ corresponding to objective values worse than (or equal to) $f_0(x^\star)$.

(see §2.4.2). The set $f_0(x^\star) + K$ can be interpreted as the set of values that are worse than, or equal to, $f_0(x^\star)$, so the condition (4.57) states that every achievable value falls in this set. This is illustrated in figure 4.7. Most vector optimization problems do not have an optimal point and an optimal value, but this does occur in some special cases.

Example 4.9 *Best linear unbiased estimator.* Suppose $y = Ax + v$, where $v \in \mathbf{R}^m$ is a measurement noise, $y \in \mathbf{R}^m$ is a vector of measurements, and $x \in \mathbf{R}^n$ is a vector to be estimated, given the measurement y. We assume that A has rank n, and that the measurement noise satisfies $\mathbf{E}\,v = 0$, $\mathbf{E}\,vv^T = I$, *i.e.*, its components are zero mean and uncorrelated.

A *linear estimator* of x has the form $\hat{x} = Fy$. The estimator is called *unbiased* if for all x we have $\mathbf{E}\,\hat{x} = x$, *i.e.*, if $FA = I$. The error covariance of an unbiased estimator is

$$\mathbf{E}(\hat{x} - x)(\hat{x} - x)^T = \mathbf{E}\,Fvv^T F^T = FF^T.$$

Our goal is to find an unbiased estimator that has a 'small' error covariance matrix. We can compare error covariances using matrix inequality, *i.e.*, with respect to $\mathbf{S}_+^n$. This has the following interpretation: Suppose $\hat{x}_1 = F_1 y$, $\hat{x}_2 = F_2 y$ are two unbiased estimators. Then the first estimator is at least as good as the second, *i.e.*, $F_1 F_1^T \preceq F_2 F_2^T$, if and only if for all c,

$$\mathbf{E}(c^T \hat{x}_1 - c^T x)^2 \le \mathbf{E}(c^T \hat{x}_2 - c^T x)^2.$$

In other words, for any linear function of x, the estimator F_1 yields at least as good an estimate as does F_2.

We can express the problem of finding an unbiased estimator for x as the vector optimization problem

$$\begin{array}{ll} \text{minimize (w.r.t. } \mathbf{S}_+^n) & FF^T \\ \text{subject to} & FA = I, \end{array} \qquad (4.58)$$

with variable $F \in \mathbf{R}^{n \times m}$. The objective FF^T is convex with respect to $\mathbf{S}_+^n$, so the problem (4.58) is a convex vector optimization problem. An easy way to see this is to observe that $v^T FF^T v = \|F^T v\|_2^2$ is a convex function of F for any fixed v.

It is a famous result that the problem (4.58) has an optimal solution, the least-squares estimator, or pseudo-inverse,

$$F^\star = A^\dagger = (A^T A)^{-1} A^T.$$

For any F with $FA = I$, we have $FF^T \succeq F^\star F^{\star T}$. The matrix

$$F^\star F^{\star T} = A^\dagger A^{\dagger T} = (A^T A)^{-1}$$

is the optimal value of the problem (4.58).

4.7.3   Pareto optimal points and values

We now consider the case (which occurs in most vector optimization problems of interest) in which the set of achievable objective values does not have a minimum element, so the problem does not have an optimal point or optimal value. In these cases *minimal* elements of the set of achievable values play an important role. We say that a feasible point x is *Pareto optimal* (or *efficient*) if $f_0(x)$ is a minimal element of the set of achievable values $\mathcal{O}$. In this case we say that $f_0(x)$ is a *Pareto optimal value* for the vector optimization problem (4.56). Thus, a point x is Pareto optimal if it is feasible and, for any feasible y, $f_0(y) \preceq_K f_0(x)$ implies $f_0(y) = f_0(x)$. In other words: any feasible point y that is better than or equal to x (*i.e.*, $f_0(y) \preceq_K f_0(x)$) has exactly the same objective value as x.

A point x is Pareto optimal if and only if it is feasible and

$$(f_0(x) - K) \cap \mathcal{O} = \{f_0(x)\} \qquad (4.59)$$

(see §2.4.2). The set $f_0(x) - K$ can be interpreted as the set of values that are better than or equal to $f_0(x)$, so the condition (4.59) states that the only achievable value better than or equal to $f_0(x)$ is $f_0(x)$ itself. This is illustrated in figure 4.8.

A vector optimization problem can have many Pareto optimal values (and points). The set of Pareto optimal values, denoted $\mathcal{P}$, satisfies

$$\mathcal{P} \subseteq \mathcal{O} \cap \mathbf{bd}\,\mathcal{O},$$

i.e., every Pareto optimal value is an achievable objective value that lies in the boundary of the set of achievable objective values (see exercise 4.52).

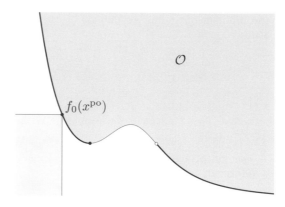

Figure 4.8 The set $\mathcal{O}$ of achievable values for a vector optimization problem with objective values in $\mathbf{R}^2$, with cone $K = \mathbf{R}_+^2$, is shown shaded. This problem does not have an optimal point or value, but it does have a set of Pareto optimal points, whose corresponding values are shown as the darkened curve on the lower left boundary of $\mathcal{O}$. The point labeled $f_0(x^{\mathrm{po}})$ is a Pareto optimal value, and x^{po} is a Pareto optimal point. The lightly shaded region is $f_0(x^{\mathrm{po}}) - K$, which is the set of all $z \in \mathbf{R}^2$ corresponding to objective values better than (or equal to) $f_0(x^{\mathrm{po}})$.

4.7.4   Scalarization

Scalarization is a standard technique for finding Pareto optimal (or optimal) points for a vector optimization problem, based on the characterization of minimum and minimal points via dual generalized inequalities given in §2.6.3. Choose any $\lambda \succ_{K^*} 0$, *i.e.*, any vector that is positive in the dual generalized inequality. Now consider the *scalar* optimization problem

$$
\begin{aligned}
\text{minimize} \quad & \lambda^T f_0(x) \\
\text{subject to} \quad & f_i(x) \leq 0, \quad i = 1, \ldots, m \\
& h_i(x) = 0, \quad i = 1, \ldots, p,
\end{aligned}
\tag{4.60}
$$

and let x be an optimal point. Then x is Pareto optimal for the vector optimization problem (4.56). This follows from the dual inequality characterization of minimal points given in §2.6.3, and is also easily shown directly. If x were not Pareto optimal, then there is a y that is feasible, satisfies $f_0(y) \preceq_K f_0(x)$, and $f_0(x) \neq f_0(y)$. Since $f_0(x) - f_0(y) \succeq_K 0$ and is nonzero, we have $\lambda^T(f_0(x) - f_0(y)) > 0$, *i.e.*, $\lambda^T f_0(x) > \lambda^T f_0(y)$. This contradicts the assumption that x is optimal for the scalar problem (4.60).

Using scalarization, we can find Pareto optimal points for *any* vector optimization problem by solving the ordinary scalar optimization problem (4.60). The vector λ, which is sometimes called the *weight vector*, must satisfy $\lambda \succ_{K^*} 0$. The weight vector is a free parameter; by varying it we obtain (possibly) different Pareto optimal solutions of the vector optimization problem (4.56). This is illustrated in figure 4.9. The figure also shows an example of a Pareto optimal point that cannot

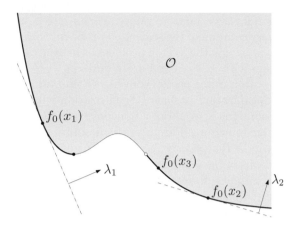

Figure 4.9 *Scalarization.* The set $\mathcal{O}$ of achievable values for a vector optimization problem with cone $K = \mathbf{R}_+^2$. Three Pareto optimal values $f_0(x_1)$, $f_0(x_2)$, $f_0(x_3)$ are shown. The first two values can be obtained by scalarization: $f_0(x_1)$ minimizes $\lambda_1^T u$ over all $u \in \mathcal{O}$ and $f_0(x_2)$ minimizes $\lambda_2^T u$, where $\lambda_1, \lambda_2 \succ 0$. The value $f_0(x_3)$ is Pareto optimal, but cannot be found by scalarization.

be obtained via scalarization, for any value of the weight vector $\lambda \succ_{K^*} 0$.

The method of scalarization can be interpreted geometrically. A point x is optimal for the scalarized problem, *i.e.*, minimizes $\lambda^T f_0$ over the feasible set, if and only if $\lambda^T(f_0(y) - f_0(x)) \geq 0$ for all feasible y. But this is the same as saying that $\{u \mid -\lambda^T(u - f_0(x)) = 0\}$ is a supporting hyperplane to the set of achievable objective values $\mathcal{O}$ at the point $f_0(x)$; in particular

$$\{u \mid \lambda^T(u - f_0(x)) < 0\} \cap \mathcal{O} = \emptyset. \tag{4.61}$$

(See figure 4.9.) Thus, when we find an optimal point for the scalarized problem, we not only find a Pareto optimal point for the original vector optimization problem; we also find an entire halfspace in $\mathbf{R}^q$, given by (4.61), of objective values that cannot be achieved.

Scalarization of convex vector optimization problems

Now suppose the vector optimization problem (4.56) is convex. Then the scalarized problem (4.60) is also convex, since $\lambda^T f_0$ is a (scalar-valued) convex function (by the results in §3.6). This means that we can find Pareto optimal points of a convex vector optimization problem by solving a convex scalar optimization problem. For each choice of the weight vector $\lambda \succ_{K^*} 0$ we get a (usually different) Pareto optimal point.

For convex vector optimization problems we have a partial converse: For every Pareto optimal point x^{po}, there is some nonzero $\lambda \succeq_{K^*} 0$ such that x^{po} is a solution of the scalarized problem (4.60). So, roughly speaking, for convex problems the method of scalarization yields all Pareto optimal points, as the weight vector λ

varies over the K^*-nonnegative, nonzero values. We have to be careful here, because it is *not* true that every solution of the scalarized problem, with $\lambda \succeq_{K^*} 0$ and $\lambda \neq 0$, is a Pareto optimal point for the vector problem. (In contrast, *every* solution of the scalarized problem with $\lambda \succ_{K^*} 0$ is Pareto optimal.)

In some cases we can use this partial converse to find *all* Pareto optimal points of a convex vector optimization problem. Scalarization with $\lambda \succ_{K^*} 0$ gives a set of Pareto optimal points (as it would in a nonconvex vector optimization problem as well). To find the remaining Pareto optimal solutions, we have to consider nonzero weight vectors λ that satisfy $\lambda \succeq_{K^*} 0$. For each such weight vector, we first identify all solutions of the scalarized problem. Then among these solutions we must check which are, in fact, Pareto optimal for the vector optimization problem. These 'extreme' Pareto optimal points can also be found as the limits of the Pareto optimal points obtained from positive weight vectors.

To establish this partial converse, we consider the set

$$\mathcal{A} = \mathcal{O} + K = \{t \in \mathbf{R}^q \mid f_0(x) \preceq_K t \text{ for some feasible } x\}, \qquad (4.62)$$

which consists of all values that are worse than or equal to (with respect to $\preceq_K$) some achievable objective value. While the set $\mathcal{O}$ of achievable objective values need not be convex, the set $\mathcal{A}$ is convex, when the problem is convex. Moreover, the minimal elements of $\mathcal{A}$ are exactly the same as the minimal elements of the set $\mathcal{O}$ of achievable values, *i.e.*, they are the same as the Pareto optimal values. (See exercise 4.53.) Now we use the results of §2.6.3 to conclude that any minimal element of $\mathcal{A}$ minimizes $\lambda^T z$ over $\mathcal{A}$ for some nonzero $\lambda \succeq_{K^*} 0$. This means that every Pareto optimal point for the vector optimization problem is optimal for the scalarized problem, for some nonzero weight $\lambda \succeq_{K^*} 0$.

Example 4.10 *Minimal upper bound on a set of matrices.* We consider the (convex) vector optimization problem, with respect to the positive semidefinite cone,

$$\begin{array}{ll} \text{minimize (w.r.t. } \mathbf{S}^n_+) & X \\ \text{subject to} & X \succeq A_i, \quad i = 1, \ldots, m, \end{array} \qquad (4.63)$$

where $A_i \in \mathbf{S}^n$, $i = 1, \ldots, m$, are given. The constraints mean that X is an upper bound on the given matrices $A_1, \ldots, A_m$; a Pareto optimal solution of (4.63) is a *minimal upper bound* on the matrices.

To find a Pareto optimal point, we apply scalarization: we choose any $W \in \mathbf{S}^n_{++}$ and form the problem

$$\begin{array}{ll} \text{minimize} & \mathbf{tr}(WX) \\ \text{subject to} & X \succeq A_i, \quad i = 1, \ldots, m, \end{array} \qquad (4.64)$$

which is an SDP. Different choices for W will, in general, give different minimal solutions.

The partial converse tells us that if X is Pareto optimal for the vector problem (4.63) then it is optimal for the SDP (4.64), for some nonzero weight matrix $W \succeq 0$. (In this case, however, not every solution of (4.64) is Pareto optimal for the vector optimization problem.)

We can give a simple geometric interpretation for this problem. We associate with each $A \in \mathbf{S}^n_{++}$ an ellipsoid centered at the origin, given by

$$\mathcal{E}_A = \{u \mid u^T A^{-1} u \leq 1\},$$

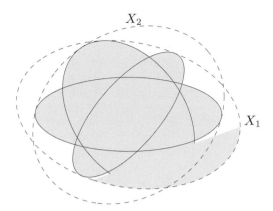

Figure 4.10 Geometric interpretation of the problem (4.63). The three shaded ellipsoids correspond to the data A_1, A_2, $A_3 \in \mathbf{S}^2_{++}$; the Pareto optimal points correspond to minimal ellipsoids that contain them. The two ellipsoids, with boundaries labeled X_1 and X_2, show two minimal ellipsoids obtained by solving the SDP (4.64) for two different weight matrices W_1 and W_2.

so that $A \preceq B$ if and only if $\mathcal{E}_A \subseteq \mathcal{E}_B$. A Pareto optimal point X for the problem (4.63) corresponds to a minimal ellipsoid that contains the ellipsoids associated with $A_1, \ldots, A_m$. An example is shown in figure 4.10.

4.7.5   Multicriterion optimization

When a vector optimization problem involves the cone $K = \mathbf{R}^q_+$, it is called a *multicriterion* or *multi-objective* optimization problem. The components of f_0, say, $F_1, \ldots, F_q$, can be interpreted as q different scalar objectives, each of which we would like to minimize. We refer to F_i as the *ith objective* of the problem. A multicriterion optimization problem is convex if $f_1, \ldots, f_m$ are convex, $h_1, \ldots, h_p$ are affine, and the objectives $F_1, \ldots, F_q$ are convex.

Since multicriterion problems are vector optimization problems, all of the material of §4.7.1–§4.7.4 applies. For multicriterion problems, though, we can be a bit more specific in the interpretations. If x is feasible, we can think of $F_i(x)$ as its score or value, according to the ith objective. If x and y are both feasible, $F_i(x) \leq F_i(y)$ means that x is at least as good as y, according to the ith objective; $F_i(x) < F_i(y)$ means that x is better than y, or x beats y, according to the ith objective. If x and y are both feasible, we say that x is *better than* y, or x *dominates* y, if $F_i(x) \leq F_i(y)$ for $i = 1, \ldots, q$, and for at least one j, $F_j(x) < F_j(y)$. Roughly speaking, x is better than y if x meets or beats y on all objectives, and beats it in at least one objective.

In a multicriterion problem, an optimal point $x^\star$ satisfies

$$F_i(x^\star) \leq F_i(y), \quad i = 1, \ldots, q,$$

for every feasible y. In other words, $x^\star$ is simultaneously optimal for each of the scalar problems

$$
\begin{array}{ll}
\text{minimize} & F_j(x) \\
\text{subject to} & f_i(x) \le 0, \quad i = 1, \dots, m \\
& h_i(x) = 0, \quad i = 1, \dots, p,
\end{array}
$$

for $j = 1, \dots, q$. When there is an optimal point, we say that the objectives are *noncompeting*, since no compromises have to be made among the objectives; each objective is as small as it could be made, even if the others were ignored.

A Pareto optimal point x^{po} satisfies the following: if y is feasible and $F_i(y) \le F_i(x^{\text{po}})$ for $i = 1, \dots, q$, then $F_i(x^{\text{po}}) = F_i(y)$, $i = 1, \dots, q$. This can be restated as: a point is Pareto optimal if and only if it is feasible and there is no better feasible point. In particular, if a feasible point is not Pareto optimal, there is at least one other feasible point that is better. In searching for good points, then, we can clearly limit our search to Pareto optimal points.

Trade-off analysis

Now suppose that x and y are Pareto optimal points with, say,

$$
\begin{array}{ll}
F_i(x) < F_i(y), & i \in A \\
F_i(x) = F_i(y), & i \in B \\
F_i(x) > F_i(y), & i \in C,
\end{array}
$$

where $A \cup B \cup C = \{1, \dots, q\}$. In other words, A is the set of (indices of) objectives for which x beats y, B is the set of objectives for which the points x and y are tied, and C is the set of objectives for which y beats x. If A and C are empty, then the two points x and y have exactly the same objective values. If this is not the case, then both A and C must be nonempty. In other words, when comparing two Pareto optimal points, they either obtain the same performance (*i.e.*, all objectives equal), or, each beats the other in at least one objective.

In comparing the point x to y, we say that we have *traded* or *traded off* better objective values for $i \in A$ for worse objective values for $i \in C$. *Optimal trade-off analysis* (or just trade-off analysis) is the study of how much worse we must do in one or more objectives in order to do better in some other objectives, or more generally, the study of what sets of objective values are achievable.

As an example, consider a bi-criterion (*i.e.*, two criterion) problem. Suppose x is a Pareto optimal point, with objectives $F_1(x)$ and $F_2(x)$. We might ask how much larger $F_2(z)$ would have to be, in order to obtain a feasible point z with $F_1(z) \le F_1(x) - a$, where $a > 0$ is some constant. Roughly speaking, we are asking how much we must pay in the second objective to obtain an improvement of a in the first objective. If a large increase in F_2 must be accepted to realize a small decrease in F_1, we say that there is a *strong trade-off* between the objectives, near the Pareto optimal value $(F_1(x), F_2(x))$. If, on the other hand, a large decrease in F_1 can be obtained with only a small increase in F_2, we say that the trade-off between the objectives is *weak* (near the Pareto optimal value $(F_1(x), F_2(x))$).

We can also consider the case in which we trade worse performance in the first objective for an improvement in the second. Here we find how much smaller $F_2(z)$

can be made, to obtain a feasible point z with $F_1(z) \leq F_1(x) + a$, where $a > 0$ is some constant. In this case we receive a benefit in the second objective, *i.e.*, a reduction in F_2 compared to $F_2(x)$. If this benefit is large (*i.e.*, by increasing F_1 a small amount we obtain a large reduction in F_2), we say the objectives exhibit a strong trade-off. If it is small, we say the objectives trade off weakly (near the Pareto optimal value $(F_1(x), F_2(x))$).

Optimal trade-off surface

The set of Pareto optimal values for a multicriterion problem is called the *optimal trade-off surface* (in general, when $q > 2$) or the *optimal trade-off curve* (when $q = 2$). (Since it would be foolish to accept any point that is not Pareto optimal, we can restrict our trade-off analysis to Pareto optimal points.) Trade-off analysis is also sometimes called *exploring the optimal trade-off surface*. (The optimal trade-off surface is usually, but not always, a surface in the usual sense. If the problem has an optimal point, for example, the optimal trade-off surface consists of a single point, the optimal value.)

An optimal trade-off curve is readily interpreted. An example is shown in figure 4.11, on page 185, for a (convex) bi-criterion problem. From this curve we can easily visualize and understand the trade-offs between the two objectives.

- The endpoint at the right shows the smallest possible value of F_2, without any consideration of F_1.

- The endpoint at the left shows the smallest possible value of F_1, without any consideration of F_2.

- By finding the intersection of the curve with a vertical line at $F_1 = \alpha$, we can see how large F_2 must be to achieve $F_1 \leq \alpha$.

- By finding the intersection of the curve with a horizontal line at $F_2 = \beta$, we can see how large F_1 must be to achieve $F_2 \leq \beta$.

- The slope of the optimal trade-off curve at a point on the curve (*i.e.*, a Pareto optimal value) shows the *local* optimal trade-off between the two objectives. Where the slope is steep, small changes in F_1 are accompanied by large changes in F_2.

- A point of large curvature is one where small decreases in one objective can only be accomplished by a large increase in the other. This is the prover- bial *knee of the trade-off curve*, and in many applications represents a good compromise solution.

All of these have simple extensions to a trade-off surface, although visualizing a surface with more than three objectives is difficult.

Scalarizing multicriterion problems

When we scalarize a multicriterion problem by forming the weighted sum objective

$$\lambda^T f_0(x) = \sum_{i=1}^{q} \lambda_i F_i(x),$$

where $\lambda \succ 0$, we can interpret λ_i as the *weight* we attach to the ith objective. The weight λ_i can be thought of as quantifying our desire to make F_i small (or our objection to having F_i large). In particular, we should take λ_i large if we want F_i to be small; if we care much less about F_i, we can take λ_i small. We can interpret the ratio λ_i/λ_j as the *relative weight* or relative importance of the ith objective compared to the jth objective. Alternatively, we can think of λ_i/λ_j as *exchange rate* between the two objectives, since in the weighted sum objective a decrease (say) in F_i by α is considered the same as an increase in F_j in the amount $(\lambda_i/\lambda_j)\alpha$.

These interpretations give us some intuition about how to set or change the weights while exploring the optimal trade-off surface. Suppose, for example, that the weight vector $\lambda \succ 0$ yields the Pareto optimal point x^{po}, with objective values $F_1(x^{\mathrm{po}}), \ldots, F_q(x^{\mathrm{po}})$. To find a (possibly) new Pareto optimal point which trades off a better kth objective value (say), for (possibly) worse objective values for the other objectives, we form a new weight vector $\tilde{\lambda}$ with

$$\tilde{\lambda}_k > \lambda_k, \qquad \tilde{\lambda}_j = \lambda_j, \quad j \neq k, \quad j = 1, \ldots, q,$$

i.e., we increase the weight on the kth objective. This yields a new Pareto optimal point $\tilde{x}^{\mathrm{po}}$ with $F_k(\tilde{x}^{\mathrm{po}}) \leq F_k(x^{\mathrm{po}})$ (and usually, $F_k(\tilde{x}^{\mathrm{po}}) < F_k(x^{\mathrm{po}})$), *i.e.*, a new Pareto optimal point with an improved kth objective.

We can also see that at any point where the optimal trade-off surface is smooth, λ gives the inward normal to the surface at the associated Pareto optimal point. In particular, when we choose a weight vector λ and apply scalarization, we obtain a Pareto optimal point where λ gives the local trade-offs among objectives.

In practice, optimal trade-off surfaces are explored by ad hoc adjustment of the weights, based on the intuitive ideas above. We will see later (in chapter 5) that the basic idea of scalarization, *i.e.*, minimizing a weighted sum of objectives, and then adjusting the weights to obtain a suitable solution, is the essence of duality.

4.7.6   Examples

Regularized least-squares

We are given $A \in \mathbf{R}^{m \times n}$ and $b \in \mathbf{R}^m$, and want to choose $x \in \mathbf{R}^n$ taking into account two quadratic objectives:

- $F_1(x) = \|Ax - b\|_2^2 = x^T A^T A x - 2b^T A x + b^T b$ is a measure of the misfit between Ax and b,

- $F_2(x) = \|x\|_2^2 = x^T x$ is a measure of the size of x.

Our goal is to find x that gives a good fit (*i.e.*, small F_1) and that is not large (*i.e.*, small F_2). We can formulate this problem as a vector optimization problem with respect to the cone $\mathbf{R}_+^2$, *i.e.*, a bi-criterion problem (with no constraints):

$$\text{minimize (w.r.t. } \mathbf{R}_+^2) \quad f_0(x) = (F_1(x), F_2(x)).$$

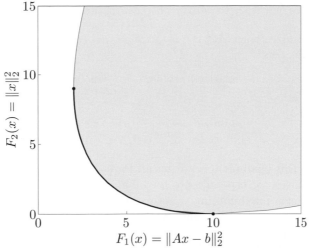

Figure 4.11 Optimal trade-off curve for a regularized least-squares problem. The shaded set is the set of achievable values $(\|Ax-b\|_2^2, \|x\|_2^2)$. The optimal trade-off curve, shown darker, is the lower left part of the boundary.

We can scalarize this problem by taking $\lambda_1 > 0$ and $\lambda_2 > 0$ and minimizing the scalar weighted sum objective

$$
\begin{aligned}
\lambda^T f_0(x) &= \lambda_1 F_1(x) + \lambda_2 F_2(x) \\
&= x^T(\lambda_1 A^T A + \lambda_2 I)x - 2\lambda_1 b^T A x + \lambda_1 b^T b,
\end{aligned}
$$

which yields

$$
x(\mu) = (\lambda_1 A^T A + \lambda_2 I)^{-1}\lambda_1 A^T b = (A^T A + \mu I)^{-1}A^T b,
$$

where $\mu = \lambda_2/\lambda_1$. For any $\mu > 0$, this point is Pareto optimal for the bi-criterion problem. We can interpret $\mu = \lambda_2/\lambda_1$ as the relative weight we assign F_2 compared to F_1.

This method produces all Pareto optimal points, except two, associated with the extremes $\mu \to \infty$ and $\mu \to 0$. In the first case we have the Pareto optimal solution $x = 0$, which would be obtained by scalarization with $\lambda = (0, 1)$. At the other extreme we have the Pareto optimal solution $A^\dagger b$, where $A^\dagger$ is the pseudo-inverse of A. This Pareto optimal solution is obtained as the limit of the optimal solution of the scalarized problem as $\mu \to 0$, *i.e.*, as $\lambda \to (1, 0)$. (We will encounter the regularized least-squares problem again in §6.3.2.)

Figure 4.11 shows the optimal trade-off curve and the set of achievable values for a regularized least-squares problem with problem data $A \in \mathbf{R}^{100\times 10}$, $b \in \mathbf{R}^{100}$. (See exercise 4.50 for more discussion.)

Risk-return trade-off in portfolio optimization

The classical Markowitz portfolio optimization problem described on page 155 is naturally expressed as a bi-criterion problem, where the objectives are the negative mean return (since we wish to *maximize* mean return) and the variance of the return:

$$\begin{array}{ll} \text{minimize (w.r.t. } \mathbf{R}_+^2) & (F_1(x), F_2(x)) = (-\bar{p}^T x, x^T \Sigma x) \\ \text{subject to} & \mathbf{1}^T x = 1, \quad x \succeq 0. \end{array}$$

In forming the associated scalarized problem, we can (without loss of generality) take $\lambda_1 = 1$ and $\lambda_2 = \mu > 0$:

$$\begin{array}{ll} \text{minimize} & -\bar{p}^T x + \mu x^T \Sigma x \\ \text{subject to} & \mathbf{1}^T x = 1, \quad x \succeq 0, \end{array}$$

which is a QP. In this example too, we get all Pareto optimal portfolios except for the two limiting cases corresponding to $\mu \to 0$ and $\mu \to \infty$. Roughly speaking, in the first case we get a maximum mean return, without regard for return variance; in the second case we form a minimum variance return, without regard for mean return. Assuming that $\bar{p}_k > \bar{p}_i$ for $i \neq k$, *i.e.*, that asset k is the unique asset with maximum mean return, the portfolio allocation $x = e_k$ is the only one corresponding to $\mu \to 0$. (In other words, we concentrate the portfolio entirely in the asset that has maximum mean return.) In many portfolio problems asset n corresponds to a *risk-free* investment, with (deterministic) return r_{rf}. Assuming that Σ, with its last row and column (which are zero) removed, is full rank, then the other extreme Pareto optimal portfolio is $x = e_n$, *i.e.*, the portfolio is concentrated entirely in the risk-free asset.

As a specific example, we consider a simple portfolio optimization problem with 4 assets, with price change mean and standard deviations given in the following table.

Asset	$\bar{p}_i$	$\Sigma_{ii}^{1/2}$
1	12%	20%
2	10%	10%
3	7%	5%
4	3%	0%

Asset 4 is a risk-free asset, with a (certain) 3% return. Assets 3, 2, and 1 have increasing mean returns, ranging from 7% to 12%, as well as increasing standard deviations, which range from 5% to 20%. The correlation coefficients between the assets are $\rho_{12} = 30\%$, $\rho_{13} = -40\%$, and $\rho_{23} = 0\%$.

Figure 4.12 shows the optimal trade-off curve for this portfolio optimization problem. The plot is given in the conventional way, with the horizontal axis showing standard deviation (*i.e.*, squareroot of variance) and the vertical axis showing expected return. The lower plot shows the optimal asset allocation vector x for each Pareto optimal point.

The results in this simple example agree with our intuition. For small risk, the optimal allocation consists mostly of the risk-free asset, with a mixture of the other assets in smaller quantities. Note that a mixture of asset 3 and asset 1, which are negatively correlated, gives some hedging, *i.e.*, lowers variance for a given level of mean return. At the other end of the trade-off curve, we see that aggressive growth portfolios (*i.e.*, those with large mean returns) concentrate the allocation in assets 1 and 2, the ones with the largest mean returns (and variances).

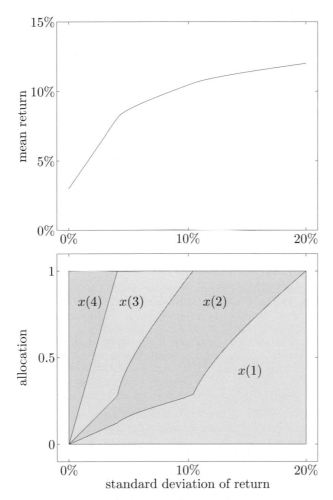

Figure 4.12 *Top.* Optimal risk-return trade-off curve for a simple portfolio optimization problem. The lefthand endpoint corresponds to putting all resources in the risk-free asset, and so has zero standard deviation. The righthand endpoint corresponds to putting all resources in asset 1, which has highest mean return. *Bottom.* Corresponding optimal allocations.

Bibliography

Linear programming has been studied extensively since the 1940s, and is the subject of many excellent books, including Dantzig [Dan63], Luenberger [Lue84], Schrijver [Sch86], Papadimitriou and Steiglitz [PS98], Bertsimas and Tsitsiklis [BT97], Vanderbei [Van96], and Roos, Terlaky, and Vial [RTV97]. Dantzig and Schrijver also provide detailed accounts of the history of linear programming. For a recent survey, see Todd [Tod02].

Schaible [Sch82, Sch83] gives an overview of fractional programming, which includes linear-fractional problems and extensions such as convex-concave fractional problems (see exercise 4.7). The model of a growing economy in example 4.7 appears in von Neumann [vN46].

Research on quadratic programming began in the 1950s (see, *e.g.*, Frank and Wolfe [FW56], Markowitz [Mar56], Hildreth [Hil57]), and was in part motivated by the portfolio optimization problem discussed on page 155 (Markowitz [Mar52]), and the LP with random cost discussed on page 154 (see Freund [Fre56]).

Interest in second-order cone programming is more recent, and started with Nesterov and Nemirovski [NN94, §6.2.3]. The theory and applications of SOCPs are surveyed by Alizadeh and Goldfarb [AG03], Ben-Tal and Nemirovski [BTN01, lecture 3] (where the problem is referred to as *conic quadratic programming*), and Lobo, Vandenberghe, Boyd, and Lebret [LVBL98].

Robust linear programming, and robust convex optimization in general, originated with Ben-Tal and Nemirovski [BTN98, BTN99] and El Ghaoui and Lebret [EL97]. Goldfarb and Iyengar [GI03a, GI03b] discuss robust QCQPs and applications in portfolio optimization. El Ghaoui, Oustry, and Lebret [EOL98] focus on robust semidefinite programming.

Geometric programming has been known since the 1960s. Its use in engineering design was first advocated by Duffin, Peterson, and Zener [DPZ67] and Zener [Zen71]. Peterson [Pet76] and Ecker [Eck80] describe the progress made during the 1970s. These articles and books also include examples of engineering applications, in particular in chemical and civil engineering. Fishburn and Dunlop [FD85], Sapatnekar, Rao, Vaidya, and Kang [SRVK93], and Hershenson, Boyd, and Lee [HBL01]) apply geometric programming to problems in integrated circuit design. The cantilever beam design example (page 163) is from Vanderplaats [Van84, page 147]. The variational characterization of the Perron-Frobenius eigenvalue (page 165) is proved in Berman and Plemmons [BP94, page 31].

Nesterov and Nemirovski [NN94, chapter 4] introduced the conic form problem (4.49) as a standard problem format in nonlinear convex optimization. The cone programming approach is further developed in Ben-Tal and Nemirovski [BTN01], who also describe numerous applications.

Alizadeh [Ali91] and Nesterov and Nemirovski [NN94, §6.4] were the first to make a systematic study of semidefinite programming, and to point out the wide variety of applications in convex optimization. Subsequent research in semidefinite programming during the 1990s was driven by applications in combinatorial optimization (Goemans and Williamson [GW95]), control (Boyd, El Ghaoui, Feron, and Balakrishnan [BEFB94], Scherer, Gahinet, and Chilali [SGC97], Dullerud and Paganini [DP00]), communications and signal processing (Luo [Luo03], Davidson, Luo, Wong, and Ma [DLW00, MDW+02]), and other areas of engineering. The book edited by Wolkowicz, Saigal, and Vandenberghe [WSV00] and the articles by Todd [Tod01], Lewis and Overton [LO96], and Vandenberghe and Boyd [VB95] provide overviews and extensive bibliographies. Connections between SDP and moment problems, of which we give a simple example on page 170, are explored in detail by Bertsimas and Sethuraman [BS00], Nesterov [Nes00], and Lasserre [Las02]. The fastest mixing Markov chain problem is from Boyd, Diaconis, and Xiao [BDX04].

Multicriterion optimization and Pareto optimality are fundamental tools in economics; see Pareto [Par71], Debreu [Deb59] and Luenberger [Lue95]. The result in example 4.9 is known as the Gauss-Markov theorem (Kailath, Sayed, and Hassibi [KSH00, page 97]).

Exercises

Basic terminology and optimality conditions

4.1 Consider the optimization problem

$$\begin{array}{ll} \text{minimize} & f_0(x_1, x_2) \\ \text{subject to} & 2x_1 + x_2 \geq 1 \\ & x_1 + 3x_2 \geq 1 \\ & x_1 \geq 0, \quad x_2 \geq 0. \end{array}$$

Make a sketch of the feasible set. For each of the following objective functions, give the optimal set and the optimal value.

(a) $f_0(x_1, x_2) = x_1 + x_2$.

(b) $f_0(x_1, x_2) = -x_1 - x_2$.

(c) $f_0(x_1, x_2) = x_1$.

(d) $f_0(x_1, x_2) = \max\{x_1, x_2\}$.

(e) $f_0(x_1, x_2) = x_1^2 + 9x_2^2$.

4.2 Consider the optimization problem

$$\text{minimize} \quad f_0(x) = -\sum_{i=1}^{m} \log(b_i - a_i^T x)$$

with domain $\mathbf{dom}\, f_0 = \{x \mid Ax \prec b\}$, where $A \in \mathbf{R}^{m \times n}$ (with rows a_i^T). We assume that $\mathbf{dom}\, f_0$ is nonempty.

Prove the following facts (which include the results quoted without proof on page 141).

(a) $\mathbf{dom}\, f_0$ is unbounded if and only if there exists a $v \neq 0$ with $Av \preceq 0$.

(b) f_0 is unbounded below if and only if there exists a v with $Av \preceq 0$, $Av \neq 0$. *Hint.* There exists a v such that $Av \preceq 0$, $Av \neq 0$ if and only if there exists no $z \succ 0$ such that $A^T z = 0$. This follows from the theorem of alternatives in example 2.21, page 50.

(c) If f_0 is bounded below then its minimum is attained, *i.e.*, there exists an x that satisfies the optimality condition (4.23).

(d) The optimal set is affine: $X_{\text{opt}} = \{x^\star + v \mid Av = 0\}$, where $x^\star$ is any optimal point.

4.3 Prove that $x^\star = (1, 1/2, -1)$ is optimal for the optimization problem

$$\begin{array}{ll} \text{minimize} & (1/2)x^T P x + q^T x + r \\ \text{subject to} & -1 \leq x_i \leq 1, \quad i = 1, 2, 3, \end{array}$$

where

$$P = \begin{bmatrix} 13 & 12 & -2 \\ 12 & 17 & 6 \\ -2 & 6 & 12 \end{bmatrix}, \qquad q = \begin{bmatrix} -22.0 \\ -14.5 \\ 13.0 \end{bmatrix}, \qquad r = 1.$$

4.4 [P. Parrilo] *Symmetries and convex optimization.* Suppose $\mathcal{G} = \{Q_1, \ldots, Q_k\} \subseteq \mathbf{R}^{n \times n}$ is a group, *i.e.*, closed under products and inverse. We say that the function $f : \mathbf{R}^n \to \mathbf{R}$ is $\mathcal{G}$-*invariant*, or *symmetric with respect to* $\mathcal{G}$, if $f(Q_i x) = f(x)$ holds for all x and $i = 1, \ldots, k$. We define $\bar{x} = (1/k) \sum_{i=1}^{k} Q_i x$, which is the average of x over its $\mathcal{G}$-orbit. We define the *fixed subspace* of $\mathcal{G}$ as

$$\mathcal{F} = \{x \mid Q_i x = x, \ i = 1, \ldots, k\}.$$

(a) Show that for any $x \in \mathbf{R}^n$, we have $\bar{x} \in \mathcal{F}$.

(b) Show that if $f : \mathbf{R}^n \to \mathbf{R}$ is convex and $\mathcal{G}$-invariant, then $f(\overline{x}) \leq f(x)$.

(c) We say the optimization problem

$$
\begin{array}{ll}
\text{minimize} & f_0(x) \\
\text{subject to} & f_i(x) \leq 0, \quad i = 1, \dots, m
\end{array}
$$

is $\mathcal{G}$-invariant if the objective f_0 is $\mathcal{G}$-invariant, and the feasible set is $\mathcal{G}$-invariant, which means

$$
f_1(x) \leq 0, \dots, f_m(x) \leq 0 \implies f_1(Q_i x) \leq 0, \dots, f_m(Q_i x) \leq 0,
$$

for $i = 1, \dots, k$. Show that if the problem is convex and $\mathcal{G}$-invariant, and there exists an optimal point, then there exists an optimal point in $\mathcal{F}$. In other words, we can adjoin the equality constraints $x \in \mathcal{F}$ to the problem, without loss of generality.

(d) As an example, suppose f is convex and symmetric, *i.e.*, $f(Px) = f(x)$ for every permutation P. Show that if f has a minimizer, then it has a minimizer of the form $\alpha\mathbf{1}$. (This means to minimize f over $x \in \mathbf{R}^n$, we can just as well minimize $f(t\mathbf{1})$ over $t \in \mathbf{R}$.)

4.5 *Equivalent convex problems.* Show that the following three convex problems are equivalent. Carefully explain how the solution of each problem is obtained from the solution of the other problems. The problem data are the matrix $A \in \mathbf{R}^{m \times n}$ (with rows a_i^T), the vector $b \in \mathbf{R}^m$, and the constant $M > 0$.

(a) The *robust least-squares problem*

$$
\text{minimize} \quad \sum_{i=1}^m \phi(a_i^T x - b_i),
$$

with variable $x \in \mathbf{R}^n$, where $\phi : \mathbf{R} \to \mathbf{R}$ is defined as

$$
\phi(u) = \begin{cases} u^2 & |u| \leq M \\ M(2|u| - M) & |u| > M. \end{cases}
$$

(This function is known as the *Huber penalty function*; see §6.1.2.)

(b) The *least-squares problem with variable weights*

$$
\begin{array}{ll}
\text{minimize} & \sum_{i=1}^m (a_i^T x - b_i)^2 / (w_i + 1) + M^2 \mathbf{1}^T w \\
\text{subject to} & w \succeq 0,
\end{array}
$$

with variables $x \in \mathbf{R}^n$ and $w \in \mathbf{R}^m$, and domain $\mathcal{D} = \{(x, w) \in \mathbf{R}^n \times \mathbf{R}^m \mid w \succ -1\}$.

Hint. Optimize over w assuming x is fixed, to establish a relation with the problem in part (a).

(This problem can be interpreted as a weighted least-squares problem in which we are allowed to adjust the weight of the ith residual. The weight is one if $w_i = 0$, and decreases if we increase w_i. The second term in the objective penalizes large values of w, *i.e.*, large adjustments of the weights.)

(c) The *quadratic program*

$$
\begin{array}{ll}
\text{minimize} & \sum_{i=1}^m (u_i^2 + 2M v_i) \\
\text{subject to} & -u - v \preceq Ax - b \preceq u + v \\
& 0 \preceq u \preceq M\mathbf{1} \\
& v \succeq 0.
\end{array}
$$

4.6 *Handling convex equality constraints.* A convex optimization problem can have only *linear* equality constraint functions. In some special cases, however, it is possible to handle convex equality constraint functions, *i.e.*, constraints of the form $h(x) = 0$, where h is convex. We explore this idea in this problem.

Consider the optimization problem

$$
\begin{array}{ll}
\text{minimize} & f_0(x) \\
\text{subject to} & f_i(x) \le 0, \quad i = 1, \ldots, m \\
& h(x) = 0,
\end{array}
\tag{4.65}
$$

where f_i and h are convex functions with domain $\mathbf{R}^n$. Unless h is affine, this is *not* a convex optimization problem. Consider the related problem

$$
\begin{array}{ll}
\text{minimize} & f_0(x) \\
\text{subject to} & f_i(x) \le 0, \quad i = 1, \ldots, m, \\
& h(x) \le 0,
\end{array}
\tag{4.66}
$$

where the convex equality constraint has been relaxed to a convex inequality. This problem is, of course, convex.

Now suppose we can guarantee that at any optimal solution $x^\star$ of the convex problem (4.66), we have $h(x^\star) = 0$, *i.e.*, the inequality $h(x) \le 0$ is always active at the solution. Then we can solve the (nonconvex) problem (4.65) by solving the convex problem (4.66). Show that this is the case if there is an index r such that

- f_0 is monotonically increasing in x_r
- $f_1, \ldots, f_m$ arc nondecreasing in x_r
- h is monotonically decreasing in x_r.

We will see specific examples in exercises 4.31 and 4.58.

4.7 *Convex-concave fractional problems.* Consider a problem of the form

$$
\begin{array}{ll}
\text{minimize} & f_0(x)/(c^T x + d) \\
\text{subject to} & f_i(x) \le 0, \quad i = 1, \ldots, m \\
& Ax = b
\end{array}
$$

where $f_0, f_1, \ldots, f_m$ are convex, and the domain of the objective function is defined as $\{x \in \mathbf{dom}\, f_0 \mid c^T x + d > 0\}$.

(a) Show that this is a quasiconvex optimization problem.

(b) Show that the problem is equivalent to

$$
\begin{array}{ll}
\text{minimize} & g_0(y, t) \\
\text{subject to} & g_i(y, t) \le 0, \quad i = 1, \ldots, m \\
& Ay = bt \\
& c^T y + dt = 1,
\end{array}
$$

where g_i is the perspective of f_i (see §3.2.6). The variables are $y \in \mathbf{R}^n$ and $t \in \mathbf{R}$. Show that this problem is convex.

(c) Following a similar argument, derive a convex formulation for the *convex-concave fractional problem*

$$
\begin{array}{ll}
\text{minimize} & f_0(x)/h(x) \\
\text{subject to} & f_i(x) \le 0, \quad i = 1, \ldots, m \\
& Ax = b
\end{array}
$$

where $f_0, f_1, \ldots, f_m$ are convex, h is concave, the domain of the objective function is defined as $\{x \in \mathbf{dom}\, f_0 \cap \mathbf{dom}\, h \mid h(x) > 0\}$ and $f_0(x) \geq 0$ everywhere.

As an example, apply your technique to the (unconstrained) problem with

$$f_0(x) = (\mathbf{tr}\, F(x))/m, \qquad h(x) = (\det(F(x))^{1/m},$$

with $\mathbf{dom}(f_0/h) = \{x \mid F(x) \succ 0\}$, where $F(x) = F_0 + x_1 F_1 + \cdots + x_n F_n$ for given $F_i \in \mathbf{S}^m$. In this problem, we minimize the ratio of the arithmetic mean over the geometric mean of the eigenvalues of an affine matrix function $F(x)$.

Linear optimization problems

4.8 *Some simple LPs.* Give an explicit solution of each of the following LPs.

(a) *Minimizing a linear function over an affine set.*

$$\begin{array}{ll} \text{minimize} & c^T x \\ \text{subject to} & Ax = b. \end{array}$$

(b) *Minimizing a linear function over a halfspace.*

$$\begin{array}{ll} \text{minimize} & c^T x \\ \text{subject to} & a^T x \leq b, \end{array}$$

where $a \neq 0$.

(c) *Minimizing a linear function over a rectangle.*

$$\begin{array}{ll} \text{minimize} & c^T x \\ \text{subject to} & l \preceq x \preceq u, \end{array}$$

where l and u satisfy $l \preceq u$.

(d) *Minimizing a linear function over the probability simplex.*

$$\begin{array}{ll} \text{minimize} & c^T x \\ \text{subject to} & \mathbf{1}^T x = 1, \quad x \succeq 0. \end{array}$$

What happens if the equality constraint is replaced by an inequality $\mathbf{1}^T x \leq 1$?

We can interpret this LP as a simple portfolio optimization problem. The vector x represents the allocation of our total budget over different assets, with x_i the fraction invested in asset i. The return of each investment is fixed and given by $-c_i$, so our total return (which we want to maximize) is $-c^T x$. If we replace the budget constraint $\mathbf{1}^T x = 1$ with an inequality $\mathbf{1}^T x \leq 1$, we have the option of not investing a portion of the total budget.

(e) *Minimizing a linear function over a unit box with a total budget constraint.*

$$\begin{array}{ll} \text{minimize} & c^T x \\ \text{subject to} & \mathbf{1}^T x = \alpha, \quad 0 \preceq x \preceq \mathbf{1}, \end{array}$$

where α is an integer between 0 and n. What happens if α is not an integer (but satisfies $0 \leq \alpha \leq n$)? What if we change the equality to an inequality $\mathbf{1}^T x \leq \alpha$?

(f) *Minimizing a linear function over a unit box with a weighted budget constraint.*

$$\begin{array}{ll} \text{minimize} & c^T x \\ \text{subject to} & d^T x = \alpha, \quad 0 \preceq x \preceq \mathbf{1}, \end{array}$$

with $d \succ 0$, and $0 \leq \alpha \leq \mathbf{1}^T d$.

4.9 *Square LP.* Consider the LP

$$\begin{array}{ll} \text{minimize} & c^T x \\ \text{subject to} & Ax \preceq b \end{array}$$

with A square and nonsingular. Show that the optimal value is given by

$$p^\star = \left\{ \begin{array}{ll} c^T A^{-1} b & A^{-T} c \preceq 0 \\ -\infty & \text{otherwise.} \end{array} \right.$$

4.10 *Converting general LP to standard form.* Work out the details on page 147 of §4.3. Explain in detail the relation between the feasible sets, the optimal solutions, and the optimal values of the standard form LP and the original LP.

4.11 *Problems involving ℓ_1- and ℓ_∞-norms.* Formulate the following problems as LPs. Explain in detail the relation between the optimal solution of each problem and the solution of its equivalent LP.

    (a) Minimize $\|Ax - b\|_\infty$ (ℓ_∞-norm approximation).

    (b) Minimize $\|Ax - b\|_1$ (ℓ_1-norm approximation).

    (c) Minimize $\|Ax - b\|_1$ subject to $\|x\|_\infty \leq 1$.

    (d) Minimize $\|x\|_1$ subject to $\|Ax - b\|_\infty \leq 1$.

    (e) Minimize $\|Ax - b\|_1 + \|x\|_\infty$.

In each problem, $A \in \mathbf{R}^{m \times n}$ and $b \in \mathbf{R}^m$ are given. (See §6.1 for more problems involving approximation and constrained approximation.)

4.12 *Network flow problem.* Consider a network of n nodes, with directed links connecting each pair of nodes. The variables in the problem are the flows on each link: x_{ij} will denote the flow from node i to node j. The cost of the flow along the link from node i to node j is given by $c_{ij} x_{ij}$, where c_{ij} are given constants. The total cost across the network is

$$C = \sum_{i,j=1}^{n} c_{ij} x_{ij}.$$

Each link flow x_{ij} is also subject to a given lower bound l_{ij} (usually assumed to be nonnegative) and an upper bound u_{ij}.

The external supply at node i is given by b_i, where $b_i > 0$ means an external flow enters the network at node i, and $b_i < 0$ means that at node i, an amount $|b_i|$ flows out of the network. We assume that $\mathbf{1}^T b = 0$, *i.e.*, the total external supply equals total external demand. At each node we have conservation of flow: the total flow into node i along links and the external supply, minus the total flow out along the links, equals zero.

The problem is to minimize the total cost of flow through the network, subject to the constraints described above. Formulate this problem as an LP.

4.13 *Robust LP with interval coefficients.* Consider the problem, with variable $x \in \mathbf{R}^n$,

$$\begin{array}{ll} \text{minimize} & c^T x \\ \text{subject to} & Ax \preceq b \text{ for all } A \in \mathcal{A}, \end{array}$$

where $\mathcal{A} \subseteq \mathbf{R}^{m \times n}$ is the set

$$\mathcal{A} = \{A \in \mathbf{R}^{m \times n} \mid \bar{A}_{ij} - V_{ij} \leq A_{ij} \leq \bar{A}_{ij} + V_{ij}, \ i = 1, \ldots, m, \ j = 1, \ldots, n\}.$$

(The matrices $\bar{A}$ and V are given.) This problem can be interpreted as an LP where each coefficient of A is only known to lie in an interval, and we require that x must satisfy the constraints for all possible values of the coefficients.

Express this problem as an LP. The LP you construct should be efficient, *i.e.*, it should not have dimensions that grow exponentially with n or m.

4.14 *Approximating a matrix in infinity norm.* The ℓ_∞-norm induced norm of a matrix $A \in \mathbf{R}^{m \times n}$, denoted $\|A\|_\infty$, is given by

$$\|A\|_\infty = \sup_{x \neq 0} \frac{\|Ax\|_\infty}{\|x\|_\infty} = \max_{i=1,\ldots,m} \sum_{j=1}^{n} |a_{ij}|.$$

This norm is sometimes called the max-row-sum norm, for obvious reasons (see §A.1.5).

Consider the problem of approximating a matrix, in the max-row-sum norm, by a linear combination of other matrices. That is, we are given $k+1$ matrices $A_0, \ldots, A_k \in \mathbf{R}^{m \times n}$, and need to find $x \in \mathbf{R}^k$ that minimizes

$$\|A_0 + x_1 A_1 + \cdots + x_k A_k\|_\infty.$$

Express this problem as a linear program. Explain the significance of any extra variables in your LP. Carefully explain how your LP formulation solves this problem, *e.g.*, what is the relation between the feasible set for your LP and this problem?

4.15 *Relaxation of Boolean LP.* In a *Boolean linear program*, the variable x is constrained to have components equal to zero or one:

$$\begin{array}{ll} \text{minimize} & c^T x \\ \text{subject to} & Ax \preceq b \\ & x_i \in \{0,1\}, \quad i = 1,\ldots,n. \end{array} \qquad (4.67)$$

In general, such problems are very difficult to solve, even though the feasible set is finite (containing at most 2^n points).

In a general method called *relaxation*, the constraint that x_i be zero or one is replaced with the linear inequalities $0 \le x_i \le 1$:

$$\begin{array}{ll} \text{minimize} & c^T x \\ \text{subject to} & Ax \preceq b \\ & 0 \le x_i \le 1, \quad i = 1,\ldots,n. \end{array} \qquad (4.68)$$

We refer to this problem as the *LP relaxation* of the Boolean LP (4.67). The LP relaxation is far easier to solve than the original Boolean LP.

(a) Show that the optimal value of the LP relaxation (4.68) is a lower bound on the optimal value of the Boolean LP (4.67). What can you say about the Boolean LP if the LP relaxation is infeasible?

(b) It sometimes happens that the LP relaxation has a solution with $x_i \in \{0,1\}$. What can you say in this case?

4.16 *Minimum fuel optimal control.* We consider a linear dynamical system with state $x(t) \in \mathbf{R}^n$, $t = 0, \ldots, N$, and actuator or input signal $u(t) \in \mathbf{R}$, for $t = 0, \ldots, N-1$. The dynamics of the system is given by the linear recurrence

$$x(t+1) = Ax(t) + bu(t), \quad t = 0, \ldots, N-1,$$

where $A \in \mathbf{R}^{n \times n}$ and $b \in \mathbf{R}^n$ are given. We assume that the initial state is zero, *i.e.*, $x(0) = 0$.

The *minimum fuel optimal control problem* is to choose the inputs $u(0), \ldots, u(N-1)$ so as to minimize the total fuel consumed, which is given by

$$F = \sum_{t=0}^{N-1} f(u(t)),$$

subject to the constraint that $x(N) = x_{\mathrm{des}}$, where N is the (given) time horizon, and $x_{\mathrm{des}} \in \mathbf{R}^n$ is the (given) desired final or target state. The function $f : \mathbf{R} \to \mathbf{R}$ is the *fuel use map* for the actuator, and gives the amount of fuel used as a function of the actuator signal amplitude. In this problem we use

$$f(a) = \begin{cases} |a| & |a| \leq 1 \\ 2|a| - 1 & |a| > 1. \end{cases}$$

This means that fuel use is proportional to the absolute value of the actuator signal, for actuator signals between -1 and 1; for larger actuator signals the marginal fuel efficiency is half.

Formulate the minimum fuel optimal control problem as an LP.

4.17 *Optimal activity levels.* We consider the selection of n nonnegative activity levels, denoted $x_1, \ldots, x_n$. These activities consume m resources, which are limited. Activity j consumes $A_{ij} x_j$ of resource i, where A_{ij} are given. The total resource consumption is additive, so the total of resource i consumed is $c_i = \sum_{j=1}^n A_{ij} x_j$. (Ordinarily we have $A_{ij} \geq 0$, i.e., activity j consumes resource i. But we allow the possibility that $A_{ij} < 0$, which means that activity j actually *generates* resource i as a by-product.) Each resource consumption is limited: we must have $c_i \leq c_i^{\max}$, where $c_i^{\max}$ are given. Each activity generates revenue, which is a piecewise-linear concave function of the activity level:

$$r_j(x_j) = \begin{cases} p_j x_j & 0 \leq x_j \leq q_j \\ p_j q_j + p_j^{\mathrm{disc}} (x_j - q_j) & x_j \geq q_j. \end{cases}$$

Here $p_j > 0$ is the basic price, $q_j > 0$ is the quantity discount level, and p_j^{disc} is the quantity discount price, for (the product of) activity j. (We have $0 < p_j^{\mathrm{disc}} < p_j$.) The total revenue is the sum of the revenues associated with each activity, i.e., $\sum_{j=1}^n r_j(x_j)$. The goal is to choose activity levels that maximize the total revenue while respecting the resource limits. Show how to formulate this problem as an LP.

4.18 *Separating hyperplanes and spheres.* Suppose you are given two sets of points in $\mathbf{R}^n$, $\{v^1, v^2, \ldots, v^K\}$ and $\{w^1, w^2, \ldots, w^L\}$. Formulate the following two problems as LP feasibility problems.

(a) Determine a hyperplane that separates the two sets, i.e., find $a \in \mathbf{R}^n$ and $b \in \mathbf{R}$ with $a \neq 0$ such that

$$a^T v^i \leq b, \quad i = 1, \ldots, K, \qquad a^T w^i \geq b, \quad i = 1, \ldots, L.$$

Note that we require $a \neq 0$, so you have to make sure that your formulation excludes the trivial solution $a = 0$, $b = 0$. You can assume that

$$\mathbf{rank} \begin{bmatrix} v^1 & v^2 & \cdots & v^K & w^1 & w^2 & \cdots & w^L \\ 1 & 1 & \cdots & 1 & 1 & 1 & \cdots & 1 \end{bmatrix} = n + 1$$

(*i.e.*, the affine hull of the $K + L$ points has dimension n).

(b) Determine a sphere separating the two sets of points, i.e., find $x_c \in \mathbf{R}^n$ and $R \geq 0$ such that

$$\|v^i - x_c\|_2 \leq R, \quad i = 1, \ldots, K, \qquad \|w^i - x_c\|_2 \geq R, \quad i = 1, \ldots, L.$$

(Here x_c is the center of the sphere; R is its radius.)

(See chapter 8 for more on separating hyperplanes, separating spheres, and related topics.)

4.19 Consider the problem

$$
\begin{array}{ll}
\text{minimize} & \|Ax - b\|_1/(c^T x + d) \\
\text{subject to} & \|x\|_\infty \leq 1,
\end{array}
$$

where $A \in \mathbf{R}^{m \times n}$, $b \in \mathbf{R}^m$, $c \in \mathbf{R}^n$, and $d \in \mathbf{R}$. We assume that $d > \|c\|_1$, which implies that $c^T x + d > 0$ for all feasible x.

(a) Show that this is a quasiconvex optimization problem.

(b) Show that it is equivalent to the convex optimization problem

$$
\begin{array}{ll}
\text{minimize} & \|Ay - bt\|_1 \\
\text{subject to} & \|y\|_\infty \leq t \\
& c^T y + dt = 1,
\end{array}
$$

with variables $y \in \mathbf{R}^n$, $t \in \mathbf{R}$.

4.20 *Power assignment in a wireless communication system.* We consider n transmitters with powers $p_1, \ldots, p_n \geq 0$, transmitting to n receivers. These powers are the optimization variables in the problem. We let $G \in \mathbf{R}^{n \times n}$ denote the matrix of *path gains* from the transmitters to the receivers; $G_{ij} \geq 0$ is the path gain from transmitter j to receiver i. The *signal power* at receiver i is then $S_i = G_{ii} p_i$, and the *interference power* at receiver i is $I_i = \sum_{k \neq i} G_{ik} p_k$. The *signal to interference plus noise ratio*, denoted SINR, at receiver i, is given by $S_i/(I_i + \sigma_i)$, where $\sigma_i > 0$ is the (self-) noise power in receiver i. The objective in the problem is to maximize the minimum SINR ratio, over all receivers, *i.e.*, to maximize

$$
\min_{i=1,\ldots,n} \frac{S_i}{I_i + \sigma_i}.
$$

There are a number of constraints on the powers that must be satisfied, in addition to the obvious one $p_i \geq 0$. The first is a maximum allowable power for each transmitter, *i.e.*, $p_i \leq P_i^{\max}$, where $P_i^{\max} > 0$ is given. In addition, the transmitters are partitioned into groups, with each group sharing the same power supply, so there is a total power constraint for each group of transmitter powers. More precisely, we have subsets $K_1, \ldots, K_m$ of $\{1, \ldots, n\}$ with $K_1 \cup \cdots \cup K_m = \{1, \ldots, n\}$, and $K_j \cap K_l = 0$ if $j \neq l$. For each group K_l, the total associated transmitter power cannot exceed $P_l^{\mathrm{gp}} > 0$:

$$
\sum_{k \in K_l} p_k \leq P_l^{\mathrm{gp}}, \quad l = 1, \ldots, m.
$$

Finally, we have a limit $P_k^{\mathrm{rc}} > 0$ on the total received power at each receiver:

$$
\sum_{k=1}^n G_{ik} p_k \leq P_i^{\mathrm{rc}}, \quad i = 1, \ldots, n.
$$

(This constraint reflects the fact that the receivers will saturate if the total received power is too large.)

Formulate the SINR maximization problem as a generalized linear-fractional program.

Quadratic optimization problems

4.21 *Some simple QCQPs.* Give an explicit solution of each of the following QCQPs.

(a) *Minimizing a linear function over an ellipsoid centered at the origin.*

$$
\begin{array}{ll}
\text{minimize} & c^T x \\
\text{subject to} & x^T A x \leq 1,
\end{array}
$$

where $A \in \mathbf{S}_{++}^n$ and $c \neq 0$. What is the solution if the problem is not convex ($A \notin \mathbf{S}_+^n$)?

(b) *Minimizing a linear function over an ellipsoid.*

$$\begin{array}{ll} \text{minimize} & c^T x \\ \text{subject to} & (x - x_c)^T A (x - x_c) \le 1, \end{array}$$

where $A \in \mathbf{S}^n_{++}$ and $c \ne 0$.

(c) *Minimizing a quadratic form over an ellipsoid centered at the origin.*

$$\begin{array}{ll} \text{minimize} & x^T B x \\ \text{subject to} & x^T A x \le 1, \end{array}$$

where $A \in \mathbf{S}^n_{++}$ and $B \in \mathbf{S}^n_+$. Also consider the nonconvex extension with $B \notin \mathbf{S}^n_+$. (See §B.1.)

4.22 Consider the QCQP

$$\begin{array}{ll} \text{minimize} & (1/2)x^T P x + q^T x + r \\ \text{subject to} & x^T x \le 1, \end{array}$$

with $P \in \mathbf{S}^n_{++}$. Show that $x^\star = -(P + \lambda I)^{-1} q$ where $\lambda = \max\{0, \bar{\lambda}\}$ and $\bar{\lambda}$ is the largest solution of the nonlinear equation

$$q^T (P + \lambda I)^{-2} q = 1.$$

4.23 ℓ_4-*norm approximation via QCQP.* Formulate the ℓ_4-norm approximation problem

$$\text{minimize} \quad \|Ax - b\|_4 = \left(\sum_{i=1}^m (a_i^T x - b_i)^4 \right)^{1/4}$$

as a QCQP. The matrix $A \in \mathbf{R}^{m \times n}$ (with rows a_i^T) and the vector $b \in \mathbf{R}^m$ are given.

4.24 *Complex ℓ_1-, ℓ_2- and ℓ_∞-norm approximation.* Consider the problem

$$\text{minimize} \quad \|Ax - b\|_p,$$

where $A \in \mathbf{C}^{m \times n}$, $b \in \mathbf{C}^m$, and the variable is $x \in \mathbf{C}^n$. The complex ℓ_p-norm is defined by

$$\|y\|_p = \left(\sum_{i=1}^m |y_i|^p \right)^{1/p}$$

for $p \ge 1$, and $\|y\|_\infty = \max_{i=1,\dots,m} |y_i|$. For $p = 1$, 2, and ∞, express the complex ℓ_p-norm approximation problem as a QCQP or SOCP with real variables and data.

4.25 *Linear separation of two sets of ellipsoids.* Suppose we are given $K + L$ ellipsoids

$$\mathcal{E}_i = \{P_i u + q_i \mid \|u\|_2 \le 1\}, \quad i = 1, \dots, K + L,$$

where $P_i \in \mathbf{S}^n$. We are interested in finding a hyperplane that strictly separates $\mathcal{E}_1, \dots, \mathcal{E}_K$ from $\mathcal{E}_{K+1}, \dots, \mathcal{E}_{K+L}$, i.e., we want to compute $a \in \mathbf{R}^n$, $b \in \mathbf{R}$ such that

$$a^T x + b > 0 \text{ for } x \in \mathcal{E}_1 \cup \cdots \cup \mathcal{E}_K, \qquad a^T x + b < 0 \text{ for } x \in \mathcal{E}_{K+1} \cup \cdots \cup \mathcal{E}_{K+L},$$

or prove that no such hyperplane exists. Express this problem as an SOCP feasibility problem.

4.26 *Hyperbolic constraints as SOC constraints.* Verify that $x \in \mathbf{R}^n$, $y, z \in \mathbf{R}$ satisfy

$$x^T x \le yz, \qquad y \ge 0, \qquad z \ge 0$$

if and only if

$$\left\| \begin{bmatrix} 2x \\ y - z \end{bmatrix} \right\|_2 \le y + z, \qquad y \ge 0, \qquad z \ge 0.$$

Use this observation to cast the following problems as SOCPs.

(a) *Maximizing harmonic mean.*

$$\text{maximize} \quad \left(\sum_{i=1}^{m} 1/(a_i^T x - b_i) \right)^{-1},$$

with domain $\{x \mid Ax \succ b\}$, where a_i^T is the ith row of A.

(b) *Maximizing geometric mean.*

$$\text{maximize} \quad \left(\prod_{i=1}^{m} (a_i^T x - b_i) \right)^{1/m},$$

with domain $\{x \mid Ax \succeq b\}$, where a_i^T is the ith row of A.

4.27 *Matrix fractional minimization via SOCP.* Express the following problem as an SOCP:

$$\begin{array}{ll} \text{minimize} & (Ax + b)^T (I + B \, \mathbf{diag}(x) B^T)^{-1} (Ax + b) \\ \text{subject to} & x \succeq 0, \end{array}$$

with $A \in \mathbf{R}^{m \times n}$, $b \in \mathbf{R}^m$, $B \in \mathbf{R}^{m \times n}$. The variable is $x \in \mathbf{R}^n$.
Hint. First show that the problem is equivalent to

$$\begin{array}{ll} \text{minimize} & v^T v + w^T \, \mathbf{diag}(x)^{-1} w \\ \text{subject to} & v + Bw = Ax + b \\ & x \succeq 0, \end{array}$$

with variables $v \in \mathbf{R}^m$, $w, x \in \mathbf{R}^n$. (If $x_i = 0$ we interpret w_i^2/x_i as zero if $w_i = 0$ and as ∞ otherwise.) Then use the results of exercise 4.26.

4.28 *Robust quadratic programming.* In §4.4.2 we discussed robust linear programming as an application of second-order cone programming. In this problem we consider a similar robust variation of the (convex) *quadratic* program

$$\begin{array}{ll} \text{minimize} & (1/2)x^T P x + q^T x + r \\ \text{subject to} & Ax \preceq b. \end{array}$$

For simplicity we assume that only the matrix P is subject to errors, and the other parameters (q, r, A, b) are exactly known. The robust quadratic program is defined as

$$\begin{array}{ll} \text{minimize} & \sup_{P \in \mathcal{E}} ((1/2)x^T P x + q^T x + r) \\ \text{subject to} & Ax \preceq b \end{array}$$

where $\mathcal{E}$ is the set of possible matrices P.

For each of the following sets $\mathcal{E}$, express the robust QP as a convex problem. Be as specific as you can. If the problem can be expressed in a standard form (*e.g.*, QP, QCQP, SOCP, SDP), say so.

(a) A finite set of matrices: $\mathcal{E} = \{P_1, \ldots, P_K\}$, where $P_i \in \mathbf{S}_+^n$, $i = 1, \ldots, K$.

(b) A set specified by a nominal value $P_0 \in \mathbf{S}_+^n$ plus a bound on the eigenvalues of the deviation $P - P_0$:

$$\mathcal{E} = \{P \in \mathbf{S}^n \mid -\gamma I \preceq P - P_0 \preceq \gamma I\}$$

where $\gamma \in \mathbf{R}$ and $P_0 \in \mathbf{S}_+^n$,

(c) An ellipsoid of matrices:

$$\mathcal{E} = \left\{ P_0 + \sum_{i=1}^{K} P_i u_i \;\middle|\; \|u\|_2 \leq 1 \right\}.$$

You can assume $P_i \in \mathbf{S}_+^n$, $i = 0, \ldots, K$.

4.29 *Maximizing probability of satisfying a linear inequality.* Let c be a random variable in $\mathbf{R}^n$, normally distributed with mean $\bar{c}$ and covariance matrix R. Consider the problem

$$\begin{array}{ll} \text{maximize} & \mathbf{prob}(c^T x \geq \alpha) \\ \text{subject to} & Fx \preceq g, \quad Ax = b. \end{array}$$

Assuming there exists a feasible point $\tilde{x}$ for which $\bar{c}^T \tilde{x} \geq \alpha$, show that this problem is equivalent to a convex or quasiconvex optimization problem. Formulate the problem as a QP, QCQP, or SOCP (if the problem is convex), or explain how you can solve it by solving a sequence of QP, QCQP, or SOCP feasibility problems (if the problem is quasiconvex).

Geometric programming

4.30 A heated fluid at temperature T (degrees above ambient temperature) flows in a pipe with fixed length and circular cross section with radius r. A layer of insulation, with thickness $w \ll r$, surrounds the pipe to reduce heat loss through the pipe walls. The design variables in this problem are T, r, and w.

The heat loss is (approximately) proportional to Tr/w, so over a fixed lifetime, the energy cost due to heat loss is given by $\alpha_1 Tr/w$. The cost of the pipe, which has a fixed wall thickness, is approximately proportional to the total material, *i.e.*, it is given by $\alpha_2 r$. The cost of the insulation is also approximately proportional to the total insulation material, *i.e.*, $\alpha_3 rw$ (using $w \ll r$). The total cost is the sum of these three costs.

The heat flow down the pipe is entirely due to the flow of the fluid, which has a fixed velocity, *i.e.*, it is given by $\alpha_4 Tr^2$. The constants α_i are all positive, as are the variables T, r, and w.

Now the problem: maximize the total heat flow down the pipe, subject to an upper limit $C_{\max}$ on total cost, and the constraints

$$T_{\min} \leq T \leq T_{\max}, \qquad r_{\min} \leq r \leq r_{\max}, \qquad w_{\min} \leq w \leq w_{\max}, \quad w \leq 0.1r.$$

Express this problem as a geometric program.

4.31 *Recursive formulation of optimal beam design problem.* Show that the GP (4.46) is equivalent to the GP

$$\begin{array}{ll} \text{minimize} & \sum_{i=1}^{N} w_i h_i \\ \text{subject to} & w_i/w_{\max} \leq 1, \quad w_{\min}/w_i \leq 1, \quad i = 1, \ldots, N \\ & h_i/h_{\max} \leq 1, \quad h_{\min}/h_i \leq 1, \quad i = 1, \ldots, N \\ & h_i/(w_i S_{\max}) \leq 1, \quad S_{\min} w_i/h_i \leq 1, \quad i = 1, \ldots, N \\ & 6iF/(\sigma_{\max} w_i h_i^2) \leq 1, \quad i = 1, \ldots, N \\ & (2i-1)d_i/v_i + v_{i+1}/v_i \leq 1, \quad i = 1, \ldots, N \\ & (i-1/3)d_i/y_i + v_{i+1}/y_i + y_{i+1}/y_i \leq 1, \quad i = 1, \ldots, N \\ & y_1/y_{\max} \leq 1 \\ & Ew_i h_i^3 d_i/(6F) = 1, \quad i = 1, \ldots, N. \end{array}$$

The variables are w_i, h_i, v_i, d_i, y_i for $i = 1, \ldots, N$.

4.32 *Approximating a function as a monomial.* Suppose the function $f : \mathbf{R}^n \to \mathbf{R}$ is differentiable at a point $x_0 \succ 0$, with $f(x_0) > 0$. How would you find a monomial function $\hat{f} : \mathbf{R}^n \to \mathbf{R}$ such that $f(x_0) = \hat{f}(x_0)$ and for x near x_0, $\hat{f}(x)$ is very near $f(x)$?

4.33 Express the following problems as convex optimization problems.

(a) Minimize $\max\{p(x), q(x)\}$, where p and q are posynomials.

(b) Minimize $\exp(p(x)) + \exp(q(x))$, where p and q are posynomials.

(c) Minimize $p(x)/(r(x) - q(x))$, subject to $r(x) > q(x)$, where p, q are posynomials, and r is a monomial.

4.34 *Log-convexity of Perron-Frobenius eigenvalue.* Let $A \in \mathbf{R}^{n \times n}$ be an elementwise positive matrix, *i.e.*, $A_{ij} > 0$. (The results of this problem hold for irreducible nonnegative matrices as well.) Let $\lambda_{\mathrm{pf}}(A)$ denotes its Perron-Frobenius eigenvalue, *i.e.*, its eigenvalue of largest magnitude. (See the definition and the example on page 165.) Show that $\log \lambda_{\mathrm{pf}}(A)$ is a convex function of $\log A_{ij}$. This means, for example, that we have the inequality

$$\lambda_{\mathrm{pf}}(C) \le (\lambda_{\mathrm{pf}}(A)\lambda_{\mathrm{pf}}(B))^{1/2},$$

where $C_{ij} = (A_{ij}B_{ij})^{1/2}$, and A and B are elementwise positive matrices.

Hint. Use the characterization of the Perron-Frobenius eigenvalue given in (4.47), or, alternatively, use the characterization

$$\log \lambda_{\mathrm{pf}}(A) = \lim_{k \to \infty} (1/k) \log(\mathbf{1}^T A^k \mathbf{1}).$$

4.35 *Signomial and geometric programs.* A *signomial* is a linear combination of monomials of some positive variables $x_1, \dots, x_n$. Signomials are more general than posynomials, which are signomials with all positive coefficients. A *signomial program* is an optimization problem of the form

$$
\begin{array}{ll}
\text{minimize} & f_0(x) \\
\text{subject to} & f_i(x) \le 0, \quad i = 1, \dots, m \\
& h_i(x) = 0, \quad i = 1, \dots, p,
\end{array}
$$

where $f_0, \dots, f_m$ and $h_1, \dots, h_p$ are signomials. In general, signomial programs are very difficult to solve.

Some signomial programs can be transformed to GPs, and therefore solved efficiently. Show how to do this for a signomial program of the following form:

- The objective signomial f_0 is a posynomial, *i.e.*, its terms have only positive coefficients.

- Each inequality constraint signomial $f_1, \dots, f_m$ has exactly one term with a negative coefficient: $f_i = p_i - q_i$ where p_i is posynomial, and q_i is monomial.

- Each equality constraint signomial $h_1, \dots, h_p$ has exactly one term with a positive coefficient and one term with a negative coefficient: $h_i = r_i - s_i$ where r_i and s_i are monomials.

4.36 Explain how to reformulate a general GP as an equivalent GP in which every posynomial (in the objective and constraints) has at most two monomial terms. *Hint.* Express each sum (of monomials) as a sum of sums, each with two terms.

4.37 *Generalized posynomials and geometric programming.* Let $x_1, \dots, x_n$ be positive variables, and suppose the functions $f_i : \mathbf{R}^n \to \mathbf{R}$, $i = 1, \dots, k$, are posynomials of $x_1, \dots, x_n$. If $\phi : \mathbf{R}^k \to \mathbf{R}$ is a polynomial with nonnegative coefficients, then the composition

$$h(x) = \phi(f_1(x), \dots, f_k(x)) \tag{4.69}$$

is a posynomial, since posynomials are closed under products, sums, and multiplication by nonnegative scalars. For example, suppose f_1 and f_2 are posynomials, and consider the polynomial $\phi(z_1, z_2) = 3z_1^2 z_2 + 2z_1 + 3z_2^3$ (which has nonnegative coefficients). Then $h = 3f_1^2 f_2 + 2f_1 + f_2^3$ is a posynomial.

In this problem we consider a generalization of this idea, in which ϕ is allowed to be a posynomial, *i.e.*, can have fractional exponents. Specifically, assume that $\phi : \mathbf{R}^k \to \mathbf{R}$ is a posynomial, with all its exponents nonnegative. In this case we will call the function h defined in (4.69) a *generalized posynomial*. As an example, suppose f_1 and f_2 are posynomials, and consider the posynomial (with nonnegative exponents) $\phi(z_1, z_2) = 2z_1^{0.3} z_2^{1.2} + z_1 z_2^{0.5} + 2$. Then the function

$$h(x) = 2f_1(x)^{0.3} f_2(x)^{1.2} + f_1(x)f_2(x)^{0.5} + 2$$

is a generalized posynomial. Note that it is *not* a posynomial, however (unless f_1 and f_2 are monomials or constants).

A *generalized geometric program* (GGP) is an optimization problem of the form

$$
\begin{aligned}
\text{minimize} \quad & h_0(x) \\
\text{subject to} \quad & h_i(x) \leq 1, \quad i = 1, \ldots, m \\
& g_i(x) = 1, \quad i = 1, \ldots, p,
\end{aligned}
\tag{4.70}
$$

where $g_1, \ldots, g_p$ are monomials, and $h_0, \ldots, h_m$ are generalized posynomials.

Show how to express this generalized geometric program as an equivalent geometric program. Explain any new variables you introduce, and explain how your GP is equivalent to the GGP (4.70).

Semidefinite programming and conic form problems

4.38 *LMIs and SDPs with one variable.* The *generalized eigenvalues* of a matrix pair (A, B), where $A, B \in \mathbf{S}^n$, are defined as the roots of the polynomial $\det(\lambda B - A)$ (see §A.5.3). Suppose B is nonsingular, and that A and B can be simultaneously diagonalized by a congruence, *i.e.*, there exists a nonsingular $R \in \mathbf{R}^{n \times n}$ such that

$$
R^T A R = \mathbf{diag}(a), \qquad R^T B R = \mathbf{diag}(b),
$$

where $a, b \in \mathbf{R}^n$. (A sufficient condition for this to hold is that there exists t_1, t_2 such that $t_1 A + t_2 B \succ 0$.)

(a) Show that the generalized eigenvalues of (A, B) are real, and given by $\lambda_i = a_i / b_i$, $i = 1, \ldots, n$.

(b) Express the solution of the SDP

$$
\begin{aligned}
\text{minimize} \quad & ct \\
\text{subject to} \quad & tB \preceq A,
\end{aligned}
$$

with variable $t \in \mathbf{R}$, in terms of a and b.

4.39 *SDPs and congruence transformations.* Consider the SDP

$$
\begin{aligned}
\text{minimize} \quad & c^T x \\
\text{subject to} \quad & x_1 F_1 + x_2 F_2 + \cdots + x_n F_n + G \preceq 0,
\end{aligned}
$$

with $F_i, G \in \mathbf{S}^k$, $c \in \mathbf{R}^n$.

(a) Suppose $R \in \mathbf{R}^{k \times k}$ is nonsingular. Show that the SDP is equivalent to the SDP

$$
\begin{aligned}
\text{minimize} \quad & c^T x \\
\text{subject to} \quad & x_1 \tilde{F}_1 + x_2 \tilde{F}_2 + \cdots + x_n \tilde{F}_n + \tilde{G} \preceq 0,
\end{aligned}
$$

where $\tilde{F}_i = R^T F_i R$, $\tilde{G} = R^T G R$.

(b) Suppose there exists a nonsingular R such that $\tilde{F}_i$ and $\tilde{G}$ are diagonal. Show that the SDP is equivalent to an LP.

(c) Suppose there exists a nonsingular R such that $\tilde{F}_i$ and $\tilde{G}$ have the form

$$
\tilde{F}_i = \begin{bmatrix} \alpha_i I & a_i \\ a_i^T & \alpha_i \end{bmatrix}, \quad i = 1, \ldots, n, \qquad
\tilde{G} = \begin{bmatrix} \beta I & b \\ b^T & \beta \end{bmatrix},
$$

where $\alpha_i, \beta \in \mathbf{R}$, $a_i, b \in \mathbf{R}^{k-1}$. Show that the SDP is equivalent to an SOCP with a single second-order cone constraint.

4.40 *LPs, QPs, QCQPs, and SOCPs as SDPs.* Express the following problems as SDPs.

(a) The LP (4.27).

(b) The QP (4.34), the QCQP (4.35) and the SOCP (4.36). *Hint.* Suppose $A \in \mathbf{S}^r_{++}$, $C \in \mathbf{S}^s$, and $B \in \mathbf{R}^{r \times s}$. Then

$$\begin{bmatrix} A & B \\ B^T & C \end{bmatrix} \succeq 0 \iff C - B^T A^{-1} B \succeq 0.$$

For a more complete statement, which applies also to singular A, and a proof, see §A.5.5.

(c) The matrix fractional optimization problem

$$\text{minimize} \quad (Ax + b)^T F(x)^{-1} (Ax + b)$$

where $A \in \mathbf{R}^{m \times n}$, $b \in \mathbf{R}^m$,

$$F(x) = F_0 + x_1 F_1 + \cdots + x_n F_n,$$

with $F_i \in \mathbf{S}^m$, and we take the domain of the objective to be $\{x \mid F(x) \succ 0\}$. You can assume the problem is feasible (there exists at least one x with $F(x) \succ 0$).

4.41 *LMI tests for copositive matrices and P_0-matrices.* A matrix $A \in \mathbf{S}^n$ is said to be *copositive* if $x^T A x \geq 0$ for all $x \succeq 0$ (see exercise 2.35). A matrix $A \in \mathbf{R}^{n \times n}$ is said to be a P_0-*matrix* if $\max_{i=1,\ldots,n} x_i (Ax)_i \geq 0$ for all x. Checking whether a matrix is copositive or a P_0-matrix is very difficult in general. However, there exist useful sufficient conditions that can be verified using semidefinite programming.

(a) Show that A is copositive if it can be decomposed as a sum of a positive semidefinite and an elementwise nonnegative matrix:

$$A = B + C, \qquad B \succeq 0, \qquad C_{ij} \geq 0, \quad i, j = 1, \ldots, n. \tag{4.71}$$

Express the problem of finding B and C that satisfy (4.71) as an SDP feasibility problem.

(b) Show that A is a P_0-matrix if there exists a positive diagonal matrix D such that

$$DA + A^T D \succeq 0. \tag{4.72}$$

Express the problem of finding a D that satisfies (4.72) as an SDP feasibility problem.

4.42 *Complex LMIs and SDPs.* A complex LMI has the form

$$x_1 F_1 + \cdots + x_n F_n + G \preceq 0$$

where $F_1, \ldots, F_n$, G are complex $n \times n$ Hermitian matrices, *i.e.*, $F_i^H = F_i$, $G^H = G$, and $x \in \mathbf{R}^n$ is a real variable. A complex SDP is the problem of minimizing a (real) linear function of x subject to a complex LMI constraint.

Complex LMIs and SDPs can be transformed to real LMIs and SDPs, using the fact that

$$X \succeq 0 \iff \begin{bmatrix} \Re X & -\Im X \\ \Im X & \Re X \end{bmatrix} \succeq 0,$$

where $\Re X \in \mathbf{R}^{n \times n}$ is the real part of the complex Hermitian matrix X, and $\Im X \in \mathbf{R}^{n \times n}$ is the imaginary part of X.

Verify this result, and show how to pose a complex SDP as a real SDP.

4.43 *Eigenvalue optimization via SDP.* Suppose $A : \mathbf{R}^n \to \mathbf{S}^m$ is affine, *i.e.*,

$$A(x) = A_0 + x_1 A_1 + \cdots + x_n A_n$$

where $A_i \in \mathbf{S}^m$. Let $\lambda_1(x) \geq \lambda_2(x) \geq \cdots \geq \lambda_m(x)$ denote the eigenvalues of $A(x)$. Show how to pose the following problems as SDPs.

(a) Minimize the maximum eigenvalue $\lambda_1(x)$.

(b) Minimize the spread of the eigenvalues, $\lambda_1(x) - \lambda_m(x)$.

(c) Minimize the condition number of $A(x)$, subject to $A(x) \succ 0$. The condition number is defined as $\kappa(A(x)) = \lambda_1(x)/\lambda_m(x)$, with domain $\{x \mid A(x) \succ 0\}$. You may assume that $A(x) \succ 0$ for at least one x.

Hint. You need to minimize λ/γ, subject to

$$0 \prec \gamma I \preceq A(x) \preceq \lambda I.$$

Change variables to $y = x/\gamma$, $t = \lambda/\gamma$, $s = 1/\gamma$.

(d) Minimize the sum of the absolute values of the eigenvalues, $|\lambda_1(x)| + \cdots + |\lambda_m(x)|$.
Hint. Express $A(x)$ as $A(x) = A_+ - A_-$, where $A_+ \succeq 0$, $A_- \succeq 0$.

4.44 *Optimization over polynomials.* Pose the following problem as an SDP. Find the polynomial $p : \mathbf{R} \to \mathbf{R}$,

$$p(t) = x_1 + x_2 t + \cdots + x_{2k+1} t^{2k},$$

that satisfies given bounds $l_i \leq p(t_i) \leq u_i$, at m specified points t_i, and, of all the polynomials that satisfy these bounds, has the greatest minimum value:

$$\begin{array}{ll} \text{maximize} & \inf_t p(t) \\ \text{subject to} & l_i \leq p(t_i) \leq u_i, \quad i = 1, \ldots, m. \end{array}$$

The variables are $x \in \mathbf{R}^{2k+1}$.
Hint. Use the LMI characterization of nonnegative polynomials derived in exercise 2.37, part (b).

4.45 [Nes00, Par00] *Sum-of-squares representation via LMIs.* Consider a polynomial $p : \mathbf{R}^n \to \mathbf{R}$ of degree $2k$. The polynomial is said to be positive semidefinite (PSD) if $p(x) \geq 0$ for all $x \in \mathbf{R}^n$. Except for special cases (*e.g.*, $n = 1$ or $k = 1$), it is extremely difficult to determine whether or not a given polynomial is PSD, let alone solve an optimization problem, with the coefficients of p as variables, with the constraint that p be PSD.

A famous sufficient condition for a polynomial to be PSD is that it have the form

$$p(x) = \sum_{i=1}^{r} q_i(x)^2,$$

for some polynomials q_i, with degree no more than k. A polynomial p that has this sum-of-squares form is called SOS.

The condition that a polynomial p be SOS (viewed as a constraint on its coefficients) turns out to be equivalent to an LMI, and therefore a variety of optimization problems, with SOS constraints, can be posed as SDPs. You will explore these ideas in this problem.

(a) Let $f_1, \ldots, f_s$ be all monomials of degree k or less. (Here we mean monomial in the standard sense, *i.e.*, $x_1^{m_1} \cdots x_n^{m_n}$, where $m_i \in \mathbf{Z}_+$, and not in the sense used in geometric programming.) Show that if p can be expressed as a positive semidefinite quadratic form $p = f^T V f$, with $V \in \mathbf{S}_+^s$, then p is SOS. Conversely, show that if p is SOS, then it can be expressed as a positive semidefinite quadratic form in the monomials, *i.e.*, $p = f^T V f$, for some $V \in \mathbf{S}_+^s$.

(b) Show that the condition $p = f^T V f$ is a set of linear equality constraints relating the coefficients of p and the matrix V. Combined with part (a) above, this shows that the condition that p be SOS is equivalent to a set of linear equalities relating V and the coefficients of p, and the matrix inequality $V \succeq 0$.

(c) Work out the LMI conditions for SOS explicitly for the case where p is polynomial of degree four in two variables.

4.46 *Multidimensional moments.* The moments of a random variable t on $\mathbf{R}^2$ are defined as $\mu_{ij} = \mathbf{E}\, t_1^i t_2^j$, where i, j are nonnegative integers. In this problem we derive necessary conditions for a set of numbers μ_{ij}, $0 \le i, j \le 2k$, $i + j \le 2k$, to be the moments of a distribution on $\mathbf{R}^2$.

Let $p : \mathbf{R}^2 \to \mathbf{R}$ be a polynomial of degree k with coefficients c_{ij},

$$p(t) = \sum_{i=0}^{k} \sum_{j=0}^{k-i} c_{ij} t_1^i t_2^j,$$

and let t be a random variable with moments μ_{ij}. Suppose $c \in \mathbf{R}^{(k+1)(k+2)/2}$ contains the coefficients c_{ij} in some specific order, and $\mu \in \mathbf{R}^{(k+1)(2k+1)}$ contains the moments μ_{ij} in the same order. Show that $\mathbf{E}\, p(t)^2$ can be expressed as a quadratic form in c:

$$\mathbf{E}\, p(t)^2 = c^T H(\mu) c,$$

where $H : \mathbf{R}^{(k+1)(2k+1)} \to \mathbf{S}^{(k+1)(k+2)/2}$ is a linear function of μ. From this, conclude that μ must satisfy the LMI $H(\mu) \succeq 0$.

Remark: For random variables on $\mathbf{R}$, the matrix H can be taken as the Hankel matrix defined in (4.52). In this case, $H(\mu) \succeq 0$ is a necessary and sufficient condition for μ to be the moments of a distribution, or the limit of a sequence of moments. On $\mathbf{R}^2$, however, the LMI is only a necessary condition.

4.47 *Maximum determinant positive semidefinite matrix completion.* We consider a matrix $A \in \mathbf{S}^n$, with some entries specified, and the others not specified. The *positive semidefinite matrix completion problem* is to determine values of the unspecified entries of the matrix so that $A \succeq 0$ (or to determine that such a completion does not exist).

(a) Explain why we can assume without loss of generality that the diagonal entries of A are specified.

(b) Show how to formulate the positive semidefinite completion problem as an SDP feasibility problem.

(c) Assume that A has at least one completion that is positive definite, and the diagonal entries of A are specified (*i.e.*, fixed). The positive definite completion with largest determinant is called the *maximum determinant completion*. Show that the maximum determinant completion is unique. Show that if $A^\star$ is the maximum determinant completion, then $(A^\star)^{-1}$ has zeros in all the entries of the original matrix that were not specified. *Hint.* The gradient of the function $f(X) = \log \det X$ is $\nabla f(X) = X^{-1}$ (see §A.4.1).

(d) Suppose A is specified on its tridiagonal part, *i.e.*, we are given $A_{11}, \ldots, A_{nn}$ and $A_{12}, \ldots, A_{n-1,n}$. Show that if there exists a positive definite completion of A, then there is a positive definite completion whose inverse is tridiagonal.

4.48 *Generalized eigenvalue minimization.* Recall (from example 3.37, or §A.5.3) that the largest generalized eigenvalue of a pair of matrices $(A, B) \in \mathbf{S}^k \times \mathbf{S}^k_{++}$ is given by

$$\lambda_{\max}(A, B) = \sup_{u \ne 0} \frac{u^T A u}{u^T B u} = \max\{\lambda \mid \det(\lambda B - A) = 0\}.$$

As we have seen, this function is quasiconvex (if we take $\mathbf{S}^k \times \mathbf{S}^k_{++}$ as its domain).

We consider the problem

$$\text{minimize} \quad \lambda_{\max}(A(x), B(x)) \tag{4.73}$$

where $A, B : \mathbf{R}^n \to \mathbf{S}^k$ are affine functions, defined as

$$A(x) = A_0 + x_1 A_1 + \cdots + x_n A_n, \qquad B(x) = B_0 + x_1 B_1 + \cdots + x_n B_n.$$

with $A_i, B_i \in \mathbf{S}^k$.

(a) Give a family of convex functions $\phi_t : \mathbf{S}^k \times \mathbf{S}^k \to \mathbf{R}$, that satisfy

$$\lambda_{\max}(A, B) \le t \iff \phi_t(A, B) \le 0$$

for all $(A, B) \in \mathbf{S}^k \times \mathbf{S}^k_{++}$. Show that this allows us to solve (4.73) by solving a sequence of convex feasibility problems.

(b) Give a family of matrix-convex functions $\Phi_t : \mathbf{S}^k \times \mathbf{S}^k \to \mathbf{S}^k$ that satisfy

$$\lambda_{\max}(A, B) \le t \iff \Phi_t(A, B) \preceq 0$$

for all $(A, B) \in \mathbf{S}^k \times \mathbf{S}^k_{++}$. Show that this allows us to solve (4.73) by solving a sequence of convex feasibility problems with LMI constraints.

(c) Suppose $B(x) = (a^T x + b)I$, with $a \ne 0$. Show that (4.73) is equivalent to the convex problem

$$\begin{array}{ll}
\text{minimize} & \lambda_{\max}(s A_0 + y_1 A_1 + \cdots + y_n A_n) \\
\text{subject to} & a^T y + bs = 1 \\
& s \ge 0,
\end{array}$$

with variables $y \in \mathbf{R}^n$, $s \in \mathbf{R}$.

4.49 *Generalized fractional programming.* Let $K \in \mathbf{R}^m$ be a proper cone. Show that the function $f_0 : \mathbf{R}^n \to \mathbf{R}^m$, defined by

$$f_0(x) = \inf\{t \mid Cx + d \preceq_K t(Fx + g)\}, \qquad \mathbf{dom}\, f_0 = \{x \mid Fx + g \succ_K 0\},$$

with $C, F \in \mathbf{R}^{m \times n}$, $d, g \in \mathbf{R}^m$, is quasiconvex.

A quasiconvex optimization problem with objective function of this form is called a *generalized fractional program*. Express the generalized linear-fractional program of page 152 and the generalized eigenvalue minimization problem (4.73) as generalized fractional programs.

Vector and multicriterion optimization

4.50 *Bi-criterion optimization.* Figure 4.11 shows the optimal trade-off curve and the set of achievable values for the bi-criterion optimization problem

$$\text{minimize (w.r.t. } \mathbf{R}^2_+) \quad (\|Ax - b\|^2, \|x\|^2_2),$$

for some $A \in \mathbf{R}^{100 \times 10}$, $b \in \mathbf{R}^{100}$. Answer the following questions using information from the plot. We denote by x_{ls} the solution of the least-squares problem

$$\text{minimize} \quad \|Ax - b\|^2_2.$$

(a) What is $\|x_{\mathrm{ls}}\|_2$?

(b) What is $\|Ax_{\mathrm{ls}} - b\|_2$?

(c) What is $\|b\|_2$?

(d) Give the optimal value of the problem

$$
\begin{array}{ll}
\text{minimize} & \|Ax - b\|_2^2 \\
\text{subject to} & \|x\|_2^2 = 1.
\end{array}
$$

(e) Give the optimal value of the problem

$$
\begin{array}{ll}
\text{minimize} & \|Ax - b\|_2^2 \\
\text{subject to} & \|x\|_2^2 \leq 1.
\end{array}
$$

(f) Give the optimal value of the problem

$$
\text{minimize} \ \|Ax - b\|_2^2 + \|x\|_2^2.
$$

(g) What is the rank of A?

4.51 *Monotone transformation of objective in vector optimization.* Consider the vector optimization problem (4.56). Suppose we form a new vector optimization problem by replacing the objective f_0 with $\phi \circ f_0$, where $\phi : \mathbf{R}^q \to \mathbf{R}^q$ satisfies

$$
u \preceq_K v, \ u \neq v \implies \phi(u) \preceq_K \phi(v), \ \phi(u) \neq \phi(v).
$$

Show that a point x is Pareto optimal (or optimal) for one problem if and only if it is Pareto optimal (optimal) for the other, so the two problems are equivalent. In particular, composing each objective in a multicriterion problem with an increasing function does not affect the Pareto optimal points.

4.52 *Pareto optimal points and the boundary of the set of achievable values.* Consider a vector optimization problem with cone K. Let $\mathcal{P}$ denote the set of Pareto optimal values, and let $\mathcal{O}$ denote the set of achievable objective values. Show that $\mathcal{P} \subseteq \mathcal{O} \cap \mathbf{bd}\,\mathcal{O}$, *i.e.*, every Pareto optimal value is an achievable objective value that lies in the boundary of the set of achievable objective values.

4.53 Suppose the vector optimization problem (4.56) is convex. Show that the set

$$
\mathcal{A} = \mathcal{O} + K = \{t \in \mathbf{R}^q \mid f_0(x) \preceq_K t \text{ for some feasible } x\},
$$

is convex. Also show that the minimal elements of $\mathcal{A}$ are the same as the minimal points of $\mathcal{O}$.

4.54 *Scalarization and optimal points.* Suppose a (not necessarily convex) vector optimization problem has an optimal point $x^\star$. Show that $x^\star$ is a solution of the associated scalarized problem for any choice of $\lambda \succ_{K^*} 0$. Also show the converse: If a point x is a solution of the scalarized problem for any choice of $\lambda \succ_{K^*} 0$, then it is an optimal point for the (not necessarily convex) vector optimization problem.

4.55 *Generalization of weighted-sum scalarization.* In §4.7.4 we showed how to obtain Pareto optimal solutions of a vector optimization problem by replacing the vector objective $f_0 : \mathbf{R}^n \to \mathbf{R}^q$ with the scalar objective $\lambda^T f_0$, where $\lambda \succ_{K^*} 0$. Let $\psi : \mathbf{R}^q \to \mathbf{R}$ be a K-increasing function, *i.e.*, satisfying

$$
u \preceq_K v, \ u \neq v \implies \psi(u) < \psi(v).
$$

Show that any solution of the problem

$$
\begin{array}{ll}
\text{minimize} & \psi(f_0(x)) \\
\text{subject to} & f_i(x) \leq 0, \quad i = 1, \dots, m \\
& h_i(x) = 0, \quad i = 1, \dots, p
\end{array}
$$

is Pareto optimal for the vector optimization problem

$$
\begin{aligned}
\text{minimize (w.r.t. } K) \quad & f_0(x) \\
\text{subject to} \quad & f_i(x) \le 0, \quad i = 1, \dots, m \\
& h_i(x) = 0, \quad i = 1, \dots, p.
\end{aligned}
$$

Note that $\psi(u) = \lambda^T u$, where $\lambda \succ_{K^*} 0$, is a special case.

As a related example, show that in a multicriterion optimization problem (*i.e.*, a vector optimization problem with $f_0 = F : \mathbf{R}^n \to \mathbf{R}^q$, and $K = \mathbf{R}^q_+$), a *unique* solution of the scalar optimization problem

$$
\begin{aligned}
\text{minimize} \quad & \max_{i=1,\dots,q} F_i(x) \\
\text{subject to} \quad & f_i(x) \le 0, \quad i = 1, \dots, m \\
& h_i(x) = 0, \quad i = 1, \dots, p,
\end{aligned}
$$

is Pareto optimal.

Miscellaneous problems

4.56 [P. Parrilo] We consider the problem of minimizing the convex function $f_0 : \mathbf{R}^n \to \mathbf{R}$ over the convex hull of the union of some convex sets, $\mathbf{conv}\left(\bigcup_{i=1}^q C_i\right)$. These sets are described via convex inequalities,

$$
C_i = \{x \mid f_{ij}(x) \le 0, \ j = 1, \dots, k_i\},
$$

where $f_{ij} : \mathbf{R}^n \to \mathbf{R}$ are convex. Our goal is to formulate this problem as a convex optimization problem.

The obvious approach is to introduce variables $x_1, \dots, x_q \in \mathbf{R}^n$, with $x_i \in C_i$, $\theta \in \mathbf{R}^q$ with $\theta \succeq 0$, $\mathbf{1}^T \theta = 1$, and a variable $x \in \mathbf{R}^n$, with $x = \theta_1 x_1 + \cdots + \theta_q x_q$. This equality constraint is not affine in the variables, so this approach does not yield a convex problem. A more sophisticated formulation is given by

$$
\begin{aligned}
\text{minimize} \quad & f_0(x) \\
\text{subject to} \quad & s_i f_{ij}(z_i/s_i) \le 0, \quad i = 1, \dots, q, \quad j = 1, \dots, k_i \\
& \mathbf{1}^T s = 1, \quad s \succeq 0 \\
& x = z_1 + \cdots + z_q,
\end{aligned}
$$

with variables $z_1, \dots, z_q \in \mathbf{R}^n$, $x \in \mathbf{R}^n$, and $s_1, \dots, s_q \in \mathbf{R}$. (When $s_i = 0$, we take $s_i f_{ij}(z_i/s_i)$ to be 0 if $z_i = 0$ and ∞ if $z_i \ne 0$.) Explain why this problem is convex, and equivalent to the original problem.

4.57 *Capacity of a communication channel.* We consider a communication channel, with input $X(t) \in \{1, \dots, n\}$, and output $Y(t) \in \{1, \dots, m\}$, for $t = 1, 2, \dots$ (in seconds, say). The relation between the input and the output is given statistically:

$$
p_{ij} = \mathbf{prob}(Y(t) = i \mid X(t) = j), \quad i = 1, \dots, m, \quad j = 1, \dots, n.
$$

The matrix $P \in \mathbf{R}^{m \times n}$ is called the *channel transition matrix*, and the channel is called a *discrete memoryless channel*.

A famous result of Shannon states that information can be sent over the communication channel, with arbitrarily small probability of error, at any rate less than a number C, called the *channel capacity*, in bits per second. Shannon also showed that the capacity of a discrete memoryless channel can be found by solving an optimization problem. Assume that X has a probability distribution denoted $x \in \mathbf{R}^n$, *i.e.*,

$$
x_j = \mathbf{prob}(X = j), \quad j = 1, \dots, n.
$$

The *mutual information* between X and Y is given by

$$I(X;Y) = \sum_{i=1}^{m} \sum_{j=1}^{n} x_j p_{ij} \log_2 \frac{p_{ij}}{\sum_{k=1}^{n} x_k p_{ik}}.$$

Then the channel capacity C is given by

$$C = \sup_x I(X;Y),$$

where the supremum is over all possible probability distributions for the input X, *i.e.*, over $x \succeq 0$, $\mathbf{1}^T x = 1$.

Show how the channel capacity can be computed using convex optimization.

Hint. Introduce the variable $y = Px$, which gives the probability distribution of the output Y, and show that the mutual information can be expressed as

$$I(X;Y) = c^T x - \sum_{i=1}^{m} y_i \log_2 y_i,$$

where $c_j = \sum_{i=1}^{m} p_{ij} \log_2 p_{ij}$, $j = 1, \ldots, n$.

4.58 *Optimal consumption.* In this problem we consider the optimal way to consume (or spend) an initial amount of money (or other asset) k_0 over time. The variables are $c_0, \ldots, c_T$, where $c_t \geq 0$ denotes the *consumption* in period t. The utility derived from a consumption level c is given by $u(c)$, where $u : \mathbf{R} \to \mathbf{R}$ is an increasing concave function. The present value of the utility derived from the consumption is given by

$$U = \sum_{t=0}^{T} \beta^t u(c_t),$$

where $0 < \beta < 1$ is a *discount factor*.

Let k_t denote the amount of money available for investment in period t. We assume that it earns an investment return given by $f(k_t)$, where $f : \mathbf{R} \to \mathbf{R}$ is an increasing, concave *investment return function*, which satisfies $f(0) = 0$. For example if the funds earn simple interest at rate R percent per period, we have $f(a) = (R/100)a$. The amount to be consumed, *i.e.*, c_t, is withdrawn at the end of the period, so we have the recursion

$$k_{t+1} = k_t + f(k_t) - c_t, \quad t = 0, \ldots, T.$$

The initial sum $k_0 > 0$ is given. We require $k_t \geq 0$, $t = 1, \ldots, T+1$ (but more sophisticated models, which allow $k_t < 0$, can be considered).

Show how to formulate the problem of maximizing U as a convex optimization problem. Explain how the problem you formulate is equivalent to this one, and exactly how the two are related.

Hint. Show that we can replace the recursion for k_t given above with the inequalities

$$k_{t+1} \leq k_t + f(k_t) - c_t, \quad t = 0, \ldots, T.$$

(Interpretation: the inequalities give you the option of throwing money away in each period.) For a more general version of this trick, see exercise 4.6.

4.59 *Robust optimization.* In some optimization problems there is uncertainty or variation in the objective and constraint functions, due to parameters or factors that are either beyond our control or unknown. We can model this situation by making the objective and constraint functions $f_0, \ldots, f_m$ functions of the optimization variable $x \in \mathbf{R}^n$ and a parameter vector $u \in \mathbf{R}^k$ that is unknown, or varies. In the *stochastic optimization* approach, the parameter vector u is modeled as a random variable with a known distribution, and we work with the expected values $\mathbf{E}_u f_i(x, u)$. In the *worst-case analysis* approach, we are given a set U that u is known to lie in, and we work with the maximum or worst-case values $\sup_{u \in U} f_i(x, u)$. To simplify the discussion, we assume there are no equality constraints.

(a) *Stochastic optimization.* We consider the problem

$$
\begin{array}{ll}
\text{minimize} & \mathbf{E} f_0(x, u) \\
\text{subject to} & \mathbf{E} f_i(x, u) \le 0, \quad i = 1, \ldots, m,
\end{array}
$$

where the expectation is with respect to u. Show that if f_i are convex in x for each u, then this stochastic optimization problem is convex.

(b) *Worst-case optimization.* We consider the problem

$$
\begin{array}{ll}
\text{minimize} & \sup_{u \in U} f_0(x, u) \\
\text{subject to} & \sup_{u \in U} f_i(x, u) \le 0, \quad i = 1, \ldots, m.
\end{array}
$$

Show that if f_i are convex in x for each u, then this worst-case optimization problem is convex.

(c) *Finite set of possible parameter values.* The observations made in parts (a) and (b) are most useful when we have analytical or easily evaluated expressions for the expected values $\mathbf{E} f_i(x, u)$ or the worst-case values $\sup_{u \in U} f_i(x, u)$.

Suppose we are given the set of possible values of the parameter is finite, *i.e.*, we have $u \in \{u_1, \ldots, u_N\}$. For the stochastic case, we are also given the probabilities of each value: $\mathbf{prob}(u = u_i) = p_i$, where $p \in \mathbf{R}^N$, $p \succeq 0$, $\mathbf{1}^T p = 1$. In the worst-case formulation, we simply take $U \in \{u_1, \ldots, u_N\}$.

Show how to set up the worst-case and stochastic optimization problems explicitly (*i.e.*, give explicit expressions for $\sup_{u \in U} f_i$ and $\mathbf{E}_u f_i$).

4.60 *Log-optimal investment strategy.* We consider a portfolio problem with n assets held over N periods. At the beginning of each period, we re-invest our total wealth, redistributing it over the n assets using a fixed, constant, allocation strategy $x \in \mathbf{R}^n$, where $x \succ 0$, $\mathbf{1}^T x = 1$. In other words, if $W(t-1)$ is our wealth at the beginning of period t, then during period t we invest $x_i W(t-1)$ in asset i. We denote by $\lambda(t)$ the total return during period t, *i.e.*, $\lambda(t) = W(t)/W(t-1)$. At the end of the N periods our wealth has been multiplied by the factor $\prod_{t=1}^{N} \lambda(t)$. We call

$$
\frac{1}{N} \sum_{t=1}^{N} \log \lambda(t)
$$

the *growth rate* of the investment over the N periods. We are interested in determining an allocation strategy x that maximizes growth of our total wealth for large N.

We use a discrete stochastic model to account for the uncertainty in the returns. We assume that during each period there are m possible scenarios, with probabilities π_j, $j = 1, \ldots, m$. In scenario j, the return for asset i over one period is given by p_{ij}. Therefore, the return $\lambda(t)$ of our portfolio during period t is a random variable, with m possible values $p_1^T x, \ldots, p_m^T x$, and distribution

$$
\pi_j = \mathbf{prob}(\lambda(t) = p_j^T x), \quad j = 1, \ldots, m.
$$

We assume the same scenarios for each period, with (identical) independent distributions. Using the law of large numbers, we have

$$
\lim_{N \to \infty} \frac{1}{N} \log \left(\frac{W(N)}{W(0)} \right) = \lim_{N \to \infty} \frac{1}{N} \sum_{t=1}^{N} \log \lambda(t) = \mathbf{E} \log \lambda(t) = \sum_{j=1}^{m} \pi_j \log(p_j^T x).
$$

In other words, with investment strategy x, the long term growth rate is given by

$$R_{\mathrm{lt}} = \sum_{j=1}^{m} \pi_j \log(p_j^T x).$$

The investment strategy x that maximizes this quantity is called the *log-optimal invest-ment strategy*, and can be found by solving the optimization problem

$$\begin{array}{ll} \text{maximize} & \sum_{j=1}^{m} \pi_j \log(p_j^T x) \\ \text{subject to} & x \succeq 0, \quad \mathbf{1}^T x = 1, \end{array}$$

with variable $x \in \mathbf{R}^n$.

Show that this is a convex optimization problem.

4.61 *Optimization with logistic model.* A random variable $X \in \{0, 1\}$ satisfies

$$\mathbf{prob}(X = 1) = p = \frac{\exp(a^T x + b)}{1 + \exp(a^T x + b)},$$

where $x \in \mathbf{R}^n$ is a vector of variables that affect the probability, and a and b are known parameters. We can think of $X = 1$ as the event that a consumer buys a product, and x as a vector of variables that affect the probability, *e.g.*, advertising effort, retail price, discounted price, packaging expense, and other factors. The variable x, which we are to optimize over, is subject to a set of linear constraints, $Fx \preceq g$.

Formulate the following problems as convex optimization problems.

(a) *Maximizing buying probability.* The goal is to choose x to maximize p.

(b) *Maximizing expected profit.* Let $c^T x + d$ be the profit derived from selling the product, which we assume is positive for all feasible x. The goal is to maximize the expected profit, which is $p(c^T x + d)$.

4.62 *Optimal power and bandwidth allocation in a Gaussian broadcast channel.* We consider a communication system in which a central node transmits messages to n receivers. ('Gaussian' refers to the type of noise that corrupts the transmissions.) Each receiver channel is characterized by its (transmit) power level $P_i \geq 0$ and its bandwidth $W_i \geq 0$. The power and bandwidth of a receiver channel determine its *bit rate* R_i (the rate at which information can be sent) via

$$R_i = \alpha_i W_i \log(1 + \beta_i P_i / W_i),$$

where α_i and β_i are known positive constants. For $W_i = 0$, we take $R_i = 0$ (which is what you get if you take the limit as $W_i \to 0$).

The powers must satisfy a total power constraint, which has the form

$$P_1 + \cdots + P_n = P_{\mathrm{tot}},$$

where $P_{\mathrm{tot}} > 0$ is a given total power available to allocate among the channels. Similarly, the bandwidths must satisfy

$$W_1 + \cdots + W_n = W_{\mathrm{tot}},$$

where $W_{\mathrm{tot}} > 0$ is the (given) total available bandwidth. The optimization variables in this problem are the powers and bandwidths, *i.e.*, $P_1, \ldots, P_n, W_1, \ldots, W_n$.

The objective is to maximize the total utility,

$$\sum_{i=1}^{n} u_i(R_i),$$

where $u_i : \mathbf{R} \to \mathbf{R}$ is the utility function associated with the ith receiver. (You can think of $u_i(R_i)$ as the revenue obtained for providing a bit rate R_i to receiver i, so the objective is to maximize the total revenue.) You can assume that the utility functions u_i are nondecreasing and concave.

Pose this problem as a convex optimization problem.

4.63 *Optimally balancing manufacturing cost and yield.* The vector $x \in \mathbf{R}^n$ denotes the nominal parameters in a manufacturing process. The yield of the process, *i.e.*, the fraction of manufactured goods that is acceptable, is given by $Y(x)$. We assume that Y is log-concave (which is often the case; see example 3.43). The cost per unit to manufacture the product is given by $c^T x$, where $c \in \mathbf{R}^n$. The cost per acceptable unit is $c^T x / Y(x)$. We want to minimize $c^T x / Y(x)$, subject to some convex constraints on x such as a linear inequalities $Ax \preceq b$. (You can assume that over the feasible set we have $c^T x > 0$ and $Y(x) > 0$.)

This problem is *not* a convex or quasiconvex optimization problem, but it can be solved using convex optimization and a one-dimensional search. The basic ideas are given below; you must supply all details and justification.

(a) Show that the function $f : \mathbf{R} \to \mathbf{R}$ given by

$$f(a) = \sup\{Y(x) \mid Ax \preceq b, \ c^T x = a\},$$

which gives the maximum yield versus cost, is log-concave. This means that by solving a convex optimization problem (in x) we can evaluate the function f.

(b) Suppose that we evaluate the function f for enough values of a to give a good approximation over the range of interest. Explain how to use these data to (approximately) solve the problem of minimizing cost per good product.

4.64 *Optimization with recourse.* In an optimization problem with recourse, also called *two-stage optimization*, the cost function and constraints depend not only on our choice of variables, but also on a discrete random variable $s \in \{1, \ldots, S\}$, which is interpreted as specifying which of S *scenarios* occurred. The scenario random variable s has known probability distribution π, with $\pi_i = \mathbf{prob}(s = i)$, $i = 1, \ldots, S$.

In two-stage optimization, we are to choose the values of two variables, $x \in \mathbf{R}^n$ and $z \in \mathbf{R}^q$. The variable x must be chosen *before* the particular scenario s is known; the variable z, however, is chosen *after* the value of the scenario random variable is known. In other words, z is a function of the scenario random variable s. To describe our choice z, we list the values we would choose under the different scenarios, *i.e.*, we list the vectors

$$z_1, \ldots, z_S \in \mathbf{R}^q.$$

Here z_3 is our choice of z when $s = 3$ occurs, and so on. The set of values

$$x \in \mathbf{R}^n, \qquad z_1, \ldots, z_S \in \mathbf{R}^q$$

is called the *policy*, since it tells us what choice to make for x (independent of which scenario occurs), and also, what choice to make for z in each possible scenario.

The variable z is called the *recourse variable* (or *second-stage variable*), since it allows us to take some action or make a choice after we know which scenario occurred. In contrast, our choice of x (which is called the *first-stage variable*) must be made without any knowledge of the scenario.

For simplicity we will consider the case with no constraints. The cost function is given by

$$f : \mathbf{R}^n \times \mathbf{R}^q \times \{1, \ldots, S\} \to \mathbf{R},$$

where $f(x, z, i)$ gives the cost when the first-stage choice x is made, second-stage choice z is made, and scenario i occurs. We will take as the overall objective, to be minimized over all policies, the expected cost

$$\mathbf{E}\, f(x, z_s, s) = \sum_{i=1}^{S} \pi_i f(x, z_i, i).$$

Suppose that f is a convex function of (x, z), for each scenario $i = 1, \dots, S$. Explain how to find an optimal policy, *i.e.*, one that minimizes the expected cost over all possible policies, using convex optimization.

4.65 *Optimal operation of a hybrid vehicle.* A hybrid vehicle has an internal combustion engine, a motor/generator connected to a storage battery, and a conventional (friction) brake. In this exercise we consider a (highly simplified) model of a *parallel hybrid vehicle*, in which both the motor/generator and the engine are directly connected to the drive wheels. The engine can provide power to the wheels, and the brake can take power from the wheels, turning it into heat. The motor/generator can act as a motor, when it uses energy stored in the battery to deliver power to the wheels, or as a generator, when it takes power from the wheels or engine, and uses the power to charge the battery. When the generator takes power from the wheels and charges the battery, it is called *regenerative braking*; unlike ordinary friction braking, the energy taken from the wheels is *stored*, and can be used later. The vehicle is judged by driving it over a known, fixed test track to evaluate its fuel efficiency.

A diagram illustrating the power flow in the hybrid vehicle is shown below. The arrows indicate the direction in which the power flow is considered positive. The engine power p_{eng}, for example, is positive when it is delivering power; the brake power p_{br} is positive when it is taking power from the wheels. The power p_{req} is the required power at the wheels. It is positive when the wheels require power (*e.g.*, when the vehicle accelerates, climbs a hill, or cruises on level terrain). The required wheel power is negative when the vehicle must decelerate rapidly, or descend a hill.

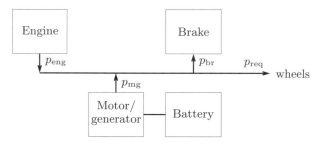

All of these powers are functions of time, which we discretize in one second intervals, with $t = 1, 2, \dots, T$. The required wheel power $p_{\mathrm{req}}(1), \dots, p_{\mathrm{req}}(T)$ is given. (The speed of the vehicle on the track is specified, so together with known road slope information, and known aerodynamic and other losses, the power required at the wheels can be calculated.) Power is conserved, which means we have

$$p_{\mathrm{req}}(t) = p_{\mathrm{eng}}(t) + p_{\mathrm{mg}}(t) - p_{\mathrm{br}}(t), \quad t = 1, \dots, T.$$

The brake can only dissipate power, so we have $p_{\mathrm{br}}(t) \geq 0$ for each t. The engine can only provide power, and only up to a given limit $P_{\mathrm{eng}}^{\mathrm{max}}$, *i.e.*, we have

$$0 \leq p_{\mathrm{eng}}(t) \leq P_{\mathrm{eng}}^{\mathrm{max}}, \quad t = 1, \dots, T.$$

The motor/generator power is also limited: p_{mg} must satisfy

$$P_{\mathrm{mg}}^{\mathrm{min}} \leq p_{\mathrm{mg}}(t) \leq P_{\mathrm{mg}}^{\mathrm{max}}, \quad t = 1, \dots, T.$$

Here $P_{\mathrm{mg}}^{\mathrm{max}} > 0$ is the maximum motor power, and $-P_{\mathrm{mg}}^{\mathrm{min}} > 0$ is the maximum generator power.

The battery charge or energy at time t is denoted $E(t)$, $t = 1, \dots, T + 1$. The battery energy satisfies

$$E(t + 1) = E(t) - p_{\mathrm{mg}}(t) - \eta |p_{\mathrm{mg}}(t)|, \quad t = 1, \dots, T,$$

where $\eta > 0$ is a known parameter. (The term $-p_{\mathrm{mg}}(t)$ represents the energy removed or added the battery by the motor/generator, ignoring any losses. The term $-\eta|p_{\mathrm{mg}}(t)|$ represents energy lost through inefficiencies in the battery or motor/generator.)

The battery charge must be between 0 (empty) and its limit $E_{\mathrm{batt}}^{\max}$ (full), at all times. (If $E(t) = 0$, the battery is fully discharged, and no more energy can be extracted from it; when $E(t) = E_{\mathrm{batt}}^{\max}$, the battery is full and cannot be charged.) To make the comparison with non-hybrid vehicles fair, we fix the initial battery charge to equal the final battery charge, so the net energy change is zero over the track: $E(1) = E(T+1)$. We do not specify the value of the initial (and final) energy.

The objective in the problem is the total fuel consumed by the engine, which is

$$F_{\mathrm{total}} = \sum_{t=1}^{T} F(p_{\mathrm{eng}}(t)),$$

where $F : \mathbf{R} \to \mathbf{R}$ is the *fuel use characteristic* of the engine. We assume that F is positive, increasing, and convex.

Formulate this problem as a convex optimization problem, with variables $p_{\mathrm{eng}}(t)$, $p_{\mathrm{mg}}(t)$, and $p_{\mathrm{br}}(t)$ for $t = 1, \ldots, T$, and $E(t)$ for $t = 1, \ldots, T+1$. Explain why your formulation is equivalent to the problem described above.

Chapter 5

Duality

5.1 The Lagrange dual function

5.1.1 The Lagrangian

We consider an optimization problem in the standard form (4.1):

$$
\begin{array}{ll}
\text{minimize} & f_0(x) \\
\text{subject to} & f_i(x) \le 0, \quad i = 1, \dots, m \\
& h_i(x) = 0, \quad i = 1, \dots, p,
\end{array}
\qquad (5.1)
$$

with variable $x \in \mathbf{R}^n$. We assume its domain $\mathcal{D} = \bigcap_{i=0}^{m} \mathbf{dom}\, f_i \cap \bigcap_{i=1}^{p} \mathbf{dom}\, h_i$ is nonempty, and denote the optimal value of (5.1) by $p^\star$. We do not assume the problem (5.1) is convex.

The basic idea in Lagrangian duality is to take the constraints in (5.1) into account by augmenting the objective function with a weighted sum of the constraint functions. We define the *Lagrangian* $L : \mathbf{R}^n \times \mathbf{R}^m \times \mathbf{R}^p \to \mathbf{R}$ associated with the problem (5.1) as

$$
L(x, \lambda, \nu) = f_0(x) + \sum_{i=1}^{m} \lambda_i f_i(x) + \sum_{i=1}^{p} \nu_i h_i(x),
$$

with $\mathbf{dom}\, L = \mathcal{D} \times \mathbf{R}^m \times \mathbf{R}^p$. We refer to λ_i as the *Lagrange multiplier* associated with the ith inequality constraint $f_i(x) \le 0$; similarly we refer to ν_i as the Lagrange multiplier associated with the ith equality constraint $h_i(x) = 0$. The vectors λ and ν are called the *dual variables* or *Lagrange multiplier vectors* associated with the problem (5.1).

5.1.2   The Lagrange dual function

We define the *Lagrange dual function* (or just *dual function*) $g : \mathbf{R}^m \times \mathbf{R}^p \to \mathbf{R}$ as the minimum value of the Lagrangian over x: for $\lambda \in \mathbf{R}^m$, $\nu \in \mathbf{R}^p$,

$$g(\lambda, \nu) = \inf_{x \in \mathcal{D}} L(x, \lambda, \nu) = \inf_{x \in \mathcal{D}} \left(f_0(x) + \sum_{i=1}^{m} \lambda_i f_i(x) + \sum_{i=1}^{p} \nu_i h_i(x) \right).$$

When the Lagrangian is unbounded below in x, the dual function takes on the value $-\infty$. Since the dual function is the pointwise infimum of a family of affine functions of (λ, ν), it is concave, even when the problem (5.1) is not convex.

5.1.3   Lower bounds on optimal value

The dual function yields lower bounds on the optimal value $p^\star$ of the problem (5.1): For any $\lambda \succeq 0$ and any ν we have

$$g(\lambda, \nu) \leq p^\star. \tag{5.2}$$

This important property is easily verified. Suppose $\tilde{x}$ is a feasible point for the problem (5.1), *i.e.*, $f_i(\tilde{x}) \leq 0$ and $h_i(\tilde{x}) = 0$, and $\lambda \succeq 0$. Then we have

$$\sum_{i=1}^{m} \lambda_i f_i(\tilde{x}) + \sum_{i=1}^{p} \nu_i h_i(\tilde{x}) \leq 0,$$

since each term in the first sum is nonpositive, and each term in the second sum is zero, and therefore

$$L(\tilde{x}, \lambda, \nu) = f_0(\tilde{x}) + \sum_{i=1}^{m} \lambda_i f_i(\tilde{x}) + \sum_{i=1}^{p} \nu_i h_i(\tilde{x}) \leq f_0(\tilde{x}).$$

Hence
$$g(\lambda, \nu) = \inf_{x \in \mathcal{D}} L(x, \lambda, \nu) \leq L(\tilde{x}, \lambda, \nu) \leq f_0(\tilde{x}).$$

Since $g(\lambda, \nu) \leq f_0(\tilde{x})$ holds for every feasible point $\tilde{x}$, the inequality (5.2) follows. The lower bound (5.2) is illustrated in figure 5.1, for a simple problem with $x \in \mathbf{R}$ and one inequality constraint.

The inequality (5.2) holds, but is vacuous, when $g(\lambda, \nu) = -\infty$. The dual function gives a nontrivial lower bound on $p^\star$ only when $\lambda \succeq 0$ and $(\lambda, \nu) \in \mathbf{dom}\, g$, *i.e.*, $g(\lambda, \nu) > -\infty$. We refer to a pair (λ, ν) with $\lambda \succeq 0$ and $(\lambda, \nu) \in \mathbf{dom}\, g$ as *dual feasible*, for reasons that will become clear later.

5.1.4   Linear approximation interpretation

The Lagrangian and lower bound property can be given a simple interpretation, based on a linear approximation of the indicator functions of the sets $\{0\}$ and $-\mathbf{R}_+$.

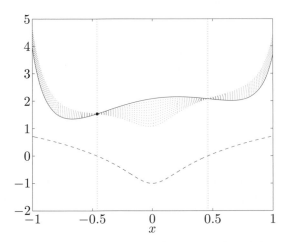

Figure 5.1 *Lower bound from a dual feasible point.* The solid curve shows the objective function f_0, and the dashed curve shows the constraint function f_1. The feasible set is the interval $[-0.46, 0.46]$, which is indicated by the two dotted vertical lines. The optimal point and value are $x^\star = -0.46$, $p^\star = 1.54$ (shown as a circle). The dotted curves show $L(x, \lambda)$ for $\lambda = 0.1, 0.2, \ldots, 1.0$. Each of these has a minimum value smaller than $p^\star$, since on the feasible set (and for $\lambda \geq 0$) we have $L(x, \lambda) \leq f_0(x)$.

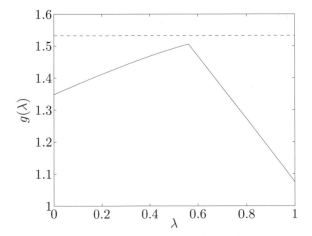

Figure 5.2 The dual function g for the problem in figure 5.1. Neither f_0 nor f_1 is convex, but the dual function is concave. The horizontal dashed line shows $p^\star$, the optimal value of the problem.

We first rewrite the original problem (5.1) as an unconstrained problem,

$$\text{minimize} \quad f_0(x) + \sum_{i=1}^{m} I_-(f_i(x)) + \sum_{i=1}^{p} I_0(h_i(x)), \qquad (5.3)$$

where $I_- : \mathbf{R} \to \mathbf{R}$ is the indicator function for the nonpositive reals,

$$I_-(u) = \begin{cases} 0 & u \le 0 \\ \infty & u > 0, \end{cases}$$

and similarly, I_0 is the indicator function of $\{0\}$. In the formulation (5.3), the function $I_-(u)$ can be interpreted as expressing our irritation or displeasure associated with a constraint function value $u = f_i(x)$: It is zero if $f_i(x) \le 0$, and infinite if $f_i(x) > 0$. In a similar way, $I_0(u)$ gives our displeasure for an equality constraint value $u = h_i(x)$. We can think of I_- as a "brick wall" or "infinitely hard" displeasure function; our displeasure rises from zero to infinite as $f_i(x)$ transitions from nonpositive to positive.

Now suppose in the formulation (5.3) we replace the function $I_-(u)$ with the linear function $\lambda_i u$, where $\lambda_i \ge 0$, and the function $I_0(u)$ with $\nu_i u$. The objective becomes the Lagrangian function $L(x, \lambda, \nu)$, and the dual function value $g(\lambda, \nu)$ is the optimal value of the problem

$$\text{minimize} \quad L(x, \lambda, \nu) = f_0(x) + \sum_{i=1}^{m} \lambda_i f_i(x) + \sum_{i=1}^{p} \nu_i h_i(x). \qquad (5.4)$$

In this formulation, we use a linear or "soft" displeasure function in place of I_- and I_0. For an inequality constraint, our displeasure is zero when $f_i(x) = 0$, and is positive when $f_i(x) > 0$ (assuming $\lambda_i > 0$); our displeasure grows as the constraint becomes "more violated". Unlike the original formulation, in which any nonpositive value of $f_i(x)$ is acceptable, in the soft formulation we actually derive pleasure from constraints that have margin, *i.e.*, from $f_i(x) < 0$.

Clearly the approximation of the indicator function $I_-(u)$ with a linear function $\lambda_i u$ is rather poor. But the linear function is at least an *underestimator* of the indicator function. Since $\lambda_i u \le I_-(u)$ and $\nu_i u \le I_0(u)$ for all u, we see immediately that the dual function yields a lower bound on the optimal value of the original problem.

The idea of replacing the "hard" constraints with "soft" versions will come up again when we consider interior-point methods (§11.2.1).

5.1.5   Examples

In this section we give some examples for which we can derive an analytical expression for the Lagrange dual function.

Least-squares solution of linear equations

We consider the problem

$$\begin{array}{ll} \text{minimize} & x^T x \\ \text{subject to} & Ax = b, \end{array} \qquad (5.5)$$

where $A \in \mathbf{R}^{p \times n}$. This problem has no inequality constraints and p (linear) equality constraints. The Lagrangian is $L(x, \nu) = x^T x + \nu^T (Ax - b)$, with domain $\mathbf{R}^n \times$

$\mathbf{R}^p$. The dual function is given by $g(\nu) = \inf_x L(x, \nu)$. Since $L(x, \nu)$ is a convex quadratic function of x, we can find the minimizing x from the optimality condition

$$\nabla_x L(x, \nu) = 2x + A^T \nu = 0,$$

which yields $x = -(1/2)A^T \nu$. Therefore the dual function is

$$g(\nu) = L(-(1/2)A^T \nu, \nu) = -(1/4)\nu^T AA^T \nu - b^T \nu,$$

which is a concave quadratic function, with domain $\mathbf{R}^p$. The lower bound property (5.2) states that for any $\nu \in \mathbf{R}^p$, we have

$$-(1/4)\nu^T AA^T \nu - b^T \nu \leq \inf\{x^T x \mid Ax = b\}.$$

Standard form LP

Consider an LP in standard form,

$$\begin{array}{ll} \text{minimize} & c^T x \\ \text{subject to} & Ax = b \\ & x \succeq 0, \end{array} \qquad (5.6)$$

which has inequality constraint functions $f_i(x) = -x_i$, $i = 1, \ldots, n$. To form the Lagrangian we introduce multipliers λ_i for the n inequality constraints and multipliers ν_i for the equality constraints, and obtain

$$L(x, \lambda, \nu) = c^T x - \sum_{i=1}^{n} \lambda_i x_i + \nu^T (Ax - b) = -b^T \nu + (c + A^T \nu - \lambda)^T x.$$

The dual function is

$$g(\lambda, \nu) = \inf_x L(x, \lambda, \nu) = -b^T \nu + \inf_x (c + A^T \nu - \lambda)^T x,$$

which is easily determined analytically, since a linear function is bounded below only when it is identically zero. Thus, $g(\lambda, \nu) = -\infty$ except when $c + A^T \nu - \lambda = 0$, in which case it is $-b^T \nu$:

$$g(\lambda, \nu) = \begin{cases} -b^T \nu & A^T \nu - \lambda + c = 0 \\ -\infty & \text{otherwise.} \end{cases}$$

Note that the dual function g is finite only on a proper affine subset of $\mathbf{R}^m \times \mathbf{R}^p$. We will see that this is a common occurrence.

The lower bound property (5.2) is nontrivial only when λ and ν satisfy $\lambda \succeq 0$ and $A^T \nu - \lambda + c = 0$. When this occurs, $-b^T \nu$ is a lower bound on the optimal value of the LP (5.6).

Two-way partitioning problem

We consider the (nonconvex) problem

$$\begin{array}{ll} \text{minimize} & x^T W x \\ \text{subject to} & x_i^2 = 1, \quad i = 1, \ldots, n, \end{array} \qquad (5.7)$$

where $W \in \mathbf{S}^n$. The constraints restrict the values of x_i to 1 or -1, so the problem is equivalent to finding the vector with components ± 1 that minimizes $x^T W x$. The feasible set here is finite (it contains 2^n points) so this problem can in principle be solved by simply checking the objective value of each feasible point. Since the number of feasible points grows exponentially, however, this is possible only for small problems (say, with $n \leq 30$). In general (and for n larger than, say, 50) the problem (5.7) is very difficult to solve.

We can interpret the problem (5.7) as a two-way partitioning problem on a set of n elements, say, $\{1, \ldots, n\}$: A feasible x corresponds to the partition

$$\{1, \ldots, n\} \;=\; \{i \mid x_i = -1\} \;\cup\; \{i \mid x_i = 1\}.$$

The matrix coefficient W_{ij} can be interpreted as the cost of having the elements i and j in the same partition, and $-W_{ij}$ is the cost of having i and j in different partitions. The objective in (5.7) is the total cost, over all pairs of elements, and the problem (5.7) is to find the partition with least total cost.

We now derive the dual function for this problem. The Lagrangian is

$$
\begin{aligned}
L(x, \nu) \;&=\; x^T W x + \sum_{i=1}^n \nu_i (x_i^2 - 1) \\
&=\; x^T (W + \mathbf{diag}(\nu)) x - \mathbf{1}^T \nu.
\end{aligned}
$$

We obtain the Lagrange dual function by minimizing over x:

$$
\begin{aligned}
g(\nu) \;&=\; \inf_x x^T (W + \mathbf{diag}(\nu)) x - \mathbf{1}^T \nu \\
&=\; \begin{cases} -\mathbf{1}^T \nu & W + \mathbf{diag}(\nu) \succeq 0 \\ -\infty & \text{otherwise,} \end{cases}
\end{aligned}
$$

where we use the fact that the infimum of a quadratic form is either zero (if the form is positive semidefinite) or $-\infty$ (if the form is not positive semidefinite).

This dual function provides lower bounds on the optimal value of the difficult problem (5.7). For example, we can take the specific value of the dual variable

$$\nu = -\lambda_{\min}(W) \mathbf{1},$$

which is dual feasible, since

$$W + \mathbf{diag}(\nu) = W - \lambda_{\min}(W) I \succeq 0.$$

This yields the bound on the optimal value $p^\star$

$$p^\star \geq -\mathbf{1}^T \nu = n \lambda_{\min}(W). \tag{5.8}$$

Remark 5.1 This lower bound on $p^\star$ can also be obtained without using the Lagrange dual function. First, we replace the constraints $x_1^2 = 1, \ldots, x_n^2 = 1$ with $\sum_{i=1}^n x_i^2 = n$, to obtain the modified problem

$$
\begin{array}{ll}
\text{minimize} & x^T W x \\
\text{subject to} & \sum_{i=1}^n x_i^2 = n.
\end{array} \tag{5.9}
$$

The constraints of the original problem (5.7) imply the constraint here, so the optimal value of the problem (5.9) is a lower bound on $p^\star$, the optimal value of (5.7). But the modified problem (5.9) is easily solved as an eigenvalue problem, with optimal value $n\lambda_{\min}(W)$.

5.1.6   The Lagrange dual function and conjugate functions

Recall from §3.3 that the conjugate f^* of a function $f : \mathbf{R}^n \to \mathbf{R}$ is given by

$$f^*(y) = \sup_{x \in \mathbf{dom} f} \left(y^T x - f(x)\right).$$

The conjugate function and Lagrange dual function are closely related. To see one simple connection, consider the problem

$$\begin{array}{ll} \text{minimize} & f(x) \\ \text{subject to} & x = 0 \end{array}$$

(which is not very interesting, and solvable by inspection). This problem has Lagrangian $L(x, \nu) = f(x) + \nu^T x$, and dual function

$$g(\nu) = \inf_x \left(f(x) + \nu^T x\right) = -\sup_x \left((-\nu)^T x - f(x)\right) = -f^*(-\nu).$$

More generally (and more usefully), consider an optimization problem with linear inequality and equality constraints,

$$\begin{array}{ll} \text{minimize} & f_0(x) \\ \text{subject to} & Ax \preceq b \\ & Cx = d. \end{array} \tag{5.10}$$

Using the conjugate of f_0 we can write the dual function for the problem (5.10) as

$$\begin{aligned} g(\lambda, \nu) &= \inf_x \left(f_0(x) + \lambda^T(Ax - b) + \nu^T(Cx - d)\right) \\ &= -b^T \lambda - d^T \nu + \inf_x \left(f_0(x) + (A^T \lambda + C^T \nu)^T x\right) \\ &= -b^T \lambda - d^T \nu - f_0^*(-A^T \lambda - C^T \nu). \end{aligned} \tag{5.11}$$

The domain of g follows from the domain of f_0^*:

$$\mathbf{dom}\, g = \{(\lambda, \nu) \mid -A^T \lambda - C^T \nu \in \mathbf{dom}\, f_0^*\}.$$

Let us illustrate this with a few examples.

Equality constrained norm minimization

Consider the problem

$$\begin{array}{ll} \text{minimize} & \|x\| \\ \text{subject to} & Ax = b, \end{array} \tag{5.12}$$

where $\|\cdot\|$ is any norm. Recall (from example 3.26 on page 93) that the conjugate of $f_0 = \|\cdot\|$ is given by

$$f_0^*(y) = \begin{cases} 0 & \|y\|_* \le 1 \\ \infty & \text{otherwise,} \end{cases}$$

the indicator function of the dual norm unit ball.

Using the result (5.11) above, the dual function for the problem (5.12) is given by

$$g(\nu) = -b^T\nu - f_0^*(-A^T\nu) = \begin{cases} -b^T\nu & \|A^T\nu\|_* \le 1 \\ -\infty & \text{otherwise.} \end{cases}$$

Entropy maximization

Consider the entropy maximization problem

$$\begin{array}{ll} \text{minimize} & f_0(x) = \sum_{i=1}^n x_i \log x_i \\ \text{subject to} & Ax \preceq b \\ & \mathbf{1}^T x = 1 \end{array} \qquad (5.13)$$

where $\mathbf{dom}\, f_0 = \mathbf{R}_{++}^n$. The conjugate of the negative entropy function $u \log u$, with scalar variable u, is e^{v-1} (see example 3.21 on page 91). Since f_0 is a sum of negative entropy functions of different variables, we conclude that its conjugate is

$$f_0^*(y) = \sum_{i=1}^n e^{y_i - 1},$$

with $\mathbf{dom}\, f_0^* = \mathbf{R}^n$. Using the result (5.11) above, the dual function of (5.13) is given by

$$g(\lambda, \nu) = -b^T\lambda - \nu - \sum_{i=1}^n e^{-a_i^T\lambda - \nu - 1} = -b^T\lambda - \nu - e^{-\nu-1}\sum_{i=1}^n e^{-a_i^T\lambda}$$

where a_i is the ith column of A.

Minimum volume covering ellipsoid

Consider the problem with variable $X \in \mathbf{S}^n$,

$$\begin{array}{ll} \text{minimize} & f_0(X) = \log\det X^{-1} \\ \text{subject to} & a_i^T X a_i \le 1, \quad i = 1, \dots, m, \end{array} \qquad (5.14)$$

where $\mathbf{dom}\, f_0 = \mathbf{S}_{++}^n$. The problem (5.14) has a simple geometric interpretation. With each $X \in \mathbf{S}_{++}^n$ we associate the ellipsoid, centered at the origin,

$$\mathcal{E}_X = \{z \mid z^T X z \le 1\}.$$

The volume of this ellipsoid is proportional to $\left(\det X^{-1}\right)^{1/2}$, so the objective of (5.14) is, except for a constant and a factor of two, the logarithm of the volume of $\mathcal{E}_X$. The constraints of the problem (5.14) are that $a_i \in \mathcal{E}_X$. Thus the problem (5.14) is to determine the minimum volume ellipsoid, centered at the origin, that includes the points $a_1, \ldots, a_m$.

The inequality constraints in problem (5.14) are affine; they can be expressed as

$$\mathbf{tr}\left((a_i a_i^T) X\right) \leq 1.$$

In example 3.23 (page 92) we found that the conjugate of f_0 is

$$f_0^*(Y) = \log \det(-Y)^{-1} - n,$$

with $\mathbf{dom}\, f_0^* = -\mathbf{S}_{++}^n$. Applying the result (5.11) above, the dual function for the problem (5.14) is given by

$$g(\lambda) = \begin{cases} \log \det \left(\sum_{i=1}^m \lambda_i a_i a_i^T\right) - \mathbf{1}^T \lambda + n & \sum_{i=1}^m \lambda_i a_i a_i^T \succ 0 \\ -\infty & \text{otherwise.} \end{cases} \tag{5.15}$$

Thus, for any $\lambda \succeq 0$ with $\sum_{i=1}^m \lambda_i a_i a_i^T \succ 0$, the number

$$\log \det \left(\sum_{i=1}^m \lambda_i a_i a_i^T\right) - \mathbf{1}^T \lambda + n$$

is a lower bound on the optimal value of the problem (5.14).

5.2 The Lagrange dual problem

For each pair (λ, ν) with $\lambda \succeq 0$, the Lagrange dual function gives us a lower bound on the optimal value $p^\star$ of the optimization problem (5.1). Thus we have a lower bound that depends on some parameters λ, ν. A natural question is: What is the *best* lower bound that can be obtained from the Lagrange dual function?

This leads to the optimization problem

$$\begin{array}{ll} \text{maximize} & g(\lambda, \nu) \\ \text{subject to} & \lambda \succeq 0. \end{array} \tag{5.16}$$

This problem is called the *Lagrange dual problem* associated with the problem (5.1). In this context the original problem (5.1) is sometimes called the *primal problem*. The term *dual feasible*, to describe a pair (λ, ν) with $\lambda \succeq 0$ and $g(\lambda, \nu) > -\infty$, now makes sense. It means, as the name implies, that (λ, ν) is feasible for the dual problem (5.16). We refer to $(\lambda^\star, \nu^\star)$ as *dual optimal* or *optimal Lagrange multipliers* if they are optimal for the problem (5.16).

The Lagrange dual problem (5.16) is a convex optimization problem, since the objective to be maximized is concave and the constraint is convex. This is the case whether or not the primal problem (5.1) is convex.

5.2.1   Making dual constraints explicit

The examples above show that it is not uncommon for the domain of the dual function,

$$\mathbf{dom}\,g = \{(\lambda, \nu) \mid g(\lambda, \nu) > -\infty\},$$

to have dimension smaller than $m + p$. In many cases we can identify the affine hull of $\mathbf{dom}\,g$, and describe it as a set of linear equality constraints. Roughly speaking, this means we can identify the equality constraints that are 'hidden' or 'implicit' in the objective g of the dual problem (5.16). In this case we can form an equivalent problem, in which these equality constraints are given explicitly as constraints. The following examples demonstrate this idea.

Lagrange dual of standard form LP

On page 219 we found that the Lagrange dual function for the standard form LP

$$\begin{array}{ll}
\text{minimize} & c^T x \\
\text{subject to} & Ax = b \\
& x \succeq 0
\end{array} \tag{5.17}$$

is given by

$$g(\lambda, \nu) = \left\{ \begin{array}{ll} -b^T \nu & A^T \nu - \lambda + c = 0 \\ -\infty & \text{otherwise.} \end{array} \right.$$

Strictly speaking, the Lagrange dual problem of the standard form LP is to maximize this dual function g subject to $\lambda \succeq 0$, *i.e.*,

$$\begin{array}{ll}
\text{maximize} & g(\lambda, \nu) = \left\{ \begin{array}{ll} -b^T \nu & A^T \nu - \lambda + c = 0 \\ -\infty & \text{otherwise} \end{array} \right. \\
\text{subject to} & \lambda \succeq 0.
\end{array} \tag{5.18}$$

Here g is finite only when $A^T \nu - \lambda + c = 0$. We can form an equivalent problem by making these equality constraints explicit:

$$\begin{array}{ll}
\text{maximize} & -b^T \nu \\
\text{subject to} & A^T \nu - \lambda + c = 0 \\
& \lambda \succeq 0.
\end{array} \tag{5.19}$$

This problem, in turn, can be expressed as

$$\begin{array}{ll}
\text{maximize} & -b^T \nu \\
\text{subject to} & A^T \nu + c \succeq 0,
\end{array} \tag{5.20}$$

which is an LP in inequality form.

Note the subtle distinctions between these three problems. The Lagrange dual of the standard form LP (5.17) is the problem (5.18), which is equivalent to (but not the same as) the problems (5.19) and (5.20). With some abuse of terminology, we refer to the problem (5.19) or the problem (5.20) as the Lagrange dual of the standard form LP (5.17).

Lagrange dual of inequality form LP

In a similar way we can find the Lagrange dual problem of a linear program in inequality form

$$\begin{array}{ll} \text{minimize} & c^T x \\ \text{subject to} & Ax \preceq b. \end{array} \tag{5.21}$$

The Lagrangian is

$$L(x, \lambda) = c^T x + \lambda^T (Ax - b) = -b^T \lambda + (A^T \lambda + c)^T x,$$

so the dual function is

$$g(\lambda) = \inf_x L(x, \lambda) = -b^T \lambda + \inf_x (A^T \lambda + c)^T x.$$

The infimum of a linear function is $-\infty$, except in the special case when it is identically zero, so the dual function is

$$g(\lambda) = \left\{ \begin{array}{ll} -b^T \lambda & A^T \lambda + c = 0 \\ -\infty & \text{otherwise.} \end{array} \right.$$

The dual variable λ is dual feasible if $\lambda \succeq 0$ and $A^T \lambda + c = 0$.

The Lagrange dual of the LP (5.21) is to maximize g over all $\lambda \succeq 0$. Again we can reformulate this by explicitly including the dual feasibility conditions as constraints, as in

$$\begin{array}{ll} \text{maximize} & -b^T \lambda \\ \text{subject to} & A^T \lambda + c = 0 \\ & \lambda \succeq 0, \end{array} \tag{5.22}$$

which is an LP in standard form.

Note the interesting symmetry between the standard and inequality form LPs and their duals: The dual of a standard form LP is an LP with only inequality constraints, and vice versa. One can also verify that the Lagrange dual of (5.22) is (equivalent to) the primal problem (5.21).

5.2.2   Weak duality

The optimal value of the Lagrange dual problem, which we denote $d^\star$, is, by definition, the best lower bound on $p^\star$ that can be obtained from the Lagrange dual function. In particular, we have the simple but important inequality

$$d^\star \leq p^\star, \tag{5.23}$$

which holds even if the original problem is not convex. This property is called *weak duality*.

The weak duality inequality (5.23) holds when $d^\star$ and $p^\star$ are infinite. For example, if the primal problem is unbounded below, so that $p^\star = -\infty$, we must have $d^\star = -\infty$, *i.e.*, the Lagrange dual problem is infeasible. Conversely, if the dual problem is unbounded above, so that $d^\star = \infty$, we must have $p^\star = \infty$, *i.e.*, the primal problem is infeasible.

We refer to the difference $p^\star - d^\star$ as the *optimal duality gap* of the original problem, since it gives the gap between the optimal value of the primal problem and the best (*i.e.*, greatest) lower bound on it that can be obtained from the Lagrange dual function. The optimal duality gap is always nonnegative.

The bound (5.23) can sometimes be used to find a lower bound on the optimal value of a problem that is difficult to solve, since the dual problem is always convex, and in many cases can be solved efficiently, to find $d^\star$. As an example, consider the two-way partitioning problem (5.7) described on page 219. The dual problem is an SDP,

$$\begin{array}{ll} \text{maximize} & -\mathbf{1}^T \nu \\ \text{subject to} & W + \mathbf{diag}(\nu) \succeq 0, \end{array}$$

with variable $\nu \in \mathbf{R}^n$. This problem can be solved efficiently, even for relatively large values of n, such as $n = 1000$. Its optimal value is a lower bound on the optimal value of the two-way partitioning problem, and is always at least as good as the lower bound (5.8) based on $\lambda_{\min}(W)$.

5.2.3   Strong duality and Slater's constraint qualification

If the equality

$$d^\star = p^\star \qquad\qquad (5.24)$$

holds, *i.e.*, the optimal duality gap is zero, then we say that *strong duality* holds. This means that the best bound that can be obtained from the Lagrange dual function is tight.

Strong duality does not, in general, hold. But if the primal problem (5.1) is convex, *i.e.*, of the form

$$\begin{array}{ll} \text{minimize} & f_0(x) \\ \text{subject to} & f_i(x) \le 0, \quad i = 1, \ldots, m, \\ & Ax = b, \end{array} \qquad (5.25)$$

with $f_0, \ldots, f_m$ convex, we usually (but not always) have strong duality. There are many results that establish conditions on the problem, beyond convexity, under which strong duality holds. These conditions are called *constraint qualifications*.

One simple constraint qualification is *Slater's condition*: There exists an $x \in \mathbf{relint}\,\mathcal{D}$ such that

$$f_i(x) < 0, \quad i = 1, \ldots, m, \qquad Ax = b. \qquad (5.26)$$

Such a point is sometimes called *strictly feasible*, since the inequality constraints hold with strict inequalities. Slater's theorem states that strong duality holds, if Slater's condition holds (and the problem is convex).

Slater's condition can be refined when some of the inequality constraint functions f_i are affine. If the first k constraint functions $f_1, \ldots, f_k$ are affine, then strong duality holds provided the following weaker condition holds: There exists an $x \in \mathbf{relint}\,\mathcal{D}$ with

$$f_i(x) \le 0, \quad i = 1, \ldots, k, \qquad f_i(x) < 0, \quad i = k+1, \ldots, m, \qquad Ax = b. \quad (5.27)$$

In other words, the affine inequalities do not need to hold with strict inequality. Note that the refined Slater condition (5.27) reduces to feasibility when the constraints are all linear equalities and inequalities, and $\mathbf{dom}\, f_0$ is open.

Slater's condition (and the refinement (5.27)) not only implies strong duality for convex problems. It also implies that the dual optimal value is attained when $d^\star > -\infty$, *i.e.*, there exists a dual feasible $(\lambda^\star, \nu^\star)$ with $g(\lambda^\star, \nu^\star) = d^\star = p^\star$. We will prove that strong duality obtains, when the primal problem is convex and Slater's condition holds, in §5.3.2.

5.2.4 Examples

Least-squares solution of linear equations

Recall the problem (5.5):
$$\begin{array}{ll} \text{minimize} & x^T x \\ \text{subject to} & Ax = b. \end{array}$$

The associated dual problem is
$$\text{maximize} \quad -(1/4)\nu^T A A^T \nu - b^T \nu,$$

which is an unconstrained concave quadratic maximization problem.

Slater's condition is simply that the primal problem is feasible, so $p^\star = d^\star$ provided $b \in \mathcal{R}(A)$, *i.e.*, $p^\star < \infty$. In fact for this problem we always have strong duality, even when $p^\star = \infty$. This is the case when $b \notin \mathcal{R}(A)$, so there is a z with $A^T z = 0$, $b^T z \neq 0$. It follows that the dual function is unbounded above along the line $\{tz \mid t \in \mathbf{R}\}$, so $d^\star = \infty$ as well.

Lagrange dual of LP

By the weaker form of Slater's condition, we find that strong duality holds for any LP (in standard or inequality form) provided the primal problem is feasible. Applying this result to the duals, we conclude that strong duality holds for LPs if the dual is feasible. This leaves only one possible situation in which strong duality for LPs can fail: both the primal and dual problems are infeasible. This pathological case can, in fact, occur; see exercise 5.23.

Lagrange dual of QCQP

We consider the QCQP
$$\begin{array}{ll} \text{minimize} & (1/2)x^T P_0 x + q_0^T x + r_0 \\ \text{subject to} & (1/2)x^T P_i x + q_i^T x + r_i \leq 0, \quad i = 1, \dots, m, \end{array} \tag{5.28}$$

with $P_0 \in \mathbf{S}_{++}^n$, and $P_i \in \mathbf{S}_+^n$, $i = 1, \dots, m$. The Lagrangian is
$$L(x, \lambda) = (1/2)x^T P(\lambda)x + q(\lambda)^T x + r(\lambda),$$

where
$$P(\lambda) = P_0 + \sum_{i=1}^m \lambda_i P_i, \qquad q(\lambda) = q_0 + \sum_{i=1}^m \lambda_i q_i, \qquad r(\lambda) = r_0 + \sum_{i=1}^m \lambda_i r_i.$$

It is possible to derive an expression for $g(\lambda)$ for general λ, but it is quite complicated. If $\lambda \succeq 0$, however, we have $P(\lambda) \succ 0$ and

$$g(\lambda) = \inf_x L(x, \lambda) = -(1/2)q(\lambda)^T P(\lambda)^{-1}q(\lambda) + r(\lambda).$$

We can therefore express the dual problem as

$$
\begin{array}{ll}
\text{maximize} & -(1/2)q(\lambda)^T P(\lambda)^{-1}q(\lambda) + r(\lambda) \\
\text{subject to} & \lambda \succeq 0.
\end{array}
\tag{5.29}
$$

The Slater condition says that strong duality between (5.29) and (5.28) holds if the quadratic inequality constraints are strictly feasible, *i.e.*, there exists an x with

$$(1/2)x^T P_i x + q_i^T x + r_i < 0, \quad i = 1, \dots, m.$$

Entropy maximization

Our next example is the entropy maximization problem (5.13):

$$
\begin{array}{ll}
\text{minimize} & \sum_{i=1}^n x_i \log x_i \\
\text{subject to} & Ax \preceq b \\
& \mathbf{1}^T x = 1,
\end{array}
$$

with domain $\mathcal{D} = \mathbf{R}_+^n$. The Lagrange dual function was derived on page 222; the dual problem is

$$
\begin{array}{ll}
\text{maximize} & -b^T \lambda - \nu - e^{-\nu-1} \sum_{i=1}^n e^{-a_i^T \lambda} \\
\text{subject to} & \lambda \succeq 0,
\end{array}
\tag{5.30}
$$

with variables $\lambda \in \mathbf{R}^m$, $\nu \in \mathbf{R}$. The (weaker) Slater condition for (5.13) tells us that the optimal duality gap is zero if there exists an $x \succ 0$ with $Ax \preceq b$ and $\mathbf{1}^T x = 1$.

We can simplify the dual problem (5.30) by maximizing over the dual variable ν analytically. For fixed λ, the objective function is maximized when the derivative with respect to ν is zero, *i.e.*,

$$\nu = \log \sum_{i=1}^n e^{-a_i^T \lambda} - 1.$$

Substituting this optimal value of ν into the dual problem gives

$$
\begin{array}{ll}
\text{maximize} & -b^T \lambda - \log\left(\sum_{i=1}^n e^{-a_i^T \lambda}\right) \\
\text{subject to} & \lambda \succeq 0,
\end{array}
$$

which is a geometric program (in convex form) with nonnegativity constraints.

Minimum volume covering ellipsoid

We consider the problem (5.14):

$$
\begin{array}{ll}
\text{minimize} & \log \det X^{-1} \\
\text{subject to} & a_i^T X a_i \leq 1, \quad i = 1, \dots, m,
\end{array}
$$

with domain $\mathcal{D} = \mathbf{S}_{++}^n$. The Lagrange dual function is given by (5.15), so the dual problem can be expressed as

$$\begin{array}{ll}
\text{maximize} & \log\det\left(\sum_{i=1}^m \lambda_i a_i a_i^T\right) - \mathbf{1}^T\lambda + n \\
\text{subject to} & \lambda \succeq 0
\end{array} \qquad (5.31)$$

where we take $\log\det X = -\infty$ if $X \not\succ 0$.

The (weaker) Slater condition for the problem (5.14) is that there exists an $X \in \mathbf{S}_{++}^n$ with $a_i^T X a_i \le 1$, for $i = 1, \ldots, m$. This is always satisfied, so strong duality always obtains between (5.14) and the dual problem (5.31).

A nonconvex quadratic problem with strong duality

On rare occasions strong duality obtains for a *nonconvex* problem. As an important example, we consider the problem of minimizing a nonconvex quadratic function over the unit ball,

$$\begin{array}{ll}
\text{minimize} & x^T A x + 2b^T x \\
\text{subject to} & x^T x \le 1,
\end{array} \qquad (5.32)$$

where $A \in \mathbf{S}^n$, $A \not\succeq 0$, and $b \in \mathbf{R}^n$. Since $A \not\succeq 0$, this is not a convex problem. This problem is sometimes called the *trust region problem*, and arises in minimizing a second-order approximation of a function over the unit ball, which is the region in which the approximation is assumed to be approximately valid.

The Lagrangian is

$$L(x, \lambda) = x^T A x + 2b^T x + \lambda(x^T x - 1) = x^T(A + \lambda I)x + 2b^T x - \lambda,$$

so the dual function is given by

$$g(\lambda) = \begin{cases} -b^T(A + \lambda I)^\dagger b - \lambda & A + \lambda I \succeq 0, \quad b \in \mathcal{R}(A + \lambda I) \\ -\infty & \text{otherwise}, \end{cases}$$

where $(A + \lambda I)^\dagger$ is the pseudo-inverse of $A + \lambda I$. The Lagrange dual problem is thus

$$\begin{array}{ll}
\text{maximize} & -b^T(A + \lambda I)^\dagger b - \lambda \\
\text{subject to} & A + \lambda I \succeq 0, \quad b \in \mathcal{R}(A + \lambda I),
\end{array} \qquad (5.33)$$

with variable $\lambda \in \mathbf{R}$. Although it is not obvious from this expression, this is a convex optimization problem. In fact, it is readily solved since it can be expressed as

$$\begin{array}{ll}
\text{maximize} & -\sum_{i=1}^n (q_i^T b)^2/(\lambda_i + \lambda) - \lambda \\
\text{subject to} & \lambda \ge -\lambda_{\min}(A),
\end{array}$$

where λ_i and q_i are the eigenvalues and corresponding (orthonormal) eigenvectors of A, and we interpret $(q_i^T b)^2/0$ as 0 if $q_i^T b = 0$ and as ∞ otherwise.

Despite the fact that the original problem (5.32) is not convex, we always have zero optimal duality gap for this problem: The optimal values of (5.32) and (5.33) are always the same. In fact, a more general result holds: strong duality holds for any optimization problem with quadratic objective and one quadratic inequality constraint, provided Slater's condition holds; see §B.1.

5.2.5   Mixed strategies for matrix games

In this section we use strong duality to derive a basic result for zero-sum matrix games. We consider a game with two players. Player 1 makes a choice (or *move*) $k \in \{1, \ldots, n\}$, and player 2 makes a choice $l \in \{1, \ldots, m\}$. Player 1 then makes a payment of P_{kl} to player 2, where $P \in \mathbf{R}^{n \times m}$ is the *payoff matrix* for the game. The goal of player 1 is to make the payment as small as possible, while the goal of player 2 is to maximize it.

The players use randomized or *mixed strategies*, which means that each player makes his or her choice randomly and independently of the other player's choice, according to a probability distribution:

$$\mathbf{prob}(k = i) = u_i, \quad i = 1, \ldots, n, \qquad \mathbf{prob}(l = i) = v_i, \quad i = 1, \ldots, m.$$

Here u and v give the probability distributions of the choices of the two players, *i.e.*, their associated strategies. The expected payoff from player 1 to player 2 is then

$$\sum_{k=1}^{n} \sum_{l=1}^{m} u_k v_l P_{kl} = u^T P v.$$

Player 1 wishes to choose u to minimize $u^T P v$, while player 2 wishes to choose v to maximize $u^T P v$.

Let us first analyze the game from the point of view of player 1, assuming her strategy u is known to player 2 (which clearly gives an advantage to player 2). Player 2 will choose v to maximize $u^T P v$, which results in the expected payoff

$$\sup\{u^T P v \mid v \succeq 0, \ \mathbf{1}^T v = 1\} = \max_{i=1,\ldots,m} (P^T u)_i.$$

The best thing player 1 can do is to choose u to minimize this worst-case payoff to player 2, *i.e.*, to choose a strategy u that solves the problem

$$\begin{array}{ll} \text{minimize} & \max_{i=1,\ldots,m} (P^T u)_i \\ \text{subject to} & u \succeq 0, \quad \mathbf{1}^T u = 1, \end{array} \qquad (5.34)$$

which is a piecewise-linear convex optimization problem. We will denote the optimal value of this problem as $p_1^\star$. This is the smallest expected payoff player 1 can arrange to have, assuming that player 2 knows the strategy of player 1, and plays to his own maximum advantage.

In a similar way we can consider the situation in which v, the strategy of player 2, is known to player 1 (which gives an advantage to player 1). In this case player 1 chooses u to minimize $u^T P v$, which results in an expected payoff of

$$\inf\{u^T P v \mid u \succeq 0, \ \mathbf{1}^T u = 1\} = \min_{i=1,\ldots,n} (P v)_i.$$

Player 2 chooses v to maximize this, *i.e.*, chooses a strategy v that solves the problem

$$\begin{array}{ll} \text{maximize} & \min_{i=1,\ldots,n} (P v)_i \\ \text{subject to} & v \succeq 0, \quad \mathbf{1}^T v = 1, \end{array} \qquad (5.35)$$

which is another convex optimization problem, with piecewise-linear (concave) objective. We will denote the optimal value of this problem as $p_2^\star$. This is the largest expected payoff player 2 can guarantee getting, assuming that player 1 knows the strategy of player 2.

It is intuitively obvious that knowing your opponent's strategy gives an advantage (or at least, cannot hurt), and indeed, it is easily shown that we always have $p_1^\star \ge p_2^\star$. We can interpret the difference, $p_1^\star - p_2^\star$, which is nonnegative, as the advantage conferred on a player by knowing the opponent's strategy.

Using duality, we can establish a result that is at first surprising: $p_1^\star = p_2^\star$. In other words, in a matrix game with mixed strategies, there is *no* advantage to knowing your opponent's strategy. We will establish this result by showing that the two problems (5.34) and (5.35) are Lagrange dual problems, for which strong duality obtains.

We start by formulating (5.34) as an LP,

$$
\begin{array}{ll}
\text{minimize} & t \\
\text{subject to} & u \succeq 0, \quad \mathbf{1}^T u = 1 \\
& P^T u \preceq t\mathbf{1},
\end{array}
$$

with extra variable $t \in \mathbf{R}$. Introducing the multiplier λ for $P^T u \preceq t\mathbf{1}$, μ for $u \succeq 0$, and ν for $\mathbf{1}^T u = 1$, the Lagrangian is

$$
t + \lambda^T (P^T u - t\mathbf{1}) - \mu^T u + \nu(1 - \mathbf{1}^T u) = \nu + (1 - \mathbf{1}^T \lambda)t + (P\lambda - \nu\mathbf{1} - \mu)^T u,
$$

so the dual function is

$$
g(\lambda, \mu, \nu) = \left\{
\begin{array}{ll}
\nu & \mathbf{1}^T \lambda = 1, \quad P\lambda - \nu\mathbf{1} = \mu \\
-\infty & \text{otherwise.}
\end{array}
\right.
$$

The dual problem is then

$$
\begin{array}{ll}
\text{maximize} & \nu \\
\text{subject to} & \lambda \succeq 0, \quad \mathbf{1}^T \lambda = 1, \quad \mu \succeq 0 \\
& P\lambda - \nu\mathbf{1} = \mu.
\end{array}
$$

Eliminating μ we obtain the following Lagrange dual of (5.34):

$$
\begin{array}{ll}
\text{maximize} & \nu \\
\text{subject to} & \lambda \succeq 0, \quad \mathbf{1}^T \lambda = 1 \\
& P\lambda \succeq \nu\mathbf{1},
\end{array}
$$

with variables λ, ν. But this is clearly equivalent to (5.35). Since the LPs are feasible, we have strong duality; the optimal values of (5.34) and (5.35) are equal.

5.3   Geometric interpretation

5.3.1   Weak and strong duality via set of values

We can give a simple geometric interpretation of the dual function in terms of the set

$$\mathcal{G} = \{(f_1(x), \ldots, f_m(x), h_1(x), \ldots, h_p(x), f_0(x)) \in \mathbf{R}^m \times \mathbf{R}^p \times \mathbf{R} \mid x \in \mathcal{D}\}, \quad (5.36)$$

which is the set of values taken on by the constraint and objective functions. The optimal value $p^\star$ of (5.1) is easily expressed in terms of $\mathcal{G}$ as

$$p^\star = \inf\{t \mid (u, v, t) \in \mathcal{G}, \ u \preceq 0, \ v = 0\}.$$

To evaluate the dual function at (λ, ν), we minimize the affine function

$$(\lambda, \nu, 1)^T(u, v, t) = \sum_{i=1}^m \lambda_i u_i + \sum_{i=1}^p \nu_i v_i + t$$

over $(u, v, t) \in \mathcal{G}$, $i.e.$, we have

$$g(\lambda, \nu) = \inf\{(\lambda, \nu, 1)^T(u, v, t) \mid (u, v, t) \in \mathcal{G}\}.$$

In particular, we see that if the infimum is finite, then the inequality

$$(\lambda, \nu, 1)^T(u, v, t) \geq g(\lambda, \nu)$$

defines a supporting hyperplane to $\mathcal{G}$. This is sometimes referred to as a *nonvertical* supporting hyperplane, because the last component of the normal vector is nonzero.

Now suppose $\lambda \succeq 0$. Then, obviously, $t \geq (\lambda, \nu, 1)^T(u, v, t)$ if $u \preceq 0$ and $v = 0$. Therefore

$$
\begin{aligned}
p^\star &= \inf\{t \mid (u, v, t) \in \mathcal{G}, \ u \preceq 0, \ v = 0\} \\
&\geq \inf\{(\lambda, \nu, 1)^T(u, v, t) \mid (u, v, t) \in \mathcal{G}, \ u \preceq 0, \ v = 0\} \\
&\geq \inf\{(\lambda, \nu, 1)^T(u, v, t) \mid (u, v, t) \in \mathcal{G}\} \\
&= g(\lambda, \nu),
\end{aligned}
$$

i.e., we have weak duality. This interpretation is illustrated in figures 5.3 and 5.4, for a simple problem with one inequality constraint.

Epigraph variation

In this section we describe a variation on the geometric interpretation of duality in terms of $\mathcal{G}$, which explains why strong duality obtains for (most) convex problems. We define the set $\mathcal{A} \subseteq \mathbf{R}^m \times \mathbf{R}^p \times \mathbf{R}$ as

$$\mathcal{A} = \mathcal{G} + (\mathbf{R}_+^m \times \{0\} \times \mathbf{R}_+), \quad (5.37)$$

or, more explicitly,

$$
\begin{aligned}
\mathcal{A} = \{(u, v, t) \mid &\exists x \in \mathcal{D}, \ f_i(x) \leq u_i, \ i = 1, \ldots, m, \\
&h_i(x) = v_i, \ i = 1, \ldots, p, \ f_0(x) \leq t\},
\end{aligned}
$$

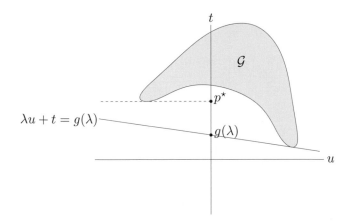

Figure 5.3 Geometric interpretation of dual function and lower bound $g(\lambda) \leq p^\star$, for a problem with one (inequality) constraint. Given λ, we minimize $(\lambda, 1)^T(u, t)$ over $\mathcal{G} = \{(f_1(x), f_0(x)) \mid x \in \mathcal{D}\}$. This yields a supporting hyperplane with slope $-\lambda$. The intersection of this hyperplane with the $u = 0$ axis gives $g(\lambda)$.

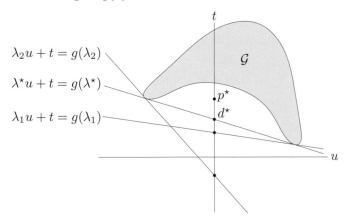

Figure 5.4 Supporting hyperplanes corresponding to three dual feasible values of λ, including the optimum $\lambda^\star$. Strong duality does not hold; the optimal duality gap $p^\star - d^\star$ is positive.

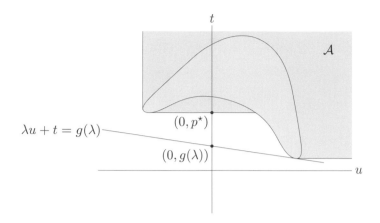

Figure 5.5 Geometric interpretation of dual function and lower bound $g(\lambda) \leq p^\star$, for a problem with one (inequality) constraint. Given λ, we minimize $(\lambda, 1)^T (u, t)$ over $\mathcal{A} = \{(u, t) \mid \exists x \in \mathcal{D}, \ f_0(x) \leq t, \ f_1(x) \leq u\}$. This yields a supporting hyperplane with slope $-\lambda$. The intersection of this hyperplane with the $u = 0$ axis gives $g(\lambda)$.

We can think of $\mathcal{A}$ as a sort of epigraph form of $\mathcal{G}$, since $\mathcal{A}$ includes all the points in $\mathcal{G}$, as well as points that are 'worse', *i.e.*, those with larger objective or inequality constraint function values.

We can express the optimal value in terms of $\mathcal{A}$ as

$$p^\star = \inf\{t \mid (0, 0, t) \in \mathcal{A}\}.$$

To evaluate the dual function at a point (λ, ν) with $\lambda \succeq 0$, we can minimize the affine function $(\lambda, \nu, 1)^T (u, v, t)$ over $\mathcal{A}$: If $\lambda \succeq 0$, then

$$g(\lambda, \nu) = \inf\{(\lambda, \nu, 1)^T (u, v, t) \mid (u, v, t) \in \mathcal{A}\}.$$

If the infimum is finite, then

$$(\lambda, \nu, 1)^T (u, v, t) \geq g(\lambda, \nu)$$

defines a nonvertical supporting hyperplane to $\mathcal{A}$.

In particular, since $(0, 0, p^\star) \in \mathbf{bd}\,\mathcal{A}$, we have

$$p^\star = (\lambda, \nu, 1)^T (0, 0, p^\star) \geq g(\lambda, \nu), \tag{5.38}$$

the weak duality lower bound. Strong duality holds if and only if we have equality in (5.38) for some dual feasible (λ, ν), *i.e.*, there exists a nonvertical supporting hyperplane to $\mathcal{A}$ at its boundary point $(0, 0, p^\star)$.

This second interpretation is illustrated in figure 5.5.

5.3.2   Proof of strong duality under constraint qualification

In this section we prove that Slater's constraint qualification guarantees strong duality (and that the dual optimum is attained) for a convex problem. We consider the primal problem (5.25), with $f_0, \ldots, f_m$ convex, and assume Slater's condition holds: There exists $\tilde{x} \in \mathbf{relint}\, \mathcal{D}$ with $f_i(\tilde{x}) < 0$, $i = 1, \ldots, m$, and $A\tilde{x} = b$. In order to simplify the proof, we make two additional assumptions: first that $\mathcal{D}$ has nonempty interior (hence, $\mathbf{relint}\, \mathcal{D} = \mathbf{int}\, \mathcal{D}$) and second, that $\mathbf{rank}\, A = p$. We assume that $p^\star$ is finite. (Since there is a feasible point, we can only have $p^\star = -\infty$ or $p^\star$ finite; if $p^\star = -\infty$, then $d^\star = -\infty$ by weak duality.)

The set $\mathcal{A}$ defined in (5.37) is readily shown to be convex if the underlying problem is convex. We define a second convex set $\mathcal{B}$ as

$$\mathcal{B} = \{(0, 0, s) \in \mathbf{R}^m \times \mathbf{R}^p \times \mathbf{R} \mid s < p^\star\}.$$

The sets $\mathcal{A}$ and $\mathcal{B}$ do not intersect. To see this, suppose $(u, v, t) \in \mathcal{A} \cap \mathcal{B}$. Since $(u, v, t) \in \mathcal{B}$ we have $u = 0$, $v = 0$, and $t < p^\star$. Since $(u, v, t) \in \mathcal{A}$, there exists an x with $f_i(x) \leq 0$, $i = 1, \ldots, m$, $Ax - b = 0$, and $f_0(x) \leq t < p^\star$, which is impossible since $p^\star$ is the optimal value of the primal problem.

By the separating hyperplane theorem of §2.5.1 there exists $(\tilde{\lambda}, \tilde{\nu}, \mu) \neq 0$ and α such that

$$(u, v, t) \in \mathcal{A} \implies \tilde{\lambda}^T u + \tilde{\nu}^T v + \mu t \geq \alpha, \tag{5.39}$$

and

$$(u, v, t) \in \mathcal{B} \implies \tilde{\lambda}^T u + \tilde{\nu}^T v + \mu t \leq \alpha. \tag{5.40}$$

From (5.39) we conclude that $\tilde{\lambda} \succeq 0$ and $\mu \geq 0$. (Otherwise $\tilde{\lambda}^T u + \mu t$ is unbounded below over $\mathcal{A}$, contradicting (5.39).) The condition (5.40) simply means that $\mu t \leq \alpha$ for all $t < p^\star$, and hence, $\mu p^\star \leq \alpha$. Together with (5.39) we conclude that for any $x \in \mathcal{D}$,

$$\sum_{i=1}^m \tilde{\lambda}_i f_i(x) + \tilde{\nu}^T(Ax - b) + \mu f_0(x) \geq \alpha \geq \mu p^\star. \tag{5.41}$$

Assume that $\mu > 0$. In that case we can divide (5.41) by μ to obtain

$$L(x, \tilde{\lambda}/\mu, \tilde{\nu}/\mu) \geq p^\star$$

for all $x \in \mathcal{D}$, from which it follows, by minimizing over x, that $g(\lambda, \nu) \geq p^\star$, where we define

$$\lambda = \tilde{\lambda}/\mu, \qquad \nu = \tilde{\nu}/\mu.$$

By weak duality we have $g(\lambda, \nu) \leq p^\star$, so in fact $g(\lambda, \nu) = p^\star$. This shows that strong duality holds, and that the dual optimum is attained, at least in the case when $\mu > 0$.

Now consider the case $\mu = 0$. From (5.41), we conclude that for all $x \in \mathcal{D}$,

$$\sum_{i=1}^m \tilde{\lambda}_i f_i(x) + \tilde{\nu}^T(Ax - b) \geq 0. \tag{5.42}$$

Applying this to the point $\tilde{x}$ that satisfies the Slater condition, we have

$$\sum_{i=1}^m \tilde{\lambda}_i f_i(\tilde{x}) \geq 0.$$

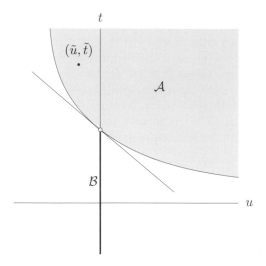

Figure 5.6 Illustration of strong duality proof, for a convex problem that satisfies Slater's constraint qualification. The set $\mathcal{A}$ is shown shaded, and the set $\mathcal{B}$ is the thick vertical line segment, not including the point $(0, p^\star)$, shown as a small open circle. The two sets are convex and do not intersect, so they can be separated by a hyperplane. Slater's constraint qualification guarantees that any separating hyperplane must be nonvertical, since it must pass to the left of the point $(\tilde{u}, \tilde{t}) = (f_1(\tilde{x}), f_0(\tilde{x}))$, where $\tilde{x}$ is strictly feasible.

Since $f_i(\tilde{x}) < 0$ and $\tilde{\lambda}_i \geq 0$, we conclude that $\tilde{\lambda} = 0$. From $(\tilde{\lambda}, \tilde{\nu}, \mu) \neq 0$ and $\tilde{\lambda} = 0$, $\mu = 0$, we conclude that $\tilde{\nu} \neq 0$. Then (5.42) implies that for all $x \in \mathcal{D}$, $\tilde{\nu}^T(Ax - b) \geq 0$. But $\tilde{x}$ satisfies $\tilde{\nu}^T(A\tilde{x} - b) = 0$, and since $\tilde{x} \in \mathbf{int}\,\mathcal{D}$, there are points in $\mathcal{D}$ with $\tilde{\nu}^T(Ax - b) < 0$ unless $A^T\tilde{\nu} = 0$. This, of course, contradicts our assumption that $\mathbf{rank}\,A = p$.

The geometric idea behind the proof is illustrated in figure 5.6, for a simple problem with one inequality constraint. The hyperplane separating $\mathcal{A}$ and $\mathcal{B}$ defines a supporting hyperplane to $\mathcal{A}$ at $(0, p^\star)$. Slater's constraint qualification is used to establish that the hyperplane must be nonvertical (*i.e.*, has a normal vector of the form $(\lambda^\star, 1)$). (For a simple example of a convex problem with one inequality constraint for which strong duality fails, see exercise 5.21.)

5.3.3   Multicriterion interpretation

There is a natural connection between Lagrange duality for a problem without equality constraints,

$$
\begin{array}{ll}
\text{minimize} & f_0(x) \\
\text{subject to} & f_i(x) \leq 0, \quad i = 1, \dots, m,
\end{array}
\tag{5.43}
$$

and the scalarization method for the (unconstrained) multicriterion problem

$$\text{minimize (w.r.t. } \mathbf{R}_+^{m+1}) \quad F(x) = (f_1(x), \dots, f_m(x), f_0(x)) \tag{5.44}$$

(see §4.7.4). In scalarization, we choose a positive vector $\tilde{\lambda}$, and minimize the scalar function $\tilde{\lambda}^T F(x)$; any minimizer is guaranteed to be Pareto optimal. Since we can scale $\tilde{\lambda}$ by a positive constant, without affecting the minimizers, we can, without loss of generality, take $\tilde{\lambda} = (\lambda, 1)$. Thus, in scalarization we minimize the function

$$\tilde{\lambda}^T F(x) = f_0(x) + \sum_{i=1}^m \lambda_i f_i(x),$$

which is exactly the Lagrangian for the problem (5.43).

   To establish that every Pareto optimal point of a convex multicriterion problem minimizes the function $\tilde{\lambda}^T F(x)$ for some nonnegative weight vector $\tilde{\lambda}$, we considered the set $\mathcal{A}$, defined in (4.62),

$$\mathcal{A} = \{t \in \mathbf{R}^{m+1} \mid \exists x \in \mathcal{D}, \ f_i(x) \le t_i, \ i = 0, \dots, m\},$$

which is exactly the same as the set $\mathcal{A}$ defined in (5.37), that arises in Lagrange duality. Here too we constructed the required weight vector as a supporting hyperplane to the set, at an arbitrary Pareto optimal point. In multicriterion optimization, we interpret the components of the weight vector as giving the relative weights between the objective functions. When we fix the last component of the weight vector (associated with f_0) to be one, the other weights have the interpretation of the cost relative to f_0, *i.e.*, the cost relative to the objective.

5.4   Saddle-point interpretation

In this section we give several interpretations of Lagrange duality. The material of this section will not be used in the sequel.

5.4.1   Max-min characterization of weak and strong duality

It is possible to express the primal and the dual optimization problems in a form that is more symmetric. To simplify the discussion we assume there are no equality constraints; the results are easily extended to cover them.

   First note that

$$\sup_{\lambda \succeq 0} L(x, \lambda) = \sup_{\lambda \succeq 0} \left(f_0(x) + \sum_{i=1}^m \lambda_i f_i(x) \right)$$

$$= \begin{cases} f_0(x) & f_i(x) \le 0, \quad i = 1, \dots, m \\ \infty & \text{otherwise.} \end{cases}$$

Indeed, suppose x is not feasible, and $f_i(x) > 0$ for some i. Then $\sup_{\lambda \succeq 0} L(x, \lambda) = \infty$, as can be seen by choosing $\lambda_j = 0$, $j \neq i$, and $\lambda_i \to \infty$. On the other hand, if $f_i(x) \leq 0$, $i = 1, \dots, m$, then the optimal choice of λ is $\lambda = 0$ and $\sup_{\lambda \succeq 0} L(x, \lambda) = f_0(x)$. This means that we can express the optimal value of the primal problem as

$$p^\star = \inf_x \sup_{\lambda \succeq 0} L(x, \lambda).$$

By the definition of the dual function, we also have

$$d^\star = \sup_{\lambda \succeq 0} \inf_x L(x, \lambda).$$

Thus, weak duality can be expressed as the inequality

$$\sup_{\lambda \succeq 0} \inf_x L(x, \lambda) \leq \inf_x \sup_{\lambda \succeq 0} L(x, \lambda), \tag{5.45}$$

and strong duality as the equality

$$\sup_{\lambda \succeq 0} \inf_x L(x, \lambda) = \inf_x \sup_{\lambda \succeq 0} L(x, \lambda).$$

Strong duality means that the order of the minimization over x and the maximization over $\lambda \succeq 0$ can be switched without affecting the result.

In fact, the inequality (5.45) does not depend on any properties of L: We have

$$\sup_{z \in Z} \inf_{w \in W} f(w, z) \leq \inf_{w \in W} \sup_{z \in Z} f(w, z) \tag{5.46}$$

for any $f : \mathbf{R}^n \times \mathbf{R}^m \to \mathbf{R}$ (and any $W \subseteq \mathbf{R}^n$ and $Z \subseteq \mathbf{R}^m$). This general inequality is called the *max-min inequality*. When equality holds, *i.e.*,

$$\sup_{z \in Z} \inf_{w \in W} f(w, z) = \inf_{w \in W} \sup_{z \in Z} f(w, z) \tag{5.47}$$

we say that f (and W and Z) satisfy the *strong max-min property* or the *saddle-point* property. Of course the strong max-min property holds only in special cases, for example, when $f : \mathbf{R}^n \times \mathbf{R}^m \to \mathbf{R}$ is the Lagrangian of a problem for which strong duality obtains, $W = \mathbf{R}^n$, and $Z = \mathbf{R}^m_+$.

5.4.2   Saddle-point interpretation

We refer to a pair $\tilde{w} \in W$, $\tilde{z} \in Z$ as a *saddle-point* for f (and W and Z) if

$$f(\tilde{w}, z) \leq f(\tilde{w}, \tilde{z}) \leq f(w, \tilde{z})$$

for all $w \in W$ and $z \in Z$. In other words, $\tilde{w}$ minimizes $f(w, \tilde{z})$ (over $w \in W$) and $\tilde{z}$ maximizes $f(\tilde{w}, z)$ (over $z \in Z$):

$$f(\tilde{w}, \tilde{z}) = \inf_{w \in W} f(w, \tilde{z}), \qquad f(\tilde{w}, \tilde{z}) = \sup_{z \in Z} f(\tilde{w}, z).$$

This implies that the strong max-min property (5.47) holds, and that the common value is $f(\tilde{w}, \tilde{z})$.

Returning to our discussion of Lagrange duality, we see that if $x^\star$ and $\lambda^\star$ are primal and dual optimal points for a problem in which strong duality obtains, they form a saddle-point for the Lagrangian. The converse is also true: If (x, λ) is a saddle-point of the Lagrangian, then x is primal optimal, λ is dual optimal, and the optimal duality gap is zero.

5.4.3   Game interpretation

We can interpret the max-min inequality (5.46), the max-min equality (5.47), and the saddle-point property, in terms of a continuous *zero-sum game*. If the first player chooses $w \in W$, and the second player selects $z \in Z$, then player 1 pays an amount $f(w, z)$ to player 2. Player 1 therefore wants to minimize f, while player 2 wants to maximize f. (The game is called continuous since the choices are vectors, and not discrete.)

Suppose that player 1 makes his choice first, and then player 2, after learning the choice of player 1, makes her selection. Player 2 wants to maximize the payoff $f(w, z)$, and so will choose $z \in Z$ to maximize $f(w, z)$. The resulting payoff will be $\sup_{z \in Z} f(w, z)$, which depends on w, the choice of the first player. (We assume here that the supremum is achieved; if not the optimal payoff can be arbitrarily close to $\sup_{z \in Z} f(w, z)$.) Player 1 knows (or assumes) that player 2 will follow this strategy, and so will choose $w \in W$ to make this worst-case payoff to player 2 as small as possible. Thus player 1 chooses

$$\operatorname*{argmin}_{w \in W} \sup_{z \in Z} f(w, z),$$

which results in the payoff

$$\inf_{w \in W} \sup_{z \in Z} f(w, z)$$

from player 1 to player 2.

Now suppose the order of play is reversed: Player 2 must choose $z \in Z$ first, and then player 1 chooses $w \in W$ (with knowledge of z). Following a similar argument, if the players follow the optimal strategy, player 2 should choose $z \in Z$ to maximize $\inf_{w \in W} f(w, z)$, which results in the payoff of

$$\sup_{z \in Z} \inf_{w \in W} f(w, z)$$

from player 1 to player 2.

The max-min inequality (5.46) states the (intuitively obvious) fact that it is better for a player to go second, or more precisely, for a player to know his or her opponent's choice before choosing. In other words, the payoff to player 2 will be larger if player 1 must choose first. When the saddle-point property (5.47) holds, there is no advantage to playing second.

If $(\tilde{w}, \tilde{z})$ is a saddle-point for f (and W and Z), then it is called a *solution* of the game; $\tilde{w}$ is called the optimal choice or strategy for player 1, and $\tilde{z}$ is called the optimal choice or strategy for player 2. In this case there is no advantage to playing second.

Now consider the special case where the payoff function is the Lagrangian, $W = \mathbf{R}^n$ and $Z = \mathbf{R}_+^m$. Here player 1 chooses the primal variable x, while player 2 chooses the dual variable $\lambda \succeq 0$. By the argument above, the optimal choice for player 2, if she must choose first, is any $\lambda^\star$ which is dual optimal, which results in a payoff to player 2 of $d^\star$. Conversely, if player 1 must choose first, his optimal choice is any primal optimal $x^\star$, which results in a payoff of $p^\star$.

The optimal duality gap for the problem is exactly equal to the advantage afforded the player who goes second, *i.e.*, the player who has the advantage of knowing his or her opponent's choice before choosing. If strong duality holds, then there is no advantage to the players of knowing their opponent's choice.

5.4.4   Price or tax interpretation

Lagrange duality has an interesting economic interpretation. Suppose the variable x denotes how an enterprise operates and $f_0(x)$ denotes the cost of operating at x, *i.e.*, $-f_0(x)$ is the profit (say, in dollars) made at the operating condition x. Each constraint $f_i(x) \le 0$ represents some limit, such as a limit on resources (*e.g.*, warehouse space, labor) or a regulatory limit (*e.g.*, environmental). The operating condition that maximizes profit while respecting the limits can be found by solving the problem

$$\begin{array}{ll} \text{minimize} & f_0(x) \\ \text{subject to} & f_i(x) \le 0, \quad i = 1, \ldots, m. \end{array}$$

The resulting optimal profit is $-p^\star$.

Now imagine a second scenario in which the limits can be violated, by paying an additional cost which is linear in the amount of violation, measured by f_i. Thus the payment made by the enterprise for the ith limit or constraint is $\lambda_i f_i(x)$. Payments are also made *to* the firm for constraints that are not tight; if $f_i(x) < 0$, then $\lambda_i f_i(x)$ represents a payment to the firm. The coefficient λ_i has the interpretation of the price for violating $f_i(x) \le 0$; its units are dollars per unit violation (as measured by f_i). For the same price the enterprise can sell any 'unused' portion of the ith constraint. We assume $\lambda_i \ge 0$, *i.e.*, the firm must pay for violations (and receives income if a constraint is not tight).

As an example, suppose the first constraint in the original problem, $f_1(x) \le 0$, represents a limit on warehouse space (say, in square meters). In this new arrangement, we open the possibility that the firm can rent extra warehouse space at a cost of λ_1 dollars per square meter and also rent out unused space, at the same rate.

The total cost to the firm, for operating condition x, and constraint prices λ_i, is $L(x, \lambda) = f_0(x) + \sum_{i=1}^m \lambda_i f_i(x)$. The firm will obviously operate so as to minimize its total cost $L(x, \lambda)$, which yields a cost $g(\lambda)$. The dual function therefore represents the optimal cost to the firm, as a function of the constraint price vector λ. The optimal dual value, $d^\star$, is the optimal cost to the enterprise under the least favorable set of prices.

Using this interpretation we can paraphrase weak duality as follows: The optimal cost to the firm in the second scenario (in which constraint violations can be bought and sold) is less than or equal to the cost in the original situation (which has constraints that cannot be violated), even with the most unfavorable prices. This is obvious: If $x^\star$ is optimal in the first scenario, then the operating cost of $x^\star$ in the second scenario will be lower than $f_0(x^\star)$, since some income can be derived from the constraints that are not tight. The optimal duality gap is then the minimum possible advantage to the enterprise of being allowed to pay for constraint violations (and receive payments for nontight constraints).

Now suppose strong duality holds, and the dual optimum is attained. We can interpret a dual optimal $\lambda^\star$ as a set of prices for which there is no advantage to the firm in being allowed to pay for constraint violations (or receive payments for nontight constraints). For this reason a dual optimal $\lambda^\star$ is sometimes called a set of *shadow prices* for the original problem.

5.5   Optimality conditions

We remind the reader that we do not assume the problem (5.1) is convex, unless explicitly stated.

5.5.1   Certificate of suboptimality and stopping criteria

If we can find a dual feasible (λ, ν), we establish a lower bound on the optimal value of the primal problem: $p^\star \geq g(\lambda, \nu)$. Thus a dual feasible point (λ, ν) provides a *proof* or *certificate* that $p^\star \geq g(\lambda, \nu)$. Strong duality means there exist arbitrarily good certificates.

Dual feasible points allow us to bound how suboptimal a given feasible point is, without knowing the exact value of $p^\star$. Indeed, if x is primal feasible and (λ, ν) is dual feasible, then

$$f_0(x) - p^\star \leq f_0(x) - g(\lambda, \nu).$$

In particular, this establishes that x is ϵ-suboptimal, with $\epsilon = f_0(x) - g(\lambda, \nu)$. (It also establishes that (λ, ν) is ϵ-suboptimal for the dual problem.)

We refer to the gap between primal and dual objectives,

$$f_0(x) - g(\lambda, \nu),$$

as the *duality gap* associated with the primal feasible point x and dual feasible point (λ, ν). A primal dual feasible pair x, (λ, ν) localizes the optimal value of the primal (and dual) problems to an interval:

$$p^\star \in [g(\lambda, \nu), f_0(x)], \qquad d^\star \in [g(\lambda, \nu), f_0(x)],$$

the width of which is the duality gap.

If the duality gap of the primal dual feasible pair x, (λ, ν) is zero, i.e., $f_0(x) = g(\lambda, \nu)$, then x is primal optimal and (λ, ν) is dual optimal. We can think of (λ, ν)

as a certificate that proves x is optimal (and, similarly, we can think of x as a certificate that proves (λ, ν) is dual optimal).

These observations can be used in optimization algorithms to provide nonheuristic stopping criteria. Suppose an algorithm produces a sequence of primal feasible $x^{(k)}$ and dual feasible $(\lambda^{(k)}, \nu^{(k)})$, for $k = 1, 2, \ldots$, and $\epsilon_{\mathrm{abs}} > 0$ is a given required absolute accuracy. Then the stopping criterion (*i.e.*, the condition for terminating the algorithm)

$$f_0(x^{(k)}) - g(\lambda^{(k)}, \nu^{(k)}) \le \epsilon_{\mathrm{abs}}$$

guarantees that when the algorithm terminates, $x^{(k)}$ is ϵ_{abs}-suboptimal. Indeed, $(\lambda^{(k)}, \nu^{(k)})$ is a certificate that proves it. (Of course strong duality must hold if this method is to work for arbitrarily small tolerances ϵ_{abs}.)

A similar condition can be used to guarantee a given relative accuracy $\epsilon_{\mathrm{rel}} > 0$. If

$$g(\lambda^{(k)}, \nu^{(k)}) > 0, \qquad \frac{f_0(x^{(k)}) - g(\lambda^{(k)}, \nu^{(k)})}{g(\lambda^{(k)}, \nu^{(k)})} \le \epsilon_{\mathrm{rel}}$$

holds, or

$$f_0(x^{(k)}) < 0, \qquad \frac{f_0(x^{(k)}) - g(\lambda^{(k)}, \nu^{(k)})}{-f_0(x^{(k)})} \le \epsilon_{\mathrm{rel}}$$

holds, then $p^\star \ne 0$ and the relative error

$$\frac{f_0(x^{(k)}) - p^\star}{|p^\star|}$$

is guaranteed to be less than or equal to ϵ_{rel}.

5.5.2   Complementary slackness

Suppose that the primal and dual optimal values are attained and equal (so, in particular, strong duality holds). Let $x^\star$ be a primal optimal and $(\lambda^\star, \nu^\star)$ be a dual optimal point. This means that

$$
\begin{aligned}
f_0(x^\star) &= g(\lambda^\star, \nu^\star) \\
&= \inf_x \left(f_0(x) + \sum_{i=1}^m \lambda_i^\star f_i(x) + \sum_{i=1}^p \nu_i^\star h_i(x) \right) \\
&\le f_0(x^\star) + \sum_{i=1}^m \lambda_i^\star f_i(x^\star) + \sum_{i=1}^p \nu_i^\star h_i(x^\star) \\
&\le f_0(x^\star).
\end{aligned}
$$

The first line states that the optimal duality gap is zero, and the second line is the definition of the dual function. The third line follows since the infimum of the Lagrangian over x is less than or equal to its value at $x = x^\star$. The last inequality follows from $\lambda_i^\star \ge 0$, $f_i(x^\star) \le 0$, $i = 1, \ldots, m$, and $h_i(x^\star) = 0$, $i = 1, \ldots, p$. We conclude that the two inequalities in this chain hold with equality.

We can draw several interesting conclusions from this. For example, since the inequality in the third line is an equality, we conclude that $x^\star$ minimizes $L(x, \lambda^\star, \nu^\star)$ over x. (The Lagrangian $L(x, \lambda^\star, \nu^\star)$ can have other minimizers; $x^\star$ is simply *a* minimizer.)

Another important conclusion is that

$$\sum_{i=1}^{m} \lambda_i^\star f_i(x^\star) = 0.$$

Since each term in this sum is nonpositive, we conclude that

$$\lambda_i^\star f_i(x^\star) = 0, \quad i = 1, \dots, m. \tag{5.48}$$

This condition is known as *complementary slackness*; it holds for any primal optimal $x^\star$ and any dual optimal $(\lambda^\star, \nu^\star)$ (when strong duality holds). We can express the complementary slackness condition as

$$\lambda_i^\star > 0 \implies f_i(x^\star) = 0,$$

or, equivalently,

$$f_i(x^\star) < 0 \implies \lambda_i^\star = 0.$$

Roughly speaking, this means the ith optimal Lagrange multiplier is zero unless the ith constraint is active at the optimum.

5.5.3   KKT optimality conditions

We now assume that the functions $f_0, \dots, f_m, h_1, \dots, h_p$ are differentiable (and therefore have open domains), but we make no assumptions yet about convexity.

KKT conditions for nonconvex problems

As above, let $x^\star$ and $(\lambda^\star, \nu^\star)$ be any primal and dual optimal points with zero duality gap. Since $x^\star$ minimizes $L(x, \lambda^\star, \nu^\star)$ over x, it follows that its gradient must vanish at $x^\star$, i.e.,

$$\nabla f_0(x^\star) + \sum_{i=1}^{m} \lambda_i^\star \nabla f_i(x^\star) + \sum_{i=1}^{p} \nu_i^\star \nabla h_i(x^\star) = 0.$$

Thus we have

$$\begin{array}{rcll}
f_i(x^\star) & \leq & 0, & i = 1, \dots, m \\
h_i(x^\star) & = & 0, & i = 1, \dots, p \\
\lambda_i^\star & \geq & 0, & i = 1, \dots, m \\
\lambda_i^\star f_i(x^\star) & = & 0, & i = 1, \dots, m \\
\nabla f_0(x^\star) + \sum_{i=1}^{m} \lambda_i^\star \nabla f_i(x^\star) + \sum_{i=1}^{p} \nu_i^\star \nabla h_i(x^\star) & = & 0,
\end{array} \tag{5.49}$$

which are called the *Karush-Kuhn-Tucker* (KKT) conditions.

To summarize, for *any* optimization problem with differentiable objective and constraint functions for which strong duality obtains, any pair of primal and dual optimal points must satisfy the KKT conditions (5.49).

KKT conditions for convex problems

When the primal problem is convex, the KKT conditions are also sufficient for the points to be primal and dual optimal. In other words, if f_i are convex and h_i are affine, and $\tilde{x}$, $\tilde{\lambda}$, $\tilde{\nu}$ are any points that satisfy the KKT conditions

$$
\begin{array}{rcll}
f_i(\tilde{x}) & \leq & 0, & i = 1, \dots, m \\
h_i(\tilde{x}) & = & 0, & i = 1, \dots, p \\
\tilde{\lambda}_i & \geq & 0, & i = 1, \dots, m \\
\tilde{\lambda}_i f_i(\tilde{x}) & = & 0, & i = 1, \dots, m \\
\nabla f_0(\tilde{x}) + \sum_{i=1}^{m} \tilde{\lambda}_i \nabla f_i(\tilde{x}) + \sum_{i=1}^{p} \tilde{\nu}_i \nabla h_i(\tilde{x}) & = & 0, &
\end{array}
$$

then $\tilde{x}$ and $(\tilde{\lambda}, \tilde{\nu})$ are primal and dual optimal, with zero duality gap.

To see this, note that the first two conditions state that $\tilde{x}$ is primal feasible. Since $\tilde{\lambda}_i \geq 0$, $L(x, \tilde{\lambda}, \tilde{\nu})$ is convex in x; the last KKT condition states that its gradient with respect to x vanishes at $x = \tilde{x}$, so it follows that $\tilde{x}$ minimizes $L(x, \tilde{\lambda}, \tilde{\nu})$ over x. From this we conclude that

$$
\begin{aligned}
g(\tilde{\lambda}, \tilde{\nu}) & = L(\tilde{x}, \tilde{\lambda}, \tilde{\nu}) \\
& = f_0(\tilde{x}) + \sum_{i=1}^{m} \tilde{\lambda}_i f_i(\tilde{x}) + \sum_{i=1}^{p} \tilde{\nu}_i h_i(\tilde{x}) \\
& = f_0(\tilde{x}),
\end{aligned}
$$

where in the last line we use $h_i(\tilde{x}) = 0$ and $\tilde{\lambda}_i f_i(\tilde{x}) = 0$. This shows that $\tilde{x}$ and $(\tilde{\lambda}, \tilde{\nu})$ have zero duality gap, and therefore are primal and dual optimal. In summary, for any *convex* optimization problem with differentiable objective and constraint functions, any points that satisfy the KKT conditions are primal and dual optimal, and have zero duality gap.

If a convex optimization problem with differentiable objective and constraint functions satisfies Slater's condition, then the KKT conditions provide necessary and sufficient conditions for optimality: Slater's condition implies that the optimal duality gap is zero and the dual optimum is attained, so x is optimal if and only if there are (λ, ν) that, together with x, satisfy the KKT conditions.

The KKT conditions play an important role in optimization. In a few special cases it is possible to solve the KKT conditions (and therefore, the optimization problem) analytically. More generally, many algorithms for convex optimization are conceived as, or can be interpreted as, methods for solving the KKT conditions.

Example 5.1 *Equality constrained convex quadratic minimization.* We consider the problem

$$
\begin{array}{ll}
\text{minimize} & (1/2)x^T P x + q^T x + r \\
\text{subject to} & A x = b,
\end{array}
\tag{5.50}
$$

where $P \in \mathbf{S}_+^n$. The KKT conditions for this problem are

$$
A x^\star = b, \qquad P x^\star + q + A^T \nu^\star = 0,
$$

which we can write as

$$
\begin{bmatrix} P & A^T \\ A & 0 \end{bmatrix} \begin{bmatrix} x^\star \\ \nu^\star \end{bmatrix} = \begin{bmatrix} -q \\ b \end{bmatrix}.
$$

Solving this set of $m + n$ equations in the $m + n$ variables $x^\star$, $\nu^\star$ gives the optimal primal and dual variables for (5.50).

Example 5.2 *Water-filling.* We consider the convex optimization problem

$$\begin{array}{ll} \text{minimize} & -\sum_{i=1}^{n} \log(\alpha_i + x_i) \\ \text{subject to} & x \succeq 0, \quad \mathbf{1}^T x = 1, \end{array}$$

where $\alpha_i > 0$. This problem arises in information theory, in allocating power to a set of n communication channels. The variable x_i represents the transmitter power allocated to the ith channel, and $\log(\alpha_i + x_i)$ gives the capacity or communication rate of the channel, so the problem is to allocate a total power of one to the channels, in order to maximize the total communication rate.

Introducing Lagrange multipliers $\lambda^\star \in \mathbf{R}^n$ for the inequality constraints $x^\star \succeq 0$, and a multiplier $\nu^\star \in \mathbf{R}$ for the equality constraint $\mathbf{1}^T x = 1$, we obtain the KKT conditions

$$x^\star \succeq 0, \qquad \mathbf{1}^T x^\star = 1, \qquad \lambda^\star \succeq 0, \qquad \lambda_i^\star x_i^\star = 0, \quad i = 1, \ldots, n,$$

$$-1/(\alpha_i + x_i^\star) - \lambda_i^\star + \nu^\star = 0, \quad i = 1, \ldots, n.$$

We can directly solve these equations to find $x^\star$, $\lambda^\star$, and $\nu^\star$. We start by noting that $\lambda^\star$ acts as a slack variable in the last equation, so it can be eliminated, leaving

$$x^\star \succeq 0, \qquad \mathbf{1}^T x^\star = 1, \qquad x_i^\star \left(\nu^\star - 1/(\alpha_i + x_i^\star)\right) = 0, \quad i = 1, \ldots, n,$$

$$\nu^\star \geq 1/(\alpha_i + x_i^\star), \quad i = 1, \ldots, n.$$

If $\nu^\star < 1/\alpha_i$, this last condition can only hold if $x_i^\star > 0$, which by the third condition implies that $\nu^\star = 1/(\alpha_i + x_i^\star)$. Solving for $x_i^\star$, we conclude that $x_i^\star = 1/\nu^\star - \alpha_i$ if $\nu^\star < 1/\alpha_i$. If $\nu^\star \geq 1/\alpha_i$, then $x_i^\star > 0$ is impossible, because it would imply $\nu^\star \geq 1/\alpha_i > 1/(\alpha_i + x_i^\star)$, which violates the complementary slackness condition. Therefore, $x_i^\star = 0$ if $\nu^\star \geq 1/\alpha_i$. Thus we have

$$x_i^\star = \begin{cases} 1/\nu^\star - \alpha_i & \nu^\star < 1/\alpha_i \\ 0 & \nu^\star \geq 1/\alpha_i, \end{cases}$$

or, put more simply, $x_i^\star = \max\{0, 1/\nu^\star - \alpha_i\}$. Substituting this expression for $x_i^\star$ into the condition $\mathbf{1}^T x^\star = 1$ we obtain

$$\sum_{i=1}^{n} \max\{0, 1/\nu^\star - \alpha_i\} = 1.$$

The lefthand side is a piecewise-linear increasing function of $1/\nu^\star$, with breakpoints at α_i, so the equation has a unique solution which is readily determined.

This solution method is called *water-filling* for the following reason. We think of α_i as the ground level above patch i, and then flood the region with water to a depth $1/\nu$, as illustrated in figure 5.7. The total amount of water used is then $\sum_{i=1}^{n} \max\{0, 1/\nu^\star - \alpha_i\}$. We then increase the flood level until we have used a total amount of water equal to one. The depth of water above patch i is then the optimal value $x_i^\star$.

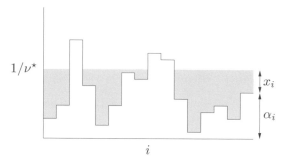

Figure 5.7 Illustration of water-filling algorithm. The height of each patch is given by α_i. The region is flooded to a level $1/\nu^\star$ which uses a total quantity of water equal to one. The height of the water (shown shaded) above each patch is the optimal value of $x_i^\star$.

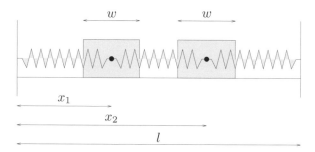

Figure 5.8 Two blocks connected by springs to each other, and the left and right walls. The blocks have width $w > 0$, and cannot penetrate each other or the walls.

5.5.4   Mechanics interpretation of KKT conditions

The KKT conditions can be given a nice interpretation in mechanics (which indeed, was one of Lagrange's primary motivations). We illustrate the idea with a simple example. The system shown in figure 5.8 consists of two blocks attached to each other, and to walls at the left and right, by three springs. The position of the blocks are given by $x \in \mathbf{R}^2$, where x_1 is the displacement of the (middle of the) left block, and x_2 is the displacement of the right block. The left wall is at position 0, and the right wall is at position l.

The potential energy in the springs, as a function of the block positions, is given by

$$f_0(x_1, x_2) = \frac{1}{2}k_1 x_1^2 + \frac{1}{2}k_2(x_2 - x_1)^2 + \frac{1}{2}k_3(l - x_2)^2,$$

where $k_i > 0$ are the stiffness constants of the three springs. The equilibrium position $x^\star$ is the position that minimizes the potential energy subject to the inequalities

$$w/2 - x_1 \leq 0, \qquad w + x_1 - x_2 \leq 0, \qquad w/2 - l + x_2 \leq 0. \qquad (5.51)$$

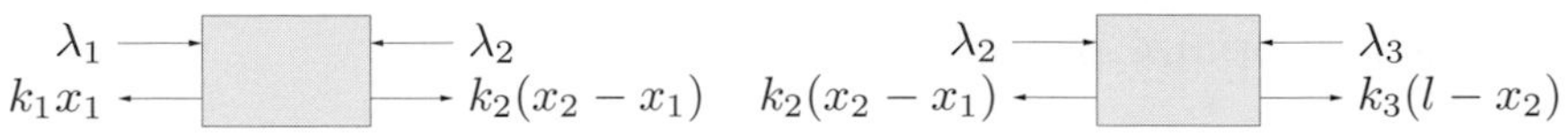

Figure 5.9 Force analysis of the block-spring system. The total force on each block, due to the springs and also to contact forces, must be zero. The Lagrange multipliers, shown on top, are the contact forces between the walls and blocks. The spring forces are shown at bottom.

These constraints are called *kinematic constraints*, and express the fact that the blocks have width $w > 0$, and cannot penetrate each other or the walls. The equilibrium position is therefore given by the solution of the optimization problem

$$
\begin{aligned}
\text{minimize} \quad & (1/2)\left(k_1 x_1^2 + k_2(x_2 - x_1)^2 + k_3(l - x_2)^2\right) \\
\text{subject to} \quad & w/2 - x_1 \le 0 \\
& w + x_1 - x_2 \le 0 \\
& w/2 - l + x_2 \le 0,
\end{aligned}
\tag{5.52}
$$

which is a QP.

With λ_1, λ_2, λ_3 as Lagrange multipliers, the KKT conditions for this problem consist of the kinematic constraints (5.51), the nonnegativity constraints $\lambda_i \ge 0$, the complementary slackness conditions

$$
\lambda_1(w/2 - x_1) = 0, \qquad \lambda_2(w - x_2 + x_1) = 0, \qquad \lambda_3(w/2 - l + x_2) = 0, \tag{5.53}
$$

and the zero gradient condition

$$
\begin{bmatrix} k_1 x_1 - k_2(x_2 - x_1) \\ k_2(x_2 - x_1) - k_3(l - x_2) \end{bmatrix} + \lambda_1 \begin{bmatrix} -1 \\ 0 \end{bmatrix} + \lambda_2 \begin{bmatrix} 1 \\ -1 \end{bmatrix} + \lambda_3 \begin{bmatrix} 0 \\ 1 \end{bmatrix} = 0. \tag{5.54}
$$

The equation (5.54) can be interpreted as the force balance equations for the two blocks, provided we interpret the Lagrange multipliers as *contact forces* that act between the walls and blocks, as illustrated in figure 5.9. The first equation states that the sum of the forces on the first block is zero: The term $-k_1 x_1$ is the force exerted on the left block by the left spring, the term $k_2(x_2 - x_1)$ is the force exerted by the middle spring, λ_1 is the force exerted by the left wall, and $-\lambda_2$ is the force exerted by the right block. The contact forces must point away from the contact surface (as expressed by the constraints $\lambda_1 \ge 0$ and $-\lambda_2 \le 0$), and are nonzero only when there is contact (as expressed by the first two complementary slackness conditions (5.53)). In a similar way, the second equation in (5.54) is the force balance for the second block, and the last condition in (5.53) states that λ_3 is zero unless the right block touches the wall.

In this example, the potential energy and kinematic constraint functions are convex, and (the refined form of) Slater's constraint qualification holds provided $2w \le l$, *i.e.*, there is enough room between the walls to fit the two blocks, so we can conclude that the energy formulation of the equilibrium given by (5.52), gives the same result as the force balance formulation, given by the KKT conditions.

5.5.5   Solving the primal problem via the dual

We mentioned at the beginning of §5.5.3 that if strong duality holds and a dual optimal solution $(\lambda^\star, \nu^\star)$ exists, then any primal optimal point is also a minimizer of $L(x, \lambda^\star, \nu^\star)$. This fact sometimes allows us to compute a primal optimal solution from a dual optimal solution.

More precisely, suppose we have strong duality and an optimal $(\lambda^\star, \nu^\star)$ is known. Suppose that the minimizer of $L(x, \lambda^\star, \nu^\star)$, *i.e.*, the solution of

$$\text{minimize} \quad f_0(x) + \sum_{i=1}^{m} \lambda_i^\star f_i(x) + \sum_{i=1}^{p} \nu_i^\star h_i(x), \qquad (5.55)$$

is unique. (For a convex problem this occurs, for example, if $L(x, \lambda^\star, \nu^\star)$ is a strictly convex function of x.) Then if the solution of (5.55) is primal feasible, it must be primal optimal; if it is not primal feasible, then no primal optimal point can exist, *i.e.*, we can conclude that the primal optimum is not attained. This observation is interesting when the dual problem is easier to solve than the primal problem, for example, because it can be solved analytically, or has some special structure that can be exploited.

Example 5.3 *Entropy maximization.* We consider the entropy maximization problem

$$\begin{array}{ll} \text{minimize} & f_0(x) = \sum_{i=1}^{n} x_i \log x_i \\ \text{subject to} & Ax \preceq b \\ & \mathbf{1}^T x = 1 \end{array}$$

with domain $\mathbf{R}_{++}^n$, and its dual problem

$$\begin{array}{ll} \text{maximize} & -b^T \lambda - \nu - e^{-\nu-1} \sum_{i=1}^{n} e^{-a_i^T \lambda} \\ \text{subject to} & \lambda \succeq 0 \end{array}$$

where a_i are the columns of A (see pages 222 and 228). We assume that the weak form of Slater's condition holds, *i.e.*, there exists an $x \succ 0$ with $Ax \preceq b$ and $\mathbf{1}^T x = 1$, so strong duality holds and an optimal solution $(\lambda^\star, \nu^\star)$ exists.

Suppose we have solved the dual problem. The Lagrangian at $(\lambda^\star, \nu^\star)$ is

$$L(x, \lambda^\star, \nu^\star) = \sum_{i=1}^{n} x_i \log x_i + \lambda^{\star T}(Ax - b) + \nu^\star(\mathbf{1}^T x - 1)$$

which is strictly convex on $\mathcal{D}$ and bounded below, so it has a unique solution $x^\star$, given by

$$x_i^\star = 1/\exp(a_i^T \lambda^\star + \nu^\star + 1), \quad i = 1, \ldots, n.$$

If $x^\star$ is primal feasible, it must be the optimal solution of the primal problem (5.13). If $x^\star$ is not primal feasible, then we can conclude that the primal optimum is not attained.

Example 5.4 *Minimizing a separable function subject to an equality constraint.* We consider the problem

$$\begin{array}{ll} \text{minimize} & f_0(x) = \sum_{i=1}^{n} f_i(x_i) \\ \text{subject to} & a^T x = b, \end{array}$$

where $a \in \mathbf{R}^n$, $b \in \mathbf{R}$, and $f_i : \mathbf{R} \to \mathbf{R}$ are differentiable and strictly convex. The objective function is called *separable* since it is a sum of functions of the individual variables $x_1, \ldots, x_n$. We assume that the domain of f_0 intersects the constraint set, i.e., there exists a point $x_0 \in \mathbf{dom}\, f_0$ with $a^T x_0 = b$. This implies the problem has a unique optimal point $x^\star$.

The Lagrangian is

$$L(x, \nu) = \sum_{i=1}^{n} f_i(x_i) + \nu(a^T x - b) = -b\nu + \sum_{i=1}^{n}(f_i(x_i) + \nu a_i x_i),$$

which is also separable, so the dual function is

$$
\begin{aligned}
g(\nu) &= -b\nu + \inf_{x}\left(\sum_{i=1}^{n}(f_i(x_i) + \nu a_i x_i)\right) \\
&= -b\nu + \sum_{i=1}^{n} \inf_{x_i}(f_i(x_i) + \nu a_i x_i) \\
&= -b\nu - \sum_{i=1}^{n} f_i^*(-\nu a_i).
\end{aligned}
$$

The dual problem is thus

$$\text{maximize} \quad -b\nu - \sum_{i=1}^{n} f_i^*(-\nu a_i),$$

with (scalar) variable $\nu \in \mathbf{R}$.

Now suppose we have found an optimal dual variable $\nu^\star$. (There are several simple methods for solving a convex problem with one scalar variable, such as the bisection method.) Since each f_i is strictly convex, the function $L(x, \nu^\star)$ is strictly convex in x, and so has a unique minimizer $\tilde{x}$. But we also know that $x^\star$ minimizes $L(x, \nu^\star)$, so we must have $\tilde{x} = x^\star$. We can recover $x^\star$ from $\nabla_x L(x, \nu^\star) = 0$, i.e., by solving the equations $f_i'(x_i^\star) = -\nu^\star a_i$.

5.6   Perturbation and sensitivity analysis

When strong duality obtains, the optimal dual variables give very useful information about the sensitivity of the optimal value with respect to perturbations of the constraints.

5.6.1   The perturbed problem

We consider the following perturbed version of the original optimization problem (5.1):

$$
\begin{aligned}
\text{minimize} \quad & f_0(x) \\
\text{subject to} \quad & f_i(x) \le u_i, \quad i = 1, \ldots, m \\
& h_i(x) = v_i, \quad i = 1, \ldots, p,
\end{aligned}
\tag{5.56}
$$

with variable $x \in \mathbf{R}^n$. This problem coincides with the original problem (5.1) when $u = 0$, $v = 0$. When u_i is positive it means that we have relaxed the ith inequality constraint; when u_i is negative, it means that we have tightened the constraint. Thus the perturbed problem (5.56) results from the original problem (5.1) by tightening or relaxing each inequality constraint by u_i, and changing the righthand side of the equality constraints by v_i.

We define $p^\star(u, v)$ as the optimal value of the perturbed problem (5.56):

$$p^\star(u, v) = \inf\{f_0(x) \mid \exists x \in \mathcal{D}, \ f_i(x) \le u_i, \ i = 1, \dots, m,$$
$$h_i(x) = v_i, \ i = 1, \dots, p\}.$$

We can have $p^\star(u, v) = \infty$, which corresponds to perturbations of the constraints that result in infeasibility. Note that $p^\star(0, 0) = p^\star$, the optimal value of the unperturbed problem (5.1). (We hope this slight abuse of notation will cause no confusion.) Roughly speaking, the function $p^\star : \mathbf{R}^m \times \mathbf{R}^p \to \mathbf{R}$ gives the optimal value of the problem as a function of perturbations to the righthand sides of the constraints.

When the original problem is convex, the function $p^\star$ is a convex function of u and v; indeed, its epigraph is precisely the closure of the set $\mathcal{A}$ defined in (5.37) (see exercise 5.32).

5.6.2   A global inequality

Now we assume that strong duality holds, and that the dual optimum is attained. (This is the case if the original problem is convex, and Slater's condition is satisfied). Let $(\lambda^\star, \nu^\star)$ be optimal for the dual (5.16) of the unperturbed problem. Then for all u and v we have

$$p^\star(u, v) \ge p^\star(0, 0) - \lambda^{\star T}u - \nu^{\star T}v. \tag{5.57}$$

To establish this inequality, suppose that x is any feasible point for the perturbed problem, i.e., $f_i(x) \le u_i$ for $i = 1, \dots, m$, and $h_i(x) = v_i$ for $i = 1, \dots, p$. Then we have, by strong duality,

$$p^\star(0, 0) = g(\lambda^\star, \nu^\star) \quad \le \quad f_0(x) + \sum_{i=1}^m \lambda_i^\star f_i(x) + \sum_{i=1}^p \nu_i^\star h_i(x)$$
$$\le \quad f_0(x) + \lambda^{\star T}u + \nu^{\star T}v.$$

(The first inequality follows from the definition of $g(\lambda^\star, \nu^\star)$; the second follows since $\lambda^\star \succeq 0$.) We conclude that for any x feasible for the perturbed problem, we have

$$f_0(x) \ge p^\star(0, 0) - \lambda^{\star T}u - \nu^{\star T}v,$$

from which (5.57) follows.

Sensitivity interpretations

When strong duality holds, various sensitivity interpretations of the optimal Lagrange variables follow directly from the inequality (5.57). Some of the conclusions are:

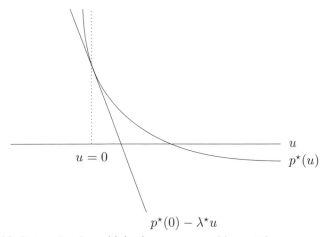

Figure 5.10 Optimal value $p^\star(u)$ of a convex problem with one constraint $f_1(x) \le u$, as a function of u. For $u = 0$, we have the original unperturbed problem; for $u < 0$ the constraint is tightened, and for $u > 0$ the constraint is loosened. The affine function $p^\star(0) - \lambda^\star u$ is a lower bound on $p^\star$.

- If $\lambda_i^\star$ is large and we tighten the ith constraint (*i.e.*, choose $u_i < 0$), then the optimal value $p^\star(u, v)$ is guaranteed to increase greatly.

- If $\nu_i^\star$ is large and positive and we take $v_i < 0$, or if $\nu_i^\star$ is large and negative and we take $v_i > 0$, then the optimal value $p^\star(u, v)$ is guaranteed to increase greatly.

- If $\lambda_i^\star$ is small, and we loosen the ith constraint ($u_i > 0$), then the optimal value $p^\star(u, v)$ will not decrease too much.

- If $\nu_i^\star$ is small and positive, and $v_i > 0$, or if $\nu_i^\star$ is small and negative and $v_i < 0$, then the optimal value $p^\star(u, v)$ will not decrease too much.

The inequality (5.57), and the conclusions listed above, give a *lower bound* on the perturbed optimal value, but no upper bound. For this reason the results are *not* symmetric with respect to loosening or tightening a constraint. For example, suppose that $\lambda_i^\star$ is large, and we loosen the ith constraint a bit (*i.e.*, take u_i small and positive). In this case the inequality (5.57) is not useful; it does not, for example, imply that the optimal value will decrease considerably.

The inequality (5.57) is illustrated in figure 5.10 for a convex problem with one inequality constraint. The inequality states that the affine function $p^\star(0) - \lambda^\star u$ is a lower bound on the convex function $p^\star$.

5.6.3 Local sensitivity analysis

Suppose now that $p^\star(u, v)$ is differentiable at $u = 0$, $v = 0$. Then, provided strong duality holds, the optimal dual variables $\lambda^\star$, $\nu^\star$ are related to the gradient of $p^\star$ at

$u = 0$, $v = 0$:

$$\lambda_i^\star = -\frac{\partial p^\star(0,0)}{\partial u_i}, \qquad \nu_i^\star = -\frac{\partial p^\star(0,0)}{\partial v_i}. \tag{5.58}$$

This property can be seen in the example shown in figure 5.10, where $-\lambda^\star$ is the slope of $p^\star$ near $u = 0$.

Thus, when $p^\star(u, v)$ is differentiable at $u = 0$, $v = 0$, and strong duality holds, the optimal Lagrange multipliers are exactly the local sensitivities of the optimal value with respect to constraint perturbations. In contrast to the nondifferentiable case, this interpretation *is* symmetric: Tightening the ith inequality constraint a small amount (*i.e.*, taking u_i small and negative) yields an increase in $p^\star$ of approximately $-\lambda_i^\star u_i$; loosening the ith constraint a small amount (*i.e.*, taking u_i small and positive) yields a decrease in $p^\star$ of approximately $\lambda_i^\star u_i$.

To show (5.58), suppose $p^\star(u, v)$ is differentiable and strong duality holds. For the perturbation $u = te_i$, $v = 0$, where e_i is the ith unit vector, we have

$$\lim_{t \to 0} \frac{p^\star(te_i, 0) - p^\star}{t} = \frac{\partial p^\star(0,0)}{\partial u_i}.$$

The inequality (5.57) states that for $t > 0$,

$$\frac{p^\star(te_i, 0) - p^\star}{t} \geq -\lambda_i^\star,$$

while for $t < 0$ we have the opposite inequality. Taking the limit $t \to 0$, with $t > 0$, yields

$$\frac{\partial p^\star(0,0)}{\partial u_i} \geq -\lambda_i^\star,$$

while taking the limit with $t < 0$ yields the opposite inequality, so we conclude that

$$\frac{\partial p^\star(0,0)}{\partial u_i} = -\lambda_i^\star.$$

The same method can be used to establish

$$\frac{\partial p^\star(0,0)}{\partial v_i} = -\nu_i^\star.$$

The local sensitivity result (5.58) gives us a quantitative measure of how active a constraint is at the optimum $x^\star$. If $f_i(x^\star) < 0$, then the constraint is inactive, and it follows that the constraint can be tightened or loosened a small amount without affecting the optimal value. By complementary slackness, the associated optimal Lagrange multiplier must be zero. But now suppose that $f_i(x^\star) = 0$, *i.e.*, the ith constraint is active at the optimum. The ith optimal Lagrange multiplier tells us how active the constraint is: If $\lambda_i^\star$ is small, it means that the constraint can be loosened or tightened a bit without much effect on the optimal value; if $\lambda_i^\star$ is large, it means that if the constraint is loosened or tightened a bit, the effect on the optimal value will be great.

Shadow price interpretation

We can also give a simple geometric interpretation of the result (5.58) in terms of economics. We consider (for simplicity) a convex problem with no equality constraints, which satisfies Slater's condition. The variable $x \in \mathbf{R}^m$ determines how a firm operates, and the objective f_0 is the cost, *i.e.*, $-f_0$ is the profit. Each constraint $f_i(x) \leq 0$ represents a limit on some resource such as labor, steel, or warehouse space. The (negative) perturbed optimal cost function $-p^\star(u)$ tells us how much more or less profit could be made if more, or less, of each resource were made available to the firm. If it is differentiable near $u = 0$, then we have

$$\lambda_i^\star = -\frac{\partial p^\star(0)}{\partial u_i}.$$

In other words, $\lambda_i^\star$ tells us approximately how much more profit the firm could make, for a small increase in availability of resource i.

It follows that $\lambda_i^\star$ would be the natural or equilibrium *price* for resource i, if it were possible for the firm to buy or sell it. Suppose, for example, that the firm can buy or sell resource i, at a price that is less than $\lambda_i^\star$. In this case it would certainly buy some of the resource, which would allow it to operate in a way that increases its profit more than the cost of buying the resource. Conversely, if the price exceeds $\lambda_i^\star$, the firm would sell some of its allocation of resource i, and obtain a net gain since its income from selling some of the resource would be larger than its drop in profit due to the reduction in availability of the resource.

5.7 Examples

In this section we show by example that simple equivalent reformulations of a problem can lead to very different dual problems. We consider the following types of reformulations:

- Introducing new variables and associated equality constraints.
- Replacing the objective with an increasing function of the original objective.
- Making explicit constraints implicit, *i.e.*, incorporating them into the domain of the objective.

5.7.1 Introducing new variables and equality constraints

Consider an unconstrained problem of the form

$$\text{minimize} \quad f_0(Ax + b). \tag{5.59}$$

Its Lagrange dual function is the constant $p^\star$. So while we do have strong duality, *i.e.*, $p^\star = d^\star$, the Lagrangian dual is neither useful nor interesting.

Now let us reformulate the problem (5.59) as

$$\begin{array}{ll}\text{minimize} & f_0(y) \\ \text{subject to} & Ax + b = y.\end{array} \qquad (5.60)$$

Here we have introduced new variables y, as well as new equality constraints $Ax + b = y$. The problems (5.59) and (5.60) are clearly equivalent.

The Lagrangian of the reformulated problem is

$$L(x, y, \nu) = f_0(y) + \nu^T (Ax + b - y).$$

To find the dual function we minimize L over x and y. Minimizing over x we find that $g(\nu) = -\infty$ unless $A^T \nu = 0$, in which case we are left with

$$g(\nu) = b^T \nu + \inf_y (f_0(y) - \nu^T y) = b^T \nu - f_0^*(\nu),$$

where f_0^* is the conjugate of f_0. The dual problem of (5.60) can therefore be expressed as

$$\begin{array}{ll}\text{maximize} & b^T \nu - f_0^*(\nu) \\ \text{subject to} & A^T \nu = 0.\end{array} \qquad (5.61)$$

Thus, the dual of the reformulated problem (5.60) is considerably more useful than the dual of the original problem (5.59).

Example 5.5 *Unconstrained geometric program.* Consider the unconstrained geometric program

$$\text{minimize} \quad \log \left(\sum_{i=1}^m \exp(a_i^T x + b_i) \right).$$

We first reformulate it by introducing new variables and equality constraints:

$$\begin{array}{ll}\text{minimize} & f_0(y) = \log \left(\sum_{i=1}^m \exp y_i \right) \\ \text{subject to} & Ax + b = y,\end{array}$$

where a_i^T are the rows of A. The conjugate of the log-sum-exp function is

$$f_0^*(\nu) = \begin{cases} \sum_{i=1}^m \nu_i \log \nu_i & \nu \succeq 0, \ \mathbf{1}^T \nu = 1 \\ \infty & \text{otherwise} \end{cases}$$

(example 3.25, page 93), so the dual of the reformulated problem can be expressed as

$$\begin{array}{ll}\text{maximize} & b^T \nu - \sum_{i=1}^m \nu_i \log \nu_i \\ \text{subject to} & \mathbf{1}^T \nu = 1 \\ & A^T \nu = 0 \\ & \nu \succeq 0,\end{array} \qquad (5.62)$$

which is an entropy maximization problem.

Example 5.6 *Norm approximation problem.* We consider the unconstrained norm approximation problem

$$\text{minimize} \quad \|Ax - b\|, \qquad (5.63)$$

where $\|\cdot\|$ is any norm. Here too the Lagrange dual function is constant, equal to the optimal value of (5.63), and therefore not useful.

Once again we reformulate the problem as

$$\begin{array}{ll} \text{minimize} & \|y\| \\ \text{subject to} & Ax - b = y. \end{array}$$

The Lagrange dual problem is, following (5.61),

$$\begin{array}{ll} \text{maximize} & b^T \nu \\ \text{subject to} & \|\nu\|_* \leq 1 \\ & A^T \nu = 0, \end{array} \qquad (5.64)$$

where we use the fact that the conjugate of a norm is the indicator function of the dual norm unit ball (example 3.26, page 93).

The idea of introducing new equality constraints can be applied to the constraint functions as well. Consider, for example, the problem

$$\begin{array}{ll} \text{minimize} & f_0(A_0 x + b_0) \\ \text{subject to} & f_i(A_i x + b_i) \leq 0, \quad i = 1, \dots, m, \end{array} \qquad (5.65)$$

where $A_i \in \mathbf{R}^{k_i \times n}$ and $f_i : \mathbf{R}^{k_i} \to \mathbf{R}$ are convex. (For simplicity we do not include equality constraints here.) We introduce a new variable $y_i \in \mathbf{R}^{k_i}$, for $i = 0, \dots, m$, and reformulate the problem as

$$\begin{array}{ll} \text{minimize} & f_0(y_0) \\ \text{subject to} & f_i(y_i) \leq 0, \quad i = 1, \dots, m \\ & A_i x + b_i = y_i, \quad i = 0, \dots, m. \end{array} \qquad (5.66)$$

The Lagrangian for this problem is

$$L(x, y_0, \dots, y_m, \lambda, \nu_0, \dots, \nu_m) = f_0(y_0) + \sum_{i=1}^m \lambda_i f_i(y_i) + \sum_{i=0}^m \nu_i^T (A_i x + b_i - y_i).$$

To find the dual function we minimize over x and y_i. The minimum over x is $-\infty$ unless

$$\sum_{i=0}^m A_i^T \nu_i = 0,$$

in which case we have, for $\lambda \succ 0$,

$$g(\lambda, \nu_0, \dots, \nu_m)$$
$$= \sum_{i=0}^m \nu_i^T b_i + \inf_{y_0, \dots, y_m} \left(f_0(y_0) + \sum_{i=1}^m \lambda_i f_i(y_i) - \sum_{i=0}^m \nu_i^T y_i \right)$$
$$= \sum_{i=0}^m \nu_i^T b_i + \inf_{y_0} \left(f_0(y_0) - \nu_0^T y_0 \right) + \sum_{i=1}^m \lambda_i \inf_{y_i} \left(f_i(y_i) - (\nu_i/\lambda_i)^T y_i \right)$$
$$= \sum_{i=0}^m \nu_i^T b_i - f_0^*(\nu_0) - \sum_{i=1}^m \lambda_i f_i^*(\nu_i/\lambda_i).$$

The last expression involves the perspective of the conjugate function, and is therefore concave in the dual variables. Finally, we address the question of what happens when $\lambda \succeq 0$, but some λ_i are zero. If $\lambda_i = 0$ and $\nu_i \neq 0$, then the dual function is $-\infty$. If $\lambda_i = 0$ and $\nu_i = 0$, however, the terms involving y_i, ν_i, and λ_i are all zero. Thus, the expression above for g is valid for all $\lambda \succeq 0$, if we take $\lambda_i f_i^*(\nu_i/\lambda_i) = 0$ when $\lambda_i = 0$ and $\nu_i = 0$, and $\lambda_i f_i^*(\nu_i/\lambda_i) = \infty$ when $\lambda_i = 0$ and $\nu_i \neq 0$.

Therefore we can express the dual of the problem (5.66) as

$$
\begin{aligned}
\text{maximize} \quad & \textstyle\sum_{i=0}^m \nu_i^T b_i - f_0^*(\nu_0) - \sum_{i=1}^m \lambda_i f_i^*(\nu_i/\lambda_i) \\
\text{subject to} \quad & \lambda \succeq 0 \\
& \textstyle\sum_{i=0}^m A_i^T \nu_i = 0.
\end{aligned}
\tag{5.67}
$$

Example 5.7 *Inequality constrained geometric program.* The inequality constrained geometric program

$$
\begin{aligned}
\text{minimize} \quad & \log\left(\textstyle\sum_{k=1}^{K_0} e^{a_{0k}^T x + b_{0k}}\right) \\
\text{subject to} \quad & \log\left(\textstyle\sum_{k=1}^{K_i} e^{a_{ik}^T x + b_{ik}}\right) \leq 0, \quad i = 1, \ldots, m
\end{aligned}
$$

is of the form (5.65) with $f_i : \mathbf{R}^{K_i} \to \mathbf{R}$ given by $f_i(y) = \log\left(\sum_{k=1}^{K_i} e^{y_k}\right)$. The conjugate of this function is

$$
f_i^*(\nu) = \begin{cases} \sum_{k=1}^{K_i} \nu_k \log \nu_k & \nu \succeq 0, \quad \mathbf{1}^T \nu = 1 \\ \infty & \text{otherwise.} \end{cases}
$$

Using (5.67) we can immediately write down the dual problem as

$$
\begin{aligned}
\text{maximize} \quad & b_0^T \nu_0 - \textstyle\sum_{k=1}^{K_0} \nu_{0k} \log \nu_{0k} + \sum_{i=1}^m \left(b_i^T \nu_i - \sum_{k=1}^{K_i} \nu_{ik} \log(\nu_{ik}/\lambda_i)\right) \\
\text{subject to} \quad & \nu_0 \succeq 0, \quad \mathbf{1}^T \nu_0 = 1 \\
& \nu_i \succeq 0, \quad \mathbf{1}^T \nu_i = \lambda_i, \quad i = 1, \ldots, m \\
& \lambda_i \geq 0, \quad i = 1, \ldots, m \\
& \textstyle\sum_{i=0}^m A_i^T \nu_i = 0,
\end{aligned}
$$

which further simplifies to

$$
\begin{aligned}
\text{maximize} \quad & b_0^T \nu_0 - \textstyle\sum_{k=1}^{K_0} \nu_{0k} \log \nu_{0k} + \sum_{i=1}^m \left(b_i^T \nu_i - \sum_{k=1}^{K_i} \nu_{ik} \log(\nu_{ik}/\mathbf{1}^T \nu_i)\right) \\
\text{subject to} \quad & \nu_i \succeq 0, \quad i = 0, \ldots, m \\
& \mathbf{1}^T \nu_0 = 1 \\
& \textstyle\sum_{i=0}^m A_i^T \nu_i = 0.
\end{aligned}
$$

5.7.2 Transforming the objective

If we replace the objective f_0 by an increasing function of f_0, the resulting problem is clearly equivalent (see §4.1.3). The dual of this equivalent problem, however, can be very different from the dual of the original problem.

Example 5.8 We consider again the minimum norm problem

$$
\text{minimize} \quad \|Ax - b\|,
$$

where $\|\cdot\|$ is some norm. We reformulate this problem as

$$\begin{array}{ll}\text{minimize} & (1/2)\|y\|^2 \\ \text{subject to} & Ax - b = y.\end{array}$$

Here we have introduced new variables, and replaced the objective by half its square. Evidently it is equivalent to the original problem.

The dual of the reformulated problem is

$$\begin{array}{ll}\text{maximize} & -(1/2)\|\nu\|_*^2 + b^T\nu \\ \text{subject to} & A^T\nu = 0,\end{array}$$

where we use the fact that the conjugate of $(1/2)\|\cdot\|^2$ is $(1/2)\|\cdot\|_*^2$ (see example 3.27, page 93).

Note that this dual problem is not the same as the dual problem (5.64) derived earlier.

5.7.3   Implicit constraints

The next simple reformulation we study is to include some of the constraints in the objective function, by modifying the objective function to be infinite when the constraint is violated.

Example 5.9 *Linear program with box constraints.* We consider the linear program

$$\begin{array}{ll}\text{minimize} & c^T x \\ \text{subject to} & Ax = b \\ & l \preceq x \preceq u\end{array} \qquad (5.68)$$

where $A \in \mathbf{R}^{p \times n}$ and $l \prec u$. The constraints $l \preceq x \preceq u$ are sometimes called *box constraints* or *variable bounds*.

We can, of course, derive the dual of this linear program. The dual will have a Lagrange multiplier ν associated with the equality constraint, λ_1 associated with the inequality constraint $x \preceq u$, and λ_2 associated with the inequality constraint $l \preceq x$. The dual is

$$\begin{array}{ll}\text{maximize} & -b^T\nu - \lambda_1^T u + \lambda_2^T l \\ \text{subject to} & A^T\nu + \lambda_1 - \lambda_2 + c = 0 \\ & \lambda_1 \succeq 0, \quad \lambda_2 \succeq 0.\end{array} \qquad (5.69)$$

Instead, let us first reformulate the problem (5.68) as

$$\begin{array}{ll}\text{minimize} & f_0(x) \\ \text{subject to} & Ax = b,\end{array} \qquad (5.70)$$

where we define

$$f_0(x) = \left\{ \begin{array}{ll} c^T x & l \preceq x \preceq u \\ \infty & \text{otherwise.} \end{array} \right.$$

The problem (5.70) is clearly equivalent to (5.68); we have merely made the explicit box constraints implicit.

The dual function for the problem (5.70) is

$$
\begin{aligned}
g(\nu) &= \inf_{l \preceq x \preceq u} \left(c^T x + \nu^T (Ax - b) \right) \\
&= -b^T \nu - u^T (A^T \nu + c)^- + l^T (A^T \nu + c)^+
\end{aligned}
$$

where $y_i^+ = \max\{y_i, 0\}$, $y_i^- = \max\{-y_i, 0\}$. So here we are able to derive an analytical formula for g, which is a concave piecewise-linear function.

The dual problem is the unconstrained problem

$$
\text{maximize} \quad -b^T \nu - u^T (A^T \nu + c)^- + l^T (A^T \nu + c)^+, \tag{5.71}
$$

which has a quite different form from the dual of the original problem.

(The problems (5.69) and (5.71) are closely related, in fact, equivalent; see exercise 5.8.)

5.8   Theorems of alternatives

5.8.1   Weak alternatives via the dual function

In this section we apply Lagrange duality theory to the problem of determining feasibility of a system of inequalities and equalities

$$
f_i(x) \leq 0, \quad i = 1, \dots, m, \qquad h_i(x) = 0, \quad i = 1, \dots, p. \tag{5.72}
$$

We assume the domain of the inequality system (5.72), $\mathcal{D} = \bigcap_{i=1}^m \operatorname{\mathbf{dom}} f_i \cap \bigcap_{i=1}^p \operatorname{\mathbf{dom}} h_i$, is nonempty. We can think of (5.72) as the standard problem (5.1), with objective $f_0 = 0$, i.e.,

$$
\begin{array}{ll}
\text{minimize} & 0 \\
\text{subject to} & f_i(x) \leq 0, \quad i = 1, \dots, m \\
& h_i(x) = 0, \quad i = 1, \dots, p.
\end{array} \tag{5.73}
$$

This problem has optimal value

$$
p^\star = \begin{cases} 0 & \text{(5.72) is feasible} \\ \infty & \text{(5.72) is infeasible,} \end{cases} \tag{5.74}
$$

so solving the optimization problem (5.73) is the same as solving the inequality system (5.72).

The dual function

We associate with the inequality system (5.72) the dual function

$$
g(\lambda, \nu) = \inf_{x \in \mathcal{D}} \left(\sum_{i=1}^m \lambda_i f_i(x) + \sum_{i=1}^p \nu_i h_i(x) \right),
$$

which is the same as the dual function for the optimization problem (5.73). Since $f_0 = 0$, the dual function is positive homogeneous in (λ, ν): For $\alpha > 0$, $g(\alpha\lambda, \alpha\nu) = \alpha g(\lambda, \nu)$. The dual problem associated with (5.73) is to maximize $g(\lambda, \nu)$ subject to $\lambda \succeq 0$. Since g is homogeneous, the optimal value of this dual problem is given by

$$d^\star = \begin{cases} \infty & \lambda \succeq 0, \ g(\lambda, \nu) > 0 \text{ is feasible} \\ 0 & \lambda \succeq 0, \ g(\lambda, \nu) > 0 \text{ is infeasible.} \end{cases} \tag{5.75}$$

Weak duality tells us that $d^\star \leq p^\star$. Combining this fact with (5.74) and (5.75) yields the following: If the inequality system

$$\lambda \succeq 0, \qquad g(\lambda, \nu) > 0 \tag{5.76}$$

is feasible (which means $d^\star = \infty$), then the inequality system (5.72) is infeasible (since we then have $p^\star = \infty$). Indeed, we can interpret any solution (λ, ν) of the inequalities (5.76) as a *proof* or *certificate* of infeasibility of the system (5.72).

We can restate this implication in terms of feasibility of the original system: If the original inequality system (5.72) is feasible, then the inequality system (5.76) must be infeasible. We can interpret an x which satisfies (5.72) as a certificate establishing infeasibility of the inequality system (5.76).

Two systems of inequalities (and equalities) are called *weak alternatives* if at most one of the two is feasible. Thus, the systems (5.72) and (5.76) are weak alternatives. This is true whether or not the inequalities (5.72) are convex (*i.e.*, f_i convex, h_i affine); moreover, the alternative inequality system (5.76) is always convex (*i.e.*, g is concave and the constraints $\lambda_i \geq 0$ are convex).

Strict inequalities

We can also study feasibility of the *strict* inequality system

$$f_i(x) < 0, \quad i = 1, \dots, m, \qquad h_i(x) = 0, \quad i = 1, \dots, p. \tag{5.77}$$

With g defined as for the nonstrict inequality system, we have the alternative inequality system

$$\lambda \succeq 0, \qquad \lambda \neq 0, \qquad g(\lambda, \nu) \geq 0. \tag{5.78}$$

We can show directly that (5.77) and (5.78) are weak alternatives. Suppose there exists an $\tilde{x}$ with $f_i(\tilde{x}) < 0$, $h_i(\tilde{x}) = 0$. Then for any $\lambda \succeq 0$, $\lambda \neq 0$, and ν,

$$\lambda_1 f_1(\tilde{x}) + \cdots + \lambda_m f_m(\tilde{x}) + \nu_1 h_1(\tilde{x}) + \cdots + \nu_p h_p(\tilde{x}) < 0.$$

It follows that

$$\begin{aligned} g(\lambda, \nu) &= \inf_{x \in \mathcal{D}} \left(\sum_{i=1}^m \lambda_i f_i(x) + \sum_{i=1}^p \nu_i h_i(x) \right) \\ &\leq \sum_{i=1}^m \lambda_i f_i(\tilde{x}) + \sum_{i=1}^p \nu_i h_i(\tilde{x}) \\ &< 0. \end{aligned}$$

Therefore, feasibility of (5.77) implies that there does not exist (λ, ν) satisfying (5.78).

Thus, we can prove infeasibility of (5.77) by producing a solution of the system (5.78); we can prove infeasibility of (5.78) by producing a solution of the system (5.77).

5.8.2   Strong alternatives

When the original inequality system is convex, *i.e.*, f_i are convex and h_i are affine, and some type of constraint qualification holds, then the pairs of weak alternatives described above are *strong alternatives*, which means that *exactly one* of the two alternatives holds. In other words, each of the inequality systems is feasible if and only if the other is infeasible.

In this section we assume that f_i are convex and h_i are affine, so the inequality system (5.72) can be expressed as

$$f_i(x) \le 0, \quad i = 1, \ldots, m, \qquad Ax = b,$$

where $A \in \mathbf{R}^{p \times n}$.

Strict inequalities

We first study the strict inequality system

$$f_i(x) < 0, \quad i = 1, \ldots, m, \qquad Ax = b, \tag{5.79}$$

and its alternative

$$\lambda \succeq 0, \qquad \lambda \ne 0, \qquad g(\lambda, \nu) \ge 0. \tag{5.80}$$

We need one technical condition: There exists an $x \in \mathbf{relint}\, \mathcal{D}$ with $Ax = b$. In other words we not only assume that the linear equality constraints are consistent, but also that they have a solution in $\mathbf{relint}\, \mathcal{D}$. (Very often $\mathcal{D} = \mathbf{R}^n$, so the condition is satisfied if the equality constraints are consistent.) Under this condition, exactly one of the inequality systems (5.79) and (5.80) is feasible. In other words, the inequality systems (5.79) and (5.80) are strong alternatives.

We will establish this result by considering the related optimization problem

$$\begin{array}{ll} \text{minimize} & s \\ \text{subject to} & f_i(x) - s \le 0, \quad i = 1, \ldots, m \\ & Ax = b \end{array} \tag{5.81}$$

with variables x, s, and domain $\mathcal{D} \times \mathbf{R}$. The optimal value $p^\star$ of this problem is negative if and only if there exists a solution to the strict inequality system (5.79).

The Lagrange dual function for the problem (5.81) is

$$\inf_{x \in \mathcal{D},\, s} \left(s + \sum_{i=1}^{m} \lambda_i (f_i(x) - s) + \nu^T (Ax - b) \right) = \begin{cases} g(\lambda, \nu) & \mathbf{1}^T \lambda = 1 \\ -\infty & \text{otherwise.} \end{cases}$$

Therefore we can express the dual problem of (5.81) as

$$
\begin{array}{ll}
\text{maximize} & g(\lambda, \nu) \\
\text{subject to} & \lambda \succeq 0, \quad \mathbf{1}^T \lambda = 1.
\end{array}
$$

Now we observe that Slater's condition holds for the problem (5.81). By the hypothesis there exists an $\tilde{x} \in \mathbf{relint}\, \mathcal{D}$ with $A\tilde{x} = b$. Choosing any $\tilde{s} > \max_i f_i(\tilde{x})$ yields a point $(\tilde{x}, \tilde{s})$ which is strictly feasible for (5.81). Therefore we have $d^\star = p^\star$, and the dual optimum $d^\star$ is attained. In other words, there exist $(\lambda^\star, \nu^\star)$ such that

$$
g(\lambda^\star, \nu^\star) = p^\star, \qquad \lambda^\star \succeq 0, \qquad \mathbf{1}^T \lambda^\star = 1. \tag{5.82}
$$

Now suppose that the strict inequality system (5.79) is infeasible, which means that $p^\star \geq 0$. Then $(\lambda^\star, \nu^\star)$ from (5.82) satisfy the alternate inequality system (5.80). Similarly, if the alternate inequality system (5.80) is feasible, then $d^\star = p^\star \geq 0$, which shows that the strict inequality system (5.79) is infeasible. Thus, the inequality systems (5.79) and (5.80) are strong alternatives; each is feasible if and only if the other is not.

Nonstrict inequalities

We now consider the nonstrict inequality system

$$
f_i(x) \leq 0, \quad i = 1, \dots, m, \qquad Ax = b, \tag{5.83}
$$

and its alternative

$$
\lambda \succeq 0, \qquad g(\lambda, \nu) > 0. \tag{5.84}
$$

We will show these are strong alternatives, provided the following conditions hold: There exists an $x \in \mathbf{relint}\, \mathcal{D}$ with $Ax = b$, and the optimal value $p^\star$ of (5.81) is attained. This holds, for example, if $\mathcal{D} = \mathbf{R}^n$ and $\max_i f_i(x) \to \infty$ as $x \to \infty$. With these assumptions we have, as in the strict case, that $p^\star = d^\star$, and that both the primal and dual optimal values are attained. Now suppose that the nonstrict inequality system (5.83) is infeasible, which means that $p^\star > 0$. (Here we use the assumption that the primal optimal value is attained.) Then $(\lambda^\star, \nu^\star)$ from (5.82) satisfy the alternate inequality system (5.84). Thus, the inequality systems (5.83) and (5.84) are strong alternatives; each is feasible if and only if the other is not.

5.8.3   Examples

Linear inequalities

Consider the system of linear inequalities $Ax \preceq b$. The dual function is

$$
g(\lambda) = \inf_x \lambda^T (Ax - b) = \begin{cases} -b^T \lambda & A^T \lambda = 0 \\ -\infty & \text{otherwise.} \end{cases}
$$

The alternative inequality system is therefore

$$
\lambda \succeq 0, \qquad A^T \lambda = 0, \qquad b^T \lambda < 0.
$$

These are, in fact, strong alternatives. This follows since the optimum in the related problem (5.81) is achieved, unless it is unbounded below.

We now consider the system of strict linear inequalities $Ax \prec b$, which has the strong alternative system

$$\lambda \succeq 0, \qquad \lambda \neq 0, \qquad A^T \lambda = 0, \qquad b^T \lambda \leq 0.$$

In fact we have encountered (and proved) this result before, in §2.5.1; see (2.17) and (2.18) (on page 50).

Intersection of ellipsoids

We consider m ellipsoids, described as

$$\mathcal{E}_i = \{x \mid f_i(x) \leq 0\},$$

with $f_i(x) = x^T A_i x + 2b_i^T x + c_i$, $i = 1, \ldots, m$, where $A_i \in \mathbf{S}_{++}^n$. We ask when the intersection of these ellipsoids has nonempty interior. This is equivalent to feasibility of the set of strict quadratic inequalities

$$f_i(x) = x^T A_i x + 2b_i^T x + c_i < 0, \quad i = 1, \ldots, m. \tag{5.85}$$

The dual function g is

$$\begin{aligned}
g(\lambda) &= \inf_x \left(x^T A(\lambda) x + 2b(\lambda)^T x + c(\lambda) \right) \\
&= \begin{cases} -b(\lambda)^T A(\lambda)^\dagger b(\lambda) + c(\lambda) & A(\lambda) \succeq 0, \quad b(\lambda) \in \mathcal{R}(A(\lambda)) \\ -\infty & \text{otherwise,} \end{cases}
\end{aligned}$$

where

$$A(\lambda) = \sum_{i=1}^m \lambda_i A_i, \qquad b(\lambda) = \sum_{i=1}^m \lambda_i b_i, \qquad c(\lambda) = \sum_{i=1}^m \lambda_i c_i.$$

Note that for $\lambda \succeq 0$, $\lambda \neq 0$, we have $A(\lambda) \succ 0$, so we can simplify the expression for the dual function as

$$g(\lambda) = -b(\lambda)^T A(\lambda)^{-1} b(\lambda) + c(\lambda).$$

The strong alternative of the system (5.85) is therefore

$$\lambda \succeq 0, \qquad \lambda \neq 0, \qquad -b(\lambda)^T A(\lambda)^{-1} b(\lambda) + c(\lambda) \geq 0. \tag{5.86}$$

We can give a simple geometric interpretation of this pair of strong alternatives. For any nonzero $\lambda \succeq 0$, the (possibly empty) ellipsoid

$$\mathcal{E}_\lambda = \{x \mid x^T A(\lambda) x + 2b(\lambda)^T x + c(\lambda) \leq 0\}$$

contains $\mathcal{E}_1 \cap \cdots \cap \mathcal{E}_m$, since $f_i(x) \leq 0$ implies $\sum_{i=1}^m \lambda_i f_i(x) \leq 0$. Now, $\mathcal{E}_\lambda$ has empty interior if and only if

$$\inf_x \left(x^T A(\lambda) x + 2b(\lambda)^T x + c(\lambda) \right) = -b(\lambda)^T A(\lambda)^{-1} b(\lambda) + c(\lambda) \geq 0.$$

Therefore the alternative system (5.86) means that $\mathcal{E}_\lambda$ has empty interior.

Weak duality is obvious: If (5.86) holds, then $\mathcal{E}_\lambda$ contains the intersection $\mathcal{E}_1 \cap \cdots \cap \mathcal{E}_m$, and has empty interior, so naturally the intersection has empty interior. The fact that these are strong alternatives states the (not obvious) fact that if the intersection $\mathcal{E}_1 \cap \cdots \cap \mathcal{E}_m$ has empty interior, then we can construct an ellipsoid $\mathcal{E}_\lambda$ that contains the intersection and has empty interior.

Farkas' lemma

In this section we describe a pair of strong alternatives for a mixture of strict and nonstrict linear inequalities, known as *Farkas' lemma*: The system of inequalities

$$Ax \preceq 0, \qquad c^T x < 0, \tag{5.87}$$

where $A \in \mathbf{R}^{m \times n}$ and $c \in \mathbf{R}^n$, and the system of equalities and inequalities

$$A^T y + c = 0, \qquad y \succeq 0, \tag{5.88}$$

are strong alternatives.

We can prove Farkas' lemma directly, using LP duality. Consider the LP

$$\begin{array}{ll} \text{minimize} & c^T x \\ \text{subject to} & Ax \preceq 0, \end{array} \tag{5.89}$$

and its dual

$$\begin{array}{ll} \text{maximize} & 0 \\ \text{subject to} & A^T y + c = 0 \\ & y \succeq 0. \end{array} \tag{5.90}$$

The primal LP (5.89) is homogeneous, and so has optimal value 0, if (5.87) is not feasible, and optimal value $-\infty$, if (5.87) is feasible. The dual LP (5.90) has optimal value 0, if (5.88) is feasible, and optimal value $-\infty$, if (5.88) is infeasible.

Since $x = 0$ is feasible in (5.89), we can rule out the one case in which strong duality can fail for LPs, so we must have $p^\star = d^\star$. Combined with the remarks above, this shows that (5.87) and (5.88) are strong alternatives.

Example 5.10 *Arbitrage-free bounds on price.* We consider a set of n assets, with prices at the beginning of an investment period $p_1, \ldots, p_n$, respectively. At the end of the investment period, the value of the assets is $v_1, \ldots, v_n$. If $x_1, \ldots, x_n$ represents the initial investment in each asset (with $x_j < 0$ meaning a short position in asset j), the cost of the initial investment is $p^T x$, and the final value of the investment is $v^T x$.

The value of the assets at the end of the investment period, v, is uncertain. We will assume that only m possible scenarios, or outcomes, are possible. If outcome i occurs, the final value of the assets is $v^{(i)}$, and therefore, the overall value of the investments is $v^{(i)T} x$.

If there is an investment vector x with $p^T x < 0$, and in all possible scenarios, the final value is nonnegative, i.e., $v^{(i)T} x \geq 0$ for $i = 1, \ldots, m$, then an *arbitrage* is said to exist. The condition $p^T x < 0$ means you are *paid* to accept the investment mix, and the condition $v^{(i)T} x \geq 0$ for $i = 1, \ldots, m$ means that no matter what outcome occurs, the final value is nonnegative, so an arbitrage corresponds to a guaranteed money-making investment strategy. It is generally assumed that the prices and values are such that no arbitrage exists. This means that the inequality system

$$Vx \succeq 0, \qquad p^T x < 0$$

is infeasible, where $V_{ij} = v_j^{(i)}$.

Using Farkas' lemma, we have no arbitrage if and only if there exists y such that

$$-V^T y + p = 0, \qquad y \succeq 0.$$

We can use this characterization of arbitrage-free prices and values to solve several interesting problems.

Suppose, for example, that the values V are known, and all prices except the last one, p_n, are known. The set of prices p_n that are consistent with the no-arbitrage assumption is an interval, which can be found by solving a pair of LPs. The optimal value of the LP

$$
\begin{array}{ll}
\text{minimize} & p_n \\
\text{subject to} & V^T y = p, \quad y \succeq 0,
\end{array}
$$

with variables p_n and y, gives the smallest possible arbitrage-free price for asset n. Solving the same LP with maximization instead of minimization yields the largest possible price for asset n. If the two values are equal, *i.e.*, the no-arbitrage assumption leads us to a unique price for asset n, we say the market is *complete*. For an example, see exercise 5.38.

This method can be used to find bounds on the price of a derivative or option that is based on the final value of other underlying assets, *i.e.*, when the value or payoff of asset n is a function of the values of the other assets.

5.9   Generalized inequalities

In this section we examine how Lagrange duality extends to a problem with generalized inequality constraints

$$
\begin{array}{ll}
\text{minimize} & f_0(x) \\
\text{subject to} & f_i(x) \preceq_{K_i} 0, \quad i = 1, \dots, m \\
& h_i(x) = 0, \quad i = 1, \dots, p,
\end{array} \tag{5.91}
$$

where $K_i \subseteq \mathbf{R}^{k_i}$ are proper cones. For now, we do not assume convexity of the problem (5.91). We assume the domain of (5.91), $\mathcal{D} = \bigcap_{i=0}^{m} \mathbf{dom}\, f_i \cap \bigcap_{i=1}^{p} \mathbf{dom}\, h_i$, is nonempty.

5.9.1   The Lagrange dual

With each generalized inequality $f_i(x) \preceq_{K_i} 0$ in (5.91) we associate a Lagrange multiplier *vector* $\lambda_i \in \mathbf{R}^{k_i}$ and define the associated Lagrangian as

$$
L(x, \lambda, \nu) = f_0(x) + \lambda_1^T f_1(x) + \cdots + \lambda_m^T f_m(x) + \nu_1 h_1(x) + \cdots + \nu_p h_p(x),
$$

where $\lambda = (\lambda_1, \dots, \lambda_m)$ and $\nu = (\nu_1, \dots, \nu_p)$. The dual function is defined exactly as in a problem with scalar inequalities:

$$
g(\lambda, \nu) = \inf_{x \in \mathcal{D}} L(x, \lambda, \nu) = \inf_{x \in \mathcal{D}} \left(f_0(x) + \sum_{i=1}^{m} \lambda_i^T f_i(x) + \sum_{i=1}^{p} \nu_i h_i(x) \right).
$$

Since the Lagrangian is affine in the dual variables (λ, ν), and the dual function is a pointwise infimum of the Lagrangian, the dual function is concave.

As in a problem with scalar inequalities, the dual function gives lower bounds on $p^\star$, the optimal value of the primal problem (5.91). For a problem with scalar inequalities, we require $\lambda_i \geq 0$. Here the nonnegativity requirement on the dual variables is replaced by the condition

$$\lambda_i \succeq_{K_i^*} 0, \quad i = 1, \ldots, m,$$

where K_i^* denotes the dual cone of K_i. In other words, the Lagrange multipliers associated with inequalities must be *dual* nonnegative.

Weak duality follows immediately from the definition of dual cone. If $\lambda_i \succeq_{K_i^*} 0$ and $f_i(\tilde{x}) \preceq_{K_i} 0$, then $\lambda_i^T f_i(\tilde{x}) \leq 0$. Therefore for any primal feasible point $\tilde{x}$ and any $\lambda_i \succeq_{K_i^*} 0$, we have

$$f_0(\tilde{x}) + \sum_{i=1}^m \lambda_i^T f_i(\tilde{x}) + \sum_{i=1}^p \nu_i h_i(\tilde{x}) \leq f_0(\tilde{x}).$$

Taking the infimum over $\tilde{x}$ yields $g(\lambda, \nu) \leq p^\star$.

The Lagrange dual optimization problem is

$$\begin{array}{ll} \text{maximize} & g(\lambda, \nu) \\ \text{subject to} & \lambda_i \succeq_{K_i^*} 0, \quad i = 1, \ldots, m. \end{array} \qquad (5.92)$$

We always have *weak duality*, i.e., $d^\star \leq p^\star$, where $d^\star$ denotes the optimal value of the dual problem (5.92), whether or not the primal problem (5.91) is convex.

Slater's condition and strong duality

As might be expected, *strong* duality ($d^\star = p^\star$) holds when the primal problem is convex and satisfies an appropriate constraint qualification. For example, a generalized version of Slater's condition for the problem

$$\begin{array}{ll} \text{minimize} & f_0(x) \\ \text{subject to} & f_i(x) \preceq_{K_i} 0, \quad i = 1, \ldots, m \\ & Ax = b, \end{array}$$

where f_0 is convex and f_i is K_i-convex, is that there exists an $x \in \mathbf{relint}\, \mathcal{D}$ with $Ax = b$ and $f_i(x) \prec_{K_i} 0$, $i = 1, \ldots, m$. This condition implies strong duality (and also, that the dual optimum is attained).

Example 5.11 *Lagrange dual of semidefinite program.* We consider a semidefinite program in inequality form,

$$\begin{array}{ll} \text{minimize} & c^T x \\ \text{subject to} & x_1 F_1 + \cdots + x_n F_n + G \preceq 0 \end{array} \qquad (5.93)$$

where $F_1, \ldots, F_n, G \in \mathbf{S}^k$. (Here f_1 is affine, and K_1 is $\mathbf{S}_+^k$, the positive semidefinite cone.)

We associate with the constraint a dual variable or multiplier $Z \in \mathbf{S}^k$, so the Lagrangian is

$$\begin{aligned} L(x, Z) &= c^T x + \mathbf{tr}\left((x_1 F_1 + \cdots + x_n F_n + G)\, Z\right) \\ &= x_1 (c_1 + \mathbf{tr}(F_1 Z)) + \cdots + x_n (c_n + \mathbf{tr}(F_n Z)) + \mathbf{tr}(GZ), \end{aligned}$$

which is affine in x. The dual function is given by

$$g(Z) = \inf_x L(x, Z) = \begin{cases} \mathbf{tr}(GZ) & \mathbf{tr}(F_iZ) + c_i = 0, \quad i = 1, \ldots, n \\ -\infty & \text{otherwise.} \end{cases}$$

The dual problem can therefore be expressed as

$$\begin{array}{ll} \text{maximize} & \mathbf{tr}(GZ) \\ \text{subject to} & \mathbf{tr}(F_iZ) + c_i = 0, \quad i = 1, \ldots, n \\ & Z \succeq 0. \end{array}$$

(We use the fact that $\mathbf{S}_+^k$ is self-dual, $i.e.$, $(\mathbf{S}_+^k)^* = \mathbf{S}_+^k$; see §2.6.)

Strong duality obtains if the semidefinite program (5.93) is strictly feasible, $i.e.$, there exists an x with

$$x_1F_1 + \cdots + x_nF_n + G \prec 0.$$

Example 5.12 *Lagrange dual of cone program in standard form.* We consider the cone program

$$\begin{array}{ll} \text{minimize} & c^Tx \\ \text{subject to} & Ax = b \\ & x \succeq_K 0, \end{array}$$

where $A \in \mathbf{R}^{m \times n}$, $b \in \mathbf{R}^m$, and $K \subseteq \mathbf{R}^n$ is a proper cone. We associate with the equality constraint a multiplier $\nu \in \mathbf{R}^m$, and with the nonnegativity constraint a multiplier $\lambda \in \mathbf{R}^n$. The Lagrangian is

$$L(x, \lambda, \nu) = c^Tx - \lambda^Tx + \nu^T(Ax - b),$$

so the dual function is

$$g(\lambda, \nu) = \inf_x L(x, \lambda, \nu) = \begin{cases} -b^T\nu & A^T\nu - \lambda + c = 0 \\ -\infty & \text{otherwise.} \end{cases}$$

The dual problem can be expressed as

$$\begin{array}{ll} \text{maximize} & -b^T\nu \\ \text{subject to} & A^T\nu + c = \lambda \\ & \lambda \succeq_{K^*} 0. \end{array}$$

By eliminating λ and defining $y = -\nu$, this problem can be simplified to

$$\begin{array}{ll} \text{maximize} & b^Ty \\ \text{subject to} & A^Ty \preceq_{K^*} c, \end{array}$$

which is a cone program in inequality form, involving the dual generalized inequality.

Strong duality obtains if the Slater condition holds, $i.e.$, there is an $x \succ_K 0$ with $Ax = b$.

5.9.2   Optimality conditions

The optimality conditions of §5.5 are readily extended to problems with generalized inequalities. We first derive the complementary slackness conditions.

Complementary slackness

Assume that the primal and dual optimal values are equal, and attained at the optimal points $x^\star$, $\lambda^\star$, $\nu^\star$. As in §5.5.2, the complementary slackness conditions follow directly from the equality $f_0(x^\star) = g(\lambda^\star, \nu^\star)$, along with the definition of g. We have

$$
\begin{aligned}
f_0(x^\star) &= g(\lambda^\star, \nu^\star) \\
&\leq f_0(x^\star) + \sum_{i=1}^{m} \lambda_i^{\star T} f_i(x^\star) + \sum_{i=1}^{p} \nu_i^\star h_i(x^\star) \\
&\leq f_0(x^\star),
\end{aligned}
$$

and therefore we conclude that $x^\star$ minimizes $L(x, \lambda^\star, \nu^\star)$, and also that the two sums in the second line are zero. Since the second sum is zero (since $x^\star$ satisfies the equality constraints), we have $\sum_{i=1}^{m} \lambda_i^{\star T} f_i(x^\star) = 0$. Since each term in this sum is nonpositive, we conclude that

$$
\lambda_i^{\star T} f_i(x^\star) = 0, \quad i = 1, \ldots, m, \tag{5.94}
$$

which generalizes the complementary slackness condition (5.48). From (5.94) we can conclude that

$$
\lambda_i^\star \succ_{K_i^\star} 0 \implies f_i(x^\star) = 0, \qquad f_i(x^\star) \prec_{K_i} 0, \implies \lambda_i^\star = 0.
$$

However, in contrast to problems with scalar inequalities, it is possible to satisfy (5.94) with $\lambda_i^\star \neq 0$ and $f_i(x^\star) \neq 0$.

KKT conditions

Now we add the assumption that the functions f_i, h_i are differentiable, and generalize the KKT conditions of §5.5.3 to problems with generalized inequalities. Since $x^\star$ minimizes $L(x, \lambda^\star, \nu^\star)$, its gradient with respect to x vanishes at $x^\star$:

$$
\nabla f_0(x^\star) + \sum_{i=1}^{m} Df_i(x^\star)^T \lambda_i^\star + \sum_{i=1}^{p} \nu_i^\star \nabla h_i(x^\star) = 0,
$$

where $Df_i(x^\star) \in \mathbf{R}^{k_i \times n}$ is the derivative of f_i evaluated at $x^\star$ (see §A.4.1). Thus, if strong duality holds, any primal optimal $x^\star$ and any dual optimal $(\lambda^\star, \nu^\star)$ must satisfy the optimality conditions (or KKT conditions)

$$
\begin{aligned}
f_i(x^\star) &\preceq_{K_i} 0, & i &= 1, \ldots, m \\
h_i(x^\star) &= 0, & i &= 1, \ldots, p \\
\lambda_i^\star &\succeq_{K_i^\star} 0, & i &= 1, \ldots, m \\
\lambda_i^{\star T} f_i(x^\star) &= 0, & i &= 1, \ldots, m \\
\nabla f_0(x^\star) + \sum_{i=1}^{m} Df_i(x^\star)^T \lambda_i^\star + \sum_{i=1}^{p} \nu_i^\star \nabla h_i(x^\star) &= 0.
\end{aligned}
\tag{5.95}
$$

If the primal problem is convex, the converse also holds, *i.e.*, the conditions (5.95) are sufficient conditions for optimality of $x^\star$, $(\lambda^\star, \nu^\star)$.

5.9.3   Perturbation and sensitivity analysis

The results of §5.6 can be extended to problems involving generalized inequalities. We consider the associated perturbed version of the problem,

$$
\begin{array}{ll}
\text{minimize} & f_0(x) \\
\text{subject to} & f_i(x) \preceq_{K_i} u_i, \quad i = 1, \ldots, m \\
& h_i(x) = v_i, \quad i = 1, \ldots, p,
\end{array}
$$

where $u_i \in \mathbf{R}^{k_i}$, and $v \in \mathbf{R}^p$. We define $p^\star(u, v)$ as the optimal value of the perturbed problem. As in the case with scalar inequalities, $p^\star$ is a convex function when the original problem is convex.

Now let $(\lambda^\star, \nu^\star)$ be optimal for the dual of the original (unperturbed) problem, which we assume has zero duality gap. Then for all u and v we have

$$
p^\star(u, v) \geq p^\star - \sum_{i=1}^m \lambda_i^{\star T} u_i - \nu^{\star T} v,
$$

the analog of the global sensitivity inequality (5.57). The local sensitivity result holds as well: If $p^\star(u, v)$ is differentiable at $u = 0$, $v = 0$, then the optimal dual variables $\lambda_i^\star$ satisfies

$$
\lambda_i^\star = -\nabla_{u_i} p^\star(0, 0),
$$

the analog of (5.58).

Example 5.13 *Semidefinite program in inequality form.* We consider a semidefinite program in inequality form, as in example 5.11. The primal problem is

$$
\begin{array}{ll}
\text{minimize} & c^T x \\
\text{subject to} & F(x) = x_1 F_1 + \cdots + x_n F_n + G \preceq 0,
\end{array}
$$

with variable $x \in \mathbf{R}^n$ (and $F_1, \ldots, F_n, G \in \mathbf{S}^k$), and the dual problem is

$$
\begin{array}{ll}
\text{maximize} & \mathbf{tr}(GZ) \\
\text{subject to} & \mathbf{tr}(F_i Z) + c_i = 0, \quad i = 1, \ldots, n \\
& Z \succeq 0,
\end{array}
$$

with variable $Z \in \mathbf{S}^k$.

Suppose that $x^\star$ and $Z^\star$ are primal and dual optimal, respectively, with zero duality gap. The complementary slackness condition is $\mathbf{tr}(F(x^\star)Z^\star) = 0$. Since $F(x^\star) \preceq 0$ and $Z^\star \succeq 0$, we can conclude that $F(x^\star)Z^\star = 0$. Thus, the complementary slackness condition can be expressed as

$$
\mathcal{R}(F(x^\star)) \perp \mathcal{R}(Z^\star),
$$

i.e., the ranges of the primal and dual matrices are orthogonal.

Let $p^\star(U)$ denote the optimal value of the perturbed SDP

$$
\begin{array}{ll}
\text{minimize} & c^T x \\
\text{subject to} & F(x) = x_1 F_1 + \cdots + x_n F_n + G \preceq U.
\end{array}
$$

Then we have, for all U, $p^\star(U) \geq p^\star - \mathbf{tr}(Z^\star U)$. If $p^\star(U)$ is differentiable at $U = 0$, then we have
$$\nabla p^\star(0) = -Z^\star.$$
This means that for U small, the optimal value of the perturbed SDP is very close to (the lower bound) $p^\star - \mathbf{tr}(Z^\star U)$.

5.9.4 Theorems of alternatives

We can derive theorems of alternatives for systems of generalized inequalities and equalities

$$f_i(x) \preceq_{K_i} 0, \quad i = 1, \ldots, m, \qquad h_i(x) = 0, \quad i = 1, \ldots, p, \qquad (5.96)$$

where $K_i \subseteq \mathbf{R}^{k_i}$ are proper cones. We will also consider systems with strict inequalities,

$$f_i(x) \prec_{K_i} 0, \quad i = 1, \ldots, m, \qquad h_i(x) = 0, \quad i = 1, \ldots, p. \qquad (5.97)$$

We assume that $\mathcal{D} = \bigcap_{i=0}^m \mathbf{dom}\, f_i \cap \bigcap_{i=1}^p \mathbf{dom}\, h_i$ is nonempty.

Weak alternatives

We associate with the systems (5.96) and (5.97) the dual function

$$g(\lambda, \nu) = \inf_{x \in \mathcal{D}} \left(\sum_{i=1}^m \lambda_i^T f_i(x) + \sum_{i=1}^p \nu_i h_i(x) \right)$$

where $\lambda = (\lambda_1, \ldots, \lambda_m)$ with $\lambda_i \in \mathbf{R}^{k_i}$ and $\nu \in \mathbf{R}^p$. In analogy with (5.76), we claim that

$$\lambda_i \succeq_{K_i^*} 0, \quad i = 1, \ldots, m, \qquad g(\lambda, \nu) > 0 \qquad (5.98)$$

is a weak alternative to the system (5.96). To verify this, suppose there exists an x satisfying (5.96) and (λ, ν) satisfying (5.98). Then we have a contradiction:

$$0 < g(\lambda, \nu) \leq \lambda_1^T f_1(x) + \cdots + \lambda_m^T f_m(x) + \nu_1 h_1(x) + \cdots + \nu_p h_p(x) \leq 0.$$

Therefore at least one of the two systems (5.96) and (5.98) must be infeasible, *i.e.*, the two systems are weak alternatives.

In a similar way, we can prove that (5.97) and the system

$$\lambda_i \succeq_{K_i^*} 0, \quad i = 1, \ldots, m, \qquad \lambda \neq 0, \qquad g(\lambda, \nu) \geq 0.$$

form a pair of weak alternatives.

Strong alternatives

We now assume that the functions f_i are K_i-convex, and the functions h_i are affine. We first consider a system with strict inequalities

$$f_i(x) \prec_{K_i} 0, \quad i = 1, \ldots, m, \qquad Ax = b, \qquad (5.99)$$

and its alternative

$$\lambda_i \succeq_{K_i^*} 0, \quad i = 1, \ldots, m, \qquad \lambda \neq 0, \qquad g(\lambda, \nu) \geq 0. \tag{5.100}$$

We have already seen that (5.99) and (5.100) are weak alternatives. They are also strong alternatives provided the following constraint qualification holds: There exists an $\tilde{x} \in \mathbf{relint}\, \mathcal{D}$ with $A\tilde{x} = b$. To prove this, we select a set of vectors $e_i \succ_{K_i} 0$, and consider the problem

$$
\begin{array}{ll}
\text{minimize} & s \\
\text{subject to} & f_i(x) \preceq_{K_i} se_i, \quad i = 1, \ldots, m \\
& Ax = b
\end{array}
\tag{5.101}
$$

with variables x and $s \in \mathbf{R}$. Slater's condition holds since $(\tilde{x}, \tilde{s})$ satisfies the strict inequalities $f_i(\tilde{x}) \prec_{K_i} \tilde{s}e_i$ provided $\tilde{s}$ is large enough.

The dual of (5.101) is

$$
\begin{array}{ll}
\text{maximize} & g(\lambda, \nu) \\
\text{subject to} & \lambda_i \succeq_{K_i^*} 0, \quad i = 1, \ldots, m \\
& \sum_{i=1}^m e_i^T \lambda_i = 1
\end{array}
\tag{5.102}
$$

with variables $\lambda = (\lambda_1, \ldots, \lambda_m)$ and ν.

Now suppose the system (5.99) is infeasible. Then the optimal value of (5.101) is nonnegative. Since Slater's condition is satisfied, we have strong duality and the dual optimum is attained. Therefore there exist $(\tilde{\lambda}, \tilde{\nu})$ that satisfy the constraints of (5.102) and $g(\tilde{\lambda}, \tilde{\nu}) \geq 0$, i.e., the system (5.100) has a solution.

As we noted in the case of scalar inequalities, existence of an $x \in \mathbf{relint}\, \mathcal{D}$ with $Ax = b$ is not sufficient for the system of nonstrict inequalities

$$f_i(x) \preceq_{K_i} 0, \quad i = 1, \ldots, m, \qquad Ax = b$$

and its alternative

$$\lambda_i \succeq_{K_i^*} 0, \quad i = 1, \ldots, m, \qquad g(\lambda, \nu) > 0$$

to be strong alternatives. An additional condition is required, e.g., that the optimal value of (5.101) is attained.

Example 5.14 *Feasibility of a linear matrix inequality.* The following systems are strong alternatives:

$$F(x) = x_1 F_1 + \cdots + x_n F_n + G \prec 0,$$

where $F_i, G \in \mathbf{S}^k$, and

$$Z \succeq 0, \qquad Z \neq 0, \qquad \mathbf{tr}(GZ) \geq 0, \qquad \mathbf{tr}(F_i Z) = 0, \quad i = 1, \ldots, n,$$

where $Z \in \mathbf{S}^k$. This follows from the general result, if we take for K the positive semidefinite cone $\mathbf{S}_+^k$, and

$$
g(Z) = \inf_x \left(\mathbf{tr}(F(x)Z) \right) = \begin{cases} \mathbf{tr}(GZ) & \mathbf{tr}(F_i Z) = 0, \quad i = 1, \ldots, n \\ -\infty & \text{otherwise.} \end{cases}
$$

The nonstrict inequality case is slightly more involved, and we need an extra assumption on the matrices F_i to have strong alternatives. One such condition is

$$\sum_{i=1}^{n} v_i F_i \succeq 0 \Longrightarrow \sum_{i=1}^{n} v_i F_i = 0.$$

If this condition holds, the following systems are strong alternatives:

$$F(x) = x_1 F_1 + \cdots + x_n F_n + G \preceq 0$$

and

$$Z \succeq 0, \qquad \mathbf{tr}(GZ) > 0, \qquad \mathbf{tr}(F_i Z) = 0, \quad i = 1, \ldots, n$$

(see exercise 5.44).

Bibliography

Lagrange duality is covered in detail by Luenberger [Lue69, chapter 8], Rockafellar [Roc70, part VI], Whittle [Whi71], Hiriart-Urruty and Lemaréchal [HUL93], and Bertsekas, Nedić, and Ozdaglar [Ber03]. The name is derived from Lagrange's method of multipliers for optimization problems with equality constraints; see Courant and Hilbert [CH53, chapter IV].

The max-min result for matrix games in §5.2.5 predates linear programming duality. It is proved via a theorem of alternatives by von Neuman and Morgenstern [vNM53, page 153]. The strong duality result for linear programming on page 227 is due to von Neumann [vN63] and Gale, Kuhn, and Tucker [GKT51]. Strong duality for the nonconvex quadratic problem (5.32) is a fundamental result in the literature on trust region methods for nonlinear optimization (Nocedal and Wright [NW99, page 78]). It is also related to the S-procedure in control theory, discussed in appendix §B.1. For an extension of the proof of strong duality of §5.3.2 to the refined Slater condition (5.27), see Rockafellar [Roc70, page 277].

Conditions that guarantee the saddle-point property (5.47) can be found in Rockafellar [Roc70, part VII] and Bertsekas, Nedić, and Ozdaglar [Ber03, chapter 2]; see also exercise 5.25.

The KKT conditions are named after Karush (whose unpublished 1939 Master's thesis is summarized in Kuhn [Kuh76]), Kuhn, and Tucker [KT51]. Related optimality conditions were also derived by John [Joh85]. The water-filling algorithm in example 5.2 has applications in information theory and communications (Cover and Thomas [CT91, page 252]).

Farkas' lemma was published by Farkas [Far02]. It is the best known theorem of alternatives for systems of linear inequalities and equalities, but many variants exist; see Mangasarian [Man94, §2.4]. The application of Farkas' lemma to asset pricing (example 5.10) is discussed by Bertsimas and Tsitsiklis [BT97, page 167] and Ross [Ros99].

The extension of Lagrange duality to problems with generalized inequalities appears in Isii [Isi64], Luenberger [Lue69, chapter 8], Berman [Ber73], and Rockafellar [Roc89, page 47]. It is discussed in the context of cone programming in Nesterov and Nemirovski [NN94, §4.2] and Ben-Tal and Nemirovski [BTN01, lecture 2]. Theorems of alternatives for generalized inequalities were studied by Ben-Israel [BI69], Berman and Ben-Israel [BBI71], and Craven and Kohila [CK77]. Bellman and Fan [BF63], Wolkowicz [Wol81], and Lasserre [Las95] give extensions of Farkas' lemma to linear matrix inequalities.

Exercises

Basic definitions

5.1 *A simple example.* Consider the optimization problem

$$\begin{array}{ll} \text{minimize} & x^2 + 1 \\ \text{subject to} & (x-2)(x-4) \le 0, \end{array}$$

with variable $x \in \mathbf{R}$.

   (a) *Analysis of primal problem.* Give the feasible set, the optimal value, and the optimal solution.

   (b) *Lagrangian and dual function.* Plot the objective $x^2 + 1$ versus x. On the same plot, show the feasible set, optimal point and value, and plot the Lagrangian $L(x, \lambda)$ versus x for a few positive values of λ. Verify the lower bound property ($p^\star \ge \inf_x L(x, \lambda)$ for $\lambda \ge 0$). Derive and sketch the Lagrange dual function g.

   (c) *Lagrange dual problem.* State the dual problem, and verify that it is a concave maximization problem. Find the dual optimal value and dual optimal solution $\lambda^\star$. Does strong duality hold?

   (d) *Sensitivity analysis.* Let $p^\star(u)$ denote the optimal value of the problem

$$\begin{array}{ll} \text{minimize} & x^2 + 1 \\ \text{subject to} & (x-2)(x-4) \le u, \end{array}$$

     as a function of the parameter u. Plot $p^\star(u)$. Verify that $dp^\star(0)/du = -\lambda^\star$.

5.2 *Weak duality for unbounded and infeasible problems.* The weak duality inequality, $d^\star \le p^\star$, clearly holds when $d^\star = -\infty$ or $p^\star = \infty$. Show that it holds in the other two cases as well: If $p^\star = -\infty$, then we must have $d^\star = -\infty$, and also, if $d^\star = \infty$, then we must have $p^\star = \infty$.

5.3 *Problems with one inequality constraint.* Express the dual problem of

$$\begin{array}{ll} \text{minimize} & c^T x \\ \text{subject to} & f(x) \le 0, \end{array}$$

with $c \ne 0$, in terms of the conjugate f^*. Explain why the problem you give is convex. We do not assume f is convex.

Examples and applications

5.4 *Interpretation of LP dual via relaxed problems.* Consider the inequality form LP

$$\begin{array}{ll} \text{minimize} & c^T x \\ \text{subject to} & Ax \preceq b, \end{array}$$

with $A \in \mathbf{R}^{m \times n}$, $b \in \mathbf{R}^m$. In this exercise we develop a simple geometric interpretation of the dual LP (5.22).

Let $w \in \mathbf{R}_+^m$. If x is feasible for the LP, *i.e.*, satisfies $Ax \preceq b$, then it also satisfies the inequality

$$w^T A x \le w^T b.$$

Geometrically, for any $w \succeq 0$, the halfspace $H_w = \{x \mid w^T A x \le w^T b\}$ contains the feasible set for the LP. Therefore if we minimize the objective $c^T x$ over the halfspace H_w we get a lower bound on $p^\star$.

(a) Derive an expression for the minimum value of $c^T x$ over the halfspace H_w (which will depend on the choice of $w \succeq 0$).

(b) Formulate the problem of finding the best such bound, by maximizing the lower bound over $w \succeq 0$.

(c) Relate the results of (a) and (b) to the Lagrange dual of the LP, given by (5.22).

5.5 *Dual of general LP.* Find the dual function of the LP

$$\begin{array}{ll} \text{minimize} & c^T x \\ \text{subject to} & Gx \preceq h \\ & Ax = b. \end{array}$$

Give the dual problem, and make the implicit equality constraints explicit.

5.6 *Lower bounds in Chebyshev approximation from least-squares.* Consider the Chebyshev or ℓ_∞-norm approximation problem

$$\text{minimize} \quad \|Ax - b\|_\infty, \tag{5.103}$$

where $A \in \mathbf{R}^{m \times n}$ and $\mathbf{rank}\, A = n$. Let x_{ch} denote an optimal solution (there may be multiple optimal solutions; x_{ch} denotes one of them).

The Chebyshev problem has no closed-form solution, but the corresponding least-squares problem does. Define

$$x_{\mathrm{ls}} = \mathrm{argmin}\, \|Ax - b\|_2 = (A^T A)^{-1} A^T b.$$

We address the following question. Suppose that for a particular A and b we have computed the least-squares solution x_{ls} (but not x_{ch}). How suboptimal is x_{ls} for the Chebyshev problem? In other words, how much larger is $\|Ax_{\mathrm{ls}} - b\|_\infty$ than $\|Ax_{\mathrm{ch}} - b\|_\infty$?

(a) Prove the lower bound

$$\|Ax_{\mathrm{ls}} - b\|_\infty \le \sqrt{m} \, \|Ax_{\mathrm{ch}} - b\|_\infty,$$

using the fact that for all $z \in \mathbf{R}^m$,

$$\frac{1}{\sqrt{m}} \|z\|_2 \le \|z\|_\infty \le \|z\|_2.$$

(b) In example 5.6 (page 254) we derived a dual for the general norm approximation problem. Applying the results to the ℓ_∞-norm (and its dual norm, the ℓ_1-norm), we can state the following dual for the Chebyshev approximation problem:

$$\begin{array}{ll} \text{maximize} & b^T \nu \\ \text{subject to} & \|\nu\|_1 \le 1 \\ & A^T \nu = 0. \end{array} \tag{5.104}$$

Any feasible ν corresponds to a lower bound $b^T \nu$ on $\|Ax_{\mathrm{ch}} - b\|_\infty$.

Denote the least-squares residual as $r_{\mathrm{ls}} = b - Ax_{\mathrm{ls}}$. Assuming $r_{\mathrm{ls}} \ne 0$, show that

$$\hat{\nu} = -r_{\mathrm{ls}}/\|r_{\mathrm{ls}}\|_1, \qquad \tilde{\nu} = r_{\mathrm{ls}}/\|r_{\mathrm{ls}}\|_1,$$

are both feasible in (5.104). By duality $b^T \hat{\nu}$ and $b^T \tilde{\nu}$ are lower bounds on $\|Ax_{\mathrm{ch}} - b\|_\infty$. Which is the better bound? How do these bounds compare with the bound derived in part (a)?

5.7 *Piecewise-linear minimization.* We consider the convex piecewise-linear minimization problem

$$\text{minimize} \quad \max_{i=1,\ldots,m}(a_i^T x + b_i) \qquad (5.105)$$

with variable $x \in \mathbf{R}^n$.

(a) Derive a dual problem, based on the Lagrange dual of the equivalent problem

$$\begin{array}{ll} \text{minimize} & \max_{i=1,\ldots,m} y_i \\ \text{subject to} & a_i^T x + b_i = y_i, \quad i = 1, \ldots, m, \end{array}$$

with variables $x \in \mathbf{R}^n$, $y \in \mathbf{R}^m$.

(b) Formulate the piecewise-linear minimization problem (5.105) as an LP, and form the dual of the LP. Relate the LP dual to the dual obtained in part (a).

(c) Suppose we approximate the objective function in (5.105) by the smooth function

$$f_0(x) = \log\left(\sum_{i=1}^m \exp(a_i^T x + b_i)\right),$$

and solve the unconstrained geometric program

$$\text{minimize} \quad \log\left(\sum_{i=1}^m \exp(a_i^T x + b_i)\right). \qquad (5.106)$$

A dual of this problem is given by (5.62). Let $p_{\text{pwl}}^\star$ and $p_{\text{gp}}^\star$ be the optimal values of (5.105) and (5.106), respectively. Show that

$$0 \le p_{\text{gp}}^\star - p_{\text{pwl}}^\star \le \log m.$$

(d) Derive similar bounds for the difference between $p_{\text{pwl}}^\star$ and the optimal value of

$$\text{minimize} \quad (1/\gamma)\log\left(\sum_{i=1}^m \exp(\gamma(a_i^T x + b_i))\right),$$

where $\gamma > 0$ is a parameter. What happens as we increase γ?

5.8 Relate the two dual problems derived in example 5.9 on page 257.

5.9 *Suboptimality of a simple covering ellipsoid.* Recall the problem of determining the minimum volume ellipsoid, centered at the origin, that contains the points $a_1, \ldots, a_m \in \mathbf{R}^n$ (problem (5.14), page 222):

$$\begin{array}{ll} \text{minimize} & f_0(X) = \log\det(X^{-1}) \\ \text{subject to} & a_i^T X a_i \le 1, \quad i = 1, \ldots, m, \end{array}$$

with $\mathbf{dom}\, f_0 = \mathbf{S}_{++}^n$. We assume that the vectors $a_1, \ldots, a_m$ span $\mathbf{R}^n$ (which implies that the problem is bounded below).

(a) Show that the matrix

$$X_{\text{sim}} = \left(\sum_{k=1}^m a_k a_k^T\right)^{-1},$$

is feasible. *Hint.* Show that

$$\begin{bmatrix} \sum_{k=1}^m a_k a_k^T & a_i \\ a_i^T & 1 \end{bmatrix} \succeq 0,$$

and use Schur complements (§A.5.5) to prove that $a_i^T X a_i \le 1$ for $i = 1, \ldots, m$.

(b) Now we establish a bound on how suboptimal the feasible point X_{sim} is, via the dual problem,

$$
\begin{array}{ll}
\text{maximize} & \log\det\left(\sum_{i=1}^{m}\lambda_i a_i a_i^T\right) - \mathbf{1}^T\lambda + n \\
\text{subject to} & \lambda \succeq 0,
\end{array}
$$

with the implicit constraint $\sum_{i=1}^{m}\lambda_i a_i a_i^T \succ 0$. (This dual is derived on page 222.) To derive a bound, we restrict our attention to dual variables of the form $\lambda = t\mathbf{1}$, where $t > 0$. Find (analytically) the optimal value of t, and evaluate the dual objective at this λ. Use this to prove that the volume of the ellipsoid $\{u \mid u^T X_{\text{sim}} u \leq 1\}$ is no more than a factor $(m/n)^{n/2}$ more than the volume of the minimum volume ellipsoid.

5.10 *Optimal experiment design.* The following problems arise in experiment design (see §7.5).

(a) *D-optimal design.*

$$
\begin{array}{ll}
\text{minimize} & \log\det\left(\sum_{i=1}^{p} x_i v_i v_i^T\right)^{-1} \\
\text{subject to} & x \succeq 0, \quad \mathbf{1}^T x = 1.
\end{array}
$$

(b) *A-optimal design.*

$$
\begin{array}{ll}
\text{minimize} & \mathbf{tr}\left(\sum_{i=1}^{p} x_i v_i v_i^T\right)^{-1} \\
\text{subject to} & x \succeq 0, \quad \mathbf{1}^T x = 1.
\end{array}
$$

The domain of both problems is $\{x \mid \sum_{i=1}^{p} x_i v_i v_i^T \succ 0\}$. The variable is $x \in \mathbf{R}^p$; the vectors $v_1, \ldots, v_p \in \mathbf{R}^n$ are given.

Derive dual problems by first introducing a new variable $X \in \mathbf{S}^n$ and an equality constraint $X = \sum_{i=1}^{p} x_i v_i v_i^T$, and then applying Lagrange duality. Simplify the dual problems as much as you can.

5.11 Derive a dual problem for

$$
\text{minimize} \quad \sum_{i=1}^{N}\|A_i x + b_i\|_2 + (1/2)\|x - x_0\|_2^2.
$$

The problem data are $A_i \in \mathbf{R}^{m_i \times n}$, $b_i \in \mathbf{R}^{m_i}$, and $x_0 \in \mathbf{R}^n$. First introduce new variables $y_i \in \mathbf{R}^{m_i}$ and equality constraints $y_i = A_i x + b_i$.

5.12 *Analytic centering.* Derive a dual problem for

$$
\text{minimize} \quad -\sum_{i=1}^{m}\log(b_i - a_i^T x)
$$

with domain $\{x \mid a_i^T x < b_i, \ i = 1, \ldots, m\}$. First introduce new variables y_i and equality constraints $y_i = b_i - a_i^T x$.

(The solution of this problem is called the *analytic center* of the linear inequalities $a_i^T x \leq b_i$, $i = 1, \ldots, m$. Analytic centers have geometric applications (see §8.5.3), and play an important role in barrier methods (see chapter 11).)

5.13 *Lagrangian relaxation of Boolean LP.* A *Boolean linear program* is an optimization problem of the form

$$
\begin{array}{ll}
\text{minimize} & c^T x \\
\text{subject to} & Ax \preceq b \\
& x_i \in \{0, 1\}, \quad i = 1, \ldots, n,
\end{array}
$$

and is, in general, very difficult to solve. In exercise 4.15 we studied the LP relaxation of this problem,

$$
\begin{array}{ll}
\text{minimize} & c^T x \\
\text{subject to} & Ax \preceq b \\
& 0 \leq x_i \leq 1, \quad i = 1, \ldots, n,
\end{array}
\tag{5.107}
$$

which is far easier to solve, and gives a lower bound on the optimal value of the Boolean LP. In this problem we derive another lower bound for the Boolean LP, and work out the relation between the two lower bounds.

(a) *Lagrangian relaxation.* The Boolean LP can be reformulated as the problem

$$\begin{array}{ll} \text{minimize} & c^T x \\ \text{subject to} & Ax \preceq b \\ & x_i(1 - x_i) = 0, \quad i = 1, \dots, n, \end{array}$$

which has quadratic equality constraints. Find the Lagrange dual of this problem. The optimal value of the dual problem (which is convex) gives a lower bound on the optimal value of the Boolean LP. This method of finding a lower bound on the optimal value is called *Lagrangian relaxation.*

(b) Show that the lower bound obtained via Lagrangian relaxation, and via the LP relaxation (5.107), are the same. *Hint.* Derive the dual of the LP relaxation (5.107).

5.14 *A penalty method for equality constraints.* We consider the problem

$$\begin{array}{ll} \text{minimize} & f_0(x) \\ \text{subject to} & Ax = b, \end{array} \tag{5.108}$$

where $f_0 : \mathbf{R}^n \to \mathbf{R}$ is convex and differentiable, and $A \in \mathbf{R}^{m \times n}$ with $\mathbf{rank}\, A = m$. In a *quadratic penalty method*, we form an auxiliary function

$$\phi(x) = f_0(x) + \alpha \|Ax - b\|_2^2,$$

where $\alpha > 0$ is a parameter. This auxiliary function consists of the objective plus the *penalty term* $\alpha\|Ax - b\|_2^2$. The idea is that a minimizer of the auxiliary function, $\tilde{x}$, should be an approximate solution of the original problem. Intuition suggests that the larger the penalty weight α, the better the approximation $\tilde{x}$ to a solution of the original problem. Suppose $\tilde{x}$ is a minimizer of ϕ. Show how to find, from $\tilde{x}$, a dual feasible point for (5.108). Find the corresponding lower bound on the optimal value of (5.108).

5.15 Consider the problem

$$\begin{array}{ll} \text{minimize} & f_0(x) \\ \text{subject to} & f_i(x) \leq 0, \quad i = 1, \dots, m, \end{array} \tag{5.109}$$

where the functions $f_i : \mathbf{R}^n \to \mathbf{R}$ are differentiable and convex. Let $h_1, \dots, h_m : \mathbf{R} \to \mathbf{R}$ be increasing differentiable convex functions. Show that

$$\phi(x) = f_0(x) + \sum_{i=1}^m h_i(f_i(x))$$

is convex. Suppose $\tilde{x}$ minimizes ϕ. Show how to find from $\tilde{x}$ a feasible point for the dual of (5.109). Find the corresponding lower bound on the optimal value of (5.109).

5.16 *An exact penalty method for inequality constraints.* Consider the problem

$$\begin{array}{ll} \text{minimize} & f_0(x) \\ \text{subject to} & f_i(x) \leq 0, \quad i = 1, \dots, m, \end{array} \tag{5.110}$$

where the functions $f_i : \mathbf{R}^n \to \mathbf{R}$ are differentiable and convex. In an exact penalty method, we solve the auxiliary problem

$$\text{minimize} \quad \phi(x) = f_0(x) + \alpha \max_{i=1,\dots,m} \max\{0, f_i(x)\}, \tag{5.111}$$

where $\alpha > 0$ is a parameter. The second term in ϕ penalizes deviations of x from feasibility. The method is called an *exact* penalty method if for sufficiently large α, solutions of the auxiliary problem (5.111) also solve the original problem (5.110).

(a) Show that ϕ is convex.

(b) The auxiliary problem can be expressed as

$$\begin{array}{ll} \text{minimize} & f_0(x) + \alpha y \\ \text{subject to} & f_i(x) \le y, \quad i = 1, \dots, m \\ & 0 \le y \end{array}$$

where the variables are x and $y \in \mathbf{R}$. Find the Lagrange dual of this problem, and express it in terms of the Lagrange dual function g of (5.110).

(c) Use the result in (b) to prove the following property. Suppose $\lambda^\star$ is an optimal solution of the Lagrange dual of (5.110), and that strong duality holds. If $\alpha > \mathbf{1}^T \lambda^\star$, then any solution of the auxiliary problem (5.111) is also an optimal solution of (5.110).

5.17 *Robust linear programming with polyhedral uncertainty.* Consider the robust LP

$$\begin{array}{ll} \text{minimize} & c^T x \\ \text{subject to} & \sup_{a \in \mathcal{P}_i} a^T x \le b_i, \quad i = 1, \dots, m, \end{array}$$

with variable $x \in \mathbf{R}^n$, where $\mathcal{P}_i = \{a \mid C_i a \preceq d_i\}$. The problem data are $c \in \mathbf{R}^n$, $C_i \in \mathbf{R}^{m_i \times n}$, $d_i \in \mathbf{R}^{m_i}$, and $b \in \mathbf{R}^m$. We assume the polyhedra $\mathcal{P}_i$ are nonempty.

Show that this problem is equivalent to the LP

$$\begin{array}{ll} \text{minimize} & c^T x \\ \text{subject to} & d_i^T z_i \le b_i, \quad i = 1, \dots, m \\ & C_i^T z_i = x, \quad i = 1, \dots, m \\ & z_i \succeq 0, \quad i = 1, \dots, m \end{array}$$

with variables $x \in \mathbf{R}^n$ and $z_i \in \mathbf{R}^{m_i}$, $i = 1, \dots, m$. *Hint.* Find the dual of the problem of maximizing $a_i^T x$ over $a_i \in \mathcal{P}_i$ (with variable a_i).

5.18 *Separating hyperplane between two polyhedra.* Formulate the following problem as an LP or an LP feasibility problem. Find a separating hyperplane that strictly separates two polyhedra

$$\mathcal{P}_1 = \{x \mid Ax \preceq b\}, \qquad \mathcal{P}_2 = \{x \mid Cx \preceq d\},$$

i.e., find a vector $a \in \mathbf{R}^n$ and a scalar γ such that

$$a^T x > \gamma \text{ for } x \in \mathcal{P}_1, \qquad a^T x < \gamma \text{ for } x \in \mathcal{P}_2.$$

You can assume that $\mathcal{P}_1$ and $\mathcal{P}_2$ do not intersect.

Hint. The vector a and scalar γ must satisfy

$$\inf_{x \in \mathcal{P}_1} a^T x > \gamma > \sup_{x \in \mathcal{P}_2} a^T x.$$

Use LP duality to simplify the infimum and supremum in these conditions.

5.19 *The sum of the largest elements of a vector.* Define $f : \mathbf{R}^n \to \mathbf{R}$ as

$$f(x) = \sum_{i=1}^{r} x_{[i]},$$

where r is an integer between 1 and n, and $x_{[1]} \ge x_{[2]} \ge \cdots \ge x_{[r]}$ are the components of x sorted in decreasing order. In other words, $f(x)$ is the sum of the r largest elements of x. In this problem we study the constraint

$$f(x) \le \alpha.$$

As we have seen in chapter 3, page 80, this is a convex constraint, and equivalent to a set of $n!/(r!(n-r)!)$ linear inequalities

$$x_{i_1} + \cdots + x_{i_r} \le \alpha, \quad 1 \le i_1 < i_2 < \cdots < i_r \le n.$$

The purpose of this problem is to derive a more compact representation.

(a) Given a vector $x \in \mathbf{R}^n$, show that $f(x)$ is equal to the optimal value of the LP

$$
\begin{array}{ll}
\text{maximize} & x^T y \\
\text{subject to} & 0 \preceq y \preceq \mathbf{1} \\
& \mathbf{1}^T y = r
\end{array}
$$

with $y \in \mathbf{R}^n$ as variable.

(b) Derive the dual of the LP in part (a). Show that it can be written as

$$
\begin{array}{ll}
\text{minimize} & rt + \mathbf{1}^T u \\
\text{subject to} & t\mathbf{1} + u \succeq x \\
& u \succeq 0,
\end{array}
$$

where the variables are $t \in \mathbf{R}$, $u \in \mathbf{R}^n$. By duality this LP has the same optimal value as the LP in (a), *i.e.*, $f(x)$. We therefore have the following result: x satisfies $f(x) \leq \alpha$ if and only if there exist $t \in \mathbf{R}$, $u \in \mathbf{R}^n$ such that

$$
rt + \mathbf{1}^T u \leq \alpha, \qquad t\mathbf{1} + u \succeq x, \qquad u \succeq 0.
$$

These conditions form a set of $2n+1$ linear inequalities in the $2n+1$ variables x, u, t.

(c) As an application, we consider an extension of the classical Markowitz portfolio optimization problem

$$
\begin{array}{ll}
\text{minimize} & x^T \Sigma x \\
\text{subject to} & \bar{p}^T x \geq r_{\min} \\
& \mathbf{1}^T x = 1, \quad x \succeq 0
\end{array}
$$

discussed in chapter 4, page 155. The variable is the portfolio $x \in \mathbf{R}^n$; $\bar{p}$ and Σ are the mean and covariance matrix of the price change vector p.

Suppose we add a *diversification constraint*, requiring that no more than 80% of the total budget can be invested in any 10% of the assets. This constraint can be expressed as

$$
\sum_{i=1}^{\lfloor 0.1n \rfloor} x_{[i]} \leq 0.8.
$$

Formulate the portfolio optimization problem with diversification constraint as a QP.

5.20 *Dual of channel capacity problem.* Derive a dual for the problem

$$
\begin{array}{ll}
\text{minimize} & -c^T x + \sum_{i=1}^{m} y_i \log y_i \\
\text{subject to} & Px = y \\
& x \succeq 0, \quad \mathbf{1}^T x = 1,
\end{array}
$$

where $P \in \mathbf{R}^{m \times n}$ has nonnegative elements, and its columns add up to one (*i.e.*, $P^T \mathbf{1} = 1$). The variables are $x \in \mathbf{R}^n$, $y \in \mathbf{R}^m$. (For $c_j = \sum_{i=1}^{m} p_{ij} \log p_{ij}$, the optimal value is, up to a factor $\log 2$, the negative of the capacity of a discrete memoryless channel with channel transition probability matrix P; see exercise 4.57.)

Simplify the dual problem as much as possible.

Strong duality and Slater's condition

5.21 *A convex problem in which strong duality fails.* Consider the optimization problem

$$
\begin{array}{ll}
\text{minimize} & e^{-x} \\
\text{subject to} & x^2/y \le 0
\end{array}
$$

with variables x and y, and domain $\mathcal{D} = \{(x, y) \mid y > 0\}$.

    (a) Verify that this is a convex optimization problem. Find the optimal value.

    (b) Give the Lagrange dual problem, and find the optimal solution $\lambda^\star$ and optimal value $d^\star$ of the dual problem. What is the optimal duality gap?

    (c) Does Slater's condition hold for this problem?

    (d) What is the optimal value $p^\star(u)$ of the perturbed problem

$$
\begin{array}{ll}
\text{minimize} & e^{-x} \\
\text{subject to} & x^2/y \le u
\end{array}
$$

    as a function of u? Verify that the global sensitivity inequality

$$
p^\star(u) \ge p^\star(0) - \lambda^\star u
$$

    does not hold.

5.22 *Geometric interpretation of duality.* For each of the following optimization problems, draw a sketch of the sets

$$
\begin{aligned}
\mathcal{G} &= \{(u, t) \mid \exists x \in \mathcal{D}, \ f_0(x) = t, \ f_1(x) = u\}, \\
\mathcal{A} &= \{(u, t) \mid \exists x \in \mathcal{D}, \ f_0(x) \le t, \ f_1(x) \le u\},
\end{aligned}
$$

give the dual problem, and solve the primal and dual problems. Is the problem convex? Is Slater's condition satisfied? Does strong duality hold?

The domain of the problem is $\mathbf{R}$ unless otherwise stated.

    (a) Minimize x subject to $x^2 \le 1$.

    (b) Minimize x subject to $x^2 \le 0$.

    (c) Minimize x subject to $|x| \le 0$.

    (d) Minimize x subject to $f_1(x) \le 0$ where

$$
f_1(x) = \begin{cases} -x + 2 & x \ge 1 \\ x & -1 \le x \le 1 \\ -x - 2 & x \le -1. \end{cases}
$$

    (e) Minimize x^3 subject to $-x + 1 \le 0$.

    (f) Minimize x^3 subject to $-x + 1 \le 0$ with domain $\mathcal{D} = \mathbf{R}_+$.

5.23 *Strong duality in linear programming.* We prove that strong duality holds for the LP

$$
\begin{array}{ll}
\text{minimize} & c^T x \\
\text{subject to} & Ax \preceq b
\end{array}
$$

and its dual

$$
\begin{array}{ll}
\text{maximize} & -b^T z \\
\text{subject to} & A^T z + c = 0, \quad z \succeq 0,
\end{array}
$$

provided at least one of the problems is feasible. In other words, the only possible exception to strong duality occurs when $p^\star = \infty$ and $d^\star = -\infty$.

(a) Suppose $p^\star$ is finite and $x^\star$ is an optimal solution. (If finite, the optimal value of an LP is attained.) Let $I \subseteq \{1, 2, \ldots, m\}$ be the set of active constraints at $x^\star$:

$$a_i^T x^\star = b_i, \quad i \in I, \qquad a_i^T x^\star < b_i, \quad i \notin I.$$

Show that there exists a $z \in \mathbf{R}^m$ that satisfies

$$z_i \geq 0, \quad i \in I, \qquad z_i = 0, \quad i \notin I, \qquad \sum_{i \in I} z_i a_i + c = 0.$$

Show that z is dual optimal with objective value $c^T x^\star$.

Hint. Assume there exists no such z, i.e., $-c \notin \{\sum_{i \in I} z_i a_i \mid z_i \geq 0\}$. Reduce this to a contradiction by applying the strict separating hyperplane theorem of example 2.20, page 49. Alternatively, you can use Farkas' lemma (see §5.8.3).

(b) Suppose $p^\star = \infty$ and the dual problem is feasible. Show that $d^\star = \infty$. *Hint.* Show that there exists a nonzero $v \in \mathbf{R}^m$ such that $A^T v = 0$, $v \succeq 0$, $b^T v < 0$. If the dual is feasible, it is unbounded in the direction v.

(c) Consider the example

$$
\begin{array}{ll}
\text{minimize} & x \\
\text{subject to} & \begin{bmatrix} 0 \\ 1 \end{bmatrix} x \preceq \begin{bmatrix} -1 \\ 1 \end{bmatrix}.
\end{array}
$$

Formulate the dual LP, and solve the primal and dual problems. Show that $p^\star = \infty$ and $d^\star = -\infty$.

5.24 *Weak max-min inequality.* Show that the weak max-min inequality

$$\sup_{z \in Z} \inf_{w \in W} f(w, z) \leq \inf_{w \in W} \sup_{z \in Z} f(w, z)$$

always holds, with no assumptions on $f : \mathbf{R}^n \times \mathbf{R}^m \to \mathbf{R}$, $W \subseteq \mathbf{R}^n$, or $Z \subseteq \mathbf{R}^m$.

5.25 [BL00, page 95] *Convex-concave functions and the saddle-point property.* We derive conditions under which the saddle-point property

$$\sup_{z \in Z} \inf_{w \in W} f(w, z) = \inf_{w \in W} \sup_{z \in Z} f(w, z) \tag{5.112}$$

holds, where $f : \mathbf{R}^n \times \mathbf{R}^m \to \mathbf{R}$, $W \times Z \subseteq \operatorname{\mathbf{dom}} f$, and W and Z are nonempty. We will assume that the function

$$g_z(w) = \begin{cases} f(w, z) & w \in W \\ \infty & \text{otherwise} \end{cases}$$

is closed and convex for all $z \in Z$, and the function

$$h_w(z) = \begin{cases} -f(w, z) & z \in Z \\ \infty & \text{otherwise} \end{cases}$$

is closed and convex for all $w \in W$.

(a) The righthand side of (5.112) can be expressed as $p(0)$, where

$$p(u) = \inf_{w \in W} \sup_{z \in Z} (f(w, z) + u^T z).$$

Show that p is a convex function.

(b) Show that the conjugate of p is given by

$$p^*(v) = \begin{cases} -\inf_{w \in W} f(w,v) & v \in Z \\ \infty & \text{otherwise.} \end{cases}$$

(c) Show that the conjugate of p^* is given by

$$p^{**}(u) = \sup_{z \in Z} \inf_{w \in W} (f(w,z) + u^T z).$$

Combining this with (a), we can express the max-min equality (5.112) as $p^{**}(0) = p(0)$.

(d) From exercises 3.28 and 3.39 (d), we know that $p^{**}(0) = p(0)$ if $0 \in \mathbf{int\,dom}\,p$. Conclude that this is the case if W and Z are bounded.

(e) As another consequence of exercises 3.28 and 3.39, we have $p^{**}(0) = p(0)$ if $0 \in \mathbf{dom}\,p$ and p is closed. Show that p is closed if the sublevel sets of g_z are bounded.

Optimality conditions

5.26 Consider the QCQP

$$\begin{array}{ll} \text{minimize} & x_1^2 + x_2^2 \\ \text{subject to} & (x_1 - 1)^2 + (x_2 - 1)^2 \le 1 \\ & (x_1 - 1)^2 + (x_2 + 1)^2 \le 1 \end{array}$$

with variable $x \in \mathbf{R}^2$.

(a) Sketch the feasible set and level sets of the objective. Find the optimal point x^* and optimal value p^*.

(b) Give the KKT conditions. Do there exist Lagrange multipliers λ_1^* and λ_2^* that prove that x^* is optimal?

(c) Derive and solve the Lagrange dual problem. Does strong duality hold?

5.27 *Equality constrained least-squares.* Consider the equality constrained least-squares problem

$$\begin{array}{ll} \text{minimize} & \|Ax - b\|_2^2 \\ \text{subject to} & Gx = h \end{array}$$

where $A \in \mathbf{R}^{m \times n}$ with $\mathbf{rank}\,A = n$, and $G \in \mathbf{R}^{p \times n}$ with $\mathbf{rank}\,G = p$.

Give the KKT conditions, and derive expressions for the primal solution x^* and the dual solution ν^*.

5.28 Prove (without using any linear programming code) that the optimal solution of the LP

$$\begin{array}{ll} \text{minimize} & 47x_1 + 93x_2 + 17x_3 - 93x_4 \\ \text{subject to} & \begin{bmatrix} -1 & -6 & 1 & 3 \\ -1 & -2 & 7 & 1 \\ 0 & 3 & -10 & -1 \\ -6 & -11 & -2 & 12 \\ 1 & 6 & -1 & -3 \end{bmatrix} \begin{bmatrix} x_1 \\ x_2 \\ x_3 \\ x_4 \end{bmatrix} \preceq \begin{bmatrix} -3 \\ 5 \\ -8 \\ -7 \\ 4 \end{bmatrix} \end{array}$$

is unique, and given by $x^* = (1, 1, 1, 1)$.

5.29 The problem

$$\begin{array}{ll} \text{minimize} & -3x_1^2 + x_2^2 + 2x_3^2 + 2(x_1 + x_2 + x_3) \\ \text{subject to} & x_1^2 + x_2^2 + x_3^2 = 1, \end{array}$$

is a special case of (5.32), so strong duality holds even though the problem is not convex. Derive the KKT conditions. Find all solutions x, ν that satisfy the KKT conditions. Which pair corresponds to the optimum?

5.30 Derive the KKT conditions for the problem

$$\begin{array}{ll}\text{minimize} & \operatorname{\mathbf{tr}} X - \log\det X \\ \text{subject to} & Xs = y,\end{array}$$

with variable $X \in \mathbf{S}^n$ and domain $\mathbf{S}^n_{++}$. $y \in \mathbf{R}^n$ and $s \in \mathbf{R}^n$ are given, with $s^T y = 1$. Verify that the optimal solution is given by

$$X^\star = I + yy^T - \frac{1}{s^T s}ss^T.$$

5.31 *Supporting hyperplane interpretation of KKT conditions.* Consider a convex problem with no equality constraints,

$$\begin{array}{ll}\text{minimize} & f_0(x) \\ \text{subject to} & f_i(x) \le 0, \quad i = 1,\ldots,m.\end{array}$$

Assume that $x^\star \in \mathbf{R}^n$ and $\lambda^\star \in \mathbf{R}^m$ satisfy the KKT conditions

$$\begin{array}{rcll} f_i(x^\star) & \le & 0, & i = 1,\ldots,m \\ \lambda_i^\star & \ge & 0, & i = 1,\ldots,m \\ \lambda_i^\star f_i(x^\star) & = & 0, & i = 1,\ldots,m \\ \nabla f_0(x^\star) + \sum_{i=1}^m \lambda_i^\star \nabla f_i(x^\star) & = & 0. \end{array}$$

Show that

$$\nabla f_0(x^\star)^T(x - x^\star) \ge 0$$

for all feasible x. In other words the KKT conditions imply the simple optimality criterion of §4.2.3.

Perturbation and sensitivity analysis

5.32 *Optimal value of perturbed problem.* Let $f_0, f_1, \ldots, f_m : \mathbf{R}^n \to \mathbf{R}$ be convex. Show that the function

$$p^\star(u,v) = \inf\{f_0(x) \mid \exists x \in \mathcal{D}, \ f_i(x) \le u_i, \ i = 1,\ldots,m, \ Ax - b = v\}$$

is convex. This function is the optimal cost of the perturbed problem, as a function of the perturbations u and v (see §5.6.1).

5.33 *Parametrized ℓ_1-norm approximation.* Consider the ℓ_1-norm minimization problem

$$\text{minimize} \quad \|Ax + b + \epsilon d\|_1$$

with variable $x \in \mathbf{R}^3$, and

$$A = \begin{bmatrix} -2 & 7 & 1 \\ -5 & -1 & 3 \\ -7 & 3 & -5 \\ -1 & 4 & -4 \\ 1 & 5 & 5 \\ 2 & -5 & -1 \end{bmatrix}, \qquad b = \begin{bmatrix} -4 \\ 3 \\ 9 \\ 0 \\ -11 \\ 5 \end{bmatrix}, \qquad d = \begin{bmatrix} -10 \\ -13 \\ -27 \\ -10 \\ -7 \\ 14 \end{bmatrix}.$$

We denote by $p^\star(\epsilon)$ the optimal value as a function of ϵ.

(a) Suppose $\epsilon = 0$. Prove that $x^\star = \mathbf{1}$ is optimal. Are there any other optimal points?

(b) Show that $p^\star(\epsilon)$ is affine on an interval that includes $\epsilon = 0$.

5.34 Consider the pair of primal and dual LPs

$$
\begin{aligned}
\text{minimize} \quad & (c + \epsilon d)^T x \\
\text{subject to} \quad & Ax \preceq b + \epsilon f
\end{aligned}
$$

and

$$
\begin{aligned}
\text{maximize} \quad & -(b + \epsilon f)^T z \\
\text{subject to} \quad & A^T z + c + \epsilon d = 0 \\
& z \succeq 0
\end{aligned}
$$

where

$$
A = \begin{bmatrix}
-4 & 12 & -2 & 1 \\
-17 & 12 & 7 & 11 \\
1 & 0 & -6 & 1 \\
3 & 3 & 22 & -1 \\
-11 & 2 & -1 & -8
\end{bmatrix}, \quad
b = \begin{bmatrix}
8 \\ 13 \\ -4 \\ 27 \\ -18
\end{bmatrix}, \quad
f = \begin{bmatrix}
6 \\ 15 \\ -13 \\ 48 \\ 8
\end{bmatrix},
$$

$c = (49, -34, -50, -5)$, $d = (3, 8, 21, 25)$, and ϵ is a parameter.

(a) Prove that $x^\star = (1, 1, 1, 1)$ is optimal when $\epsilon = 0$, by constructing a dual optimal point $z^\star$ that has the same objective value as $x^\star$. Are there any other primal or dual optimal solutions?

(b) Give an explicit expression for the optimal value $p^\star(\epsilon)$ as a function of ϵ on an interval that contains $\epsilon = 0$. Specify the interval on which your expression is valid. Also give explicit expressions for the primal solution $x^\star(\epsilon)$ and the dual solution $z^\star(\epsilon)$ as a function of ϵ, on the same interval.

Hint. First calculate $x^\star(\epsilon)$ and $z^\star(\epsilon)$, assuming that the primal and dual constraints that are active at the optimum for $\epsilon = 0$, remain active at the optimum for values of ϵ around 0. Then verify that this assumption is correct.

5.35 *Sensitivity analysis for GPs.* Consider a GP

$$
\begin{aligned}
\text{minimize} \quad & f_0(x) \\
\text{subject to} \quad & f_i(x) \le 1, \quad i = 1, \ldots, m \\
& h_i(x) = 1, \quad i = 1, \ldots, p,
\end{aligned}
$$

where $f_0, \ldots, f_m$ are posynomials, $h_1, \ldots, h_p$ are monomials, and the domain of the problem is $\mathbf{R}^n_{++}$. We define the perturbed GP as

$$
\begin{aligned}
\text{minimize} \quad & f_0(x) \\
\text{subject to} \quad & f_i(x) \le e^{u_i}, \quad i = 1, \ldots, m \\
& h_i(x) = e^{v_i}, \quad i = 1, \ldots, p,
\end{aligned}
$$

and we denote the optimal value of the perturbed GP as $p^\star(u, v)$. We can think of u_i and v_i as relative, or fractional, perturbations of the constraints. For example, $u_1 = -0.01$ corresponds to tightening the first inequality constraint by (approximately) 1%.

Let $\lambda^\star$ and $\nu^\star$ be optimal dual variables for the convex form GP

$$
\begin{aligned}
\text{minimize} \quad & \log f_0(y) \\
\text{subject to} \quad & \log f_i(y) \le 0, \quad i = 1, \ldots, m \\
& \log h_i(y) = 0, \quad i = 1, \ldots, p,
\end{aligned}
$$

with variables $y_i = \log x_i$. Assuming that $p^\star(u, v)$ is differentiable at $u = 0$, $v = 0$, relate $\lambda^\star$ and $\nu^\star$ to the derivatives of $p^\star(u, v)$ at $u = 0$, $v = 0$. Justify the statement "Relaxing the ith constraint by α percent will give an improvement in the objective of around $\alpha \lambda_i^\star$ percent, for α small."

Theorems of alternatives

5.36 *Alternatives for linear equalities.* Consider the linear equations $Ax = b$, where $A \in \mathbf{R}^{m \times n}$. From linear algebra we know that this equation has a solution if and only $b \in \mathcal{R}(A)$, which occurs if and only if $b \perp \mathcal{N}(A^T)$. In other words, $Ax = b$ has a solution if and only if there exists no $y \in \mathbf{R}^m$ such that $A^T y = 0$ and $b^T y \neq 0$.

Derive this result from the theorems of alternatives in §5.8.2.

5.37 [BT97] *Existence of equilibrium distribution in finite state Markov chain.* Let $P \in \mathbf{R}^{n \times n}$ be a matrix that satisfies

$$p_{ij} \geq 0, \quad i, j = 1, \ldots, n, \qquad P^T \mathbf{1} = \mathbf{1},$$

i.e., the coefficients are nonnegative and the columns sum to one. Use Farkas' lemma to prove there exists a $y \in \mathbf{R}^n$ such that

$$Py = y, \qquad y \succeq 0, \qquad \mathbf{1}^T y = 1.$$

(We can interpret y as an equilibrium distribution of the Markov chain with n states and transition probability matrix P.)

5.38 [BT97] *Option pricing.* We apply the results of example 5.10, page 263, to a simple problem with three assets: a riskless asset with fixed return $r > 1$ over the investment period of interest (for example, a bond), a stock, and an option on the stock. The option gives us the right to purchase the stock at the end of the period, for a predetermined price K.

We consider two scenarios. In the first scenario, the price of the stock goes up from S at the beginning of the period, to Su at the end of the period, where $u > r$. In this scenario, we exercise the option only if $Su > K$, in which case we make a profit of $Su - K$. Otherwise, we do not exercise the option, and make zero profit. The value of the option at the end of the period, in the first scenario, is therefore $\max\{0, Su - K\}$.

In the second scenario, the price of the stock goes down from S to Sd, where $d < 1$. The value at the end of the period is $\max\{0, Sd - K\}$.

In the notation of example 5.10,

$$V = \begin{bmatrix} r & uS & \max\{0, Su - K\} \\ r & dS & \max\{0, Sd - K\} \end{bmatrix}, \qquad p_1 = 1, \qquad p_2 = S, \qquad p_3 = C,$$

where C is the price of the option.

Show that for given r, S, K, u, d, the option price C is uniquely determined by the no-arbitrage condition. In other words, the market for the option is complete.

Generalized inequalities

5.39 *SDP relaxations of two-way partitioning problem.* We consider the two-way partitioning problem (5.7), described on page 219,

$$\begin{array}{ll} \text{minimize} & x^T W x \\ \text{subject to} & x_i^2 = 1, \quad i = 1, \ldots, n, \end{array} \qquad (5.113)$$

with variable $x \in \mathbf{R}^n$. The Lagrange dual of this (nonconvex) problem is given by the SDP

$$\begin{array}{ll} \text{maximize} & -\mathbf{1}^T \nu \\ \text{subject to} & W + \mathbf{diag}(\nu) \succeq 0 \end{array} \qquad (5.114)$$

with variable $\nu \in \mathbf{R}^n$. The optimal value of this SDP gives a lower bound on the optimal value of the partitioning problem (5.113). In this exercise we derive another SDP that gives a lower bound on the optimal value of the two-way partitioning problem, and explore the connection between the two SDPs.

(a) *Two-way partitioning problem in matrix form.* Show that the two-way partitioning problem can be cast as

$$\begin{array}{ll} \text{minimize} & \mathbf{tr}(WX) \\ \text{subject to} & X \succeq 0, \quad \mathbf{rank}\, X = 1 \\ & X_{ii} = 1, \quad i = 1, \dots, n, \end{array}$$

with variable $X \in \mathbf{S}^n$. *Hint.* Show that if X is feasible, then it has the form $X = xx^T$, where $x \in \mathbf{R}^n$ satisfies $x_i \in \{-1, 1\}$ (and vice versa).

(b) *SDP relaxation of two-way partitioning problem.* Using the formulation in part (a), we can form the relaxation

$$\begin{array}{ll} \text{minimize} & \mathbf{tr}(WX) \\ \text{subject to} & X \succeq 0 \\ & X_{ii} = 1, \quad i = 1, \dots, n, \end{array} \qquad (5.115)$$

with variable $X \in \mathbf{S}^n$. This problem is an SDP, and therefore can be solved efficiently. Explain why its optimal value gives a lower bound on the optimal value of the two-way partitioning problem (5.113). What can you say if an optimal point $X^\star$ for this SDP has rank one?

(c) We now have two SDPs that give a lower bound on the optimal value of the two-way partitioning problem (5.113): the SDP relaxation (5.115) found in part (b), and the Lagrange dual of the two-way partitioning problem, given in (5.114). What is the relation between the two SDPs? What can you say about the lower bounds found by them? *Hint:* Relate the two SDPs via duality.

5.40 *E-optimal experiment design.* A variation on the two optimal experiment design problems of exercise 5.10 is the *E-optimal design* problem

$$\begin{array}{ll} \text{minimize} & \lambda_{\max} \left(\sum_{i=1}^{p} x_i v_i v_i^T \right)^{-1} \\ \text{subject to} & x \succeq 0, \quad \mathbf{1}^T x = 1. \end{array}$$

(See also §7.5.) Derive a dual for this problem, by first reformulating it as

$$\begin{array}{ll} \text{minimize} & 1/t \\ \text{subject to} & \sum_{i=1}^{p} x_i v_i v_i^T \succeq tI \\ & x \succeq 0, \quad \mathbf{1}^T x = 1, \end{array}$$

with variables $t \in \mathbf{R}$, $x \in \mathbf{R}^p$ and domain $\mathbf{R}_{++} \times \mathbf{R}^p$, and applying Lagrange duality. Simplify the dual problem as much as you can.

5.41 *Dual of fastest mixing Markov chain problem.* On page 174, we encountered the SDP

$$\begin{array}{ll} \text{minimize} & t \\ \text{subject to} & -tI \preceq P - (1/n)\mathbf{1}\mathbf{1}^T \preceq tI \\ & P\mathbf{1} = \mathbf{1} \\ & P_{ij} \geq 0, \quad i, j = 1, \dots, n \\ & P_{ij} = 0 \text{ for } (i, j) \notin \mathcal{E}, \end{array}$$

with variables $t \in \mathbf{R}$, $P \in \mathbf{S}^n$.

Show that the dual of this problem can be expressed as

$$\begin{array}{ll} \text{maximize} & \mathbf{1}^T z - (1/n)\mathbf{1}^T Y \mathbf{1} \\ \text{subject to} & \|Y\|_{2*} \leq 1 \\ & (z_i + z_j) \leq Y_{ij} \text{ for } (i, j) \in \mathcal{E} \end{array}$$

with variables $z \in \mathbf{R}^n$ and $Y \in \mathbf{S}^n$. The norm $\| \cdot \|_{2*}$ is the dual of the spectral norm on $\mathbf{S}^n$: $\|Y\|_{2*} = \sum_{i=1}^{n} |\lambda_i(Y)|$, the sum of the absolute values of the eigenvalues of Y. (See §A.1.6, page 637.)

5.42 *Lagrange dual of conic form problem in inequality form.* Find the Lagrange dual problem of the conic form problem in inequality form

$$
\begin{array}{ll}
\text{minimize} & c^T x \\
\text{subject to} & Ax \preceq_K b
\end{array}
$$

where $A \in \mathbf{R}^{m \times n}$, $b \in \mathbf{R}^m$, and K is a proper cone in $\mathbf{R}^m$. Make any implicit equality constraints explicit.

5.43 *Dual of SOCP.* Show that the dual of the SOCP

$$
\begin{array}{ll}
\text{minimize} & f^T x \\
\text{subject to} & \|A_i x + b_i\|_2 \le c_i^T x + d_i, \quad i = 1, \ldots, m,
\end{array}
$$

with variables $x \in \mathbf{R}^n$, can be expressed as

$$
\begin{array}{ll}
\text{maximize} & \sum_{i=1}^{m} (b_i^T u_i - d_i v_i) \\
\text{subject to} & \sum_{i=1}^{m} (A_i^T u_i - c_i v_i) + f = 0 \\
& \|u_i\|_2 \le v_i, \quad i = 1, \ldots, m,
\end{array}
$$

with variables $u_i \in \mathbf{R}^{n_i}$, $v_i \in \mathbf{R}$, $i = 1, \ldots, m$. The problem data are $f \in \mathbf{R}^n$, $A_i \in \mathbf{R}^{n_i \times n}$, $b_i \in \mathbf{R}^{n_i}$, $c_i \in \mathbf{R}$ and $d_i \in \mathbf{R}$, $i = 1, \ldots, m$.
Derive the dual in the following two ways.

(a) Introduce new variables $y_i \in \mathbf{R}^{n_i}$ and $t_i \in \mathbf{R}$ and equalities $y_i = A_i x + b_i$, $t_i = c_i^T x + d_i$, and derive the Lagrange dual.

(b) Start from the conic formulation of the SOCP and use the conic dual. Use the fact that the second-order cone is self-dual.

5.44 *Strong alternatives for nonstrict LMIs.* In example 5.14, page 270, we mentioned that the system

$$
Z \succeq 0, \qquad \mathbf{tr}(GZ) > 0, \qquad \mathbf{tr}(F_i Z) = 0, \quad i = 1, \ldots, n, \qquad (5.116)
$$

is a strong alternative for the nonstrict LMI

$$
F(x) = x_1 F_1 + \cdots + x_n F_n + G \preceq 0, \qquad (5.117)
$$

if the matrices F_i satisfy

$$
\sum_{i=1}^{n} v_i F_i \succeq 0 \implies \sum_{i=1}^{n} v_i F_i = 0. \qquad (5.118)
$$

In this exercise we prove this result, and give an example to illustrate that the systems are not always strong alternatives.

(a) Suppose (5.118) holds, and that the optimal value of the auxiliary SDP

$$
\begin{array}{ll}
\text{minimize} & s \\
\text{subject to} & F(x) \preceq sI
\end{array}
$$

is positive. Show that the optimal value is attained. If follows from the discussion in §5.9.4 that the systems (5.117) and (5.116) are strong alternatives.
Hint. The proof simplifies if you assume, without loss of generality, that the matrices $F_1, \ldots, F_n$ are independent, so (5.118) may be replaced by $\sum_{i=1}^{n} v_i F_i \succeq 0 \Rightarrow v = 0$.

(b) Take $n = 1$, and

$$
G = \begin{bmatrix} 0 & 1 \\ 1 & 0 \end{bmatrix}, \qquad F_1 = \begin{bmatrix} 0 & 0 \\ 0 & 1 \end{bmatrix}.
$$

Show that (5.117) and (5.116) are both infeasible.

Part II

Applications

Chapter 6

Approximation and fitting

6.1 Norm approximation

6.1.1 Basic norm approximation problem

The simplest *norm approximation problem* is an unconstrained problem of the form

$$\text{minimize} \quad \|Ax - b\| \tag{6.1}$$

where $A \in \mathbf{R}^{m \times n}$ and $b \in \mathbf{R}^m$ are problem data, $x \in \mathbf{R}^n$ is the variable, and $\|\cdot\|$ is a norm on $\mathbf{R}^m$. A solution of the norm approximation problem is sometimes called an *approximate solution* of $Ax \approx b$, in the norm $\|\cdot\|$. The vector

$$r = Ax - b$$

is called the *residual* for the problem; its components are sometimes called the individual *residuals* associated with x.

The norm approximation problem (6.1) is a convex problem, and is solvable, *i.e.*, there is always at least one optimal solution. Its optimal value is zero if and only if $b \in \mathcal{R}(A)$; the problem is more interesting and useful, however, when $b \notin \mathcal{R}(A)$. We can assume without loss of generality that the columns of A are independent; in particular, that $m \geq n$. When $m = n$ the optimal point is simply $A^{-1}b$, so we can assume that $m > n$.

Approximation interpretation

By expressing Ax as

$$Ax = x_1 a_1 + \cdots + x_n a_n,$$

where $a_1, \ldots, a_n \in \mathbf{R}^m$ are the columns of A, we see that the goal of the norm approximation problem is to fit or approximate the vector b by a linear combination of the columns of A, as closely as possible, with deviation measured in the norm $\|\cdot\|$.

The approximation problem is also called the *regression problem*. In this context the vectors $a_1, \ldots, a_n$ are called the *regressors*, and the vector $x_1 a_1 + \cdots + x_n a_n$, where x is an optimal solution of the problem, is called the *regression of b* (onto the regressors).

Estimation interpretation

A closely related interpretation of the norm approximation problem arises in the problem of estimating a parameter vector on the basis of an imperfect linear vector measurement. We consider a linear measurement model

$$y = Ax + v,$$

where $y \in \mathbf{R}^m$ is a vector measurement, $x \in \mathbf{R}^n$ is a vector of parameters to be estimated, and $v \in \mathbf{R}^m$ is some measurement error that is unknown, but presumed to be small (in the norm $\| \cdot \|$). The estimation problem is to make a sensible guess as to what x is, given y.

If we guess that x has the value $\hat{x}$, then we are implicitly making the guess that v has the value $y - A\hat{x}$. Assuming that smaller values of v (measured by $\| \cdot \|$) are more plausible than larger values, the most plausible guess for x is

$$\hat{x} = \operatorname{argmin}_z \|Az - y\|.$$

(These ideas can be expressed more formally in a statistical framework; see chapter 7.)

Geometric interpretation

We consider the subspace $\mathcal{A} = \mathcal{R}(A) \subseteq \mathbf{R}^m$, and a point $b \in \mathbf{R}^m$. A *projection* of the point b onto the subspace $\mathcal{A}$, in the norm $\| \cdot \|$, is any point in $\mathcal{A}$ that is closest to b, *i.e.*, any optimal point for the problem

$$\begin{array}{ll} \text{minimize} & \|u - b\| \\ \text{subject to} & u \in \mathcal{A}. \end{array}$$

Parametrizing an arbitrary element of $\mathcal{R}(A)$ as $u = Ax$, we see that solving the norm approximation problem (6.1) is equivalent to computing a projection of b onto $\mathcal{A}$.

Design interpretation

We can interpret the norm approximation problem (6.1) as a problem of optimal design. The n variables $x_1, \ldots, x_n$ are *design variables* whose values are to be determined. The vector $y = Ax$ gives a vector of m *results*, which we assume to be linear functions of the design variables x. The vector b is a vector of *target* or *desired results*. The goal is to choose a vector of design variables that achieves, as closely as possible, the desired results, *i.e.*, $Ax \approx b$. We can interpret the residual vector r as the deviation between the actual results (*i.e.*, Ax) and the desired or target results (*i.e.*, b). If we measure the quality of a design by the norm of the deviation between the actual results and the desired results, then the norm approximation problem (6.1) is the problem of finding the best design.

Weighted norm approximation problems

An extension of the norm approximation problem is the *weighted norm approximation problem*

$$\text{minimize} \quad \|W(Ax - b)\|$$

where the problem data $W \in \mathbf{R}^{m \times m}$ is called the *weighting matrix*. The weighting matrix is often diagonal, in which case it gives different relative emphasis to different components of the residual vector $r = Ax - b$.

The weighted norm problem can be considered as a norm approximation problem with norm $\|\cdot\|$, and data $\tilde{A} = WA$, $\tilde{b} = Wb$, and therefore treated as a standard norm approximation problem (6.1). Alternatively, the weighted norm approximation problem can be considered a norm approximation problem with data A and b, and the W-*weighted norm* defined by

$$\|z\|_W = \|Wz\|$$

(assuming here that W is nonsingular).

Least-squares approximation

The most common norm approximation problem involves the Euclidean or ℓ_2-norm. By squaring the objective, we obtain an equivalent problem which is called the *least-squares approximation problem*,

$$\text{minimize} \quad \|Ax - b\|_2^2 = r_1^2 + r_2^2 + \cdots + r_m^2,$$

where the objective is the sum of squares of the residuals. This problem can be solved analytically by expressing the objective as the convex quadratic function

$$f(x) = x^T A^T A x - 2b^T A x + b^T b.$$

A point x minimizes f if and only if

$$\nabla f(x) = 2A^T A x - 2A^T b = 0,$$

i.e., if and only if x satisfies the so-called *normal equations*

$$A^T A x = A^T b,$$

which always have a solution. Since we assume the columns of A are independent, the least-squares approximation problem has the unique solution $x = (A^T A)^{-1} A^T b$.

Chebyshev or minimax approximation

When the ℓ_∞-norm is used, the norm approximation problem

$$\text{minimize} \quad \|Ax - b\|_\infty = \max\{|r_1|, \ldots, |r_m|\}$$

is called the *Chebyshev approximation problem*, or *minimax approximation problem*, since we are to minimize the maximum (absolute value) residual. The Chebyshev approximation problem can be cast as an LP

$$\begin{array}{ll} \text{minimize} & t \\ \text{subject to} & -t\mathbf{1} \preceq Ax - b \preceq t\mathbf{1}, \end{array}$$

with variables $x \in \mathbf{R}^n$ and $t \in \mathbf{R}$.

Sum of absolute residuals approximation

When the ℓ_1-norm is used, the norm approximation problem

$$\text{minimize} \quad \|Ax - b\|_1 = |r_1| + \cdots + |r_m|$$

is called the sum of (absolute) residuals approximation problem, or, in the context of estimation, a *robust estimator* (for reasons that will be clear soon). Like the Chebyshev approximation problem, the ℓ_1-norm approximation problem can be cast as an LP

$$\begin{array}{ll} \text{minimize} & \mathbf{1}^T t \\ \text{subject to} & -t \preceq Ax - b \preceq t, \end{array}$$

with variables $x \in \mathbf{R}^n$ and $t \in \mathbf{R}^m$.

6.1.2   Penalty function approximation

In ℓ_p-norm approximation, for $1 \leq p < \infty$, the objective is

$$(|r_1|^p + \cdots + |r_m|^p)^{1/p}.$$

As in least-squares problems, we can consider the equivalent problem with objective

$$|r_1|^p + \cdots + |r_m|^p,$$

which is a separable and symmetric function of the residuals. In particular, the objective depends only on the *amplitude distribution* of the residuals, *i.e.*, the residuals in sorted order.

We will consider a useful generalization of the ℓ_p-norm approximation problem, in which the objective depends only on the amplitude distribution of the residuals. The *penalty function approximation problem* has the form

$$\begin{array}{ll} \text{minimize} & \phi(r_1) + \cdots + \phi(r_m) \\ \text{subject to} & r = Ax - b, \end{array} \tag{6.2}$$

where $\phi : \mathbf{R} \to \mathbf{R}$ is called the (residual) *penalty function*. We assume that ϕ is convex, so the penalty function approximation problem is a convex optimization problem. In many cases, the penalty function ϕ is symmetric, nonnegative, and satisfies $\phi(0) = 0$, but we will not use these properties in our analysis.

Interpretation

We can interpret the penalty function approximation problem (6.2) as follows. For the choice x, we obtain the approximation Ax of b, which has the associated residual vector r. A penalty function assesses a cost or penalty for each component of residual, given by $\phi(r_i)$; the total penalty is the sum of the penalties for each residual, *i.e.*, $\phi(r_1) + \cdots + \phi(r_m)$. Different choices of x lead to different resulting residuals, and therefore, different total penalties. In the penalty function approximation problem, we minimize the total penalty incurred by the residuals.

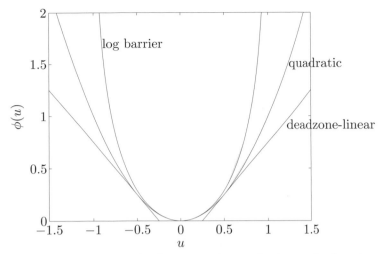

Figure 6.1 Some common penalty functions: the quadratic penalty function $\phi(u) = u^2$, the deadzone-linear penalty function with deadzone width $a = 1/4$, and the log barrier penalty function with limit $a = 1$.

Example 6.1 *Some common penalty functions and associated approximation problems.*

- By taking $\phi(u) = |u|^p$, where $p \geq 1$, the penalty function approximation problem is equivalent to the ℓ_p-norm approximation problem. In particular, the quadratic penalty function $\phi(u) = u^2$ yields least-squares or Euclidean norm approximation, and the absolute value penalty function $\phi(u) = |u|$ yields ℓ_1-norm approximation.

- The *deadzone-linear* penalty function (with deadzone width $a > 0$) is given by

$$\phi(u) = \left\{ \begin{array}{ll} 0 & |u| \leq a \\ |u| - a & |u| > a. \end{array} \right.$$

  The deadzone-linear function assesses no penalty for residuals smaller than a.

- The *log barrier* penalty function (with limit $a > 0$) has the form

$$\phi(u) = \left\{ \begin{array}{ll} -a^2 \log(1 - (u/a)^2) & |u| < a \\ \infty & |u| \geq a. \end{array} \right.$$

  The log barrier penalty function assesses an infinite penalty for residuals larger than a.

A deadzone-linear, log barrier, and quadratic penalty function are plotted in figure 6.1. Note that the log barrier function is very close to the quadratic penalty for $|u/a| \leq 0.25$ (see exercise 6.1).

Scaling the penalty function by a positive number does not affect the solution of the penalty function approximation problem, since this merely scales the objective function. But the *shape* of the penalty function has a large effect on the solution of the penalty function approximation problem. Roughly speaking, $\phi(u)$ is a measure of our dislike of a residual of value u. If ϕ is very small (or even zero) for small values of u, it means we care very little (or not at all) if residuals have these values. If $\phi(u)$ grows rapidly as u becomes large, it means we have a strong dislike for large residuals; if ϕ becomes infinite outside some interval, it means that residuals outside the interval are unacceptable. This simple interpretation gives insight into the solution of a penalty function approximation problem, as well as guidelines for choosing a penalty function.

As an example, let us compare ℓ_1-norm and ℓ_2-norm approximation, associated with the penalty functions $\phi_1(u) = |u|$ and $\phi_2(u) = u^2$, respectively. For $|u| = 1$, the two penalty functions assign the same penalty. For small u we have $\phi_1(u) \gg \phi_2(u)$, so ℓ_1-norm approximation puts relatively larger emphasis on small residuals compared to ℓ_2-norm approximation. For large u we have $\phi_2(u) \gg \phi_1(u)$, so ℓ_1-norm approximation puts less weight on large residuals, compared to ℓ_2-norm approximation. This difference in relative weightings for small and large residuals is reflected in the solutions of the associated approximation problems. The amplitude distribution of the optimal residual for the ℓ_1-norm approximation problem will tend to have more zero and very small residuals, compared to the ℓ_2-norm approximation solution. In contrast, the ℓ_2-norm solution will tend to have relatively fewer large residuals (since large residuals incur a much larger penalty in ℓ_2-norm approximation than in ℓ_1-norm approximation).

Example

An example will illustrate these ideas. We take a matrix $A \in \mathbf{R}^{100 \times 30}$ and vector $b \in \mathbf{R}^{100}$ (chosen at random, but the results are typical), and compute the ℓ_1-norm and ℓ_2-norm approximate solutions of $Ax \approx b$, as well as the penalty function approximations with a deadzone-linear penalty (with $a = 0.5$) and log barrier penalty (with $a = 1$). Figure 6.2 shows the four associated penalty functions, and the amplitude distributions of the optimal residuals for these four penalty approximations. From the plots of the penalty functions we note that

- The ℓ_1-norm penalty puts the most weight on small residuals and the least weight on large residuals.

- The ℓ_2-norm penalty puts very small weight on small residuals, but strong weight on large residuals.

- The deadzone-linear penalty function puts no weight on residuals smaller than 0.5, and relatively little weight on large residuals.

- The log barrier penalty puts weight very much like the ℓ_2-norm penalty for small residuals, but puts very strong weight on residuals larger than around 0.8, and infinite weight on residuals larger than 1.

Several features are clear from the amplitude distributions:

- For the ℓ_1-optimal solution, many residuals are either zero or very small. The ℓ_1-optimal solution also has relatively more large residuals.

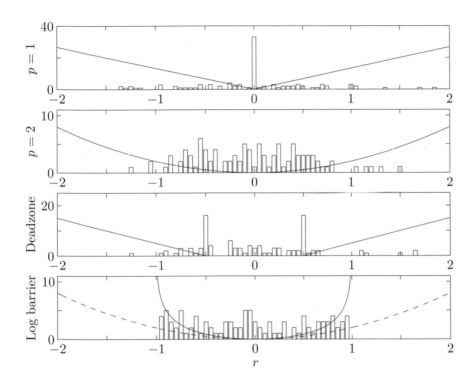

Figure 6.2 Histogram of residual amplitudes for four penalty functions, with the (scaled) penalty functions also shown for reference. For the log barrier plot, the quadratic penalty is also shown, in dashed curve.

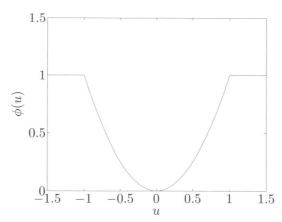

Figure 6.3 A (nonconvex) penalty function that assesses a fixed penalty to residuals larger than a threshold (which in this example is one): $\phi(u) = u^2$ if $|u| \leq 1$ and $\phi(u) = 1$ if $|u| > 1$. As a result, penalty approximation with this function would be relatively insensitive to outliers.

- The ℓ_2-norm approximation has many modest residuals, and relatively few larger ones.

- For the deadzone-linear penalty, we see that many residuals have the value ± 0.5, right at the edge of the 'free' zone, for which no penalty is assessed.

- For the log barrier penalty, we see that no residuals have a magnitude larger than 1, but otherwise the residual distribution is similar to the residual distribution for ℓ_2-norm approximation.

Sensitivity to outliers or large errors

In the estimation or regression context, an *outlier* is a measurement $y_i = a_i^T x + v_i$ for which the noise v_i is relatively large. This is often associated with faulty data or a flawed measurement. When outliers occur, any estimate of x will be associated with a residual vector with some large components. Ideally we would like to guess which measurements are outliers, and either remove them from the estimation process or greatly lower their weight in forming the estimate. (We cannot, however, assign zero penalty for very large residuals, because then the optimal point would likely make all residuals large, which yields a total penalty of zero.) This could be accomplished using penalty function approximation, with a penalty function such as

$$\phi(u) = \left\{ \begin{array}{ll} u^2 & |u| \leq M \\ M^2 & |u| > M, \end{array} \right. \tag{6.3}$$

shown in figure 6.3. This penalty function agrees with least-squares for any residual smaller than M, but puts a fixed weight on any residual larger than M, no matter how much larger it is. In other words, residuals larger than M are ignored; they are assumed to be associated with outliers or bad data. Unfortunately, the penalty

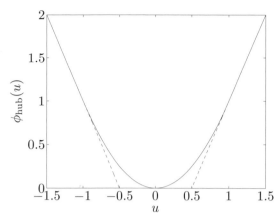

Figure 6.4 The solid line is the robust least-squares or Huber penalty function ϕ_{hub}, with $M = 1$. For $|u| \leq M$ it is quadratic, and for $|u| > M$ it grows linearly.

function (6.3) is not convex, and the associated penalty function approximation problem becomes a hard combinatorial optimization problem.

The sensitivity of a penalty function based estimation method to outliers depends on the (relative) value of the penalty function for large residuals. If we restrict ourselves to convex penalty functions (which result in convex optimization problems), the ones that are least sensitive are those for which $\phi(u)$ grows linearly, *i.e.*, like $|u|$, for large u. Penalty functions with this property are sometimes called *robust*, since the associated penalty function approximation methods are much less sensitive to outliers or large errors than, for example, least-squares.

One obvious example of a robust penalty function is $\phi(u) = |u|$, corresponding to ℓ_1-norm approximation. Another example is the *robust least-squares* or *Huber penalty function*, given by

$$\phi_{\mathrm{hub}}(u) = \begin{cases} u^2 & |u| \leq M \\ M(2|u| - M) & |u| > M, \end{cases} \tag{6.4}$$

shown in figure 6.4. This penalty function agrees with the least-squares penalty function for residuals smaller than M, and then reverts to ℓ_1-like linear growth for larger residuals. The Huber penalty function can be considered a convex approximation of the outlier penalty function (6.3), in the following sense: They agree for $|u| \leq M$, and for $|u| > M$, the Huber penalty function is the convex function closest to the outlier penalty function (6.3).

Example 6.2 *Robust regression.* Figure 6.5 shows 42 points (t_i, y_i) in a plane, with two obvious outliers (one at the upper left, and one at lower right). The dashed line shows the least-squares approximation of the points by a straight line $f(t) = \alpha + \beta t$. The coefficients α and β are obtained by solving the least-squares problem

$$\text{minimize} \quad \sum_{i=1}^{42} (y_i - \alpha - \beta t_i)^2,$$

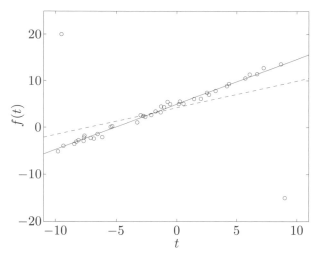

Figure 6.5 The 42 circles show points that can be well approximated by
an affine function, except for the two outliers at upper left and lower right.
The dashed line is the least-squares fit of a straight line $f(t) = \alpha + \beta t$
to the points, and is rotated away from the main locus of points, toward
the outliers. The solid line shows the robust least-squares fit, obtained by
minimizing Huber's penalty function with $M = 1$. This gives a far better fit
to the non-outlier data.

with variables α and β. The least-squares approximation is clearly rotated away from
the main locus of the points, toward the two outliers.

The solid line shows the robust least-squares approximation, obtained by minimizing
the Huber penalty function

$$\text{minimize}\quad \textstyle\sum_{i=1}^{42} \phi_{\text{hub}}(y_i - \alpha - \beta t_i),$$

with $M = 1$. This approximation is far less affected by the outliers.

Since ℓ_1-norm approximation is among the (convex) penalty function approximation methods that are most robust to outliers, ℓ_1-norm approximation is sometimes called *robust estimation* or *robust regression*. The robustness property of
ℓ_1-norm estimation can also be understood in a statistical framework; see page 353.

Small residuals and ℓ_1-norm approximation

We can also focus on small residuals. Least-squares approximation puts very small
weight on small residuals, since $\phi(u) = u^2$ is very small when u is small. Penalty
functions such as the deadzone-linear penalty function put *zero* weight on small
residuals. For penalty functions that are very small for small residuals, we expect
the optimal residuals to be small, but not very small. Roughly speaking, there is
little or no incentive to drive small residuals smaller.

In contrast, penalty functions that put relatively large weight on small residuals,
such as $\phi(u) = |u|$, corresponding to ℓ_1-norm approximation, tend to produce

optimal residuals many of which are very small, or even exactly zero. This means that in ℓ_1-norm approximation, we typically find that many of the equations are satisfied exactly, *i.e.*, we have $a_i^T x = b_i$ for many i. This phenomenon can be seen in figure 6.2.

.1.3   Approximation with constraints

It is possible to add constraints to the basic norm approximation problem (6.1). When these constraints are convex, the resulting problem is convex. Constraints arise for a variety of reasons.

- In an approximation problem, constraints can be used to rule out certain unacceptable approximations of the vector b, or to ensure that the approximator Ax satisfies certain properties.

- In an estimation problem, the constraints arise as prior knowledge of the vector x to be estimated, or from prior knowledge of the estimation error v.

- Constraints arise in a geometric setting in determining the projection of a point b on a set more complicated than a subspace, for example, a cone or polyhedron.

Some examples will make these clear.

Nonnegativity constraints on variables

We can add the constraint $x \succeq 0$ to the basic norm approximation problem:

$$
\begin{array}{ll}
\text{minimize} & \|Ax - b\| \\
\text{subject to} & x \succeq 0.
\end{array}
$$

In an estimation setting, nonnegativity constraints arise when we estimate a vector x of parameters known to be nonnegative, *e.g.*, powers, intensities, or rates. The geometric interpretation is that we are determining the projection of a vector b onto the cone generated by the columns of A. We can also interpret this problem as approximating b using a nonnegative linear (*i.e.*, conic) combination of the columns of A.

Variable bounds

Here we add the constraint $l \preceq x \preceq u$, where $l, \ u \in \mathbf{R}^n$ are problem parameters:

$$
\begin{array}{ll}
\text{minimize} & \|Ax - b\| \\
\text{subject to} & l \preceq x \preceq u.
\end{array}
$$

In an estimation setting, variable bounds arise as prior knowledge of intervals in which each variable lies. The geometric interpretation is that we are determining the projection of a vector b onto the image of a box under the linear mapping induced by A.

Probability distribution

We can impose the constraint that x satisfy $x \succeq 0$, $\mathbf{1}^T x = 1$:

$$
\begin{array}{ll}
\text{minimize} & \|Ax - b\| \\
\text{subject to} & x \succeq 0, \quad \mathbf{1}^T x = 1.
\end{array}
$$

This would arise in the estimation of proportions or relative frequencies, which are nonnegative and sum to one. It can also be interpreted as approximating b by a convex combination of the columns of A. (We will have much more to say about estimating probabilities in §7.2.)

Norm ball constraint

We can add to the basic norm approximation problem the constraint that x lie in a norm ball:

$$
\begin{array}{ll}
\text{minimize} & \|Ax - b\| \\
\text{subject to} & \|x - x_0\| \leq d,
\end{array}
$$

where x_0 and d are problem parameters. Such a constraint can be added for several reasons.

- In an estimation setting, x_0 is a prior guess of what the parameter x is, and d is the maximum plausible deviation of our estimate from our prior guess. Our estimate of the parameter x is the value $\hat{x}$ which best matches the measured data (*i.e.*, minimizes $\|Az - b\|$) among all plausible candidates (*i.e.*, z that satisfy $\|z - x_0\| \leq d$).

- The constraint $\|x - x_0\| \leq d$ can denote a *trust region*. Here the linear relation $y = Ax$ is only an approximation of some nonlinear relation $y = f(x)$ that is valid when x is near some point x_0, specifically $\|x - x_0\| \leq d$. The problem is to minimize $\|Ax - b\|$ but only over those x for which the model $y = Ax$ is trusted.

These ideas also come up in the context of regularization; see §6.3.2.

6.2   Least-norm problems

The basic *least-norm problem* has the form

$$
\begin{array}{ll}
\text{minimize} & \|x\| \\
\text{subject to} & Ax = b
\end{array}
\tag{6.5}
$$

where the data are $A \in \mathbf{R}^{m \times n}$ and $b \in \mathbf{R}^m$, the variable is $x \in \mathbf{R}^n$, and $\|\cdot\|$ is a norm on $\mathbf{R}^n$. A solution of the problem, which always exists if the linear equations $Ax = b$ have a solution, is called a *least-norm solution* of $Ax = b$. The least-norm problem is, of course, a convex optimization problem.

We can assume without loss of generality that the rows of A are independent, so $m \leq n$. When $m = n$, the only feasible point is $x = A^{-1}b$; the least-norm problem is interesting only when $m < n$, *i.e.*, when the equation $Ax = b$ is underdetermined.

Reformulation as norm approximation problem

The least-norm problem (6.5) can be formulated as a norm approximation problem by eliminating the equality constraint. Let x_0 be any solution of $Ax = b$, and let $Z \in \mathbf{R}^{n \times k}$ be a matrix whose columns are a basis for the nullspace of A. The general solution of $Ax = b$ can then be expressed as $x_0 + Zu$ where $u \in \mathbf{R}^k$. The least-norm problem (6.5) can be expressed as

$$\text{minimize} \quad \|x_0 + Zu\|,$$

with variable $u \in \mathbf{R}^k$, which is a norm approximation problem. In particular, our analysis and discussion of norm approximation problems applies to least-norm problems as well (when interpreted correctly).

Control or design interpretation

We can interpret the least-norm problem (6.5) as a problem of optimal design or optimal control. The n variables $x_1, \dots, x_n$ are *design variables* whose values are to be determined. In a control setting, the variables $x_1, \dots, x_n$ represent inputs, whose values we are to choose. The vector $y = Ax$ gives m attributes or results of the design x, which we assume to be linear functions of the design variables x. The $m < n$ equations $Ax = b$ represent m *specifications* or *requirements* on the design. Since $m < n$, the design is underspecified; there are $n - m$ degrees of freedom in the design (assuming A is rank m).

Among all the designs that satisfy the specifications, the least-norm problem chooses the smallest design, as measured by the norm $\| \cdot \|$. This can be thought of as the most efficient design, in the sense that it achieves the specifications $Ax = b$, with the smallest possible x.

Estimation interpretation

We assume that x is a vector of parameters to be estimated. We have $m < n$ perfect (noise free) linear measurements, given by $Ax = b$. Since we have fewer measurements than parameters to estimate, our measurements do not completely determine x. Any parameter vector x that satisfies $Ax = b$ is consistent with our measurements.

To make a good guess about what x is, without taking further measurements, we must use prior information. Suppose our prior information, or assumption, is that x is more likely to be small (as measured by $\| \cdot \|$) than large. The least-norm problem chooses as our estimate of the parameter vector x the one that is smallest (hence, most plausible) among all parameter vectors that are consistent with the measurements $Ax = b$. (For a statistical interpretation of the least-norm problem, see page 359.)

Geometric interpretation

We can also give a simple geometric interpretation of the least-norm problem (6.5). The feasible set $\{x \mid Ax = b\}$ is affine, and the objective is the distance (measured by the norm $\| \cdot \|$) between x and the point 0. The least-norm problem finds the point in the affine set with minimum distance to 0, *i.e.*, it determines the projection of the point 0 on the affine set $\{x \mid Ax = b\}$.

Least-squares solution of linear equations

The most common least-norm problem involves the Euclidean or ℓ_2-norm. By squaring the objective we obtain the equivalent problem

$$\begin{array}{ll} \text{minimize} & \|x\|_2^2 \\ \text{subject to} & Ax = b, \end{array}$$

the unique solution of which is called the *least-squares solution* of the equations $Ax = b$. Like the least-squares approximation problem, this problem can be solved analytically. Introducing the dual variable $\nu \in \mathbf{R}^m$, the optimality conditions are

$$2x^\star + A^T \nu^\star = 0, \qquad Ax^\star = b,$$

which is a pair of linear equations, and readily solved. From the first equation we obtain $x^\star = -(1/2)A^T\nu^\star$; substituting this into the second equation we obtain $-(1/2)AA^T\nu^\star = b$, and conclude

$$\nu^\star = -2(AA^T)^{-1}b, \qquad x^\star = A^T(AA^T)^{-1}b.$$

(Since $\mathbf{rank}\, A = m < n$, the matrix AA^T is invertible.)

Least-penalty problems

A useful variation on the least-norm problem (6.5) is the *least-penalty problem*

$$\begin{array}{ll} \text{minimize} & \phi(x_1) + \cdots + \phi(x_n) \\ \text{subject to} & Ax = b, \end{array} \tag{6.6}$$

where $\phi : \mathbf{R} \to \mathbf{R}$ is convex, nonnegative, and satisfies $\phi(0) = 0$. The penalty function value $\phi(u)$ quantifies our dislike of a component of x having value u; the least-penalty problem then finds x that has least total penalty, subject to the constraint $Ax = b$.

All of the discussion and interpretation of penalty functions in penalty function approximation can be transposed to the least-penalty problem, by substituting the amplitude distribution of x (in the least-penalty problem) for the amplitude distribution of the residual r (in the penalty approximation problem).

Sparse solutions via least ℓ_1-norm

Recall from the discussion on page 300 that ℓ_1-norm approximation gives relatively large weight to small residuals, and therefore results in many optimal residuals small, or even zero. A similar effect occurs in the least-norm context. The least ℓ_1-norm problem,

$$\begin{array}{ll} \text{minimize} & \|x\|_1 \\ \text{subject to} & Ax = b, \end{array}$$

tends to produce a solution x with a large number of components equal to zero. In other words, the least ℓ_1-norm problem tends to produce *sparse* solutions of $Ax = b$, often with m nonzero components.

It is easy to find solutions of $Ax = b$ that have only m nonzero components. Choose any set of m indices (out of $1, \ldots, n$) which are to be the nonzero components of x. The equation $Ax = b$ reduces to $\tilde{A}\tilde{x} = b$, where $\tilde{A}$ is the $m \times m$ submatrix of A obtained by selecting only the chosen columns, and $\tilde{x} \in \mathbf{R}^m$ is the subvector of x containing the m selected components. If $\tilde{A}$ is nonsingular, then we can take $\tilde{x} = \tilde{A}^{-1}b$, which gives a feasible solution x with m or less nonzero components. If $\tilde{A}$ is singular and $b \notin \mathcal{R}(\tilde{A})$, the equation $\tilde{A}\tilde{x} = b$ is unsolvable, which means there is no feasible x with the chosen set of nonzero components. If $\tilde{A}$ is singular and $b \in \mathcal{R}(\tilde{A})$, there is a feasible solution with fewer than m nonzero components.

This approach can be used to find the smallest x with m (or fewer) nonzero entries, but in general requires examining and comparing all $n!/(m!(n-m)!)$ choices of m nonzero coefficients of the n coefficients in x. Solving the least ℓ_1-norm problem, on the other hand, gives a good heuristic for finding a sparse, and small, solution of $Ax = b$.

6.3   Regularized approximation

6.3.1   Bi-criterion formulation

In the basic form of regularized approximation, the goal is to find a vector x that is small (if possible), and also makes the residual $Ax - b$ small. This is naturally described as a (convex) vector optimization problem with two objectives, $\|Ax - b\|$ and $\|x\|$:

$$\text{minimize (w.r.t. } \mathbf{R}_+^2) \quad (\|Ax - b\|, \|x\|). \tag{6.7}$$

The two norms can be different: the first, used to measure the size of the residual, is on $\mathbf{R}^m$; the second, used to measure the size of x, is on $\mathbf{R}^n$.

The optimal trade-off between the two objectives can be found using several methods. The optimal trade-off curve of $\|Ax - b\|$ versus $\|x\|$, which shows how large one of the objectives must be made to have the other one small, can then be plotted. One endpoint of the optimal trade-off curve between $\|Ax - b\|$ and $\|x\|$ is easy to describe. The minimum value of $\|x\|$ is zero, and is achieved only when $x = 0$. For this value of x, the residual norm has the value $\|b\|$.

The other endpoint of the trade-off curve is more complicated to describe. Let C denote the set of minimizers of $\|Ax - b\|$ (with no constraint on $\|x\|$). Then any minimum norm point in C is Pareto optimal, corresponding to the other endpoint of the trade-off curve. In other words, Pareto optimal points at this endpoint are given by minimum norm minimizers of $\|Ax - b\|$. If both norms are Euclidean, this Pareto optimal point is unique, and given by $x = A^\dagger b$, where $A^\dagger$ is the pseudo-inverse of A. (See §4.7.6, page 184, and §A.5.4.)

6.3.2   Regularization

Regularization is a common scalarization method used to solve the bi-criterion problem (6.7). One form of regularization is to minimize the weighted sum of the objectives:

$$\text{minimize} \quad \|Ax - b\| + \gamma\|x\|, \tag{6.8}$$

where $\gamma > 0$ is a problem parameter. As γ varies over $(0, \infty)$, the solution of (6.8) traces out the optimal trade-off curve.

Another common method of regularization, especially when the Euclidean norm is used, is to minimize the weighted sum of squared norms, *i.e.*,

$$\text{minimize} \quad \|Ax - b\|^2 + \delta\|x\|^2, \tag{6.9}$$

for a variety of values of $\delta > 0$.

These regularized approximation problems each solve the bi-criterion problem of making both $\|Ax - b\|$ and $\|x\|$ small, by adding an extra term or penalty associated with the norm of x.

Interpretations

Regularization is used in several contexts. In an estimation setting, the extra term penalizing large $\|x\|$ can be interpreted as our prior knowledge that $\|x\|$ is not too large. In an optimal design setting, the extra term adds the cost of using large values of the design variables to the cost of missing the target specifications.

The constraint that $\|x\|$ be small can also reflect a modeling issue. It might be, for example, that $y = Ax$ is only a good approximation of the true relationship $y = f(x)$ between x and y. In order to have $f(x) \approx b$, we want $Ax \approx b$, and also need x small in order to ensure that $f(x) \approx Ax$.

We will see in §6.4.1 and §6.4.2 that regularization can be used to take into account variation in the matrix A. Roughly speaking, a large x is one for which variation in A causes large variation in Ax, and hence should be avoided.

Regularization is also used when the matrix A is square, and the goal is to solve the linear equations $Ax = b$. In cases where A is poorly conditioned, or even singular, regularization gives a compromise between solving the equations (*i.e.*, making $\|Ax - b\|$ zero) and keeping x of reasonable size.

Regularization comes up in a statistical setting; see §7.1.2.

Tikhonov regularization

The most common form of regularization is based on (6.9), with Euclidean norms, which results in a (convex) quadratic optimization problem:

$$\text{minimize} \quad \|Ax - b\|_2^2 + \delta\|x\|_2^2 = x^T(A^TA + \delta I)x - 2b^TAx + b^Tb. \tag{6.10}$$

This *Tikhonov regularization problem* has the analytical solution

$$x = (A^TA + \delta I)^{-1}A^Tb.$$

Since $A^TA + \delta I \succ 0$ for any $\delta > 0$, the Tikhonov regularized least-squares solution requires no rank (or dimension) assumptions on the matrix A.

Smoothing regularization

The idea of regularization, *i.e.*, adding to the objective a term that penalizes large
x, can be extended in several ways. In one useful extension we add a regularization
term of the form $\|Dx\|$, in place of $\|x\|$. In many applications, the matrix D
represents an approximate differentiation or second-order differentiation operator,
so $\|Dx\|$ represents a measure of the variation or smoothness of x.

For example, suppose that the vector $x \in \mathbf{R}^n$ represents the value of some
continuous physical parameter, say, temperature, along the interval $[0,1]$: x_i is
the temperature at the point i/n. A simple approximation of the gradient or
first derivative of the parameter near i/n is given by $n(x_{i+1} - x_i)$, and a simple
approximation of its second derivative is given by the second difference

$$n\left(n(x_{i+1} - x_i) - n(x_i - x_{i-1})\right) = n^2(x_{i+1} - 2x_i + x_{i-1}).$$

If Δ is the (tridiagonal, Toeplitz) matrix

$$\Delta = n^2 \begin{bmatrix} 1 & -2 & 1 & 0 & \cdots & 0 & 0 & 0 & 0 \\ 0 & 1 & -2 & 1 & \cdots & 0 & 0 & 0 & 0 \\ 0 & 0 & 1 & -2 & \cdots & 0 & 0 & 0 & 0 \\ \vdots & \vdots & \vdots & \vdots & & \vdots & \vdots & \vdots & \vdots \\ 0 & 0 & 0 & 0 & \cdots & -2 & 1 & 0 & 0 \\ 0 & 0 & 0 & 0 & \cdots & 1 & -2 & 1 & 0 \\ 0 & 0 & 0 & 0 & \cdots & 0 & 1 & -2 & 1 \end{bmatrix} \in \mathbf{R}^{(n-2)\times n},$$

then Δx represents an approximation of the second derivative of the parameter, so
$\|\Delta x\|_2^2$ represents a measure of the mean-square curvature of the parameter over
the interval $[0,1]$.

The Tikhonov regularized problem

$$\text{minimize} \quad \|Ax - b\|_2^2 + \delta\|\Delta x\|_2^2$$

can be used to trade off the objective $\|Ax - b\|^2$, which might represent a measure
of fit, or consistency with experimental data, and the objective $\|\Delta x\|^2$, which is
(approximately) the mean-square curvature of the underlying physical parameter.
The parameter δ is used to control the amount of regularization required, or to
plot the optimal trade-off curve of fit versus smoothness.

We can also add several regularization terms. For example, we can add terms
associated with smoothness and size, as in

$$\text{minimize} \quad \|Ax - b\|_2^2 + \delta\|\Delta x\|_2^2 + \eta\|x\|_2^2.$$

Here, the parameter $\delta \geq 0$ is used to control the smoothness of the approximate
solution, and the parameter $\eta \geq 0$ is used to control its size.

Example 6.3 *Optimal input design.* We consider a dynamical system with scalar
input sequence $u(0), u(1), \ldots, u(N)$, and scalar output sequence $y(0), y(1), \ldots, y(N)$,
related by convolution:

$$y(t) = \sum_{\tau=0}^{t} h(\tau)u(t - \tau), \quad t = 0, 1, \ldots, N.$$

The sequence $h(0)$, $h(1), \ldots, h(N)$ is called the *convolution kernel* or *impulse response* of the system.

Our goal is to choose the input sequence u to achieve several goals.

- *Output tracking.* The primary goal is that the output y should track, or follow, a desired target or reference signal y_{des}. We measure output tracking error by the quadratic function

$$J_{\text{track}} = \frac{1}{N+1} \sum_{t=0}^{N} (y(t) - y_{\text{des}}(t))^2.$$

- *Small input.* The input should not be large. We measure the magnitude of the input by the quadratic function

$$J_{\text{mag}} = \frac{1}{N+1} \sum_{t=0}^{N} u(t)^2.$$

- *Small input variations.* The input should not vary rapidly. We measure the magnitude of the input variations by the quadratic function

$$J_{\text{der}} = \frac{1}{N} \sum_{t=0}^{N-1} (u(t+1) - u(t))^2.$$

By minimizing a weighted sum

$$J_{\text{track}} + \delta J_{\text{der}} + \eta J_{\text{mag}},$$

where $\delta > 0$ and $\eta > 0$, we can trade off the three objectives.

Now we consider a specific example, with $N = 200$, and impulse response

$$h(t) = \frac{1}{9}(0.9)^t (1 - 0.4 \cos(2t)).$$

Figure 6.6 shows the optimal input, and corresponding output (along with the desired trajectory y_{des}), for three values of the regularization parameters δ and η. The top row shows the optimal input and corresponding output for $\delta = 0$, $\eta = 0.005$. In this case we have some regularization for the magnitude of the input, but no regularization for its variation. While the tracking is good (*i.e.*, we have J_{track} is small), the input required is large, and rapidly varying. The second row corresponds to $\delta = 0$, $\eta = 0.05$. In this case we have more magnitude regularization, but still no regularization for variation in u. The corresponding input is indeed smaller, at the cost of a larger tracking error. The bottom row shows the results for $\delta = 0.3$, $\eta = 0.05$. In this case we have added some regularization for the variation. The input variation is substantially reduced, with not much increase in output tracking error.

ℓ_1-norm regularization

Regularization with an ℓ_1-norm can be used as a heuristic for finding a sparse solution. For example, consider the problem

$$\text{minimize} \quad \|Ax - b\|_2 + \gamma \|x\|_1, \tag{6.11}$$

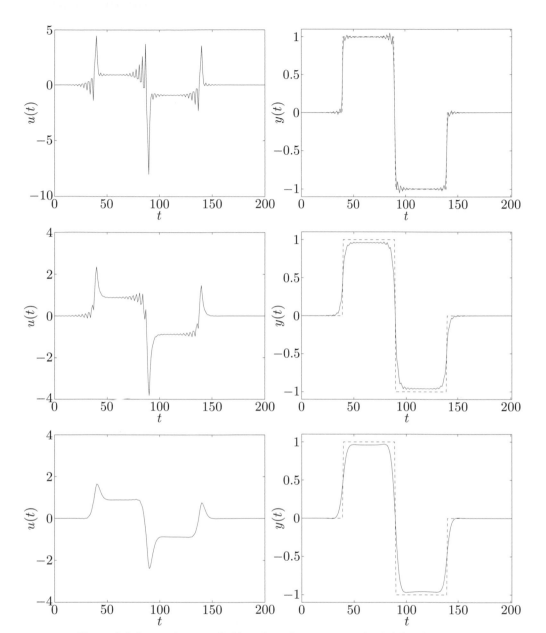

Figure 6.6 Optimal inputs (left) and resulting outputs (right) for three values of the regularization parameters δ (which corresponds to input variation) and η (which corresponds to input magnitude). The dashed line in the righthand plots shows the desired output y_{des}. Top row: $\delta = 0$, $\eta = 0.005$; middle row: $\delta = 0$, $\eta = 0.05$; bottom row: $\delta = 0.3$, $\eta = 0.05$.

in which the residual is measured with the Euclidean norm and the regularization is done with an ℓ_1-norm. By varying the parameter γ we can sweep out the optimal trade-off curve between $\|Ax - b\|_2$ and $\|x\|_1$, which serves as an approximation of the optimal trade-off curve between $\|Ax - b\|_2$ and the sparsity or cardinality $\mathbf{card}(x)$ of the vector x, $i.e.$, the number of nonzero elements. The problem (6.11) can be recast and solved as an SOCP.

Example 6.4 *Regressor selection problem.*   We are given a matrix $A \in \mathbf{R}^{m \times n}$, whose columns are potential regressors, and a vector $b \in \mathbf{R}^m$ that is to be fit by a linear combination of $k < n$ columns of A. The problem is to choose the subset of k regressors to be used, and the associated coefficients. We can express this problem as

$$\begin{array}{ll} \text{minimize} & \|Ax - b\|_2 \\ \text{subject to} & \mathbf{card}(x) \leq k. \end{array}$$

In general, this is a hard combinatorial problem.

One straightforward approach is to check every possible sparsity pattern in x with k nonzero entries. For a fixed sparsity pattern, we can find the optimal x by solving a least-squares problem, $i.e.$, minimizing $\|\tilde{A}\tilde{x} - b\|_2$, where $\tilde{A}$ denotes the submatrix of A obtained by keeping the columns corresponding to the sparsity pattern, and $\tilde{x}$ is the subvector with the nonzero components of x. This is done for each of the $n!/(k!(n-k)!)$ sparsity patterns with k nonzeros.

A good heuristic approach is to solve the problem (6.11) for different values of γ, finding the smallest value of γ that results in a solution with $\mathbf{card}(x) = k$. We then fix this sparsity pattern and find the value of x that minimizes $\|Ax - b\|_2$.

Figure 6.7 illustrates a numerical example with $A \in \mathbf{R}^{10 \times 20}$, $x \in \mathbf{R}^{20}$, $b \in \mathbf{R}^{10}$. The circles on the dashed curve are the (globally) Pareto optimal values for the trade-off between $\mathbf{card}(x)$ (vertical axis) and the residual $\|Ax - b\|_2$ (horizontal axis). For each k, the Pareto optimal point was obtained by enumerating all possible sparsity patterns with k nonzero entries, as described above. The circles on the solid curve were obtained with the heuristic approach, by using the sparsity patterns of the solutions of problem (6.11) for different values of γ. Note that for $\mathbf{card}(x) = 1$, the heuristic method actually finds the global optimum.

This idea will come up again in *basis pursuit* (§6.5.4).

6.3.3   Reconstruction, smoothing, and de-noising

In this section we describe an important special case of the bi-criterion approximation problem described above, and give some examples showing how different regularization methods perform. In *reconstruction problems*, we start with a *signal* represented by a vector $x \in \mathbf{R}^n$. The coefficients x_i correspond to the value of some function of time, evaluated (or *sampled*, in the language of signal processing) at evenly spaced points. It is usually assumed that the signal does not vary too rapidly, which means that usually, we have $x_i \approx x_{i+1}$. (In this section we consider signals in one dimension, $e.g.$, audio signals, but the same ideas can be applied to signals in two or more dimensions, $e.g.$, images or video.)

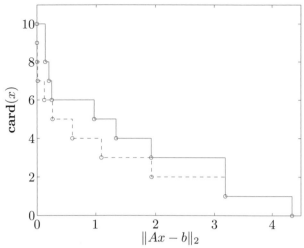

Figure 6.7 Sparse regressor selection with a matrix $A \in \mathbf{R}^{10 \times 20}$. The circles on the dashed line are the Pareto optimal values for the trade-off between the residual $\|Ax - b\|_2$ and the number of nonzero elements $\mathbf{card}(x)$. The points indicated by circles on the solid line are obtained via the ℓ_1-norm regularized heuristic.

The signal x is corrupted by an additive noise v:

$$x_{\mathrm{cor}} = x + v.$$

The noise can be modeled in many different ways, but here we simply assume that it is unknown, small, and, unlike the signal, rapidly varying. The goal is to form an estimate $\hat{x}$ of the original signal x, given the corrupted signal x_{cor}. This process is called *signal reconstruction* (since we are trying to reconstruct the original signal from the corrupted version) or *de-noising* (since we are trying to remove the noise from the corrupted signal). Most reconstruction methods end up performing some sort of smoothing operation on x_{cor} to produce $\hat{x}$, so the process is also called *smoothing*.

One simple formulation of the reconstruction problem is the bi-criterion problem

$$\text{minimize (w.r.t. } \mathbf{R}_+^2)\quad (\|\hat{x} - x_{\mathrm{cor}}\|_2, \phi(\hat{x})), \tag{6.12}$$

where $\hat{x}$ is the variable and x_{cor} is a problem parameter. The function $\phi : \mathbf{R}^n \to \mathbf{R}$ is convex, and is called the *regularization function* or *smoothing objective*. It is meant to measure the roughness, or lack of smoothness, of the estimate $\hat{x}$. The reconstruction problem (6.12) seeks signals that are close (in ℓ_2-norm) to the corrupted signal, and that are smooth, *i.e.*, for which $\phi(\hat{x})$ is small. The reconstruction problem (6.12) is a convex bi-criterion problem. We can find the Pareto optimal points by scalarization, and solving a (scalar) convex optimization problem.

Quadratic smoothing

The simplest reconstruction method uses the quadratic smoothing function

$$\phi_{\mathrm{quad}}(x) = \sum_{i=1}^{n-1}(x_{i+1} - x_i)^2 = \|Dx\|_2^2,$$

where $D \in \mathbf{R}^{(n-1)\times n}$ is the bidiagonal matrix

$$D = \begin{bmatrix} -1 & 1 & 0 & \cdots & 0 & 0 & 0 \\ 0 & -1 & 1 & \cdots & 0 & 0 & 0 \\ \vdots & \vdots & \vdots & & \vdots & \vdots & \vdots \\ 0 & 0 & 0 & \cdots & -1 & 1 & 0 \\ 0 & 0 & 0 & \cdots & 0 & -1 & 1 \end{bmatrix}.$$

We can obtain the optimal trade-off between $\|\hat{x} - x_{\mathrm{cor}}\|_2$ and $\|D\hat{x}\|_2$ by minimizing

$$\|\hat{x} - x_{\mathrm{cor}}\|_2^2 + \delta\|D\hat{x}\|_2^2,$$

where $\delta > 0$ parametrizes the optimal trade-off curve. The solution of this quadratic problem,

$$\hat{x} = (I + \delta D^T D)^{-1} x_{\mathrm{cor}},$$

can be computed very efficiently since $I + \delta D^T D$ is tridiagonal; see appendix C.

Quadratic smoothing example

Figure 6.8 shows a signal $x \in \mathbf{R}^{4000}$ (top) and the corrupted signal x_{cor} (bottom). The optimal trade-off curve between the objectives $\|\hat{x} - x_{\mathrm{cor}}\|_2$ and $\|D\hat{x}\|_2$ is shown in figure 6.9. The extreme point on the left of the trade-off curve corresponds to $\hat{x} = x_{\mathrm{cor}}$, and has objective value $\|Dx_{\mathrm{cor}}\|_2 = 4.4$. The extreme point on the right corresponds to $\hat{x} = 0$, for which $\|\hat{x} - x_{\mathrm{cor}}\|_2 = \|x_{\mathrm{cor}}\|_2 = 16.2$. Note the clear knee in the trade-off curve near $\|\hat{x} - x_{\mathrm{cor}}\|_2 \approx 3$.

Figure 6.10 shows three smoothed signals on the optimal trade-off curve, corresponding to $\|\hat{x} - x_{\mathrm{cor}}\|_2 = 8$ (top), 3 (middle), and 1 (bottom). Comparing the reconstructed signals with the original signal x, we see that the best reconstruction is obtained for $\|\hat{x} - x_{\mathrm{cor}}\|_2 = 3$, which corresponds to the knee of the trade-off curve. For higher values of $\|\hat{x} - x_{\mathrm{cor}}\|_2$, there is too much smoothing; for smaller values there is too little smoothing.

Total variation reconstruction

Simple quadratic smoothing works well as a reconstruction method when the original signal is very smooth, and the noise is rapidly varying. But any rapid variations in the original signal will, obviously, be attenuated or removed by quadratic smoothing. In this section we describe a reconstruction method that can remove much of the noise, while still preserving occasional rapid variations in the original signal. The method is based on the smoothing function

$$\phi_{\mathrm{tv}}(\hat{x}) = \sum_{i=1}^{n-1}|\hat{x}_{i+1} - \hat{x}_i| = \|D\hat{x}\|_1,$$

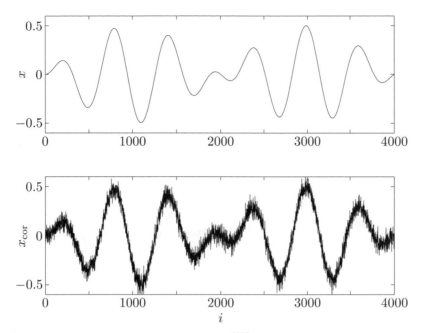

Figure 6.8 *Top:* the original signal $x \in \mathbf{R}^{4000}$. *Bottom:* the corrupted signal x_{cor}.

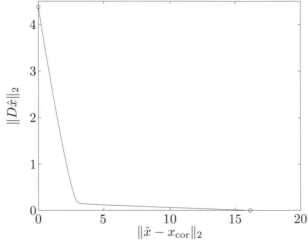

Figure 6.9 Optimal trade-off curve between $\|D\hat{x}\|_2$ and $\|\hat{x} - x_{\mathrm{cor}}\|_2$. The curve has a clear knee near $\|\hat{x} - x_{\mathrm{cor}}\| \approx 3$.

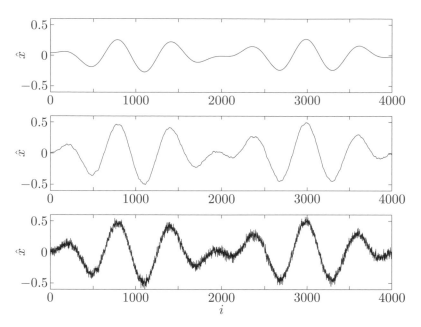

Figure 6.10 Three smoothed or reconstructed signals $\hat{x}$. The top one corresponds to $\|\hat{x} - x_{\mathrm{cor}}\|_2 = 8$, the middle one to $\|\hat{x} - x_{\mathrm{cor}}\|_2 = 3$, and the bottom one to $\|\hat{x} - x_{\mathrm{cor}}\|_2 = 1$.

which is called the *total variation* of $x \in \mathbf{R}^n$. Like the quadratic smoothness measure ϕ_{quad}, the total variation function assigns large values to rapidly varying $\hat{x}$. The total variation measure, however, assigns relatively less penalty to large values of $|x_{i+1} - x_i|$.

Total variation reconstruction example

Figure 6.11 shows a signal $x \in \mathbf{R}^{2000}$ (in the top plot), and the signal corrupted with noise x_{cor}. The signal is mostly smooth, but has several rapid variations or jumps in value; the noise is rapidly varying.

We first use quadratic smoothing. Figure 6.12 shows three smoothed signals on the optimal trade-off curve between $\|D\hat{x}\|_2$ and $\|\hat{x} - x_{\mathrm{cor}}\|_2$. In the first two signals, the rapid variations in the original signal are also smoothed. In the third signal the steep edges in the signal are better preserved, but there is still a significant amount of noise left.

Now we demonstrate total variation reconstruction. Figure 6.13 shows the optimal trade-off curve between $\|D\hat{x}\|_1$ and $\|\hat{x} - x_{\mathrm{corr}}\|_2$. Figure 6.14 shows the reconstructed signals on the optimal trade-off curve, for $\|D\hat{x}\|_1 = 5$ (top), $\|D\hat{x}\|_1 = 8$ (middle), and $\|D\hat{x}\|_1 = 10$ (bottom). We observe that, unlike quadratic smoothing, total variation reconstruction preserves the sharp transitions in the signal.

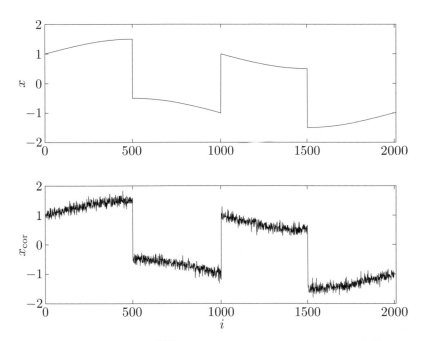

Figure 6.11 A signal $x \in \mathbf{R}^{2000}$, and the corrupted signal $x_{\mathrm{cor}} \in \mathbf{R}^{2000}$. The noise is rapidly varying, and the signal is mostly smooth, with a few rapid variations.

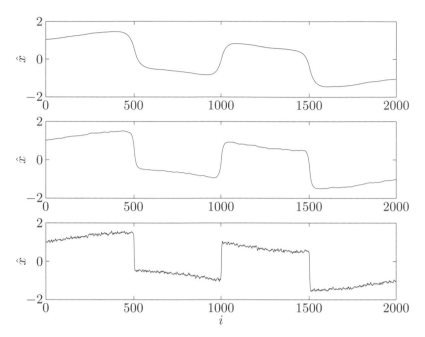

Figure 6.12 Three quadratically smoothed signals $\hat{x}$. The top one corresponds to $\|\hat{x} - x_{\mathrm{cor}}\|_2 = 10$, the middle one to $\|\hat{x} - x_{\mathrm{cor}}\|_2 = 7$, and the bottom one to $\|\hat{x} - x_{\mathrm{cor}}\|_2 = 4$. The top one greatly reduces the noise, but also excessively smooths out the rapid variations in the signal. The bottom smoothed signal does not give enough noise reduction, and still smooths out the rapid variations in the original signal. The middle smoothed signal gives the best compromise, but still smooths out the rapid variations.

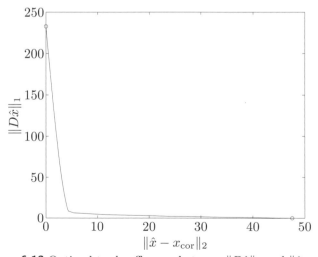

Figure 6.13 Optimal trade-off curve between $\|D\hat{x}\|_1$ and $\|\hat{x} - x_{\mathrm{cor}}\|_2$.

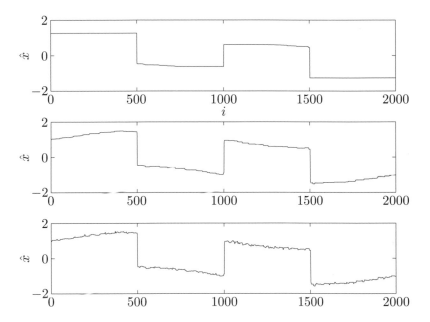

Figure 6.14 Three reconstructed signals $\hat{x}$, using total variation reconstruction. The top one corresponds to $\|D\hat{x}\|_1 = 5$, the middle one to $\|D\hat{x}\|_1 = 8$, and the bottom one to $\|D\hat{x}\|_1 = 10$. The bottom one does not give quite enough noise reduction, while the top one eliminates some of the slowly varying parts of the signal. Note that in total variation reconstruction, unlike quadratic smoothing, the sharp changes in the signal are preserved.

6.4   Robust approximation

6.4.1   Stochastic robust approximation

We consider an approximation problem with basic objective $\|Ax - b\|$, but also wish to take into account some uncertainty or possible variation in the data matrix A. (The same ideas can be extended to handle the case where there is uncertainty in both A and b.) In this section we consider some statistical models for the variation in A.

We assume that A is a random variable taking values in $\mathbf{R}^{m \times n}$, with mean $\bar{A}$, so we can describe A as

$$A = \bar{A} + U,$$

where U is a random matrix with zero mean. Here, the constant matrix $\bar{A}$ gives the average value of A, and U describes its statistical variation.

It is natural to use the expected value of $\|Ax - b\|$ as the objective:

$$\text{minimize} \quad \mathbf{E} \, \|Ax - b\|. \tag{6.13}$$

We refer to this problem as the *stochastic robust approximation problem*. It is always a convex optimization problem, but usually not tractable since in most cases it is very difficult to evaluate the objective or its derivatives.

One simple case in which the stochastic robust approximation problem (6.13) can be solved occurs when A assumes only a finite number of values, *i.e.*,

$$\mathbf{prob}(A = A_i) = p_i, \quad i = 1, \dots, k,$$

where $A_i \in \mathbf{R}^{m \times n}$, $\mathbf{1}^T p = 1$, $p \succeq 0$. In this case the problem (6.13) has the form

$$\text{minimize} \quad p_1 \|A_1 x - b\| + \cdots + p_k \|A_k x - b\|,$$

which is often called a *sum-of-norms problem*. It can be expressed as

$$
\begin{aligned}
\text{minimize} \quad & p^T t \\
\text{subject to} \quad & \|A_i x - b\| \le t_i, \quad i = 1, \dots, k,
\end{aligned}
$$

where the variables are $x \in \mathbf{R}^n$ and $t \in \mathbf{R}^k$. If the norm is the Euclidean norm, this sum-of-norms problem is an SOCP. If the norm is the ℓ_1- or ℓ_∞-norm, the sum-of-norms problem can be expressed as an LP; see exercise 6.8.

Some variations on the statistical robust approximation problem (6.13) are tractable. As an example, consider the statistical robust least-squares problem

$$\text{minimize} \quad \mathbf{E} \, \|Ax - b\|_2^2,$$

where the norm is the Euclidean norm. We can express the objective as

$$
\begin{aligned}
\mathbf{E} \, \|Ax - b\|_2^2 &= \mathbf{E}(\bar{A}x - b + Ux)^T (\bar{A}x - b + Ux) \\
&= (\bar{A}x - b)^T (\bar{A}x - b) + \mathbf{E} \, x^T U^T U x \\
&= \|\bar{A}x - b\|_2^2 + x^T P x,
\end{aligned}
$$

where $P = \mathbf{E}\, U^T U$. Therefore the statistical robust approximation problem has the form of a regularized least-squares problem

$$\text{minimize} \quad \|\bar{A}x - b\|_2^2 + \|P^{1/2}x\|_2^2,$$

with solution

$$x = (\bar{A}^T \bar{A} + P)^{-1} \bar{A}^T b.$$

This makes perfect sense: when the matrix A is subject to variation, the vector Ax will have more variation the larger x is, and Jensen's inequality tells us that variation in Ax will increase the average value of $\|Ax - b\|_2$. So we need to balance making $\bar{A}x - b$ small with the desire for a small x (to keep the variation in Ax small), which is the essential idea of regularization.

This observation gives us another interpretation of the Tikhonov regularized least-squares problem (6.10), as a robust least-squares problem, taking into account possible variation in the matrix A. The solution of the Tikhonov regularized least-squares problem (6.10) minimizes $\mathbf{E}\, \|(A + U)x - b\|^2$, where U_{ij} are zero mean, uncorrelated random variables, with variance δ/m (and here, A is deterministic).

6.4.2   Worst-case robust approximation

It is also possible to model the variation in the matrix A using a set-based, worst-case approach. We describe the uncertainty by a set of possible values for A:

$$A \in \mathcal{A} \subseteq \mathbf{R}^{m \times n},$$

which we assume is nonempty and bounded. We define the associated *worst-case error* of a candidate approximate solution $x \in \mathbf{R}^n$ as

$$e_{\mathrm{wc}}(x) = \sup\{\|Ax - b\| \mid A \in \mathcal{A}\},$$

which is always a convex function of x. The (worst-case) *robust approximation problem* is to minimize the worst-case error:

$$\text{minimize} \quad e_{\mathrm{wc}}(x) = \sup\{\|Ax - b\| \mid A \in \mathcal{A}\}, \tag{6.14}$$

where the variable is x, and the problem data are b and the set $\mathcal{A}$. When $\mathcal{A}$ is the singleton $\mathcal{A} = \{A\}$, the robust approximation problem (6.14) reduces to the basic norm approximation problem (6.1). The robust approximation problem is always a convex optimization problem, but its tractability depends on the norm used and the description of the uncertainty set $\mathcal{A}$.

Example 6.5   *Comparison of stochastic and worst-case robust approximation.*   To illustrate the difference between the stochastic and worst-case formulations of the robust approximation problem, we consider the least-squares problem

$$\text{minimize} \quad \|A(u)x - b\|_2^2,$$

where $u \in \mathbf{R}$ is an uncertain parameter and $A(u) = A_0 + uA_1$. We consider a specific instance of the problem, with $A(u) \in \mathbf{R}^{20 \times 10}$, $\|A_0\| = 10$, $\|A_1\| = 1$, and u

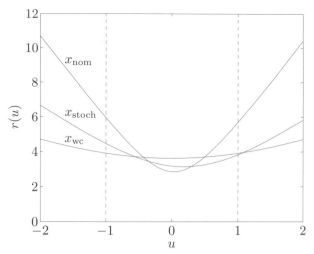

Figure 6.15 The residual $r(u) = \|A(u)x - b\|_2$ as a function of the uncertain parameter u for three approximate solutions x: (1) the nominal least-squares solution x_{nom}; (2) the solution of the stochastic robust approximation problem x_{stoch} (assuming u is uniformly distributed on $[-1, 1]$); and (3) the solution of the worst-case robust approximation problem x_{wc}, assuming the parameter u lies in the interval $[-1, 1]$. The nominal solution achieves the smallest residual when $u = 0$, but gives much larger residuals as u approaches -1 or 1. The worst-case solution has a larger residual when $u = 0$, but its residuals do not rise much as the parameter u varies over the interval $[-1, 1]$.

in the interval $[-1, 1]$. (So, roughly speaking, the variation in the matrix A is around $\pm 10\%$.)

We find three approximate solutions:

- *Nominal optimal.* The optimal solution x_{nom} is found, assuming $A(u)$ has its nominal value A_0.

- *Stochastic robust approximation.* We find x_{stoch}, which minimizes $\mathbf{E}\,\|A(u)x - b\|_2^2$, assuming the parameter u is uniformly distributed on $[-1, 1]$.

- *Worst-case robust approximation.* We find x_{wc}, which minimizes

$$\sup_{-1 \le u \le 1} \|A(u)x - b\|_2 = \max\{\|(A_0 - A_1)x - b\|_2, \|(A_0 + A_1)x - b\|_2\}.$$

For each of these three values of x, we plot the residual $r(u) = \|A(u)x - b\|_2$ as a function of the uncertain parameter u, in figure 6.15. These plots show how sensitive an approximate solution can be to variation in the parameter u. The nominal solution achieves the smallest residual when $u = 0$, but is quite sensitive to parameter variation: it gives much larger residuals as u deviates from 0, and approaches -1 or 1. The worst-case solution has a larger residual when $u = 0$, but its residuals do not rise much as u varies over the interval $[-1, 1]$. The stochastic robust approximate solution is in between.

The robust approximation problem (6.14) arises in many contexts and applications. In an estimation setting, the set $\mathcal{A}$ gives our uncertainty in the linear relation between the vector to be estimated and our measurement vector. Sometimes the noise term v in the model $y = Ax + v$ is called *additive noise* or *additive error*, since it is added to the 'ideal' measurement Ax. In contrast, the variation in A is called *multiplicative error*, since it multiplies the variable x.

In an optimal design setting, the variation can represent uncertainty (arising in manufacture, say) of the linear equations that relate the design variables x to the results vector Ax. The robust approximation problem (6.14) is then interpreted as the robust design problem: find design variables x that minimize the worst possible mismatch between Ax and b, over all possible values of A.

Finite set

Here we have $\mathcal{A} = \{A_1, \ldots, A_k\}$, and the robust approximation problem is

$$\text{minimize} \quad \max_{i=1,\ldots,k} \|A_i x - b\|.$$

This problem is equivalent to the robust approximation problem with the polyhedral set $\mathcal{A} = \mathbf{conv}\{A_1, \ldots, A_k\}$:

$$\text{minimize} \quad \sup\left\{\|Ax - b\| \mid A \in \mathbf{conv}\{A_1, \ldots, A_k\}\right\}.$$

We can cast the problem in epigraph form as

$$\begin{aligned} \text{minimize} \quad & t \\ \text{subject to} \quad & \|A_i x - b\| \le t, \quad i = 1, \ldots, k, \end{aligned}$$

which can be solved in a variety of ways, depending on the norm used. If the norm is the Euclidean norm, this is an SOCP. If the norm is the ℓ_1- or ℓ_∞-norm, we can express it as an LP.

Norm bound error

Here the uncertainty set $\mathcal{A}$ is a norm ball, $\mathcal{A} = \{\bar{A} + U \mid \|U\| \le a\}$, where $\|\cdot\|$ is a norm on $\mathbf{R}^{m \times n}$. In this case we have

$$e_{\text{wc}}(x) = \sup\{\|\bar{A}x - b + Ux\| \mid \|U\| \le a\},$$

which must be carefully interpreted since the first norm appearing is on $\mathbf{R}^m$ (and is used to measure the size of the residual) and the second one appearing is on $\mathbf{R}^{m \times n}$ (used to define the norm ball $\mathcal{A}$).

This expression for $e_{\text{wc}}(x)$ can be simplified in several cases. As an example, let us take the Euclidean norm on $\mathbf{R}^n$ and the associated induced norm on $\mathbf{R}^{m \times n}$, i.e., the maximum singular value. If $\bar{A}x - b \ne 0$ and $x \ne 0$, the supremum in the expression for $e_{\text{wc}}(x)$ is attained for $U = auv^T$, with

$$u = \frac{\bar{A}x - b}{\|\bar{A}x - b\|_2}, \qquad v = \frac{x}{\|x\|_2},$$

and the resulting worst-case error is

$$e_{\text{wc}}(x) = \|\bar{A}x - b\|_2 + a\|x\|_2.$$

(It is easily verified that this expression is also valid if x or $\bar{A}x - b$ is zero.) The robust approximation problem (6.14) then becomes

$$\text{minimize} \quad \|\bar{A}x - b\|_2 + a\|x\|_2,$$

which is a regularized norm problem, solvable as the SOCP

$$\begin{array}{ll} \text{minimize} & t_1 + at_2 \\ \text{subject to} & \|\bar{A}x - b\|_2 \leq t_1, \quad \|x\|_2 \leq t_2. \end{array}$$

Since the solution of this problem is the same as the solution of the regularized least-squares problem

$$\text{minimize} \quad \|\bar{A}x - b\|_2^2 + \delta\|x\|_2^2$$

for some value of the regularization parameter δ, we have another interpretation of the regularized least-squares problem as a worst-case robust approximation problem.

Uncertainty ellipsoids

We can also describe the variation in A by giving an ellipsoid of possible values for each row:

$$\mathcal{A} = \{[a_1 \ \cdots \ a_m]^T \mid a_i \in \mathcal{E}_i, \ i = 1, \dots, m\},$$

where

$$\mathcal{E}_i = \{\bar{a}_i + P_i u \mid \|u\|_2 \leq 1\}.$$

The matrix $P_i \in \mathbf{R}^{n \times n}$ describes the variation in a_i. We allow P_i to have a nontrivial nullspace, in order to model the situation when the variation in a_i is restricted to a subspace. As an extreme case, we take $P_i = 0$ if there is no uncertainty in a_i.

With this ellipsoidal uncertainty description, we can give an explicit expression for the worst-case magnitude of each residual:

$$\begin{aligned} \sup_{a_i \in \mathcal{E}_i} |a_i^T x - b_i| &= \sup\{|\bar{a}_i^T x - b_i + (P_i u)^T x| \mid \|u\|_2 \leq 1\} \\ &= |\bar{a}_i^T x - b_i| + \|P_i^T x\|_2. \end{aligned}$$

Using this result we can solve several robust approximation problems. For example, the robust ℓ_2-norm approximation problem

$$\text{minimize} \quad e_{\text{wc}}(x) = \sup\{\|Ax - b\|_2 \mid a_i \in \mathcal{E}_i, \ i = 1, \dots, m\}$$

can be reduced to an SOCP, as follows. An explicit expression for the worst-case error is given by

$$e_{\text{wc}}(x) = \left(\sum_{i=1}^m \left(\sup_{a_i \in \mathcal{E}_i} |a_i^T x - b_i| \right)^2 \right)^{1/2} = \left(\sum_{i=1}^m (|\bar{a}_i^T x - b_i| + \|P_i^T x\|_2)^2 \right)^{1/2}.$$

To minimize $e_{\text{wc}}(x)$ we can solve

$$\begin{array}{ll} \text{minimize} & \|t\|_2 \\ \text{subject to} & |\bar{a}_i^T x - b_i| + \|P_i^T x\|_2 \leq t_i, \quad i = 1, \dots, m, \end{array}$$

where we introduced new variables $t_1, \ldots, t_m$. This problem can be formulated as

$$
\begin{array}{ll}
\text{minimize} & \|t\|_2 \\
\text{subject to} & \bar{a}_i^T x - b_i + \|P_i^T x\|_2 \leq t_i, \quad i = 1, \ldots, m \\
& -\bar{a}_i^T x + b_i + \|P_i^T x\|_2 \leq t_i, \quad i = 1, \ldots, m,
\end{array}
$$

which becomes an SOCP when put in epigraph form.

Norm bounded error with linear structure

As a generalization of the norm bound description $\mathcal{A} = \{\bar{A} + U \mid \|U\| \leq a\}$, we can define $\mathcal{A}$ as the image of a norm ball under an affine transformation:

$$
\mathcal{A} = \{\bar{A} + u_1 A_1 + u_2 A_2 + \cdots + u_p A_p \mid \|u\| \leq 1\},
$$

where $\| \cdot \|$ is a norm on $\mathbf{R}^p$, and the $p + 1$ matrices $\bar{A}, A_1, \ldots, A_p \in \mathbf{R}^{m \times n}$ are given. The worst-case error can be expressed as

$$
\begin{aligned}
e_{\text{wc}}(x) &= \sup_{\|u\| \leq 1} \|(\bar{A} + u_1 A_1 + \cdots + u_p A_p)x - b\| \\
&= \sup_{\|u\| \leq 1} \|P(x)u + q(x)\|,
\end{aligned}
$$

where P and q are defined as

$$
P(x) = \begin{bmatrix} A_1 x & A_2 x & \cdots & A_p x \end{bmatrix} \in \mathbf{R}^{m \times p}, \qquad q(x) = \bar{A}x - b \in \mathbf{R}^m.
$$

As a first example, we consider the robust Chebyshev approximation problem

$$
\text{minimize} \quad e_{\text{wc}}(x) = \sup_{\|u\|_\infty \leq 1} \|(\bar{A} + u_1 A_1 + \cdots + u_p A_p)x - b\|_\infty.
$$

In this case we can derive an explicit expression for the worst-case error. Let $p_i(x)^T$ denote the ith row of $P(x)$. We have

$$
\begin{aligned}
e_{\text{wc}}(x) &= \sup_{\|u\|_\infty \leq 1} \|P(x)u + q(x)\|_\infty \\
&= \max_{i=1,\ldots,m} \sup_{\|u\|_\infty \leq 1} |p_i(x)^T u + q_i(x)| \\
&= \max_{i=1,\ldots,m} (\|p_i(x)\|_1 + |q_i(x)|).
\end{aligned}
$$

The robust Chebyshev approximation problem can therefore be cast as an LP

$$
\begin{array}{ll}
\text{minimize} & t \\
\text{subject to} & -y_0 \preceq \bar{A}x - b \preceq y_0 \\
& -y_k \preceq A_k x \preceq y_k, \quad k = 1, \ldots, p \\
& y_0 + \sum_{k=1}^{p} y_k \preceq t\mathbf{1},
\end{array}
$$

with variables $x \in \mathbf{R}^n$, $y_k \in \mathbf{R}^m$, $t \in \mathbf{R}$.

As another example, we consider the robust least-squares problem

$$
\text{minimize} \quad e_{\text{wc}}(x) = \sup_{\|u\|_2 \leq 1} \|(\bar{A} + u_1 A_1 + \cdots + u_p A_p)x - b\|_2.
$$

Here we use Lagrange duality to evaluate e_{wc}. The worst-case error $e_{\text{wc}}(x)$ is the squareroot of the optimal value of the (nonconvex) quadratic optimization problem

$$\begin{array}{ll} \text{maximize} & \|P(x)u + q(x)\|_2^2 \\ \text{subject to} & u^T u \leq 1, \end{array}$$

with u as variable. The Lagrange dual of this problem can be expressed as the SDP

$$\begin{array}{ll} \text{minimize} & t + \lambda \\ \text{subject to} & \begin{bmatrix} I & P(x) & q(x) \\ P(x)^T & \lambda I & 0 \\ q(x)^T & 0 & t \end{bmatrix} \succeq 0 \end{array} \qquad (6.15)$$

with variables t, $\lambda \in \mathbf{R}$. Moreover, as mentioned in §5.2 and §B.1 (and proved in §B.4), strong duality holds for this pair of primal and dual problems. In other words, for fixed x, we can compute $e_{\text{wc}}(x)^2$ by solving the SDP (6.15) with variables t and λ. Optimizing jointly over t, λ, and x is equivalent to minimizing $e_{\text{wc}}(x)^2$. We conclude that the robust least-squares problem is equivalent to the SDP (6.15) with x, λ, t as variables.

Example 6.6 *Comparison of worst-case robust, Tikhonov regularized, and nominal least-squares solutions.* We consider an instance of the robust approximation problem

$$\text{minimize} \quad \sup_{\|u\|_2 \leq 1} \|(\bar{A} + u_1 A_1 + u_2 A_2)x - b\|_2, \qquad (6.16)$$

with dimensions $m = 50$, $n = 20$. The matrix $\bar{A}$ has norm 10, and the two matrices A_1 and A_2 have norm 1, so the variation in the matrix A is, roughly speaking, around 10%. The uncertainty parameters u_1 and u_2 lie in the unit disk in $\mathbf{R}^2$.

We compute the optimal solution of the robust least-squares problem (6.16) x_{rls}, as well as the solution of the nominal least-squares problem x_{ls} (*i.e.*, assuming $u = 0$), and also the Tikhonov regularized solution x_{tik}, with $\delta = 1$.

To illustrate the sensitivity of each of these approximate solutions to the parameter u, we generate 10^5 parameter vectors, uniformly distributed on the unit disk, and evaluate the residual

$$\|(A_0 + u_1 A_1 + u_2 A_2)x - b\|_2$$

for each parameter value. The distributions of the residuals are shown in figure 6.16.

We can make several observations. First, the residuals of the nominal least-squares solution are widely spread, from a smallest value around 0.52 to a largest value around 4.9. In particular, the least-squares solution is very sensitive to parameter variation. In contrast, both the robust least-squares and Tikhonov regularized solutions exhibit far smaller variation in residual as the uncertainty parameter varies over the unit disk. The robust least-squares solution, for example, achieves a residual between 2.0 and 2.6 for all parameters in the unit disk.

6.5 Function fitting and interpolation

In function fitting problems, we select a member of a finite-dimensional subspace of functions that best fits some given data or requirements. For simplicity we

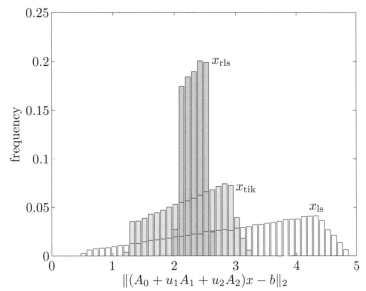

Figure 6.16 Distribution of the residuals for the three solutions of a least-squares problem (6.16): x_{ls}, the least-squares solution assuming $u = 0$; x_{tik}, the Tikhonov regularized solution with $\delta = 1$; and x_{rls}, the robust least-squares solution. The histograms were obtained by generating 10^5 values of the uncertain parameter vector u from a uniform distribution on the unit disk in $\mathbf{R}^2$. The bins have width 0.1.

consider real-valued functions; the ideas are readily extended to handle vector-valued functions as well.

6.5.1   Function families

We consider a family of functions $f_1, \ldots, f_n : \mathbf{R}^k \to \mathbf{R}$, with common domain $\mathbf{dom}\, f_i = D$. With each $x \in \mathbf{R}^n$ we associate the function $f : \mathbf{R}^k \to \mathbf{R}$ given by

$$f(u) = x_1 f_1(u) + \cdots + x_n f_n(u) \tag{6.17}$$

with $\mathbf{dom}\, f = D$. The family $\{f_1, \ldots, f_n\}$ is sometimes called the set of *basis functions* (for the fitting problem) even when the functions are not independent. The vector $x \in \mathbf{R}^n$, which parametrizes the subspace of functions, is our optimization variable, and is sometimes called the *coefficient vector*. The basis functions generate a subspace $\mathcal{F}$ of functions on D.

In many applications the basis functions are specially chosen, using prior knowledge or experience, in order to reasonably model functions of interest with the finite-dimensional subspace of functions. In other cases, more generic function families are used. We describe a few of these below.

Polynomials

One common subspace of functions on $\mathbf{R}$ consists of polynomials of degree less than n. The simplest basis consists of the powers, *i.e.*, $f_i(t) = t^{i-1}$, $i = 1, \ldots, n$. In many applications, the same subspace is described using a different basis, for example, a set of polynomials $f_1, \ldots, f_n$, of degree less than n, that are orthonormal with respect to some positive function (or measure) $\phi : \mathbf{R}^n \to \mathbf{R}_+$, *i.e.*,

$$\int f_i(t) f_j(t) \phi(t)\, dt = \left\{ \begin{array}{ll} 1 & i = j \\ 0 & i \neq j. \end{array} \right.$$

Another common basis for polynomials is the *Lagrange basis* $f_1, \ldots, f_n$ associated with distinct points $t_1, \ldots, t_n$, which satisfy

$$f_i(t_j) = \left\{ \begin{array}{ll} 1 & i = j \\ 0 & i \neq j. \end{array} \right.$$

We can also consider polynomials on $\mathbf{R}^k$, with a maximum total degree, or a maximum degree for each variable.

As a related example, we have *trigonometric polynomials* of degree less than n, with basis

$$\sin kt, \quad k = 1, \ldots, n-1, \qquad \cos kt, \quad k = 0, \ldots, n-1.$$

Piecewise-linear functions

We start with a *triangularization* of the domain D, which means the following. We have a set of *mesh* or *grid points* $g_1, \ldots, g_n \in \mathbf{R}^k$, and a partition of D into a set of simplexes:

$$D = S_1 \cup \cdots \cup S_m, \qquad \mathbf{int}(S_i \cap S_j) = \emptyset \text{ for } i \neq j.$$

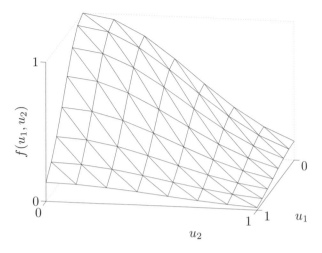

Figure 6.17 A piecewise-linear function of two variables, on the unit square. The triangulation consists of 98 simplexes, and a uniform grid of 64 points in the unit square.

Each simplex is the convex hull of $k+1$ grid points, and we require that each grid point is a vertex of any simplex it lies in.

Given a triangularization, we can construct a piecewise-linear (or more precisely, piecewise-affine) function f by assigning function values $f(g_i) = x_i$ to the grid points, and then extending the function affinely on each simplex. The function f can be expressed as (6.17) where the basis functions f_i are affine on each simplex and are defined by the conditions

$$f_i(g_j) = \begin{cases} 1 & i = j \\ 0 & i \neq j. \end{cases}$$

By construction, such a function is continuous.

Figure 6.17 shows an example for $k = 2$.

Piecewise polynomials and splines

The idea of piecewise-affine functions on a triangulated domain is readily extended to piecewise polynomials and other functions.

Piecewise polynomials are defined as polynomials (of some maximum degree) on each simplex of the triangulation, which are continuous, *i.e.*, the polynomials agree at the boundaries between simplexes. By further restricting the piecewise polynomials to have continuous derivatives up to a certain order, we can define various classes of *spline functions*. Figure 6.18 shows an example of a cubic spline, *i.e.*, a piecewise polynomial of degree 3 on $\mathbf{R}$, with continuous first and second derivatives.

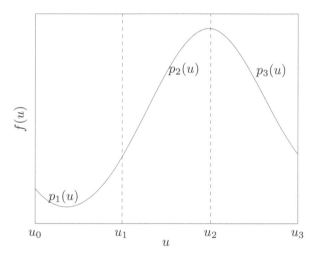

Figure 6.18 *Cubic spline.* A cubic spline is a piecewise polynomial, with continuous first and second derivatives. In this example, the cubic spline f is formed from the three cubic polynomials p_1 (on $[u_0, u_1]$), p_2 (on $[u_1, u_2]$), and p_3 (on $[u_2, u_3]$). Adjacent polynomials have the same function value, and equal first and second derivatives, at the boundary points u_1 and u_2. In this example, the dimension of the family of functions is $n = 6$, since we have 12 polynomial coefficients (4 per cubic polynomial), and 6 equality constraints (3 each at u_1 and u_2).

6.5.2   Constraints

In this section we describe some constraints that can be imposed on the function f, and therefore, on the variable $x \in \mathbf{R}^n$.

Function value interpolation and inequalities

Let v be a point in D. The value of f at v,

$$f(v) = \sum_{i=1}^{n} x_i f_i(v),$$

is a linear function of x. Therefore *interpolation conditions*

$$f(v_j) = z_j, \quad j = 1, \ldots, m,$$

which require the function f to have the values $z_j \in \mathbf{R}$ at specified points $v_j \in D$, form a set of linear equalities in x. More generally, inequalities on the function value at a given point, as in $l \leq f(v) \leq u$, are linear inequalities on the variable x. There are many other interesting convex constraints on f (hence, x) that involve the function values at a finite set of points $v_1, \ldots, v_N$. For example, the Lipschitz constraint

$$|f(v_j) - f(v_k)| \leq L \|v_j - v_k\|, \quad j, \ k = 1, \ldots, m,$$

forms a set of linear inequalities in x.

We can also impose inequalities on the function values at an infinite number of points. As an example, consider the nonnegativity constraint

$$f(u) \geq 0 \text{ for all } u \in D.$$

This is a convex constraint on x (since it is the intersection of an infinite number of halfspaces), but may not lead to a tractable problem except in special cases that exploit the particular structure of the functions. One simple example occurs when the functions are piecewise-linear. In this case, if the function values are nonnegative at the grid points, the function is nonnegative everywhere, so we obtain a simple (finite) set of linear inequalities.

As a less trivial example, consider the case when the functions are polynomials on $\mathbf{R}$, with even maximum degree $2k$ (*i.e.*, $n = 2k + 1$), and $D = \mathbf{R}$. As shown in exercise 2.37, page 65, the nonnegativity constraint

$$p(u) = x_1 + x_2 u + \cdots + x_{2k+1} u^{2k} \geq 0 \quad \text{for all } u \in \mathbf{R},$$

is equivalent to

$$x_i = \sum_{m+n=i+1} Y_{mn}, \quad i = 1, \ldots, 2k+1, \qquad Y \succeq 0,$$

where $Y \in \mathbf{S}^{k+1}$ is an auxiliary variable.

Derivative constraints

Suppose the basis functions f_i are differentiable at a point $v \in D$. The gradient

$$\nabla f(v) = \sum_{i=1}^{n} x_i \nabla f_i(v),$$

is a linear function of x, so interpolation conditions on the derivative of f at v reduce to linear equality constraints on x. Requiring that the norm of the gradient at v not exceed a given limit,

$$\|\nabla f(v)\| = \left\| \sum_{i=1}^{n} x_i \nabla f_i(v) \right\| \leq M,$$

is a convex constraint on x. The same idea extends to higher derivatives. For example, if f is twice differentiable at v, the requirement that

$$lI \preceq \nabla^2 f(v) \preceq uI$$

is a linear matrix inequality in x, hence convex.

We can also impose constraints on the derivatives at an infinite number of points. For example, we can require that f is monotone:

$$f(u) \geq f(v) \text{ for all } u, \ v \in D, \ u \succeq v.$$

This is a convex constraint in x, but may not lead to a tractable problem except in special cases. When f is piecewise affine, for example, the monotonicity constraint is equivalent to the condition $\nabla f(v) \succeq 0$ inside each of the simplexes. Since the gradient is a linear function of the grid point values, this leads to a simple (finite) set of linear inequalities.

As another example, we can require that the function be convex, $i.e.$, satisfy

$$f((u+v)/2) \leq (f(u) + f(v))/2 \text{ for all } u, \ v \in D$$

(which is enough to ensure convexity when f is continuous). This is a convex constraint, which has a tractable representation in some cases. One obvious example is when f is quadratic, in which case the convexity constraint reduces to the requirement that the quadratic part of f be nonnegative, which is an LMI. Another example in which a convexity constraint leads to a tractable problem is described in more detail in §6.5.5.

Integral constraints

Any linear functional $\mathcal{L}$ on the subspace of functions can be expressed as a linear function of x, $i.e.$, we have $\mathcal{L}(f) = c^T x$. Evaluation of f (or a derivative) at a point is just a special case. As another example, the linear functional

$$\mathcal{L}(f) = \int_D \phi(u) f(u) \, du,$$

where $\phi : \mathbf{R}^k \to \mathbf{R}$, can be expressed as $\mathcal{L}(f) = c^T x$, where

$$c_i = \int_D \phi(u) f_i(u) \, du.$$

Thus, a constraint of the form $\mathcal{L}(f) = a$ is a linear equality constraint on x. One example of such a constraint is the *moment constraint*

$$\int_D t^m f(t) \, dt = a$$

(where $f : \mathbf{R} \to \mathbf{R}$).

6.5.3   Fitting and interpolation problems

Minimum norm function fitting

In a fitting problem, we are given data

$$(u_1, y_1), \quad \dots, \quad (u_m, y_m)$$

with $u_i \in D$ and $y_i \in \mathbf{R}$, and seek a function $f \in \mathcal{F}$ that matches this data as closely as possible. For example in least-squares fitting we consider the problem

$$\text{minimize} \quad \sum_{i=1}^m (f(u_i) - y_i)^2,$$

which is a simple least-squares problem in the variable x. We can add a variety of constraints, for example linear inequalities that must be satisfied by f at various points, constraints on the derivatives of f, monotonicity constraints, or moment constraints.

Example 6.7 *Polynomial fitting.* We are given data $u_1, \dots, u_m \in \mathbf{R}$ and $v_1, \dots, v_m \in \mathbf{R}$, and hope to approximately fit a polynomial of the form

$$p(u) = x_1 + x_2 u + \cdots + x_n u^{n-1}$$

to the data. For each x we form the vector of errors,

$$e = (p(u_1) - v_1, \dots, p(u_m) - v_m).$$

To find the polynomial that minimizes the norm of the error, we solve the norm approximation problem

$$\text{minimize} \quad \|e\| = \|Ax - v\|$$

with variable $x \in \mathbf{R}^n$, where $A_{ij} = u_i^{j-1}$, $i = 1, \dots, m$, $j = 1, \dots, n$.

Figure 6.19 shows an example with $m = 40$ data points and $n = 6$ (*i.e.*, polynomials of maximum degree 5), for the ℓ_2- and ℓ_∞-norms.

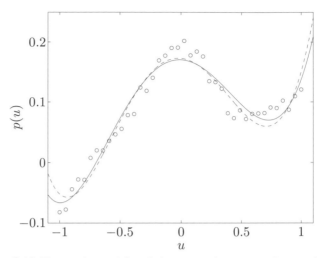

Figure 6.19 Two polynomials of degree 5 that approximate the 40 data points shown as circles. The polynomial shown as a solid line minimizes the ℓ_2-norm of the error; the polynomial shown as a dashed line minimizes the ℓ_∞-norm.

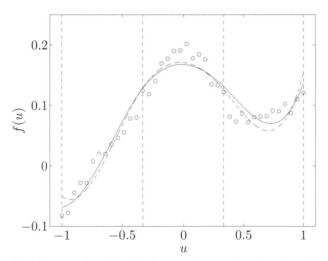

Figure 6.20 Two cubic splines that approximate the 40 data points shown as circles (which are the same as the data in figure 6.19). The spline shown as a solid line minimizes the ℓ_2-norm of the error; the spline shown as a dashed line minimizes the ℓ_∞-norm. As in the polynomial approximation shown in figure 6.19, the dimension of the subspace of fitting functions is 6.

Example 6.8 *Spline fitting.* Figure 6.20 shows the same data as in example 6.7, and two optimal fits with cubic splines. The interval $[-1, 1]$ is divided into three equal intervals, and we consider piecewise polynomials, with maximum degree 3, with continuous first and second derivatives. The dimension of this subspace of functions is 6, the same as the dimension of polynomials with maximum degree 5, considered in example 6.7.

In the simplest forms of function fitting, we have $m \gg n$, *i.e.*, the number of data points is much larger than the dimension of the subspace of functions. Smoothing is accomplished automatically, since all members of the subspace are smooth.

Least-norm interpolation

In another variation of function fitting, we have fewer data points than the dimension of the subspace of functions. In the simplest case, we require that the function we choose must satisfy the interpolation conditions

$$f(u_i) = y_i, \quad i = 1, \ldots, m,$$

which are linear equality constraints on x. Among the functions that satisfy these interpolation conditions, we might seek one that is smoothest, or smallest. These lead to least-norm problems.

In the most general function fitting problem, we can optimize an objective (such as some measure of the error e), subject to a variety of convex constraints that represent our prior knowledge of the underlying function.

Interpolation, extrapolation, and bounding

By evaluating the optimal function fit $\hat{f}$ at a point v not in the original data set, we obtain a guess of what the value of the underlying function is, at the point v. This is called *interpolation* when v is between or near the given data points (*e.g.*, $v \in \mathbf{conv}\{v_1, \ldots, v_m\}$), and *extrapolation* otherwise.

We can also produce an interval in which the value $f(v)$ can lie, by maximizing and minimizing (the linear function) $f(v)$, subject to the constraints. We can use the function fit to help identify faulty data or outliers. Here we might use, for example, an ℓ_1-norm fit, and look for data points with large errors.

6.5.4   Sparse descriptions and basis pursuit

In *basis pursuit*, there is a very large number of basis functions, and the goal is to find a good fit of the given data as a linear combination of a small number of the basis functions. (In this context the function family is linearly dependent, and is sometimes referred to as an *over-complete basis* or *dictionary*.) This is called basis pursuit since we are selecting a much smaller basis, from the given over-complete basis, to model the data.

Thus we seek a function $f \in \mathcal{F}$ that fits the data well,

$$f(u_i) \approx y_i, \quad i = 1, \ldots, m,$$

with a sparse coefficient vector x, *i.e.*, $\mathbf{card}(x)$ small. In this case we refer to

$$f = x_1 f_1 + \cdots + x_n f_n = \sum_{i \in \mathcal{B}} x_i f_i,$$

where $\mathcal{B} = \{i \mid x_i \neq 0\}$ is the set of indices of the chosen basis elements, as a *sparse description* of the data. Mathematically, basis pursuit is the same as the regressor selection problem (see §6.4), but the interpretation (and scale) of the optimization problem are different.

Sparse descriptions and basis pursuit have many uses. They can be used for de-noising or smoothing, or data compression for efficient transmission or storage of a signal. In data compression, the sender and receiver both know the dictionary, or basis elements. To send a signal to the receiver, the sender first finds a sparse representation of the signal, and then sends to the receiver only the nonzero coefficients (to some precision). Using these coefficients, the receiver can reconstruct (an approximation of) the original signal.

One common approach to basis pursuit is the same as the method for regressor selection described in §6.4, and based on ℓ_1-norm regularization as a heuristic for finding sparse descriptions. We first solve the convex problem

$$\text{minimize} \quad \sum_{i=1}^{m}(f(u_i) - y_i)^2 + \gamma\|x\|_1, \qquad (6.18)$$

where $\gamma > 0$ is a parameter used to trade off the quality of the fit to the data, and the sparsity of the coefficient vector. The solution of this problem can be used directly, or followed by a refinement step, in which the best fit is found, using the sparsity pattern of the solution of (6.18). In other words, we first solve (6.18), to obtain $\hat{x}$. We then set $\mathcal{B} = \{i \mid \hat{x}_i \neq 0\}$, *i.e.*, the set of indices corresponding to nonzero coefficients. Then we solve the least-squares problem

$$\text{minimize} \quad \sum_{i=1}^{m}(f(u_i) - y_i)^2$$

with variables x_i, $i \in \mathcal{B}$, and $x_i = 0$ for $i \notin \mathcal{B}$.

In basis pursuit and sparse description applications it is not uncommon to have a very large dictionary, with n on the order of 10^4 or much more. To be effective, algorithms for solving (6.18) must exploit problem structure, which derives from the structure of the dictionary signals.

Time-frequency analysis via basis pursuit

In this section we illustrate basis pursuit and sparse representation with a simple example. We consider functions (or signals) on $\mathbf{R}$, with the range of interest $[0, 1]$. We think of the independent variable as time, so we use t (instead of u) to denote it.

We first describe the basis functions in the dictionary. Each basis function is a *Gaussian sinusoidal pulse*, or *Gabor function*, with form

$$e^{-(t-\tau)^2/\sigma^2}\cos(\omega t + \phi),$$

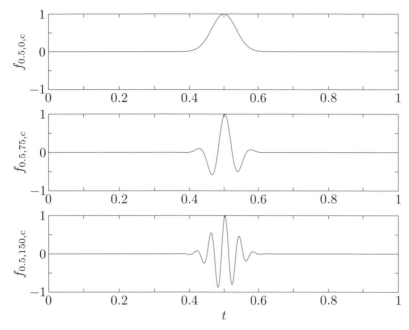

Figure 6.21 Three of the basis elements in the dictionary, all with center time $\tau = 0.5$ and cosine phase. The top signal has frequency $\omega = 0$, the middle one has frequency $\omega = 75$, and the bottom one has frequency $\omega = 150$.

where $\sigma > 0$ gives the width of the pulse, τ is the time of (the center of) the pulse, $\omega \geq 0$ is the frequency, and ϕ is the phase angle. All of the basis functions have width $\sigma = 0.05$. The pulse times and frequencies are

$$\tau = 0.002k, \quad k = 0, \ldots, 500, \qquad \omega = 5k, \quad k = 0, \ldots, 30.$$

For each time τ, there is one basis element with frequency zero (and phase $\phi = 0$), and 2 basis elements (cosine and sine, *i.e.*, phase $\phi = 0$ and $\phi = \pi/2$) for each of 30 remaining frequencies, so all together there are $501 \times 61 = 30561$ basis elements. The basis elements are naturally indexed by time, frequency, and phase (cosine or sine), so we denote them as

$$\begin{aligned} f_{\tau,\omega,\mathrm{c}}, \qquad & \tau = 0, 0.002, \ldots, 1, \qquad \omega = 0, 5, \ldots, 150, \\ f_{\tau,\omega,\mathrm{s}}, \qquad & \tau = 0, 0.002, \ldots, 1, \qquad \omega = 5, \ldots, 150. \end{aligned}$$

Three of these basis functions (all with time $\tau = 0.5$) are shown in figure 6.21.

Basis pursuit with this dictionary can be thought of as a *time-frequency analysis* of the data. If a basis element $f_{\tau,\omega,\mathrm{c}}$ or $f_{\tau,\omega,\mathrm{s}}$ appears in the sparse representation of a signal (*i.e.*, with a nonzero coefficient), we can interpret this as meaning that the data contains the frequency ω at time τ.

We will use basis pursuit to find a sparse approximation of the signal

$$y(t) = a(t) \sin \theta(t)$$

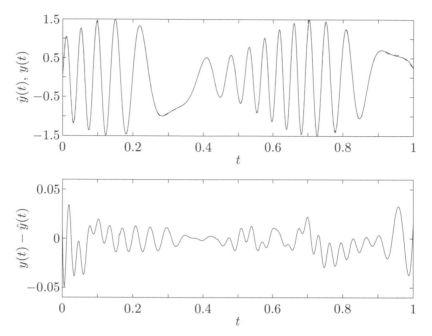

Figure 6.22 *Top.* The original signal (solid line) and approximation $\hat{y}$ obtained by basis pursuit (dashed line) are almost indistinguishable. *Bottom.* The approximation error $y(t) - \hat{y}(t)$, with different vertical scale.

where
$$a(t) = 1 + 0.5\sin(11t), \qquad \theta(t) = 30\sin(5t).$$

(This signal is chosen only because it is simple to describe, and exhibits noticeable changes in its spectral content over time.) We can interpret $a(t)$ as the signal amplitude, and $\theta(t)$ as its total phase. We can also interpret

$$\omega(t) = \left|\frac{d\theta}{dt}\right| - 150|\cos(5t)|$$

as the *instantaneous frequency* of the signal at time t. The data are given as 501 uniformly spaced samples over the interval $[0, 1]$, *i.e.*, we are given 501 pairs (t_k, y_k) with

$$t_k = 0.005k, \quad y_k = y(t_k), \quad k = 0, \dots, 500.$$

We first solve the ℓ_1-norm regularized least-squares problem (6.18), with $\gamma = 1$. The resulting optimal coefficient vector is very sparse, with only 42 nonzero coefficients out of 30561. We then find the least-squares fit of the original signal using these 42 basis vectors. The result $\hat{y}$ is compared with the original signal y in figure 6.22. The top figure shows the approximated signal (in dashed line) and, almost indistinguishable, the original signal $y(t)$ (in solid line). The bottom figure shows the error $y(t) - \hat{y}(t)$. As is clear from the figure, we have obtained an

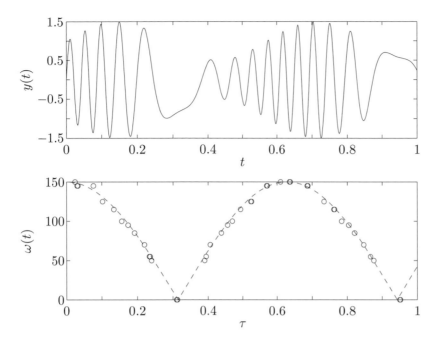

Figure 6.23 *Top: Original signal. Bottom: Time-frequency plot.* The dashed curve shows the instantaneous frequency $\omega(t) = 150|\cos(5t)|$ of the original signal. Each circle corresponds to a chosen basis element in the approximation obtained by basis pursuit. The horizontal axis shows the time index τ, and the vertical axis shows the frequency index ω of the basis element.

approximation $\hat{y}$ with a very good relative fit. The relative error is

$$\frac{(1/501)\sum_{i=1}^{501}(y(t_i) - \hat{y}(t_i))^2}{(1/501)\sum_{i=1}^{501} y(t_i)^2} = 2.6 \cdot 10^{-4}.$$

By plotting the pattern of nonzero coefficients versus time and frequency, we obtain a time-frequency analysis of the original data. Such a plot is shown in figure 6.23, along with the instantaneous frequency. The plot shows that the nonzero components closely track the instantaneous frequency.

6.5.5   Interpolation with convex functions

In some special cases we can solve interpolation problems involving an infinite-dimensional set of functions, using finite-dimensional convex optimization. In this section we describe an example.

We start with the following question: When does there exist a convex function $f : \mathbf{R}^k \to \mathbf{R}$, with $\mathbf{dom}\, f = \mathbf{R}^k$, that satisfies the interpolation conditions

$$f(u_i) = y_i, \quad i = 1, \ldots, m,$$

at given points $u_i \in \mathbf{R}^k$? (Here we do not restrict f to lie in any finite-dimensional subspace of functions.) The answer is: if and only if there exist $g_1, \ldots, g_m$ such that

$$y_j \geq y_i + g_i^T(u_j - u_i), \quad i, \, j = 1, \ldots, m. \tag{6.19}$$

To see this, first suppose that f is convex, $\mathbf{dom}\, f = \mathbf{R}^k$, and $f(u_i) = y_i$, $i = 1, \ldots, m$. At each u_i we can find a vector g_i such that

$$f(z) \geq f(u_i) + g_i^T(z - u_i) \tag{6.20}$$

for all z. If f is differentiable, we can take $g_i = \nabla f(u_i)$; in the more general case, we can construct g_i by finding a supporting hyperplane to $\mathbf{epi}\, f$ at (u_i, y_i). (The vectors g_i are called *subgradients*.) By applying (6.20) to $z = u_j$, we obtain (6.19).

Conversely, suppose $g_1, \ldots, g_m$ satisfy (6.19). Define f as

$$f(z) = \max_{i=1,\ldots,m} \, (y_i + g_i^T(z - u_i))$$

for all $z \in \mathbf{R}^k$. Clearly, f is a (piecewise-linear) convex function. The inequalities (6.19) imply that $f(u_i) = y_i$, for $i = 1, \ldots, m$.

We can use this result to solve several problems involving interpolation, approximation, or bounding, with convex functions.

Fitting a convex function to given data

Perhaps the simplest application is to compute the least-squares fit of a convex function to given data (u_i, y_i), $i = 1, \ldots, m$:

$$
\begin{array}{ll}
\text{minimize} & \sum_{i=1}^m (y_i - f(u_i))^2 \\
\text{subject to} & f : \mathbf{R}^k \to \mathbf{R} \text{ is convex}, \quad \mathbf{dom}\, f = \mathbf{R}^k.
\end{array}
$$

This is an infinite-dimensional problem, since the variable is f, which is in the space of continuous real-valued functions on $\mathbf{R}^k$. Using the result above, we can formulate this problem as

$$
\begin{array}{ll}
\text{minimize} & \sum_{i=1}^m (y_i - \hat{y}_i)^2 \\
\text{subject to} & \hat{y}_j \geq \hat{y}_i + g_i^T(u_j - u_i), \quad i, \, j = 1, \ldots, m,
\end{array}
$$

which is a QP with variables $\hat{y} \in \mathbf{R}^m$ and $g_1, \ldots, g_m \in \mathbf{R}^k$. The optimal value of this problem is zero if and only if the given data can be interpolated by a convex function, *i.e.*, if there is a convex function that satisfies $f(u_i) = y_i$. An example is shown in figure 6.24.

Bounding values of an interpolating convex function

As another simple example, suppose that we are given data (u_i, y_i), $i = 1, \ldots, m$, which can be interpolated by a convex function. We would like to determine the range of possible values of $f(u_0)$, where u_0 is another point in $\mathbf{R}^k$, and f is any convex function that interpolates the given data. To find the smallest possible value of $f(u_0)$ we solve the LP

$$
\begin{array}{ll}
\text{minimize} & y_0 \\
\text{subject to} & y_j \geq y_i + g_i^T(u_j - u_i), \quad i, \, j = 0, \ldots, m,
\end{array}
$$

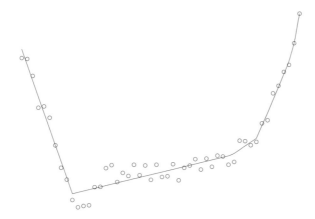

Figure 6.24 Least-squares fit of a convex function to data, shown as circles. The (piecewise-linear) function shown minimizes the sum of squared fitting error, over all convex functions.

which is an LP with variables $y_0 \in \mathbf{R}$, $g_0, \ldots, g_m \in \mathbf{R}^k$. By maximizing y_0 (which is also an LP) we find the largest possible value of $f(u_0)$ for a convex function that interpolates the given data.

Interpolation with monotone convex functions

As an extension of convex interpolation, we can consider interpolation with a convex and monotone nondecreasing function. It can be shown that there exists a convex function $f : \mathbf{R}^k \to \mathbf{R}$, with $\mathbf{dom}\, f = \mathbf{R}^k$, that satisfies the interpolation conditions

$$f(u_i) = y_i, \quad i = 1, \ldots, m,$$

and is monotone nondecreasing (*i.e.*, $f(u) \geq f(v)$ whenever $u \succeq v$), if and only if there exist $g_1, \ldots, g_m \in \mathbf{R}^k$, such that

$$g_i \succeq 0, \quad i = 1, \ldots, m, \qquad y_j \geq y_i + g_i^T (u_j - u_i), \quad i, j = 1, \ldots, m. \qquad (6.21)$$

In other words, we add to the convex interpolation conditions (6.19), the condition that the subgradients g_i are all nonnegative. (See exercise 6.12.)

Bounding consumer preference

As an application, we consider a problem of predicting consumer preferences. We consider different *baskets of goods*, consisting of different amounts of n consumer goods. A goods basket is specified by a vector $x \in [0, 1]^n$ where x_i denotes the amount of consumer good i. We assume the amounts are normalized so that $0 \leq x_i \leq 1$, *i.e.*, $x_i = 0$ is the minimum and $x_i = 1$ is the maximum possible amount of good i. Given two baskets of goods x and $\tilde{x}$, a consumer can either prefer x to $\tilde{x}$, or prefer $\tilde{x}$ to x, or consider x and $\tilde{x}$ equally attractive. We consider one model consumer, whose choices are repeatable.

We model consumer preference in the following way. We assume there is an underlying *utility function* $u : \mathbf{R}^n \to \mathbf{R}$, with domain $[0, 1]^n$; $u(x)$ gives a measure of the utility derived by the consumer from the goods basket x. Given a choice between two baskets of goods, the consumer chooses the one that has larger utility, and will be ambivalent when the two baskets have equal utility. It is reasonable to assume that u is monotone nondecreasing. This means that the consumer always prefers to have more of any good, with the amounts of all other goods the same. It is also reasonable to assume that u is concave. This models *satiation*, or decreasing marginal utility as we increase the amount of goods.

Now suppose we are given some consumer preference data, but we do not know the underlying utility function u. Specifically, we have a set of goods baskets $a_1, \ldots, a_m \in [0, 1]^n$, and some information about preferences among them:

$$u(a_i) > u(a_j) \text{ for } (i, j) \in \mathcal{P}, \qquad u(a_i) \geq u(a_j) \text{ for } (i, j) \in \mathcal{P}_{\text{weak}}, \qquad (6.22)$$

where $\mathcal{P}, \mathcal{P}_{\text{weak}} \subseteq \{1, \ldots, m\} \times \{1, \ldots, m\}$ are given. Here $\mathcal{P}$ gives the set of known preferences: $(i, j) \in \mathcal{P}$ means that basket a_i is known to be preferred to basket a_j. The set $\mathcal{P}_{\text{weak}}$ gives the set of known weak preferences: $(i, j) \in \mathcal{P}_{\text{weak}}$ means that basket a_i is preferred to basket a_j, or that the two baskets are equally attractive.

We first consider the following question: How can we determine if the given data are consistent, *i.e.*, whether or not there exists a concave nondecreasing utility function u for which (6.22) holds? This is equivalent to solving the feasibility problem

$$
\begin{array}{ll}
\text{find} & u \\
\text{subject to} & u : \mathbf{R}^n \to \mathbf{R} \text{ concave and nondecreasing} \\
& u(a_i) > u(a_j), \quad (i, j) \in \mathcal{P} \\
& u(a_i) \geq u(a_j), \quad (i, j) \in \mathcal{P}_{\text{weak}},
\end{array}
\qquad (6.23)
$$

with the function u as the (infinite-dimensional) optimization variable. Since the constraints in (6.23) are all homogeneous, we can express the problem in the equivalent form

$$
\begin{array}{ll}
\text{find} & u \\
\text{subject to} & u : \mathbf{R}^n \to \mathbf{R} \text{ concave and nondecreasing} \\
& u(a_i) \geq u(a_j) + 1, \quad (i, j) \in \mathcal{P} \\
& u(a_i) \geq u(a_j), \quad (i, j) \in \mathcal{P}_{\text{weak}},
\end{array}
\qquad (6.24)
$$

which uses only nonstrict inequalities. (It is clear that if u satisfies (6.24), then it must satisfy (6.23); conversely, if u satisfies (6.23), then it can be scaled to satisfy (6.24).) This problem, in turn, can be cast as a (finite-dimensional) linear programming feasibility problem, using the interpolation result on page 339:

$$
\begin{array}{ll}
\text{find} & u_1, \ldots, u_m, \ g_1, \ldots, g_m \\
\text{subject to} & g_i \succeq 0, \quad i = 1, \ldots, m \\
& u_j \leq u_i + g_i^T (a_j - a_i), \quad i, j = 1, \ldots, m \\
& u_i \geq u_j + 1, \quad (i, j) \in \mathcal{P} \\
& u_i \geq u_j, \quad (i, j) \in \mathcal{P}_{\text{weak}}.
\end{array}
\qquad (6.25)
$$

By solving this linear programming feasibility problem, we can determine whether there exists a concave, nondecreasing utility function that is consistent with the given sets of strict and nonstrict preferences. If (6.25) is feasible, there is at least one such utility function (and indeed, we can construct one that is piecewise-linear, from a feasible $u_1, \ldots, u_m, g_1, \ldots, g_m$). If (6.25) is not feasible, we can conclude that there is no concave increasing utility function that is consistent with the given sets of strict and nonstrict preferences.

As an example, suppose that $\mathcal{P}$ and $\mathcal{P}_{\text{weak}}$ are consumer preferences that are known to be consistent with at least one concave increasing utility function. Consider a pair (k, l) that is not in $\mathcal{P}$ or $\mathcal{P}_{\text{weak}}$, *i.e.*, consumer preference between baskets k and l is not known. In some cases we can conclude that a preference holds between basket k and l, even without knowing the underlying preference function. To do this we augment the known preferences (6.22) with the inequality $u(a_k) \leq u(a_l)$, which means that basket l is preferred to basket k, or they are equally attractive. We then solve the feasibility linear program (6.25), including the extra weak preference $u(a_k) \leq u(a_l)$. If the augmented set of preferences is infeasible, it means that any concave nondecreasing utility function that is consistent with the original given consumer preference data must also satisfy $u(a_k) > u(a_l)$. In other words, we can conclude that basket k is preferred to basket l, without knowing the underlying utility function.

Example 6.9 Here we give a simple numerical example that illustrates the discussion above. We consider baskets of two goods (so we can easily plot the goods baskets). To generate the consumer preference data $\mathcal{P}$, we compute 40 random points in $[0, 1]^2$, and then compare them using the utility function

$$u(x_1, x_2) = (1.1x_1^{1/2} + 0.8x_2^{1/2})/1.9.$$

These goods baskets, and a few level curves of the utility function u, are shown in figure 6.25.

We now use the consumer preference data (but not, of course, the true utility function u) to compare each of these 40 goods baskets to the basket $a_0 = (0.5, 0.5)$. For each original basket a_i, we solve the linear programming feasibility problem described above, to see if we can conclude that basket a_0 is preferred to basket a_i. Similarly, we check whether we can conclude that basket a_i is preferred to basket a_0. For each basket a_i, there are three possible outcomes: we can conclude that a_0 is definitely preferred to a_i, that a_i is definitely preferred to a_0, or (if both LP feasibility problems are feasible) that no conclusion is possible. (Here, *definitely preferred* means that the preference holds for any concave nondecreasing utility function that is consistent with the original given data.)

We find that 21 of the baskets are definitely rejected in favor of $(0.5, 0.5)$, and 14 of the baskets are definitely preferred. We cannot make any conclusion, from the consumer preference data, about the remaining 5 baskets. These results are shown in figure 6.26. Note that goods baskets below and to the left of $(0.5, 0.5)$ will definitely be rejected in favor of $(0.5, 0.5)$, using only the monotonicity property of the utility function, and similarly, those points that are above and to the right of $(0.5, 0.5)$ must be preferred. So for these 17 points, there is no need to solve the feasibility LP (6.25). Classifying the 23 points in the other two quadrants, however, requires the concavity assumption, and solving the feasibility LP (6.25).

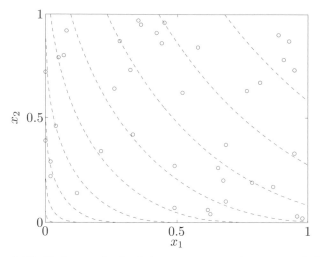

Figure 6.25 Forty goods baskets $a_1, \dots, a_{40}$, shown as circles.   The $0.1, 0.2, \dots, 0.9$ level curves of the true utility function u are shown as dashed lines. This utility function is used to find the consumer preference data $\mathcal{P}$ among the 40 baskets.

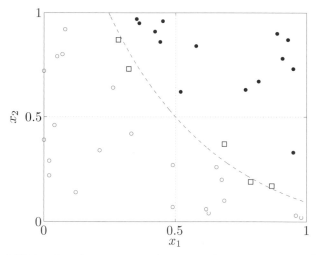

Figure 6.26 Results of consumer preference analysis using the LP (6.25), for a new goods basket $a_0 = (0.5, 0.5)$. The original baskets are displayed as open circles if they are definitely rejected ($u(a_k) < u(a_0)$), as solid black circles if they are definitely preferred ($u(a_k) > u(a_0)$), and as squares when no conclusion can be made. The level curve of the underlying utility function, that passes through $(0.5, 0.5)$, is shown as a dashed curve. The vertical and horizontal lines passing through $(0.5, 0.5)$ divide $[0, 1]^2$ into four quadrants. Points in the upper right quadrant must be preferred to $(0.5, 0.5)$, by the monotonicity assumption on u. Similarly, $(0.5, 0.5)$ must be preferred to the points in the lower left quadrant. For the points in the other two quadrants, the results are not obvious.

Bibliography

The robustness properties of approximations with different penalty functions were analyzed by Huber [Hub64, Hub81], who also proposed the penalty function (6.4). The log-barrier penalty function arises in control theory, where it is applied to the system closed-loop frequency response, and has several names, *e.g.*, *central* $\mathbf{H}_\infty$, or *risk-averse* control; see Boyd and Barratt [BB91] and the references therein.

Regularized approximation is covered in many books, including Tikhonov and Arsenin [TA77] and Hansen [Han98]. Tikhonov regularization is sometimes called *ridge regression* (Golub and Van Loan [GL89, page 564]). Least-squares approximation with ℓ_1-norm regularization is also known under the name *lasso* (Tibshirani [Tib96]). Other least-squares regularization and regressor selection techniques are discussed and compared in Hastie, Tibshirani, and Friedman [HTF01, §3.4].

Total variation denoising was introduced for image reconstruction by Rudin, Osher, and Fatemi [ROF92].

The robust least-squares problem with norm bounded uncertainty (page 321) was introduced by El Ghaoui and Lebret [EL97], and Chandrasekaran, Golub, Gu, and Sayed [CGGS98]. El Ghaoui and Lebret also give the SDP formulation of the robust least-squares problem with structured uncertainty (page 323).

Chen, Donoho, and Saunders [CDS01] discuss basis pursuit via linear programming. They refer to the ℓ_1-norm regularized problem (6.18) as *basis pursuit denoising*. Meyer and Pratt [MP68] is an early paper on the problem of bounding utility functions.

Exercises

Norm approximation and least-norm problems

6.1 *Quadratic bounds for log barrier penalty.* Let $\phi : \mathbf{R} \to \mathbf{R}$ be the log barrier penalty function with limit $a > 0$:

$$\phi(u) = \begin{cases} -a^2 \log(1 - (u/a)^2) & |u| < a \\ \infty & \text{otherwise.} \end{cases}$$

Show that if $u \in \mathbf{R}^m$ satisfies $\|u\|_\infty < a$, then

$$\|u\|_2^2 \leq \sum_{i=1}^m \phi(u_i) \leq \frac{\phi(\|u\|_\infty)}{\|u\|_\infty^2} \|u\|_2^2.$$

This means that $\sum_{i=1}^m \phi(u_i)$ is well approximated by $\|u\|_2^2$ if $\|u\|_\infty$ is small compared to a. For example, if $\|u\|_\infty / a = 0.25$, then

$$\|u\|_2^2 \leq \sum_{i=1}^m \phi(u_i) \leq 1.033 \cdot \|u\|_2^2.$$

6.2 ℓ_1-, ℓ_2-, *and* ℓ_∞-*norm approximation by a constant vector.* What is the solution of the norm approximation problem with one scalar variable $x \in \mathbf{R}$,

$$\text{minimize} \quad \|x\mathbf{1} - b\|,$$

for the ℓ_1-, ℓ_2-, and ℓ_∞-norms?

6.3 Formulate the following approximation problems as LPs, QPs, SOCPs, or SDPs. The problem data are $A \in \mathbf{R}^{m \times n}$ and $b \in \mathbf{R}^m$. The rows of A are denoted a_i^T.

(a) *Deadzone-linear penalty approximation:* minimize $\sum_{i=1}^m \phi(a_i^T x - b_i)$, where

$$\phi(u) = \begin{cases} 0 & |u| \leq a \\ |u| - a & |u| > a, \end{cases}$$

where $a > 0$.

(b) *Log-barrier penalty approximation:* minimize $\sum_{i=1}^m \phi(a_i^T x - b_i)$, where

$$\phi(u) = \begin{cases} -a^2 \log(1 - (u/a)^2) & |u| < a \\ \infty & |u| \geq a, \end{cases}$$

with $a > 0$.

(c) *Huber penalty approximation:* minimize $\sum_{i=1}^m \phi(a_i^T x - b_i)$, where

$$\phi(u) = \begin{cases} u^2 & |u| \leq M \\ M(2|u| - M) & |u| > M, \end{cases}$$

with $M > 0$.

(d) *Log-Chebyshev approximation:* minimize $\max_{i=1,\ldots,m} |\log(a_i^T x) - \log b_i|$. We assume $b \succ 0$. An equivalent convex form is

$$\begin{aligned} &\text{minimize} && t \\ &\text{subject to} && 1/t \leq a_i^T x / b_i \leq t, \quad i = 1, \ldots, m, \end{aligned}$$

with variables $x \in \mathbf{R}^n$ and $t \in \mathbf{R}$, and domain $\mathbf{R}^n \times \mathbf{R}_{++}$.

(e) Minimizing the sum of the largest k residuals:

$$
\begin{array}{ll}
\text{minimize} & \sum_{i=1}^{k} |r|_{[i]} \\
\text{subject to} & r = Ax - b,
\end{array}
$$

where $|r|_{[1]} \geq |r|_{[2]} \geq \cdots \geq |r|_{[m]}$ are the numbers $|r_1|, |r_2|, \ldots, |r_m|$ sorted in decreasing order. (For $k = 1$, this reduces to ℓ_∞-norm approximation; for $k = m$, it reduces to ℓ_1-norm approximation.) *Hint.* See exercise 5.19.

6.4 *A differentiable approximation of ℓ_1-norm approximation.* The function $\phi(u) = (u^2+\epsilon)^{1/2}$, with parameter $\epsilon > 0$, is sometimes used as a differentiable approximation of the absolute value function $|u|$. To approximately solve the ℓ_1-norm approximation problem

$$\text{minimize} \quad \|Ax - b\|_1, \tag{6.26}$$

where $A \in \mathbf{R}^{m \times n}$, we solve instead the problem

$$\text{minimize} \quad \sum_{i=1}^{m} \phi(a_i^T x - b_i), \tag{6.27}$$

where a_i^T is the ith row of A. We assume **rank** $A = n$.

Let $p^\star$ denote the optimal value of the ℓ_1-norm approximation problem (6.26). Let $\hat{x}$ denote the optimal solution of the approximate problem (6.27), and let $\hat{r}$ denote the associated residual, $\hat{r} = A\hat{x} - b$.

(a) Show that $p^\star \geq \sum_{i=1}^{m} \hat{r}_i^2 / (\hat{r}_i^2 + \epsilon)^{1/2}$.

(b) Show that

$$
\|A\hat{x} - b\|_1 \leq p^\star + \sum_{i=1}^{m} |\hat{r}_i| \left(1 - \frac{|\hat{r}_i|}{(\hat{r}_i^2 + \epsilon)^{1/2}} \right).
$$

(By evaluating the righthand side after computing $\hat{x}$, we obtain a bound on how suboptimal $\hat{x}$ is for the ℓ_1-norm approximation problem.)

6.5 *Minimum length approximation.* Consider the problem

$$
\begin{array}{ll}
\text{minimize} & \text{length}(x) \\
\text{subject to} & \|Ax - b\| \leq \epsilon,
\end{array}
$$

where $\text{length}(x) = \min\{k \mid x_i = 0 \text{ for } i > k\}$. The problem variable is $x \in \mathbf{R}^n$; the problem parameters are $A \in \mathbf{R}^{m \times n}$, $b \in \mathbf{R}^m$, and $\epsilon > 0$. In a regression context, we are asked to find the minimum number of columns of A, taken in order, that can approximate the vector b within ϵ.

Show that this is a quasiconvex optimization problem.

6.6 *Duals of some penalty function approximation problems.* Derive a Lagrange dual for the problem

$$
\begin{array}{ll}
\text{minimize} & \sum_{i=1}^{m} \phi(r_i) \\
\text{subject to} & r = Ax - b,
\end{array}
$$

for the following penalty functions $\phi : \mathbf{R} \to \mathbf{R}$. The variables are $x \in \mathbf{R}^n$, $r \in \mathbf{R}^m$.

(a) *Deadzone-linear penalty* (with deadzone width $a = 1$),

$$
\phi(u) = \left\{
\begin{array}{ll}
0 & |u| \leq 1 \\
|u| - 1 & |u| > 1.
\end{array}
\right.
$$

(b) *Huber penalty* (with $M = 1$),

$$
\phi(u) = \left\{
\begin{array}{ll}
u^2 & |u| \leq 1 \\
2|u| - 1 & |u| > 1.
\end{array}
\right.
$$

(c) *Log-barrier* (with limit $a = 1$),

$$\phi(u) = -\log(1 - u^2), \qquad \mathbf{dom}\,\phi = (-1, 1).$$

(d) *Relative deviation from one,*

$$\phi(u) = \max\{u, 1/u\} = \begin{cases} u & u \geq 1 \\ 1/u & u \leq 1, \end{cases}$$

with $\mathbf{dom}\,\phi = \mathbf{R}_{++}$.

Regularization and robust approximation

6.7 *Bi-criterion optimization with Euclidean norms.* We consider the bi-criterion optimization problem

$$\text{minimize (w.r.t. } \mathbf{R}_+^2) \quad (\|Ax - b\|_2^2, \|x\|_2^2),$$

where $A \in \mathbf{R}^{m \times n}$ has rank r, and $b \in \mathbf{R}^m$. Show how to find the solution of each of the following problems from the singular value decomposition of A,

$$A = U\,\mathbf{diag}(\sigma)V^T = \sum_{i=1}^r \sigma_i u_i v_i^T$$

(see §A.5.4).

(a) *Tikhonov regularization:* minimize $\|Ax - b\|_2^2 + \delta\|x\|_2^2$.
(b) Minimize $\|Ax - b\|_2^2$ subject to $\|x\|_2^2 = \gamma$.
(c) Maximize $\|Ax - b\|_2^2$ subject to $\|x\|_2^2 = \gamma$.

Here δ and γ are positive parameters.

Your results provide efficient methods for computing the optimal trade-off curve and the set of achievable values of the bi-criterion problem.

6.8 Formulate the following robust approximation problems as LPs, QPs, SOCPs, or SDPs. For each subproblem, consider the ℓ_1-, ℓ_2-, and the ℓ_∞-norms.

(a) *Stochastic robust approximation with a finite set of parameter values,* i.e., the sum-of-norms problem

$$\text{minimize} \quad \sum_{i=1}^k p_i \|A_i x - b\|$$

where $p \succeq 0$ and $\mathbf{1}^T p = 1$. (See §6.4.1.)

(b) *Worst-case robust approximation with coefficient bounds:*

$$\text{minimize} \quad \sup_{A \in \mathcal{A}} \|Ax - b\|$$

where

$$\mathcal{A} = \{A \in \mathbf{R}^{m \times n} \mid l_{ij} \leq a_{ij} \leq u_{ij},\ i = 1, \dots, m,\ j = 1, \dots, n\}.$$

Here the uncertainty set is described by giving upper and lower bounds for the components of A. We assume $l_{ij} < u_{ij}$.

(c) *Worst-case robust approximation with polyhedral uncertainty:*

$$\text{minimize} \quad \sup_{A \in \mathcal{A}} \|Ax - b\|$$

where

$$\mathcal{A} = \{[a_1 \ \cdots \ a_m]^T \mid C_i a_i \preceq d_i,\ i = 1, \dots, m\}.$$

The uncertainty is described by giving a polyhedron $\mathcal{P}_i = \{a_i \mid C_i a_i \preceq d_i\}$ of possible values for each row. The parameters $C_i \in \mathbf{R}^{p_i \times n}$, $d_i \in \mathbf{R}^{p_i}$, $i = 1, \dots, m$, are given. We assume that the polyhedra $\mathcal{P}_i$ are nonempty and bounded.

Function fitting and interpolation

6.9 *Minimax rational function fitting.* Show that the following problem is quasiconvex:

$$\text{minimize} \quad \max_{i=1,\ldots,k} \left| \frac{p(t_i)}{q(t_i)} - y_i \right|$$

where

$$p(t) = a_0 + a_1 t + a_2 t^2 + \cdots + a_m t^m, \qquad q(t) = 1 + b_1 t + \cdots + b_n t^n,$$

and the domain of the objective function is defined as

$$D = \{(a, b) \in \mathbf{R}^{m+1} \times \mathbf{R}^n \mid q(t) > 0, \ \alpha \le t \le \beta\}.$$

In this problem we fit a rational function $p(t)/q(t)$ to given data, while constraining the denominator polynomial to be positive on the interval $[\alpha, \beta]$. The optimization variables are the numerator and denominator coefficients a_i, b_i. The interpolation points $t_i \in [\alpha, \beta]$, and desired function values y_i, $i = 1, \ldots, k$, are given.

6.10 *Fitting data with a concave nonnegative nondecreasing quadratic function.* We are given the data

$$x_1, \ldots, x_N \in \mathbf{R}^n, \qquad y_1, \ldots, y_N \in \mathbf{R},$$

and wish to fit a quadratic function of the form

$$f(x) = (1/2)x^T P x + q^T x + r,$$

where $P \in \mathbf{S}^n$, $q \in \mathbf{R}^n$, and $r \in \mathbf{R}$ are the parameters in the model (and, therefore, the variables in the fitting problem).

Our model will be used only on the box $\mathcal{B} = \{x \in \mathbf{R}^n \mid l \preceq x \preceq u\}$. You can assume that $l \prec u$, and that the given data points x_i are in this box.

We will use the simple sum of squared errors objective,

$$\sum_{i=1}^{N} (f(x_i) - y_i)^2,$$

as the criterion for the fit. We also impose several constraints on the function f. First, it must be concave. Second, it must be nonnegative on $\mathcal{B}$, *i.e.*, $f(z) \ge 0$ for all $z \in \mathcal{B}$. Third, f must be nondecreasing on $\mathcal{B}$, *i.e.*, whenever $z, \tilde{z} \in \mathcal{B}$ satisfy $z \preceq \tilde{z}$, we have $f(z) \le f(\tilde{z})$.

Show how to formulate this fitting problem as a convex problem. Simplify your formulation as much as you can.

6.11 *Least-squares direction interpolation.* Suppose $F_1, \ldots, F_n : \mathbf{R}^k \to \mathbf{R}^p$, and we form the linear combination $F : \mathbf{R}^k \to \mathbf{R}^p$,

$$F(u) = x_1 F_1(u) + \cdots + x_n F_n(u),$$

where x is the variable in the interpolation problem.

In this problem we require that $\angle(F(v_j), q_j) = 0$, $j = 1, \ldots, m$, where q_j are given vectors in $\mathbf{R}^p$, which we assume satisfy $\|q_j\|_2 = 1$. In other words, we require the direction of F to take on specified values at the points v_j. To ensure that $F(v_j)$ is not zero (which makes the angle undefined), we impose the minimum length constraints $\|F(v_j)\|_2 \ge \epsilon$, $j = 1, \ldots, m$, where $\epsilon > 0$ is given.

Show how to find x that minimizes $\|x\|^2$, and satisfies the direction (and minimum length) conditions above, using convex optimization.

6.12 *Interpolation with monotone functions.* A function $f : \mathbf{R}^k \to \mathbf{R}$ is monotone nondecreasing (with respect to $\mathbf{R}^k_+$) if $f(u) \ge f(v)$ whenever $u \succeq v$.

(a) Show that there exists a monotone nondecreasing function $f : \mathbf{R}^k \to \mathbf{R}$, that satisfies $f(u_i) = y_i$ for $i = 1, \dots, m$, if and only if

$$y_i \geq y_j \text{ whenever } u_i \succeq u_j, \quad i, \ j = 1, \dots, m.$$

(b) Show that there exists a convex monotone nondecreasing function $f : \mathbf{R}^k \to \mathbf{R}$, with $\mathbf{dom}\, f = \mathbf{R}^k$, that satisfies $f(u_i) = y_i$ for $i = 1, \dots, m$, if and only if there exist $g_i \in \mathbf{R}^k$, $i = 1, \dots, m$, such that

$$g_i \succeq 0, \quad i = 1, \dots, m, \qquad y_j \geq y_i + g_i^T(u_j - u_i), \quad i, \ j = 1, \dots, m.$$

6.13 *Interpolation with quasiconvex functions.* Show that there exists a quasiconvex function $f : \mathbf{R}^k \to \mathbf{R}$, that satisfies $f(u_i) = y_i$ for $i = 1, \dots, m$, if and only if there exist $g_i \in \mathbf{R}^k$, $i = 1, \dots, m$, such that

$$g_i^T(u_j - u_i) \leq -1 \text{ whenever } y_j < y_i, \quad i, \ j = 1, \dots, m.$$

6.14 [Nes00] *Interpolation with positive-real functions.* Suppose $z_1, \dots, z_n \in \mathbf{C}$ are n distinct points with $|z_i| > 1$. We define K_{np} as the set of vectors $y \in \mathbf{C}^n$ for which there exists a function $f : \mathbf{C} \to \mathbf{C}$ that satisfies the following conditions.

- f is *positive-real*, which means it is analytic outside the unit circle (*i.e.*, for $|z| > 1$), and its real part is nonnegative outside the unit circle ($\Re f(z) \geq 0$ for $|z| > 1$).

- f satisfies the *interpolation conditions*

$$f(z_1) = y_1, \qquad f(z_2) = y_2, \qquad \dots, \qquad f(z_n) = y_n.$$

If we denote the set of positive-real functions as $\mathcal{F}$, then we can express K_{np} as

$$K_{\mathrm{np}} = \{ y \in \mathbf{C}^n \mid \exists f \in \mathcal{F}, \ y_k = f(z_k), \ k = 1, \dots, n \}.$$

(a) It can be shown that f is positive-real if and only if there exists a nondecreasing function ρ such that for all z with $|z| > 1$,

$$f(z) = i\Im f(\infty) + \int_0^{2\pi} \frac{e^{i\theta} + z^{-1}}{e^{i\theta} - z^{-1}}\, d\rho(\theta),$$

where $i = \sqrt{-1}$ (see [KN77, page 389]). Use this representation to show that K_{np} is a closed convex cone.

(b) We will use the inner product $\Re(x^H y)$ between vectors $x, y \in \mathbf{C}^n$, where x^H denotes the complex conjugate transpose of x. Show that the dual cone of K_{np} is given by

$$K_{\mathrm{np}}^* = \left\{ x \in \mathbf{C}^n \ \middle|\ \Im(\mathbf{1}^T x) = 0, \ \Re\left(\sum_{l=1}^n x_l \frac{e^{-i\theta} + \bar{z}_l^{-1}}{e^{-i\theta} - \bar{z}_l^{-1}} \right) \geq 0 \ \forall \theta \in [0, 2\pi] \right\}.$$

(c) Show that

$$K_{\mathrm{np}}^* = \left\{ x \in \mathbf{C}^n \ \middle|\ \exists Q \in \mathbf{H}_+^n, \ x_l = \sum_{k=1}^n \frac{Q_{kl}}{1 - z_k^{-1} \bar{z}_l^{-1}}, \ l = 1, \dots, n \right\}$$

where $\mathbf{H}_+^n$ denotes the set of positive semidefinite Hermitian matrices of size $n \times n$. Use the following result (known as *Riesz-Fejér theorem*; see [KN77, page 60]). A function of the form

$$\sum_{k=0}^n (y_k e^{-ik\theta} + \bar{y}_k e^{ik\theta})$$

is nonnegative for all θ if and only if there exist $a_0, \ldots, a_n \in \mathbf{C}$ such that

$$\sum_{k=0}^{n} (y_k e^{-ik\theta} + \bar{y}_k e^{ik\theta}) = \left| \sum_{k=0}^{n} a_k e^{ik\theta} \right|^2 .$$

(d) Show that $K_{\mathrm{np}} = \{y \in \mathbf{C}^n \mid P(y) \succeq 0\}$ where $P(y) \in \mathbf{H}^n$ is defined as

$$P(y)_{kl} = \frac{y_k + \bar{y}_l}{1 - z_k^{-1} \bar{z}_l^{-1}}, \quad l, k = 1, \ldots, n.$$

The matrix $P(y)$ is called the *Nevanlinna-Pick matrix* associated with the points z_k, y_k.

Hint. As we noted in part (a), K_{np} is a closed convex cone, so $K_{\mathrm{np}} = K_{\mathrm{np}}^{**}$.

(e) As an application, pose the following problem as a convex optimization problem:

$$\begin{array}{ll} \text{minimize} & \sum_{k=1}^{n} |f(z_k) - w_k|^2 \\ \text{subject to} & f \in \mathcal{F}. \end{array}$$

The problem data are n points z_k with $|z_k| > 1$ and n complex numbers $w_1, \ldots, w_n$. We optimize over all positive-real functions f.

Chapter 7

Statistical estimation

7.1 Parametric distribution estimation

7.1.1 Maximum likelihood estimation

We consider a family of probability distributions on $\mathbf{R}^m$, indexed by a vector $x \in \mathbf{R}^n$, with densities $p_x(\cdot)$. When considered as a function of x, for fixed $y \in \mathbf{R}^m$, the function $p_x(y)$ is called the *likelihood function*. It is more convenient to work with its logarithm, which is called the *log-likelihood function*, and denoted l:

$$l(x) = \log p_x(y).$$

There are often constraints on the values of the parameter x, which can represent prior knowledge about x, or the domain of the likelihood function. These constraints can be explicitly given, or incorporated into the likelihood function by assigning $p_x(y) = 0$ (for all y) whenever x does not satisfy the prior information constraints. (Thus, the log-likelihood function can be assigned the value $-\infty$ for parameters x that violate the prior information constraints.)

Now consider the problem of estimating the value of the parameter x, based on observing one sample y from the distribution. A widely used method, called *maximum likelihood (ML) estimation*, is to estimate x as

$$\hat{x}_{\mathrm{ml}} = \mathrm{argmax}_x p_x(y) = \mathrm{argmax}_x l(x),$$

i.e., to choose as our estimate a value of the parameter that maximizes the likelihood (or log-likelihood) function for the observed value of y. If we have prior information about x, such as $x \in C \subseteq \mathbf{R}^n$, we can add the constraint $x \in C$ explicitly, or impose it implicitly, by redefining $p_x(y)$ to be zero for $x \notin C$.

The problem of finding a maximum likelihood estimate of the parameter vector x can be expressed as

$$
\begin{array}{ll}
\text{maximize} & l(x) = \log p_x(y) \\
\text{subject to} & x \in C,
\end{array}
\tag{7.1}
$$

where $x \in C$ gives the prior information or other constraints on the parameter vector x. In this optimization problem, the vector $x \in \mathbf{R}^n$ (which is the parameter in the probability density) is the variable, and the vector $y \in \mathbf{R}^m$ (which is the observed sample) is a problem parameter.

The maximum likelihood estimation problem (7.1) is a convex optimization problem if the log-likelihood function l is concave for each value of y, and the set C can be described by a set of linear equality and convex inequality constraints, a situation which occurs in many estimation problems. For these problems we can compute an ML estimate using convex optimization.

Linear measurements with IID noise

We consider a linear measurement model,

$$y_i = a_i^T x + v_i, \quad i = 1, \dots, m,$$

where $x \in \mathbf{R}^n$ is a vector of parameters to be estimated, $y_i \in \mathbf{R}$ are the measured or observed quantities, and v_i are the measurement errors or noise. We assume that v_i are independent, identically distributed (IID), with density p on $\mathbf{R}$. The likelihood function is then

$$p_x(y) = \prod_{i=1}^{m} p(y_i - a_i^T x),$$

so the log-likelihood function is

$$l(x) = \log p_x(y) = \sum_{i=1}^{m} \log p(y_i - a_i^T x).$$

The ML estimate is any optimal point for the problem

$$\text{maximize} \quad \sum_{i=1}^{m} \log p(y_i - a_i^T x), \tag{7.2}$$

with variable x. If the density p is log-concave, this problem is convex, and has the form of a penalty approximation problem ((6.2), page 294), with penalty function $-\log p$.

Example 7.1 *ML estimation for some common noise densities.*

- *Gaussian noise.* When v_i are Gaussian with zero mean and variance σ^2, the density is $p(z) = (2\pi\sigma^2)^{-1/2} e^{-z^2/2\sigma^2}$, and the log-likelihood function is

  $$l(x) = -(m/2) \log(2\pi\sigma^2) - \frac{1}{2\sigma^2} \|Ax - y\|_2^2,$$

  where A is the matrix with rows $a_1^T, \dots, a_m^T$. Therefore the ML estimate of x is $x_{\mathrm{ml}} = \operatorname{argmin}_x \|Ax - y\|_2^2$, the solution of a least-squares approximation problem.

- *Laplacian noise.* When v_i are Laplacian, *i.e.*, have density $p(z) = (1/2a)e^{-|z|/a}$ (where $a > 0$), the ML estimate is $\hat{x} = \operatorname{argmin}_x \|Ax - y\|_1$, the solution of the ℓ_1-norm approximation problem.

- *Uniform noise.* When v_i are uniformly distributed on $[-a, a]$, we have $p(z) = 1/(2a)$ on $[-a, a]$, and an ML estimate is any x satisfying $\|Ax - y\|_\infty \le a$.

ML interpretation of penalty function approximation

Conversely, we can interpret any penalty function approximation problem

$$\text{minimize} \quad \sum_{i=1}^{m} \phi(b_i - a_i^T x)$$

as a maximum likelihood estimation problem, with noise density

$$p(z) = \frac{e^{-\phi(z)}}{\int e^{-\phi(u)} \, du},$$

and measurements b. This observation gives a statistical interpretation of the penalty function approximation problem. Suppose, for example, that the penalty function ϕ grows very rapidly for large values, which means that we attach a very large cost or penalty to large residuals. The corresponding noise density function p will have very small tails, and the ML estimator will avoid (if possible) estimates with any large residuals because these correspond to very unlikely events.

We can also understand the robustness of ℓ_1-norm approximation to large errors in terms of maximum likelihood estimation. We interpret ℓ_1-norm approximation as maximum likelihood estimation with a noise density that is Laplacian; ℓ_2-norm approximation is maximum likelihood estimation with a Gaussian noise density. The Laplacian density has larger tails than the Gaussian, *i.e.*, the probability of a very large v_i is far larger with a Laplacian than a Gaussian density. As a result, the associated maximum likelihood method expects to see greater numbers of large residuals.

Counting problems with Poisson distribution

In a wide variety of problems the random variable y is nonnegative integer valued, with a Poisson distribution with mean $\mu > 0$:

$$\mathbf{prob}(y = k) = \frac{e^{-\mu}\mu^k}{k!}.$$

Often y represents the count or number of events (such as photon arrivals, traffic accidents, etc.) of a Poisson process over some period of time.

In a simple statistical model, the mean μ is modeled as an affine function of a vector $u \in \mathbf{R}^n$:

$$\mu = a^T u + b.$$

Here u is called the vector of *explanatory variables*, and the vector $a \in \mathbf{R}^n$ and number $b \in \mathbf{R}$ are called the *model parameters*. For example, if y is the number of traffic accidents in some region over some period, u_1 might be the total traffic flow through the region during the period, u_2 the rainfall in the region during the period, and so on.

We are given a number of observations which consist of pairs (u_i, y_i), $i = 1, \dots, m$, where y_i is the observed value of y for which the value of the explanatory variable is $u_i \in \mathbf{R}^n$. Our job is to find a maximum likelihood estimate of the model parameters $a \in \mathbf{R}^n$ and $b \in \mathbf{R}$ from these data.

The likelihood function has the form

$$\prod_{i=1}^{m} \frac{(a^T u_i + b)^{y_i} \exp(-(a^T u_i + b))}{y_i!},$$

so the log-likelihood function is

$$l(a, b) = \sum_{i=1}^{m} (y_i \log(a^T u_i + b) - (a^T u_i + b) - \log(y_i!)).$$

We can find an ML estimate of a and b by solving the convex optimization problem

$$\text{maximize} \quad \sum_{i=1}^{m} (y_i \log(a^T u_i + b) - (a^T u_i + b)),$$

where the variables are a and b.

Logistic regression

We consider a random variable $y \in \{0, 1\}$, with

$$\mathbf{prob}(y = 1) = p, \qquad \mathbf{prob}(y = 0) = 1 - p,$$

where $p \in [0, 1]$, and is assumed to depend on a vector of explanatory variables $u \in \mathbf{R}^n$. For example, $y = 1$ might mean that an individual in a population acquires a certain disease. The probability of acquiring the disease is p, which is modeled as a function of some explanatory variables u, which might represent weight, age, height, blood pressure, and other medically relevant variables.

The *logistic model* has the form

$$p = \frac{\exp(a^T u + b)}{1 + \exp(a^T u + b)}, \tag{7.3}$$

where $a \in \mathbf{R}^n$ and $b \in \mathbf{R}$ are the model parameters that determine how the probability p varies as a function of the explanatory variable u.

Now suppose we are given some data consisting of a set of values of the explanatory variables $u_1, \ldots, u_m \in \mathbf{R}^n$ along with the corresponding outcomes $y_1, \ldots, y_m \in \{0, 1\}$. Our job is to find a maximum likelihood estimate of the model parameters $a \in \mathbf{R}^n$ and $b \in \mathbf{R}$. Finding an ML estimate of a and b is sometimes called *logistic regression*.

We can re-order the data so for $u_1, \ldots, u_q$, the outcome is $y = 1$, and for $u_{q+1}, \ldots, u_m$ the outcome is $y = 0$. The likelihood function then has the form

$$\prod_{i=1}^{q} p_i \prod_{i=q+1}^{m} (1 - p_i),$$

where p_i is given by the logistic model with explanatory variable u_i. The log-likelihood function has the form

$$l(a, b) = \sum_{i=1}^{q} \log p_i + \sum_{i=q+1}^{m} \log(1 - p_i)$$

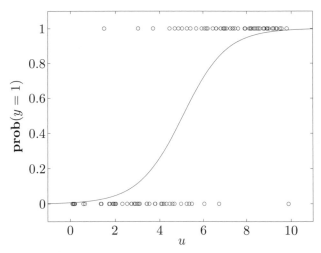

Figure 7.1 *Logistic regression.* The circles show 50 points (u_i, y_i), where $u_i \in \mathbf{R}$ is the explanatory variable, and $y_i \in \{0, 1\}$ is the outcome. The data suggest that for $u < 5$ or so, the outcome is more likely to be $y = 0$, while for $u > 5$ or so, the outcome is more likely to be $y = 1$. The data also suggest that for $u < 2$ or so, the outcome is very likely to be $y = 0$, and for $u > 8$ or so, the outcome is very likely to be $y = 1$. The solid curve shows $\mathbf{prob}(y = 1) = \exp(au + b)/(1 + \exp(au + b))$ for the maximum likelihood parameters a, b. This maximum likelihood model is consistent with our informal observations about the data set.

$$
= \sum_{i=1}^{q} \log \frac{\exp(a^T u_i + b)}{1 + \exp(a^T u_i + b)} + \sum_{i=q+1}^{m} \log \frac{1}{1 + \exp(a^T u_i + b)}
$$

$$
= \sum_{i=1}^{q} (a^T u_i + b) - \sum_{i=1}^{m} \log(1 + \exp(a^T u_i + b)).
$$

Since l is a concave function of a and b, the logistic regression problem can be solved as a convex optimization problem. Figure 7.1 shows an example with $u \in \mathbf{R}$.

Covariance estimation for Gaussian variables

Suppose $y \in \mathbf{R}^n$ is a Gaussian random variable with zero mean and covariance matrix $R = \mathbf{E}\, yy^T$, so its density is

$$
p_R(y) = (2\pi)^{-n/2} \det(R)^{-1/2} \exp(-y^T R^{-1} y/2),
$$

where $R \in \mathbf{S}^n_{++}$. We want to estimate the covariance matrix R based on N independent samples $y_1, \ldots, y_N \in \mathbf{R}^n$ drawn from the distribution, and using prior knowledge about R.

The log-likelihood function has the form

$$
l(R) \quad = \quad \log p_R(y_1, \ldots, y_N)
$$

$$= -(Nn/2)\log(2\pi) - (N/2)\log\det R - (1/2)\sum_{k=1}^{N} y_k^T R^{-1} y_k$$

$$= -(Nn/2)\log(2\pi) - (N/2)\log\det R - (N/2)\operatorname{\mathbf{tr}}(R^{-1}Y),$$

where

$$Y = \frac{1}{N}\sum_{k=1}^{N} y_k y_k^T$$

is the sample covariance of $y_1, \ldots, y_N$. This log-likelihood function is *not* a concave function of R (although it is concave on a subset of its domain $\mathbf{S}_{++}^n$; see exercise 7.4), but a change of variable yields a concave log-likelihood function. Let S denote the inverse of the covariance matrix, $S = R^{-1}$ (which is called the *information matrix*). Using S in place of R as a new parameter, the log-likelihood function has the form

$$l(S) = -(Nn/2)\log(2\pi) + (N/2)\log\det S - (N/2)\operatorname{\mathbf{tr}}(SY),$$

which *is* a concave function of S.

Therefore the ML estimate of S (hence, R) is found by solving the problem

$$\begin{array}{ll}\text{maximize} & \log\det S - \operatorname{\mathbf{tr}}(SY) \\ \text{subject to} & S \in \mathcal{S}\end{array} \qquad (7.4)$$

where $\mathcal{S}$ is our prior knowledge of $S = R^{-1}$. (We also have the implicit constraint that $S \in \mathbf{S}_{++}^n$.) Since the objective function is concave, this is a convex problem if the set $\mathcal{S}$ can be described by a set of linear equality and convex inequality constraints.

First we examine the case in which no prior assumptions are made on R (hence, S), other than $R \succ 0$. In this case the problem (7.4) can be solved analytically. The gradient of the objective is $S^{-1} - Y$, so the optimal S satisfies $S^{-1} = Y$ if $Y \in \mathbf{S}_{++}^n$. (If $Y \notin \mathbf{S}_{++}^n$, the log-likelihood function is unbounded above.) Therefore, when we have no prior assumptions about R, the maximum likelihood estimate of the covariance is, simply, the sample covariance: $\hat{R}_{\mathrm{ml}} = Y$.

Now we consider some examples of constraints on R that can be expressed as convex constraints on the information matrix S. We can handle lower and upper (matrix) bounds on R, of the form

$$L \preceq R \preceq U,$$

where L and U are symmetric and positive definite, as

$$U^{-1} \preceq R^{-1} \preceq L^{-1}.$$

A condition number constraint on R,

$$\lambda_{\max}(R) \le \kappa_{\max}\lambda_{\min}(R),$$

can be expressed as

$$\lambda_{\max}(S) \le \kappa_{\max}\lambda_{\min}(S).$$

This is equivalent to the existence of $u > 0$ such that $uI \preceq S \preceq \kappa_{\max} uI$. We can therefore solve the ML problem, with the condition number constraint on R, by solving the convex problem

$$
\begin{array}{ll}
\text{maximize} & \log \det S - \mathbf{tr}(SY) \\
\text{subject to} & uI \preceq S \preceq \kappa_{\max} uI
\end{array}
\tag{7.5}
$$

where the variables are $S \in \mathbf{S}^n$ and $u \in \mathbf{R}$.

As another example, suppose we are given bounds on the variance of some linear functions of the underlying random vector y,

$$
\mathbf{E}(c_i^T y)^2 \le \alpha_i, \quad i = 1, \ldots, K.
$$

These prior assumptions can be expressed as

$$
\mathbf{E}(c_i^T y)^2 = c_i^T R c_i = c_i^T S^{-1} c_i \le \alpha_i, \quad i = 1, \ldots, K.
$$

Since $c_i^T S^{-1} c_i$ is a convex function of S (provided $S \succ 0$, which holds here), these bounds can be imposed in the ML problem.

7.1.2 Maximum a posteriori probability estimation

Maximum a posteriori probability (MAP) estimation can be considered a Bayesian version of maximum likelihood estimation, with a prior probability density on the underlying parameter x. We assume that x (the vector to be estimated) and y (the observation) are random variables with a joint probability density $p(x, y)$. This is in contrast to the statistical estimation setup, where x is a parameter, not a random variable.

The *prior density* of x is given by

$$
p_x(x) = \int p(x, y) \, dy.
$$

This density represents our prior information about what the values of the vector x might be, before we observe the vector y. Similarly, the prior density of y is given by

$$
p_y(y) = \int p(x, y) \, dx.
$$

This density represents the prior information about what the measurement or observation vector y will be.

The conditional density of y, given x, is given by

$$
p_{y|x}(x, y) = \frac{p(x, y)}{p_x(x)}.
$$

In the MAP estimation method, $p_{y|x}$ plays the role of the parameter dependent density p_x in the maximum likelihood estimation setup. The conditional density of x, given y, is given by

$$
p_{x|y}(x, y) = \frac{p(x, y)}{p_y(y)} = p_{y|x}(x, y) \frac{p_x(x)}{p_y(y)}.
$$

When we substitute the observed value y into $p_{x|y}$, we obtain the *posterior density* of x. It represents our knowledge of x after the observation.

In the MAP estimation method, our estimate of x, given the observation y, is given by

$$
\begin{aligned}
\hat{x}_{\text{map}} &= \operatorname{argmax}_x p_{x|y}(x, y) \\
&= \operatorname{argmax}_x p_{y|x}(x, y) p_x(x) \\
&= \operatorname{argmax}_x p(x, y).
\end{aligned}
$$

In other words, we take as estimate of x the value that maximizes the conditional density of x, given the observed value of y. The only difference between this estimate and the maximum likelihood estimate is the second term, $p_x(x)$, appearing here. This term can be interpreted as taking our prior knowledge of x into account. Note that if the prior density of x is uniform over a set C, then finding the MAP estimate is the same as maximizing the likelihood function subject to $x \in C$, which is the ML estimation problem (7.1).

Taking logarithms, we can express the MAP estimate as

$$
\hat{x}_{\text{map}} = \operatorname{argmax}_x (\log p_{y|x}(x, y) + \log p_x(x)). \tag{7.6}
$$

The first term is essentially the same as the log-likelihood function; the second term penalizes choices of x that are unlikely, according to the prior density (*i.e.*, x with $p_x(x)$ small).

Brushing aside the philosophical differences in setup, the only difference between finding the MAP estimate (via (7.6)) and the ML estimate (via (7.1)) is the presence of an extra term in the optimization problem, associated with the prior density of x. Therefore, for any maximum likelihood estimation problem with concave log-likelihood function, we can add a prior density for x that is log-concave, and the resulting MAP estimation problem will be convex.

Linear measurements with IID noise

Suppose that $x \in \mathbf{R}^n$ and $y \in \mathbf{R}^m$ are related by

$$
y_i = a_i^T x + v_i, \quad i = 1, \dots, m,
$$

where v_i are IID with density p_v on $\mathbf{R}$, and x has prior density p_x on $\mathbf{R}^n$. The joint density of x and y is then

$$
p(x, y) = p_x(x) \prod_{i=1}^m p_v(y_i - a_i^T x),
$$

and the MAP estimate can be found by solving the optimization problem

$$
\text{maximize} \quad \log p_x(x) + \textstyle\sum_{i=1}^m \log p_v(y_i - a_i^T x). \tag{7.7}
$$

If p_x and p_v are log-concave, this problem is convex. The only difference between the MAP estimation problem (7.7) and the associated ML estimation problem (7.2) is the extra term $\log p_x(x)$.

For example, if v_i are uniform on $[-a, a]$, and the prior distribution of x is Gaussian with mean $\bar{x}$ and covariance Σ, the MAP estimate is found by solving the QP

$$\begin{array}{ll}\text{minimize} & (x - \bar{x})^T \Sigma^{-1}(x - \bar{x}) \\ \text{subject to} & \|Ax - y\|_\infty \le a,\end{array}$$

with variable x.

MAP with perfect linear measurements

Suppose $x \in \mathbf{R}^n$ is a vector of parameters to be estimated, with prior density p_x. We have m perfect (noise free, deterministic) linear measurements, given by $y = Ax$. In other words, the conditional distribution of y, given x, is a point mass with value one at the point Ax. The MAP estimate can be found by solving the problem

$$\begin{array}{ll}\text{maximize} & \log p_x(x) \\ \text{subject to} & Ax = y.\end{array}$$

If p_x is log-concave, this is a convex problem.

If under the prior distribution, the parameters x_i are IID with density p on $\mathbf{R}$, then the MAP estimation problem has the form

$$\begin{array}{ll}\text{maximize} & \sum_{i=1}^n \log p(x_i) \\ \text{subject to} & Ax = y,\end{array}$$

which is a least-penalty problem ((6.6), page 304), with penalty function $\phi(u) = -\log p(u)$.

Conversely, we can interpret any least-penalty problem,

$$\begin{array}{ll}\text{minimize} & \phi(x_1) + \cdots + \phi(x_n) \\ \text{subject to} & Ax = b\end{array}$$

as a MAP estimation problem, with m perfect linear measurements (*i.e.*, $Ax = b$) and x_i IID with density

$$p(z) = \frac{e^{-\phi(z)}}{\int e^{-\phi(u)}\, du}.$$

7.2   Nonparametric distribution estimation

We consider a random variable X with values in the finite set $\{\alpha_1, \ldots, \alpha_n\} \subseteq \mathbf{R}$. (We take the values to be in $\mathbf{R}$ for simplicity; the same ideas can be applied when the values are in $\mathbf{R}^k$, for example.) The distribution of X is characterized by $p \in \mathbf{R}^n$, with $\mathbf{prob}(X = \alpha_k) = p_k$. Clearly, p satisfies $p \succeq 0$, $\mathbf{1}^T p = 1$. Conversely, if $p \in \mathbf{R}^n$ satisfies $p \succeq 0$, $\mathbf{1}^T p = 1$, then it defines a probability distribution for a random variable X, defined as $\mathbf{prob}(X = \alpha_k) = p_k$. Thus, the probability simplex

$$\{p \in \mathbf{R}^n \mid p \succeq 0,\ \mathbf{1}^T p = 1\}$$

is in one-to-one correspondence with all possible probability distributions for a random variable X taking values in $\{\alpha_1, \ldots, \alpha_n\}$.

In this section we discuss methods used to estimate the distribution p based on a combination of prior information and, possibly, observations and measurements.

Prior information

Many types of prior information about p can be expressed in terms of linear equality constraints or inequalities. If $f : \mathbf{R} \to \mathbf{R}$ is any function, then

$$\mathbf{E} f(X) = \sum_{i=1}^{n} p_i f(\alpha_i)$$

is a linear function of p. As a special case, if $C \subseteq \mathbf{R}$, then $\mathbf{prob}(X \in C)$ is a linear function of p:

$$\mathbf{prob}(X \in C) = c^T p, \qquad c_i = \begin{cases} 1 & \alpha_i \in C \\ 0 & \alpha_i \notin C. \end{cases}$$

It follows that known expected values of certain functions (*e.g.*, moments) or known probabilities of certain sets can be incorporated as linear equality constraints on $p \in \mathbf{R}^n$. Inequalities on expected values or probabilities can be expressed as linear inequalities on $p \in \mathbf{R}^n$.

For example, suppose we know that X has mean $\mathbf{E} X = \alpha$, second moment $\mathbf{E} X^2 = \beta$, and $\mathbf{prob}(X \geq 0) \leq 0.3$. This prior information can be expressed as

$$\mathbf{E} X = \sum_{i=1}^{n} \alpha_i p_i = \alpha, \qquad \mathbf{E} X^2 = \sum_{i=1}^{n} \alpha_i^2 p_i = \beta, \qquad \sum_{\alpha_i \geq 0} p_i \leq 0.3,$$

which are two linear equalities and one linear inequality in p.

We can also include some prior constraints that involve nonlinear functions of p. As an example, the variance of X is given by

$$\mathbf{var}(X) = \mathbf{E} X^2 - (\mathbf{E} X)^2 = \sum_{i=1}^{n} \alpha_i^2 p_i - \left(\sum_{i=1}^{n} \alpha_i p_i \right)^2.$$

The first term is a linear function of p and the second term is concave quadratic in p, so the variance of X is a concave function of p. It follows that a *lower bound* on the variance of X can be expressed as a convex quadratic inequality on p.

As another example, suppose A and B are subsets of $\mathbf{R}$, and consider the conditional probability of A given B:

$$\mathbf{prob}(X \in A | X \in B) = \frac{\mathbf{prob}(X \in A \cap B)}{\mathbf{prob}(X \in B)}.$$

This function is linear-fractional in $p \in \mathbf{R}^n$: it can be expressed as

$$\mathbf{prob}(X \in A | X \in B) = c^T p / d^T p,$$

where

$$c_i = \begin{cases} 1 & \alpha_i \in A \cap B \\ 0 & \alpha_i \notin A \cap B \end{cases}, \qquad d_i = \begin{cases} 1 & \alpha_i \in B \\ 0 & \alpha_i \notin B. \end{cases}$$

Therefore we can express the prior constraints

$$l \le \mathbf{prob}(X \in A | X \in B) \le u$$

as the linear inequality constraints on p

$$ld^T p \le c^T p \le ud^T p.$$

Several other types of prior information can be expressed in terms of nonlinear convex inequalities. For example, the entropy of X, given by

$$-\sum_{i=1}^{n} p_i \log p_i,$$

is a concave function of p, so we can impose a minimum value of entropy as a convex inequality on p. If q represents another distribution, *i.e.*, $q \succeq 0$, $\mathbf{1}^T q = 1$, then the Kullback-Leibler divergence between the distribution q and the distribution p is given by

$$\sum_{i=1}^{n} p_i \log(p_i / q_i),$$

which is convex in p (and q as well; see example 3.19, page 90). It follows that we can impose a maximum Kullback-Leibler divergence between p and a given distribution q, as a convex inequality on p.

In the next few paragraphs we express the prior information about the distribution p as $p \in \mathcal{P}$. We assume that $\mathcal{P}$ can be described by a set of linear equalities and convex inequalities. We include in the prior information $\mathcal{P}$ the basic constraints $p \succeq 0$, $\mathbf{1}^T p = 1$.

Bounding probabilities and expected values

Given prior information about the distribution, say $p \in \mathcal{P}$, we can compute upper or lower bounds on the expected value of a function, or probability of a set. For example to determine a lower bound on $\mathbf{E} f(X)$ over all distributions that satisfy the prior information $p \in \mathcal{P}$, we solve the convex problem

$$\begin{array}{ll} \text{minimize} & \sum_{i=1}^{n} f(\alpha_i) p_i \\ \text{subject to} & p \in \mathcal{P}. \end{array}$$

Maximum likelihood estimation

We can use maximum likelihood estimation to estimate p based on observations from the distribution. Suppose we observe N independent samples $x_1, \ldots, x_N$ from the distribution. Let k_i denote the number of these samples with value α_i, so that $k_1 + \cdots + k_n = N$, the total number of observed samples. The log-likelihood function is then

$$l(p) = \sum_{i=1}^{n} k_i \log p_i,$$

which is a concave function of p. The maximum likelihood estimate of p can be found by solving the convex problem

$$\begin{array}{ll} \text{maximize} & l(p) = \sum_{i=1}^{n} k_i \log p_i \\ \text{subject to} & p \in \mathcal{P}, \end{array}$$

with variable p.

Maximum entropy

The maximum entropy distribution consistent with the prior assumptions can be found by solving the convex problem

$$\begin{array}{ll} \text{minimize} & \sum_{i=1}^{n} p_i \log p_i \\ \text{subject to} & p \in \mathcal{P}. \end{array}$$

Enthusiasts describe the maximum entropy distribution as the most equivocal or most random, among those consistent with the prior information.

Minimum Kullback-Leibler divergence

We can find the distribution p that has minimum Kullback-Leibler divergence from a given prior distribution q, among those consistent with prior information, by solving the convex problem

$$\begin{array}{ll} \text{minimize} & \sum_{i=1}^{n} p_i \log(p_i/q_i) \\ \text{subject to} & p \in \mathcal{P}, \end{array}$$

Note that when the prior distribution is the uniform distribution, *i.e.*, $q = (1/n)\mathbf{1}$, this problem reduces to the maximum entropy problem.

Example 7.2 We consider a probability distribution on 100 equidistant points α_i in the interval $[-1, 1]$. We impose the following prior assumptions:

$$\begin{array}{rcl} \mathbf{E}\,X & \in & [-0.1, 0.1] \\ \mathbf{E}\,X^2 & \in & [0.5, 0.6] \\ \mathbf{E}(3X^3 - 2X) & \in & [-0.3, -0.2] \\ \mathbf{prob}(X < 0) & \in & [0.3, 0.4]. \end{array} \qquad (7.8)$$

Along with the constraints $\mathbf{1}^T p = 1$, $p \succeq 0$, these constraints describe a polyhedron of probability distributions.

Figure 7.2 shows the maximum entropy distribution that satisfies these constraints. The maximum entropy distribution satisfies

$$\begin{array}{rcl} \mathbf{E}\,X & = & 0.056 \\ \mathbf{E}\,X^2 & = & 0.5 \\ \mathbf{E}(3X^3 - 2X) & = & -0.2 \\ \mathbf{prob}(X < 0) & = & 0.4. \end{array}$$

To illustrate bounding probabilities, we compute upper and lower bounds on the cumulative distribution $\mathbf{prob}(X \leq \alpha_i)$, for $i = 1, \ldots, 100$. For each value of i,

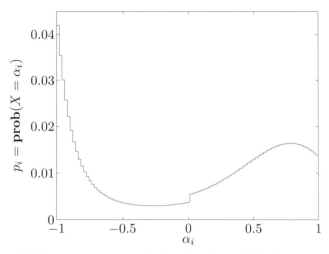

Figure 7.2 Maximum entropy distribution that satisfies the constraints (7.8).

we solve two LPs: one that maximizes $\mathbf{prob}(X \leq \alpha_i)$, and one that minimizes $\mathbf{prob}(X \leq \alpha_i)$, over all distributions consistent with the prior assumptions (7.8). The results are shown in figure 7.3. The upper and lower curves show the upper and lower bounds, respectively; the middle curve shows the cumulative distribution of the maximum entropy distribution.

Example 7.3 *Bounding risk probability with known marginal distributions.* Suppose X and Y are two random variables that give the return on two investments. We assume that X takes values in $\{\alpha_1, \ldots, \alpha_n\} \subseteq \mathbf{R}$ and Y takes values in $\{\beta_1, \ldots, \beta_m\} \subseteq \mathbf{R}$, with $p_{ij} = \mathbf{prob}(X = \alpha_i, Y = \beta_j)$. The marginal distributions of the two returns X and Y are known, *i.e.*,

$$\sum_{j=1}^{m} p_{ij} = r_i, \quad i = 1, \ldots, n, \qquad \sum_{i=1}^{n} p_{ij} = q_j, \quad j = 1, \ldots, m, \qquad (7.9)$$

but otherwise nothing is known about the joint distribution p. This defines a polyhedron of joint distributions consistent with the given marginals.

Now suppose we make both investments, so our total return is the random variable $X + Y$. We are interested in computing an upper bound on the probability of some level of loss, or low return, *i.e.*, $\mathbf{prob}(X + Y < \gamma)$. We can compute a tight upper bound on this probability by solving the LP

$$\begin{array}{ll} \text{maximize} & \sum \{p_{ij} \mid \alpha_i + \beta_j < \gamma\} \\ \text{subject to} & (7.9), \quad p_{ij} \geq 0, \quad i = 1, \ldots n, \quad j = 1, \ldots, m. \end{array}$$

The optimal value of this LP is the maximum probability of loss. The optimal solution $p^\star$ is the joint distribution, consistent with the given marginal distributions, that maximizes the probability of the loss.

The same method can be applied to a derivative of the two investments. Let $R(X, Y)$ be the return of the derivative, where $R : \mathbf{R}^2 \to \mathbf{R}$. We can compute sharp lower

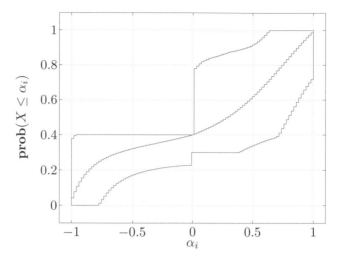

Figure 7.3 The top and bottom curves show the maximum and minimum possible values of the cumulative distribution function, $\mathbf{prob}(X \leq \alpha_i)$, over all distributions that satisfy (7.8). The middle curve is the cumulative distribution of the maximum entropy distribution that satisfies (7.8).

and upper bounds on $\mathbf{prob}(R < \gamma)$ by solving a similar LP, with objective function

$$\sum \{p_{ij} \mid R(\alpha_i, \beta_j) < \gamma\},$$

which we can minimize and maximize.

7.3   Optimal detector design and hypothesis testing

Suppose X is a random variable with values in $\{1, \ldots, n\}$, with a distribution that depends on a parameter $\theta \in \{1, \ldots, m\}$. The distributions of X, for the m possible values of θ, can be represented by a matrix $P \in \mathbf{R}^{n \times m}$, with elements

$$p_{kj} = \mathbf{prob}(X = k \mid \theta = j).$$

The jth column of P gives the probability distribution associated with the parameter value $\theta = j$.

We consider the problem of estimating θ, based on an observed sample of X. In other words, the sample X is generated from one of the m possible distributions, and we are to guess which one. The m values of θ are called *hypotheses*, and guessing which hypothesis is correct (*i.e.*, which distribution generated the observed sample X) is called *hypothesis testing*. In many cases one of the hypotheses corresponds to some normal situation, and each of the other hypotheses corresponds to some abnormal event. In this case hypothesis testing can be interpreted as observing a value of X, and then guessing whether or not an abnormal event has occurred, and if so, which one. For this reason hypothesis testing is also called *detection*.

In most cases there is no significance to the ordering of the hypotheses; they are simply m different hypotheses, arbitrarily labeled $\theta = 1, \ldots, m$. If $\hat{\theta} = \theta$, where $\hat{\theta}$ denotes the estimate of θ, then we have correctly guessed the parameter value θ. If $\hat{\theta} \neq \theta$, then we have (incorrectly) guessed the parameter value θ; we have mistaken $\hat{\theta}$ for θ. In other cases, there is significance in the ordering of the hypotheses. In this case, an event such as $\hat{\theta} > \theta$, *i.e.*, the event that we overestimate θ, is meaningful.

It is also possible to parametrize θ by values other than $\{1, \ldots, m\}$, say as $\theta \in \{\theta_1, \ldots, \theta_m\}$, where θ_i are (distinct) values. These values could be real numbers, or vectors, for example, specifying the mean and variance of the kth distribution. In this case, a quantity such as $\|\hat{\theta} - \theta\|$, which is the norm of the parameter estimation error, is meaningful.

7.3.1 Deterministic and randomized detectors

A (deterministic) *estimator* or *detector* is a function ψ from $\{1, \ldots, n\}$ (the set of possible observed values) into $\{1, \ldots, m\}$ (the set of hypotheses). If X is observed to have value k, then our guess for the value of θ is $\hat{\theta} = \psi(k)$. One obvious deterministic detector is the *maximum likelihood detector*, given by

$$\hat{\theta} = \psi_{\mathrm{ml}}(k) = \operatorname*{argmax}_{j} p_{kj}. \tag{7.10}$$

When we observe the value $X = k$, the maximum likelihood estimate of θ is a value that maximizes the probability of observing $X = k$, over the set of possible distributions.

We will consider a generalization of the deterministic detector, in which the estimate of θ, given an observed value of X, is random. A *randomized detector* of θ is a random variable $\hat{\theta} \in \{1, \ldots, m\}$, with a distribution that depends on the observed value of X. A randomized detector can be defined in terms of a matrix $T \in \mathbf{R}^{m \times n}$ with elements

$$t_{ik} = \mathbf{prob}(\hat{\theta} = i \mid X = k).$$

The interpretation is as follows: if we observe $X = k$, then the detector gives $\hat{\theta} = i$ with probability t_{ik}. The kth column of T, which we will denote t_k, gives the probability distribution of $\hat{\theta}$, when we observe $X = k$. If each column of T is a unit vector, then the randomized detector is a deterministic detector, *i.e.*, $\hat{\theta}$ is a (deterministic) function of the observed value of X.

At first glance, it seems that intentionally introducing additional randomization into the estimation or detection process can only make the estimator worse. But we will see below examples in which a randomized detector outperforms all deterministic estimators.

We are interested in designing the matrix T that defines the randomized detector. Obviously the columns t_k of T must satisfy the (linear equality and inequality) constraints

$$t_k \succeq 0, \qquad \mathbf{1}^T t_k = 1. \tag{7.11}$$

7.3.2   Detection probability matrix

For the randomized detector defined by the matrix T, we define the *detection probability matrix* as $D = TP$. We have

$$D_{ij} = (TP)_{ij} = \mathbf{prob}(\hat{\theta} = i \mid \theta = j),$$

so D_{ij} is the probability of guessing $\hat{\theta} = i$, when in fact $\theta = j$. The $m \times m$ detection probability matrix D characterizes the performance of the randomized detector defined by T. The diagonal entry D_{ii} is the probability of guessing $\hat{\theta} = i$ when $\theta = i$, i.e., the probability of correctly detecting that $\theta = i$. The off-diagonal entry D_{ij} (with $i \neq j$) is the probability of mistaking $\theta = i$ for $\theta = j$, i.e., the probability that our guess is $\hat{\theta} = i$, when in fact $\theta = j$. If $D = I$, the detector is perfect: no matter what the parameter θ is, we correctly guess $\hat{\theta} = \theta$.

The diagonal entries of D, arranged in a vector, are called the *detection probabilities*, and denoted P^{d}:

$$P_i^{\mathrm{d}} = D_{ii} = \mathbf{prob}(\hat{\theta} = i \mid \theta = i).$$

The *error probabilities* are the complements, and are denoted P^{e}:

$$P_i^{\mathrm{e}} = 1 - D_{ii} = \mathbf{prob}(\hat{\theta} \neq i \mid \theta = i).$$

Since the columns of the detection probability matrix D add up to one, we can express the error probabilities as

$$P_i^{\mathrm{e}} = \sum_{j \neq i} D_{ji}.$$

7.3.3   Optimal detector design

In this section we show that a wide variety of objectives for detector design are linear, affine, or convex piecewise-linear functions of D, and therefore also of T (which is the optimization variable). Similarly, a variety of constraints for detector design can be expressed in terms of linear inequalities in D. It follows that a wide variety of optimal detector design problems can be expressed as LPs. We will see in §7.3.4 that some of these LPs have simple solutions; in this section we simply formulate the problem.

Limits on errors and detection probabilities

We can impose a lower bound on the probability of correctly detecting the jth hypothesis,

$$P_j^{\mathrm{d}} = D_{jj} \geq L_j,$$

which is a linear inequality in D (hence, T). Similarly, we can impose a maximum allowable probability for mistaking $\theta = i$ for $\theta = j$:

$$D_{ij} \leq U_{ij},$$

which are also linear constraints on T. We can take any of the detection probabilities as an objective to be maximized, or any of the error probabilities as an objective to be minimized.

Minimax detector design

We can take as objective (to be minimized) the *minimax error probability*, $\max_j P_j^{\mathrm{e}}$, which is a piecewise-linear convex function of D (hence, also of T). With this as the only objective, we have the problem of minimizing the maximum probability of detection error,

$$\begin{array}{ll} \text{minimize} & \max_j P_j^{\mathrm{e}} \\ \text{subject to} & t_k \succeq 0, \quad \mathbf{1}^T t_k = 1, \quad k = 1, \ldots, n, \end{array}$$

where the variables are $t_1, \ldots, t_n \in \mathbf{R}^m$. This can be reformulated as an LP. The minimax detector minimizes the worst-case (largest) probability of error over all m hypotheses.

We can, of course, add further constraints to the minimax detector design problem.

Bayes detector design

In Bayes detector design, we have a prior distribution for the hypotheses, given by $q \in \mathbf{R}^m$, where

$$q_i = \mathbf{prob}(\theta = i).$$

In this case, the probabilities p_{ij} are interpreted as conditional probabilities of X, given θ. The probability of error for the detector is then given by $q^T P^{\mathrm{e}}$, which is an affine function of T. The Bayes optimal detector is the solution of the LP

$$\begin{array}{ll} \text{minimize} & q^T P^{\mathrm{e}} \\ \text{subject to} & t_k \succeq 0, \quad \mathbf{1}^T t_k = 1, \quad k = 1, \ldots, n. \end{array}$$

We will see in §7.3.4 that this problem has a simple analytical solution.

One special case is when $q = (1/m)\mathbf{1}$. In this case the Bayes optimal detector minimizes the average probability of error, where the (unweighted) average is over the hypotheses. In §7.3.4 we will see that the maximum likelihood detector (7.10) is optimal for this problem.

Bias, mean-square error, and other quantities

In this section we assume that the ordering of the values of θ have some significance, *i.e.*, that the value $\theta = i$ can be interpreted as a larger value of the parameter than $\theta = j$, when $i > j$. This might be the case, for example, when $\theta = i$ corresponds to the hypothesis that i events have occurred. Here we may be interested in quantities such as

$$\mathbf{prob}(\hat{\theta} > \theta \mid \theta = i),$$

which is the probability that we overestimate θ when $\theta = i$. This is an affine function of D:

$$\mathbf{prob}(\hat{\theta} > \theta \mid \theta = i) = \sum_{j > i} D_{ji},$$

so a maximum allowable value for this probability can be expressed as a linear inequality on D (hence, T). As another example, the probability of misclassifying θ by more than one, when $\theta = i$,

$$\mathbf{prob}(|\hat{\theta} - \theta| > 1 \mid \theta = i) = \sum_{|j-i|>1} D_{ji},$$

is also a linear function of D.

We now suppose that the parameters have values $\{\theta_1, \ldots, \theta_m\} \subseteq \mathbf{R}$. The estimation or detection (parameter) error is then given by $\hat{\theta} - \theta$, and a number of quantities of interest are given by linear functions of D. Examples include:

- *Bias.* The bias of the detector, when $\theta = \theta_i$, is given by the linear function

$$\mathbf{E}_i(\hat{\theta} - \theta) = \sum_{j=1}^m (\theta_j - \theta_i) D_{ji},$$

  where the subscript on $\mathbf{E}$ means the expectation is with respect to the distribution of the hypothesis $\theta = \theta_i$.

- *Mean square error.* The mean square error of the detector, when $\theta = \theta_i$, is given by the linear function

$$\mathbf{E}_i(\hat{\theta} - \theta)^2 = \sum_{j=1}^m (\theta_j - \theta_i)^2 D_{ji}.$$

- *Average absolute error.* The average absolute error of the detector, when $\theta = \theta_i$, is given by the linear function

$$\mathbf{E}_i |\hat{\theta} - \theta| = \sum_{j=1}^m |\theta_j - \theta_i| D_{ji}.$$

7.3.4   Multicriterion formulation and scalarization

The optimal detector design problem can be considered a multicriterion problem, with the constraints (7.11), and the $m(m-1)$ objectives given by the off-diagonal entries of D, which are the probabilities of the different types of detection error:

$$\begin{array}{lll} \text{minimize (w.r.t. } \mathbf{R}_+^{m(m-1)}) & D_{ij}, \quad i, \ j = 1, \ldots, m, \quad i \neq j \\ \text{subject to} & t_k \succeq 0, \quad \mathbf{1}^T t_k = 1, \quad k = 1, \ldots, n, \end{array} \qquad (7.12)$$

with variables $t_1, \ldots, t_n \in \mathbf{R}^m$. Since each objective D_{ij} is a linear function of the variables, this is a multicriterion linear program.

We can scalarize this multicriterion problem by forming the weighted sum objective

$$\sum_{i,j=1}^m W_{ij} D_{ij} = \mathbf{tr}(W^T D)$$

where the weight matrix $W \in \mathbf{R}^{m \times m}$ satisfies

$$W_{ii} = 0, \quad i = 1, \ldots, m, \qquad W_{ij} > 0, \quad i, j = 1, \ldots, m, \quad i \neq j.$$

This objective is a weighted sum of the $m(m-1)$ error probabilities, with weight W_{ij} associated with the error of guessing $\hat{\theta} = i$ when in fact $\theta = j$. The weight matrix is sometimes called the *loss matrix*.

To find a Pareto optimal point for the multicriterion problem (7.12), we form the scalar optimization problem

$$
\begin{array}{ll}
\text{minimize} & \mathbf{tr}(W^T D) \\
\text{subject to} & t_k \succeq 0, \quad \mathbf{1}^T t_k = 1, \quad k = 1, \ldots, n,
\end{array}
\tag{7.13}
$$

which is an LP. This LP is separable in the variables $t_1, \ldots, t_n$. The objective can be expressed as a sum of (linear) functions of t_k:

$$\mathbf{tr}(W^T D) = \mathbf{tr}(W^T T P) = \mathbf{tr}(P W^T T) = \sum_{k=1}^{n} c_k^T t_k,$$

where c_k is the kth column of $W P^T$. The constraints are separable (*i.e.*, we have separate constraints on each t_i). Therefore we can solve the LP (7.13) by separately solving

$$
\begin{array}{ll}
\text{minimize} & c_k^T t_k \\
\text{subject to} & t_k \succeq 0, \quad \mathbf{1}^T t_k = 1,
\end{array}
$$

for $k = 1, \ldots, n$. Each of these LPs has a simple analytical solution (see exercise 4.8). We first find an index q such that $c_{kq} = \min_j c_{kj}$. Then we take $t_k^\star = e_q$. This optimal point corresponds to a deterministic detector: when $X = k$ is observed, our estimate is

$$\hat{\theta} = \operatorname*{argmin}_{j} (W P^T)_{jk}.
\tag{7.14}$$

Thus, for every weight matrix W with positive off-diagonal elements we can find a deterministic detector that minimizes the weighted sum objective. This seems to suggest that randomized detectors are not needed, but we will see this is not the case. The Pareto optimal trade-off surface for the multicriterion LP (7.12) is piecewise-linear; the deterministic detectors of the form (7.14) correspond to the vertices on the Pareto optimal surface.

MAP and ML detectors

Consider a Bayes detector design with prior distribution q. The mean probability of error is

$$q^T P^{\mathrm{e}} = \sum_{j=1}^{m} q_j \sum_{i \neq j} D_{ij} = \sum_{i,j=1}^{m} W_{ij} D_{ij},$$

if we define the weight matrix W as

$$W_{ij} = q_j, \quad i, j = 1, \ldots, m, \quad i \neq j, \qquad W_{ii} = 0, \quad i = 1, \ldots, m.$$

Thus, a Bayes optimal detector is given by the deterministic detector (7.14), with

$$(WP^T)_{jk} = \sum_{i \neq j} q_i p_{ki} = \sum_{i=1}^{m} q_i p_{ki} - q_j p_{kj}.$$

The first term is independent of j, so the optimal detector is simply

$$\hat{\theta} = \operatorname*{argmax}_{j}(p_{kj}q_j),$$

when $X = k$ is observed. The solution has a simple interpretation: Since $p_{kj}q_j$ gives the probability that $\theta = j$ and $X = k$, this detector is a maximum a posteriori probability (MAP) detector.

For the special case $q = (1/m)\mathbf{1}$, *i.e.*, a uniform prior distribution on θ, this MAP detector reduces to a maximum likelihood (ML) detector:

$$\hat{\theta} = \operatorname*{argmax}_{j} p_{kj}.$$

Thus, a maximum likelihood detector minimizes the (unweighted) average or mean probability of error.

7.3.5 Binary hypothesis testing

As an illustration, we consider the special case $m = 2$, which is called *binary hypothesis testing*. The random variable X is generated from one of two distributions, which we denote $p \in \mathbf{R}^n$ and $q \in \mathbf{R}^n$, to simplify the notation. Often the hypothesis $\theta = 1$ corresponds to some normal situation, and the hypothesis $\theta = 2$ corresponds to some abnormal event that we are trying to detect. If $\hat{\theta} = 1$, we say the test is negative (*i.e.*, we guess that the event did not occur); if $\hat{\theta} = 2$, we say the test is positive (*i.e.*, we guess that the event did occur).

The detection probability matrix $D \in \mathbf{R}^{2 \times 2}$ is traditionally expressed as

$$D = \begin{bmatrix} 1 - P_{\mathrm{fp}} & P_{\mathrm{fn}} \\ P_{\mathrm{fp}} & 1 - P_{\mathrm{fn}} \end{bmatrix}.$$

Here P_{fn} is the probability of a *false negative* (*i.e.*, the test is negative when in fact the event has occurred) and P_{fp} is the probability of a *false positive* (*i.e.*, the test is positive when in fact the event has not occurred), which is also called the *false alarm probability*. The optimal detector design problem is a bi-criterion problem, with objectives P_{fn} and P_{fp}.

The optimal trade-off curve between P_{fn} and P_{fp} is called the *receiver operating characteristic* (ROC), and is determined by the distributions p and q. The ROC can be found by scalarizing the bi-criterion problem, as described in §7.3.4. For the weight matrix W, an optimal detector (7.14) is

$$\hat{\theta} = \begin{cases} 1 & W_{21}p_k > W_{12}q_k \\ 2 & W_{21}p_k \leq W_{12}q_k \end{cases}$$

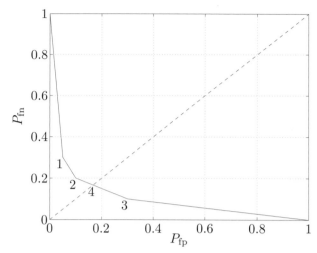

Figure 7.4 Optimal trade-off curve between probability of a false negative, and probability of a false positive test result, for the matrix P given in (7.15). The vertices of the trade-off curve, labeled 1–3, correspond to deterministic detectors; the point labeled 4, which is a randomized detector, is the minimax detector. The dashed line shows $P_{\mathrm{fn}} = P_{\mathrm{fp}}$, the points where the error probabilities are equal.

when $X = k$ is observed. This is called a *likelihood ratio threshold test*: if the ratio p_k/q_k is more than the threshold W_{12}/W_{21}, the test is negative (*i.e.*, $\hat\theta = 1$); otherwise the test is positive. By choosing different values of the threshold, we obtain (deterministic) Pareto optimal detectors that give different levels of false positive versus false negative error probabilities. This result is known as the Neyman-Pearson lemma.

The likelihood ratio detectors do not give all the Pareto optimal detectors; they are the vertices of the optimal trade-off curve, which is piecewise-linear.

Example 7.4 We consider a binary hypothesis testing example with $n = 4$, and

$$P = \begin{bmatrix} 0.70 & 0.10 \\ 0.20 & 0.10 \\ 0.05 & 0.70 \\ 0.05 & 0.10 \end{bmatrix}. \tag{7.15}$$

The optimal trade-off curve between P_{fn} and P_{fp}, *i.e.*, the receiver operating curve, is shown in figure 7.4. The left endpoint corresponds to the detector which is always negative, independent of the observed value of X; the right endpoint corresponds to the detector that is always positive. The vertices labeled 1, 2, and 3 correspond to the deterministic detectors

$$T^{(1)} = \begin{bmatrix} 1 & 1 & 0 & 1 \\ 0 & 0 & 1 & 0 \end{bmatrix},$$

$$T^{(2)} = \begin{bmatrix} 1 & 1 & 0 & 0 \\ 0 & 0 & 1 & 1 \end{bmatrix},$$

$$T^{(3)} = \begin{bmatrix} 1 & 0 & 0 & 0 \\ 0 & 1 & 1 & 1 \end{bmatrix},$$

respectively. The point labeled 4 corresponds to the nondeterministic detector

$$T^{(4)} = \begin{bmatrix} 1 & 2/3 & 0 & 0 \\ 0 & 1/3 & 1 & 1 \end{bmatrix},$$

which is the minimax detector. This minimax detector yields equal probability of a false positive and false negative, which in this case is $1/6$. Every deterministic detector has either a false positive or false negative probability that exceeds $1/6$, so this is an example where a randomized detector outperforms every deterministic detector.

7.3.6   Robust detectors

So far we have assumed that P, which gives the distribution of the observed variable X, for each value of the parameter θ, is known. In this section we consider the case where these distributions are not known, but certain prior information about them is given. We assume that $P \in \mathcal{P}$, where $\mathcal{P}$ is the set of possible distributions. With a randomized detector characterized by T, the detection probability matrix D now depends on the particular value of P. We will judge the error probabilities by their worst-case values, over $P \in \mathcal{P}$. We define the *worst-case detection probability matrix* D^{wc} as

$$D_{ij}^{\mathrm{wc}} = \sup_{P \in \mathcal{P}} D_{ij}, \quad i, j = 1, \ldots, m, \quad i \neq j$$

and

$$D_{ii}^{\mathrm{wc}} = \inf_{P \in \mathcal{P}} D_{ii}, \quad i = 1, \ldots, m.$$

The off-diagonal entries give the largest possible probability of errors, and the diagonal entries give the smallest possible probability of detection, over $P \in \mathcal{P}$. Note that $\sum_{i=1}^{n} D_{ij}^{\mathrm{wc}} \neq 1$ in general, *i.e.*, the columns of a worst-case detection probability matrix do not necessarily add up to one.

We define the worst-case probability of error as

$$P_i^{\mathrm{wce}} = 1 - D_{ii}^{\mathrm{wc}}.$$

Thus, P_i^{wce} is the largest probability of error, when $\theta = i$, over all possible distributions in $\mathcal{P}$.

Using the worst-case detection probability matrix, or the worst-case probability of error vector, we can develop various robust versions of detector design problems. In the rest of this section we concentrate on the robust minimax detector design problem, as a generic example that illustrates the ideas.

We define the *robust minimax detector* as the detector that minimizes the worst-case probability of error, over all hypotheses, *i.e.*, minimizes the objective

$$\max_i P_i^{\mathrm{wce}} = \max_{i=1,\ldots,m} \sup_{P \in \mathcal{P}} (1 - (TP)_{ii}) = 1 - \min_{i=1,\ldots,m} \inf_{P \in \mathcal{P}} (TP)_{ii}.$$

The robust minimax detector minimizes the worst possible probability of error, over all m hypotheses, and over all $P \in \mathcal{P}$.

Robust minimax detector for finite $\mathcal{P}$

When the set of possible distributions is finite, the robust minimax detector design problem is readily formulated as an LP. With $\mathcal{P} = \{P_1, \ldots, P_k\}$, we can find the robust minimax detector by solving

$$
\begin{aligned}
\text{maximize} \quad & \min_{i=1,\ldots,m} \inf_{P \in \mathcal{P}} (TP)_{ii} = \min_{i=1,\ldots,m} \min_{j=1,\ldots,k} (TP_j)_{ii} \\
\text{subject to} \quad & t_i \succeq 0, \quad \mathbf{1}^T t_i = 1, \quad i = 1, \ldots, n,
\end{aligned}
$$

The objective is piecewise-linear and concave, so this problem can be expressed as an LP. Note that we can just as well consider $\mathcal{P}$ to be the polyhedron $\mathbf{conv}\,\mathcal{P}$; the associated worst-case detection matrix, and robust minimax detector, are the same.

Robust minimax detector for polyhedral $\mathcal{P}$

It is also possible to efficiently formulate the robust minimax detector problem as an LP when $\mathcal{P}$ is a polyhedron described by linear equality and inequality constraints. This formulation is less obvious, and relies on a dual representation of $\mathcal{P}$.

To simplify the discussion, we assume that $\mathcal{P}$ has the form

$$
\mathcal{P} = \left\{ P = [p_1 \ \cdots \ p_m] \mid A_k p_k = b_k, \ \mathbf{1}^T p_k = 1, \ p_k \succeq 0 \right\}. \tag{7.16}
$$

In other words, for each distribution p_k, we are given some expected values $A_k p_k = b_k$. (These might represent known moments, probabilities, etc.) The extension to the case where we are given inequalities on expected values is straightforward.

The robust minimax design problem is

$$
\begin{aligned}
\text{maximize} \quad & \gamma \\
\text{subject to} \quad & \inf\{\tilde{t}_i^T p \mid A_i p = b_i, \ \mathbf{1}^T p = 1, \ p \succeq 0\} \geq \gamma, \quad i = 1, \ldots, m \\
& t_i \succeq 0, \quad \mathbf{1}^T t_i = 1, \quad i = 1, \ldots, n,
\end{aligned}
$$

where $\tilde{t}_i^T$ denotes the ith row of T (so that $(TP)_{ii} = \tilde{t}_i^T p_i$). By LP duality,

$$
\inf\{\tilde{t}_i^T p \mid A_i p = b_i, \ \mathbf{1}^T p = 1, \ p \succeq 0\} = \sup\{\nu^T b_i + \mu \mid A_i^T \nu + \mu \mathbf{1} \preceq \tilde{t}_i\}.
$$

Using this, the robust minimax detector design problem can be expressed as the LP

$$
\begin{aligned}
\text{maximize} \quad & \gamma \\
\text{subject to} \quad & \nu_i^T b_i + \mu_i \geq \gamma, \quad i = 1, \ldots, m \\
& A_i^T \nu_i + \mu_i \mathbf{1} \preceq \tilde{t}_i, \quad i = 1, \ldots, m \\
& t_i \succeq 0, \quad \mathbf{1}^T t_i = 1, \quad i = 1, \ldots, n,
\end{aligned}
$$

with variables $\nu_1, \ldots, \nu_m$, $\mu_1, \ldots, \mu_n$, and T (which has columns t_i and rows $\tilde{t}_i^T$).

Example 7.5 *Robust binary hypothesis testing.* Suppose $m = 2$ and the set $\mathcal{P}$ in (7.16) is defined by

$$
A_1 = A_2 = A = \begin{bmatrix} a_1 & a_2 & \cdots & a_n \\ a_1^2 & a_2^2 & \cdots & a_n^2 \end{bmatrix}, \qquad b_1 = \begin{bmatrix} \alpha_1 \\ \alpha_2 \end{bmatrix}, \qquad b_2 = \begin{bmatrix} \beta_1 \\ \beta_2 \end{bmatrix}.
$$

Designing a robust minimax detector for this set $\mathcal{P}$ can be interpreted as a binary hypothesis testing problem: based on an observation of a random variable $X \in \{a_1, \ldots, a_n\}$, choose between the following two hypotheses:

1. $\mathbf{E}\,X = \alpha_1$, $\mathbf{E}\,X^2 = \alpha_2$

2. $\mathbf{E}\,X = \beta_1$, $\mathbf{E}\,X^2 = \beta_2$.

Let $\tilde{t}^T$ denote the first row of T (and so, $(\mathbf{1}-\tilde{t})^T$ is the second row). For given $\tilde{t}$, the worst-case probabilities of correct detection are

$$
D_{11}^{\mathrm{wc}} = \inf\left\{\tilde{t}^T p \;\middle|\; \sum_{i=1}^n a_i p_i = \alpha_1,\; \sum_{i=1}^n a_i^2 p_i = \alpha_2,\; \mathbf{1}^T p = 1,\; p \succeq 0\right\}
$$

$$
D_{22}^{\mathrm{wc}} = \inf\left\{(\mathbf{1}-\tilde{t})^T p \;\middle|\; \sum_{i=1}^n a_i p_i = \beta_1,\; \sum_{i=1}^n a_i^2 p_i = \beta_2,\; \mathbf{1}^T p = 1,\; p \succeq 0\right\}.
$$

Using LP duality we can express D_{11}^{wc} as the optimal value of the LP

$$
\begin{array}{ll}
\text{maximize} & z_0 + z_1\alpha_1 + z_2\alpha_2 \\
\text{subject to} & z_0 + a_i z_1 + a_i^2 z_2 \le \tilde{t}_i, \quad i = 1,\dots,n,
\end{array}
$$

with variables $z_0, z_1, z_2 \in \mathbf{R}$. Similarly D_{22}^{wc} is the optimal value of the LP

$$
\begin{array}{ll}
\text{maximize} & w_0 + w_1\beta_1 + w_2\beta_2 \\
\text{subject to} & w_0 + a_i w_1 + a_i^2 w_2 \le 1-\tilde{t}_i, \quad i = 1,\dots,n,
\end{array}
$$

with variables $w_0, w_1, w_2 \in \mathbf{R}$. To obtain the minimax detector, we have to maximize the minimum of D_{11}^{wc} and D_{22}^{wc}, $i.e.$, solve the LP

$$
\begin{array}{ll}
\text{maximize} & \gamma \\
\text{subject to} & z_0 + z_1\alpha_2 + z_2\alpha_2 \ge \gamma \\
& w_0 + \beta_1 w_1 + \beta_2 w_2 \ge \gamma \\
& z_0 + z_1 a_i + z_2 a_i^2 \le \tilde{t}_i, \quad i = 1,\dots,n \\
& w_0 + w_1 a_i + w_2 a_i^2 \le 1-\tilde{t}_i, \quad i = 1,\dots,n \\
& 0 \preceq \tilde{t} \preceq \mathbf{1}.
\end{array}
$$

The variables are $z_0, z_1, z_2, w_0, w_1, w_2$ and $\tilde{t}$.

7.4   Chebyshev and Chernoff bounds

In this section we consider two types of classical bounds on the probability of a set, and show that generalizations of each can be cast as convex optimization problems. The original classical bounds correspond to simple convex optimization problems with analytical solutions; the convex optimization formulation of the general cases allow us to compute better bounds, or bounds for more complex situations.

7.4.1   Chebyshev bounds

Chebyshev bounds give an upper bound on the probability of a set based on known expected values of certain functions ($e.g.$, mean and variance). The simplest example is Markov's inequality: If X is a random variable on $\mathbf{R}_+$ with $\mathbf{E}\,X = \mu$, then we have $\mathbf{prob}(X \geq 1) \leq \mu$, no matter what the distribution of X is. Another simple example is Chebyshev's bound: If X is a random variable on $\mathbf{R}$ with $\mathbf{E}\,X = \mu$ and $\mathbf{E}(X - \mu)^2 = \sigma^2$, then we have $\mathbf{prob}(|X - \mu| \geq 1) \leq \sigma^2$, again no matter what the distribution of X is. The idea behind these simple bounds can be generalized to a setting in which convex optimization is used to compute a bound on the probability.

Let X be a random variable on $S \subseteq \mathbf{R}^m$, and $C \subseteq S$ be the set for which we want to bound $\mathbf{prob}(X \in C)$. Let 1_C denote the 0-1 indicator function of the set C, i.e., $1_C(z) = 1$ if $z \in C$ and $1_C(z) = 0$ if $z \notin C$.

Our prior knowledge of the distribution consists of known expected values of some functions:

$$\mathbf{E}\,f_i(X) = a_i, \quad i = 1, \dots, n,$$

where $f_i : \mathbf{R}^m \to \mathbf{R}$. We take f_0 to be the constant function with value one, for which we always have $\mathbf{E}\,f_0(X) = a_0 = 1$. Consider a linear combination of the functions f_i, given by

$$f(z) = \sum_{i=0}^{n} x_i f_i(z),$$

where $x_i \in \mathbf{R}$, $i = 0, \dots, n$. From our knowledge of $\mathbf{E}\,f_i(X)$, we have $\mathbf{E}\,f(X) = a^T x$.

Now suppose that f satisfies the condition $f(z) \geq 1_C(z)$ for all $z \in S$, i.e., f is pointwise greater than or equal to the indicator function of C (on S). Then we have

$$\mathbf{E}\,f(X) = a^T x \geq \mathbf{E}\,1_C(X) = \mathbf{prob}(X \in C).$$

In other words, $a^T x$ is an upper bound on $\mathbf{prob}(X \in C)$, valid for all distributions supported on S, with $\mathbf{E}\,f_i(X) = a_i$.

We can search for the best such upper bound on $\mathbf{prob}(X \in C)$, by solving the problem

$$
\begin{array}{ll}
\text{minimize} & x_0 + a_1 x_1 + \cdots + a_n x_n \\
\text{subject to} & f(z) = \sum_{i=0}^{n} x_i f_i(z) \geq 1 \text{ for } z \in C \\
& f(z) = \sum_{i=0}^{n} x_i f_i(z) \geq 0 \text{ for } z \in S, \ z \notin C,
\end{array}
\qquad (7.17)
$$

with variable $x \in \mathbf{R}^{n+1}$. This problem is always convex, since the constraints can be expressed as

$$g_1(x) = 1 - \inf_{z \in C} f(z) \leq 0, \qquad g_2(x) = -\inf_{z \in S \setminus C} f(z) \leq 0$$

(g_1 and g_2 are convex). The problem (7.17) can also be thought of as a semi-infinite linear program, i.e., an optimization problem with a linear objective and an infinite number of linear inequalities, one for each $z \in S$.

In simple cases we can solve the problem (7.17) analytically. As an example, we take $S = \mathbf{R}_+$, $C = [1, \infty)$, $f_0(z) = 1$, and $f_1(z) = z$, with $\mathbf{E}\,f_1(X) = \mathbf{E}\,X = \mu \leq 1$ as our prior information. The constraint $f(z) \geq 0$ for $z \in S$ reduces to $x_0 \geq 0$, $x_1 \geq 0$. The constraint $f(z) \geq 1$ for $z \in C$, i.e., $x_0 + x_1 z \geq 1$ for all $z \geq 1$, reduces to $x_0 + x_1 \geq 1$. The problem (7.17) is then

$$
\begin{array}{ll}
\text{minimize} & x_0 + \mu x_1 \\
\text{subject to} & x_0 \geq 0, \quad x_1 \geq 0 \\
& x_0 + x_1 \geq 1.
\end{array}
$$

Since $0 \leq \mu \leq 1$, the optimal point for this simple LP is $x_0 = 0$, $x_1 = 1$. This gives the classical Markov bound $\mathbf{prob}(X \geq 1) \leq \mu$.

In other cases we can solve the problem (7.17) using convex optimization.

Remark 7.1 *Duality and the Chebyshev bound problem.* The Chebyshev bound problem (7.17) determines a bound on $\mathbf{prob}(X \in C)$ for all probability measures that satisfy the given expected value constraints. Thus we can think of the Chebyshev bound problem (7.17) as producing a bound on the optimal value of the infinite-dimensional problem

$$
\begin{array}{ll}
\text{maximize} & \int_C \pi(dz) \\
\text{subject to} & \int_S f_i(z)\pi(dz) = a_i, \quad i = 1, \dots, n \\
& \int_S \pi(dz) = 1 \\
& \pi \geq 0,
\end{array}
\tag{7.18}
$$

where the variable is the measure π, and $\pi \geq 0$ means that the measure is nonnegative.

Since the Chebyshev problem (7.17) produces a bound on the problem (7.18), it should not be a surprise that they are related by duality. While semi-infinite and infinite-dimensional problems are beyond the scope of this book, we can still formally construct a dual of the problem (7.17), introducing a Lagrange multiplier *function* $p : S \to \mathbf{R}$, with $p(z)$ the Lagrange multiplier associated with the inequality $f(z) \geq 1$ (for $z \in C$) or $f(z) \geq 0$ (for $z \in S \backslash C$). Using an integral over z where we would have a sum in the finite-dimensional case, we arrive at the formal dual

$$
\begin{array}{ll}
\text{maximize} & \int_C p(z)\, dz \\
\text{subject to} & \int_S f_i(z)p(z)\, dz = a_i, \quad i = 1, \dots, n \\
& \int_S p(z)\, dz = 1 \\
& p(z) \geq 0 \text{ for all } z \in S,
\end{array}
$$

where the optimization variable is the *function* p. This is, essentially, the same as (7.18).

Probability bounds with known first and second moments

As an example, suppose that $S = \mathbf{R}^m$, and that we are given the first and second moments of the random variable X:

$$
\mathbf{E}\, X = a \in \mathbf{R}^m, \qquad \mathbf{E}\, X X^T = \Sigma \in \mathbf{S}^m.
$$

In other words, we are given the expected value of the m functions z_i, $i = 1, \dots, m$, and the $m(m+1)/2$ functions $z_i z_j$, $i, j = 1, \dots, m$, but no other information about the distribution.

In this case we can express f as the general quadratic function

$$
f(z) = z^T P z + 2q^T z + r,
$$

where the variables (*i.e.*, the vector x in the discussion above) are $P \in \mathbf{S}^m$, $q \in \mathbf{R}^m$, and $r \in \mathbf{R}$. From our knowledge of the first and second moments, we find that

$$
\begin{aligned}
\mathbf{E}\, f(X) &= \mathbf{E}(X^T P X + 2q^T X + r) \\
&= \mathbf{E}\,\mathbf{tr}(P X X^T) + 2\,\mathbf{E}\, q^T X + r \\
&= \mathbf{tr}(\Sigma P) + 2q^T a + r.
\end{aligned}
$$

The constraint that $f(z) \geq 0$ for all z can be expressed as the linear matrix inequality

$$\begin{bmatrix} P & q \\ q^T & r \end{bmatrix} \succeq 0.$$

In particular, we have $P \succeq 0$.

Now suppose that the set C is the complement of an open polyhedron,

$$C = \mathbf{R}^m \setminus \mathcal{P}, \qquad \mathcal{P} = \{z \mid a_i^T z < b_i, \; i = 1, \ldots, k\}.$$

The condition that $f(z) \geq 1$ for all $z \in C$ is the same as requiring that

$$a_i^T z \geq b_i \implies z^T P z + 2q^T z + r \geq 1$$

for $i = 1, \ldots, k$. This, in turn, can be expressed as: there exist $\tau_1, \ldots, \tau_k \geq 0$ such that

$$\begin{bmatrix} P & q \\ q^T & r-1 \end{bmatrix} \succeq \tau_i \begin{bmatrix} 0 & a_i/2 \\ a_i^T/2 & -b_i \end{bmatrix}, \qquad i = 1, \ldots, k.$$

(See §B.2.)

Putting it all together, the Chebyshev bound problem (7.17) can be expressed as

$$
\begin{aligned}
\text{minimize} \quad & \mathbf{tr}(\Sigma P) + 2q^T a + r \\
\text{subject to} \quad & \begin{bmatrix} P & q \\ q^T & r-1 \end{bmatrix} \succeq \tau_i \begin{bmatrix} 0 & a_i/2 \\ a_i^T/2 & -b_i \end{bmatrix}, \quad i = 1, \ldots, k \\
& \tau_i \geq 0, \quad i = 1, \ldots, k \\
& \begin{bmatrix} P & q \\ q^T & r \end{bmatrix} \succeq 0,
\end{aligned}
\tag{7.19}
$$

which is a semidefinite program in the variables P, q, r, and $\tau_1, \ldots, \tau_k$. The optimal value, say α, is an upper bound on $\mathbf{prob}(X \in C)$ over all distributions with mean a and second moment Σ. Or, turning it around, $1 - \alpha$ is a lower bound on $\mathbf{prob}(X \in \mathcal{P})$.

Remark 7.2 *Duality and the Chebyshev bound problem.* The dual SDP associated with (7.19) can be expressed as

$$
\begin{aligned}
\text{maximize} \quad & \sum_{i=1}^{k} \lambda_i \\
\text{subject to} \quad & a_i^T z_i \geq b \lambda_i, \quad i = 1, \ldots, k \\
& \sum_{i=1}^{k} \begin{bmatrix} Z_i & z_i \\ z_i^T & \lambda_i \end{bmatrix} \preceq \begin{bmatrix} \Sigma & a \\ a^T & 1 \end{bmatrix} \\
& \begin{bmatrix} Z_i & z_i \\ z_i^T & \lambda_i \end{bmatrix} \succeq 0, \quad i = 1, \ldots, k.
\end{aligned}
$$

The variables are $Z_i \in \mathbf{S}^m$, $z_i \in \mathbf{R}^m$, and $\lambda_i \in \mathbf{R}$, for $i = 1, \ldots, k$. Since the SDP (7.19) is strictly feasible, strong duality holds and the dual optimum is attained.

We can give an interesting probability interpretation to the dual problem. Suppose Z_i, z_i, λ_i are dual feasible and that the first r components of λ are positive, and the rest are zero. For simplicity we also assume that $\sum_{i=1}^{k} \lambda_i < 1$. We define

$$
\begin{aligned}
x_i &= (1/\lambda_i) z_i, \quad i = 1, \ldots, r, \\
w_0 &= \frac{1}{\mu} \left(a - \sum_{i=1}^{r} \lambda_i x_i \right), \\
W &= \frac{1}{\mu} \left(\Sigma - \sum_{i=1}^{r} \lambda_i x_i x_i^T \right),
\end{aligned}
$$

where $\mu = 1 - \sum_{i=1}^{k} \lambda_i$. With these definitions the dual feasibility constraints can be expressed as

$$
a_i^T x_i \geq b_i, \quad i = 1, \ldots, r
$$

and

$$
\sum_{i=1}^{r} \lambda_i \begin{bmatrix} x_i x_i^T & x_i \\ x_i^T & 1 \end{bmatrix} + \mu \begin{bmatrix} W & w_0 \\ w_0^T & 1 \end{bmatrix} = \begin{bmatrix} \Sigma & a \\ a^T & 1 \end{bmatrix}.
$$

Moreover, from dual feasibility,

$$
\begin{aligned}
\mu \begin{bmatrix} W & w_0 \\ w_0^T & 1 \end{bmatrix}
&= \begin{bmatrix} \Sigma & a \\ a^T & 1 \end{bmatrix} - \sum_{i=1}^{r} \lambda_i \begin{bmatrix} x_i x_i^T & x_i \\ x_i^T & 1 \end{bmatrix} \\
&= \begin{bmatrix} \Sigma & a \\ a^T & 1 \end{bmatrix} - \sum_{i=1}^{r} \begin{bmatrix} (1/\lambda_i) z_i z_i^T & z_i \\ z_i^T & \lambda_i \end{bmatrix} \\
&\succeq \begin{bmatrix} \Sigma & a \\ a^T & 1 \end{bmatrix} - \sum_{i=1}^{r} \begin{bmatrix} Z_i & z_i \\ z_i^T & \lambda_i \end{bmatrix} \\
&\succeq 0.
\end{aligned}
$$

Therefore, $W \succeq w_0 w_0^T$, so it can be factored as $W - w_0 w_0^T = \sum_{i=1}^{s} w_i w_i^T$. Now consider a discrete random variable X with the following distribution. If $s \geq 1$, we take

$$
\begin{array}{lll}
X = x_i & \text{with probability } \lambda_i, & i = 1, \ldots, r \\
X = w_0 + \sqrt{s}\, w_i & \text{with probability } \mu/(2s), & i = 1, \ldots, s \\
X = w_0 - \sqrt{s}\, w_i & \text{with probability } \mu/(2s), & i = 1, \ldots, s.
\end{array}
$$

If $s = 0$, we take

$$
\begin{array}{ll}
X = x_i & \text{with probability } \lambda_i, \quad i = 1, \ldots, r \\
X = w_0 & \text{with probability } \mu.
\end{array}
$$

It is easily verified that $\mathbf{E}\, X = a$ and $\mathbf{E}\, X X^T = \Sigma$, i.e., the distribution matches the given moments. Furthermore, since $x_i \in C$,

$$
\mathbf{prob}(X \in C) \geq \sum_{i=1}^{r} \lambda_i.
$$

In particular, by applying this interpretation to the dual optimal solution, we can construct a distribution that satisfies the Chebyshev bound from (7.19) with equality, which shows that the Chebyshev bound is sharp for this case.

7.4.2   Chernoff bounds

Let X be a random variable on $\mathbf{R}$. The Chernoff bound states that

$$\mathbf{prob}(X \geq u) \leq \inf_{\lambda \geq 0} \mathbf{E}\, e^{\lambda(X-u)},$$

which can be expressed as

$$\log \mathbf{prob}(X \geq u) \leq \inf_{\lambda \geq 0}\{-\lambda u + \log \mathbf{E}\, e^{\lambda X}\}. \qquad (7.20)$$

Recall (from example 3.41, page 106) that the righthand term, $\log \mathbf{E}\, e^{\lambda X}$, is called the cumulant generating function of the distribution, and is always convex, so the function to be minimized is convex. The bound (7.20) is most useful in cases when the cumulant generating function has an analytical expression, and the minimization over λ can be carried out analytically.

For example, if X is Gaussian with zero mean and unit variance, the cumulant generating function is

$$\log \mathbf{E}\, e^{\lambda X} = \lambda^2/2,$$

and the infimum over $\lambda \geq 0$ of $-\lambda u + \lambda^2/2$ occurs with $\lambda = u$ (if $u \geq 0$), so the Chernoff bound is (for $u \geq 0$)

$$\mathbf{prob}(X \geq u) \leq e^{-u^2/2}.$$

The idea behind the Chernoff bound can be extended to a more general setting, in which convex optimization is used to compute a bound on the probability of a set in $\mathbf{R}^m$. Let $C \subseteq \mathbf{R}^m$, and as in the description of Chebyshev bounds above, let 1_C denote the 0-1 indicator function of C. We will derive an upper bound on $\mathbf{prob}(X \in C)$. (In principle we can compute $\mathbf{prob}(X \in C)$, for example by Monte Carlo simulation, or numerical integration, but either of these can be a daunting computational task, and neither method produces guaranteed bounds.)

Let $\lambda \in \mathbf{R}^m$ and $\mu \in \mathbf{R}$, and consider the function $f : \mathbf{R}^m \rightarrow \mathbf{R}$ given by

$$f(z) = e^{\lambda^T z + \mu}.$$

As in the development of Chebyshev bounds, if f satisfies $f(z) \geq 1_C(z)$ for all z, then we can conclude that

$$\mathbf{prob}(X \in C) = \mathbf{E}\, 1_C(X) \leq \mathbf{E}\, f(X).$$

Clearly we have $f(z) \geq 0$ for all z; to have $f(z) \geq 1$ for $z \in C$ is the same as $\lambda^T z + \mu \geq 0$ for all $z \in C$, i.e., $-\lambda^T z \leq \mu$ for all $z \in C$. Thus, if $-\lambda^T z \leq \mu$ for all $z \in C$, we have the bound

$$\mathbf{prob}(X \in C) \leq \mathbf{E}\exp(\lambda^T X + \mu),$$

or, taking logarithms,

$$\log \mathbf{prob}(X \in C) \leq \mu + \log \mathbf{E}\exp(\lambda^T X).$$

From this we obtain a general form of Chernoff's bound:

$$
\begin{aligned}
\log \mathbf{prob}(X \in C) \;\le\;& \inf\{\mu + \log \mathbf{E}\exp(\lambda^T X) \mid -\lambda^T z \le \mu \text{ for all } z \in C\} \\
=\;& \inf_\lambda \left(\sup_{z \in C}(-\lambda^T z) + \log \mathbf{E}\exp(\lambda^T X) \right) \\
=\;& \inf \left(S_C(-\lambda) + \log \mathbf{E}\exp(\lambda^T X) \right),
\end{aligned}
$$

where S_C is the support function of C. Note that the second term, $\log \mathbf{E}\exp(\lambda^T X)$, is the cumulant generating function of the distribution, and is always convex (see example 3.41, page 106). Evaluating this bound is, in general, a convex optimization problem.

Chernoff bound for a Gaussian variable on a polyhedron

As a specific example, suppose that X is a Gaussian random vector on $\mathbf{R}^m$ with zero mean and covariance I, so its cumulant generating function is

$$
\log \mathbf{E}\exp(\lambda^T X) = \lambda^T \lambda / 2.
$$

We take C to be a polyhedron described by inequalities:

$$
C = \{x \mid Ax \preceq b\},
$$

which we assume is nonempty.

For use in the Chernoff bound, we use a dual characterization of the support function S_C:

$$
\begin{aligned}
S_C(y) \;=\;& \sup\{y^T x \mid Ax \preceq b\} \\
=\;& -\inf\{-y^T x \mid Ax \preceq b\} \\
=\;& -\sup\{-b^T u \mid A^T u = y,\; u \succeq 0\} \\
=\;& \inf\{b^T u \mid A^T u = y,\; u \succeq 0\}
\end{aligned}
$$

where in the third line we use LP duality:

$$
\inf\{c^T x \mid Ax \preceq b\} = \sup\{-b^T u \mid A^T u + c = 0,\; u \succeq 0\}
$$

with $c = -y$. Using this expression for S_C in the Chernoff bound we obtain

$$
\begin{aligned}
\log \mathbf{prob}(X \in C) \;\le\;& \inf_\lambda \left(S_C(-\lambda) + \log \mathbf{E}\exp(\lambda^T X) \right) \\
=\;& \inf_\lambda \inf_u \{b^T u + \lambda^T \lambda / 2 \mid u \succeq 0,\; A^T u + \lambda = 0\}.
\end{aligned}
$$

Thus, the Chernoff bound on $\mathbf{prob}(X \in C)$ is the exponential of the optimal value of the QP

$$
\begin{array}{ll}
\text{minimize} & b^T u + \lambda^T \lambda / 2 \\
\text{subject to} & u \succeq 0, \quad A^T u + \lambda = 0,
\end{array}
\tag{7.21}
$$

where the variables are u and λ.

This problem has an interesting geometric interpretation. It is equivalent to

$$\text{minimize} \quad b^T u + (1/2)\|A^T u\|_2^2$$
$$\text{subject to} \quad u \succeq 0,$$

which is the dual of

$$\text{maximize} \quad -(1/2)\|x\|_2^2$$
$$\text{subject to} \quad Ax \preceq b.$$

In other words, the Chernoff bound is

$$\mathbf{prob}(X \in C) \leq \exp(-\mathbf{dist}(0, C)^2/2), \tag{7.22}$$

where $\mathbf{dist}(0, C)$ is the Euclidean distance of the origin to C.

Remark 7.3 The bound (7.22) can also be derived without using Chernoff's inequality. If the distance between 0 and C is d, then there is a halfspace $\mathcal{H} = \{z \mid a^T z \geq d\}$, with $\|a\|_2 = 1$, that contains C. The random variable $a^T X$ is $\mathcal{N}(0, 1)$, so

$$\mathbf{prob}(X \in C) \leq \mathbf{prob}(X \in \mathcal{H}) = \Phi(-d),$$

where Φ is the cumulative distribution function of a zero mean, unit variance Gaussian. Since $\Phi(-d) \leq e^{-d^2/2}$ for $d \geq 0$, this bound is at least as sharp as the Chernoff bound (7.22).

7.4.3   Example

In this section we illustrate the Chebyshev and Chernoff probability bounding methods with a detection example. We have a set of m possible symbols or signals $s \in \{s_1, s_2, \ldots, s_m\} \subseteq \mathbf{R}^n$, which is called the *signal constellation*. One of these signals is transmitted over a noisy channel. The received signal is $x = s + v$, where v is a noise, modeled as a random variable. We assume that $\mathbf{E}\,v = 0$ and $\mathbf{E}\,vv^T = \sigma^2 I$, i.e., the noise components $v_1, \ldots, v_n$ are zero mean, uncorrelated, and have variance σ^2. The receiver must estimate which signal was sent on the basis of the received signal $x = s + v$. The *minimum distance detector* chooses as estimate the symbol s_k closest (in Euclidean norm) to x. (If the noise v is Gaussian, then minimum distance decoding is the same as maximum likelihood decoding.)

If the signal s_k is transmitted, correct detection occurs if s_k is the estimate, given x. This occurs when the signal s_k is closer to x than the other signals, i.e.,

$$\|x - s_k\|_2 < \|x - s_j\|_2, \quad j \neq k.$$

Thus, correct detection of symbol s_k occurs if the random variable v satisfies the linear inequalities

$$2(s_j - s_k)^T(s_k + v) < \|s_j\|_2^2 - \|s_k\|_2^2, \quad j \neq k.$$

These inequalities define the *Voronoi region* V_k of s_k in the signal constellation, i.e., the set of points closer to s_k than any other signal in the constellation. The probability of correct detection of s_k is $\mathbf{prob}(s_k + v \in V_k)$.

Figure 7.5 shows a simple example with $m = 7$ signals, with dimension $n = 2$.

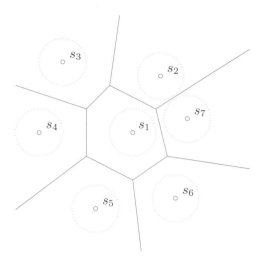

Figure 7.5 A constellation of 7 signals $s_1, \ldots, s_7 \in \mathbf{R}^2$, shown as small circles. The line segments show the boundaries of the corresponding Voronoi regions. The minimum distance detector selects symbol s_k when the received signal lies closer to s_k than to any of the other points, *i.e.*, if the received signal is in the interior of the Voronoi region around symbol s_k. The circles around each point have radius one, to show the scale.

Chebyshev bounds

The SDP bound (7.19) provides a lower bound on the probability of correct detection, and is plotted in figure 7.6, as a function of the noise standard deviation σ, for the three symbols s_1, s_2, and s_3. These bounds hold for *any* noise distribution with zero mean and covariance $\sigma^2 I$. They are tight in the sense that there exists a noise distribution with zero mean and covariance $\Sigma = \sigma^2 I$, for which the probability of error is equal to the lower bound. This is illustrated in figure 7.7, for the first Voronoi set, and $\sigma = 1$.

Chernoff bounds

We use the same example to illustrate the Chernoff bound. Here we assume that the noise is Gaussian, *i.e.*, $v \sim \mathcal{N}(0, \sigma^2 I)$. If symbol s_k is transmitted, the probability of correct detection is the probability that $s_k + v \in V_k$. To find a lower bound for this probability, we use the QP (7.21) to compute upper bounds on the probability that the ML detector selects symbol i, $i = 1, \ldots, m$, $i \neq k$. (Each of these upper bounds is related to the distance of s_k to the Voronoi set V_i.) Adding these upper bounds on the probabilities of mistaking s_k for s_i, we obtain an upper bound on the probability of error, and therefore, a lower bound on the probability of correct detection of symbol s_k. The resulting lower bound, for s_1, is shown in figure 7.8, along with an estimate of the probability of correct detection obtained using Monte Carlo analysis.

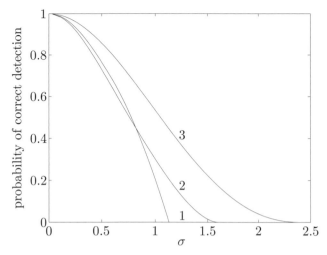

Figure 7.6 Chebyshev lower bounds on the probability of correct detection for symbols s_1, s_2, and s_3. These bounds are valid for any noise distribution that has zero mean and covariance $\sigma^2 I$.

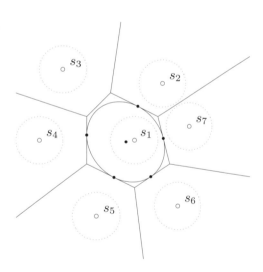

Figure 7.7 The Chebyshev lower bound on the probability of correct detection of symbol 1 is equal to 0.2048 when $\sigma = 1$. This bound is achieved by the discrete distribution illustrated in the figure. The solid circles are the possible values of the received signal $s_1 + v$. The point in the center of the ellipse has probability 0.2048. The five points on the boundary have a total probability 0.7952. The ellipse is defined by $x^T P x + 2q^T x + r = 1$, where P, q, and r are the optimal solution of the SDP (7.19).

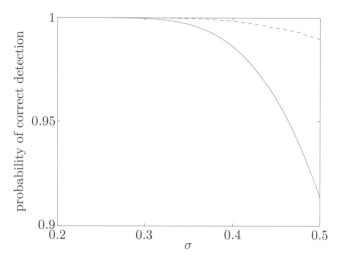

Figure 7.8 The Chernoff lower bound (solid line) and a Monte Carlo estimate (dashed line) of the probability of correct detection of symbol s_1, as a function of σ. In this example the noise is Gaussian with zero mean and covariance $\sigma^2 I$.

7.5   Experiment design

We consider the problem of estimating a vector $x \in \mathbf{R}^n$ from measurements or experiments

$$y_i = a_i^T x + w_i, \quad i = 1, \ldots, m,$$

where w_i is measurement noise. We assume that w_i are independent Gaussian random variables with zero mean and unit variance, and that the measurement vectors $a_1, \ldots, a_m$ span $\mathbf{R}^n$. The maximum likelihood estimate of x, which is the same as the minimum variance estimate, is given by the least-squares solution

$$\hat{x} = \left(\sum_{i=1}^m a_i a_i^T \right)^{-1} \sum_{i=1}^m y_i a_i.$$

The associated estimation error $e = \hat{x} - x$ has zero mean and covariance matrix

$$E = \mathbf{E}\, ee^T = \left(\sum_{i=1}^m a_i a_i^T \right)^{-1}.$$

The matrix E characterizes the accuracy of the estimation, or the informativeness of the experiments. For example the α-confidence level ellipsoid for x is given by

$$\mathcal{E} = \{ z \mid (z - \hat{x})^T E^{-1} (z - \hat{x}) \leq \beta \},$$

where β is a constant that depends on n and α.

   We suppose that the vectors $a_1, \ldots, a_m$, which characterize the measurements, can be chosen among p possible test vectors $v_1, \ldots, v_p \in \mathbf{R}^n$, *i.e.*, each a_i is one of the v_j. The goal of *experiment design* is to choose the vectors a_i, from among the possible choices, so that the error covariance E is small (in some sense). In other words, each of m experiments or measurements can be chosen from a fixed menu of p possible experiments; our job is to find a set of measurements that (together) are maximally informative.

Let m_j denote the number of experiments for which a_i is chosen to have the value v_j, so we have

$$m_1 + \cdots + m_p = m.$$

We can express the error covariance matrix as

$$E = \left(\sum_{i=1}^{m} a_i a_i^T \right)^{-1} = \left(\sum_{j=1}^{p} m_j v_j v_j^T \right)^{-1}.$$

This shows that the error covariance depends only on the numbers of each type of experiment chosen (*i.e.*, $m_1, \ldots, m_p$).

The basic experiment design problem is as follows. Given the menu of possible choices for experiments, *i.e.*, $v_1, \ldots, v_p$, and the total number m of experiments to be carried out, choose the numbers of each type of experiment, *i.e.*, $m_1, \ldots, m_p$, to make the error covariance E small (in some sense). The variables $m_1, \ldots, m_p$ must, of course, be integers and sum to m, the given total number of experiments. This leads to the optimization problem

$$
\begin{aligned}
\text{minimize (w.r.t. } \mathbf{S}_+^n) \quad & E = \left(\textstyle\sum_{j=1}^{p} m_j v_j v_j^T \right)^{-1} \\
\text{subject to} \quad & m_i \geq 0, \quad m_1 + \cdots + m_p = m \\
& m_i \in \mathbf{Z},
\end{aligned}
\tag{7.23}
$$

where the variables are the integers $m_1, \ldots, m_p$.

The basic experiment design problem (7.23) is a vector optimization problem over the positive semidefinite cone. If one experiment design results in E, and another in $\tilde{E}$, with $E \preceq \tilde{E}$, then certainly the first experiment design is as good as or better than the second. For example, the confidence ellipsoid for the first experiment design (translated to the origin for comparison) is contained in the confidence ellipsoid of the second. We can also say that the first experiment design allows us to estimate $q^T x$ better (*i.e.*, with lower variance) than the second experiment design, for any vector q, since the variance of our estimate of $q^T x$ is given by $q^T E q$ for the first experiment design and $q^T \tilde{E} q$ for the second. We will see below several common scalarizations for the problem.

7.5.1   The relaxed experiment design problem

The basic experiment design problem (7.23) can be a hard combinatorial problem when m, the total number of experiments, is comparable to n, since in this case the m_i are all small integers. In the case when m is large compared to n, however, a good approximate solution of (7.23) can be found by ignoring, or relaxing, the constraint that the m_i are integers. Let $\lambda_i = m_i/m$, which is the fraction of the total number of experiments for which $a_j = v_i$, or the relative frequency of experiment i. We can express the error covariance in terms of λ_i as

$$E = \frac{1}{m}\left(\sum_{i=1}^{p}\lambda_i v_i v_i^T\right)^{-1}. \qquad (7.24)$$

The vector $\lambda \in \mathbf{R}^p$ satisfies $\lambda \succeq 0$, $\mathbf{1}^T\lambda = 1$, and also, each λ_i is an integer multiple of $1/m$. By ignoring this last constraint, we arrive at the problem

$$\begin{array}{ll} \text{minimize (w.r.t. } \mathbf{S}_+^n) & E = (1/m)\left(\sum_{i=1}^{p}\lambda_i v_i v_i^T\right)^{-1} \\ \text{subject to} & \lambda \succeq 0, \quad \mathbf{1}^T\lambda = 1, \end{array} \qquad (7.25)$$

with variable $\lambda \in \mathbf{R}^p$. To distinguish this from the original combinatorial experiment design problem (7.23), we refer to it as the *relaxed experiment design problem*. The relaxed experiment design problem (7.25) is a convex optimization problem, since the objective E is an $\mathbf{S}_+^n$-convex function of λ.

Several statements can be made about the relation between the (combinatorial) experiment design problem (7.23) and the relaxed problem (7.25). Clearly the optimal value of the relaxed problem provides a lower bound on the optimal value of the combinatorial one, since the combinatorial problem has an additional constraint. From a solution of the relaxed problem (7.25) we can construct a suboptimal solution of the combinatorial problem (7.23) as follows. First, we apply simple rounding to get

$$m_i = \mathbf{round}(m\lambda_i), \quad i = 1,\dots,p.$$

Corresponding to this choice of $m_1,\dots,m_p$ is the vector $\tilde{\lambda}$,

$$\tilde{\lambda}_i = (1/m)\mathbf{round}(m\lambda_i), \quad i = 1,\dots,p.$$

The vector $\tilde{\lambda}$ satisfies the constraint that each entry is an integer multiple of $1/m$. Clearly we have $|\lambda_i - \tilde{\lambda}_i| \leq 1/(2m)$, so for m large, we have $\lambda \approx \tilde{\lambda}$. This implies that the constraint $\mathbf{1}^T\tilde{\lambda} = 1$ is nearly satisfied, for large m, and also that the error covariance matrices associated with $\tilde{\lambda}$ and λ are close.

We can also give an alternative interpretation of the relaxed experiment design problem (7.25). We can interpret the vector $\lambda \in \mathbf{R}^p$ as defining a probability distribution on the experiments $v_1,\dots,v_p$. Our choice of λ corresponds to a *random experiment*: each experiment a_i takes the form v_j with probability λ_j.

In the rest of this section, we consider only the relaxed experiment design problem, so we drop the qualifier 'relaxed' in our discussion.

7.5.2 Scalarizations

Several scalarizations have been proposed for the experiment design problem (7.25), which is a vector optimization problem over the positive semidefinite cone.

D-optimal design

The most widely used scalarization is called D-*optimal design*, in which we minimize the determinant of the error covariance matrix E. This corresponds to designing the experiment to minimize the volume of the resulting confidence ellipsoid (for a fixed confidence level). Ignoring the constant factor $1/m$ in E, and taking the logarithm of the objective, we can pose this problem as

$$
\begin{array}{ll}
\text{minimize} & \log \det \left(\sum_{i=1}^{p} \lambda_i v_i v_i^T \right)^{-1} \\
\text{subject to} & \lambda \succeq 0, \quad \mathbf{1}^T \lambda = 1,
\end{array}
\tag{7.26}
$$

which is a convex optimization problem.

E-optimal design

In E-*optimal design*, we minimize the norm of the error covariance matrix, *i.e.*, the maximum eigenvalue of E. Since the diameter (twice the longest semi-axis) of the confidence ellipsoid $\mathcal{E}$ is proportional to $\|E\|_2^{1/2}$, minimizing $\|E\|_2$ can be interpreted geometrically as minimizing the diameter of the confidence ellipsoid. E-optimal design can also be interpreted as minimizing the maximum variance of $q^T e$, over all q with $\|q\|_2 = 1$.

The E-optimal experiment design problem is

$$
\begin{array}{ll}
\text{minimize} & \left\| \left(\sum_{i=1}^{p} \lambda_i v_i v_i^T \right)^{-1} \right\|_2 \\
\text{subject to} & \lambda \succeq 0, \quad \mathbf{1}^T \lambda = 1.
\end{array}
$$

The objective is a convex function of λ, so this is a convex problem.

The E-optimal experiment design problem can be cast as an SDP

$$
\begin{array}{ll}
\text{maximize} & t \\
\text{subject to} & \sum_{i=1}^{p} \lambda_i v_i v_i^T \succeq tI \\
& \lambda \succeq 0, \quad \mathbf{1}^T \lambda = 1,
\end{array}
\tag{7.27}
$$

with variables $\lambda \in \mathbf{R}^p$ and $t \in \mathbf{R}$.

A-optimal design

In A-*optimal experiment design*, we minimize $\mathbf{tr}\, E$, the trace of the covariance matrix. This objective is simply the mean of the norm of the error squared:

$$
\mathbf{E} \, \|e\|_2^2 = \mathbf{E} \, \mathbf{tr}(ee^T) = \mathbf{tr}\, E.
$$

The A-optimal experiment design problem is

$$
\begin{array}{ll}
\text{minimize} & \mathbf{tr} \left(\sum_{i=1}^{p} \lambda_i v_i v_i^T \right)^{-1} \\
\text{subject to} & \lambda \succeq 0, \quad \mathbf{1}^T \lambda = 1.
\end{array}
\tag{7.28}
$$

This, too, is a convex problem. Like the E-optimal experiment design problem, it can be cast as an SDP:

$$
\begin{array}{ll}
\text{minimize} & \mathbf{1}^T u \\
\text{subject to} & \begin{bmatrix} \sum_{i=1}^{p} \lambda_i v_i v_i^T & e_k \\ e_k^T & u_k \end{bmatrix} \succeq 0, \quad k = 1, \dots, n \\
& \lambda \succeq 0, \quad \mathbf{1}^T \lambda = 1,
\end{array}
$$

where the variables are $u \in \mathbf{R}^n$ and $\lambda \in \mathbf{R}^p$, and here, e_k is the kth unit vector.

Optimal experiment design and duality

The Lagrange duals of the three scalarizations have an interesting geometric meaning.

The dual of the D-optimal experiment design problem (7.26) can be expressed as

$$
\begin{array}{ll}
\text{maximize} & \log \det W + n \log n \\
\text{subject to} & v_i^T W v_i \leq 1, \quad i = 1, \ldots, p,
\end{array}
$$

with variable $W \in \mathbf{S}^n$ and domain $\mathbf{S}_{++}^n$ (see exercise 5.10). This dual problem has a simple interpretation: The optimal solution $W^\star$ determines the minimum volume ellipsoid, centered at the origin, given by $\{x \mid x^T W^\star x \leq 1\}$, that contains the points $v_1, \ldots, v_p$. (See also the discussion of problem (5.14) on page 222.) By complementary slackness,

$$
\lambda_i^\star (1 - v_i^T W^\star v_i) = 0, \quad i = 1, \ldots, p, \tag{7.29}
$$

i.e., the optimal experiment design only uses the experiments v_i which lie on the surface of the minimum volume ellipsoid.

The duals of the E-optimal and A-optimal design problems can be given a similar interpretation. The duals of problems (7.27) and (7.28) can be expressed as

$$
\begin{array}{ll}
\text{maximize} & \mathbf{tr}\, W \\
\text{subject to} & v_i^T W v_i \leq 1, \quad i = 1, \ldots, p \\
& W \succeq 0,
\end{array} \tag{7.30}
$$

and

$$
\begin{array}{ll}
\text{maximize} & (\mathbf{tr}\, W^{1/2})^2 \\
\text{subject to} & v_i^T W v_i \leq 1, \quad i = 1, \ldots, p,
\end{array} \tag{7.31}
$$

respectively. The variable in both problems is $W \in \mathbf{S}^n$. In the second problem there is an implicit constraint $W \in \mathbf{S}_+^n$. (See exercises 5.40 and 5.10.)

As for the D-optimal design, the optimal solution $W^\star$ determines a minimal ellipsoid $\{x \mid x^T W^\star x \leq 1\}$ that contains the points $v_1, \ldots, v_p$. Moreover $W^\star$ and $\lambda^\star$ satisfy the complementary slackness conditions (7.29), *i.e.*, the optimal design only uses experiments v_i that lie on the surface of the ellipsoid defined by $W^\star$.

Experiment design example

We consider a problem with $x \in \mathbf{R}^2$, and $p = 20$. The 20 candidate measurement vectors a_i are shown as circles in figure 7.9. The origin is indicated with a cross. The D-optimal experiment has only two nonzero λ_i, indicated as solid circles in figure 7.9. The E-optimal experiment has two nonzero λ_i, indicated as solid circles in figure 7.10. The A-optimal experiment has three nonzero λ_i, indicated as solid circles in figure 7.11. We also show the three ellipsoids $\{x \mid x^T W^\star x \leq 1\}$ associated with the dual optimal solutions $W^\star$. The resulting 90% confidence ellipsoids are shown in figure 7.12, along with the confidence ellipsoid for the 'uniform' design, with equal weight $\lambda_i = 1/p$ on all experiments.

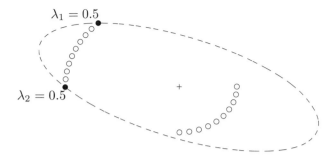

Figure 7.9 Experiment design example. The 20 candidate measurement vectors are indicated with circles. The D-optimal design uses the two measurement vectors indicated with solid circles, and puts an equal weight $\lambda_i = 0.5$ on each of them. The ellipsoid is the minimum volume ellipsoid centered at the origin, that contains the points v_i.

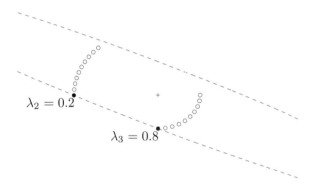

Figure 7.10 The E-optimal design uses two measurement vectors. The dashed lines are (part of) the boundary of the ellipsoid $\{x \mid x^T W^\star x \le 1\}$ where $W^\star$ is the solution of the dual problem (7.30).

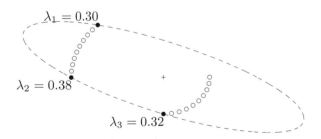

Figure 7.11 The A-optimal design uses three measurement vectors. The dashed line shows the ellipsoid $\{x \mid x^T W^\star x \le 1\}$ associated with the solution of the dual problem (7.31).

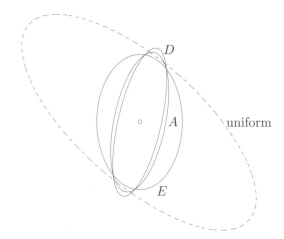

Figure 7.12 Shape of the 90% confidence ellipsoids for D-optimal, A-optimal, E-optimal, and uniform designs.

7.5.3   Extensions

Resource limits

Suppose that associated with each experiment is a cost c_i, which could represent the economic cost, or time required, to carry out an experiment with v_i. The total cost, or time required (if the experiments are carried out sequentially) is then

$$m_1 c_1 + \cdots + m_p c_p = mc^T \lambda.$$

We can add a limit on total cost by adding the linear inequality $mc^T \lambda \leq B$, where B is a budget, to the basic experiment design problem. We can add multiple linear inequalities, representing limits on multiple resources.

Multiple measurements per experiment

We can also consider a generalization in which each experiment yields multiple measurements. In other words, when we carry out an experiment using one of the possible choices, we obtain several measurements. To model this situation we can use the same notation as before, with v_i as matrices in $\mathbf{R}^{n \times k_i}$:

$$v_i = \begin{bmatrix} u_{i1} & \cdots & u_{ik_i} \end{bmatrix},$$

where k_i is the number of (scalar) measurements obtained when the experiment v_i is carried out. The error covariance matrix, in this more complicated setup, has the exact same form.

In conjunction with additional linear inequalities representing limits on cost or time, we can model discounts or time savings associated with performing groups of measurements simultaneously. Suppose, for example, that the cost of simultaneously making (scalar) measurements v_1 and v_2 is less than the sum of the costs of making them separately. We can take v_3 to be the matrix

$$v_3 = \begin{bmatrix} v_1 & v_2 \end{bmatrix}$$

and assign costs c_1, c_2, and c_3 associated with making the first measurement alone, the second measurement alone, and the two simultaneously, respectively.

When we solve the experiment design problem, λ_1 will give us the fraction of times we should carry out the first experiment alone, λ_2 will give us the fraction of times we should carry out the second experiment alone, and λ_3 will give us the fraction of times we should carry out the two experiments simultaneously. (Normally we would expect a choice to be made here; we would not expect to have $\lambda_1 > 0$, $\lambda_2 > 0$, and $\lambda_3 > 0$.)

Bibliography

ML and MAP estimation, hypothesis testing, and detection are covered in books on statistics, pattern recognition, statistical signal processing, or communications; see, for example, Bickel and Doksum [BD77], Duda, Hart, and Stork [DHS99], Scharf [Sch91], or Proakis [Pro01].

Logistic regression is discussed in Hastie, Tibshirani, and Friedman [HTF01, §4.4]. For the covariance estimation problem of page 355, see Anderson [And70].

Generalizations of Chebyshev's inequality were studied extensively in the sixties, by Isii [Isi64], Marshall and Olkin [MO60], Karlin and Studden [KS66, chapter 12], and others. The connection with semidefinite programming was made more recently by Bertsimas and Sethuraman [BS00] and Lasserre [Las02].

The terminology in §7.5 (A-, D-, and E-optimality) is standard in the literature on optimal experiment design (see, for example, Pukelsheim [Puk93]). The geometric interpretation of the dual D-optimal design problem is discussed by Titterington [Tit75].

Exercises

Estimation

7.1 *Linear measurements with exponentially distributed noise.* Show how to solve the ML estimation problem (7.2) when the noise is exponentially distributed, with density

$$p(z) = \begin{cases} (1/a)e^{-z/a} & z \geq 0 \\ 0 & z < 0, \end{cases}$$

where $a > 0$.

7.2 *ML estimation and ℓ_∞-norm approximation.* We consider the linear measurement model $y = Ax + v$ of page 352, with a uniform noise distribution of the form

$$p(z) = \begin{cases} 1/(2\alpha) & |z| \leq \alpha \\ 0 & |z| > \alpha. \end{cases}$$

As mentioned in example 7.1, page 352, any x that satisfies $\|Ax - y\|_\infty \leq \alpha$ is a ML estimate.

Now assume that the parameter α is not known, and we wish to estimate α, along with the parameters x. Show that the ML estimates of x and α are found by solving the ℓ_∞-norm approximation problem

$$\text{minimize} \quad \|Ax - y\|_\infty,$$

where a_i^T are the rows of A.

7.3 *Probit model.* Suppose $y \in \{0, 1\}$ is random variable given by

$$y = \begin{cases} 1 & a^T u + b + v \leq 0 \\ 0 & a^T u + b + v > 0, \end{cases}$$

where the vector $u \in \mathbf{R}^n$ is a vector of explanatory variables (as in the logistic model described on page 354), and v is a zero mean unit variance Gaussian variable.

Formulate the ML estimation problem of estimating a and b, given data consisting of pairs (u_i, y_i), $i = 1, \ldots, N$, as a convex optimization problem.

7.4 *Estimation of covariance and mean of a multivariate normal distribution.* We consider the problem of estimating the covariance matrix R and the mean a of a Gaussian probability density function

$$p_{R,a}(y) = (2\pi)^{-n/2} \det(R)^{-1/2} \exp(-(y-a)^T R^{-1}(y-a)/2),$$

based on N independent samples $y_1, y_2, \ldots, y_N \in \mathbf{R}^n$.

(a) We first consider the estimation problem when there are no additional constraints on R and a. Let μ and Y be the sample mean and covariance, defined as

$$\mu = \frac{1}{N} \sum_{k=1}^{N} y_k, \qquad Y = \frac{1}{N} \sum_{k=1}^{N} (y_k - \mu)(y_k - \mu)^T.$$

Show that the log-likelihood function

$$l(R, a) = -(Nn/2)\log(2\pi) - (N/2)\log\det R - (1/2)\sum_{k=1}^{N}(y_k - a)^T R^{-1}(y_k - a)$$

can be expressed as

$$l(R,a) = \frac{N}{2}\left(-n\log(2\pi) - \log\det R - \mathbf{tr}(R^{-1}Y) - (a-\mu)^T R^{-1}(a-\mu)\right).$$

Use this expression to show that if $Y \succ 0$, the ML estimates of R and a are unique, and given by

$$a_{\mathrm{ml}} = \mu, \qquad R_{\mathrm{ml}} = Y.$$

(b) The log-likelihood function includes a convex term $(-\log\det R)$, so it is not obviously concave. Show that l is concave, jointly in R and a, in the region defined by

$$R \preceq 2Y.$$

This means we can use convex optimization to compute simultaneous ML estimates of R and a, subject to convex constraints, as long as the constraints include $R \preceq 2Y$, i.e., the estimate R must not exceed twice the unconstrained ML estimate.

7.5 *Markov chain estimation.* Consider a Markov chain with n states, and transition probability matrix $P \in \mathbf{R}^{n \times n}$ defined as

$$P_{ij} = \mathbf{prob}(y(t+1) = i \mid y(t) = j).$$

The transition probabilities must satisfy $P_{ij} \geq 0$ and $\sum_{i=1}^{n} P_{ij} = 1$, $j = 1, \ldots, n$. We consider the problem of estimating the transition probabilities, given an observed sample sequence $y(1) = k_1$, $y(2) = k_2$, $\ldots$, $y(N) = k_n$.

(a) Show that if there are no other prior constraints on P_{ij}, then the ML estimates are the empirical transition frequencies: $\hat{P}_{ij}$ is the ratio of the number of times the state transitioned from j into i, divided by the number of times it was j, in the observed sample.

(b) Suppose that an equilibrium distribution p of the Markov chain is known, i.e., a vector $q \in \mathbf{R}^n_+$ satisfying $\mathbf{1}^T q = 1$ and $Pq = q$. Show that the problem of computing the ML estimate of P, given the observed sequence and knowledge of q, can be expressed as a convex optimization problem.

7.6 *Estimation of mean and variance.* Consider a random variable $x \in \mathbf{R}$ with density p, which is normalized, i.e., has zero mean and unit variance. Consider a random variable $y = (x+b)/a$ obtained by an affine transformation of x, where $a > 0$. The random variable y has mean b and variance $1/a^2$. As a and b vary over $\mathbf{R}_+$ and $\mathbf{R}$, respectively, we generate a family of densities obtained from p by scaling and shifting, uniquely parametrized by mean and variance.

Show that if p is log-concave, then finding the ML estimate of a and b, given samples $y_1, \ldots, y_n$ of y, is a convex problem.

As an example, work out an analytical solution for the ML estimates of a and b, assuming p is a normalized Laplacian density, $p(x) = e^{-2|x|}$.

7.7 *ML estimation of Poisson distributions.* Suppose x_i, $i = 1, \ldots, n$, are independent random variables with Poisson distributions

$$\mathbf{prob}(x_i = k) = \frac{e^{-\mu_i}\mu_i^k}{k!},$$

with unknown means μ_i. The variables x_i represent the number of times that one of n possible independent events occurs during a certain period. In emission tomography, for example, they might represent the number of photons emitted by n sources.

We consider an experiment designed to determine the means μ_i. The experiment involves m detectors. If event i occurs, it is detected by detector j with probability p_{ji}. We assume the probabilities p_{ji} are given (with $p_{ji} \geq 0$, $\sum_{j=1}^{m} p_{ji} \leq 1$). The total number of events recorded by detector j is denoted y_j,

$$y_j = \sum_{i=1}^{n} y_{ji}, \quad j = 1, \ldots, m.$$

Formulate the ML estimation problem of estimating the means μ_i, based on observed values of y_j, $j = 1, \ldots, m$, as a convex optimization problem.

Hint. The variables y_{ji} have Poisson distributions with means $p_{ji}\mu_i$, *i.e.*,

$$\mathbf{prob}(y_{ji} = k) = \frac{e^{-p_{ji}\mu_i}(p_{ji}\mu_i)^k}{k!}.$$

The sum of n independent Poisson variables with means $\lambda_1, \ldots, \lambda_n$ has a Poisson distribution with mean $\lambda_1 + \cdots + \lambda_n$.

7.8 *Estimation using sign measurements.* We consider the measurement setup

$$y_i = \mathbf{sign}(a_i^T x + b_i + v_i), \quad i = 1, \ldots, m,$$

where $x \in \mathbf{R}^n$ is the vector to be estimated, and $y_i \in \{-1, 1\}$ are the measurements. The vectors $a_i \in \mathbf{R}^n$ and scalars $b_i \in \mathbf{R}$ are known, and v_i are IID noises with a log-concave probability density. (You can assume that $a_i^T x + b_i + v_i = 0$ does not occur.) Show that maximum likelihood estimation of x is a convex optimization problem.

7.9 *Estimation with unknown sensor nonlinearity.* We consider the measurement setup

$$y_i = f(a_i^T x + b_i + v_i), \quad i = 1, \ldots, m,$$

where $x \in \mathbf{R}^n$ is the vector to be estimated, $y_i \in \mathbf{R}$ are the measurements, $a_i \in \mathbf{R}^n$, $b_i \in \mathbf{R}$ are known, and v_i are IID noises with log-concave probability density. The function $f : \mathbf{R} \to \mathbf{R}$, which represents a measurement nonlinearity, is *not* known. However, it is known that $f'(t) \in [l, u]$ for all t, where $0 < l < u$ are given.

Explain how to use convex optimization to find a maximum likelihood estimate of x, as well as the function f. (This is an infinite-dimensional ML estimation problem, but you can be informal in your approach and explanation.)

7.10 *Nonparametric distributions on $\mathbf{R}^k$.* We consider a random variable $x \in \mathbf{R}^k$ with values in a finite set $\{\alpha_1, \ldots, \alpha_n\}$, and with distribution

$$p_i = \mathbf{prob}(x = \alpha_i), \quad i = 1, \ldots, n.$$

Show that a lower bound on the covariance of X,

$$S \preceq \mathbf{E}(X - \mathbf{E}\,X)(X - \mathbf{E}\,X)^T,$$

is a convex constraint in p.

Optimal detector design

7.11 *Randomized detectors.* Show that every randomized detector can be expressed as a convex combination of a set of deterministic detectors: If

$$T = \begin{bmatrix} t_1 & t_2 & \cdots & t_n \end{bmatrix} \in \mathbf{R}^{m \times n}$$

satisfies $t_k \succeq 0$ and $\mathbf{1}^T t_k = 1$, then T can be expressed as

$$T = \theta_1 T_1 + \cdots + \theta_N T_N,$$

where T_i is a zero-one matrix with exactly one element equal to one per column, and $\theta_i \geq 0$, $\sum_{i=1}^{N} \theta_i = 1$. What is the maximum number of deterministic detectors N we may need?

We can interpret this convex decomposition as follows. The randomized detector can be realized as a bank of N deterministic detectors. When we observe $X = k$, the estimator chooses a random index from the set $\{1, \ldots, N\}$, with probability $\mathbf{prob}(j = i) = \theta_i$, and then uses deterministic detector T_j.

7.12 *Optimal action.* In detector design, we are given a matrix $P \in \mathbf{R}^{n \times m}$ (whose columns are probability distributions), and then design a matrix $T \in \mathbf{R}^{m \times n}$ (whose columns are probability distributions), so that $D = TP$ has large diagonal elements (and small off-diagonal elements). In this problem we study the dual problem: Given P, find a matrix $S \in \mathbf{R}^{m \times n}$ (whose columns are probability distributions), so that $\tilde{D} = PS \in \mathbf{R}^{n \times n}$ has large diagonal elements (and small off-diagonal elements). To make the problem specific, we take the objective to be maximizing the minimum element of $\tilde{D}$ on the diagonal.

We can interpret this problem as follows. There are n *outcomes*, which depend (stochastically) on which of m inputs or *actions* we take: P_{ij} is the probability that outcome i occurs, given action j. Our goal is find a (randomized) strategy that, to the extent possible, causes any specified outcome to occur. The strategy is given by the matrix S: S_{ji} is the probability that we take action j, when we want outcome i to occur. The matrix $\tilde{D}$ gives the action error probability matrix: $\tilde{D}_{ij}$ is the probability that outcome i occurs, when we want outcome j to occur. In particular, $\tilde{D}_{ii}$ is the probability that outcome i occurs, when we want it to occur.

Show that this problem has a simple analytical solution. Show that (unlike the corresponding detector problem) there is always an optimal solution that is deterministic.

Hint. Show that the problem is separable in the columns of S.

Chebyshev and Chernoff bounds

7.13 *Chebyshev-type inequalities on a finite set.* Assume X is a random variable taking values in the set $\{\alpha_1, \alpha_2, \ldots, \alpha_m\}$, and let S be a subset of $\{\alpha_1, \ldots, \alpha_m\}$. The distribution of X is unknown, but we are given the expected values of n functions f_i:

$$\mathbf{E} \, f_i(X) = b_i, \quad i = 1, \ldots, n. \tag{7.32}$$

Show that the optimal value of the LP

$$
\begin{array}{ll}
\text{minimize} & x_0 + \sum_{i=1}^{n} b_i x_i \\
\text{subject to} & x_0 + \sum_{i=1}^{n} f_i(\alpha) x_i \geq 1, \quad \alpha \in S \\
& x_0 + \sum_{i=1}^{n} f_i(\alpha) x_i \geq 0, \quad \alpha \not\in S,
\end{array}
$$

with variables $x_0, \ldots, x_n$, is an upper bound on $\mathbf{prob}(X \in S)$, valid for all distributions that satisfy (7.32). Show that there always exists a distribution that achieves the upper bound.

Chapter 8

Geometric problems

8.1 Projection on a set

The *distance* of a point $x_0 \in \mathbf{R}^n$ to a closed set $C \subseteq \mathbf{R}^n$, in the norm $\|\cdot\|$, is defined as

$$\mathbf{dist}(x_0, C) = \inf\{\|x_0 - x\| \mid x \in C\}.$$

The infimum here is always achieved. We refer to any point $z \in C$ which is closest to x_0, *i.e.*, satisfies $\|z - x_0\| = \mathbf{dist}(x_0, C)$, as a *projection* of x_0 on C. In general there can be more than one projection of x_0 on C, *i.e.*, several points in C closest to x_0.

In some special cases we can establish that the projection of a point on a set is unique. For example, if C is closed and convex, and the norm is strictly convex (*e.g.*, the Euclidean norm), then for any x_0 there is always exactly one $z \in C$ which is closest to x_0. As an interesting converse, we have the following result: If for every x_0 there is a unique Euclidean projection of x_0 on C, then C is closed and convex (see exercise 8.2).

We use the notation $P_C : \mathbf{R}^n \to \mathbf{R}^n$ to denote any function for which $P_C(x_0)$ is a projection of x_0 on C, *i.e.*, for all x_0,

$$P_C(x_0) \in C, \qquad \|x_0 - P_C(x_0)\| = \mathbf{dist}(x_0, C).$$

In other words, we have

$$P_C(x_0) = \operatorname{argmin}\{\|x - x_0\| \mid x \in C\}.$$

We refer to P_C as *projection on C*.

Example 8.1 *Projection on the unit square in $\mathbf{R}^2$*. Consider the (boundary of the) unit square in $\mathbf{R}^2$, *i.e.*, $C = \{x \in \mathbf{R}^2 \mid \|x\|_\infty = 1\}$. We take $x_0 = 0$.

In the ℓ_1-norm, the four points $(1,0)$, $(0,-1)$, $(-1,0)$, and $(0,1)$ are closest to $x_0 = 0$, with distance 1, so we have $\mathbf{dist}(x_0, C) = 1$ in the ℓ_1-norm. The same statement holds for the ℓ_2-norm.

In the ℓ_∞-norm, all points in C lie at a distance 1 from x_0, and $\mathbf{dist}(x_0, C) = 1$.

Example 8.2 *Projection onto rank-k matrices.* Consider the set of $m \times n$ matrices with rank less than or equal to k,

$$C = \{X \in \mathbf{R}^{m \times n} \mid \mathbf{rank}\, X \leq k\},$$

with $k \leq \min\{m, n\}$, and let $X_0 \in \mathbf{R}^{m \times n}$. We can find a projection of X_0 on C, in the (spectral or maximum singular value) norm $\|\cdot\|_2$, via the singular value decomposition. Let

$$X_0 = \sum_{i=1}^{r} \sigma_i u_i v_i^T$$

be the singular value decomposition of X_0, where $r = \mathbf{rank}\, X_0$. Then the matrix $Y = \sum_{i=1}^{\min\{k,r\}} \sigma_i u_i v_i^T$ is a projection of X_0 on C.

8.1.1   Projecting a point on a convex set

If C is convex, then we can compute the projection $P_C(x_0)$ and the distance $\mathbf{dist}(x_0, C)$ by solving a convex optimization problem. We represent the set C by a set of linear equalities and convex inequalities

$$Ax = b, \qquad f_i(x) \leq 0, \quad i = 1, \ldots, m, \tag{8.1}$$

and find the projection of x_0 on C by solving the problem

$$
\begin{array}{ll}
\text{minimize} & \|x - x_0\| \\
\text{subject to} & f_i(x) \leq 0, \quad i = 1, \ldots, m \\
& Ax = b,
\end{array}
\tag{8.2}
$$

with variable x. This problem is feasible if and only if C is nonempty; when it is feasible, its optimal value is $\mathbf{dist}(x_0, C)$, and any optimal point is a projection of x_0 on C.

Euclidean projection on a polyhedron

The projection of x_0 on a polyhedron described by linear inequalities $Ax \preceq b$ can be computed by solving the QP

$$
\begin{array}{ll}
\text{minimize} & \|x - x_0\|_2^2 \\
\text{subject to} & Ax \preceq b.
\end{array}
$$

Some special cases have simple analytical solutions.

- The Euclidean projection of x_0 on a hyperplane $C = \{x \mid a^T x = b\}$ is given by

$$P_C(x_0) = x_0 + (b - a^T x_0)a/\|a\|_2^2.$$

- The Euclidean projection of x_0 on a halfspace $C = \{x \mid a^T x \leq b\}$ is given by

$$P_C(x_0) = \begin{cases} x_0 + (b - a^T x_0)a/\|a\|_2^2 & a^T x_0 > b \\ x_0 & a^T x_0 \leq b. \end{cases}$$

- The Euclidean projection of x_0 on a rectangle $C = \{x \mid l \preceq x \preceq u\}$ (where $l \prec u$) is given by

$$P_C(x_0)_k = \begin{cases} l_k & x_{0k} \leq l_k \\ x_{0k} & l_k \leq x_{0k} \leq u_k \\ u_k & x_{0k} \geq u_k. \end{cases}$$

Euclidean projection on a proper cone

Let $x = P_K(x_0)$ denote the Euclidean projection of a point x_0 on a proper cone K. The KKT conditions of

$$\begin{array}{ll} \text{minimize} & \|x - x_0\|_2^2 \\ \text{subject to} & x \succeq_K 0 \end{array}$$

are given by

$$x \succeq_K 0, \qquad x - x_0 = z, \qquad z \succeq_{K^*} 0, \qquad z^T x = 0.$$

Introducing the notation $x_+ = x$ and $x_- = z$, we can express these conditions as

$$x_0 = x_+ - x_-, \qquad x_+ \succeq_K 0, \qquad x_- \succeq_{K^*} 0, \qquad x_+^T x_- = 0.$$

In other words, by projecting x_0 on the cone K, we decompose it into the difference of two orthogonal elements: one nonnegative with respect to K (and which is the projection of x_0 on K), and the other nonnegative with respect to K^*.

Some specific examples:

- For $K = \mathbf{R}_+^n$, we have $P_K(x_0)_k = \max\{x_{0k}, 0\}$. The Euclidean projection of a vector onto the nonnegative orthant is found by replacing each negative component with 0.

- For $K = \mathbf{S}_+^n$, and the Euclidean (or Frobenius) norm $\|\cdot\|_F$, we have $P_K(X_0) = \sum_{i=1}^n \max\{0, \lambda_i\} v_i v_i^T$, where $X_0 = \sum_{i=1}^n \lambda_i v_i v_i^T$ is the eigenvalue decomposition of X_0. To project a symmetric matrix onto the positive semidefinite cone, we form its eigenvalue expansion and drop terms associated with negative eigenvalues. This matrix is also the projection onto the positive semidefinite cone in the ℓ_2-, or spectral norm.

8.1.2   Separating a point and a convex set

Suppose C is a closed convex set described by the equalities and inequalities (8.1). If $x_0 \in C$, then $\mathbf{dist}(x_0, C) = 0$, and the optimal point for the problem (8.2) is x_0. If $x_0 \notin C$ then $\mathbf{dist}(x_0, C) > 0$, and the optimal value of the problem (8.2) is positive. In this case we will see that any dual optimal point provides a separating hyperplane between the point x_0 and the set C.

The link between projecting a point on a convex set and finding a hyperplane that separates them (when the point is not in the set) should not be surprising. Indeed, our proof of the separating hyperplane theorem, given in §2.5.1, relies on

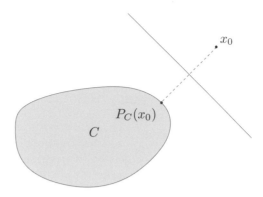

Figure 8.1 A point x_0 and its Euclidean projection $P_C(x_0)$ on a convex set C. The hyperplane midway between the two, with normal vector $P_C(x_0) - x_0$, strictly separates the point and the set. This property does not hold for general norms; see exercise 8.4.

finding the Euclidean distance between the sets. If $P_C(x_0)$ denotes the Euclidean projection of x_0 on C, where $x_0 \notin C$, then the hyperplane

$$(P_C(x_0) - x_0)^T (x - (1/2)(x_0 + P_C(x_0))) = 0$$

(strictly) separates x_0 from C, as illustrated in figure 8.1. In other norms, however, the clearest link between the projection problem and the separating hyperplane problem is via Lagrange duality.

We first express (8.2) as

$$\begin{array}{ll} \text{minimize} & \|y\| \\ \text{subject to} & f_i(x) \le 0, \quad i = 1, \dots, m \\ & Ax = b \\ & x_0 - x = y \end{array}$$

with variables x and y. The Lagrangian of this problem is

$$L(x, y, \lambda, \mu, \nu) = \|y\| + \sum_{i=1}^{m} \lambda_i f_i(x) + \nu^T (Ax - b) + \mu^T (x_0 - x - y)$$

and the dual function is

$$g(\lambda, \mu, \nu) = \begin{cases} \inf_x \left(\sum_{i=1}^{m} \lambda_i f_i(x) + \nu^T (Ax - b) + \mu^T (x_0 - x) \right) & \|\mu\|_* \le 1 \\ -\infty & \text{otherwise,} \end{cases}$$

so we obtain the dual problem

$$\begin{array}{ll} \text{maximize} & \mu^T x_0 + \inf_x \left(\sum_{i=1}^{m} \lambda_i f_i(x) + \nu^T (Ax - b) - \mu^T x \right) \\ \text{subject to} & \lambda \succeq 0 \\ & \|\mu\|_* \le 1, \end{array}$$

with variables λ, μ, ν. We can interpret the dual problem as follows. Suppose λ, μ, ν are dual feasible with a positive dual objective value, *i.e.*, $\lambda \succeq 0$, $\|\mu\|_* \le 1$, and

$$\mu^T x_0 - \mu^T x + \sum_{i=1}^m \lambda_i f_i(x) + \nu^T (Ax - b) > 0$$

for all x. This implies that $\mu^T x_0 > \mu^T x$ for $x \in C$, and therefore μ defines a strictly separating hyperplane. In particular, suppose (8.2) is strictly feasible, so strong duality holds. If $x_0 \notin C$, the optimal value is positive, and any dual optimal solution defines a strictly separating hyperplane.

Note that this construction of a separating hyperplane, via duality, works for any norm. In contrast, the simple construction described above only works for the Euclidean norm.

Separating a point from a polyhedron

The dual problem of

$$\begin{array}{ll} \text{minimize} & \|y\| \\ \text{subject to} & Ax \preceq b \\ & x_0 - x = y \end{array}$$

is

$$\begin{array}{ll} \text{maximize} & \mu^T x_0 - b^T \lambda \\ \text{subject to} & A^T \lambda = \mu \\ & \|\mu\|_* \leq 1 \\ & \lambda \succeq 0 \end{array}$$

which can be further simplified as

$$\begin{array}{ll} \text{maximize} & (Ax_0 - b)^T \lambda \\ \text{subject to} & \|A^T \lambda\|_* \leq 1 \\ & \lambda \succeq 0. \end{array}$$

It is easily verified that if the dual objective is positive, then $A^T \lambda$ is the normal vector to a separating hyperplane: If $Ax \preceq b$, then

$$(A^T \lambda)^T x = \lambda^T (Ax) \leq \lambda^T b < \lambda^T Ax_0,$$

so $\mu = A^T \lambda$ defines a separating hyperplane.

8.1.3   Projection and separation via indicator and support functions

The ideas described above in §8.1.1 and §8.1.2 can be expressed in a compact form in terms of the indicator function I_C and the support function S_C of the set C, defined as

$$S_C(x) = \sup_{y \in C} x^T y, \qquad I_C(x) = \begin{cases} 0 & x \in C \\ +\infty & x \notin C. \end{cases}$$

The problem of projecting x_0 on a closed convex set C can be expressed compactly as

$$\begin{array}{ll} \text{minimize} & \|x - x_0\| \\ \text{subject to} & I_C(x) \leq 0, \end{array}$$

or, equivalently, as

$$
\begin{array}{ll}
\text{minimize} & \|y\| \\
\text{subject to} & I_C(x) \le 0 \\
& x_0 - x = y
\end{array}
$$

where the variables are x and y. The dual function of this problem is

$$
\begin{aligned}
g(z, \lambda) &= \inf_{x,y} \left(\|y\| + \lambda I_C(x) + z^T(x_0 - x - y) \right) \\
&= \begin{cases} z^T x_0 + \inf_x \left(-z^T x + I_C(x) \right) & \|z\|_* \le 1, \quad \lambda \ge 0 \\ -\infty & \text{otherwise} \end{cases} \\
&= \begin{cases} z^T x_0 - S_C(z) & \|z\|_* \le 1, \quad \lambda \ge 0 \\ -\infty & \text{otherwise} \end{cases}
\end{aligned}
$$

so we obtain the dual problem

$$
\begin{array}{ll}
\text{maximize} & z^T x_0 - S_C(z) \\
\text{subject to} & \|z\|_* \le 1.
\end{array}
$$

If z is dual optimal with a positive objective value, then $z^T x_0 > z^T x$ for all $x \in C$, i.e., z defines a separating hyperplane.

8.2   Distance between sets

The distance between two sets C and D, in a norm $\| \cdot \|$, is defined as

$$
\mathbf{dist}(C, D) = \inf\{\|x - y\| \mid x \in C, \ y \in D\}.
$$

The two sets C and D do not intersect if $\mathbf{dist}(C, D) > 0$. They intersect if $\mathbf{dist}(C, D) = 0$ and the infimum in the definition is attained (which is the case, for example, if the sets are closed and one of the sets is bounded).

The distance between sets can be expressed in terms of the distance between a point and a set,

$$
\mathbf{dist}(C, D) = \mathbf{dist}(0, D - C),
$$

so the results of the previous section can be applied. In this section, however, we derive results specifically for problems involving distance between sets. This allows us to exploit the structure of the set $C - D$, and makes the interpretation easier.

8.2.1   Computing the distance between convex sets

Suppose C and D are described by two sets of convex inequalities

$$
C = \{x \mid f_i(x) \le 0, \ i = 1, \dots, m\}, \qquad D = \{x \mid g_i(x) \le 0, \ i = 1, \dots, p\}.
$$

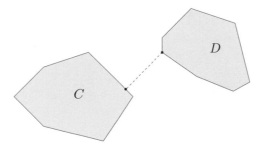

Figure 8.2 Euclidean distance between polyhedra C and D. The dashed line connects the two points in C and D, respectively, that are closest to each other in Euclidean norm. These points can be found by solving a QP.

(We can include linear equalities, but exclude them here for simplicity.) We can find $\mathbf{dist}(C, D)$ by solving the convex optimization problem

$$
\begin{array}{ll}
\text{minimize} & \|x - y\| \\
\text{subject to} & f_i(x) \le 0, \quad i = 1, \ldots, m \\
& g_i(y) \le 0, \quad i = 1, \ldots, p.
\end{array}
\tag{8.3}
$$

Euclidean distance between polyhedra

Let C and D be two polyhedra described by the sets of linear inequalities $A_1 x \preceq b_1$ and $A_2 x \preceq b_2$, respectively. The distance between C and D is the distance between the closest pair of points, one in C and the other in D, as illustrated in figure 8.2. The distance between them is the optimal value of the problem

$$
\begin{array}{ll}
\text{minimize} & \|x - y\|_2 \\
\text{subject to} & A_1 x \preceq b_1 \\
& A_2 y \preceq b_2.
\end{array}
\tag{8.4}
$$

We can square the objective to obtain an equivalent QP.

8.2.2   Separating convex sets

The dual of the problem (8.3) of finding the distance between two convex sets has an interesting geometric interpretation in terms of separating hyperplanes between the sets. We first express the problem in the following equivalent form:

$$
\begin{array}{ll}
\text{minimize} & \|w\| \\
\text{subject to} & f_i(x) \le 0, \quad i = 1, \ldots, m \\
& g_i(y) \le 0, \quad i = 1, \ldots, p \\
& x - y = w.
\end{array}
\tag{8.5}
$$

The dual function is

$$
g(\lambda, z, \mu) = \inf_{x,y,w} \left(\|w\| + \sum_{i=1}^{m} \lambda_i f_i(x) + \sum_{i=1}^{p} \mu_i g_i(y) + z^T (x - y - w) \right)
$$

$$= \begin{cases} \inf_x \left(\sum_{i=1}^m \lambda_i f_i(x) + z^T x\right) + \inf_y \left(\sum_{i=1}^p \mu_i g_i(y) - z^T y\right) & \|z\|_* \leq 1 \\ -\infty & \text{otherwise,} \end{cases}$$

which results in the dual problem

$$\begin{array}{ll} \text{maximize} & \inf_x \left(\sum_{i=1}^m \lambda_i f_i(x) + z^T x\right) + \inf_y \left(\sum_{i=1}^p \mu_i g_i(y) - z^T y\right) \\ \text{subject to} & \|z\|_* \leq 1 \\ & \lambda \succeq 0, \quad \mu \succeq 0. \end{array} \tag{8.6}$$

We can interpret this geometrically as follows. If λ, μ are dual feasible with a positive objective value, then

$$\sum_{i=1}^m \lambda_i f_i(x) + z^T x + \sum_{i=1}^p \mu_i g_i(y) - z^T y > 0$$

for all x and y. In particular, for $x \in C$ and $y \in D$, we have $z^T x - z^T y > 0$, so we see that z defines a hyperplane that strictly separates C and D.

Therefore, if strong duality holds between the two problems (8.5) and (8.6) (which is the case when (8.5) is strictly feasible), we can make the following conclusion. If the distance between the two sets is positive, then they can be strictly separated by a hyperplane.

Separating polyhedra

Applying these duality results to sets defined by linear inequalities $A_1 x \preceq b_1$ and $A_2 x \preceq b_2$, we find the dual problem

$$\begin{array}{ll} \text{maximize} & -b_1^T \lambda - b_2^T \mu \\ \text{subject to} & A_1^T \lambda + z = 0 \\ & A_2^T \mu - z = 0 \\ & \|z\|_* \leq 1 \\ & \lambda \succeq 0, \quad \mu \succeq 0. \end{array}$$

If λ, μ, and z are dual feasible, then for all $x \in C$, $y \in D$,

$$z^T x = -\lambda^T A_1 x \geq -\lambda^T b_1, \qquad z^T y = \mu^T A_2 x \leq \mu^T b_2,$$

and, if the dual objective value is positive,

$$z^T x - z^T y \geq -\lambda^T b_1 - \mu^T b_2 > 0,$$

i.e., z defines a separating hyperplane.

8.2.3   Distance and separation via indicator and support functions

The ideas described above in §8.2.1 and §8.2.2 can be expressed in a compact form using indicator and support functions. The problem of finding the distance between two convex sets can be posed as the convex problem

$$\begin{array}{ll} \text{minimize} & \|x - y\| \\ \text{subject to} & I_C(x) \leq 0 \\ & I_D(y) \leq 0, \end{array}$$

which is equivalent to

$$\begin{array}{ll} \text{minimize} & \|w\| \\ \text{subject to} & I_C(x) \le 0 \\ & I_D(y) \le 0 \\ & x - y = w. \end{array}$$

The dual of this problem is

$$\begin{array}{ll} \text{maximize} & -S_C(-z) - S_D(z) \\ \text{subject to} & \|z\|_* \le 1. \end{array}$$

If z is dual feasible with a positive objective value, then $S_D(z) < -S_C(-z)$, i.e.,

$$\sup_{x \in D} z^T x < \inf_{x \in C} z^T x.$$

In other words, z defines a hyperplane that strictly separates C and D.

8.3   Euclidean distance and angle problems

Suppose $a_1, \dots, a_n$ is a set of vectors in $\mathbf{R}^n$, which we assume (for now) have known Euclidean lengths

$$l_1 = \|a_1\|_2, \quad \dots, \quad l_n = \|a_n\|_2.$$

We will refer to the set of vectors as a *configuration*, or, when they are independent, a *basis*. In this section we consider optimization problems involving various geometric properties of the configuration, such as the Euclidean distances between pairs of the vectors, the angles between pairs of the vectors, and various geometric measures of the conditioning of the basis.

8.3.1   Gram matrix and realizability

The lengths, distances, and angles can be expressed in terms of the *Gram matrix* associated with the vectors $a_1, \dots, a_n$, given by

$$G = A^T A, \qquad A = \begin{bmatrix} a_1 & \cdots & a_n \end{bmatrix},$$

so that $G_{ij} = a_i^T a_j$. The diagonal entries of G are given by

$$G_{ii} = l_i^2, \quad i = 1, \dots, n,$$

which (for now) we assume are known and fixed. The distance d_{ij} between a_i and a_j is

$$\begin{aligned} d_{ij} &= \|a_i - a_j\|_2 \\ &= (l_i^2 + l_j^2 - 2a_i^T a_j)^{1/2} \\ &= (l_i^2 + l_j^2 - 2G_{ij})^{1/2}. \end{aligned}$$

Conversely, we can express G_{ij} in terms of d_{ij} as

$$G_{ij} = \frac{l_i^2 + l_j^2 - d_{ij}^2}{2},$$

which we note, for future reference, is an affine function of d_{ij}^2.

The *correlation coefficient* ρ_{ij} between (nonzero) a_i and a_j is given by

$$\rho_{ij} = \frac{a_i^T a_j}{\|a_i\|_2 \|a_j\|_2} = \frac{G_{ij}}{l_i l_j},$$

so that $G_{ij} = l_i l_j \rho_{ij}$ is a linear function of ρ_{ij}. The angle θ_{ij} between (nonzero) a_i and a_j is given by

$$\theta_{ij} = \cos^{-1} \rho_{ij} = \cos^{-1}(G_{ij}/(l_i l_j)),$$

where we take $\cos^{-1} \rho \in [0, \pi]$. Thus, we have $G_{ij} = l_i l_j \cos \theta_{ij}$.

The lengths, distances, and angles are invariant under orthogonal transformations: If $Q \in \mathbf{R}^{n \times n}$ is orthogonal, then the set of vectors $Qa_1, \ldots, Qa_n$ has the same Gram matrix, and therefore the same lengths, distances, and angles.

Realizability

The Gram matrix $G = A^T A$ is, of course, symmetric and positive semidefinite. The converse is a basic result of linear algebra: A matrix $G \in \mathbf{S}^n$ is the Gram matrix of a set of vectors $a_1, \ldots, a_n$ if and only if $G \succeq 0$. When $G \succeq 0$, we can construct a configuration with Gram matrix G by finding a matrix A with $A^T A = G$. One solution of this equation is the symmetric squareroot $A = G^{1/2}$. When $G \succ 0$, we can find a solution via the Cholesky factorization of G: If $LL^T = G$, then we can take $A = L^T$. Moreover, we can construct *all* configurations with the given Gram matrix G, given any one solution A, by orthogonal transformation: If $\tilde{A}^T \tilde{A} = G$ is *any* solution, then $\tilde{A} = QA$ for some orthogonal matrix Q.

Thus, a set of lengths, distances, and angles (or correlation coefficients) is *realizable*, *i.e.*, those of some configuration, if and only if the associated Gram matrix G is positive semidefinite, and has diagonal elements $l_1^2, \ldots, l_n^2$.

We can use this fact to express several geometric problems as convex optimization problems, with $G \in \mathbf{S}^n$ as the optimization variable. Realizability imposes the constraint $G \succeq 0$ and $G_{ii} = l_i^2$, $i = 1, \ldots, n$; we list below several other convex constraints and objectives.

Angle and distance constraints

We can fix an angle to have a certain value, $\theta_{ij} = \alpha$, via the linear equality constraint $G_{ij} = l_i l_j \cos \alpha$. More generally, we can impose a lower and upper bound on an angle, $\alpha \le \theta_{ij} \le \beta$, by the constraint

$$l_i l_j \cos \alpha \ge G_{ij} \ge l_i l_j \cos \beta,$$

which is a pair of linear inequalities on G. (Here we use the fact that $\cos^{-1}$ is monotone decreasing.) We can maximize or minimize a particular angle θ_{ij}, by minimizing or maximizing G_{ij} (again using monotonicity of $\cos^{-1}$).

In a similar way we can impose constraints on the distances. To require that d_{ij} lies in an interval, we use

$$d_{\min} \le d_{ij} \le d_{\max} \iff d_{\min}^2 \le d_{ij}^2 \le d_{\max}^2$$
$$\iff d_{\min}^2 \le l_i^2 + l_j^2 - 2G_{ij} \le d_{\max}^2,$$

which is a pair of linear inequalities on G. We can minimize or maximize a distance, by minimizing or maximizing its square, which is an affine function of G.

As a simple example, suppose we are given ranges (*i.e.*, an interval of possible values) for some of the angles and some of the distances. We can then find the minimum and maximum possible value of some other angle, or some other distance, over all configurations, by solving two SDPs. We can reconstruct the two extreme configurations by factoring the resulting optimal Gram matrices.

Singular value and condition number constraints

The singular values of A, $\sigma_1 \ge \cdots \ge \sigma_n$, are the squareroots of the eigenvalues $\lambda_1 \ge \cdots \ge \lambda_n$ of G. Therefore σ_1^2 is a convex function of G, and σ_n^2 is a concave function of G. Thus we can impose an upper bound on the maximum singular value of A, or minimize it; we can impose a lower bound on the minimum singular value, or maximize it. The condition number of A, σ_1/σ_n, is a quasiconvex function of G, so we can impose a maximum allowable value, or minimize it over all configurations that satisfy the other geometric constraints, by quasiconvex optimization.

Roughly speaking, the constraints we can impose as convex constraints on G are those that require $a_1, \ldots, a_n$ to be a well conditioned basis.

Dual basis

When $G \succ 0$, $a_1, \ldots, a_n$ form a basis for $\mathbf{R}^n$. The associated *dual basis* is $b_1, \ldots, b_n$, where

$$b_i^T a_j = \begin{cases} 1 & i = j \\ 0 & i \ne j. \end{cases}$$

The dual basis vectors $b_1, \ldots, b_n$ are simply the rows of the matrix A^{-1}. As a result, the Gram matrix associated with the dual basis is G^{-1}.

We can express several geometric conditions on the dual basis as convex constraints on G. The (squared) lengths of the dual basis vectors,

$$\|b_i\|_2^2 = e_i^T G^{-1} e_i,$$

are convex functions of G, and so can be minimized. The trace of G^{-1}, another convex function of G, gives the sum of the squares of the lengths of the dual basis vectors (and is another measure of a well conditioned basis).

Ellipsoid and simplex volume

The volume of the ellipsoid $\{Au \mid \|u\|_2 \le 1\}$, which gives another measure of how well conditioned the basis is, is given by

$$\gamma(\det(A^T A))^{1/2} = \gamma(\det G)^{1/2},$$

where γ is the volume of the unit ball in $\mathbf{R}^n$. The log volume is therefore $\log \gamma + (1/2) \log \det G$, which is a concave function of G. We can therefore maximize the volume of the image ellipsoid, over a convex set of configurations, by maximizing $\log \det G$.

The same holds for any set in $\mathbf{R}^n$. The volume of the image under A is its volume, multiplied by the factor $(\det G)^{1/2}$. For example, consider the image under A of the unit simplex $\mathbf{conv}\{0, e_1, \ldots, e_n\}$, *i.e.*, the simplex $\mathbf{conv}\{0, a_1, \ldots, a_n\}$. The volume of this simplex is given by $\overline{\gamma}(\det G)^{1/2}$, where $\overline{\gamma}$ is the volume of the unit simplex in $\mathbf{R}^n$. We can maximize the volume of this simplex by maximizing $\log \det G$.

8.3.2   Problems involving angles only

Suppose we only care about the angles (or correlation coefficients) between the vectors, and do not specify the lengths or distances between them. In this case it is intuitively clear that we can simply assume the vectors a_i have length $l_i = 1$. This is easily verified: The Gram matrix has the form $G = \mathbf{diag}(l) C \mathbf{diag}(l)$, where l is the vector of lengths, and C is the correlation matrix, *i.e.*, $C_{ij} = \cos \theta_{ij}$. It follows that if $G \succeq 0$ for any set of positive lengths, then $G \succeq 0$ for *all* sets of positive lengths, and in particular, this occurs if and only if $C \succeq 0$ (which is the same as assuming that all lengths are one). Thus, a set of angles $\theta_{ij} \in [0, \pi]$, $i, j = 1, \ldots, n$ is realizable if and only if $C \succeq 0$, which is a linear matrix inequality in the correlation coefficients.

As an example, suppose we are given lower and upper bounds on some of the angles (which is equivalent to imposing lower and upper bounds on the correlation coefficients). We can then find the minimum and maximum possible value of some other angle, over all configurations, by solving two SDPs.

Example 8.3 *Bounding correlation coefficients.* We consider an example in $\mathbf{R}^4$, where we are given

$$
\begin{array}{ll}
0.6 \le \rho_{12} \le 0.9, & 0.8 \le \rho_{13} \le 0.9, \\
0.5 \le \rho_{24} \le 0.7, & -0.8 \le \rho_{34} \le -0.4.
\end{array}
\tag{8.7}
$$

To find the minimum and maximum possible values of ρ_{14}, we solve the two SDPs

$$
\begin{array}{ll}
\text{minimize/maximize} & \rho_{14} \\
\text{subject to} & (8.7) \\
& \begin{bmatrix} 1 & \rho_{12} & \rho_{13} & \rho_{14} \\ \rho_{12} & 1 & \rho_{23} & \rho_{24} \\ \rho_{13} & \rho_{23} & 1 & \rho_{34} \\ \rho_{14} & \rho_{24} & \rho_{34} & 1 \end{bmatrix} \succeq 0,
\end{array}
$$

with variables $\rho_{12}, \rho_{13}, \rho_{14}, \rho_{23}, \rho_{24}, \rho_{34}$. The minimum and maximum values (to two significant digits) are -0.39 and 0.23, with corresponding correlation matrices

$$
\begin{bmatrix} 1.00 & 0.60 & 0.87 & -0.39 \\ 0.60 & 1.00 & 0.33 & 0.50 \\ 0.87 & 0.33 & 1.00 & -0.55 \\ -0.39 & 0.50 & -0.55 & 1.00 \end{bmatrix},
\quad
\begin{bmatrix} 1.00 & 0.71 & 0.80 & 0.23 \\ 0.71 & 1.00 & 0.31 & 0.59 \\ 0.80 & 0.31 & 1.00 & -0.40 \\ 0.23 & 0.59 & -0.40 & 1.00 \end{bmatrix}.
$$

8.3.3 Euclidean distance problems

In a *Euclidean distance problem*, we are concerned *only* with the distances between the vectors, d_{ij}, and do not care about the lengths of the vectors, or about the angles between them. These distances, of course, are invariant not only under orthogonal transformations, but also translation: The configuration $\tilde{a}_1 = a_1 + b, \dots, \tilde{a}_n = a_n + b$ has the same distances as the original configuration, for any $b \in \mathbf{R}^n$. In particular, for the choice

$$b = -(1/n) \sum_{i=1}^{n} a_i = -(1/n) A\mathbf{1},$$

we see that $\tilde{a}_i$ have the same distances as the original configuration, and also satisfy $\sum_{i=1}^{n} \tilde{a}_i = 0$. It follows that in a Euclidean distance problem, we can assume, without any loss of generality, that the average of the vectors $a_1, \dots, a_n$ is zero, i.e., $A\mathbf{1} = 0$.

We can solve Euclidean distance problems by considering the lengths (which cannot occur in the objective or constraints of a Euclidean distance problem) as free variables in the optimization problem. Here we rely on the fact that there is a configuration with distances $d_{ij} \geq 0$ if and only if there are lengths $l_1, \dots, l_n$ for which $G \succeq 0$, where $G_{ij} = (l_i^2 + l_j^2 - d_{ij}^2)/2$.

We define $z \in \mathbf{R}^n$ as $z_i = l_i^2$, and $D \in \mathbf{S}^n$ by $D_{ij} = d_{ij}^2$ (with, of course, $D_{ii} = 0$). The condition that $G \succeq 0$ for some choice of lengths can be expressed as

$$G = (z\mathbf{1}^T + \mathbf{1}z^T - D)/2 \succeq 0 \text{ for some } z \succeq 0, \tag{8.8}$$

which is an LMI in D and z. A matrix $D \in \mathbf{S}^n$, with nonnegative elements, zero diagonal, and which satisfies (8.8), is called a *Euclidean distance matrix*. A matrix is a Euclidean distance matrix if and only if its entries are the squares of the Euclidean distances between the vectors of some configuration. (Given a Euclidean distance matrix D and the associated length squared vector z, we can reconstruct one, or all, configurations with the given pairwise distances using the method described above.)

The condition (8.8) turns out to be equivalent to the simpler condition that D is negative semidefinite on $\mathbf{1}^\perp$, i.e.,

$$
\begin{aligned}
(8.8) \quad &\Longleftrightarrow \quad u^T D u \leq 0 \text{ for all } u \text{ with } \mathbf{1}^T u = 0 \\
&\Longleftrightarrow \quad (I - (1/n)\mathbf{1}\mathbf{1}^T) D (I - (1/n)\mathbf{1}\mathbf{1}^T) \preceq 0.
\end{aligned}
$$

This simple matrix inequality, along with $D_{ij} \geq 0$, $D_{ii} = 0$, is the classical characterization of a Euclidean distance matrix. To see the equivalence, recall that we can assume $A\mathbf{1} = 0$, which implies that $\mathbf{1}^T G\mathbf{1} = \mathbf{1}^T A^T A\mathbf{1} = 0$. It follows that $G \succeq 0$ if and only if G is positive semidefinite on $\mathbf{1}^\perp$, i.e.,

$$
\begin{aligned}
0 \quad \preceq \quad & (I - (1/n)\mathbf{1}\mathbf{1}^T) G (I - (1/n)\mathbf{1}\mathbf{1}^T) \\
= \quad & (1/2)(I - (1/n)\mathbf{1}\mathbf{1}^T)(z\mathbf{1}^T + \mathbf{1}z^T - D)(I - (1/n)\mathbf{1}\mathbf{1}^T) \\
= \quad & -(1/2)(I - (1/n)\mathbf{1}\mathbf{1}^T) D (I - (1/n)\mathbf{1}\mathbf{1}^T),
\end{aligned}
$$

which is the simplified condition.

In summary, a matrix $D \in \mathbf{S}^n$ is a Euclidean distance matrix, *i.e.*, gives the squared distances between a set of n vectors in $\mathbf{R}^n$, if and only if

$$D_{ii} = 0, \quad i = 1, \ldots, n, \qquad D_{ij} \geq 0, \quad i, j = 1, \ldots, n,$$

$$(I - (1/n)\mathbf{1}\mathbf{1}^T)D(I - (1/n)\mathbf{1}\mathbf{1}^T) \preceq 0,$$

which is a set of linear equalities, linear inequalities, and a matrix inequality in D. Therefore we can express any Euclidean distance problem that is convex in the squared distances as a convex problem with variable $D \in \mathbf{S}^n$.

8.4   Extremal volume ellipsoids

Suppose $C \subseteq \mathbf{R}^n$ is bounded and has nonempty interior. In this section we consider the problems of finding the maximum volume ellipsoid that lies inside C, and the minimum volume ellipsoid that covers C. Both problems can be formulated as convex programming problems, but are tractable only in special cases.

8.4.1   The Löwner-John ellipsoid

The minimum volume ellipsoid that contains a set C is called the *Löwner-John ellipsoid* of the set C, and is denoted $\mathcal{E}_{lj}$. To characterize $\mathcal{E}_{lj}$, it will be convenient to parametrize a general ellipsoid as

$$\mathcal{E} = \{v \mid \|Av + b\|_2 \leq 1\}, \tag{8.9}$$

i.e., the inverse image of the Euclidean unit ball under an affine mapping. We can assume without loss of generality that $A \in \mathbf{S}^n_{++}$, in which case the volume of $\mathcal{E}$ is proportional to $\det A^{-1}$. The problem of computing the minimum volume ellipsoid containing C can be expressed as

$$\begin{array}{ll} \text{minimize} & \log \det A^{-1} \\ \text{subject to} & \sup_{v \in C} \|Av + b\|_2 \leq 1, \end{array} \tag{8.10}$$

where the variables are $A \in \mathbf{S}^n$ and $b \in \mathbf{R}^n$, and there is an implicit constraint $A \succ 0$. The objective and constraint functions are both convex in A and b, so the problem (8.10) is convex. Evaluating the constraint function in (8.10), however, involves solving a convex maximization problem, and is tractable only in certain special cases.

Minimum volume ellipsoid covering a finite set

We consider the problem of finding the minimum volume ellipsoid that contains the finite set $C = \{x_1, \ldots, x_m\} \subseteq \mathbf{R}^n$. An ellipsoid covers C if and only if it covers its convex hull, so finding the minimum volume ellipsoid that covers C

is the same as finding the minimum volume ellipsoid containing the polyhedron **conv**$\{x_1, \ldots, x_m\}$. Applying (8.10), we can write this problem as

$$
\begin{array}{ll}
\text{minimize} & \log \det A^{-1} \\
\text{subject to} & \|Ax_i + b\|_2 \le 1, \quad i = 1, \ldots, m
\end{array}
\tag{8.11}
$$

where the variables are $A \in \mathbf{S}^n$ and $b \in \mathbf{R}^n$, and we have the implicit constraint $A \succ 0$. The norm constraints $\|Ax_i + b\|_2 \le 1$, $i = 1, \ldots, m$, are convex inequalities in the variables A and b. They can be replaced with the squared versions, $\|Ax_i + b\|_2^2 \le 1$, which are convex quadratic inequalities in A and b.

Minimum volume ellipsoid covering union of ellipsoids

Minimum volume covering ellipsoids can also be computed efficiently for certain sets C that are defined by quadratic inequalities. In particular, it is possible to compute the Löwner-John ellipsoid for a union or sum of ellipsoids.

As an example, consider the problem of finding the minimum volume ellipsoid $\mathcal{E}_{1j}$, that contains the ellipsoids $\mathcal{E}_1, \ldots, \mathcal{E}_m$ (and therefore, the convex hull of their union). The ellipsoids $\mathcal{E}_1$, ..., $\mathcal{E}_m$ will be described by (convex) quadratic inequalities:

$$
\mathcal{E}_i = \{x \mid x^T A_i x + 2b_i^T x + c_i \le 0\}, \quad i = 1, \ldots, m,
$$

where $A_i \in \mathbf{S}_{++}^n$. We parametrize the ellipsoid $\mathcal{E}_{1j}$ as

$$
\begin{aligned}
\mathcal{E}_{1j} &= \{x \mid \|Ax + b\|_2 \le 1\} \\
&= \{x \mid x^T A^T A x + 2(A^T b)^T x + b^T b - 1 \le 0\}
\end{aligned}
$$

where $A \in \mathbf{S}^n$ and $b \in \mathbf{R}^n$. Now we use a result from §B.2, that $\mathcal{E}_i \subseteq \mathcal{E}_{1j}$ if and only if there exists a $\tau \ge 0$ such that

$$
\begin{bmatrix} A^2 - \tau A_i & Ab - \tau b_i \\ (Ab - \tau b_i)^T & b^T b - 1 - \tau c_i \end{bmatrix} \preceq 0.
$$

The volume of $\mathcal{E}_{1j}$ is proportional to $\det A^{-1}$, so we can find the minimum volume ellipsoid that contains $\mathcal{E}_1, \ldots, \mathcal{E}_m$ by solving

$$
\begin{array}{ll}
\text{minimize} & \log \det A^{-1} \\
\text{subject to} & \tau_1 \ge 0, \ldots, \tau_m \ge 0 \\
& \begin{bmatrix} A^2 - \tau_i A_i & Ab - \tau_i b_i \\ (Ab - \tau_i b_i)^T & b^T b - 1 - \tau_i c_i \end{bmatrix} \preceq 0, \quad i = 1, \ldots, m,
\end{array}
$$

or, replacing the variable b by $\tilde{b} = Ab$,

$$
\begin{array}{ll}
\text{minimize} & \log \det A^{-1} \\
\text{subject to} & \tau_1 \ge 0, \ldots, \tau_m \ge 0 \\
& \begin{bmatrix} A^2 - \tau_i A_i & \tilde{b} - \tau_i b_i & 0 \\ (\tilde{b} - \tau_i b_i)^T & -1 - \tau_i c_i & \tilde{b}^T \\ 0 & \tilde{b} & -A^2 \end{bmatrix} \preceq 0, \quad i = 1, \ldots, m,
\end{array}
$$

which is convex in the variables $A^2 \in \mathbf{S}^n$, $\tilde{b}$, $\tau_1, \ldots, \tau_m$.

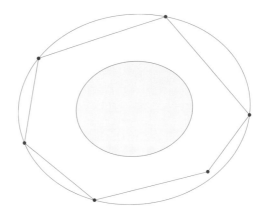

Figure 8.3 The outer ellipse is the boundary of the Löwner-John ellipsoid, *i.e.*, the minimum volume ellipsoid that encloses the points $x_1, \ldots, x_6$ (shown as dots), and therefore the polyhedron $\mathcal{P} = \mathbf{conv}\{x_1, \ldots, x_6\}$. The smaller ellipse is the boundary of the Löwner-John ellipsoid, shrunk by a factor of $n = 2$ about its center. This ellipsoid is guaranteed to lie *inside* $\mathcal{P}$.

Efficiency of Löwner-John ellipsoidal approximation

Let $\mathcal{E}_{\mathrm{lj}}$ be the Löwner-John ellipsoid of the convex set $C \subseteq \mathbf{R}^n$, which is bounded and has nonempty interior, and let x_0 be its center. If we shrink the Löwner-John ellipsoid by a factor of n, about its center, we obtain an ellipsoid that lies inside the set C:

$$x_0 + (1/n)(\mathcal{E}_{\mathrm{lj}} - x_0) \subseteq C \subseteq \mathcal{E}_{\mathrm{lj}}.$$

In other words, the Löwner-John ellipsoid approximates an arbitrary convex set, within a factor that depends only on the dimension n. Figure 8.3 shows a simple example.

The factor $1/n$ cannot be improved without additional assumptions on C. Any simplex in $\mathbf{R}^n$, for example, has the property that its Löwner-John ellipsoid must be shrunk by a factor n to fit inside it (see exercise 8.13).

We will prove this efficiency result for the special case $C = \mathbf{conv}\{x_1, \ldots, x_m\}$. We square the norm constraints in (8.11) and introduce variables $\tilde{A} = A^2$ and $\tilde{b} = Ab$, to obtain the problem

$$
\begin{array}{ll}
\text{minimize} & \log \det \tilde{A}^{-1} \\
\text{subject to} & x_i^T \tilde{A} x_i - 2\tilde{b}^T x_i + \tilde{b}^T \tilde{A}^{-1} \tilde{b} \leq 1, \quad i = 1, \ldots, m.
\end{array}
\tag{8.12}
$$

The KKT conditions for this problem are

$$\sum_{i=1}^m \lambda_i(x_i x_i^T - \tilde{A}^{-1}\tilde{b}\tilde{b}^T \tilde{A}^{-1}) = \tilde{A}^{-1}, \qquad \sum_{i=1}^m \lambda_i(x_i - \tilde{A}^{-1}\tilde{b}) = 0,$$

$$\lambda_i \geq 0, \qquad x_i^T \tilde{A} x_i - 2\tilde{b}^T x_i + \tilde{b}^T \tilde{A}^{-1}\tilde{b} \leq 1, \qquad i = 1, \ldots, m,$$

$$\lambda_i(1 - x_i^T \tilde{A} x_i + 2\tilde{b}^T x_i - \tilde{b}^T \tilde{A}^{-1}\tilde{b}) = 0, \quad i = 1, \ldots, m.$$

By a suitable affine change of coordinates, we can assume that $\tilde{A} = I$ and $\tilde{b} = 0$, *i.e.*, the minimum volume ellipsoid is the unit ball centered at the origin. The KKT conditions then simplify to

$$\sum_{i=1}^{m} \lambda_i x_i x_i^T = I, \qquad \sum_{i=1}^{m} \lambda_i x_i = 0, \qquad \lambda_i(1 - x_i^T x_i) = 0, \quad i = 1, \dots, m,$$

plus the feasibility conditions $\|x_i\|_2 \le 1$ and $\lambda_i \ge 0$. By taking the trace of both sides of the first equation, and using complementary slackness, we also have $\sum_{i=1}^{m} \lambda_i = n$.

In the new coordinates the shrunk ellipsoid is a ball with radius $1/n$, centered at the origin. We need to show that

$$\|x\|_2 \le 1/n \implies x \in C = \mathbf{conv}\{x_1, \dots, x_m\}.$$

Suppose $\|x\|_2 \le 1/n$. From the KKT conditions, we see that

$$x = \sum_{i=1}^{m} \lambda_i(x^T x_i)x_i = \sum_{i=1}^{m} \lambda_i(x^T x_i + 1/n)x_i = \sum_{i=1}^{m} \mu_i x_i, \qquad (8.13)$$

where $\mu_i = \lambda_i(x^T x_i + 1/n)$. From the Cauchy-Schwartz inequality, we note that

$$\mu_i = \lambda_i(x^T x_i + 1/n) \ge \lambda_i(-\|x\|_2\|x_i\|_2 + 1/n) \ge \lambda_i(-1/n + 1/n) = 0.$$

Furthermore

$$\sum_{i=1}^{m} \mu_i = \sum_{i=1}^{m} \lambda_i(x^T x_i + 1/n) = \sum_{i=1}^{m} \lambda_i/n = 1.$$

This, along with (8.13), shows that x is a convex combination of $x_1, \dots, x_m$, hence $x \in C$.

Efficiency of Löwner-John ellipsoidal approximation for symmetric sets

If the set C is symmetric about a point x_0, then the factor $1/n$ can be tightened to $1/\sqrt{n}$:

$$x_0 + (1/\sqrt{n})(\mathcal{E}_{\mathrm{lj}} - x_0) \subseteq C \subseteq \mathcal{E}_{\mathrm{lj}}.$$

Again, the factor $1/\sqrt{n}$ is tight. The Löwner-John ellipsoid of the cube

$$C = \{x \in \mathbf{R}^n \mid -\mathbf{1} \preceq x \preceq \mathbf{1}\}$$

is the ball with radius $\sqrt{n}$. Scaling down by $1/\sqrt{n}$ yields a ball enclosed in C, and touching the boundary at $x = \pm e_i$.

Approximating a norm by a quadratic norm

Let $\|\cdot\|$ be any norm on $\mathbf{R}^n$, and let $C = \{x \mid \|x\| \le 1\}$ be its unit ball. Let $\mathcal{E}_{\mathrm{lj}} = \{x \mid x^T A x \le 1\}$, with $A \in \mathbf{S}_{++}^n$, be the Löwner-John ellipsoid of C. Since C is symmetric about the origin, the result above tells us that $(1/\sqrt{n})\mathcal{E}_{\mathrm{lj}} \subseteq C \subseteq \mathcal{E}_{\mathrm{lj}}$. Let $\|\cdot\|_{\mathrm{lj}}$ denote the quadratic norm

$$\|z\|_{\mathrm{lj}} = (z^T A z)^{1/2},$$

whose unit ball is $\mathcal{E}_{lj}$. The inclusions $(1/\sqrt{n})\mathcal{E}_{lj} \subseteq C \subseteq \mathcal{E}_{lj}$ are equivalent to the inequalities

$$\|z\|_{lj} \leq \|z\| \leq \sqrt{n}\|z\|_{lj}$$

for all $z \in \mathbf{R}^n$. In other words, the quadratic norm $\| \cdot \|_{lj}$ approximates the norm $\| \cdot \|$ within a factor of $\sqrt{n}$. In particular, we see that any norm on $\mathbf{R}^n$ can be approximated within a factor of $\sqrt{n}$ by a quadratic norm.

8.4.2   Maximum volume inscribed ellipsoid

We now consider the problem of finding the ellipsoid of maximum volume that lies inside a convex set C, which we assume is bounded and has nonempty interior. To formulate this problem, we parametrize the ellipsoid as the image of the unit ball under an affine transformation, *i.e.*, as

$$\mathcal{E} = \{Bu + d \mid \|u\|_2 \leq 1\}.$$

Again it can be assumed that $B \in \mathbf{S}^n_{++}$, so the volume is proportional to $\det B$. We can find the maximum volume ellipsoid inside C by solving the convex optimization problem

$$\begin{array}{ll} \text{maximize} & \log \det B \\ \text{subject to} & \sup_{\|u\|_2 \leq 1} I_C(Bu + d) \leq 0 \end{array} \qquad (8.14)$$

in the variables $B \in \mathbf{S}^n$ and $d \in \mathbf{R}^n$, with implicit constraint $B \succ 0$.

Maximum volume ellipsoid in a polyhedron

We consider the case where C is a polyhedron described by a set of linear inequalities:

$$C = \{x \mid a_i^T x \leq b_i, \ i = 1, \dots, m\}.$$

To apply (8.14) we first express the constraint in a more convenient form:

$$\begin{aligned} \sup_{\|u\|_2 \leq 1} I_C(Bu + d) \leq 0 &\iff \sup_{\|u\|_2 \leq 1} a_i^T(Bu + d) \leq b_i, \quad i = 1, \dots, m \\ &\iff \|Ba_i\|_2 + a_i^T d \leq b_i, \quad i = 1, \dots, m. \end{aligned}$$

We can therefore formulate (8.14) as a convex optimization problem in the variables B and d:

$$\begin{array}{ll} \text{minimize} & \log \det B^{-1} \\ \text{subject to} & \|Ba_i\|_2 + a_i^T d \leq b_i, \quad i = 1, \dots, m. \end{array} \qquad (8.15)$$

Maximum volume ellipsoid in an intersection of ellipsoids

We can also find the maximum volume ellipsoid $\mathcal{E}$ that lies in the intersection of m ellipsoids $\mathcal{E}_1, \dots, \mathcal{E}_m$. We will describe $\mathcal{E}$ as $\mathcal{E} = \{Bu + d \mid \|u\|_2 \leq 1\}$ with $B \in \mathbf{S}^n_{++}$, and the other ellipsoids via convex quadratic inequalities,

$$\mathcal{E}_i = \{x \mid x^T A_i x + 2b_i^T x + c_i \leq 0\}, \quad i = 1, \dots, m,$$

where $A_i \in \mathbf{S}_{++}^n$. We first work out the condition under which $\mathcal{E} \subseteq \mathcal{E}_i$. This occurs if and only if

$$\sup_{\|u\|_2 \leq 1} \left((d + Bu)^T A_i (d + Bu) + 2b_i^T (d + Bu) + c_i \right)$$

$$= d^T A_i d + 2b_i^T d + c_i + \sup_{\|u\|_2 \leq 1} \left(u^T B A_i B u + 2(A_i d + b_i)^T B u \right)$$

$$\leq 0.$$

From §B.1,

$$\sup_{\|u\|_2 \leq 1} \left(u^T B A_i B u + 2(A_i d + b_i)^T B u \right) \leq -(d^T A_i d + 2b_i^T d + c_i)$$

if and only if there exists a $\lambda_i \geq 0$ such that

$$\begin{bmatrix} -\lambda_i - d^T A_i d - 2b_i^T d - c_i & (A_i d + b_i)^T B \\ B(A_i d + b_i) & \lambda_i I - B A_i B \end{bmatrix} \succeq 0.$$

The maximum volume ellipsoid contained in $\mathcal{E}_1, \ldots, \mathcal{E}_m$ can therefore be found by solving the problem

minimize $\log \det B^{-1}$

subject to $\begin{bmatrix} -\lambda_i - d^T A_i d - 2b_i^T d - c_i & (A_i d + b_i)^T B \\ B(A_i d + b_i) & \lambda_i I - B A_i B \end{bmatrix} \succeq 0, \quad i = 1, \ldots, m,$

with variables $B \in \mathbf{S}^n$, $d \in \mathbf{R}^n$, and $\lambda \in \mathbf{R}^m$, or, equivalently,

minimize $\log \det B^{-1}$

subject to $\begin{bmatrix} -\lambda_i - c_i + b_i^T A_i^{-1} b_i & 0 & (d + A_i^{-1} b_i)^T \\ 0 & \lambda_i I & B \\ d + A_i^{-1} b_i & B & A_i^{-1} \end{bmatrix} \succeq 0, \quad i = 1, \ldots, m.$

Efficiency of ellipsoidal inner approximations

Approximation efficiency results, similar to the ones for the Löwner-John ellipsoid, hold for the maximum volume inscribed ellipsoid. If $C \subseteq \mathbf{R}^n$ is convex, bounded, with nonempty interior, then the maximum volume inscribed ellipsoid, expanded by a factor of n about its center, covers the set C. The factor n can be tightened to $\sqrt{n}$ if the set C is symmetric about a point. An example is shown in figure 8.4.

8.4.3   Affine invariance of extremal volume ellipsoids

The Löwner-John ellipsoid and the maximum volume inscribed ellipsoid are both affinely invariant. If $\mathcal{E}_{\mathrm{lj}}$ is the Löwner-John ellipsoid of C, and $T \in \mathbf{R}^{n \times n}$ is nonsingular, then the Löwner-John ellipsoid of TC is $T\mathcal{E}_{\mathrm{lj}}$. A similar result holds for the maximum volume inscribed ellipsoid.

To establish this result, let $\mathcal{E}$ be any ellipsoid that covers C. Then the ellipsoid $T\mathcal{E}$ covers TC. The converse is also true: Every ellipsoid that covers TC has

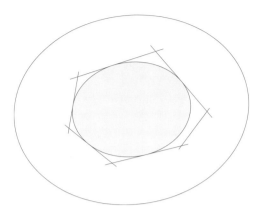

Figure 8.4 The maximum volume ellipsoid (shown shaded) inscribed in a polyhedron $\mathcal{P}$. The outer ellipse is the boundary of the inner ellipsoid, expanded by a factor $n = 2$ about its center. The expanded ellipsoid is guaranteed to cover $\mathcal{P}$.

the form $T\mathcal{E}$, where $\mathcal{E}$ is an ellipsoid that covers C. In other words, the relation $\tilde{\mathcal{E}} = T\mathcal{E}$ gives a one-to-one correspondence between the ellipsoids covering TC and the ellipsoids covering C. Moreover, the volumes of the corresponding ellipsoids are all related by the ratio $|\det T|$, so in particular, if $\mathcal{E}$ has minimum volume among ellipsoids covering C, then $T\mathcal{E}$ has minimum volume among ellipsoids covering TC.

8.5   Centering

8.5.1   Chebyshev center

Let $C \subseteq \mathbf{R}^n$ be bounded and have nonempty interior, and $x \in C$. The *depth* of a point $x \in C$ is defined as

$$\mathbf{depth}(x, C) = \mathbf{dist}(x, \mathbf{R}^n \setminus C),$$

i.e., the distance to the closest point in the exterior of C. The depth gives the radius of the largest ball, centered at x, that lies in C. A *Chebyshev center* of the set C is defined as any point of maximum depth in C:

$$x_{\mathrm{cheb}}(C) = \operatorname{argmax} \mathbf{depth}(x, C) = \operatorname{argmax} \mathbf{dist}(x, \mathbf{R}^n \setminus C).$$

A Chebyshev center is a point inside C that is farthest from the exterior of C; it is also the center of the largest ball that lies inside C. Figure 8.5 shows an example, in which C is a polyhedron, and the norm is Euclidean.

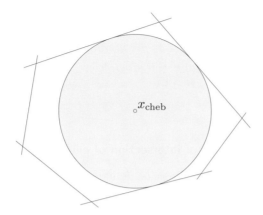

Figure 8.5 Chebyshev center of a polyhedron C, in the Euclidean norm. The center x_{cheb} is the deepest point inside C, in the sense that it is farthest from the exterior, or complement, of C. The center x_{cheb} is also the center of the largest Euclidean ball (shown lightly shaded) that lies inside C.

Chebyshev center of a convex set

When the set C is convex, the depth is a concave function for $x \in C$, so computing the Chebyshev center is a convex optimization problem (see exercise 8.5). More specifically, suppose $C \subseteq \mathbf{R}^n$ is defined by a set of convex inequalities:

$$C = \{x \mid f_1(x) \le 0, \ldots, f_m(x) \le 0\}.$$

We can find a Chebyshev center by solving the problem

$$\begin{array}{ll} \text{maximize} & R \\ \text{subject to} & g_i(x, R) \le 0, \quad i = 1, \ldots, m, \end{array} \qquad (8.16)$$

where g_i is defined as

$$g_i(x, R) = \sup_{\|u\| \le 1} f_i(x + Ru).$$

Problem (8.16) is a convex optimization problem, since each function g_i is the pointwise maximum of a family of convex functions of x and R, hence convex. However, evaluating g_i involves solving a convex *maximization* problem (either numerically or analytically), which may be very hard. In practice, we can find the Chebyshev center only in cases where the functions g_i are easy to evaluate.

Chebyshev center of a polyhedron

Suppose C is defined by a set of linear inequalities $a_i^T x \le b_i$, $i = 1, \ldots, m$. We have

$$g_i(x, R) = \sup_{\|u\| \le 1} a_i^T(x + Ru) - b_i = a_i^T x + R\|a_i\|_* - b_i$$

if $R \geq 0$, so the Chebyshev center can be found by solving the LP

$$
\begin{array}{ll}
\text{maximize} & R \\
\text{subject to} & a_i^T x + R\|a_i\|_* \leq b_i, \quad i = 1, \dots, m \\
& R \geq 0
\end{array}
$$

with variables x and R.

Euclidean Chebyshev center of intersection of ellipsoids

Let C be an intersection of m ellipsoids, defined by quadratic inequalities,

$$
C = \{x \mid x^T A_i x + 2b_i^T x + c_i \leq 0, \ i = 1, \dots, m\},
$$

where $A_i \in \mathbf{S}_{++}^n$. We have

$$
\begin{aligned}
g_i(x, R) &= \sup_{\|u\|_2 \leq 1} \left((x + Ru)^T A_i (x + Ru) + 2b_i^T (x + Ru) + c_i \right) \\
&= x^T A_i x + 2b_i^T x + c_i + \sup_{\|u\|_2 \leq 1} \left(R^2 u^T A_i u + 2R(A_i x + b_i)^T u \right).
\end{aligned}
$$

From §B.1, $g_i(x, R) \leq 0$ if and only if there exists a λ_i such that the matrix inequality

$$
\begin{bmatrix}
-x^T A_i x_i - 2b_i^T x - c_i - \lambda_i & R(A_i x + b_i)^T \\
R(A_i x + b_i) & \lambda_i I - R^2 A_i
\end{bmatrix} \succeq 0 \tag{8.17}
$$

holds. Using this result, we can express the Chebyshev centering problem as

$$
\begin{array}{ll}
\text{maximize} & R \\
\text{subject to} & \begin{bmatrix}
-\lambda_i - c_i + b_i^T A_i^{-1} b_i & 0 & (x + A_i^{-1} b_i)^T \\
0 & \lambda_i I & RI \\
x + A_i^{-1} b_i & RI & A_i^{-1}
\end{bmatrix} \succeq 0, \quad i = 1, \dots, m,
\end{array}
$$

which is an SDP with variables R, λ, and x. Note that the Schur complement of A_i^{-1} in the LMI constraint is equal to the lefthand side of (8.17).

8.5.2   Maximum volume ellipsoid center

The Chebyshev center x_{cheb} of a set $C \subseteq \mathbf{R}^n$ is the center of the largest ball that lies in C. As an extension of this idea, we define the *maximum volume ellipsoid center* of C, denoted x_{mve}, as the center of the maximum volume ellipsoid that lies in C. Figure 8.6 shows an example, where C is a polyhedron.

The maximum volume ellipsoid center is readily computed when C is defined by a set of linear inequalities, by solving the problem (8.15). (The optimal value of the variable $d \in \mathbf{R}^n$ is x_{mve}.) Since the maximum volume ellipsoid inside C is affine invariant, so is the maximum volume ellipsoid center.

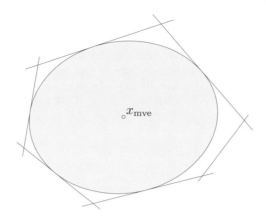

Figure 8.6 The lightly shaded ellipsoid shows the maximum volume ellipsoid contained in the set C, which is the same polyhedron as in figure 8.5. Its center x_{mve} is the maximum volume ellipsoid center of C.

8.5.3 Analytic center of a set of inequalities

The *analytic center* x_{ac} of a set of convex inequalities and linear equalities,

$$f_i(x) \le 0, \quad i = 1, \ldots, m, \qquad Fx = g$$

is defined as an optimal point for the (convex) problem

$$\begin{array}{ll}
\text{minimize} & -\sum_{i=1}^{m} \log(-f_i(x)) \\
\text{subject to} & Fx = g,
\end{array} \qquad (8.18)$$

with variable $x \in \mathbf{R}^n$ and implicit constraints $f_i(x) < 0$, $i = 1, \ldots, m$. The objective in (8.18) is called the *logarithmic barrier* associated with the set of inequalities. We assume here that the domain of the logarithmic barrier intersects the affine set defined by the equalities, *i.e.*, the strict inequality system

$$f_i(x) < 0, \quad i = 1, \ldots, m, \qquad Fx = g$$

is feasible. The logarithmic barrier is bounded below on the feasible set

$$C = \{x \mid f_i(x) < 0, \ i = 1, \ldots, m, \ Fx = g\},$$

if C is bounded.

When x is strictly feasible, *i.e.*, $Fx = g$ and $f_i(x) < 0$ for $i = 1, \ldots, m$, we can interpret $-f_i(x)$ as the margin or slack in the ith inequality. The analytic center x_{ac} is the point that maximizes the product (or geometric mean) of these slacks or margins, subject to the equality constraints $Fx = g$, and the implicit constraints $f_i(x) < 0$.

The analytic center is *not* a function of the set C described by the inequalities and equalities; two sets of inequalities and equalities can define the same set, but have different analytic centers. Still, it is not uncommon to informally use the term 'analytic center of a set C' to mean the analytic center of a particular set of equalities and inequalities that define it.

The analytic center is, however, independent of affine changes of coordinates. It is also invariant under (positive) scalings of the inequality functions, and any reparametrization of the equality constraints. In other words, if $\tilde{F}$ and $\tilde{g}$ are such that $\tilde{F}x = \tilde{g}$ if and only if $Fx = g$, and $\alpha_1, \ldots, \alpha_m > 0$, then the analytic center of

$$\alpha_i f_i(x) \leq 0, \quad i = 1, \ldots, m, \qquad \tilde{F}x = \tilde{g},$$

is the same as the analytic center of

$$f_i(x) \leq 0, \quad i = 1, \ldots, m, \qquad Fx = g$$

(see exercise 8.17).

Analytic center of a set of linear inequalities

The analytic center of a set of linear inequalities

$$a_i^T x \leq b_i, \quad i = 1, \ldots, m,$$

is the solution of the unconstrained minimization problem

$$\text{minimize} \quad -\sum_{i=1}^m \log(b_i - a_i^T x), \tag{8.19}$$

with implicit constraint $b_i - a_i^T x > 0$, $i = 1, \ldots, m$. If the polyhedron defined by the linear inequalities is bounded, then the logarithmic barrier is bounded below and strictly convex, so the analytic center is unique. (See exercise 4.2.)

We can give a geometric interpretation of the analytic center of a set of linear inequalities. Since the analytic center is independent of positive scaling of the constraint functions, we can assume without loss of generality that $\|a_i\|_2 = 1$. In this case, the slack $b_i - a_i^T x$ is the distance to the hyperplane $\mathcal{H}_i = \{x \mid a_i^T x = b_i\}$. Therefore the analytic center x_{ac} is the point that maximizes the product of distances to the defining hyperplanes.

Inner and outer ellipsoids from analytic center of linear inequalities

The analytic center of a set of linear inequalities implicitly defines an inscribed and a covering ellipsoid, defined by the Hessian of the logarithmic barrier function

$$-\sum_{i=1}^m \log(b_i - a_i^T x),$$

evaluated at the analytic center, i.e.,

$$H = \sum_{i=1}^m d_i^2 a_i a_i^T, \qquad d_i = \frac{1}{b_i - a_i^T x_{\text{ac}}}, \quad i = 1, \ldots, m.$$

We have $\mathcal{E}_{\text{inner}} \subseteq \mathcal{P} \subseteq \mathcal{E}_{\text{outer}}$, where

$$
\begin{aligned}
\mathcal{P} &= \{x \mid a_i^T x \leq b_i, \ i = 1, \ldots, m\}, \\
\mathcal{E}_{\text{inner}} &= \{x \mid (x - x_{\text{ac}})^T H (x - x_{\text{ac}}) \leq 1\}, \\
\mathcal{E}_{\text{outer}} &= \{x \mid x - x_{\text{ac}})^T H (x - x_{\text{ac}}) \leq m(m-1)\}.
\end{aligned}
$$

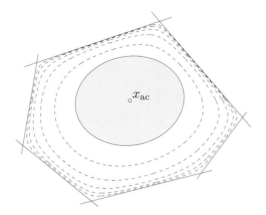

Figure 8.7 The dashed lines show five level curves of the logarithmic barrier function for the inequalities defining the polyhedron C in figure 8.5. The minimizer of the logarithmic barrier function, labeled x_{ac}, is the analytic center of the inequalities. The inner ellipsoid $\mathcal{E}_{inner} = \{x \mid (x - x_{ac})H(x - x_{ac}) \leq 1\}$, where H is the Hessian of the logarithmic barrier function at x_{ac}, is shaded.

This is a weaker result than the one for the maximum volume inscribed ellipsoid, which when scaled up by a factor of n covers the polyhedron. The inner and outer ellipsoids defined by the Hessian of the logarithmic barrier, in contrast, are related by the scale factor $(m(m-1))^{1/2}$, which is always at least n.

To show that $\mathcal{E}_{inner} \subseteq \mathcal{P}$, suppose $x \in \mathcal{E}_{inner}$, *i.e.*,

$$(x - x_{ac})^T H (x - x_{ac}) = \sum_{i=1}^{m} (d_i a_i^T (x - x_{ac}))^2 \leq 1.$$

This implies that

$$a_i^T (x - x_{ac}) \leq 1/d_i = b_i - a_i^T x_{ac}, \quad i = 1, \dots, m,$$

and therefore $a_i^T x \leq b_i$ for $i = 1, \dots, m$. (We have not used the fact that x_{ac} is the analytic center, so this result is valid if we replace x_{ac} with any strictly feasible point.)

To establish that $\mathcal{P} \subseteq \mathcal{E}_{outer}$, we will need the fact that x_{ac} is the analytic center, and therefore the gradient of the logarithmic barrier vanishes:

$$\sum_{i=1}^{m} d_i a_i = 0.$$

Now assume $x \in \mathcal{P}$. Then

$$(x - x_{ac})^T H (x - x_{ac})$$
$$= \sum_{i=1}^{m} (d_i a_i^T (x - x_{ac}))^2$$

$$= \sum_{i=1}^{m} d_i^2 (1/d_i - a_i^T(x - x_{\mathrm{ac}}))^2 - m$$

$$= \sum_{i=1}^{m} d_i^2 (b_i - a_i^T x)^2 - m$$

$$\leq \left(\sum_{i=1}^{m} d_i(b_i - a_i^T x) \right)^2 - m$$

$$= \left(\sum_{i=1}^{m} d_i(b_i - a_i^T x_{\mathrm{ac}}) + \sum_{i=1}^{m} d_i a_i^T(x_{\mathrm{ac}} - x) \right)^2 - m$$

$$= m^2 - m,$$

which shows that $x \in \mathcal{E}_{\mathrm{outer}}$. (The second equality follows from the fact that $\sum_{i=1}^{m} d_i a_i = 0$. The inequality follows from $\sum_{i=1}^{m} y_i^2 \leq \left(\sum_{i=1}^{m} y_i \right)^2$ for $y \succeq 0$. The last equality follows from $\sum_{i=1}^{m} d_i a_i = 0$, and the definition of d_i.)

Analytic center of a linear matrix inequality

The definition of analytic center can be extended to sets described by generalized inequalities with respect to a cone K, if we define a logarithm on K. For example, the analytic center of a linear matrix inequality

$$x_1 A_1 + x_2 A_2 + \cdots + x_n A_n \preceq B$$

is defined as the solution of

$$\text{minimize} \quad -\log\det(B - x_1 A_1 - \cdots - x_n A_n).$$

8.6   Classification

In pattern recognition and classification problems we are given two sets of points in $\mathbf{R}^n$, $\{x_1, \ldots, x_N\}$ and $\{y_1, \ldots, y_M\}$, and wish to find a function $f : \mathbf{R}^n \to \mathbf{R}$ (within a given family of functions) that is positive on the first set and negative on the second, *i.e.*,

$$f(x_i) > 0, \quad i = 1, \ldots, N, \qquad f(y_i) < 0, \quad i = 1, \ldots, M.$$

If these inequalities hold, we say that f, or its 0-level set $\{x \mid f(x) = 0\}$, *separates*, *classifies*, or *discriminates* the two sets of points. We sometimes also consider *weak separation*, in which the weak versions of the inequalities hold.

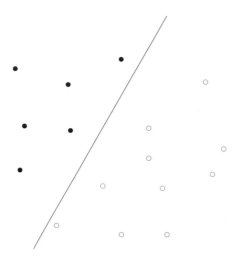

Figure 8.8 The points $x_1, \ldots, x_N$ are shown as open circles, and the points $y_1, \ldots, y_M$ are shown as filled circles. These two sets are classified by an affine function f, whose 0-level set (a line) separates them.

8.6.1   Linear discrimination

In linear discrimination, we seek an affine function $f(x) = a^T x - b$ that classifies the points, *i.e.*,

$$a^T x_i - b > 0, \quad i = 1, \ldots, N, \qquad a^T y_i - b < 0, \quad i = 1, \ldots, M. \tag{8.20}$$

Geometrically, we seek a hyperplane that separates the two sets of points. Since the strict inequalities (8.20) are homogeneous in a and b, they are feasible if and only if the set of nonstrict linear inequalities

$$a^T x_i - b \geq 1, \quad i = 1, \ldots, N, \qquad a^T y_i - b \leq -1, \quad i = 1, \ldots, M \tag{8.21}$$

(in the variables a, b) is feasible. Figure 8.8 shows a simple example of two sets of points and a linear discriminating function.

Linear discrimination alternative

The strong alternative of the set of strict inequalities (8.20) is the existence of λ, $\tilde{\lambda}$ such that

$$\lambda \succeq 0, \qquad \tilde{\lambda} \succeq 0, \qquad (\lambda, \tilde{\lambda}) \neq 0, \qquad \sum_{i=1}^{N} \lambda_i x_i = \sum_{i=1}^{M} \tilde{\lambda}_i y_i, \qquad \mathbf{1}^T \lambda = \mathbf{1}^T \tilde{\lambda} \tag{8.22}$$

(see §5.8.3). Using the third and last conditions, we can express these alternative conditions as

$$\lambda \succeq 0, \qquad \mathbf{1}^T \lambda = 1, \qquad \tilde{\lambda} \succeq 0, \qquad \mathbf{1}^T \tilde{\lambda} = 1, \qquad \sum_{i=1}^{N} \lambda_i x_i = \sum_{i=1}^{M} \tilde{\lambda}_i y_i$$

(by dividing by $\mathbf{1}^T\lambda$, which is positive, and using the same symbols for the normalized λ and $\tilde{\lambda}$). These conditions have a simple geometric interpretation: They state that there is a point in the convex hull of both $\{x_1, \ldots, x_N\}$ and $\{y_1, \ldots, y_M\}$. In other words: the two sets of points can be linearly discriminated (i.e., discriminated by an affine function) if and only if their convex hulls do not intersect. We have seen this result several times before.

Robust linear discrimination

The existence of an affine classifying function $f(x) = a^T x - b$ is equivalent to a set of linear inequalities in the variables a and b that define f. If the two sets can be linearly discriminated, then there is a polyhedron of affine functions that discriminate them, and we can choose one that optimizes some measure of robustness. We might, for example, seek the function that gives the maximum possible 'gap' between the (positive) values at the points x_i and the (negative) values at the points y_i. To do this we have to normalize a and b, since otherwise we can scale a and b by a positive constant and make the gap in the values arbitrarily large. This leads to the problem

$$
\begin{array}{ll}
\text{maximize} & t \\
\text{subject to} & a^T x_i - b \geq t, \quad i = 1, \ldots, N \\
& a^T y_i - b \leq -t, \quad i = 1, \ldots, M \\
& \|a\|_2 \leq 1,
\end{array}
\tag{8.23}
$$

with variables a, b, and t. The optimal value $t^\star$ of this convex problem (with linear objective, linear inequalities, and one quadratic inequality) is positive if and only if the two sets of points can be linearly discriminated. In this case the inequality $\|a\|_2 \leq 1$ is always tight at the optimum, i.e., we have $\|a^\star\|_2 = 1$. (See exercise 8.23.)

   We can give a simple geometric interpretation of the robust linear discrimination problem (8.23). If $\|a\|_2 = 1$ (as is the case at any optimal point), $a^T x_i - b$ is the Euclidean distance from the point x_i to the separating hyperplane $\mathcal{H} = \{z \mid a^T z = b\}$. Similarly, $b - a^T y_i$ is the distance from the point y_i to the hyperplane. Therefore the problem (8.23) finds the hyperplane that separates the two sets of points, and has maximal distance to the sets. In other words, it finds the thickest *slab* that separates the two sets.

   As suggested by the example shown in figure 8.9, the optimal value $t^\star$ (which is half the slab thickness) turns out to be half the distance between the convex hulls of the two sets of points. This can be seen clearly from the dual of the robust linear discrimination problem (8.23). The Lagrangian (for the problem of minimizing $-t$) is

$$
-t + \sum_{i=1}^N u_i (t + b - a^T x_i) + \sum_{i=1}^M v_i (t - b + a^T y_i) + \lambda(\|a\|_2 - 1).
$$

Minimizing over b and t yields the conditions $\mathbf{1}^T u = 1/2$, $\mathbf{1}^T v = 1/2$. When these hold, we have

$$
g(u, v, \lambda) = \inf_a \left(a^T \left(\sum_{i=1}^M v_i y_i - \sum_{i=1}^N u_i x_i \right) + \lambda \|a\|_2 - \lambda \right)
$$

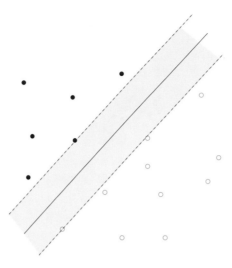

Figure 8.9 By solving the robust linear discrimination problem (8.23) we find an affine function that gives the largest gap in values between the two sets (with a normalization bound on the linear part of the function). Geometrically, we are finding the thickest slab that separates the two sets of points.

$$= \begin{cases} -\lambda & \left\| \sum_{i=1}^{M} v_i y_i - \sum_{i=1}^{N} u_i x_i \right\|_2 \le \lambda \\ -\infty & \text{otherwise.} \end{cases}$$

The dual problem can then be written as

$$\begin{array}{ll} \text{maximize} & - \left\| \sum_{i=1}^{M} v_i y_i - \sum_{i=1}^{N} u_i x_i \right\|_2 \\ \text{subject to} & u \succeq 0, \quad \mathbf{1}^T u = 1/2 \\ & v \succeq 0, \quad \mathbf{1}^T v = 1/2. \end{array}$$

We can interpret $2 \sum_{i=1}^{N} u_i x_i$ as a point in the convex hull of $\{x_1, \dots, x_N\}$ and $2 \sum_{i=1}^{M} v_i y_i$ as a point in the convex hull of $\{y_1, \dots, y_M\}$. The dual objective is to minimize (half) the distance between these two points, *i.e.*, find (half) the distance between the convex hulls of the two sets.

Support vector classifier

When the two sets of points cannot be linearly separated, we might seek an affine function that approximately classifies the points, for example, one that minimizes the number of points misclassified. Unfortunately, this is in general a difficult combinatorial optimization problem. One heuristic for approximate linear discrimination is based on *support vector classifiers*, which we describe in this section.

We start with the feasibility problem (8.21). We first relax the constraints by introducing nonnegative variables $u_1, \dots, u_N$ and $v_1, \dots, v_M$, and forming the inequalities

$$a^T x_i - b \ge 1 - u_i, \quad i = 1, \dots, N, \qquad a^T y_i - b \le -(1 - v_i), \quad i = 1, \dots, M. \quad (8.24)$$

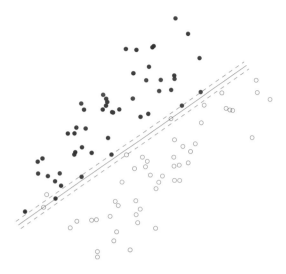

Figure 8.10 Approximate linear discrimination via linear programming. The points $x_1, \ldots, x_{50}$, shown as open circles, cannot be linearly separated from the points $y_1, \ldots, y_{50}$, shown as filled circles. The classifier shown as a solid line was obtained by solving the LP (8.25). This classifier misclassifies one point. The dashed lines are the hyperplanes $a^T z - b = \pm 1$. Four points are correctly classified, but lie in the slab defined by the dashed lines.

When $u = v = 0$, we recover the original constraints; by making u and v large enough, these inequalities can always be made feasible. We can think of u_i as a measure of how much the constraint $a^T x_i - b \geq 1$ is violated, and similarly for v_i. Our goal is to find a, b, and *sparse* nonnegative u and v that satisfy the inequalities (8.24). As a heuristic for this, we can minimize the sum of the variables u_i and v_i, by solving the LP

$$
\begin{array}{ll}
\text{minimize} & \mathbf{1}^T u + \mathbf{1}^T v \\
\text{subject to} & a^T x_i - b \geq 1 - u_i, \quad i = 1, \ldots, N \\
& a^T y_i \quad b \leq -(1 - v_i), \quad i = 1, \ldots, M \\
& u \succeq 0, \quad v \succeq 0.
\end{array}
\tag{8.25}
$$

Figure 8.10 shows an example. In this example, the affine function $a^T z - b$ misclassifies 1 out of 100 points. Note however that when $0 < u_i < 1$, the point x_i is correctly classified by the affine function $a^T z - b$, but violates the inequality $a^T x_i - b \geq 1$, and similarly for y_i. The objective function in the LP (8.25) can be interpreted as a relaxation of the number of points x_i that violate $a^T x_i - b \geq 1$ plus the number of points y_i that violate $a^T y_i - b \leq -1$. In other words, it is a relaxation of the number of points misclassified by the function $a^T z - b$, plus the number of points that are correctly classified but lie in the slab defined by $-1 < a^T z - b < 1$.

More generally, we can consider the trade-off between the number of misclassified points, and the width of the slab $\{z \mid -1 \leq a^T z - b \leq 1\}$, which is given by $2/\|a\|_2$. The standard *support vector classifier* for the sets $\{x_1, \ldots, x_N\}$,

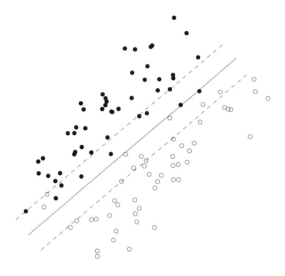

Figure 8.11 Approximate linear discrimination via support vector classifier, with $\gamma = 0.1$. The support vector classifier, shown as the solid line, misclassifies three points. Fifteen points are correctly classified but lie in the slab defined by $-1 < a^T z - b < 1$, bounded by the dashed lines.

$\{y_1, \ldots, y_M\}$ is defined as the solution of

$$
\begin{array}{ll}
\text{minimize} & \|a\|_2 + \gamma(\mathbf{1}^T u + \mathbf{1}^T v) \\
\text{subject to} & a^T x_i - b \geq 1 - u_i, \quad i = 1, \ldots, N \\
& a^T y_i - b \leq -(1 - v_i), \quad i = 1, \ldots, M \\
& u \succeq 0, \quad v \succeq 0,
\end{array}
$$

The first term is proportional to the inverse of the width of the slab defined by $-1 \leq a^T z - b \leq 1$. The second term has the same interpretation as above, *i.e.*, it is a convex relaxation for the number of misclassified points (including the points in the slab). The parameter γ, which is positive, gives the relative weight of the number of misclassified points (which we want to minimize), compared to the width of the slab (which we want to maximize). Figure 8.11 shows an example.

Approximate linear discrimination via logistic modeling

Another approach to finding an affine function that approximately classifies two sets of points that cannot be linearly separated is based on the logistic model described in §7.1.1. We start by fitting the two sets of points with a logistic model. Suppose z is a random variable with values 0 or 1, with a distribution that depends on some (deterministic) explanatory variable $u \in \mathbf{R}^n$, via a logistic model of the form

$$
\begin{aligned}
\mathbf{prob}(z = 1) &= (\exp(a^T u - b))/(1 + \exp(a^T u - b)) \\
\mathbf{prob}(z = 0) &= 1/(1 + \exp(a^T u - b)).
\end{aligned}
\tag{8.26}
$$

Now we assume that the given sets of points, $\{x_1, \ldots, x_N\}$ and $\{y_1, \ldots, y_M\}$, arise as samples from the logistic model. Specifically, $\{x_1, \ldots, x_N\}$ are the values of u for the N samples for which $z = 1$, and $\{y_1, \ldots, y_M\}$ are the values of u for the M samples for which $z = 0$. (This allows us to have $x_i = y_j$, which would rule out discrimination between the two sets. In a logistic model, it simply means that we have two samples, with the same value of explanatory variable but different outcomes.)

We can determine a and b by maximum likelihood estimation from the observed samples, by solving the convex optimization problem

$$\text{minimize} \quad -l(a, b) \tag{8.27}$$

with variables a, b, where l is the log-likelihood function

$$
\begin{aligned}
l(a, b) = {} & \sum_{i=1}^{N}(a^T x_i - b) \\
& - \sum_{i=1}^{N} \log(1 + \exp(a^T x_i - b)) - \sum_{i=1}^{M} \log(1 + \exp(a^T y_i - b))
\end{aligned}
$$

(see §7.1.1). If the two sets of points can be linearly separated, *i.e.*, if there exist a, b with $a^T x_i > b$ and $a^T y_i < b$, then the optimization problem (8.27) is unbounded below.

Once we find the maximum likelihood values of a and b, we can form a linear classifier $f(x) = a^T x - b$ for the two sets of points. This classifier has the following property: Assuming the data points are in fact generated from a logistic model with parameters a and b, it has the smallest probability of misclassification, over all linear classifiers. The hyperplane $a^T u = b$ corresponds to the points where **prob**$(z = 1) = 1/2$, *i.e.*, the two outcomes are equally likely. An example is shown in figure 8.12.

Remark 8.1 *Bayesian interpretation.* Let x and z be two random variables, taking values in $\mathbf{R}^n$ and in $\{0, 1\}$, respectively. We assume that

$$\mathbf{prob}(z = 1) = \mathbf{prob}(z = 0) = 1/2,$$

and we denote by $p_0(x)$ and $p_1(x)$ the conditional probability densities of x, given $z = 0$ and given $z = 1$, respectively. We assume that p_0 and p_1 satisfy

$$\frac{p_1(x)}{p_0(x)} = e^{a^T x - b}$$

for some a and b. Many common distributions satisfy this property. For example, p_0 and p_1 could be two normal densities on $\mathbf{R}^n$ with equal covariance matrices and different means, or they could be two exponential densities on $\mathbf{R}^n_+$.

It follows from Bayes' rule that

$$\mathbf{prob}(z = 1 \mid x = u) = \frac{p_1(u)}{p_1(u) + p_0(u)}$$

$$\mathbf{prob}(z = 0 \mid x = u) = \frac{p_0(u)}{p_1(u) + p_0(u)},$$

from which we obtain

$$\mathbf{prob}(z = 1 \mid x = u) = \frac{\exp(a^T u - b)}{1 + \exp(a^T u - b)}$$

$$\mathbf{prob}(z = 0 \mid x = u) = \frac{1}{1 + \exp(a^T u - b)}.$$

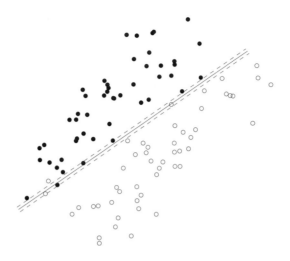

Figure 8.12 Approximate linear discrimination via logistic modeling. The points $x_1, \ldots, x_{50}$, shown as open circles, cannot be linearly separated from the points $y_1, \ldots, y_{50}$, shown as filled circles. The maximum likelihood logistic model yields the hyperplane shown as a dark line, which misclassifies only two points. The two dashed lines show $a^T u - b = \pm 1$, where the probability of each outcome, according to the logistic model, is 73%. Three points are correctly classified, but lie in between the dashed lines.

The logistic model (8.26) can therefore be interpreted as the posterior distribution of z, given that $x = u$.

8.6.2   Nonlinear discrimination

We can just as well seek a nonlinear function f, from a given subspace of functions, that is positive on one set and negative on another:

$$f(x_i) > 0, \quad i = 1, \ldots, N, \qquad f(y_i) < 0, \quad i = 1, \ldots, M.$$

Provided f is linear (or affine) in the parameters that define it, these inequalities can be solved in exactly the same way as in linear discrimination. In this section we examine some interesting special cases.

Quadratic discrimination

Suppose we take f to be quadratic: $f(x) = x^T P x + q^T x + r$. The parameters $P \in \mathbf{S}^n$, $q \in \mathbf{R}^n$, $r \in \mathbf{R}$ must satisfy the inequalities

$$x_i^T P x_i + q^T x_i + r > 0, \quad i = 1, \ldots, N$$
$$y_i^T P y_i + q^T y_i + r < 0, \quad i = 1, \ldots, M,$$

which is a set of strict linear inequalities in the variables P, q, r. As in linear discrimination, we note that f is homogeneous in P, q, and r, so we can find a solution to the strict inequalities by solving the nonstrict feasibility problem

$$x_i^T P x_i + q^T x_i + r \geq 1, \quad i = 1, \ldots, N$$
$$y_i^T P y_i + q^T y_i + r \leq -1, \quad i = 1, \ldots, M.$$

The separating surface $\{z \mid z^T P z + q^T z + r = 0\}$ is a quadratic surface, and the two classification regions

$$\{z \mid z^T P z + q^T z + r \leq 0\}, \qquad \{z \mid z^T P z + q^T z + r \geq 0\},$$

are defined by quadratic inequalities. Solving the quadratic discrimination problem, then, is the same as determining whether the two sets of points can be separated by a quadratic surface.

We can impose conditions on the shape of the separating surface or classification regions by adding constraints on P, q, and r. For example, we can require that $P \prec 0$, which means the separating surface is ellipsoidal. More specifically, it means that we seek an ellipsoid that contains all the points $x_1, \ldots, x_N$, but none of the points $y_1, \ldots, y_M$. This quadratic discrimination problem can be solved as an SDP feasibility problem

$$
\begin{array}{ll}
\text{find} & P, \ q, \ r \\
\text{subject to} & x_i^T P x_i + q^T x_i + r \geq 1, \quad i = 1, \ldots, N \\
& y_i^T P y_i + q^T y_i + r \leq -1, \quad i = 1, \ldots, M \\
& P \preceq -I,
\end{array}
$$

with variables $P \in \mathbf{S}^n$, $q \in \mathbf{R}^n$, and $r \in \mathbf{R}$. (Here we use homogeneity in P, q, r to express the constraint $P \prec 0$ as $P \preceq -I$.) Figure 8.13 shows an example.

Polynomial discrimination

We consider the set of polynomials on $\mathbf{R}^n$ with degree less than or equal to d:

$$f(x) = \sum_{i_1 + \cdots + i_n \leq d} a_{i_1 \cdots i_d} x_1^{i_1} \cdots x_n^{i_n}.$$

We can determine whether or not two sets $\{x_1, \ldots, x_N\}$ and $\{y_1, \ldots, y_M\}$ can be separated by such a polynomial by solving a set of linear inequalities in the variables $a_{i_1 \cdots i_d}$. Geometrically, we are checking whether the two sets can be separated by an algebraic surface (defined by a polynomial of degree less than or equal to d).

As an extension, the problem of determining the minimum degree polynomial on $\mathbf{R}^n$ that separates two sets of points can be solved via quasiconvex programming, since the degree of a polynomial is a quasiconvex function of the coefficients. This can be carried out by bisection on d, solving a feasibility linear program at each step. An example is shown in figure 8.14.

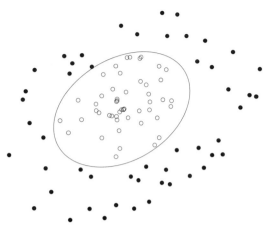

Figure 8.13 Quadratic discrimination, with the condition that $P \prec 0$. This means that we seek an ellipsoid containing all of x_i (shown as open circles) and none of the y_i (shown as filled circles). This can be solved as an SDP feasibility problem.

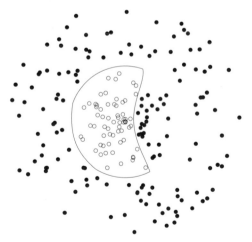

Figure 8.14 Minimum degree polynomial discrimination in $\mathbf{R}^2$. In this example, there exists no cubic polynomial that separates the points $x_1, \ldots, x_N$ (shown as open circles) from the points $y_1, \ldots, y_M$ (shown as filled circles), but they can be separated by fourth-degree polynomial, the zero level set of which is shown.

8.7   Placement and location

In this section we discuss a few variations on the following problem. We have
N points in $\mathbf{R}^2$ or $\mathbf{R}^3$, and a list of pairs of points that must be connected by
links. The positions of some of the N points are fixed; our task is to determine the
positions of the remaining points, *i.e.*, to *place* the remaining points. The objective
is to place the points so that some measure of the total interconnection length of
the links is minimized, subject to some additional constraints on the positions.
As an example application, we can think of the points as locations of plants or
warehouses of a company, and the links as the routes over which goods must be
shipped. The goal is to find locations that minimize the total transportation cost.
In another application, the points represent the position of modules or cells on an
integrated circuit, and the links represent wires that connect pairs of cells. Here
the goal might be to place the cells in such a way that the total length of wire used
to interconnect the cells is minimized.

The problem can be described in terms of an undirected graph with N nodes,
representing the N points. With each node we associate a variable $x_i \in \mathbf{R}^k$, where
$k = 2$ or $k = 3$, which represents its location or position. The problem is to
minimize

$$\sum_{(i,j)\in\mathcal{A}} f_{ij}(x_i, x_j)$$

where $\mathcal{A}$ is the set of all links in the graph, and $f_{ij} : \mathbf{R}^k \times \mathbf{R}^k \to \mathbf{R}$ is a cost
function associated with arc (i,j). (Alternatively, we can sum over all i and j, or
over $i < j$, and simply set $f_{ij} = 0$ when links i and j are not connected.) Some of
the coordinate vectors x_i are given. The optimization variables are the remaining
coordinates. Provided the functions f_{ij} are convex, this is a convex optimization
problem.

8.7.1   Linear facility location problems

In the simplest version of the problem the cost associated with arc (i,j) is the
distance between nodes i and j: $f_{ij}(x_i, x_j) = \|x_i - x_j\|$, *i.e.*, we minimize

$$\sum_{(i,j)\in\mathcal{A}} \|x_i - x_j\|.$$

We can use any norm, but the most common applications involve the Euclidean
norm or the ℓ_1-norm. For example, in circuit design it is common to route the wires
between cells along piecewise-linear paths, with each segment either horizontal or
vertical. (This is called *Manhattan routing*, since paths along the streets in a city
with a rectangular grid are also piecewise-linear, with each street aligned with one
of two orthogonal axes.) In this case, the length of wire required to connect cell i
and cell j is given by $\|x_i - x_j\|_1$.

We can include nonnegative weights that reflect differences in the cost per unit distance along different arcs:

$$\sum_{(i,j)\in\mathcal{A}} w_{ij}\|x_i - x_j\|.$$

By assigning a weight $w_{ij} = 0$ to pairs of nodes that are not connected, we can express this problem more simply using the objective

$$\sum_{i<j} w_{ij}\|x_i - x_j\|. \tag{8.28}$$

This placement problem is convex.

Example 8.4 *One free point.* Consider the case where only one point $(u, v) \in \mathbf{R}^2$ is free, and we minimize the sum of the distances to fixed points $(u_1, v_1), \ldots, (u_K, v_K)$.

- ℓ_1-*norm.* We can find a point that minimizes

$$\sum_{i=1}^{K} (|u - u_i| + |v - v_i|)$$

  analytically. An optimal point is any *median* of the fixed points. In other words, u can be taken to be any median of the points $\{u_1, \ldots, u_K\}$, and v can be taken to be any median of the points $\{v_1, \ldots, v_K\}$. (If K is odd, the minimizer is unique; if K is even, there can be a rectangle of optimal points.)

- *Euclidean norm.* The point (u, v) that minimizes the sum of the Euclidean distances,

$$\sum_{i=1}^{K} \left((u - u_i)^2 + (v - v_i)^2\right)^{1/2},$$

  is called the *Weber point* of the given fixed points.

8.7.2   Placement constraints

We now list some interesting constraints that can be added to the basic placement problem, preserving convexity. We can require some positions x_i to lie in a specified convex set, *e.g.*, a particular line, interval, square, or ellipsoid. We can constrain the relative position of one point with respect to one or more other points, for example, by limiting the distance between a pair of points. We can impose relative position constraints, *e.g.*, that one point must lie to the left of another point.

The *bounding box* of a group of points is the smallest rectangle that contains the points. We can impose a constraint that limits the points $x_1, \ldots, x_p$ (say) to lie in a bounding box with perimeter not exceeding $P_{\max}$, by adding the constraints

$$u \preceq x_i \preceq v, \quad i = 1, \ldots, p, \qquad 2\mathbf{1}^T(v - u) \leq P_{\max},$$

where u, v are additional variables.

8.7.3   Nonlinear facility location problems

More generally, we can associate a cost with each arc that is a nonlinear increasing function of the length, *i.e.*,

$$\text{minimize} \quad \sum_{i<j} w_{ij} h(\|x_i - x_j\|)$$

where h is an increasing (on $\mathbf{R}_+$) and convex function, and $w_{ij} \geq 0$. We call this a *nonlinear placement* or *nonlinear facility location problem*.

One common example uses the Euclidean norm, and the function $h(z) = z^2$, *i.e.*, we minimize

$$\sum_{i<j} w_{ij} \|x_i - x_j\|_2^2.$$

This is called a *quadratic placement problem*. The quadratic placement problem can be solved analytically when the only constraints are linear equalities; it can be solved as a QP if the constraints are linear equalities and inequalities.

Example 8.5 *One free point.* Consider the case where only one point x is free, and we minimize the sum of the squares of the Euclidean distances to fixed points $x_1, \ldots, x_K$,

$$\|x - x_1\|_2^2 + \|x - x_2\|_2^2 + \cdots + \|x - x_K\|_2^2.$$

Taking derivatives, we see that the optimal x is given by

$$\frac{1}{K}(x_1 + x_2 + \cdots + x_K),$$

i.e., the average of the fixed points.

Some other interesting possibilities are the 'deadzone' function h with deadzone width 2γ, defined as

$$h(z) = \begin{cases} 0 & |z| \leq \gamma \\ |z - \gamma| & |z| \geq \gamma, \end{cases}$$

and the 'quadratic-linear' function h, defined as

$$h(z) = \begin{cases} z^2 & |z| \leq \gamma \\ 2\gamma|z| - \gamma^2 & |z| \geq \gamma. \end{cases}$$

Example 8.6   We consider a placement problem in $\mathbf{R}^2$ with 6 free points, 8 fixed points, and 27 links. Figures 8.15–8.17 show the optimal solutions for the criteria

$$\sum_{(i,j)\in\mathcal{A}} \|x_i - x_j\|_2, \qquad \sum_{(i,j)\in\mathcal{A}} \|x_i - x_j\|_2^2, \qquad \sum_{(i,j)\in\mathcal{A}} \|x_i - x_j\|_2^4,$$

i.e., using the penalty functions $h(z) = z$, $h(z) = z^2$, and $h(z) = z^4$. The figures also show the resulting distributions of the link lengths.

Comparing the results, we see that the linear placement concentrates the free points in a small area, while the quadratic and fourth-order placements spread the points over larger areas. The linear placement includes many very short links, and a few very long ones (3 lengths under 0.2 and 2 lengths above 1.5.). The quadratic penalty function

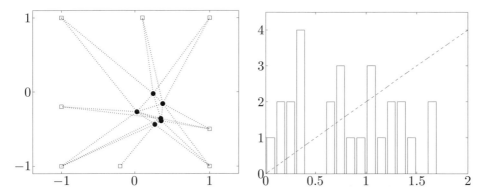

Figure 8.15 *Linear placement.* Placement problem with 6 free points (shown as dots), 8 fixed points (shown as squares), and 27 links. The coordinates of the free points minimize the sum of the Euclidean lengths of the links. The right plot is the distribution of the 27 link lengths. The dashed curve is the (scaled) penalty function $h(z) = z$.

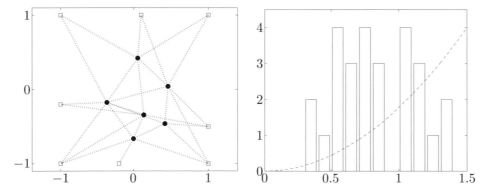

Figure 8.16 *Quadratic placement.* Placement that minimizes the sum of squares of the Euclidean lengths of the links, for the same data as in figure 8.15. The dashed curve is the (scaled) penalty function $h(z) = z^2$.

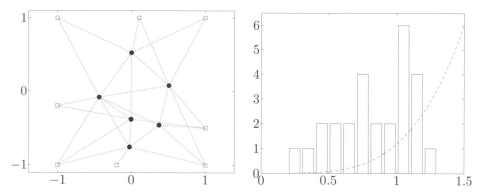

Figure 8.17 *Fourth-order placement.* Placement that minimizes the sum of the fourth powers of the Euclidean lengths of the links. The dashed curve is the (scaled) penalty function $h(z) = z^4$.

puts a higher penalty on long lengths relative to short lengths, and for lengths under 0.1, the penalty is almost negligible. As a result, the maximum length is shorter (less than 1.4), but we also have fewer short links. The fourth-order function puts an even higher penalty on long lengths, and has a wider interval (between zero and about 0.4) where it is negligible. As a result, the maximum length is shorter than for the quadratic placement, but we also have more lengths close to the maximum.

8.7.4   Location problems with path constraints

Path constraints

A p-link *path* along the points $x_1, \ldots, x_N$ is described by a sequence of nodes, $i_0, \ldots, i_p \in \{1, \ldots, N\}$. The length of the path is given by

$$\|x_{i_1} - x_{i_0}\| + \|x_{i_2} - x_{i_1}\| + \cdots + \|x_{i_p} - x_{i_{p-1}}\|,$$

which is a convex function of $x_1, \ldots, x_N$, so imposing an upper bound on the length of a path is a convex constraint. Several interesting placement problems involve path constraints, or have an objective based on path lengths. We describe one typical example, in which the objective is based on a maximum path length over a set of paths.

Minimax delay placement

We consider a directed acyclic graph with nodes $1, \ldots, N$, and arcs or links represented by a set $\mathcal{A}$ of ordered pairs: $(i, j) \in \mathcal{A}$ if and only if an arc points from i to j. We say node i is a *source node* if no arc $\mathcal{A}$ points to it; it is a *sink node* or *destination node* if no arc in $\mathcal{A}$ leaves from it. We will be interested in the maximal paths in the graph, which begin at a source node and end at a sink node.

The arcs of the graph are meant to model some kind of flow, say of goods or information, in a network with nodes at positions $x_1, \ldots, x_N$. The flow starts at a source node, then moves along a path from node to node, ending at a sink or destination node. We use the distance between successive nodes to model propagation time, or shipment time, of the goods between nodes; the total delay or propagation time of a path is (proportional to) the sum of the distances between successive nodes.

Now we can describe the minimax delay placement problem. Some of the node locations are fixed, and the others are free, *i.e.*, optimization variables. The goal is to choose the free node locations in order to minimize the maximum total delay, for any path from a source node to a sink node. Evidently this is a convex problem, since the objective

$$T_{\mathrm{max}} = \max\{\|x_{i_1} - x_{i_0}\| + \cdots + \|x_{i_p} - x_{i_{p-1}}\| \mid i_0, \ldots, i_p \text{ is a source-sink path}\} \tag{8.29}$$

is a convex function of the locations $x_1, \ldots, x_N$.

While the problem of minimizing (8.29) is convex, the number of source-sink paths can be very large, exponential in the number of nodes or arcs. There is a useful reformulation of the problem, which avoids enumerating all sink-source paths.

We first explain how we can evaluate the maximum delay T_{max} far more efficiently than by evaluating the delay for every source-sink path, and taking the maximum. Let τ_k be the maximum total delay of any path from node k to a sink node. Clearly we have $\tau_k = 0$ when k is a sink node. Consider a node k, which has outgoing arcs to nodes $j_1, \ldots, j_p$. For a path starting at node k and ending at a sink node, its first arc must lead to one of the nodes $j_1, \ldots, j_p$. If such a path first takes the arc leading to j_i, and then takes the longest path from there to a sink node, the total length is

$$\|x_{j_i} - x_k\| + \tau_{j_i},$$

i.e., the length of the arc to j_i, plus the total length of the longest path from j_i to a sink node. It follows that the maximum delay of a path starting at node k and leading to a sink node satisfies

$$\tau_k = \max\{\|x_{j_1} - x_k\| + \tau_{j_1}, \ldots, \|x_{j_p} - x_k\| + \tau_{j_p}\}. \tag{8.30}$$

(This is a simple *dynamic programming* argument.)

The equations (8.30) give a recursion for finding the maximum delay from any node: we start at the sink nodes (which have maximum delay zero), and then work backward using the equations (8.30), until we reach all source nodes. The maximum delay over any such path is then the maximum of all the τ_k, which will occur at one of the source nodes. This dynamic programming recursion shows how the maximum delay along any source-sink path can be computed recursively, without enumerating all the paths. The number of arithmetic operations required for this recursion is approximately the number of links.

Now we show how the recursion based on (8.30) can be used to formulate the minimax delay placement problem. We can express the problem as

$$
\begin{array}{ll}
\text{minimize} & \max\{\tau_k \mid k \text{ a source node}\} \\
\text{subject to} & \tau_k = 0, \quad k \text{ a sink node} \\
& \tau_k = \max\{\|x_j - x_k\| + \tau_j \mid \text{there is an arc from } k \text{ to } j\},
\end{array}
$$

with variables $\tau_1, \ldots, \tau_N$ and the free positions. This problem is not convex, but we can express it in an equivalent form that is convex, by replacing the equality constraints with inequalities. We introduce new variables $T_1, \ldots, T_N$, which will be upper bounds on $\tau_1, \ldots, \tau_N$, respectively. We will take $T_k = 0$ for all sink nodes, and in place of (8.30) we take the inequalities

$$T_k \geq \max\{\|x_{j_1} - x_k\| + T_{j_1}, \ldots, \|x_{j_p} - x_k\| + T_{j_p}\}.$$

If these inequalities are satisfied, then $T_k \geq \tau_k$. Now we form the problem

$$
\begin{array}{ll}
\text{minimize} & \max\{T_k \mid k \text{ a source node}\} \\
\text{subject to} & T_k = 0, \quad k \text{ a sink node} \\
& T_k \geq \max\{\|x_j - x_k\| + T_j \mid \text{there is an arc from } k \text{ to } j\}.
\end{array}
$$

This problem, with variables $T_1, \ldots, T_N$ and the free locations, is convex, and solves the minimax delay location problem.

8.8   Floor planning

In placement problems, the variables represent the coordinates of a number of points that are to be optimally placed. A *floor planning problem* can be considered an extension of a placement problem in two ways:

- The objects to be placed are rectangles or boxes aligned with the axes (as opposed to points), and must not overlap.

- Each rectangle or box to be placed can be reconfigured, within some limits. For example we might fix the area of each rectangle, but not the length and height separately.

The objective is usually to minimize the size (*e.g.*, area, volume, perimeter) of the *bounding box*, which is the smallest box that contains the boxes to be configured and placed.

The non-overlap constraints make the general floor planning problem a complicated combinatorial optimization problem or rectangle packing problem. However, if the *relative positioning* of the boxes is specified, several types of floor planning problems can be formulated as convex optimization problems. We explore some of these in this section. We consider the two-dimensional case, and make a few comments on extensions to higher dimensions (when they are not obvious).

We have N cells or modules $C_1, \ldots, C_N$ that are to be configured and placed in a rectangle with width W and height H, and lower left corner at the position $(0, 0)$. The geometry and position of the ith cell is specified by its width w_i and height h_i, and the coordinates (x_i, y_i) of its lower left corner. This is illustrated in figure 8.18.

The variables in the problem are x_i, y_i, w_i, h_i for $i = 1, \ldots, N$, and the width W and height H of the bounding rectangle. In all floor planning problems, we require that the cells lie inside the bounding rectangle, *i.e.*,

$$x_i \geq 0, \qquad y_i \geq 0, \qquad x_i + w_i \leq W, \qquad y_i + h_i \leq H, \qquad i = 1, \ldots, N. \quad (8.31)$$

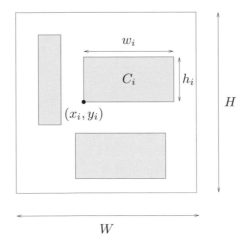

Figure 8.18 Floor planning problem. Non-overlapping rectangular cells are placed in a rectangle with width W, height H, and lower left corner at $(0, 0)$. The ith cell is specified by its width w_i, height h_i, and the coordinates of its lower left corner, (x_i, y_i).

We also require that the cells do not overlap, except possibly on their boundaries:

$$\mathbf{int}\,(C_i \cap C_j) = \emptyset \quad \text{for } i \neq j.$$

(It is also possible to require a positive minimum clearance between the cells.) The non-overlap constraint $\mathbf{int}(C_i \cap C_j) = \emptyset$ holds if and only if for $i \neq j$,

C_i is left of C_j, or C_i is right of C_j, or C_i is below C_j, or C_i is above C_j.

These four geometric conditions correspond to the inequalities

$$x_i + w_i \leq x_j, \text{ or } x_j + w_j \leq x_i, \text{ or } y_i + h_j \leq y_j, \text{ or } y_j + h_i \leq y_i, \qquad (8.32)$$

at least one of which must hold for each $i \neq j$. Note the combinatorial nature of these constraints: for each pair $i \neq j$, at least one of the four inequalities above must hold.

8.8.1   Relative positioning constraints

The idea of relative positioning constraints is to specify, for each pair of cells, one of the four possible relative positioning conditions, *i.e.*, left, right, above, or below. One simple method to specify these constraints is to give two relations on $\{1, \ldots, N\}$: $\mathcal{L}$ (meaning 'left of') and $\mathcal{B}$ (meaning 'below'). We then impose the constraint that C_i is to the left of C_j if $(i, j) \in \mathcal{L}$, and C_i is below C_j if $(i, j) \in \mathcal{B}$. This yields the constraints

$$x_i + w_i \leq x_j \text{ for } (i, j) \in \mathcal{L}, \qquad y_i + h_i \leq y_j \text{ for } (i, j) \in \mathcal{B}, \qquad (8.33)$$

for $i, j = 1, \ldots, N$. To ensure that the relations $\mathcal{L}$ and $\mathcal{B}$ specify the relative positioning of each pair of cells, we require that for each (i, j) with $i \neq j$, one of the following holds:

$$(i, j) \in \mathcal{L}, \qquad (j, i) \in \mathcal{L}, \qquad (i, j) \in \mathcal{B}, \qquad (j, i) \in \mathcal{B},$$

and that $(i, i) \notin \mathcal{L}$, $(i, i) \notin \mathcal{B}$. The inequalities (8.33) are a set of $N(N-1)/2$ linear inequalities in the variables. These inequalities imply the non-overlap inequalities (8.32), which are a set of $N(N-1)/2$ *disjunctions* of four linear inequalities.

We can assume that the relations $\mathcal{L}$ and $\mathcal{B}$ are anti-symmetric (*i.e.*, $(i, j) \in \mathcal{L} \Rightarrow (j, i) \notin \mathcal{L}$) and transitive (*i.e.*, $(i, j) \in \mathcal{L}$, $(j, k) \in \mathcal{L} \Rightarrow (i, k) \in \mathcal{L}$). (If this were not the case, the relative positioning constraints would clearly be infeasible.) Transitivity corresponds to the obvious condition that if cell C_i is to the left of cell C_j, which is to the left of cell C_k, then cell C_i must be to the left of cell C_k. In this case the inequality corresponding to $(i, k) \in \mathcal{L}$ is redundant; it is implied by the other two. By exploiting transitivity of the relations $\mathcal{L}$ and $\mathcal{B}$ we can remove redundant constraints, and obtain a compact set of relative positioning inequalities.

A minimal set of relative positioning constraints is conveniently described using two directed acyclic graphs $\mathcal{H}$ and $\mathcal{V}$ (for horizontal and vertical). Both graphs have N nodes, corresponding to the N cells in the floor planning problem. The graph $\mathcal{H}$ generates the relation $\mathcal{L}$ as follows: we have $(i, j) \in \mathcal{L}$ if and only if there is a (directed) path in $\mathcal{H}$ from i to j. Similarly, the graph $\mathcal{V}$ generates the relation $\mathcal{B}$: $(i, j) \in \mathcal{B}$ if and only if there is a (directed) path in $\mathcal{V}$ from i to j. To ensure that a relative positioning constraint is given for every pair of cells, we require that for every pair of cells, there is a directed path from one to the other in one of the graphs.

Evidently, we only need to impose the inequalities that correspond to the edges of the graphs $\mathcal{H}$ and $\mathcal{V}$; the others follow from transitivity. We arrive at the set of inequalities

$$x_i + w_i \leq x_j \text{ for } (i, j) \in \mathcal{H}, \qquad y_i + h_i \leq y_j \text{ for } (i, j) \in \mathcal{V}, \qquad (8.34)$$

which is a set of linear inequalities, one for each edge in $\mathcal{H}$ and $\mathcal{V}$. The set of inequalities (8.34) is a subset of the set of inequalities (8.33), and equivalent.

In a similar way, the $4N$ inequalities (8.31) can be reduced to a minimal, equivalent set. The constraint $x_i \geq 0$ only needs to be imposed on the left-most cells, *i.e.*, for i that are minimal in the relation $\mathcal{L}$. These correspond to the sources in the graph $\mathcal{H}$, *i.e.*, those nodes that have no edges pointing to them. Similarly, the inequalities $x_i + w_i \leq W$ only need to be imposed for the right-most cells. In the same way the vertical bounding box inequalities can be pruned to a minimal set. This yields the minimal equivalent set of bounding box inequalities

$$\begin{array}{ll} x_i \geq 0 \text{ for } i \ \mathcal{L} \text{ minimal}, & x_i + w_i \leq W \text{ for } i \ \mathcal{L} \text{ maximal}, \\ y_i \geq 0 \text{ for } i \ \mathcal{B} \text{ minimal}, & y_i + h_i \leq H \text{ for } i \ \mathcal{B} \text{ maximal}. \end{array} \qquad (8.35)$$

A simple example is shown in figure 8.19. In this example, the $\mathcal{L}$ minimal or left-most cells are C_1, C_2, and C_4, and the only right-most cell is C_5. The minimal set of inequalities specifying the horizontal relative positioning is given by

$$x_1 \geq 0, \qquad x_2 \geq 0, \qquad x_4 \geq 0, \qquad x_5 + w_5 \leq W, \qquad x_1 + w_1 \leq x_3,$$
$$x_2 + w_2 \leq x_3, \qquad x_3 + w_3 \leq x_5, \qquad x_4 + w_4 \leq x_5.$$

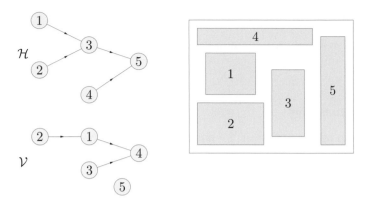

Figure 8.19 Example illustrating the horizontal and vertical graphs $\mathcal{H}$ and $\mathcal{V}$ that specify the relative positioning of the cells. If there is a path from node i to node j in $\mathcal{H}$, then cell i must be placed to the left of cell j. If there is a path from node i to node j in $\mathcal{V}$, then cell i must be placed below cell j. The floorplan shown at right satisfies the relative positioning specified by the two graphs.

The minimal set of inequalities specifying the vertical relative positioning is given by

$$y_2 \geq 0, \qquad y_3 \geq 0, \qquad y_5 \geq 0, \qquad y_4 + h_4 \leq H, \qquad y_5 + h_5 \leq H,$$
$$y_2 + h_2 \leq y_1, \qquad y_1 + h_1 \leq y_4, \qquad y_3 + h_3 \leq y_4.$$

8.8.2   Floor planning via convex optimization

In this formulation, the variables are the bounding box width and height W and H, and the cell widths, heights, and positions: w_i, h_i, x_i, and w_i, for $i = 1, \ldots, N$. We impose the bounding box constraints (8.35) and the relative positioning constraints (8.34), which are linear inequalities. As objective, we take the perimeter of the bounding box, i.e., $2(W + H)$, which is a linear function of the variables. We now list some of the constraints that can be expressed as convex inequalities or linear equalities in the variables.

Minimum spacing

We can impose a minimum spacing $\rho > 0$ between cells by changing the relative position constraints from $x_i + w_i \leq x_j$ for $(i, j) \in \mathcal{H}$, to $x_i + w_i + \rho \leq x_j$ for $(i, j) \in \mathcal{H}$, and similarly for the vertical graph. We can have a different minimum spacing associated with each edge in $\mathcal{H}$ and $\mathcal{V}$. Another possibility is to fix W and H, and maximize the minimum spacing ρ as objective.

Minimum cell area

For each cell we specify a minimum area, *i.e.*, we require that $w_i h_i \geq A_i$, where $A_i > 0$. These minimum cell area constraints can be expressed as convex inequalities in several ways, *e.g.*, $w_i \geq A_i / h_i$, $(w_i h_i)^{1/2} \geq A_i^{1/2}$, or $\log w_i + \log h_i \geq \log A_i$.

Aspect ratio constraints

We can impose upper and lower bounds on the *aspect ratio* of each cell, *i.e.*,

$$l_i \leq h_i / w_i \leq u_i.$$

Multiplying through by w_i transforms these constraints into linear inequalities. We can also fix the aspect ratio of a cell, which results in a linear equality constraint.

Alignment constraints

We can impose the constraint that two edges, or a center line, of two cells are aligned. For example, the horizontal center line of cell i aligns with the top of cell j when

$$y_i + w_i / 2 = y_j + w_j.$$

These are linear equality constraints. In a similar way we can require that a cell is flushed against the bounding box boundary.

Symmetry constraints

We can require pairs of cells to be symmetric about a vertical or horizontal axis, that can be fixed or floating (*i.e.*, whose position is fixed or not). For example, to specify that the pair of cells i and j are symmetric about the vertical axis $x = x_{\mathrm{axis}}$, we impose the linear equality constraint

$$x_{\mathrm{axis}} - (x_i + w_i / 2) = x_j + w_j / 2 - x_{\mathrm{axis}}.$$

We can require that several pairs of cells be symmetric about an unspecified vertical axis by imposing these equality constraints, and introducing x_{axis} as a new variable.

Similarity constraints

We can require that cell i be an a-scaled translate of cell j by the equality constraints $w_i = a w_j$, $h_i = a h_j$. Here the scaling factor a must be fixed. By imposing only one of these constraints, we require that the width (or height) of one cell be a given factor times the width (or height) of the other cell.

Containment constraints

We can require that a particular cell contains a given point, which imposes two linear inequalities. We can require that a particular cell lie inside a given polyhedron, again by imposing linear inequalities.

Distance constraints

We can impose a variety of constraints that limit the distance between pairs of cells. In the simplest case, we can limit the distance between the center points of cell i and j (or any other fixed points on the cells, such as lower left corners). For example, to limit the distance between the centers of cells i and j, we use the (convex) inequality

$$\|(x_i + w_i/2, y_i + h_i/2) - (x_j + w_j/2, y_j + h_j/2)\| \le D_{ij}.$$

As in placement problems, we can limit sums of distances, or use sums of distances as the objective.

We can also limit the distance $\mathbf{dist}(C_i, C_j)$ between cell i and cell j, *i.e.*, the minimum distance between a point in cell i and a point in cell j. In the general case this can be done as follows. To limit the distance between cells i and j in the norm $\| \cdot \|$, we can introduce four new variables u_i, v_i, u_j, v_j. The pair (u_i, v_i) will represent a point in C_i, and the pair (u_j, v_j) will represent a point in C_j. To ensure this we impose the linear inequalities

$$x_i \le u_i \le x_i + w_i, \qquad y_i \le v_i \le y_i + h_i,$$

and similarly for cell j. Finally, to limit $\mathbf{dist}(C_i, C_j)$, we add the convex inequality

$$\|(u_i, v_i) - (u_j, v_j)\| \le D_{ij}.$$

In many specific cases we can express these distance constraints more efficiently, by exploiting the relative positioning constraints or deriving a more explicit formulation. As an example consider the ℓ_∞-norm, and suppose cell i lies to the left of cell j (by a relative positioning constraint). The horizontal displacement between the two cells is $x_j - (x_i + w_i)$ Then we have $\mathbf{dist}(C_i, C_j) \le D_{ij}$ if and only if

$$x_j - (x_i + w_i) \le D_{ij}, \qquad y_j - (y_i + h_i) \le D_{ij}, \qquad y_i - (y_j + h_j) \le D_{ij}.$$

The first inequality states that the horizontal displacement between the right edge of cell i and the left edge of cell j does not exceed D_{ij}. The second inequality requires that the bottom of cell j is no more than D_{ij} above the top of cell i, and the third inequality requires that the bottom of cell i is no more than D_{ij} above the top of cell j. These three inequalities together are equivalent to $\mathbf{dist}(C_i, C_j) \le D_{ij}$. In this case, we do not need to introduce any new variables.

We can limit the ℓ_1- (or ℓ_2-) distance between two cells in a similar way. Here we introduce one new variable d_v, which will serve as a bound on the vertical displacement between the cells. To limit the ℓ_1-distance, we add the constraints

$$y_j - (y_i + h_i) \le d_v, \qquad y_i - (y_j + h_j) \le d_v, \qquad d_v \ge 0$$

and the constraints

$$x_j - (x_i + w_i) + d_v \le D_{ij}.$$

(The first term is the horizontal displacement and the second is an upper bound on the vertical displacement.) To limit the Euclidean distance between the cells, we replace this last constraint with

$$(x_j - (x_i + w_i))^2 + d_v^2 \le D_{ij}^2.$$

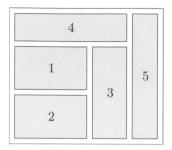

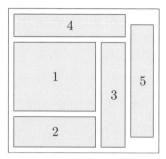
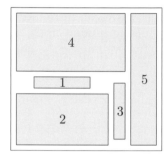

Figure 8.20 Four instances of an optimal floor plan, using the relative positioning constraints shown in figure 8.19. In each case the objective is to
minimize the perimeter, and the same minimum spacing constraint between
cells is imposed. We also require the aspect ratios to lie between 1/5 and 5.
The four cases differ in the minimum areas required for each cell. The sum
of the minimum areas is the same for each case.

Example 8.7 Figure 8.20 shows an example with 5 cells, using the ordering constraints
of figure 8.19, and four different sets of constraints. In each case we impose the
same minimum required spacing constraint, and the same aspect ratio constraint
$1/5 \le w_i/h_i \le 5$. The four cases differ in the minimum required cell areas A_i. The
values of A_i are chosen so that the total minimum required area $\sum_{i=1}^{5} A_i$ is the same
for each case.

8.8.3   Floor planning via geometric programming

The floor planning problem can also be formulated as a *geometric program* in the
variables x_i, y_i, w_i, h_i, W, H. The objectives and constraints that can be handled
in this formulation are a bit different from those that can be expressed in the convex
formulation.

First we note that the bounding box constraints (8.35) and the relative positioning constraints (8.34) are posynomial inequalities, since the lefthand sides are sums of variables, and the righthand sides are single variables, hence monomials. Dividing these inequalities by the righthand side yields standard posynomial inequalities.

In the geometric programming formulation we can minimize the bounding box area, since WH is a monomial, hence posynomial. We can also exactly specify the area of each cell, since $w_i h_i = A_i$ is a monomial equality constraint. On the other hand alignment, symmetry, and distance constraints cannot be handled in the geometric programming formulation. Similarity, however, can be; indeed it is possible to require that one cell be similar to another, without specifying the scaling ratio (which can be treated as just another variable).

Bibliography

The characterization of Euclidean distance matrices in §8.3.3 appears in Schoenberg [Sch35]; see also Gower [Gow85].

Our use of the term Löwner-John ellipsoid follows Grötschel, Lovász, and Schrijver [GLS88, page 69]. The efficiency results for ellipsoidal approximations in §8.4 were proved by John [Joh85]. Boyd, El Ghaoui, Feron, and Balakrishnan [BEFB94, §3.7] give convex formulations of several ellipsoidal approximation problems involving sets defined as unions, intersections or sums of ellipsoids.

The different centers defined in §8.5 have applications in design centering (see, for example, Seifi, Ponnambalan, and Vlach [SPV99]), and cutting-plane methods (Elzinga and Moore [EM75], Tarasov, Khachiyan, and Èrlikh [TKE88], and Ye [Ye97, chapter 8]). The inner ellipsoid defined by the Hessian of the logarithmic barrier function (page 420) is sometimes called the *Dikin ellipsoid*, and is the basis of Dikin's algorithm for linear and quadratic programming [Dik67]. The expression for the outer ellipsoid at the analytic center was given by Sonnevend [Son86]. For extensions to nonpolyhedral convex sets, see Boyd and El Ghaoui [BE93], Jarre [Jar94], and Nesterov and Nemirovski [NN94, page 34].

Convex optimization has been applied to linear and nonlinear discrimination problems since the 1960s; see Mangasarian [Man65] and Rosen [Ros65]. Standard texts that discuss pattern classification include Duda, Hart, and Stork [DHS99] and Hastie, Tibshirani, and Friedman [HTF01]. For a detailed discussion of support vector classifiers, see Vapnik [Vap00] or Schölkopf and Smola [SS01].

The Weber point defined in example 8.4 is named after Weber [Web71]. Linear and quadratic placement is used in circuit design (Kleinhaus, Sigl, Johannes, and Antreich [KSJA91, SDJ91]). Sherwani [She99] is a recent overview of algorithms for placement, layout, floor planning, and other geometric optimization problems in VLSI circuit design.

Exercises

Projection on a set

8.1 *Uniqueness of projection.* Show that if $C \subseteq \mathbf{R}^n$ is nonempty, closed and convex, and the norm $\| \cdot \|$ is strictly convex, then for every x_0 there is exactly one $x \in C$ closest to x_0. In other words the projection of x_0 on C is unique.

8.2 [Web94, Val64] *Chebyshev characterization of convexity.* A set $C \in \mathbf{R}^n$ is called a *Chebyshev set* if for every $x_0 \in \mathbf{R}^n$, there is a unique point in C closest (in Euclidean norm) to x_0. From the result in exercise 8.1, every nonempty, closed, convex set is a Chebyshev set. In this problem we show the converse, which is known as *Motzkin's theorem.*
Let $C \in \mathbf{R}^n$ be a Chebyshev set.

    (a) Show that C is nonempty and closed.

    (b) Show that P_C, the Euclidean projection on C, is continuous.

    (c) Suppose $x_0 \notin C$. Show that $P_C(x) = P_C(x_0)$ for all $x = \theta x_0 + (1-\theta)P_C(x_0)$ with $0 \le \theta \le 1$.

    (d) Suppose $x_0 \notin C$. Show that $P_C(x) = P_C(x_0)$ for all $x = \theta x_0 + (1-\theta)P_C(x_0)$ with $\theta \ge 1$.

    (e) Combining parts (c) and (d), we can conclude that all points on the ray with base $P_C(x_0)$ and direction $x_0 - P_C(x_0)$ have projection $P_C(x_0)$. Show that this implies that C is convex.

8.3 *Euclidean projection on proper cones.*

    (a) *Nonnegative orthant.* Show that Euclidean projection onto the nonnegative orthant is given by the expression on page 399.

    (b) *Positive semidefinite cone.* Show that Euclidean projection onto the positive semidefinite cone is given by the expression on page 399.

    (c) *Second-order cone.* Show that the Euclidean projection of (x_0, t_0) on the second-order cone
$$K = \{(x, t) \in \mathbf{R}^{n+1} \mid \|x\|_2 \le t\}$$
    is given by
$$P_K(x_0, t_0) = \begin{cases} 0 & \|x_0\|_2 \le -t_0 \\ (x_0, t_0) & \|x_0\|_2 \le t_0 \\ (1/2)(1 + t_0/\|x_0\|_2)(x_0, \|x_0\|_2) & \|x_0\|_2 \ge |t_0|. \end{cases}$$

8.4 The Euclidean projection of a point on a convex set yields a simple separating hyperplane
$$(P_C(x_0) - x_0)^T (x - (1/2)(x_0 + P_C(x_0))) = 0.$$

Find a counterexample that shows that this construction does not work for general norms.

8.5 [HUL93, volume 1, page 154] *Depth function and signed distance to boundary.* Let $C \subseteq \mathbf{R}^n$ be a nonempty convex set, and let $\mathbf{dist}(x, C)$ be the distance of x to C in some norm. We already know that $\mathbf{dist}(x, C)$ is a convex function of x.

    (a) Show that the depth function,
$$\mathbf{depth}(x, C) = \mathbf{dist}(x, \mathbf{R}^n \setminus C),$$
    is concave for $x \in C$.

    (b) The *signed distance* to the boundary of C is defined as
$$s(x) = \begin{cases} \mathbf{dist}(x, C) & x \notin C \\ -\mathbf{depth}(x, C) & x \in C. \end{cases}$$

    Thus, $s(x)$ is positive outside C, zero on its boundary, and negative on its interior. Show that s is a convex function.

Distance between sets

8.6 Let C, D be convex sets.

(a) Show that $\mathbf{dist}(C, x + D)$ is a convex function of x.

(b) Show that $\mathbf{dist}(tC, x + tD)$ is a convex function of (x, t) for $t > 0$.

8.7 *Separation of ellipsoids.* Let $\mathcal{E}_1$ and $\mathcal{E}_2$ be two ellipsoids defined as

$$\mathcal{E}_1 = \{x \mid (x - x_1)^T P_1^{-1}(x - x_1) \leq 1\}, \qquad \mathcal{E}_2 = \{x \mid (x - x_2)^T P_2^{-1}(x - x_2) \leq 1\},$$

where P_1, $P_2 \in \mathbf{S}_{++}^n$. Show that $\mathcal{E}_1 \cap \mathcal{E}_2 = \emptyset$ if and only if there exists an $a \in \mathbf{R}^n$ with

$$\|P_2^{1/2} a\|_2 + \|P_1^{1/2} a\|_2 < a^T(x_1 - x_2).$$

8.8 *Intersection and containment of polyhedra.* Let $\mathcal{P}_1$ and $\mathcal{P}_2$ be two polyhedra defined as

$$\mathcal{P}_1 = \{x \mid Ax \preceq b\}, \qquad \mathcal{P}_2 = \{x \mid Fx \preceq g\},$$

with $A \in \mathbf{R}^{m \times n}$, $b \in \mathbf{R}^m$, $F \in \mathbf{R}^{p \times n}$, $g \in \mathbf{R}^p$. Formulate each of the following problems as an LP feasibility problem, or a set of LP feasibility problems.

(a) Find a point in the intersection $\mathcal{P}_1 \cap \mathcal{P}_2$.

(b) Determine whether $\mathcal{P}_1 \subseteq \mathcal{P}_2$.

For each problem, derive a set of linear inequalities and equalities that forms a strong alternative, and give a geometric interpretation of the alternative.

Repeat the question for two polyhedra defined as

$$\mathcal{P}_1 = \mathbf{conv}\{v_1, \ldots, v_K\}, \qquad \mathcal{P}_2 = \mathbf{conv}\{w_1, \ldots, w_L\}.$$

Euclidean distance and angle problems

8.9 *Closest Euclidean distance matrix to given data.* We are given data $\hat{d}_{ij}$, for $i, j = 1, \ldots, n$, which are corrupted measurements of the Euclidean distances between vectors in $\mathbf{R}^k$:

$$\hat{d}_{ij} = \|x_i - x_j\|_2 + v_{ij}, \quad i, j = 1, \ldots, n,$$

where v_{ij} is some noise or error. These data satisfy $\hat{d}_{ij} \geq 0$ and $\hat{d}_{ij} = \hat{d}_{ji}$, for all i, j. The dimension k is not specified.

Show how to solve the following problem using convex optimization. Find a dimension k and $x_1, \ldots, x_n \in \mathbf{R}^k$ so that $\sum_{i,j=1}^n (d_{ij} - \hat{d}_{ij})^2$ is minimized, where $d_{ij} = \|x_i - x_j\|_2$, $i, j = 1, \ldots, n$. In other words, given some data that are approximate Euclidean distances, you are to find the closest set of actual Euclidean distances, in the least-squares sense.

8.10 *Minimax angle fitting.* Suppose that $y_1, \ldots, y_m \in \mathbf{R}^k$ are affine functions of a variable $x \in \mathbf{R}^n$:

$$y_i = A_i x + b_i, \quad i = 1, \ldots, m,$$

and $z_1, \ldots, z_m \in \mathbf{R}^k$ are given nonzero vectors. We want to choose the variable x, subject to some convex constraints, (*e.g.*, linear inequalities) to minimize the maximum angle between y_i and z_i,

$$\max\{\angle(y_1, z_1), \ldots, \angle(y_m, z_m)\}.$$

The angle between nonzero vectors is defined as usual:

$$\angle(u, v) = \cos^{-1}\left(\frac{u^T v}{\|u\|_2 \|v\|_2}\right),$$

where we take $\cos^{-1}(a) \in [0, \pi]$. We are only interested in the case when the optimal objective value does not exceed $\pi/2$.

Formulate this problem as a convex or quasiconvex optimization problem. When the constraints on x are linear inequalities, what kind of problem (or problems) do you have to solve?

8.11 *Smallest Euclidean cone containing given points.* In $\mathbf{R}^n$, we define a *Euclidean cone*, with center direction $c \neq 0$, and angular radius θ, with $0 \leq \theta \leq \pi/2$, as the set

$$\{x \in \mathbf{R}^n \mid \angle(c, x) \leq \theta\}.$$

(A Euclidean cone is a second-order cone, *i.e.*, it can be represented as the image of the second-order cone under a nonsingular linear mapping.)

Let $a_1, \ldots, a_m \in \mathbf{R}^n$. How would you find the Euclidean cone, of smallest angular radius, that contains $a_1, \ldots, a_m$? (In particular, you should explain how to solve the feasibility problem, *i.e.*, how to determine whether there is a Euclidean cone which contains the points.)

Extremal volume ellipsoids

8.12 Show that the maximum volume ellipsoid enclosed in a set is unique. Show that the Löwner-John ellipsoid of a set is unique.

8.13 *Löwner-John ellipsoid of a simplex.* In this exercise we show that the Löwner-John ellipsoid of a simplex in $\mathbf{R}^n$ must be shrunk by a factor n to fit inside the simplex. Since the Löwner-John ellipsoid is affinely invariant, it is sufficient to show the result for one particular simplex.

Derive the Löwner-John ellipsoid $\mathcal{E}_{\mathrm{lj}}$ for the simplex $C = \mathbf{conv}\{0, e_1, \ldots, e_n\}$. Show that $\mathcal{E}_{\mathrm{lj}}$ must be shrunk by a factor $1/n$ to fit inside the simplex.

8.14 *Efficiency of ellipsoidal inner approximation.* Let C be a polyhedron in $\mathbf{R}^n$ described as $C = \{x \mid Ax \preceq b\}$, and suppose that $\{x \mid Ax \prec b\}$ is nonempty.

(a) Show that the maximum volume ellipsoid enclosed in C, expanded by a factor n about its center, is an ellipsoid that contains C.

(b) Show that if C is symmetric about the origin, *i.e.*, of the form $C = \{x \mid -\mathbf{1} \preceq Ax \preceq \mathbf{1}\}$, then expanding the maximum volume inscribed ellipsoid by a factor $\sqrt{n}$ gives an ellipsoid that contains C.

8.15 *Minimum volume ellipsoid covering union of ellipsoids.* Formulate the following problem as a convex optimization problem. Find the minimum volume ellipsoid $\mathcal{E} = \{x \mid (x - x_0)^T A^{-1}(x - x_0) \leq 1\}$ that contains K given ellipsoids

$$\mathcal{E}_i = \{x \mid x^T A_i x + 2b_i^T x + c_i \leq 0\}, \quad i = 1, \ldots, K.$$

Hint. See appendix B.

8.16 *Maximum volume rectangle inside a polyhedron.* Formulate the following problem as a convex optimization problem. Find the rectangle

$$\mathcal{R} = \{x \in \mathbf{R}^n \mid l \preceq x \preceq u\}$$

of maximum volume, enclosed in a polyhedron $\mathcal{P} = \{x \mid Ax \preceq b\}$. The variables are $l, u \in \mathbf{R}^n$. Your formulation should not involve an exponential number of constraints.

Centering

8.17 *Affine invariance of analytic center.* Show that the analytic center of a set of inequalities is affine invariant. Show that it is invariant with respect to positive scaling of the inequalities.

8.18 *Analytic center and redundant inequalities.* Two sets of linear inequalities that describe the same polyhedron can have different analytic centers. Show that by adding redundant inequalities, we can make *any* interior point x_0 of a polyhedron

$$\mathcal{P} = \{x \in \mathbf{R}^n \mid Ax \preceq b\}$$

the analytic center. More specifically, suppose $A \in \mathbf{R}^{m \times n}$ and $Ax_0 \prec b$. Show that there exist $c \in \mathbf{R}^n$, $\gamma \in \mathbf{R}$, and a positive integer q, such that $\mathcal{P}$ is the solution set of the $m + q$ inequalities

$$Ax \preceq b, \qquad c^T x \leq \gamma, \qquad c^T x \leq \gamma, \qquad \ldots, \qquad c^T x \leq \gamma \qquad (8.36)$$

(where the inequality $c^T x \leq \gamma$ is added q times), and x_0 is the analytic center of (8.36).

8.19 Let x_{ac} be the analytic center of a set of linear inequalities

$$a_i^T x \leq b_i, \quad i = 1, \ldots, m,$$

and define H as the Hessian of the logarithmic barrier function at x_{ac}:

$$H = \sum_{i=1}^{m} \frac{1}{(b_i - a_i^T x_{\mathrm{ac}})^2} a_i a_i^T.$$

Show that the kth inequality is redundant (*i.e.*, it can be deleted without changing the feasible set) if

$$b_k - a_k^T x_{\mathrm{ac}} \geq m (a_k^T H^{-1} a_k)^{1/2}.$$

8.20 *Ellipsoidal approximation from analytic center of linear matrix inequality.* Let C be the solution set of the LMI

$$x_1 A_1 + x_2 A_2 + \cdots + x_n A_n \preceq B,$$

where $A_i, B \in \mathbf{S}^m$, and let x_{ac} be its analytic center. Show that

$$\mathcal{E}_{\mathrm{inner}} \subseteq C \subseteq \mathcal{E}_{\mathrm{outer}},$$

where

$$
\begin{aligned}
\mathcal{E}_{\mathrm{inner}} &= \{x \mid (x - x_{\mathrm{ac}})^T H (x - x_{\mathrm{ac}}) \leq 1\}, \\
\mathcal{E}_{\mathrm{outer}} &= \{x \mid (x - x_{\mathrm{ac}})^T H (x - x_{\mathrm{ac}}) \leq m(m-1)\},
\end{aligned}
$$

and H is the Hessian of the logarithmic barrier function

$$-\log \det(B - x_1 A_1 - x_2 A_2 - \cdots - x_n A_n)$$

evaluated at x_{ac}.

8.21 [BYT99] *Maximum likelihood interpretation of analytic center.* We use the linear measurement model of page 352,

$$y = Ax + v,$$

where $A \in \mathbf{R}^{m \times n}$. We assume the noise components v_i are IID with support $[-1, 1]$. The set of parameters x consistent with the measurements $y \in \mathbf{R}^m$ is the polyhedron defined by the linear inequalities

$$-\mathbf{1} + y \preceq Ax \preceq \mathbf{1} + y. \qquad (8.37)$$

Suppose the probability density function of v_i has the form

$$p(v) = \begin{cases} \alpha_r (1 - v^2)^r & -1 \leq v \leq 1 \\ 0 & \text{otherwise,} \end{cases}$$

where $r \geq 1$ and $\alpha_r > 0$. Show that the maximum likelihood estimate of x is the analytic center of (8.37).

8.22 *Center of gravity.* The center of gravity of a set $C \subseteq \mathbf{R}^n$ with nonempty interior is defined as

$$x_{\mathrm{cg}} = \frac{\int_C u \, du}{\int_C 1 \, du}.$$

The center of gravity is affine invariant, and (clearly) a function of the set C, and not its particular description. Unlike the centers described in the chapter, however, it is very difficult to compute the center of gravity, except in simple cases (*e.g.*, ellipsoids, balls, simplexes).

Show that the center of gravity x_{cg} is the minimizer of the convex function

$$f(x) = \int_C \|u - x\|_2^2 \, du.$$

Classification

8.23 *Robust linear discrimination.* Consider the robust linear discrimination problem given in (8.23).

(a) Show that the optimal value $t^\star$ is positive if and only if the two sets of points can be linearly separated. When the two sets of points can be linearly separated, show that the inequality $\|a\|_2 \leq 1$ is tight, *i.e.*, we have $\|a^\star\|_2 = 1$, for the optimal $a^\star$.

(b) Using the change of variables $\tilde{a} = a/t$, $\tilde{b} = b/t$, prove that the problem (8.23) is equivalent to the QP

$$
\begin{array}{ll}
\text{minimize} & \|\tilde{a}\|_2 \\
\text{subject to} & \tilde{a}^T x_i - \tilde{b} \geq 1, \quad i = 1, \dots, N \\
& \tilde{a}^T y_i - \tilde{b} \leq -1, \quad i = 1, \dots, M.
\end{array}
$$

8.24 *Linear discrimination maximally robust to weight errors.* Suppose we are given two sets of points $\{x_1, \dots, x_N\}$ and and $\{y_1, \dots, y_M\}$ in $\mathbf{R}^n$ that can be linearly separated. In §8.6.1 we showed how to find the affine function that discriminates the sets, and gives the largest gap in function values. We can also consider robustness with respect to changes in the vector a, which is sometimes called the *weight vector*. For a given a and b for which $f(x) = a^T x - b$ separates the two sets, we define the *weight error margin* as the norm of the smallest $u \in \mathbf{R}^n$ such that the affine function $(a + u)^T x - b$ no longer separates the two sets of points. In other words, the weight error margin is the maximum ρ such that

$$(a + u)^T x_i \geq b, \quad i = 1, \dots, N, \qquad (a + u)^T y_j \leq b, \quad i = 1, \dots, M,$$

holds for all u with $\|u\|_2 \leq \rho$.

Show how to find a and b that maximize the weight error margin, subject to the normalization constraint $\|a\|_2 \leq 1$.

8.25 *Most spherical separating ellipsoid.* We are given two sets of vectors $x_1, \dots, x_N \in \mathbf{R}^n$, and $y_1, \dots, y_M \in \mathbf{R}^n$, and wish to find the ellipsoid with minimum eccentricity (*i.e.*, minimum condition number of the defining matrix) that contains the points $x_1, \dots, x_N$, but not the points $y_1, \dots, y_M$. Formulate this as a convex optimization problem.

Placement and floor planning

8.26 *Quadratic placement.* We consider a placement problem in $\mathbf{R}^2$, defined by an undirected graph $\mathcal{A}$ with N nodes, and with quadratic costs:

$$\text{minimize} \quad \sum_{(i,j) \in \mathcal{A}} \|x_i - x_j\|_2^2.$$

The variables are the positions $x_i \in \mathbf{R}^2$, $i = 1, \dots, M$. The positions x_i, $i = M+1, \dots, N$ are given. We define two vectors $u, v \in \mathbf{R}^M$ by

$$u = (x_{11}, x_{21}, \dots, x_{M1}), \qquad v = (x_{12}, x_{22}, \dots, x_{M2}),$$

containing the first and second components, respectively, of the free nodes.

Show that u and v can be found by solving two sets of linear equations,

$$Cu = d_1, \qquad Cv = d_2,$$

where $C \in \mathbf{S}^M$. Give a simple expression for the coefficients of C in terms of the graph $\mathcal{A}$.

8.27 *Problems with minimum distance constraints.* We consider a problem with variables $x_1, \ldots, x_N \in \mathbf{R}^k$. The objective, $f_0(x_1, \ldots, x_N)$, is convex, and the constraints

$$f_i(x_1, \ldots, x_N) \le 0, \quad i = 1, \ldots, m,$$

are convex (*i.e.*, the functions $f_i : \mathbf{R}^{Nk} \to \mathbf{R}$ are convex). In addition, we have the *minimum distance constraints*

$$\|x_i - x_j\|_2 \ge D_{\min}, \quad i \ne j, \quad i, j = 1, \ldots, N.$$

In general, this is a hard nonconvex problem.

Following the approach taken in floorplanning, we can form a *convex restriction* of the problem, *i.e.*, a problem which is convex, but has a smaller feasible set. (Solving the restricted problem is therefore easy, and any solution is guaranteed to be feasible for the nonconvex problem.) Let $a_{ij} \in \mathbf{R}^k$, for $i < j$, $i, j = 1, \ldots, N$, satisfy $\|a_{ij}\|_2 = 1$.
Show that the restricted problem

$$\begin{array}{ll} \text{minimize} & f_0(x_1, \ldots, x_N) \\ \text{subject to} & f_i(x_1, \ldots, x_N) \le 0, \quad i = 1, \ldots, m \\ & a_{ij}^T(x_i - x_j) \ge D_{\min}, \quad i < j, \ i, j = 1, \ldots, N, \end{array}$$

is convex, and that every feasible point satisfies the minimum distance constraint.

Remark. There are many good heuristics for choosing the directions a_{ij}. One simple one starts with an approximate solution $\hat{x}_1, \ldots, \hat{x}_N$ (that need not satisfy the minimum distance constraints). We then set $a_{ij} = (\hat{x}_i - \hat{x}_j)/\|\hat{x}_i - \hat{x}_j\|_2$.

Miscellaneous problems

8.28 Let $\mathcal{P}_1$ and $\mathcal{P}_2$ be two polyhedra described as

$$\mathcal{P}_1 = \{x \mid Ax \preceq b\}, \qquad \mathcal{P}_2 = \{x \mid -\mathbf{1} \preceq Cx \preceq \mathbf{1}\},$$

where $A \in \mathbf{R}^{m \times n}$, $C \in \mathbf{R}^{p \times n}$, and $b \in \mathbf{R}^m$. The polyhedron $\mathcal{P}_2$ is symmetric about the origin. For $t \ge 0$ and $x_c \in \mathbf{R}^n$, we use the notation $t\mathcal{P}_2 + x_c$ to denote the polyhedron

$$t\mathcal{P}_2 + x_c = \{tx + x_c \mid x \in \mathcal{P}_2\},$$

which is obtained by first scaling $\mathcal{P}_2$ by a factor t about the origin, and then translating its center to x_c.

Show how to solve the following two problems, via an LP, or a set of LPs.

(a) Find the largest polyhedron $t\mathcal{P}_2 + x_c$ enclosed in $\mathcal{P}_1$, *i.e.*,

$$\begin{array}{ll} \text{maximize} & t \\ \text{subject to} & t\mathcal{P}_2 + x_c \subseteq \mathcal{P}_1 \\ & t \ge 0. \end{array}$$

(b) Find the smallest polyhedron $t\mathcal{P}_2 + x_c$ containing $\mathcal{P}_1$, *i.e.*,

$$\begin{array}{ll} \text{minimize} & t \\ \text{subject to} & \mathcal{P}_1 \subseteq t\mathcal{P}_2 + x_c \\ & t \ge 0. \end{array}$$

In both problems the variables are $t \in \mathbf{R}$ and $x_c \in \mathbf{R}^n$.

8.29 *Outer polyhedral approximations.* Let $\mathcal{P} = \{x \in \mathbf{R}^n \mid Ax \preceq b\}$ be a polyhedron, and $C \subseteq \mathbf{R}^n$ a given set (not necessarily convex). Use the support function S_C to formulate the following problem as an LP:

$$
\begin{array}{ll}
\text{minimize} & t \\
\text{subject to} & C \subseteq t\mathcal{P} + x \\
& t \geq 0.
\end{array}
$$

Here $t\mathcal{P} + x = \{tu + x \mid u \in \mathcal{P}\}$, the polyhedron $\mathcal{P}$ scaled by a factor of t about the origin, and translated by x. The variables are $t \in \mathbf{R}$ and $x \in \mathbf{R}^n$.

8.30 *Interpolation with piecewise-arc curve.* A sequence of points $a_1, \ldots, a_n \in \mathbf{R}^2$ is given. We construct a curve that passes through these points, in order, and is an arc (*i.e.*, part of a circle) or line segment (which we think of as an arc of infinite radius) between consecutive points. Many arcs connect a_i and a_{i+1}; we parameterize these arcs by giving the angle $\theta_i \in (-\pi, \pi)$ between its tangent at a_i and the line segment $[a_i, a_{i+1}]$. Thus, $\theta_i = 0$ means the arc between a_i and a_{i+1} is in fact the line segment $[a_i, a_{i+1}]$; $\theta_i = \pi/2$ means the arc between a_i and a_{i+1} is a half-circle (above the linear segment $[a_1, a_2]$); $\theta_i = -\pi/2$ means the arc between a_i and a_{i+1} is a half-circle (below the linear segment $[a_1, a_2]$). This is illustrated below.

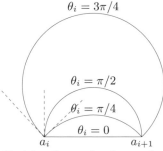

Our curve is completely specified by the angles $\theta_1, \ldots, \theta_n$, which can be chosen in the interval $(-\pi, \pi)$. The choice of θ_i affects several properties of the curve, for example, its *total arc length L*, or the *joint angle discontinuities*, which can be described as follows. At each point a_i, $i = 2, \ldots, n-1$, two arcs meet, one coming from the previous point and one going to the next point. If the tangents to these arcs exactly oppose each other, so the curve is differentiable at a_i, we say there is no joint angle discontinuity at a_i. In general, we define the joint angle discontinuity at a_i as $|\theta_{i-1} + \theta_i + \psi_i|$, where ψ_i is the angle between the line segment $[a_i, a_{i+1}]$ and the line segment $[a_{i-1}, a_i]$, *i.e.*, $\psi_i = \angle(a_i - a_{i+1}, a_{i-1} - a_i)$. This is shown below. Note that the angles ψ_i are known (since the a_i are known).

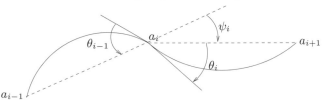

We define the *total joint angle discontinuity* as

$$
D = \sum_{i=2}^{n} |\theta_{i-1} + \theta_i + \psi_i|.
$$

Formulate the problem of minimizing total arc length length L, and total joint angle discontinuity D, as a bi-criterion convex optimization problem. Explain how you would find the extreme points on the optimal trade-off curve.

Part III

Algorithms

Chapter 9

Unconstrained minimization

9.1 Unconstrained minimization problems

In this chapter we discuss methods for solving the unconstrained optimization problem

$$\text{minimize} \quad f(x) \tag{9.1}$$

where $f : \mathbf{R}^n \to \mathbf{R}$ is convex and twice continuously differentiable (which implies that $\mathbf{dom}\, f$ is open). We will assume that the problem is solvable, i.e., there exists an optimal point $x^\star$. (More precisely, the assumptions later in the chapter will imply that $x^\star$ exists and is unique.) We denote the optimal value, $\inf_x f(x) = f(x^\star)$, as $p^\star$.

Since f is differentiable and convex, a necessary and sufficient condition for a point $x^\star$ to be optimal is

$$\nabla f(x^\star) = 0 \tag{9.2}$$

(see §4.2.3). Thus, solving the unconstrained minimization problem (9.1) is the same as finding a solution of (9.2), which is a set of n equations in the n variables $x_1, \ldots, x_n$. In a few special cases, we can find a solution to the problem (9.1) by analytically solving the optimality equation (9.2), but usually the problem must be solved by an iterative algorithm. By this we mean an algorithm that computes a sequence of points $x^{(0)}, x^{(1)}, \ldots \in \mathbf{dom}\, f$ with $f(x^{(k)}) \to p^\star$ as $k \to \infty$. Such a sequence of points is called a *minimizing sequence* for the problem (9.1). The algorithm is terminated when $f(x^{(k)}) - p^\star \le \epsilon$, where $\epsilon > 0$ is some specified tolerance.

Initial point and sublevel set

The methods described in this chapter require a suitable starting point $x^{(0)}$. The starting point must lie in $\mathbf{dom}\, f$, and in addition the sublevel set

$$S = \{x \in \mathbf{dom}\, f \mid f(x) \le f(x^{(0)})\} \tag{9.3}$$

must be closed. This condition is satisfied for all $x^{(0)} \in \mathbf{dom}\, f$ if the function f is *closed*, i.e., all its sublevel sets are closed (see §A.3.3). Continuous functions with

dom $f = \mathbf{R}^n$ are closed, so if **dom** $f = \mathbf{R}^n$, the initial sublevel set condition is satisfied by any $x^{(0)}$. Another important class of closed functions are continuous functions with open domains, for which $f(x)$ tends to infinity as x approaches **bd dom** f.

9.1.1   Examples

Quadratic minimization and least-squares

The general convex quadratic minimization problem has the form

$$\text{minimize} \quad (1/2)x^T P x + q^T x + r, \tag{9.4}$$

where $P \in \mathbf{S}^n_+$, $q \in \mathbf{R}^n$, and $r \in \mathbf{R}$. This problem can be solved via the optimality conditions, $Px^\star + q = 0$, which is a set of linear equations. When $P \succ 0$, there is a unique solution, $x^\star = -P^{-1}q$. In the more general case when P is not positive definite, any solution of $Px^\star = -q$ is optimal for (9.4); if $Px^\star = -q$ does not have a solution, then the problem (9.4) is unbounded below (see exercise 9.1). Our ability to analytically solve the quadratic minimization problem (9.4) is the basis for Newton's method, a powerful method for unconstrained minimization described in §9.5.

One special case of the quadratic minimization problem that arises very frequently is the least-squares problem

$$\text{minimize} \quad \|Ax - b\|_2^2 = x^T(A^T A)x - 2(A^T b)^T x + b^T b.$$

The optimality conditions

$$A^T A x^\star = A^T b$$

are called the *normal equations* of the least-squares problem.

Unconstrained geometric programming

As a second example, we consider an unconstrained geometric program in convex form,

$$\text{minimize} \quad f(x) = \log\left(\sum_{i=1}^m \exp(a_i^T x + b_i)\right).$$

The optimality condition is

$$\nabla f(x^\star) = \frac{1}{\sum_{j=1}^m \exp(a_j^T x^\star + b_j)} \sum_{i=1}^m \exp(a_i^T x^\star + b_i)a_i = 0,$$

which in general has no analytical solution, so here we must resort to an iterative algorithm. For this problem, **dom** $f = \mathbf{R}^n$, so any point can be chosen as the initial point $x^{(0)}$.

Analytic center of linear inequalities

We consider the optimization problem

$$\text{minimize} \quad f(x) = -\sum_{i=1}^m \log(b_i - a_i^T x), \tag{9.5}$$

where the domain of f is the open set

$$\mathbf{dom}\, f = \{x \mid a_i^T x < b_i, \ i = 1, \ldots, m\}.$$

The objective function f in this problem is called the *logarithmic barrier* for the inequalities $a_i^T x \leq b_i$. The solution of (9.5), if it exists, is called the *analytic center* of the inequalities. The initial point $x^{(0)}$ must satisfy the strict inequalities $a_i^T x^{(0)} < b_i$, $i = 1, \ldots, m$. Since f is closed, the sublevel set S for any such point is closed.

Analytic center of a linear matrix inequality

A closely related problem is

$$\text{minimize} \quad f(x) = \log \det F(x)^{-1} \tag{9.6}$$

where $F : \mathbf{R}^n \to \mathbf{S}^p$ is affine, *i.e.*,

$$F(x) = F_0 + x_1 F_1 + \cdots + x_n F_n,$$

with $F_i \in \mathbf{S}^p$. Here the domain of f is

$$\mathbf{dom}\, f = \{x \mid F(x) \succ 0\}.$$

The objective function f is called the *logarithmic barrier* for the linear matrix inequality $F(x) \succeq 0$, and the solution (if it exists) is called the analytic center of the linear matrix inequality. The initial point $x^{(0)}$ must satisfy the strict linear matrix inequality $F(x^{(0)}) \succ 0$. As in the previous example, the sublevel set of any such point will be closed, since f is closed.

9.1.2   Strong convexity and implications

In much of this chapter (with the exception of §9.6) we assume that the objective function is *strongly convex* on S, which means that there exists an $m > 0$ such that

$$\nabla^2 f(x) \succeq mI \tag{9.7}$$

for all $x \in S$. Strong convexity has several interesting consequences. For $x, y \in S$ we have

$$f(y) = f(x) + \nabla f(x)^T (y - x) + \frac{1}{2}(y - x)^T \nabla^2 f(z)(y - x)$$

for some z on the line segment $[x, y]$. By the strong convexity assumption (9.7), the last term on the righthand side is at least $(m/2)\|y - x\|_2^2$, so we have the inequality

$$f(y) \geq f(x) + \nabla f(x)^T (y - x) + \frac{m}{2}\|y - x\|_2^2 \tag{9.8}$$

for all x and y in S. When $m = 0$, we recover the basic inequality characterizing convexity; for $m > 0$ we obtain a better lower bound on $f(y)$ than follows from convexity alone.

We will first show that the inequality (9.8) can be used to bound $f(x) - p^\star$, which is the suboptimality of the point x, in terms of $\|\nabla f(x)\|_2$. The righthand side of (9.8) is a convex quadratic function of y (for fixed x). Setting the gradient with respect to y equal to zero, we find that $\tilde{y} = x - (1/m)\nabla f(x)$ minimizes the righthand side. Therefore we have

$$
\begin{aligned}
f(y) &\geq f(x) + \nabla f(x)^T (y - x) + \frac{m}{2}\|y - x\|_2^2 \\
&\geq f(x) + \nabla f(x)^T (\tilde{y} - x) + \frac{m}{2}\|\tilde{y} - x\|_2^2 \\
&= f(x) - \frac{1}{2m}\|\nabla f(x)\|_2^2.
\end{aligned}
$$

Since this holds for any $y \in S$, we have

$$
p^\star \geq f(x) - \frac{1}{2m}\|\nabla f(x)\|_2^2. \tag{9.9}
$$

This inequality shows that if the gradient is small at a point, then the point is nearly optimal. The inequality (9.9) can also be interpreted as a condition for *suboptimality* which generalizes the optimality condition (9.2):

$$
\|\nabla f(x)\|_2 \leq (2m\epsilon)^{1/2} \implies f(x) - p^\star \leq \epsilon. \tag{9.10}
$$

We can also derive a bound on $\|x - x^\star\|_2$, the distance between x and any optimal point $x^\star$, in terms of $\|\nabla f(x)\|_2$:

$$
\|x - x^\star\|_2 \leq \frac{2}{m}\|\nabla f(x)\|_2. \tag{9.11}
$$

To see this, we apply (9.8) with $y = x^\star$ to obtain

$$
\begin{aligned}
p^\star = f(x^\star) &\geq f(x) + \nabla f(x)^T (x^\star - x) + \frac{m}{2}\|x^\star - x\|_2^2 \\
&\geq f(x) - \|\nabla f(x)\|_2\|x^\star - x\|_2 + \frac{m}{2}\|x^\star - x\|_2^2,
\end{aligned}
$$

where we use the Cauchy-Schwarz inequality in the second inequality. Since $p^\star \leq f(x)$, we must have

$$
-\|\nabla f(x)\|_2\,\|x^\star - x\|_2 + \frac{m}{2}\|x^\star - x\|_2^2 \leq 0,
$$

from which (9.11) follows. One consequence of (9.11) is that the optimal point $x^\star$ is unique.

Upper bound on $\nabla^2 f(x)$

The inequality (9.8) implies that the sublevel sets contained in S are bounded, so in particular, S is bounded. Therefore the maximum eigenvalue of $\nabla^2 f(x)$, which is a continuous function of x on S, is bounded above on S, *i.e.*, there exists a constant M such that

$$
\nabla^2 f(x) \preceq MI \tag{9.12}
$$

for all $x \in S$. This upper bound on the Hessian implies for any x, $y \in S$,

$$f(y) \le f(x) + \nabla f(x)^T (y - x) + \frac{M}{2} \|y - x\|_2^2, \tag{9.13}$$

which is analogous to (9.8). Minimizing each side over y yields

$$p^\star \le f(x) - \frac{1}{2M} \|\nabla f(x)\|_2^2, \tag{9.14}$$

the counterpart of (9.9).

Condition number of sublevel sets

From the strong convexity inequality (9.7) and the inequality (9.12), we have

$$mI \preceq \nabla^2 f(x) \preceq MI \tag{9.15}$$

for all $x \in S$. The ratio $\kappa = M/m$ is thus an upper bound on the condition number of the matrix $\nabla^2 f(x)$, *i.e.*, the ratio of its largest eigenvalue to its smallest eigenvalue. We can also give a geometric interpretation of (9.15) in terms of the sublevel sets of f.

We define the *width* of a convex set $C \subseteq \mathbf{R}^n$, in the direction q, where $\|q\|_2 = 1$, as

$$W(C, q) = \sup_{z \in C} q^T z - \inf_{z \in C} q^T z.$$

The *minimum width* and *maximum width* of C are given by

$$W_{\min} = \inf_{\|q\|_2=1} W(C, q), \qquad W_{\max} = \sup_{\|q\|_2=1} W(C, q).$$

The *condition number* of the convex set C is defined as

$$\mathbf{cond}(C) = \frac{W_{\max}^2}{W_{\min}^2},$$

i.e., the square of the ratio of its maximum width to its minimum width. The condition number of C gives a measure of its *anisotropy* or *eccentricity*. If the condition number of a set C is small (say, near one) it means that the set has approximately the same width in all directions, *i.e.*, it is nearly spherical. If the condition number is large, it means that the set is far wider in some directions than in others.

Example 9.1   *Condition number of an ellipsoid.* Let $\mathcal{E}$ be the ellipsoid

$$\mathcal{E} = \{x \mid (x - x_0)^T A^{-1} (x - x_0) \le 1\},$$

where $A \in \mathbf{S}_{++}^n$. The width of $\mathcal{E}$ in the direction q is

$$\begin{aligned}
\sup_{z \in \mathcal{E}} q^T z - \inf_{z \in \mathcal{E}} q^T z &= (\|A^{1/2} q\|_2 + q^T x_0) - (-\|A^{1/2} q\|_2 + q^T x_0) \\
&= 2\|A^{1/2} q\|_2.
\end{aligned}$$

It follows that its minimum and maximum width are

$$W_{\min} = 2\lambda_{\min}(A)^{1/2}, \qquad W_{\max} = 2\lambda_{\max}(A)^{1/2},$$

and its condition number is

$$\mathbf{cond}(\mathcal{E}) = \frac{\lambda_{\max}(A)}{\lambda_{\min}(A)} = \kappa(A),$$

where $\kappa(A)$ denotes the condition number of the matrix A, *i.e.*, the ratio of its maximum singular value to its minimum singular value. Thus the condition number of the ellipsoid $\mathcal{E}$ is the same as the condition number of the matrix A that defines it.

Now suppose f satisfies $mI \preceq \nabla^2 f(x) \preceq MI$ for all $x \in S$. We will derive a bound on the condition number of the α-sublevel $C_\alpha = \{x \mid f(x) \leq \alpha\}$, where $p^\star < \alpha \leq f(x^{(0)})$. Applying (9.13) and (9.8) with $x = x^\star$, we have

$$p^\star + (M/2)\|y - x^\star\|_2^2 \geq f(y) \geq p^\star + (m/2)\|y - x^\star\|_2^2.$$

This implies that $B_{\text{inner}} \subseteq C_\alpha \subseteq B_{\text{outer}}$ where

$$
\begin{aligned}
B_{\text{inner}} &= \{y \mid \|y - x^\star\|_2 \leq (2(\alpha - p^\star)/M)^{1/2}\}, \\
B_{\text{outer}} &= \{y \mid \|y - x^\star\|_2 \leq (2(\alpha - p^\star)/m)^{1/2}\}.
\end{aligned}
$$

In other words, the α-sublevel set contains B_{inner}, and is contained in B_{outer}, which are balls with radii

$$(2(\alpha - p^\star)/M)^{1/2}, \qquad (2(\alpha - p^\star)/m)^{1/2},$$

respectively. The ratio of the radii squared gives an upper bound on the condition number of C_α:

$$\mathbf{cond}(C_\alpha) \leq \frac{M}{m}.$$

We can also give a geometric interpretation of the condition number $\kappa(\nabla^2 f(x^\star))$ of the Hessian at the optimum. From the Taylor series expansion of f around $x^\star$,

$$f(y) \approx p^\star + \frac{1}{2}(y - x^\star)^T \nabla^2 f(x^\star)(y - x^\star),$$

we see that, for α close to $p^\star$,

$$C_\alpha \approx \{y \mid (y - x^\star)^T \nabla^2 f(x^\star)(y - x^\star) \leq 2(\alpha - p^\star)\},$$

i.e., the sublevel set is well approximated by an ellipsoid with center $x^\star$. Therefore

$$\lim_{\alpha \to p^\star} \mathbf{cond}(C_\alpha) = \kappa(\nabla^2 f(x^\star)).$$

We will see that the condition number of the sublevel sets of f (which is bounded by M/m) has a strong effect on the efficiency of some common methods for unconstrained minimization.

The strong convexity constants

It must be kept in mind that the constants m and M are known only in rare cases, so the inequality (9.10) cannot be used as a practical stopping criterion. It can be considered a *conceptual* stopping criterion; it shows that if the gradient of f at x is small enough, then the difference between $f(x)$ and $p^\star$ is small. If we terminate an algorithm when $\|\nabla f(x^{(k)})\|_2 \leq \eta$, where η is chosen small enough to be (very likely) smaller than $(m\epsilon)^{1/2}$, then we have $f(x^{(k)}) - p^\star \leq \epsilon$ (very likely).

In the following sections we give convergence proofs for algorithms, which include bounds on the number of iterations required before $f(x^{(k)}) - p^\star \leq \epsilon$, where ϵ is some positive tolerance. Many of these bounds involve the (usually unknown) constants m and M, so the same comments apply. These results are at least conceptually useful; they establish that the algorithm converges, even if the bound on the number of iterations required to reach a given accuracy depends on constants that are unknown.

We will encounter one important exception to this situation. In §9.6 we will study a special class of convex functions, called *self-concordant*, for which we can provide a complete convergence analysis (for Newton's method) that does not depend on any unknown constants.

9.2 Descent methods

The algorithms described in this chapter produce a minimizing sequence $x^{(k)}$, $k = 1, \ldots$, where

$$x^{(k+1)} = x^{(k)} + t^{(k)} \Delta x^{(k)}$$

and $t^{(k)} > 0$ (except when $x^{(k)}$ is optimal). Here the concatenated symbols Δ and x that form Δx are to be read as a single entity, a vector in $\mathbf{R}^n$ called the *step* or *search direction* (even though it need not have unit norm), and $k = 0, 1, \ldots$ denotes the iteration number. The scalar $t^{(k)} \geq 0$ is called the *step size* or *step length* at iteration k (even though it is not equal to $\|x^{(k+1)} - x^{(k)}\|$ unless $\|\Delta x^{(k)}\| = 1$). The terms 'search step' and 'scale factor' are more accurate, but 'search direction' and 'step length' are the ones widely used. When we focus on one iteration of an algorithm, we sometimes drop the superscripts and use the lighter notation $x^+ = x + t\Delta x$, or $x := x + t\Delta x$, in place of $x^{(k+1)} = x^{(k)} + t^{(k)} \Delta x^{(k)}$.

All the methods we study are *descent methods*, which means that

$$f(x^{(k+1)}) < f(x^{(k)}),$$

except when $x^{(k)}$ is optimal. This implies that for all k we have $x^{(k)} \in S$, the initial sublevel set, and in particular we have $x^{(k)} \in \mathbf{dom}\, f$. From convexity we know that $\nabla f(x^{(k)})^T(y - x^{(k)}) \geq 0$ implies $f(y) \geq f(x^{(k)})$, so the search direction in a descent method must satisfy

$$\nabla f(x^{(k)})^T \Delta x^{(k)} < 0,$$

i.e., it must make an acute angle with the negative gradient. We call such a direction a *descent direction* (for f, at $x^{(k)}$).

The outline of a general descent method is as follows. It alternates between two steps: determining a descent direction Δx, and the selection of a step size t.

Algorithm 9.1 *General descent method.*

given a starting point $x \in \mathbf{dom}\, f$.

repeat
    1. Determine a descent direction Δx.
    2. *Line search.* Choose a step size $t > 0$.
    3. *Update.* $x := x + t\Delta x$.
until stopping criterion is satisfied.

The second step is called the *line search* since selection of the step size t determines where along the line $\{x + t\Delta x \mid t \in \mathbf{R}_+\}$ the next iterate will be. (A more accurate term might be *ray search*.)

A practical descent method has the same general structure, but might be organized differently. For example, the stopping criterion is often checked while, or immediately after, the descent direction Δx is computed. The stopping criterion is often of the form $\|\nabla f(x)\|_2 \le \eta$, where η is small and positive, as suggested by the suboptimality condition (9.9).

Exact line search

One line search method sometimes used in practice is *exact line search*, in which t is chosen to minimize f along the ray $\{x + t\Delta x \mid t \ge 0\}$:

$$t = \mathrm{argmin}_{s \ge 0}\ f(x + s\Delta x). \qquad (9.16)$$

An exact line search is used when the cost of the minimization problem with one variable, required in (9.16), is low compared to the cost of computing the search direction itself. In some special cases the minimizer along the ray can be found analytically, and in others it can be computed efficiently. (This is discussed in §9.7.1.)

Backtracking line search

Most line searches used in practice are *inexact*: the step length is chosen to approximately minimize f along the ray $\{x + t\Delta x \mid t \ge 0\}$, or even to just reduce f 'enough'. Many inexact line search methods have been proposed. One inexact line search method that is very simple and quite effective is called *backtracking* line search. It depends on two constants α, β with $0 < \alpha < 0.5$, $0 < \beta < 1$.

Algorithm 9.2 *Backtracking line search.*

given a descent direction Δx for f at $x \in \mathbf{dom}\, f$, $\alpha \in (0, 0.5)$, $\beta \in (0, 1)$.

$t := 1$.
while $f(x + t\Delta x) > f(x) + \alpha t \nabla f(x)^T \Delta x$, $t := \beta t$.

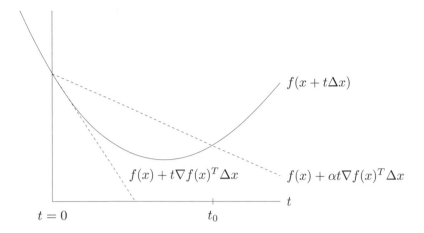

Figure 9.1 *Backtracking line search.* The curve shows f, restricted to the line over which we search. The lower dashed line shows the linear extrapolation of f, and the upper dashed line has a slope a factor of α smaller. The backtracking condition is that f lies below the upper dashed line, *i.e.*, $0 \le t \le t_0$.

The line search is called backtracking because it starts with unit step size and then reduces it by the factor β until the stopping condition $f(x + t\Delta x) \le f(x) + \alpha t \nabla f(x)^T \Delta x$ holds. Since Δx is a descent direction, we have $\nabla f(x)^T \Delta x < 0$, so for small enough t we have

$$f(x + t\Delta x) \approx f(x) + t\nabla f(x)^T \Delta x < f(x) + \alpha t \nabla f(x)^T \Delta x,$$

which shows that the backtracking line search eventually terminates. The constant α can be interpreted as the fraction of the decrease in f predicted by linear extrapolation that we will accept. (The reason for requiring α to be smaller than 0.5 will become clear later.)

The backtracking condition is illustrated in figure 9.1. This figure suggests, and it can be shown, that the backtracking exit inequality $f(x + t\Delta x) \le f(x) + \alpha t \nabla f(x)^T \Delta x$ holds for $t \ge 0$ in an interval $(0, t_0]$. It follows that the backtracking line search stops with a step length t that satisfies

$$t = 1, \qquad \text{or} \qquad t \in (\beta t_0, t_0].$$

The first case occurs when the step length $t = 1$ satisfies the backtracking condition, *i.e.*, $1 \le t_0$. In particular, we can say that the step length obtained by backtracking line search satisfies

$$t \ge \min\{1, \beta t_0\}.$$

When $\textbf{dom}\, f$ is not all of $\mathbf{R}^n$, the condition $f(x + t\Delta x) \le f(x) + \alpha t \nabla f(x)^T \Delta x$ in the backtracking line search must be interpreted carefully. By our convention that f is infinite outside its domain, the inequality implies that $x + t\Delta x \in \textbf{dom}\, f$. In a practical implementation, we first multiply t by β until $x + t\Delta x \in \textbf{dom}\, f$;

then we start to check whether the inequality $f(x + t\Delta x) \leq f(x) + \alpha t \nabla f(x)^T \Delta x$ holds.

The parameter α is typically chosen between 0.01 and 0.3, meaning that we accept a decrease in f between 1% and 30% of the prediction based on the linear extrapolation. The parameter β is often chosen to be between 0.1 (which corresponds to a very crude search) and 0.8 (which corresponds to a less crude search).

9.3 Gradient descent method

A natural choice for the search direction is the negative gradient $\Delta x = -\nabla f(x)$. The resulting algorithm is called the *gradient algorithm* or *gradient descent method*.

Algorithm 9.3 *Gradient descent method.*

given a starting point $x \in \mathbf{dom}\, f$.

repeat
    1. $\Delta x := -\nabla f(x)$.
    2. *Line search.* Choose step size t via exact or backtracking line search.
    3. *Update.* $x := x + t\Delta x$.
until stopping criterion is satisfied.

The stopping criterion is usually of the form $\|\nabla f(x)\|_2 \leq \eta$, where η is small and positive. In most implementations, this condition is checked after step 1, rather than after the update.

9.3.1 Convergence analysis

In this section we present a simple convergence analysis for the gradient method, using the lighter notation $x^+ = x + t\Delta x$ for $x^{(k+1)} = x^{(k)} + t^{(k)}\Delta x^{(k)}$, where $\Delta x = -\nabla f(x)$. We assume f is strongly convex on S, so there are positive constants m and M such that $mI \preceq \nabla^2 f(x) \preceq MI$ for all $x \in S$. Define the function $\tilde{f} : \mathbf{R} \to \mathbf{R}$ by $\tilde{f}(t) = f(x - t\nabla f(x))$, i.e., f as a function of the step length t in the negative gradient direction. In the following discussion we will only consider t for which $x - t\nabla f(x) \in S$. From the inequality (9.13), with $y = x - t\nabla f(x)$, we obtain a quadratic upper bound on $\tilde{f}$:

$$\tilde{f}(t) \leq f(x) - t\|\nabla f(x)\|_2^2 + \frac{Mt^2}{2}\|\nabla f(x)\|_2^2. \tag{9.17}$$

Analysis for exact line search

We now assume that an exact line search is used, and minimize over t both sides of the inequality (9.17). On the lefthand side we get $\tilde{f}(t_{\text{exact}})$, where t_{exact} is the step length that minimizes $\tilde{f}$. The righthand side is a simple quadratic, which is minimized by $t = 1/M$, and has minimum value $f(x) - (1/(2M))\|\nabla f(x)\|_2^2$. Therefore we have

$$f(x^+) = \tilde{f}(t_{\text{exact}}) \le f(x) - \frac{1}{2M}\|\nabla(f(x))\|_2^2.$$

Subtracting $p^\star$ from both sides, we get

$$f(x^+) - p^\star \le f(x) - p^\star - \frac{1}{2M}\|\nabla f(x)\|_2^2.$$

We combine this with $\|\nabla f(x)\|_2^2 \ge 2m(f(x) - p^\star)$ (which follows from (9.9)) to conclude

$$f(x^+) - p^\star \le (1 - m/M)(f(x) - p^\star).$$

Applying this inequality recursively, we find that

$$f(x^{(k)}) - p^\star \le c^k(f(x^{(0)}) - p^\star) \tag{9.18}$$

where $c = 1 - m/M < 1$, which shows that $f(x^{(k)})$ converges to $p^\star$ as $k \to \infty$. In particular, we must have $f(x^{(k)}) - p^\star \le \epsilon$ after at most

$$\frac{\log((f(x^{(0)}) - p^\star)/\epsilon)}{\log(1/c)} \tag{9.19}$$

iterations of the gradient method with exact line search.

This bound on the number of iterations required, even though crude, can give some insight into the gradient method. The numerator,

$$\log((f(x^{(0)}) - p^\star)/\epsilon)$$

can be interpreted as the log of the ratio of the initial suboptimality (*i.e.*, gap between $f(x^{(0)})$ and $p^\star$), to the final suboptimality (*i.e.*, less than ϵ). This term suggests that the number of iterations depends on how good the initial point is, and what the final required accuracy is.

The denominator appearing in the bound (9.19), $\log(1/c)$, is a function of M/m, which we have seen is a bound on the condition number of $\nabla^2 f(x)$ over S, or the condition number of the sublevel sets $\{z \mid f(z) \le \alpha\}$. For large condition number bound M/m, we have

$$\log(1/c) = -\log(1 - m/M) \approx m/M,$$

so our bound on the number of iterations required increases approximately linearly with increasing M/m.

We will see that the gradient method does in fact require a large number of iterations when the Hessian of f, near $x^\star$, has a large condition number. Conversely, when the sublevel sets of f are relatively isotropic, so that the condition number bound M/m can be chosen to be relatively small, the bound (9.18) shows that convergence is rapid, since c is small, or at least not too close to one.

The bound (9.18) shows that the error $f(x^{(k)}) - p^\star$ converges to zero at least as fast as a geometric series. In the context of iterative numerical methods, this is called *linear convergence*, since the error lies below a line on a log-linear plot of error versus iteration number.

Analysis for backtracking line search

Now we consider the case where a backtracking line search is used in the gradient descent method. We will show that the backtracking exit condition,

$$\tilde{f}(t) \leq f(x) - \alpha t \|\nabla f(x)\|_2^2,$$

is satisfied whenever $0 \leq t \leq 1/M$. First note that

$$0 \leq t \leq 1/M \implies -t + \frac{Mt^2}{2} \leq -t/2$$

(which follows from convexity of $-t + Mt^2/2$). Using this result and the bound (9.17), we have, for $0 \leq t \leq 1/M$,

$$
\begin{aligned}
\tilde{f}(t) &\leq f(x) - t\|\nabla f(x)\|_2^2 + \frac{Mt^2}{2}\|\nabla (f(x))\|_2^2 \\
&\leq f(x) - (t/2)\|\nabla f(x)\|_2^2 \\
&\leq f(x) - \alpha t\|\nabla f(x)\|_2^2,
\end{aligned}
$$

since $\alpha < 1/2$. Therefore the backtracking line search terminates either with $t = 1$ or with a value $t \geq \beta/M$. This provides a lower bound on the decrease in the objective function. In the first case we have

$$f(x^+) \leq f(x) - \alpha\|\nabla f(x)\|_2^2,$$

and in the second case we have

$$f(x^+) \leq f(x) - (\beta\alpha/M)\|\nabla f(x)\|_2^2.$$

Putting these together, we always have

$$f(x^+) \leq f(x) - \min\{\alpha, \beta\alpha/M\}\|\nabla f(x)\|_2^2.$$

Now we can proceed exactly as in the case of exact line search. We subtract $p^\star$ from both sides to get

$$f(x^+) - p^\star \leq f(x) - p^\star - \min\{\alpha, \beta\alpha/M\}\|\nabla f(x)\|_2^2,$$

and combine this with $\|\nabla f(x)\|_2^2 \geq 2m(f(x) - p^\star)$ to obtain

$$f(x^+) - p^\star \leq (1 - \min\{2m\alpha, 2\beta\alpha m/M\})(f(x) - p^\star).$$

From this we conclude

$$f(x^{(k)}) - p^\star \leq c^k(f(x^{(0)}) - p^\star)$$

where

$$c = 1 - \min\{2m\alpha, 2\beta\alpha m/M\} < 1.$$

In particular, $f(x^{(k)})$ converges to $p^\star$ at least as fast as a geometric series with an exponent that depends (at least in part) on the condition number bound M/m. In the terminology of iterative methods, the convergence is at least linear.

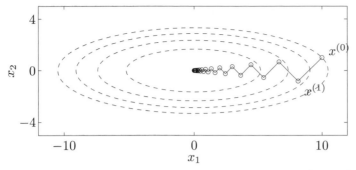

Figure 9.2 Some contour lines of the function $f(x) = (1/2)(x_1^2 + 10x_2^2)$. The condition number of the sublevel sets, which are ellipsoids, is exactly 10. The figure shows the iterates of the gradient method with exact line search, started at $x^{(0)} = (10, 1)$.

.3.2   Examples

A quadratic problem in $\mathbf{R}^2$

Our first example is very simple. We consider the quadratic objective function on $\mathbf{R}^2$

$$f(x) = \frac{1}{2}(x_1^2 + \gamma x_2^2),$$

where $\gamma > 0$. Clearly, the optimal point is $x^\star = 0$, and the optimal value is 0. The Hessian of f is constant, and has eigenvalues 1 and γ, so the condition numbers of the sublevel sets of f are all exactly

$$\frac{\max\{1, \gamma\}}{\min\{1, \gamma\}} = \max\{\gamma, 1/\gamma\}.$$

The tightest choices for the strong convexity constants m and M are

$$m = \min\{1, \gamma\}, \qquad M = \max\{1, \gamma\}.$$

We apply the gradient descent method with exact line search, starting at the point $x^{(0)} = (\gamma, 1)$. In this case we can derive the following closed-form expressions for the iterates $x^{(k)}$ and their function values (exercise 9.6):

$$x_1^{(k)} = \gamma \left(\frac{\gamma - 1}{\gamma + 1}\right)^k, \qquad x_2^{(k)} = \left(-\frac{\gamma - 1}{\gamma + 1}\right)^k,$$

and

$$f(x^{(k)}) = \frac{\gamma(\gamma + 1)}{2} \left(\frac{\gamma - 1}{\gamma + 1}\right)^{2k} = \left(\frac{\gamma - 1}{\gamma + 1}\right)^{2k} f(x^{(0)}).$$

This is illustrated in figure 9.2, for $\gamma = 10$.

For this simple example, convergence is exactly linear, *i.e.*, the error is exactly a geometric series, reduced by the factor $|(\gamma - 1)/(\gamma + 1)|^2$ at each iteration. For

$\gamma = 1$, the exact solution is found in one iteration; for γ not far from one (say, between $1/3$ and 3) convergence is rapid. The convergence is very slow for $\gamma \gg 1$ or $\gamma \ll 1$.

We can compare the convergence with the bound derived above in §9.3.1. Using the least conservative values $m = \min\{1,\gamma\}$ and $M = \max\{1,\gamma\}$, the bound (9.18) guarantees that the error in each iteration is reduced at least by the factor $c = (1 - m/M)$. We have seen that the error is in fact reduced exactly by the factor

$$\left(\frac{1 - m/M}{1 + m/M}\right)^2$$

in each iteration. For small m/M, which corresponds to large condition number, the upper bound (9.19) implies that the number of iterations required to obtain a given level of accuracy grows at most like M/m. For this example, the exact number of iterations required grows approximately like $(M/m)/4$, i.e., one quarter of the value of the bound. This shows that for this simple example, the bound on the number of iterations derived in our simple analysis is only about a factor of four conservative (using the least conservative values for m and M). In particular, the convergence rate (as well as its upper bound) is very dependent on the condition number of the sublevel sets.

A nonquadratic problem in $\mathbf{R}^2$

We now consider a nonquadratic example in $\mathbf{R}^2$, with

$$f(x_1, x_2) = e^{x_1 + 3x_2 - 0.1} + e^{x_1 - 3x_2 - 0.1} + e^{-x_1 - 0.1}. \tag{9.20}$$

We apply the gradient method with a backtracking line search, with $\alpha = 0.1$, $\beta = 0.7$. Figure 9.3 shows some level curves of f, and the iterates $x^{(k)}$ generated by the gradient method (shown as small circles). The lines connecting successive iterates show the scaled steps,

$$x^{(k+1)} - x^{(k)} = -t^{(k)} \nabla f(x^{(k)}).$$

Figure 9.4 shows the error $f(x^{(k)}) - p^\star$ versus iteration k. The plot reveals that the error converges to zero approximately as a geometric series, i.e., the convergence is approximately linear. In this example, the error is reduced from about 10 to about 10^{-7} in 20 iterations, so the error is reduced by a factor of approximately $10^{-8/20} \approx 0.4$ each iteration. This reasonably rapid convergence is predicted by our convergence analysis, since the sublevel sets of f are not too badly conditioned, which in turn means that M/m can be chosen as not too large.

To compare backtracking line search with an exact line search, we use the gradient method with an exact line search, on the same problem, and with the same starting point. The results are given in figures 9.5 and 9.4. Here too the convergence is approximately linear, about twice as fast as the gradient method with backtracking line search. With exact line search, the error is reduced by about 10^{-11} in 15 iterations, i.e., a reduction by a factor of about $10^{-11/15} \approx 0.2$ per iteration.

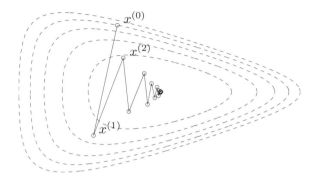

Figure 9.3 Iterates of the gradient method with backtracking line search, for the problem in $\mathbf{R}^2$ with objective f given in (9.20). The dashed curves are level curves of f, and the small circles are the iterates of the gradient method. The solid lines, which connect successive iterates, show the scaled steps $t^{(k)}\Delta x^{(k)}$.

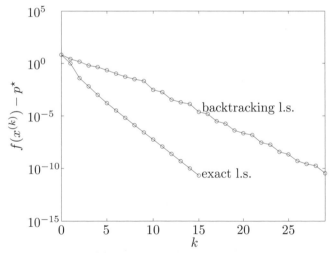

Figure 9.4 Error $f(x^{(k)}) - p^\star$ versus iteration k of the gradient method with backtracking and exact line search, for the problem in $\mathbf{R}^2$ with objective f given in (9.20). The plot shows nearly linear convergence, with the error reduced approximately by the factor 0.4 in each iteration of the gradient method with backtracking line search, and by the factor 0.2 in each iteration of the gradient method with exact line search.

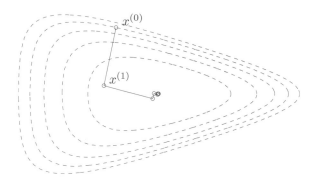

Figure 9.5 Iterates of the gradient method with exact line search for the problem in $\mathbf{R}^2$ with objective f given in (9.20).

A problem in $\mathbf{R}^{100}$

We next consider a larger example, of the form

$$f(x) = c^T x - \sum_{i=1}^{m} \log(b_i - a_i^T x), \tag{9.21}$$

with $m = 500$ terms and $n = 100$ variables.

The progress of the gradient method with backtracking line search, with parameters $\alpha = 0.1$, $\beta = 0.5$, is shown in figure 9.6. In this example we see an initial approximately linear and fairly rapid convergence for about 20 iterations, followed by a slower linear convergence. Overall, the error is reduced by a factor of around 10^6 in around 175 iterations, which gives an average error reduction by a factor of around $10^{-6/175} \approx 0.92$ per iteration. The initial convergence rate, for the first 20 iterations, is around a factor of 0.8 per iteration; the slower final convergence rate, after the first 20 iterations, is around a factor of 0.94 per iteration.

Figure 9.6 shows the convergence of the gradient method with exact line search. The convergence is again approximately linear, with an overall error reduction by approximately a factor $10^{-6/140} \approx 0.91$ per iteration. This is only a bit faster than the gradient method with backtracking line search.

Finally, we examine the influence of the backtracking line search parameters α and β on the convergence rate, by determining the number of iterations required to obtain $f(x^{(k)}) - p^\star \le 10^{-5}$. In the first experiment, we fix $\beta = 0.5$, and vary α from 0.05 to 0.5. The number of iterations required varies from about 80, for larger values of α, in the range 0.2–0.5, to about 170 for smaller values of α. This, and other experiments, suggest that the gradient method works better with fairly large α, in the range 0.2–0.5.

Similarly, we can study the effect of the choice of β by fixing $\alpha = 0.1$ and varying β from 0.05 to 0.95. Again the variation in the total number of iterations is not large, ranging from around 80 (when $\beta \approx 0.5$) to around 200 (for β small, or near 1). This experiment, and others, suggest that $\beta \approx 0.5$ is a good choice.

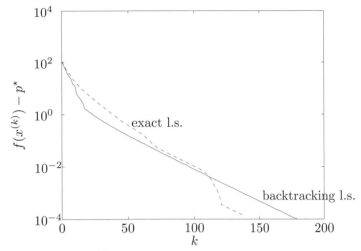

Figure 9.6 Error $f(x^{(k)}) - p^\star$ versus iteration k for the gradient method with backtracking and exact line search, for a problem in $\mathbf{R}^{100}$.

These experiments suggest that the effect of the backtracking parameters on the convergence is not large, no more than a factor of two or so.

Gradient method and condition number

Our last experiment will illustrate the importance of the condition number of $\nabla^2 f(x)$ (or the sublevel sets) on the rate of convergence of the gradient method. We start with the function given by (9.21), but replace the variable x by $x = T\bar{x}$, where

$$T = \mathbf{diag}((1, \gamma^{1/n}, \gamma^{2/n}, \dots, \gamma^{(n-1)/n})),$$

i.e., we minimize

$$\bar{f}(\bar{x}) = c^T T\bar{x} - \sum_{i=1}^{m} \log(b_i - a_i^T T\bar{x}). \tag{9.22}$$

This gives us a family of optimization problems, indexed by γ, which affects the problem condition number.

Figure 9.7 shows the number of iterations required to achieve $\bar{f}(\bar{x}^{(k)}) - \bar{p}^\star < 10^{-5}$ as a function of γ, using a backtracking line search with $\alpha = 0.3$ and $\beta = 0.7$. This plot shows that for diagonal scaling as small as $10:1$ (*i.e.*, $\gamma = 10$), the number of iterations grows to more than a thousand; for a diagonal scaling of 20 or more, the gradient method slows to essentially useless.

The condition number of the Hessian $\nabla^2 \bar{f}(\bar{x}^\star)$ at the optimum is shown in figure 9.8. For large and small γ, the condition number increases roughly as $\max\{\gamma^2, 1/\gamma^2\}$, in a very similar way as the number of iterations depends on γ. This shows again that the relation between conditioning and convergence speed is a real phenomenon, and not just an artifact of our analysis.

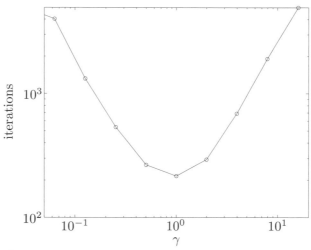

Figure 9.7 Number of iterations of the gradient method applied to problem (9.22). The vertical axis shows the number of iterations required to obtain $\bar{f}(\bar{x}^{(k)}) - \bar{p}^{\star} < 10^{-5}$. The horizontal axis shows γ, which is a parameter that controls the amount of diagonal scaling. We use a backtracking line search with $\alpha = 0.3$, $\beta = 0.7$.

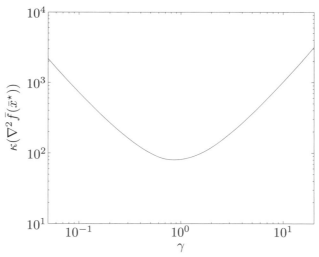

Figure 9.8 Condition number of the Hessian of the function at its minimum, as a function of γ. By comparing this plot with the one in figure 9.7, we see that the condition number has a very strong influence on convergence rate.

Conclusions

From the numerical examples shown, and others, we can make the conclusions summarized below.

- The gradient method often exhibits approximately linear convergence, *i.e.*, the error $f(x^{(k)}) - p^\star$ converges to zero approximately as a geometric series.

- The choice of backtracking parameters α, β has a noticeable but not dramatic effect on the convergence. An exact line search sometimes improves the convergence of the gradient method, but the effect is not large (and probably not worth the trouble of implementing the exact line search).

- The convergence rate depends greatly on the condition number of the Hessian, or the sublevel sets. Convergence can be very slow, even for problems that are moderately well conditioned (say, with condition number in the 100s). When the condition number is larger (say, 1000 or more) the gradient method is so slow that it is useless in practice.

The main advantage of the gradient method is its simplicity. Its main disadvantage is that its convergence rate depends so critically on the condition number of the Hessian or sublevel sets.

9.4 Steepest descent method

The first-order Taylor approximation of $f(x + v)$ around x is

$$f(x + v) \approx \widehat{f}(x + v) = f(x) + \nabla f(x)^T v.$$

The second term on the righthand side, $\nabla f(x)^T v$, is the *directional derivative* of f at x in the direction v. It gives the approximate change in f for a small step v. The step v is a descent direction if the directional derivative is negative.

We now address the question of how to choose v to make the directional derivative as negative as possible. Since the directional derivative $\nabla f(x)^T v$ is linear in v, it can be made as negative as we like by taking v large (provided v is a descent direction, *i.e.*, $\nabla f(x)^T v < 0$). To make the question sensible we have to limit the size of v, or normalize by the length of v.

Let $\| \cdot \|$ be any norm on $\mathbf{R}^n$. We define a *normalized steepest descent direction* (with respect to the norm $\| \cdot \|$) as

$$\Delta x_{\text{nsd}} = \operatorname{argmin}\{\nabla f(x)^T v \mid \|v\| = 1\}. \tag{9.23}$$

(We say 'a' steepest descent direction because there can be multiple minimizers.) A normalized steepest descent direction Δx_{nsd} is a step of unit norm that gives the largest decrease in the linear approximation of f.

A normalized steepest descent direction can be interpreted geometrically as follows. We can just as well define Δx_{nsd} as

$$\Delta x_{\text{nsd}} = \operatorname{argmin}\{\nabla f(x)^T v \mid \|v\| \leq 1\},$$

i.e., as the direction in the unit ball of $\|\cdot\|$ that extends farthest in the direction $-\nabla f(x)$.

It is also convenient to consider a steepest descent step Δx_{sd} that is *unnormalized*, by scaling the normalized steepest descent direction in a particular way:

$$\Delta x_{\mathrm{sd}} = \|\nabla f(x)\|_* \Delta x_{\mathrm{nsd}}, \qquad (9.24)$$

where $\|\cdot\|_*$ denotes the dual norm. Note that for the steepest descent step, we have

$$\nabla f(x)^T \Delta x_{\mathrm{sd}} = \|\nabla f(x)\|_* \nabla f(x)^T \Delta x_{\mathrm{nsd}} = -\|\nabla f(x)\|_*^2$$

(see exercise 9.7).

The *steepest descent method* uses the steepest descent direction as search direction.

Algorithm 9.4 *Steepest descent method.*

given a starting point $x \in \mathbf{dom}\, f$.

repeat
    1. Compute steepest descent direction Δx_{sd}.
    2. *Line search.* Choose t via backtracking or exact line search.
    3. *Update.* $x := x + t\Delta x_{\mathrm{sd}}$.
until stopping criterion is satisfied.

When exact line search is used, scale factors in the descent direction have no effect, so the normalized or unnormalized direction can be used.

9.4.1   Steepest descent for Euclidean and quadratic norms

Steepest descent for Euclidean norm

If we take the norm $\|\cdot\|$ to be the Euclidean norm we find that the steepest descent direction is simply the negative gradient, *i.e.*, $\Delta x_{\mathrm{sd}} = -\nabla f(x)$. The steepest descent method for the Euclidean norm coincides with the gradient descent method.

Steepest descent for quadratic norm

We consider the quadratic norm

$$\|z\|_P = (z^T P z)^{1/2} = \|P^{1/2} z\|_2,$$

where $P \in \mathbf{S}_{++}^n$. The normalized steepest descent direction is given by

$$\Delta x_{\mathrm{nsd}} = -\left(\nabla f(x)^T P^{-1} \nabla f(x)\right)^{-1/2} P^{-1} \nabla f(x).$$

The dual norm is given by $\|z\|_* = \|P^{-1/2} z\|_2$, so the steepest descent step with respect to $\|\cdot\|_P$ is given by

$$\Delta x_{\mathrm{sd}} = -P^{-1} \nabla f(x). \qquad (9.25)$$

The normalized steepest descent direction for a quadratic norm is illustrated in figure 9.9.

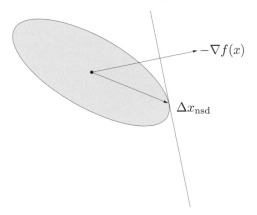

Figure 9.9 Normalized steepest descent direction for a quadratic norm. The ellipsoid shown is the unit ball of the norm, translated to the point x. The normalized steepest descent direction Δx_{nsd} at x extends as far as possible in the direction $-\nabla f(x)$ while staying in the ellipsoid. The gradient and normalized steepest descent directions are shown.

Interpretation via change of coordinates

We can give an interesting alternative interpretation of the steepest descent direction Δx_{sd} as the gradient search direction after a change of coordinates is applied to the problem. Define $\bar{u} = P^{1/2}u$, so we have $\|u\|_P = \|\bar{u}\|_2$. Using this change of coordinates, we can solve the original problem of minimizing f by solving the equivalent problem of minimizing the function $\bar{f} : \mathbf{R}^n \to \mathbf{R}$, given by

$$\bar{f}(\bar{u}) = f(P^{-1/2}\bar{u}) = f(u).$$

If we apply the gradient method to $\bar{f}$, the search direction at a point $\bar{x}$ (which corresponds to the point $x = P^{-1/2}\bar{x}$ for the original problem) is

$$\Delta \bar{x} = -\nabla \bar{f}(\bar{x}) = -P^{-1/2}\nabla f(P^{-1/2}\bar{x}) = -P^{-1/2}\nabla f(x).$$

This gradient search direction corresponds to the direction

$$\Delta x = P^{-1/2}\left(-P^{-1/2}\nabla f(x)\right) = -P^{-1}\nabla f(x)$$

for the original variable x. In other words, the steepest descent method in the quadratic norm $\|\cdot\|_P$ can be thought of as the gradient method applied to the problem after the change of coordinates $\bar{x} = P^{1/2}x$.

9.4.2   Steepest descent for ℓ_1-norm

As another example, we consider the steepest descent method for the ℓ_1-norm. A normalized steepest descent direction,

$$\Delta x_{\mathrm{nsd}} = \operatorname{argmin}\{\nabla f(x)^T v \mid \|v\|_1 \leq 1\},$$

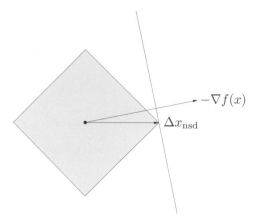

Figure 9.10 Normalized steepest descent direction for the ℓ_1-norm. The diamond is the unit ball of the ℓ_1-norm, translated to the point x. The normalized steepest descent direction can always be chosen in the direction of a standard basis vector; in this example we have $\Delta x_{\mathrm{nsd}} = e_1$.

is easily characterized. Let i be any index for which $\|\nabla f(x)\|_\infty = |(\nabla f(x))_i|$. Then a normalized steepest descent direction Δx_{nsd} for the ℓ_1-norm is given by

$$\Delta x_{\mathrm{nsd}} = -\operatorname{sign}\left(\frac{\partial f(x)}{\partial x_i}\right) e_i,$$

where e_i is the ith standard basis vector. An unnormalized steepest descent step is then

$$\Delta x_{\mathrm{sd}} = \Delta x_{\mathrm{nsd}} \|\nabla f(x)\|_\infty = -\frac{\partial f(x)}{\partial x_i} e_i.$$

Thus, the normalized steepest descent step in ℓ_1-norm can always be chosen to be a standard basis vector (or a negative standard basis vector). It is the coordinate axis direction along which the approximate decrease in f is greatest. This is illustrated in figure 9.10.

The steepest descent algorithm in the ℓ_1-norm has a very natural interpretation: At each iteration we select a component of $\nabla f(x)$ with maximum absolute value, and then decrease or increase the corresponding component of x, according to the sign of $(\nabla f(x))_i$. The algorithm is sometimes called a *coordinate-descent* algorithm, since only one component of the variable x is updated at each iteration. This can greatly simplify, or even trivialize, the line search.

Example 9.2 *Frobenius norm scaling.* In §4.5.4 we encountered the unconstrained geometric program

$$\text{minimize} \quad \sum_{i,j=1}^n M_{ij}^2 d_i^2 / d_j^2,$$

where $M \in \mathbf{R}^{n \times n}$ is given, and the variable is $d \in \mathbf{R}^n$. Using the change of variables $x_i = 2 \log d_i$ we can express this geometric program in convex form as

$$\text{minimize} \quad f(x) = \log\left(\sum_{i,j=1}^n M_{ij}^2 e^{x_i - x_j}\right).$$

It is easy to minimize f one component at a time. Keeping all components except the kth fixed, we can write $f(x) = \log(\alpha_k + \beta_k e^{-x_k} + \gamma_k e^{x_k})$, where

$$\alpha_k = M_{kk}^2 + \sum_{i,j\neq k} M_{ij}^2 e^{x_i - x_j}, \qquad \beta_k = \sum_{i\neq k} M_{ik}^2 e^{x_i}, \qquad \gamma_k = \sum_{j\neq k} M_{kj}^2 e^{-x_j}.$$

The minimum of $f(x)$, as a function of x_k, is obtained for $x_k = \log(\beta_k/\gamma_k)/2$. So for this problem an exact line search can be carried out using a simple analytical formula.

The ℓ_1-steepest descent algorithm with exact line search consists of repeating the following steps.

1. Compute the gradient

$$(\nabla f(x))_i = \frac{-\beta_i e^{-x_i} + \gamma_i e^{x_i}}{\alpha_i + \beta_i e^{-x_i} + \gamma_i e^{x_i}}, \qquad i = 1,\dots,n.$$

2. Select a largest (in absolute value) component of $\nabla f(x)$: $|\nabla f(x)|_k = \|\nabla f(x)\|_\infty$.

3. Minimize f over the scalar variable x_k, by setting $x_k = \log(\beta_k/\gamma_k)/2$.

9.4.3   Convergence analysis

In this section we extend the convergence analysis for the gradient method with backtracking line search to the steepest descent method for an arbitrary norm. We will use the fact that any norm can be bounded in terms of the Euclidean norm, so there exists constants γ, $\tilde{\gamma} \in (0,1]$ such that

$$\|x\| \geq \gamma \|x\|_2, \qquad \|x\|_* \geq \tilde{\gamma} \|x\|_2$$

(see §A.1.4).

Again we assume f is strongly convex on the initial sublevel set S. The upper bound $\nabla^2 f(x) \preceq MI$ implies an upper bound on the function $f(x + t\Delta x_{sd})$ as a function of t:

$$
\begin{aligned}
f(x + t\Delta x_{sd}) &\leq f(x) + t\nabla f(x)^T \Delta x_{sd} + \frac{M\|\Delta x_{sd}\|_2^2}{2}t^2 \\
&\leq f(x) + t\nabla f(x)^T \Delta x_{sd} + \frac{M\|\Delta x_{sd}\|^2}{2\gamma^2}t^2 \\
&= f(x) - t\|\nabla f(x)\|_*^2 + \frac{M}{2\gamma^2}t^2\|\nabla f(x)\|_*^2. \qquad (9.26)
\end{aligned}
$$

The step size $\hat{t} = \gamma^2/M$ (which minimizes the quadratic upper bound (9.26)) satisfies the exit condition for the backtracking line search:

$$f(x + \hat{t}\Delta x_{sd}) \leq f(x) - \frac{\gamma^2}{2M}\|\nabla f(x)\|_*^2 \leq f(x) + \frac{\alpha\gamma^2}{M}\nabla f(x)^T \Delta x_{sd} \qquad (9.27)$$

since $\alpha < 1/2$ and $\nabla f(x)^T \Delta x_{\mathrm{sd}} = -\|\nabla f(x)\|_*^2$. The line search therefore returns a step size $t \geq \min\{1, \beta\gamma^2/M\}$, and we have

$$
\begin{aligned}
f(x^+) = f(x + t\Delta x_{\mathrm{sd}}) &\leq f(x) - \alpha \min\{1, \beta\gamma^2/M\}\|\nabla f(x)\|_*^2 \\
&\leq f(x) - \alpha\tilde{\gamma}^2 \min\{1, \beta\gamma^2/M\}\|\nabla f(x)\|_2^2 .
\end{aligned}
$$

Subtracting $p^\star$ from both sides and using (9.9), we obtain

$$
f(x^+) - p^\star \leq c(f(x) - p^\star),
$$

where

$$
c = 1 - 2m\alpha\tilde{\gamma}^2 \min\{1, \beta\gamma^2/M\} < 1.
$$

Therefore we have

$$
f(x^{(k)}) - p^\star \leq c^k(f(x^{(0)}) - p^\star),
$$

i.e., linear convergence exactly as in the gradient method.

9.4.4 Discussion and examples

Choice of norm for steepest descent

The choice of norm used to define the steepest descent direction can have a dramatic effect on the convergence rate. For simplicity, we consider the case of steepest descent with quadratic P-norm. In §9.4.1, we showed that the steepest descent method with quadratic P-norm is the same as the gradient method applied to the problem after the change of coordinates $\bar{x} = P^{1/2}x$. We know that the gradient method works well when the condition numbers of the sublevel sets (or the Hessian near the optimal point) are moderate, and works poorly when the condition numbers are large. It follows that when the sublevel sets, after the change of coordinates $\bar{x} = P^{1/2}x$, are moderately conditioned, the steepest descent method will work well.

This observation provides a prescription for choosing P: It should be chosen so that the sublevel sets of f, transformed by $P^{-1/2}$, are well conditioned. For example if an approximation $\hat{H}$ of the Hessian at the optimal point $H(x^\star)$ were known, a very good choice of P would be $P = \hat{H}$, since the Hessian of $\tilde{f}$ at the optimum is then

$$
\hat{H}^{-1/2}\nabla^2 f(x^\star)\hat{H}^{-1/2} \approx I,
$$

and so is likely to have a low condition number.

This same idea can be described without a change of coordinates. Saying that a sublevel set has low condition number after the change of coordinates $\bar{x} = P^{1/2}x$ is the same as saying that the ellipsoid

$$
\mathcal{E} = \{x \mid x^T P x \leq 1\}
$$

approximates the shape of the sublevel set. (In other words, it gives a good approximation after appropriate scaling and translation.)

This dependence of the convergence rate on the choice of P can be viewed from two sides. The optimist's viewpoint is that for any problem, there is always a

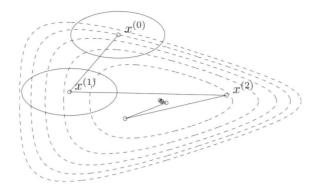

Figure 9.11 Steepest descent method with a quadratic norm $\|\cdot\|_{P_1}$. The ellipses are the boundaries of the norm balls $\{x \mid \|x - x^{(k)}\|_{P_1} \leq 1\}$ at $x^{(0)}$ and $x^{(1)}$.

choice of P for which the steepest descent method works very well. The challenge, of course, is to find such a P. The pessimist's viewpoint is that for any problem, there are a huge number of choices of P for which steepest descent works very poorly. In summary, we can say that the steepest descent method works well in cases where we can identify a matrix P for which the transformed problem has moderate condition number.

Examples

In this section we illustrate some of these ideas using the nonquadratic problem in $\mathbf{R}^2$ with objective function (9.20). We apply the steepest descent method to the problem, using the two quadratic norms defined by

$$ P_1 = \begin{bmatrix} 2 & 0 \\ 0 & 8 \end{bmatrix}, \qquad P_2 = \begin{bmatrix} 8 & 0 \\ 0 & 2 \end{bmatrix}. $$

In both cases we use a backtracking line search with $\alpha = 0.1$ and $\beta = 0.7$.

Figures 9.11 and 9.12 show the iterates for steepest descent with norm $\|\cdot\|_{P_1}$ and norm $\|\cdot\|_{P_2}$. Figure 9.13 shows the error versus iteration number for both norms. Figure 9.13 shows that the choice of norm strongly influences the convergence. With the norm $\|\cdot\|_{P_1}$, convergence is a bit more rapid than the gradient method, whereas with the norm $\|\cdot\|_{P_2}$, convergence is far slower.

This can be explained by examining the problems after the changes of coordinates $\bar{x} = P_1^{1/2} x$ and $\bar{x} = P_2^{1/2} x$, respectively. Figures 9.14 and 9.15 show the problems in the transformed coordinates. The change of variables associated with P_1 yields sublevel sets with modest condition number, so convergence is fast. The change of variables associated with P_2 yields sublevel sets that are more poorly conditioned, which explains the slower convergence.

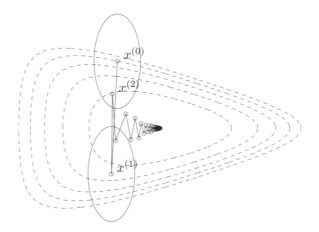

Figure 9.12 Steepest descent method, with quadratic norm $\|\cdot\|_{P_2}$.

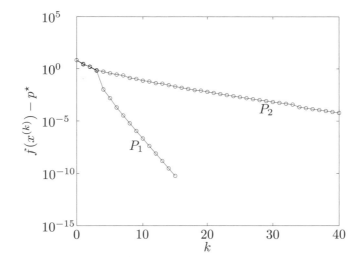

Figure 9.13 Error $f(x^{(k)}) - p^\star$ versus iteration k, for the steepest descent method with the quadratic norm $\|\cdot\|_{P_1}$ and the quadratic norm $\|\cdot\|_{P_2}$. Convergence is rapid for the norm $\|\cdot\|_{P_1}$ and very slow for $\|\cdot\|_{P_2}$.

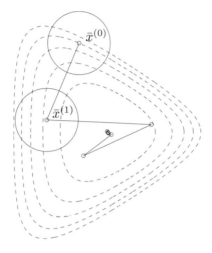

Figure 9.14 The iterates of steepest descent with norm $\| \cdot \|_{P_1}$, after the change of coordinates. This change of coordinates reduces the condition number of the sublevel sets, and so speeds up convergence.

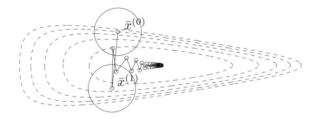

Figure 9.15 The iterates of steepest descent with norm $\| \cdot \|_{P_2}$, after the change of coordinates. This change of coordinates increases the condition number of the sublevel sets, and so slows down convergence.

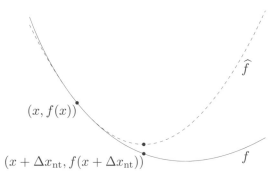

Figure 9.16 The function f (shown solid) and its second-order approximation $\widehat{f}$ at x (dashed). The Newton step Δx_{nt} is what must be added to x to give the minimizer of $\widehat{f}$.

9.5   Newton's method

9.5.1   The Newton step

For $x \in \operatorname{\mathbf{dom}} f$, the vector

$$\Delta x_{\mathrm{nt}} = -\nabla^2 f(x)^{-1} \nabla f(x)$$

is called the *Newton step* (for f, at x). Positive definiteness of $\nabla^2 f(x)$ implies that

$$\nabla f(x)^T \Delta x_{\mathrm{nt}} = -\nabla f(x)^T \nabla^2 f(x)^{-1} \nabla f(x) < 0$$

unless $\nabla f(x) = 0$, so the Newton step is a descent direction (unless x is optimal). The Newton step can be interpreted and motivated in several ways.

Minimizer of second-order approximation

The second-order Taylor approximation (or model) $\widehat{f}$ of f at x is

$$\widehat{f}(x+v) = f(x) + \nabla f(x)^T v + \frac{1}{2} v^T \nabla^2 f(x) v, \tag{9.28}$$

which is a convex quadratic function of v, and is minimized when $v = \Delta x_{\mathrm{nt}}$. Thus, the Newton step Δx_{nt} is what should be added to the point x to minimize the second-order approximation of f at x. This is illustrated in figure 9.16.

This interpretation gives us some insight into the Newton step. If the function f is quadratic, then $x + \Delta x_{\mathrm{nt}}$ is the exact minimizer of f. If the function f is nearly quadratic, intuition suggests that $x + \Delta x_{\mathrm{nt}}$ should be a very good estimate of the minimizer of f, *i.e.*, $x^\star$. Since f is twice differentiable, the quadratic model of f will be very accurate when x is near $x^\star$. It follows that when x is near $x^\star$, the point $x + \Delta x_{\mathrm{nt}}$ should be a very good estimate of $x^\star$. We will see that this intuition is correct.

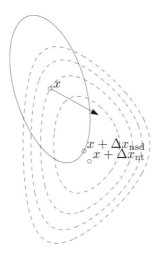

Figure 9.17 The dashed lines are level curves of a convex function. The ellipsoid shown (with solid line) is $\{x + v \mid v^T \nabla^2 f(x)v \leq 1\}$. The arrow shows $-\nabla f(x)$, the gradient descent direction. The Newton step Δx_{nt} is the steepest descent direction in the norm $\|\cdot\|_{\nabla^2 f(x)}$. The figure also shows Δx_{nsd}, the normalized steepest descent direction for the same norm.

Steepest descent direction in Hessian norm

The Newton step is also the steepest descent direction at x, for the quadratic norm defined by the Hessian $\nabla^2 f(x)$, *i.e.*,

$$\|u\|_{\nabla^2 f(x)} = (u^T \nabla^2 f(x)u)^{1/2}.$$

This gives another insight into why the Newton step should be a good search direction, and a very good search direction when x is near $x^\star$.

Recall from our discussion above that steepest descent, with quadratic norm $\|\cdot\|_P$, converges very rapidly when the Hessian, after the associated change of coordinates, has small condition number. In particular, near $x^\star$, a very good choice is $P = \nabla^2 f(x^\star)$. When x is near $x^\star$, we have $\nabla^2 f(x) \approx \nabla^2 f(x^\star)$, which explains why the Newton step is a very good choice of search direction. This is illustrated in figure 9.17.

Solution of linearized optimality condition

If we linearize the optimality condition $\nabla f(x^\star) = 0$ near x we obtain

$$\nabla f(x + v) \approx \nabla f(x) + \nabla^2 f(x)v = 0,$$

which is a linear equation in v, with solution $v = \Delta x_{\mathrm{nt}}$. So the Newton step Δx_{nt} is what must be added to x so that the linearized optimality condition holds. Again, this suggests that when x is near $x^\star$ (so the optimality conditions almost hold), the update $x + \Delta x_{\mathrm{nt}}$ should be a very good approximation of $x^\star$.

When $n = 1$, *i.e.*, $f : \mathbf{R} \to \mathbf{R}$, this interpretation is particularly simple. The solution $x^\star$ of the minimization problem is characterized by $f'(x^\star) = 0$, *i.e.*, it is

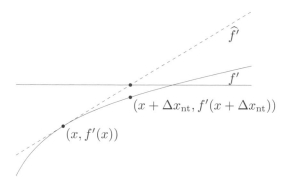

Figure 9.18 The solid curve is the derivative f' of the function f shown in figure 9.16. $\widehat{f'}$ is the linear approximation of f' at x. The Newton step Δx_{nt} is the difference between the root of $\widehat{f'}$ and the point x.

the zero-crossing of the derivative f', which is monotonically increasing since f is convex. Given our current approximation x of the solution, we form a first-order Taylor approximation of f' at x. The zero-crossing of this affine approximation is then $x + \Delta x_{\mathrm{nt}}$. This interpretation is illustrated in figure 9.18.

Affine invariance of the Newton step

An important feature of the Newton step is that it is independent of linear (or affine) changes of coordinates. Suppose $T \in \mathbf{R}^{n \times n}$ is nonsingular, and define $\bar{f}(y) = f(Ty)$. Then we have

$$\nabla \bar{f}(y) = T^T \nabla f(x), \qquad \nabla^2 \bar{f}(y) = T^T \nabla^2 f(x) T,$$

where $x = Ty$. The Newton step for $\bar{f}$ at y is therefore

$$
\begin{aligned}
\Delta y_{\mathrm{nt}} &= -\left(T^T \nabla^2 f(x) T\right)^{-1} \left(T^T \nabla f(x)\right) \\
&= -T^{-1} \nabla^2 f(x)^{-1} \nabla f(x) \\
&= T^{-1} \Delta x_{\mathrm{nt}},
\end{aligned}
$$

where Δx_{nt} is the Newton step for f at x. Hence the Newton steps of f and $\bar{f}$ are related by the same linear transformation, and

$$x + \Delta x_{\mathrm{nt}} = T(y + \Delta y_{\mathrm{nt}}).$$

The Newton decrement

The quantity

$$\lambda(x) = \left(\nabla f(x)^T \nabla^2 f(x)^{-1} \nabla f(x)\right)^{1/2}$$

is called the *Newton decrement* at x. We will see that the Newton decrement plays an important role in the analysis of Newton's method, and is also useful as a stopping criterion. We can relate the Newton decrement to the quantity $f(x) - \inf_y \widehat{f}(y)$, where $\widehat{f}$ is the second-order approximation of f at x:

$$f(x) - \inf_y \widehat{f}(y) = f(x) - \widehat{f}(x + \Delta x_{\mathrm{nt}}) = \frac{1}{2}\lambda(x)^2.$$

Thus, $\lambda^2/2$ is an estimate of $f(x) - p^\star$, based on the quadratic approximation of f at x.

We can also express the Newton decrement as

$$\lambda(x) = \left(\Delta x_{\mathrm{nt}}^T \nabla^2 f(x) \Delta x_{\mathrm{nt}}\right)^{1/2}. \tag{9.29}$$

This shows that λ is the norm of the Newton step, in the quadratic norm defined by the Hessian, *i.e.*, the norm

$$\|u\|_{\nabla^2 f(x)} = \left(u^T \nabla^2 f(x) u\right)^{1/2}.$$

The Newton decrement comes up in backtracking line search as well, since we have

$$\nabla f(x)^T \Delta x_{\mathrm{nt}} = -\lambda(x)^2. \tag{9.30}$$

This is the constant used in a backtracking line search, and can be interpreted as the directional derivative of f at x in the direction of the Newton step:

$$-\lambda(x)^2 = \nabla f(x)^T \Delta x_{\mathrm{nt}} = \left.\frac{d}{dt} f(x + \Delta x_{\mathrm{nt}} t)\right|_{t=0}.$$

Finally, we note that the Newton decrement is, like the Newton step, affine invariant. In other words, the Newton decrement of $\bar{f}(y) = f(Ty)$ at y, where T is nonsingular, is the same as the Newton decrement of f at $x = Ty$.

9.5.2 Newton's method

Newton's method, as outlined below, is sometimes called the *damped* Newton method or *guarded* Newton method, to distinguish it from the *pure* Newton method, which uses a fixed step size $t = 1$.

Algorithm 9.5 *Newton's method.*

given a starting point $x \in \mathbf{dom}\, f$, tolerance $\epsilon > 0$.

repeat
    1. *Compute the Newton step and decrement.*
        $\Delta x_{\mathrm{nt}} := -\nabla^2 f(x)^{-1} \nabla f(x); \quad \lambda^2 := \nabla f(x)^T \nabla^2 f(x)^{-1} \nabla f(x).$
    2. *Stopping criterion.* **quit** if $\lambda^2/2 \le \epsilon$.
    3. *Line search.* Choose step size t by backtracking line search.
    4. *Update.* $x := x + t\Delta x_{\mathrm{nt}}$.

This is essentially the general descent method described in §9.2, using the Newton step as search direction. The only difference (which is very minor) is that the stopping criterion is checked after computing the search direction, rather than after the update.

9.5.3 Convergence analysis

We assume, as before, that f is twice continuously differentiable, and strongly convex with constant m, i.e., $\nabla^2 f(x) \succeq mI$ for $x \in S$. We have seen that this also implies that there exists an $M > 0$ such that $\nabla^2 f(x) \preceq MI$ for all $x \in S$.

In addition, we assume that the Hessian of f is Lipschitz continuous on S with constant L, i.e.,

$$\|\nabla^2 f(x) - \nabla^2 f(y)\|_2 \le L\|x - y\|_2 \tag{9.31}$$

for all x, $y \in S$. The coefficient L, which can be interpreted as a bound on the third derivative of f, can be taken to be zero for a quadratic function. More generally L measures how well f can be approximated by a quadratic model, so we can expect the Lipschitz constant L to play a critical role in the performance of Newton's method. Intuition suggests that Newton's method will work very well for a function whose quadratic model varies slowly (i.e., has small L).

Idea and outline of convergence proof

We first give the idea and outline of the convergence proof, and the main conclusion, and then the details of the proof. We will show there are numbers η and γ with $0 < \eta \le m^2/L$ and $\gamma > 0$ such that the following hold.

- If $\|\nabla f(x^{(k)})\|_2 \ge \eta$, then

$$f(x^{(k+1)}) - f(x^{(k)}) \le -\gamma. \tag{9.32}$$

- If $\|\nabla f(x^{(k)})\|_2 < \eta$, then the backtracking line search selects $t^{(k)} = 1$ and

$$\frac{L}{2m^2}\|\nabla f(x^{(k+1)})\|_2 \le \left(\frac{L}{2m^2}\|\nabla f(x^{(k)})\|_2\right)^2. \tag{9.33}$$

Let us analyze the implications of the second condition. Suppose that it is satisfied for iteration k, i.e., $\|\nabla f(x^{(k)})\|_2 < \eta$. Since $\eta \le m^2/L$, we have $\|\nabla f(x^{(k+1)})\|_2 < \eta$, i.e., the second condition is also satisfied at iteration $k + 1$. Continuing recursively, we conclude that once the second condition holds, it will hold for all future iterates, i.e., for all $l \ge k$, we have $\|\nabla f(x^{(l)})\|_2 < \eta$. Therefore for all $l \ge k$, the algorithm takes a full Newton step $t = 1$, and

$$\frac{L}{2m^2}\|\nabla f(x^{(l+1)})\|_2 \le \left(\frac{L}{2m^2}\|\nabla f(x^{(l)})\|_2\right)^2. \tag{9.34}$$

Applying this inequality recursively, we find that for $l \ge k$,

$$\frac{L}{2m^2}\|\nabla f(x^{(l)})\|_2 \le \left(\frac{L}{2m^2}\|\nabla f(x^{(k)})\|_2\right)^{2^{l-k}} \le \left(\frac{1}{2}\right)^{2^{l-k}},$$

and hence

$$f(x^{(l)}) - p^\star \le \frac{1}{2m}\|\nabla f(x^{(l)})\|_2^2 \le \frac{2m^3}{L^2}\left(\frac{1}{2}\right)^{2^{l-k+1}}. \tag{9.35}$$

This last inequality shows that convergence is extremely rapid once the second condition is satisfied. This phenomenon is called *quadratic convergence*. Roughly speaking, the inequality (9.35) means that, after a sufficiently large number of iterations, the number of correct digits doubles at each iteration.

The iterations in Newton's method naturally fall into two stages. The second stage, which occurs once the condition $\|\nabla f(x)\|_2 \leq \eta$ holds, is called the *quadratically convergent stage*. We refer to the first stage as the *damped Newton phase*, because the algorithm can choose a step size $t < 1$. The quadratically convergent stage is also called the *pure Newton* phase, since in these iterations a step size $t = 1$ is always chosen.

Now we can estimate the total complexity. First we derive an upper bound on the number of iterations in the damped Newton phase. Since f decreases by at least γ at each iteration, the number of damped Newton steps cannot exceed

$$\frac{f(x^{(0)}) - p^\star}{\gamma},$$

since if it did, f would be less than $p^\star$, which is impossible.

We can bound the number of iterations in the quadratically convergent phase using the inequality (9.35). It implies that we must have $f(x) - p^\star \leq \epsilon$ after no more than

$$\log_2 \log_2(\epsilon_0/\epsilon)$$

iterations in the quadratically convergent phase, where $\epsilon_0 = 2m^3/L^2$.

Overall, then, the number of iterations until $f(x) - p^\star \leq \epsilon$ is bounded above by

$$\frac{f(x^{(0)}) - p^\star}{\gamma} + \log_2 \log_2(\epsilon_0/\epsilon). \tag{9.36}$$

The term $\log_2 \log_2(\epsilon_0/\epsilon)$, which bounds the number of iterations in the quadratically convergent phase, grows *extremely slowly* with required accuracy ϵ, and can be considered a constant for practical purposes, say five or six. (Six iterations of the quadratically convergent stage gives an accuracy of about $\epsilon \approx 5 \cdot 10^{-20}\epsilon_0$.)

Not quite accurately, then, we can say that the number of Newton iterations required to minimize f is bounded above by

$$\frac{f(x^{(0)}) - p^\star}{\gamma} + 6. \tag{9.37}$$

A more precise statement is that (9.37) is a bound on the number of iterations to compute an extremely good approximation of the solution.

Damped Newton phase

We now establish the inequality (9.32). Assume $\|\nabla f(x)\|_2 \geq \eta$. We first derive a lower bound on the step size selected by the line search. Strong convexity implies that $\nabla^2 f(x) \preceq MI$ on S, and therefore

$$
\begin{aligned}
f(x + t\Delta x_{\mathrm{nt}}) &\leq f(x) + t\nabla f(x)^T \Delta x_{\mathrm{nt}} + \frac{M\|\Delta x_{\mathrm{nt}}\|_2^2}{2}t^2 \\
&\leq f(x) - t\lambda(x)^2 + \frac{M}{2m}t^2\lambda(x)^2,
\end{aligned}
$$

where we use (9.30) and

$$\lambda(x)^2 = \Delta x_{\mathrm{nt}}^T \nabla^2 f(x) \Delta x_{\mathrm{nt}} \geq m \|\Delta x_{\mathrm{nt}}\|_2^2.$$

The step size $\hat{t} = m/M$ satisfies the exit condition of the line search, since

$$f(x + \hat{t}\Delta x_{\mathrm{nt}}) \leq f(x) - \frac{m}{2M}\lambda(x)^2 \leq f(x) - \alpha\hat{t}\lambda(x)^2.$$

Therefore the line search returns a step size $t \geq \beta m/M$, resulting in a decrease of the objective function

$$
\begin{aligned}
f(x^+) - f(x) &\leq -\alpha t\lambda(x)^2 \\
&\leq -\alpha\beta\frac{m}{M}\lambda(x)^2 \\
&\leq -\alpha\beta\frac{m}{M^2}\|\nabla f(x)\|_2^2 \\
&\leq -\alpha\beta\eta^2\frac{m}{M^2},
\end{aligned}
$$

where we use

$$\lambda(x)^2 = \nabla f(x)^T \nabla^2 f(x)^{-1} \nabla f(x) \geq (1/M)\|\nabla f(x)\|_2^2.$$

Therefore, (9.32) is satisfied with

$$\gamma = \alpha\beta\eta^2\frac{m}{M^2}. \tag{9.38}$$

Quadratically convergent phase

We now establish the inequality (9.33). Assume $\|\nabla f(x)\|_2 < \eta$. We first show that the backtracking line search selects unit steps, provided

$$\eta \leq 3(1 - 2\alpha)\frac{m^2}{L}.$$

By the Lipschitz condition (9.31), we have, for $t \geq 0$,

$$\|\nabla^2 f(x + t\Delta x_{\mathrm{nt}}) - \nabla^2 f(x)\|_2 \leq tL\|\Delta x_{\mathrm{nt}}\|_2,$$

and therefore

$$\left|\Delta x_{\mathrm{nt}}^T \left(\nabla^2 f(x + t\Delta x_{\mathrm{nt}}) - \nabla^2 f(x)\right)\Delta x_{\mathrm{nt}}\right| \leq tL\|\Delta x_{\mathrm{nt}}\|_2^3.$$

With $\tilde{f}(t) = f(x + t\Delta x_{\mathrm{nt}})$, we have $\tilde{f}''(t) = \Delta x_{\mathrm{nt}}^T \nabla^2 f(x + t\Delta x_{\mathrm{nt}})\Delta x_{\mathrm{nt}}$, so the inequality above is

$$|\tilde{f}''(t) - \tilde{f}''(0)| \leq tL\|\Delta x_{\mathrm{nt}}\|_2^3.$$

We will use this inequality to determine an upper bound on $\tilde{f}(t)$. We start with

$$\tilde{f}''(t) \leq \tilde{f}''(0) + tL\|\Delta x_{\mathrm{nt}}\|_2^3 \leq \lambda(x)^2 + t\frac{L}{m^{3/2}}\lambda(x)^3,$$

where we use $\tilde{f}''(0) = \lambda(x)^2$ and $\lambda(x)^2 \geq m\|\Delta x_{\mathrm{nt}}\|_2^2$. We integrate the inequality to get

$$
\begin{aligned}
\tilde{f}'(t) &\leq \tilde{f}'(0) + t\lambda(x)^2 + t^2 \frac{L}{2m^{3/2}}\lambda(x)^3 \\
&= -\lambda(x)^2 + t\lambda(x)^2 + t^2 \frac{L}{2m^{3/2}}\lambda(x)^3,
\end{aligned}
$$

using $\tilde{f}'(0) = -\lambda(x)^2$. We integrate once more to get

$$
\tilde{f}(t) \leq \tilde{f}(0) - t\lambda(x)^2 + t^2 \frac{1}{2}\lambda(x)^2 + t^3 \frac{L}{6m^{3/2}}\lambda(x)^3.
$$

Finally, we take $t = 1$ to obtain

$$
f(x + \Delta x_{\mathrm{nt}}) \leq f(x) - \frac{1}{2}\lambda(x)^2 + \frac{L}{6m^{3/2}}\lambda(x)^3. \tag{9.39}
$$

Now suppose $\|\nabla f(x)\|_2 \leq \eta \leq 3(1 - 2\alpha)m^2/L$. By strong convexity, we have

$$
\lambda(x) \leq 3(1 - 2\alpha)m^{3/2}/L,
$$

and by (9.39) we have

$$
\begin{aligned}
f(x + \Delta x_{\mathrm{nt}}) &\leq f(x) - \lambda(x)^2 \left(\frac{1}{2} - \frac{L\lambda(x)}{6m^{3/2}} \right) \\
&\leq f(x) - \alpha\lambda(x)^2 \\
&= f(x) + \alpha\nabla f(x)^T \Delta x_{\mathrm{nt}},
\end{aligned}
$$

which shows that the unit step $t = 1$ is accepted by the backtracking line search.

Let us now examine the rate of convergence. Applying the Lipschitz condition, we have

$$
\begin{aligned}
\|\nabla f(x^+)\|_2 &= \|\nabla f(x + \Delta x_{\mathrm{nt}}) - \nabla f(x) - \nabla^2 f(x)\Delta x_{\mathrm{nt}}\|_2 \\
&= \left\| \int_0^1 \left(\nabla^2 f(x + t\Delta x_{\mathrm{nt}}) - \nabla^2 f(x) \right) \Delta x_{\mathrm{nt}} \, dt \right\|_2 \\
&\leq \frac{L}{2}\|\Delta x_{\mathrm{nt}}\|_2^2 \\
&= \frac{L}{2}\|\nabla^2 f(x)^{-1}\nabla f(x)\|_2^2 \\
&\leq \frac{L}{2m^2}\|\nabla f(x)\|_2^2,
\end{aligned}
$$

i.e., the inequality (9.33).

In conclusion, the algorithm selects unit steps and satisfies the condition (9.33) if $\|\nabla f(x^{(k)})\|_2 < \eta$, where

$$
\eta = \min\{1, 3(1 - 2\alpha)\} \frac{m^2}{L}.
$$

Substituting this bound and (9.38) into (9.37), we find that the number of iterations is bounded above by

$$
6 + \frac{M^2 L^2 / m^5}{\alpha\beta \min\{1, 9(1 - 2\alpha)^2\}} (f(x^{(0)}) - p^\star). \tag{9.40}
$$

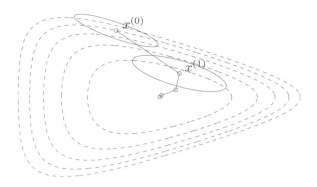

Figure 9.19 Newton's method for the problem in $\mathbf{R}^2$, with objective f given in (9.20), and backtracking line search parameters $\alpha = 0.1$, $\beta = 0.7$. Also shown are the ellipsoids $\{x \mid \|x - x^{(k)}\|_{\nabla^2 f(x^{(k)})} \leq 1\}$ at the first two iterates.

9.5.4   Examples

Example in $\mathbf{R}^2$

We first apply Newton's method with backtracking line search on the test function (9.20), with line search parameters $\alpha = 0.1$, $\beta = 0.7$. Figure 9.19 shows the Newton iterates, and also the ellipsoids

$$\{x \mid \|x - x^{(k)}\|_{\nabla^2 f(x^{(k)})} \leq 1\}$$

for the first two iterates $k = 0,\ 1$. The method works well because these ellipsoids give good approximations of the shape of the sublevel sets.

Figure 9.20 shows the error versus iteration number for the same example. This plot shows that convergence to a very high accuracy is achieved in only five iterations. Quadratic convergence is clearly apparent: The last step reduces the error from about 10^{-5} to 10^{-10}.

Example in $\mathbf{R}^{100}$

Figure 9.21 shows the convergence of Newton's method with backtracking and exact line search for a problem in $\mathbf{R}^{100}$. The objective function has the form (9.21), with the same problem data and the same starting point as was used in figure 9.6. The plot for the backtracking line search shows that a very high accuracy is attained in eight iterations. Like the example in $\mathbf{R}^2$, quadratic convergence is clearly evident after about the third iteration. The number of iterations in Newton's method with exact line search is only one smaller than with a backtracking line search. This is also typical. An exact line search usually gives a very small improvement in convergence of Newton's method. Figure 9.22 shows the step sizes for this example. After two damped steps, the steps taken by the backtracking line search are all full, i.e., $t = 1$.

Experiments with the values of the backtracking parameters α and β reveal that they have little effect on the performance of Newton's method, for this example

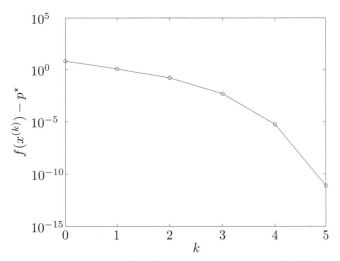

Figure 9.20 Error versus iteration k of Newton's method for the problem in $\mathbf{R}^2$. Convergence to a very high accuracy is achieved in five iterations.

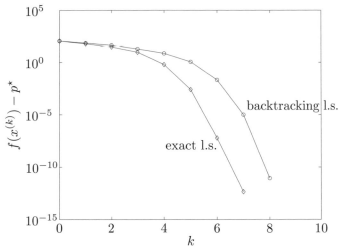

Figure 9.21 Error versus iteration for Newton's method for the problem in $\mathbf{R}^{100}$. The backtracking line search parameters are $\alpha = 0.01$, $\beta = 0.5$. Here too convergence is extremely rapid: a very high accuracy is attained in only seven or eight iterations. The convergence of Newton's method with exact line search is only one iteration faster than with backtracking line search.

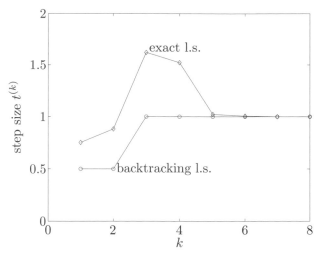

Figure 9.22 The step size t versus iteration for Newton's method with backtracking and exact line search, applied to the problem in $\mathbf{R}^{100}$. The backtracking line search takes one backtracking step in the first two iterations. After the first two iterations it always selects $t = 1$.

(and others). With α fixed at 0.01, and values of β varying between 0.2 and 1, the number of iterations required varies between 8 and 12. With β fixed at 0.5, the number of iterations is 8, for all values of α between 0.005 and 0.5. For these reasons, most practical implementations use a backtracking line search with a small value of α, such as 0.01, and a larger value of β, such as 0.5.

Example in $\mathbf{R}^{10000}$

In this last example we consider a larger problem, of the form

$$\text{minimize}\quad -\sum_{i-1}^{n} \log(1 - x_i^2) - \sum_{i=1}^{m} \log(b_i - a_i^T x)$$

with $m = 100000$ and $n = 10000$. The problem data a_i are randomly generated sparse vectors. Figure 9.23 shows the convergence of Newton's method with backtracking line search, with parameters $\alpha = 0.01$, $\beta = 0.5$. The performance is very similar to the previous convergence plots. A linearly convergent initial phase of about 13 iterations is followed by a quadratically convergent phase, that achieves a very high accuracy in 4 or 5 more iterations.

Affine invariance of Newton's method

A very important feature of Newton's method is that it is independent of linear (or affine) changes of coordinates. Let $x^{(k)}$ be the kth iterate of Newton's method, applied to $f : \mathbf{R}^n \to \mathbf{R}$. Suppose $T \in \mathbf{R}^{n \times n}$ is nonsingular, and define $\bar{f}(y) = f(Ty)$. If we use Newton's method (with the same backtracking parameters) to

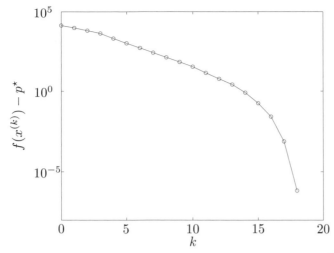

Figure 9.23 Error versus iteration of Newton's method, for a problem in $\mathbf{R}^{10000}$. A backtracking line search with parameters $\alpha = 0.01$, $\beta = 0.5$ is used. Even for this large scale problem, Newton's method requires only 18 iterations to achieve very high accuracy.

minimize $\bar{f}$, starting from $y^{(0)} = T^{-1}x^{(0)}$, then we have

$$Ty^{(k)} = x^{(k)}$$

for all k. In other words, Newton's method is the same: The iterates are related by the same change of coordinates. Even the stopping criterion is the same, since the Newton decrement for $\bar{f}$ at $y^{(k)}$ is the same as the Newton decrement for f at $x^{(k)}$. This is in stark contrast to the gradient (or steepest descent) method, which is strongly affected by changes of coordinates.

As an example, consider the family of problems given in (9.22), indexed by the parameter γ, which affects the condition number of the sublevel sets. We observed (in figures 9.7 and 9.8) that the gradient method slows to useless for values of γ smaller than 0.05 or larger than 20. In contrast, Newton's method (with $\alpha = 0.01$, $\beta = 0.5$) solves this problem (in fact, to a far higher accuracy) in nine iterations, for all values of γ between 10^{-10} and 10^{10}.

In a real implementation, with finite precision arithmetic, Newton's method is not exactly independent of affine changes of coordinates, or the condition number of the sublevel sets. But we can say that condition numbers ranging up to very large values such as 10^{10} do not adversely affect a real implementation of Newton's method. For the gradient method, a far smaller range of condition numbers can be tolerated. While choice of coordinates (or condition number of sublevel sets) is a first-order issue for gradient and steepest descent methods, it is a second-order issue for Newton's method; its only effect is in the numerical linear algebra required to compute the Newton step.

Summary

Newton's method has several very strong advantages over gradient and steepest descent methods:

- Convergence of Newton's method is rapid in general, and quadratic near $x^\star$. Once the quadratic convergence phase is reached, at most six or so iterations are required to produce a solution of very high accuracy.

- Newton's method is affine invariant. It is insensitive to the choice of coordinates, or the condition number of the sublevel sets of the objective.

- Newton's method scales well with problem size. Its performance on problems in $\mathbf{R}^{10000}$ is similar to its performance on problems in $\mathbf{R}^{10}$, with only a modest increase in the number of steps required.

- The good performance of Newton's method is not dependent on the choice of algorithm parameters. In contrast, the choice of norm for steepest descent plays a critical role in its performance.

The main disadvantage of Newton's method is the cost of forming and storing the Hessian, and the cost of computing the Newton step, which requires solving a set of linear equations. We will see in §9.7 that in many cases it is possible to exploit problem structure to substantially reduce the cost of computing the Newton step.

Another alternative is provided by a family of algorithms for unconstrained optimization called *quasi-Newton methods*. These methods require less computational effort to form the search direction, but they share some of the strong advantages of Newton methods, such as rapid convergence near $x^\star$. Since quasi-Newton methods are described in many books, and tangential to our main theme, we will not consider them in this book.

9.6   Self-concordance

There are two major shortcomings of the classical convergence analysis of Newton's method given in §9.5.3. The first is a practical one: The resulting complexity estimates involve the three constants m, M, and L, which are almost never known in practice. As a result, the bound (9.40) on the number of Newton steps required is almost never known specifically, since it depends on three constants that are, in general, not known. Of course the convergence analysis and complexity estimate are still conceptually useful.

The second shortcoming is that while Newton's method is affinely invariant, the classical analysis of Newton's method is very much dependent on the coordinate system used. If we change coordinates the constants m, M, and L all change. If for no reason other than aesthetic, we should seek an analysis of Newton's method that is, like the method itself, independent of affine changes of coordinates. In other words, we seek an alternative to the assumptions

$$mI \preceq \nabla^2 f(x) \preceq MI, \qquad \|\nabla^2 f(x) - \nabla^2 f(y)\|_2 \leq L\|x - y\|_2,$$

that is independent of affine changes of coordinates, and also allows us to analyze Newton's method.

A simple and elegant assumption that achieves this goal was discovered by Nesterov and Nemirovski, who gave the name *self-concordance* to their condition. Self-concordant functions are important for several reasons.

- They include many of the logarithmic barrier functions that play an important role in interior-point methods for solving convex optimization problems.

- The analysis of Newton's method for self-concordant functions does not depend on any unknown constants.

- Self-concordance is an affine-invariant property, *i.e.*, if we apply a linear transformation of variables to a self-concordant function, we obtain a self-concordant function. Therefore the complexity estimate that we obtain for Newton's method applied to a self-concordant function is independent of affine changes of coordinates.

9.6.1   Definition and examples

Self-concordant functions on R

We start by considering functions on $\mathbf{R}$. A convex function $f : \mathbf{R} \to \mathbf{R}$ is *self-concordant* if

$$|f'''(x)| \leq 2f''(x)^{3/2} \tag{9.41}$$

for all $x \in \mathbf{dom}\, f$. Since linear and (convex) quadratic functions have zero third derivative, they are evidently self-concordant. Some more interesting examples are given below.

Example 9.3 *Logarithm and entropy.*

- *Negative logarithm.* The function $f(x) = -\log x$ is self-concordant. Using $f''(x) = 1/x^2$, $f'''(x) = -2/x^3$, we find that

$$\frac{|f'''(x)|}{2f''(x)^{3/2}} = \frac{2/x^3}{2(1/x^2)^{3/2}} = 1,$$

so the defining inequality (9.41) holds with equality.

- *Negative entropy plus negative logarithm.* The function $f(x) = x\log x - \log x$ is self-concordant. To verify this, we use

$$f''(x) = \frac{x+1}{x^2}, \qquad f'''(x) = -\frac{x+2}{x^3}$$

to obtain

$$\frac{|f'''(x)|}{2f''(x)^{3/2}} = \frac{x+2}{2(x+1)^{3/2}}.$$

The function on the righthand side is maximized on $\mathbf{R}_+$ by $x = 0$, where its value is 1.

The negative entropy function by itself is *not* self-concordant; see exercise 11.13.

We should make two important remarks about the self-concordance definition (9.41). The first concerns the mysterious constant 2 that appears in the definition. In fact, this constant is chosen for convenience, in order to simplify the formulas later on; any other positive constant could be used instead. Suppose, for example, that the convex function $f : \mathbf{R} \to \mathbf{R}$ satisfies

$$|f'''(x)| \le k f''(x)^{3/2} \tag{9.42}$$

where k is some positive constant. Then the function $\tilde{f}(x) = (k^2/4) f(x)$ satisfies

$$
\begin{aligned}
|\tilde{f}'''(x)| &= (k^2/4)|f'''(x)| \\
&\le (k^3/4) f''(x)^{3/2} \\
&= (k^3/4)\left((4/k^2)\tilde{f}''(x)\right)^{3/2} \\
&= 2\tilde{f}''(x)^{3/2}
\end{aligned}
$$

and therefore is self-concordant. This shows that a function that satisfies (9.42) for some positive k can be scaled to satisfy the standard self-concordance inequality (9.41). So what is important is that the third derivative of the function is bounded by some multiple of the 3/2-power of its second derivative. By appropriately scaling the function, we can change the multiple to the constant 2.

The second comment is a simple calculation that shows why self-concordance is so important: it is affine invariant. Suppose we define the function $\tilde{f}$ by $\tilde{f}(y) = f(ay + b)$, where $a \ne 0$. Then $\tilde{f}$ is self-concordant if and only if f is. To see this, we substitute

$$\tilde{f}''(y) = a^2 f''(x), \qquad \tilde{f}'''(y) = a^3 f'''(x),$$

where $x = ay + b$, into the self-concordance inequality for $\tilde{f}$, *i.e.*, $|\tilde{f}'''(y)| \le 2\tilde{f}''(y)^{3/2}$, to obtain

$$|a^3 f'''(x)| \le 2(a^2 f''(x))^{3/2},$$

which (after dividing by a^3) is the self-concordance inequality for f. Roughly speaking, the self-concordance condition (9.41) is a way to limit the third derivative of a function, in a way that is independent of affine coordinate changes.

Self-concordant functions on $\mathbf{R}^n$

We now consider functions on $\mathbf{R}^n$ with $n > 1$. We say a function $f : \mathbf{R}^n \to \mathbf{R}$ is self-concordant if it is self-concordant along every line in its domain, *i.e.*, if the function $\tilde{f}(t) = f(x + tv)$ is a self-concordant function of t for all $x \in \mathbf{dom}\, f$ and for all v.

9.6.2 Self-concordant calculus

Scaling and sum

Self-concordance is preserved by scaling by a factor exceeding one: If f is self-concordant and $a \geq 1$, then af is self-concordant. Self-concordance is also preserved by addition: If f_1, f_2 are self-concordant, then $f_1 + f_2$ is self-concordant. To show this, it is sufficient to consider functions f_1, $f_2 : \mathbf{R} \to \mathbf{R}$. We have

$$
\begin{aligned}
|f_1'''(x) + f_2'''(x)| &\leq |f_1'''(x)| + |f_2'''(x)| \\
&\leq 2(f_1''(x)^{3/2} + f_2''(x)^{3/2}) \\
&\leq 2(f_1''(x) + f_2''(x))^{3/2}.
\end{aligned}
$$

In the last step we use the inequality

$$
(u^{3/2} + v^{3/2})^{2/3} \leq u + v,
$$

which holds for u, $v \geq 0$.

Composition with affine function

If $f : \mathbf{R}^n \to \mathbf{R}$ is self-concordant, and $A \in \mathbf{R}^{n \times m}$, $b \in \mathbf{R}^n$, then $f(Ax + b)$ is self-concordant.

Example 9.4 *Log barrier for linear inequalities.* The function

$$
f(x) = -\sum_{i=1}^m \log(b_i - a_i^T x),
$$

with $\mathbf{dom}\, f = \{x \mid a_i^T x < b_i,\ i = 1, \ldots, m\}$, is self-concordant. Each term $-\log(b_i - a_i^T x)$ is the composition of $-\log y$ with the affine transformation $y = b_i - a_i^T x$, and hence self-concordant. Therefore the sum is also self-concordant.

Example 9.5 *Log-determinant.* The function $f(X) = -\log \det X$ is self-concordant on $\mathbf{dom}\, f = \mathbf{S}_{++}^n$. To show this, we consider the function $\tilde{f}(t) = f(X + tV)$, where $X \succ 0$ and $V \in \mathbf{S}^n$. It can be expressed as

$$
\begin{aligned}
\tilde{f}(t) &= -\log \det(X^{1/2}(I + tX^{-1/2}VX^{-1/2})X^{1/2}) \\
&= -\log \det X - \log \det(I + tX^{-1/2}VX^{-1/2}) \\
&= -\log \det X - \sum_{i=1}^n \log(1 + t\lambda_i)
\end{aligned}
$$

where λ_i are the eigenvalues of $X^{-1/2}VX^{-1/2}$. Each term $-\log(1 + t\lambda_i)$ is a self-concordant function of t, so the sum, $\tilde{f}$, is self-concordant. It follows that f is self-concordant.

Example 9.6 *Log of concave quadratic.* The function

$$
f(x) = -\log(x^T P x + q^T x + r),
$$

where $P \in -\mathbf{S}_+^n$, is self-concordant on

$$\mathbf{dom}\, f = \{x \mid x^T P x + q^T x + r > 0\}.$$

To show this, it suffices to consider the case $n = 1$ (since by restricting f to a line, the general case reduces to the $n = 1$ case). We can then express f as

$$f(x) = -\log(px^2 + qx + r) = -\log\left(-p(x-a)(b-x)\right)$$

where $\mathbf{dom}\, f = (a, b)$ (*i.e.*, a and b are the roots of $px^2 + qx + r$). Using this expression we have

$$f(x) = -\log(-p) - \log(x - a) - \log(b - x),$$

which establishes self-concordance.

Composition with logarithm

Let $g : \mathbf{R} \to \mathbf{R}$ be a convex function with $\mathbf{dom}\, g = \mathbf{R}_{++}$, and

$$|g'''(x)| \leq 3\frac{g''(x)}{x} \tag{9.43}$$

for all x. Then

$$f(x) = -\log(-g(x)) - \log x$$

is self-concordant on $\{x \mid x > 0, \; g(x) < 0\}$. (For a proof, see exercise 9.14.)

The condition (9.43) is homogeneous and preserved under addition. It is satisfied by all (convex) quadratic functions, *i.e.*, functions of the form $ax^2 + bx + c$, where $a \geq 0$. Therefore if (9.43) holds for a function g, then it holds for the function $g(x) + ax^2 + bx + c$, where $a \geq 0$.

Example 9.7 The following functions g satisfy the condition (9.43).

- $g(x) = -x^p$ for $0 < p \leq 1$.
- $g(x) = -\log x$.
- $g(x) = x \log x$.
- $g(x) = x^p$ for $-1 \leq p \leq 0$.
- $g(x) = (ax + b)^2/x$.

It follows that in each case, the function $f(x) = -\log(-g(x)) - \log x$ is self-concordant. More generally, the function $f(x) = -\log(-g(x) - ax^2 - bx - c) - \log x$ is self-concordant on its domain,

$$\{x \mid x > 0, \; g(x) + ax^2 + bx + c < 0\},$$

provided $a \geq 0$.

Example 9.8 The composition with logarithm rule allows us to show self-concordance of the following functions.

- $f(x, y) = -\log(y^2 - x^T x)$ on $\{(x, y) \mid \|x\|_2 < y\}$.
- $f(x, y) = -2\log y - \log(y^{2/p} - x^2)$, with $p \geq 1$, on $\{(x, y) \in \mathbf{R}^2 \mid |x|^p < y\}$.
- $f(x, y) = -\log y - \log(\log y - x)$ on $\{(x, y) \mid e^x < y\}$.

We leave the details as an exercise (exercise 9.15).

9.6.3 Properties of self-concordant functions

In §9.1.2 we used strong convexity to derive bounds on the suboptimality of a point x in terms of the norm of the gradient at x. For strictly convex self-concordant functions, we can obtain similar bounds in terms of the Newton decrement

$$\lambda(x) = \left(\nabla f(x)^T \nabla^2 f(x)^{-1} \nabla f(x)\right)^{1/2}.$$

(It can be shown that the Hessian of a strictly convex self-concordant function is positive definite everywhere; see exercise 9.17.) Unlike the bounds based on the norm of the gradient, the bounds based on the Newton decrement are not affected by an affine change of coordinates.

For future reference we note that the Newton decrement can also be expressed as

$$\lambda(x) = \sup_{v \neq 0} \frac{-v^T \nabla f(x)}{(v^T \nabla^2 f(x)v)^{1/2}}$$

(see exercise 9.9). In other words, we have

$$\frac{-v^T \nabla f(x)}{(v^T \nabla^2 f(x)v)^{1/2}} \leq \lambda(x) \tag{9.44}$$

for any nonzero v, with equality for $v = \Delta x_{\rm nt}$.

Upper and lower bounds on second derivatives

Suppose $f : \mathbf{R} \to \mathbf{R}$ is a strictly convex self-concordant function. We can write the self-concordance inequality (9.41) as

$$\left| \frac{d}{dt} \left(f''(t)^{-1/2} \right) \right| \leq 1 \tag{9.45}$$

for all $t \in \mathbf{dom}\, f$ (see exercise 9.16). Assuming $t \geq 0$ and the interval between 0 and t is in $\mathbf{dom}\, f$, we can integrate (9.45) between 0 and t to obtain

$$-t \leq \int_0^t \frac{d}{d\tau} \left(f''(\tau)^{-1/2} \right) d\tau \leq t,$$

i.e., $-t \leq f''(t)^{-1/2} - f''(0)^{-1/2} \leq t$. From this we obtain lower and upper bounds on $f''(t)$:

$$\frac{f''(0)}{\left(1 + tf''(0)^{1/2}\right)^2} \leq f''(t) \leq \frac{f''(0)}{\left(1 - tf''(0)^{1/2}\right)^2}. \tag{9.46}$$

The lower bound is valid for all nonnegative $t \in \mathbf{dom}\, f$; the upper bound is valid if $t \in \mathbf{dom}\, f$ and $0 \leq t < f''(0)^{-1/2}$.

Bound on suboptimality

Let $f : \mathbf{R}^n \to \mathbf{R}$ be a strictly convex self-concordant function, and let v be a descent direction (i.e., any direction satisfying $v^T \nabla f(x) < 0$, not necessarily the

Newton direction). Define $\tilde{f} : \mathbf{R} \to \mathbf{R}$ as $\tilde{f}(t) = f(x + tv)$. By definition, the function $\tilde{f}$ is self-concordant.

Integrating the lower bound in (9.46) yields a lower bound on $\tilde{f}'(t)$:

$$\tilde{f}'(t) \geq \tilde{f}'(0) + \tilde{f}''(0)^{1/2} - \frac{\tilde{f}''(0)^{1/2}}{1 + t\tilde{f}''(0)^{1/2}}. \tag{9.47}$$

Integrating again yields a lower bound on $\tilde{f}(t)$:

$$\tilde{f}(t) \geq \tilde{f}(0) + t\tilde{f}'(0) + t\tilde{f}''(0)^{1/2} - \log(1 + t\tilde{f}''(0)^{1/2}). \tag{9.48}$$

The righthand side reaches its minimum at

$$\bar{t} = \frac{-\tilde{f}'(0)}{\tilde{f}''(0) + \tilde{f}''(0)^{1/2}\tilde{f}'(0)},$$

and evaluating at $\bar{t}$ provides a lower bound on $\tilde{f}$:

$$\begin{aligned} \inf_{t \geq 0} \tilde{f}(t) \quad &\geq \quad \tilde{f}(0) + \bar{t}\tilde{f}'(0) + \bar{t}\tilde{f}''(0)^{1/2} - \log(1 + \bar{t}\tilde{f}''(0)^{1/2}) \\ &= \quad \tilde{f}(0) - \tilde{f}'(0)\tilde{f}''(0)^{-1/2} + \log(1 + \tilde{f}'(0)\tilde{f}''(0)^{-1/2}). \end{aligned}$$

The inequality (9.44) can be expressed as

$$\lambda(x) \geq -\tilde{f}'(0)\tilde{f}''(0)^{-1/2}$$

(with equality when $v = \Delta x_{\mathrm{nt}}$), since we have

$$\tilde{f}'(0) = v^T \nabla f(x), \qquad \tilde{f}''(0) = v^T \nabla^2 f(x) v.$$

Now using the fact that $u + \log(1 - u)$ is a monotonically decreasing function of u, and the inequality above, we get

$$\inf_{t \geq 0} \tilde{f}(t) \geq \tilde{f}(0) + \lambda(x) + \log(1 - \lambda(x)).$$

This inequality holds for any descent direction v. Therefore

$$p^\star \geq f(x) + \lambda(x) + \log(1 - \lambda(x)) \tag{9.49}$$

provided $\lambda(x) < 1$. The function $-(\lambda + \log(1 - \lambda))$ is plotted in figure 9.24. It satisfies

$$-(\lambda + \log(1 - \lambda)) \approx \lambda^2 / 2,$$

for small λ, and the bound

$$-(\lambda + \log(1 - \lambda)) \leq \lambda^2$$

for $\lambda \leq 0.68$. Thus, we have the bound on suboptimality

$$p^\star \geq f(x) - \lambda(x)^2, \tag{9.50}$$

valid for $\lambda(x) \leq 0.68$.

Recall that $\lambda(x)^2/2$ is the estimate of $f(x) - p^\star$, based on the quadratic model at x; the inequality (9.50) shows that for self-concordant functions, doubling this estimate gives us a provable bound. In particular, it shows that for self-concordant functions, we can use the stopping criterion

$$\lambda(x)^2 \leq \epsilon,$$

(where $\epsilon < 0.68^2$), and guarantee that on exit $f(x) - p^\star \leq \epsilon$.

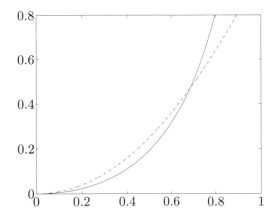

Figure 9.24 The solid line is the function $-(\lambda + \log(1-\lambda))$, which for small λ is approximately $\lambda^2/2$. The dashed line shows λ^2, which is an upper bound in the interval $0 \le \lambda \le 0.68$.

9.6.4   Analysis of Newton's method for self-concordant functions

We now analyze Newton's method with backtracking line search, when applied to a strictly convex self-concordant function f. As before, we assume that a starting point $x^{(0)}$ is known, and that the sublevel set $S = \{x \mid f(x) \le f(x^{(0)})\}$ is closed. We also assume that f is bounded below. (This implies that f has a minimizer $x^\star$; see exercise 9.19.)

The analysis is very similar to the classical analysis given in §9.5.2, except that we use self-concordance as the basic assumption instead of strong convexity and the Lipschitz condition on the Hessian, and the Newton decrement will play the role of the norm of the gradient. We will show that there are numbers η and $\gamma > 0$, with $0 < \eta \le 1/4$, that depend only on the line search parameters α and β, such that the following hold:

- If $\lambda(x^{(k)}) > \eta$, then

$$f(x^{(k+1)}) - f(x^{(k)}) \le -\gamma. \tag{9.51}$$

- If $\lambda(x^{(k)}) \le \eta$, then the backtracking line search selects $t = 1$ and

$$2\lambda(x^{(k+1)}) \le \left(2\lambda(x^{(k)})\right)^2. \tag{9.52}$$

These are the analogs of (9.32) and (9.33). As in §9.5.3, the second condition can be applied recursively, so we can conclude that for all $l \ge k$, we have $\lambda(x^{(l)}) \le \eta$, and

$$2\lambda(x^{(l)}) \le \left(2\lambda(x^{(k)})\right)^{2^{l-k}} \le (2\eta)^{2^{l-k}} \le \left(\frac{1}{2}\right)^{2^{l-k}}.$$

As a consequence, for all $l \ge k$,

$$f(x^{(l)}) - p^\star \le \lambda(x^{(l)})^2 \le \frac{1}{4}\left(\frac{1}{2}\right)^{2^{l-k+1}} \le \left(\frac{1}{2}\right)^{2^{l-k+1}},$$

and hence $f(x^{(l)}) - p^\star \le \epsilon$ if $l - k \ge \log_2 \log_2(1/\epsilon)$.

The first inequality implies that the damped phase cannot require more than $(f(x^{(0)}) - p^\star)/\gamma$ steps. Thus the total number of iterations required to obtain an accuracy $f(x) - p^\star \le \epsilon$, starting at a point $x^{(0)}$, is bounded by

$$\frac{f(x^{(0)}) - p^\star}{\gamma} + \log_2 \log_2(1/\epsilon). \tag{9.53}$$

This is the analog of the bound (9.36) in the classical analysis of Newton's method.

Damped Newton phase

Let $\tilde{f}(t) = f(x + t\Delta x_{\mathrm{nt}})$, so we have

$$\tilde{f}'(0) = -\lambda(x)^2, \qquad \tilde{f}''(0) = \lambda(x)^2.$$

If we integrate the upper bound in (9.46) twice, we obtain an upper bound for $\tilde{f}(t)$:

$$\begin{aligned}
\tilde{f}(t) &\le \tilde{f}(0) + t\tilde{f}'(0) - t\tilde{f}''(0)^{1/2} - \log\left(1 - t\tilde{f}''(0)^{1/2}\right) \\
&= \tilde{f}(0) - t\lambda(x)^2 - t\lambda(x) - \log(1 - t\lambda(x)),
\end{aligned} \tag{9.54}$$

valid for $0 \le t < 1/\lambda(x)$.

We can use this bound to show the backtracking line search always results in a step size $t \ge \beta/(1 + \lambda(x))$. To prove this we note that the point $\hat{t} = 1/(1 + \lambda(x))$ satisfies the exit condition of the line search:

$$\begin{aligned}
\tilde{f}(\hat{t}) &\le \tilde{f}(0) - \hat{t}\lambda(x)^2 - \hat{t}\lambda(x) - \log(1 - \hat{t}\lambda(x)) \\
&= \tilde{f}(0) - \lambda(x) + \log(1 + \lambda(x)) \\
&\le \tilde{f}(0) - \alpha\frac{\lambda(x)^2}{1 + \lambda(x)} \\
&= \tilde{f}(0) - \alpha\lambda(x)^2\hat{t}.
\end{aligned}$$

The second inequality follows from the fact that

$$-x + \log(1 + x) + \frac{x^2}{2(1 + x)} \le 0$$

for $x \ge 0$. Since $t \ge \beta/(1 + \lambda(x))$, we have

$$\tilde{f}(t) - \tilde{f}(0) \le -\alpha\beta\frac{\lambda(x)^2}{1 + \lambda(x)},$$

so (9.51) holds with

$$\gamma = \alpha\beta\frac{\eta^2}{1 + \eta}.$$

Quadratically convergent phase

We will show that we can take

$$\eta = (1 - 2\alpha)/4,$$

(which satisfies $0 < \eta < 1/4$, since $0 < \alpha < 1/2$), *i.e.*, if $\lambda(x^{(k)}) \le (1 - 2\alpha)/4$, then the backtracking line search accepts the unit step and (9.52) holds.

We first note that the upper bound (9.54) implies that a unit step $t = 1$ yields a point in $\textbf{dom } f$ if $\lambda(x) < 1$. Moreover, if $\lambda(x) \le (1 - 2\alpha)/2$, we have, using (9.54),

$$
\begin{aligned}
\tilde{f}(1) &\le \tilde{f}(0) - \lambda(x)^2 - \lambda(x) - \log(1 - \lambda(x)) \\
&\le \tilde{f}(0) - \frac{1}{2}\lambda(x)^2 + \lambda(x)^3 \\
&\le \tilde{f}(0) - \alpha\lambda(x)^2,
\end{aligned}
$$

so the unit step satisfies the condition of sufficient decrease. (The second line follows from the fact that $-x - \log(1 - x) \le \frac{1}{2}x^2 + x^3$ for $0 \le x \le 0.81$.)

The inequality (9.52) follows from the following fact, proved in exercise 9.18. If $\lambda(x) < 1$, and $x^+ = x - \nabla^2 f(x)^{-1} \nabla f(x)$, then

$$\lambda(x^+) \le \frac{\lambda(x)^2}{(1 - \lambda(x))^2}. \tag{9.55}$$

In particular, if $\lambda(x) \le 1/4$,

$$\lambda(x^+) \le 2\lambda(x)^2,$$

which proves that (9.52) holds when $\lambda(x^{(k)}) \le \eta$.

The final complexity bound

Putting it all together, the bound (9.53) on the number of Newton iterations becomes

$$\frac{f(x^{(0)}) - p^\star}{\gamma} + \log_2\log_2(1/\epsilon) = \frac{20 - 8\alpha}{\alpha\beta(1 - 2\alpha)^2}(f(x^{(0)}) - p^\star) + \log_2\log_2(1/\epsilon). \tag{9.56}$$

This expression depends only on the line search parameters α and β, and the final accuracy ϵ. Moreover the term involving ϵ can be safely replaced by the constant six, so the bound really depends only on α and β. For typical values of α and β, the constant that scales $f(x^{(0)}) - p^\star$ is on the order of several hundred. For example, with $\alpha = 0.1$, $\beta = 0.8$, the scaling factor is 375. With tolerance $\epsilon = 10^{-10}$, we obtain the bound

$$375(f(x^{(0)}) - p^\star) + 6. \tag{9.57}$$

We will see that this bound is fairly conservative, but does capture what appears to be the general form of the worst-case number of Newton steps required. A more refined analysis, such as the one originally given by Nesterov and Nemirovski, gives a similar bound, with a substantially smaller constant scaling $f(x^{(0)}) - p^\star$.

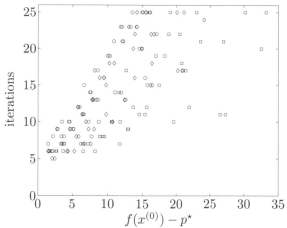

Figure 9.25 Number of Newton iterations required to minimize self-concordant functions versus $f(x^{(0)}) - p^\star$. The function f has the form $f(x) = -\sum_{i=1}^{m} \log(b_i - a_i^T x)$, where the problem data a_i and b are randomly generated. The circles show problems with $m = 100$, $n = 50$; the squares show problems with $m = 1000$, $n = 500$; and the diamonds show problems with $m = 1000$, $n = 50$. Fifty instances of each are shown.

9.6.5   Discussion and numerical examples

A family of self-concordant functions

It is interesting to compare the upper bound (9.57) with the actual number of iterations required to minimize a self-concordant function. We consider a family of problems of the form

$$f(x) = -\sum_{i=1}^{m} \log(b_i - a_i^T x).$$

The problem data a_i and b were generated as follows. For each problem instance, the coefficients of a_i were generated from independent normal distributions with mean zero and unit variance, and the coefficients b were generated from a uniform distribution on $[0, 1]$. Problem instances which were unbounded below were discarded. For each problem we first compute $x^\star$. We then generate a starting point by choosing a random direction v, and taking $x^{(0)} = x^\star + sv$, where s is chosen so that $f(x^{(0)}) - p^\star$ has a prescribed value between 0 and 35. (We should point out that starting points with values $f(x^{(0)}) - p^\star = 10$ or higher are actually very close to the boundary of the polyhedron.) We then minimize the function using Newton's method with a backtracking line search with parameters $\alpha = 0.1$, $\beta = 0.8$, and tolerance $\epsilon = 10^{-10}$.

Figure 9.25 shows the number of Newton iterations required versus $f(x^{(0)}) - p^\star$ for 150 problem instances. The circles show 50 problems with $m = 100$, $n = 50$; the squares show 50 problems with $m = 1000$, $n = 500$; and the diamonds show 50 problems with $m = 1000$, $n = 50$.

For the values of the backtracking parameters used, the complexity bound found above is

$$375(f(x^{(0)}) - p^\star) + 6, \qquad (9.58)$$

clearly a much larger value than the number of iterations required (for these 150 instances). The plot suggests that there is a valid bound of the same form, but with a much smaller constant (say, around 1.5) scaling $f(x^{(0)}) - p^\star$. Indeed, the expression

$$f(x^{(0)}) - p^\star + 6$$

is not a bad gross predictor of the number of Newton steps required, although it is clearly not the only factor. First, there are plenty of problems instances where the number of Newton steps is somewhat smaller, which correspond, we can guess, to 'lucky' starting points. Note also that for the larger problems, with 500 variables (represented by the squares), there seem to be even more cases where the number of Newton steps is unusually small.

We should mention here that the problem family we study is not just self-concordant, but in fact minimally self-concordant, by which we mean that αf is not self-concordant for $\alpha < 1$. Hence, the bound (9.58) cannot be improved by simply scaling f. (The function $f(x) = -20 \log x$ is an example of a self-concordant function which is not minimally self-concordant, since $(1/20)f$ is also self-concordant.)

Practical importance of self-concordance

We have already observed that Newton's method works in general very well for strongly convex objective functions. We can justify this vague statement empirically, and also using the classical analysis of Newton's method, which yields a complexity bound, but one that depends on several constants that are almost always unknown.

For self-concordant functions we can say somewhat more. We have a complexity bound that is completely explicit, and does not depend on any unknown constants. Empirical studies suggest that this bound can be tightened considerably, but its general form, a small constant plus a multiple of $f(x^{(0)}) - p^\star$, seems to predict, at least crudely, the number of Newton steps required to minimize an approximately minimally self-concordant function.

It is not yet clear whether self-concordant functions are in practice more easily minimized by Newton's method than non-self-concordant functions. (It is not even clear how one would make this statement precise.) At the moment, we can say that self-concordant functions are a class of functions for which we can say considerably more about the complexity of Newton's method than is the case for non-self-concordant functions.

9.7   Implementation

In this section we discuss some of the issues that arise in implementing an unconstrained minimization algorithm. We refer the reader to appendix C for more
details on numerical linear algebra.

9.7.1   Pre-computation for line searches

In the simplest implementation of a line search, $f(x + t\Delta x)$ is evaluated for each
value of t in the same way that $f(z)$ is evaluated for any $z \in \textbf{dom}\, f$. But in some
cases we can exploit the fact that f (and its derivatives, in an exact line search) are
to be evaluated at many points along the ray $\{x + t\Delta x \mid t \geq 0\}$ to reduce the total
computational effort. This usually requires some pre-computation, which is often
on the same order as computing f at any point, after which f (and its derivatives)
can be computed more efficiently along the ray.

Suppose that $x \in \textbf{dom}\, f$ and $\Delta x \in \mathbf{R}^n$, and define $\tilde{f}$ as f restricted to the line
or ray determined by x and Δx, i.e., $\tilde{f}(t) = f(x + t\Delta x)$. In a backtracking line
search we must evaluate $\tilde{f}$ for several, and possibly many, values of t; in an exact
line search method we must evaluate $\tilde{f}$ and one or more derivatives at a number of
values of t. In the simple method described above, we evaluate $\tilde{f}(t)$ by first forming
$z = x + t\Delta x$, and then evaluating $f(z)$. To evaluate $\tilde{f}'(t)$, we form $z = x + t\Delta x$,
then evaluate $\nabla f(z)$, and then compute $\tilde{f}'(t) = \nabla f(z)^T \Delta x$. In some representative
examples below we show how $\tilde{f}$ can be computed at a number of values of t more
efficiently.

Composition with an affine function

A very general case in which pre-computation can speed up the line search process
occurs when the objective has the form $f(x) = \phi(Ax + b)$, where $A \in \mathbf{R}^{p \times n}$, and ϕ
is easy to evaluate (for example, separable). To evaluate $\tilde{f}(t) = f(x + t\Delta x)$ for k
values of t using the simple approach, we form $A(x + t\Delta x) + b$ for each value of t
(which costs $2kpn$ flops), and then evaluate $\phi(A(x + t\Delta x) + b)$ for each value of t.
This can be done more efficiently by first computing $Ax + b$ and $A\Delta x$ ($4pn$ flops),
then forming $A(x + t\Delta x) + b$ for each value of t using

$$A(x + t\Delta x) + b = (Ax + b) + t(A\Delta x),$$

which costs $2kp$ flops. The total cost, keeping only the dominant terms, is $4pn + 2kp$
flops, compared to $2kpn$ for the simple method.

Analytic center of a linear matrix inequality

Here we give an example that is more specific, and more complete. We consider
the problem (9.6) of computing the analytic center of a linear matrix inequality,
i.e., minimizing $\log \det F(x)^{-1}$, where $x \in \mathbf{R}^n$ and $F : \mathbf{R}^n \to \mathbf{S}^p$ is affine. Along
the line through x with direction Δx we have

$$\tilde{f}(t) = \log \det(F(x + t\Delta x))^{-1} = -\log \det(A + tB)$$

where

$$A = F(x), \qquad B = \Delta x_1 F_1 + \cdots + \Delta x_n F_n \in \mathbf{S}^p.$$

Since $A \succ 0$, it has a Cholesky factorization $A = LL^T$, where L is lower triangular and nonsingular. Therefore we can express $\tilde{f}$ as

$$\tilde{f}(t) = -\log\det\left(L(I + tL^{-1}BL^{-T})L^T\right) = -\log\det A - \sum_{i=1}^{p}\log(1 + t\lambda_i) \quad (9.59)$$

where $\lambda_1, \ldots, \lambda_p$ are the eigenvalues of $L^{-1}BL^{-T}$. Once these eigenvalues are computed, we can evaluate $\tilde{f}(t)$, for any t, with $4p$ simple arithmetic computations, by using the formula on the right hand side of (9.59). We can evaluate $\tilde{f}'(t)$ (and similarly, any higher derivative) in $4p$ operations, using the formula

$$\tilde{f}'(t) = -\sum_{i=1}^{p}\frac{\lambda_i}{1 + t\lambda_i}.$$

Let us compare the two methods for carrying out a line search, assuming that we need to evaluate $f(x + t\Delta x)$ for k values of t. In the simple method, for each value of t we form $F(x+t\Delta x)$, and then evaluate $f(x+t\Delta x)$ as $-\log\det F(x+t\Delta x)$. For example, we can find the Cholesky factorization of $F(x + t\Delta x) = LL^T$, and then evaluate

$$-\log\det F(x + t\Delta x) = -2\sum_{i=1}^{p}\log L_{ii}.$$

The cost is np^2 to form $F(x + t\Delta x)$, plus $(1/3)p^3$ for the Cholesky factorization. Therefore the total cost of the line search is

$$k(np^2 + (1/3)p^3) = knp^2 + (1/3)kp^3.$$

Using the method outlined above, we first form A, which costs np^2, and factor it, which costs $(1/3)p^3$. We also form B (which costs np^2), and $L^{-1}BL^{-T}$, which costs $2p^3$. The eigenvalues of this matrix are then computed, at a cost of about $(4/3)p^3$ flops. This pre-computation requires a total of $2np^2 + (11/3)p^3$ flops. After finishing this pre-computation, we can now evaluate $\tilde{f}(t)$ for each value of t at a cost of $4p$ flops. The total cost is then

$$2np^2 + (11/3)p^3 + 4kp.$$

Assuming k is small compared to $p(2n + (11/3)p)$, this means the entire line search can be carried out at an effort comparable to simply evaluating f. Depending on the values of k, p, and n, the savings over the simple method can be as large as order k.

9.7.2 Computing the Newton step

In this section we briefly describe some of the issues that arise in implementing Newton's method. In most cases, the work of computing the Newton step Δx_{nt}

dominates the work involved in the line search. To compute the Newton step Δx_{nt}, we first evaluate and form the Hessian matrix $H = \nabla^2 f(x)$ and the gradient $g = \nabla f(x)$ at x. Then we solve the system of linear equations $H\Delta x_{\mathrm{nt}} = -g$ to find the Newton step. This set of equations is sometimes called the *Newton system* (since its solution gives the Newton step) or the *normal equations*, since the same type of equation arises in solving a least-squares problem (see §9.1.1).

While a general linear equation solver can be used, it is better to use methods that take advantage of the symmetry and positive definiteness of H. The most common approach is to form the Cholesky factorization of H, i.e., to compute a lower triangular matrix L that satisfies $LL^T = H$ (see §C.3.2). We then solve $Lw = -g$ by forward substitution, to obtain $w = -L^{-1}g$, and then solve $L^T \Delta x_{\mathrm{nt}} = w$ by back substitution, to obtain

$$\Delta x_{\mathrm{nt}} = L^{-T}w = -L^{-T}L^{-1}g = -H^{-1}g.$$

We can compute the Newton decrement as $\lambda^2 = -\Delta x_{\mathrm{nt}}^T g$, or use the formula

$$\lambda^2 = g^T H^{-1}g = \|L^{-1}g\|_2^2 = \|w\|_2^2.$$

If a dense (unstructured) Cholesky factorization is used, the cost of the forward and back substitution is dominated by the cost of the Cholesky factorization, which is $(1/3)n^3$ flops. The total cost of computing the Newton step Δx_{nt} is thus $F+(1/3)n^3$ flops, where F is the cost of forming H and g.

It is often possible to solve the Newton system $H\Delta x_{\mathrm{nt}} = -g$ more efficiently, by exploiting special structure in H, such as band structure or sparsity. In this context, 'structure of H' means structure that is the same for all x. For example, when we say that 'H is tridiagonal' we mean that for every $x \in \mathbf{dom}\, f$, $\nabla^2 f(x)$ is tridiagonal.

Band structure

If the Hessian H is banded with bandwidth k, i.e., $H_{ij} = 0$ for $|i - j| > k$, then the banded Cholesky factorization can be used, as well as banded forward and back substitutions. The cost of computing the Newton step $\Delta x_{\mathrm{nt}} = -H^{-1}g$ is then $F+nk^2$ flops (assuming $k \ll n$), compared to $F+(1/3)n^3$ for a dense factorization and substitution method.

The Hessian band structure condition

$$\nabla^2 f(x)_{ij} = \frac{\partial^2 f(x)}{\partial x_i \partial x_j} = 0 \quad \text{for} \quad |i - j| > k,$$

for all $x \in \mathbf{dom}\, f$, has an interesting interpretation in terms of the objective function f. Roughly speaking it means that in the objective function, each variable x_i couples nonlinearly only to the $2k + 1$ variables x_j, $j = i - k, \ldots, i + k$. This occurs when f has the partial separability form

$$f(x) = \psi_1(x_1, \ldots, x_{k+1}) + \psi_2(x_2, \ldots, x_{k+2}) + \cdots + \psi_{n-k}(x_{n-k}, \ldots, x_n),$$

where $\psi_i : \mathbf{R}^{k+1} \to \mathbf{R}$. In other words, f can be expressed as a sum of functions of k consecutive variables.

Example 9.9 Consider the problem of minimizing $f : \mathbf{R}^n \to \mathbf{R}$, which has the form

$$f(x) = \psi_1(x_1, x_2) + \psi_2(x_2, x_3) + \cdots + \psi_{n-1}(x_{n-1}, x_n),$$

where $\psi_i : \mathbf{R}^2 \to \mathbf{R}$ are convex and twice differentiable. Because of this form, the Hessian $\nabla^2 f$ is tridiagonal, since $\partial^2 f/\partial x_i \partial x_j = 0$ for $|i - j| > 1$. (And conversely, if the Hessian of a function is tridiagonal for all x, then it has this form.)

Using Cholesky factorization and forward and back substitution algorithms for tridiagonal matrices, we can solve the Newton system for this problem in order n flops. This should be compared to order n^3 flops, if the special form of f were not exploited.

Sparse structure

More generally we can exploit sparsity of the Hessian H in solving the Newton system. This sparse structure occurs whenever each variable x_i is nonlinearly coupled (in the objective) to only a few other variables, or equivalently, when the objective function can be expressed as a sum of functions, each depending on only a few variables, and each variable appearing in only a few of these functions.

To solve $H\Delta x = -g$ when H is sparse, a sparse Cholesky factorization is used to compute a permutation matrix P and lower triangular matrix L for which

$$H = PLL^T P^T.$$

The cost of this factorization depends on the particular sparsity pattern, but is often far smaller than $(1/3)n^3$, and an empirical complexity of order n (for large n) is not uncommon. The forward and back substitution are very similar to the basic method without the permutation. We solve $Lw = -P^T g$ using forward substitution, and then solve $L^T v = w$ by back substitution to obtain

$$v = L^{-T} w = -L^{-T} L^{-1} P^T g.$$

The Newton step is then $\Delta x = Pv$.

Since the sparsity pattern of H does not change as x varies (or more precisely, since we only exploit sparsity that does not change with x) we can use the same permutation matrix P for each of the Newton steps. The step of determining a good permutation matrix P, which is called the *symbolic factorization* step, can be done once, for the whole Newton process.

Diagonal plus low rank

There are many other types of structure that can be exploited in solving the Newton system $H\Delta x_{\mathrm{nt}} = -g$. Here we briefly describe one, and refer the reader to appendix C for more details. Suppose the Hessian H can be expressed as a diagonal matrix plus one of low rank, say, p. This occurs when the objective function f has the special form

$$f(x) = \sum_{i=1}^{n} \psi_i(x_i) + \psi_0(Ax + b) \tag{9.60}$$

where $A \in \mathbf{R}^{p \times n}$, $\psi_1, \ldots, \psi_n : \mathbf{R} \to \mathbf{R}$, and $\psi_0 : \mathbf{R}^p \to \mathbf{R}$. In other words, f is a separable function, plus a function that depends on a low dimensional affine function of x.

To find the Newton step Δx_{nt} for (9.60) we must solve the Newton system $H \Delta x_{\text{nt}} = -g$, with

$$H = D + A^T H_0 A.$$

Here $D = \mathbf{diag}(\psi_1''(x_1), \ldots, \psi_n''(x_n))$ is diagonal, and $H_0 = \nabla^2 \psi_0(Ax + b)$ is the Hessian of ψ_0. If we compute the Newton step without exploiting the structure, the cost of solving the Newton system is $(1/3)n^3$ flops.

Let $H_0 = L_0 L_0^T$ be the Cholesky factorization of H_0. We introduce the temporary variable $w = L_0^T A \Delta x_{\text{nt}} \in \mathbf{R}^p$, and express the Newton system as

$$D \Delta x_{\text{nt}} + A^T L_0 w = -g, \qquad w = L_0^T A \Delta x_{\text{nt}}.$$

Substituting $\Delta x_{\text{nt}} = -D^{-1}(A^T L_0 w + g)$ (from the first equation) into the second equation, we obtain

$$(I + L_0^T A D^{-1} A^T L_0) w = -L_0^T A D^{-1} g, \qquad (9.61)$$

which is a system of p linear equations.

Now we proceed as follows to compute the Newton step Δx_{nt}. First we compute the Cholesky factorization of H_0, which costs $(1/3)p^3$. We then form the dense, positive definite symmetric matrix appearing on the lefthand side of (9.61), which costs $2p^2 n$. We then solve (9.61) for w using a Cholesky factorization and a back and forward substitution, which costs $(1/3)p^3$ flops. Finally, we compute Δx_{nt} using $\Delta x_{\text{nt}} = -D^{-1}(A^T L_0 w + g)$, which costs $2np$ flops. The total cost of computing Δx_{nt} is (keeping only the dominant term) $2p^2 n$ flops, which is far smaller than $(1/3)n^3$ for $p \ll n$.

Bibliography

Dennis and Schnabel [DS96] and Ortega and Rheinboldt [OR00] are two standard references on algorithms for unconstrained minimization and nonlinear equations. The result on quadratic convergence, assuming strong convexity and Lipschitz continuity of the Hessian, is attributed to Kantorovich [Kan52]. Polyak [Pol87, §1.6] gives some insightful comments on the role of convergence results that involve unknown constants, such as the results derived in §9.5.3.

Self-concordant functions were introduced by Nesterov and Nemirovski [NN94]. All our results in §9.6 and exercises 9.14–9.20 can be found in their book, although often in a more general form or with different notation. Renegar [Ren01] gives a concise and elegant presentation of self-concordant functions and their role in the analysis of primal-dual interior-point algorithms. Peng, Roos, and Terlaky [PRT02] study interior-point methods from the viewpoint of *self-regular functions*, a class of functions that is similar, but not identical, to self-concordant functions.

References for the material in §9.7 are given at the end of appendix C.

Exercises

Unconstrained minimization

9.1 *Minimizing a quadratic function.*   Consider the problem of minimizing a quadratic function:
$$\text{minimize} \quad f(x) = (1/2)x^T P x + q^T x + r,$$
where $P \in \mathbf{S}^n$ (but we do not assume $P \succeq 0$).

    (a) Show that if $P \not\succeq 0$, *i.e.*, the objective function f is not convex, then the problem is unbounded below.

    (b) Now suppose that $P \succeq 0$ (so the objective function is convex), but the optimality condition $Px^\star = -q$ does not have a solution. Show that the problem is unbounded below.

9.2 *Minimizing a quadratic-over-linear fractional function.* Consider the problem of minimizing the function $f : \mathbf{R}^n \to \mathbf{R}$, defined as
$$f(x) = \frac{\|Ax - b\|_2^2}{c^T x + d}, \qquad \mathbf{dom}\, f = \{x \mid c^T x + d > 0\}.$$
We assume $\mathbf{rank}\, A = n$ and $b \notin \mathcal{R}(A)$.

    (a) Show that f is closed.

    (b) Show that the minimizer $x^\star$ of f is given by
$$x^\star = x_1 + t x_2$$
where $x_1 = (A^T A)^{-1} A^T b$, $x_2 = (A^T A)^{-1} c$, and $t \in \mathbf{R}$ can be calculated by solving a quadratic equation.

9.3 *Initial point and sublevel set condition.* Consider the function $f(x) = x_1^2 + x_2^2$ with domain $\mathbf{dom}\, f = \{(x_1, x_2) \mid x_1 > 1\}$.

    (a) What is $p^\star$?

    (b) Draw the sublevel set $S = \{x \mid f(x) \le f(x^{(0)})\}$ for $x^{(0)} = (2, 2)$. Is the sublevel set S closed? Is f strongly convex on S?

    (c) What happens if we apply the gradient method with backtracking line search, starting at $x^{(0)}$? Does $f(x^{(k)})$ converge to $p^\star$?

9.4 Do you agree with the following argument? The ℓ_1-norm of a vector $x \in \mathbf{R}^m$ can be expressed as
$$\|x\|_1 = (1/2) \inf_{y \succ 0} \left(\sum_{i=1}^m x_i^2 / y_i + \mathbf{1}^T y \right).$$
Therefore the ℓ_1-norm approximation problem
$$\text{minimize} \quad \|Ax - b\|_1$$
is equivalent to the minimization problem
$$\text{minimize} \quad f(x, y) = \sum_{i=1}^m (a_i^T x - b_i)^2 / y_i + \mathbf{1}^T y, \tag{9.62}$$
with $\mathbf{dom}\, f = \{(x, y) \in \mathbf{R}^n \times \mathbf{R}^m \mid y \succ 0\}$, where a_i^T is the ith row of A. Since f is twice differentiable and convex, we can solve the ℓ_1-norm approximation problem by applying Newton's method to (9.62).

9.5 *Backtracking line search.* Suppose f is strongly convex with $mI \preceq \nabla^2 f(x) \preceq MI$. Let Δx be a descent direction at x. Show that the backtracking stopping condition holds for
$$0 < t \le -\frac{\nabla f(x)^T \Delta x}{M \|\Delta x\|_2^2}.$$
Use this to give an upper bound on the number of backtracking iterations.

Gradient and steepest descent methods

9.6 *Quadratic problem in $\mathbf{R}^2$.* Verify the expressions for the iterates $x^{(k)}$ in the first example of §9.3.2.

9.7 Let Δx_{nsd} and Δx_{sd} be the normalized and unnormalized steepest descent directions at x, for the norm $\|\cdot\|$. Prove the following identities.

(a) $\nabla f(x)^T \Delta x_{\text{nsd}} = -\|\nabla f(x)\|_*$.

(b) $\nabla f(x)^T \Delta x_{\text{sd}} = -\|\nabla f(x)\|_*^2$.

(c) $\Delta x_{\text{sd}} = \operatorname{argmin}_v (\nabla f(x)^T v + (1/2)\|v\|^2)$.

9.8 *Steepest descent method in ℓ_∞-norm.* Explain how to find a steepest descent direction in the ℓ_∞-norm, and give a simple interpretation.

Newton's method

9.9 *Newton decrement.* Show that the Newton decrement $\lambda(x)$ satisfies

$$\lambda(x) = \sup_{v^T \nabla^2 f(x) v = 1} (-v^T \nabla f(x)) = \sup_{v \neq 0} \frac{-v^T \nabla f(x)}{(v^T \nabla^2 f(x) v)^{1/2}}.$$

9.10 *The pure Newton method.* Newton's method with fixed step size $t = 1$ can diverge if the initial point is not close to $x^\star$. In this problem we consider two examples.

(a) $f(x) = \log(e^x + e^{-x})$ has a unique minimizer $x^\star = 0$. Run Newton's method with fixed step size $t = 1$, starting at $x^{(0)} = 1$ and at $x^{(0)} = 1.1$.

(b) $f(x) = -\log x + x$ has a unique minimizer $x^\star = 1$. Run Newton's method with fixed step size $t = 1$, starting at $x^{(0)} = 3$.

Plot f and f', and show the first few iterates.

9.11 *Gradient and Newton methods for composition functions.* Suppose $\phi : \mathbf{R} \to \mathbf{R}$ is increasing and convex, and $f : \mathbf{R}^n \to \mathbf{R}$ is convex, so $g(x) = \phi(f(x))$ is convex. (We assume that f and g are twice differentiable.) The problems of minimizing f and minimizing g are clearly equivalent.

Compare the gradient method and Newton's method, applied to f and g. How are the search directions related? How are the methods related if an exact line search is used? *Hint.* Use the matrix inversion lemma (see §C.4.3).

9.12 *Trust region Newton method.* If $\nabla^2 f(x)$ is singular (or very ill-conditioned), the Newton step $\Delta x_{\text{nt}} = -\nabla^2 f(x)^{-1} \nabla f(x)$ is not well defined. Instead we can define a search direction Δx_{tr} as the solution of

$$\begin{array}{ll} \text{minimize} & (1/2)v^T H v + g^T v \\ \text{subject to} & \|v\|_2 \leq \gamma, \end{array}$$

where $H = \nabla^2 f(x)$, $g = \nabla f(x)$, and γ is a positive constant. The point $x + \Delta x_{\text{tr}}$ minimizes the second-order approximation of f at x, subject to the constraint that $\|(x + \Delta x_{\text{tr}}) - x\|_2 \leq \gamma$. The set $\{v \mid \|v\|_2 \leq \gamma\}$ is called the *trust region*. The parameter γ, the size of the trust region, reflects our confidence in the second-order model.

Show that Δx_{tr} minimizes

$$(1/2)v^T H v + g^T v + \hat\beta \|v\|_2^2,$$

for some $\hat\beta$. This quadratic function can be interpreted as a regularized quadratic model for f around x.

Self-concordance

9.13 *Self-concordance and the inverse barrier.*

    (a) Show that $f(x) = 1/x$ with domain $(0, 8/9)$ is self-concordant.

    (b) Show that the function

$$f(x) = \alpha \sum_{i=1}^{m} \frac{1}{b_i - a_i^T x}$$

with $\mathbf{dom}\, f = \{x \in \mathbf{R}^n \mid a_i^T x < b_i, \; i = 1, \ldots, m\}$, is self-concordant if $\mathbf{dom}\, f$ is bounded and

$$\alpha > (9/8) \max_{i=1,\ldots,m} \; \sup_{x \in \mathbf{dom}\, f} \; (b_i - a_i^T x).$$

9.14 *Composition with logarithm.* Let $g : \mathbf{R} \to \mathbf{R}$ be a convex function with $\mathbf{dom}\, g = \mathbf{R}_{++}$, and

$$|g'''(x)| \le 3 \frac{g''(x)}{x}$$

for all x. Prove that $f(x) = -\log(-g(x)) - \log x$ is self-concordant on $\{x \mid x > 0, \; g(x) < 0\}$. *Hint.* Use the inequality

$$\frac{3}{2} r p^2 + q^3 + \frac{3}{2} p^2 q + r^3 \le 1$$

which holds for $p, q, r \in \mathbf{R}_+$ with $p^2 + q^2 + r^2 = 1$.

9.15 Prove that the following functions are self-concordant. In your proof, restrict the function to a line, and apply the composition with logarithm rule.

    (a) $f(x, y) = -\log(y^2 - x^T x)$ on $\{(x, y) \mid \|x\|_2 < y\}$.

    (b) $f(x, y) = -2 \log y - \log(y^{2/p} - x^2)$, with $p \ge 1$, on $\{(x, y) \in \mathbf{R}^2 \mid |x|^p < y\}$.

    (c) $f(x, y) = -\log y - \log(\log y - x)$ on $\{(x, y) \mid e^x < y\}$.

9.16 Let $f : \mathbf{R} \to \mathbf{R}$ be a self-concordant function.

    (a) Suppose $f''(x) \ne 0$. Show that the self-concordance condition (9.41) can be expressed as

$$\left| \frac{d}{dx} \left(f''(x)^{-1/2} \right) \right| \le 1.$$

Find the 'extreme' self-concordant functions of one variable, *i.e.*, the functions f and $\tilde{f}$ that satisfy

$$\frac{d}{dx} \left(f''(x)^{-1/2} \right) = 1, \qquad \frac{d}{dx} \left(\tilde{f}''(x)^{-1/2} \right) = -1,$$

respectively.

    (b) Show that either $f''(x) = 0$ for all $x \in \mathbf{dom}\, f$, or $f''(x) > 0$ for all $x \in \mathbf{dom}\, f$.

9.17 *Upper and lower bounds on the Hessian of a self-concordant function.*

    (a) Let $f : \mathbf{R}^2 \to \mathbf{R}$ be a self-concordant function. Show that

$$\left| \frac{\partial^3 f(x)}{\partial^3 x_i} \right| \;\le\; 2 \left(\frac{\partial^2 f(x)}{\partial x_i^2} \right)^{3/2}, \qquad i = 1, 2,$$

$$\left| \frac{\partial^3 f(x)}{\partial x_i^2 \partial x_j} \right| \;\le\; 2 \frac{\partial^2 f(x)}{\partial x_i^2} \left(\frac{\partial^2 f(x)}{\partial x_j^2} \right)^{1/2}, \qquad i \ne j$$

for all $x \in \mathbf{dom}\, f$.

Hint. If $h : \mathbf{R}^2 \times \mathbf{R}^2 \times \mathbf{R}^2 \to \mathbf{R}$ is a symmetric trilinear form, *i.e.*,

$$
\begin{aligned}
h(u, v, w) \;=\; & a_1 u_1 v_1 w_1 + a_2(u_1 v_1 w_2 + u_1 v_2 w_1 + u_2 v_1 w_1) \\
& + a_3(u_1 v_2 w_2 + u_2 v_1 w_1 + u_2 v_2 w_1) + a_4 u_2 v_2 w_2,
\end{aligned}
$$

then

$$
\sup_{u,v,w\neq 0} \frac{h(u, v, w)}{\|u\|_2 \|v\|_2 \|w\|_2} = \sup_{u\neq 0} \frac{h(u, u, u)}{\|u\|_2^3}.
$$

(b) Let $f : \mathbf{R}^n \to \mathbf{R}$ be a self-concordant function. Show that the nullspace of $\nabla^2 f(x)$ is independent of x. Show that if f is strictly convex, then $\nabla^2 f(x)$ is nonsingular for all $x \in \mathbf{dom}\, f$.

Hint. Prove that if $w^T \nabla^2 f(x)w = 0$ for some $x \in \mathbf{dom}\, f$, then $w^T \nabla^2 f(y)w = 0$ for all $y \in \mathbf{dom}\, f$. To show this, apply the result in (a) to the self-concordant function $\tilde{f}(t, s) = f(x + t(y - x) + sw)$.

(c) Let $f : \mathbf{R}^n \to \mathbf{R}$ be a self-concordant function. Suppose $x \in \mathbf{dom}\, f$, $v \in \mathbf{R}^n$. Show that

$$
(1 - t\alpha)^2 \nabla^2 f(x) \preceq \nabla^2 f(x + tv) \preceq \frac{1}{(1 - t\alpha)^2} \nabla^2 f(x)
$$

for $x + tv \in \mathbf{dom}\, f$, $0 \leq t < \alpha$, where $\alpha = (v^T \nabla^2 f(x)v)^{1/2}$.

9.18 *Quadratic convergence.* Let $f : \mathbf{R}^n \to \mathbf{R}$ be a strictly convex self-concordant function. Suppose $\lambda(x) < 1$, and define $x^+ = x - \nabla^2 f(x)^{-1} \nabla f(x)$. Prove that $\lambda(x^+) \leq \lambda(x)^2/(1 - \lambda(x))^2$. *Hint.* Use the inequalities in exercise 9.17, part (c).

9.19 *Bound on the distance from the optimum.* Let $f : \mathbf{R}^n \to \mathbf{R}$ be a strictly convex self-concordant function.

(a) Suppose $\lambda(\bar{x}) < 1$ and the sublevel set $\{x \mid f(x) \leq f(\bar{x})\}$ is closed. Show that the minimum of f is attained and

$$
\left((\bar{x} - x^\star)^T \nabla^2 f(\bar{x})(\bar{x} - x^\star)\right)^{1/2} \leq \frac{\lambda(\bar{x})}{1 - \lambda(\bar{x})}.
$$

(b) Show that if f has a closed sublevel set, and is bounded below, then its minimum is attained.

9.20 *Conjugate of a self-concordant function.* Suppose $f : \mathbf{R}^n \to \mathbf{R}$ is closed, strictly convex, and self-concordant. We show that its conjugate (or Legendre transform) f^* is self-concordant.

(a) Show that for each $y \in \mathbf{dom}\, f^*$, there is a unique $x \in \mathbf{dom}\, f$ that satisfies $y = \nabla f(x)$. *Hint.* Refer to the result of exercise 9.19.

(b) Suppose $\bar{y} = \nabla f(\bar{x})$. Define

$$
g(t) = f(\bar{x} + tv), \qquad h(t) = f^*(\bar{y} + tw)
$$

where $v \in \mathbf{R}^n$ and $w = \nabla^2 f(\bar{x})v$. Show that

$$
g''(0) = h''(0), \qquad g'''(0) = -h'''(0).
$$

Use these identities to show that f^* is self-concordant.

9.21 *Optimal line search parameters.* Consider the upper bound (9.56) on the number of Newton iterations required to minimize a strictly convex self-concordant functions. What is the minimum value of the upper bound, if we minimize over α and β?

9.22 Suppose that f is strictly convex and satisfies (9.42). Give a bound on the number of Newton steps required to compute $p^\star$ within ϵ, starting at $x^{(0)}$.

Implementation

9.23 *Pre-computation for line searches.* For each of the following functions, explain how the computational cost of a line search can be reduced by a pre-computation. Give the cost of the pre-computation, and the cost of evaluating $g(t) = f(x + t\Delta x)$ and $g'(t)$ with and without the pre-computation.

(a) $f(x) = -\sum_{i=1}^{m} \log(b_i - a_i^T x)$.

(b) $f(x) = \log\left(\sum_{i=1}^{m} \exp(a_i^T x + b_i)\right)$.

(c) $f(x) = (Ax - b)^T (P_0 + x_1 P_1 + \cdots + x_n P_n)^{-1}(Ax - b)$, where $P_i \in \mathbf{S}^m$, $A \in \mathbf{R}^{m \times n}$, $b \in \mathbf{R}^m$ and $\mathbf{dom}\, f = \{x \mid P_0 + \sum_{i=1}^{n} x_i P_i \succ 0\}$.

9.24 *Exploiting block diagonal structure in the Newton system.* Suppose the Hessian $\nabla^2 f(x)$ of a convex function f is block diagonal. How do we exploit this structure when computing the Newton step? What does it mean about f?

9.25 *Smoothed fit to given data.* Consider the problem

$$\text{minimize} \quad f(x) = \sum_{i=1}^{n} \psi(x_i - y_i) + \lambda \sum_{i=1}^{n-1}(x_{i+1} - x_i)^2$$

where $\lambda > 0$ is smoothing parameter, ψ is a convex penalty function, and $x \in \mathbf{R}^n$ is the variable. We can interpret x as a smoothed fit to the vector y.

(a) What is the structure in the Hessian of f?

(b) Extend to the problem of making a smooth fit to two-dimensional data, *i.e.*, minimizing the function

$$\sum_{i,j=1}^{n} \psi(x_{ij} - y_{ij}) + \lambda \left(\sum_{i=1}^{n-1}\sum_{j=1}^{n}(x_{i+1,j} - x_{ij})^2 + \sum_{i=1}^{n}\sum_{j=1}^{n-1}(x_{i,j+1} - x_{ij})^2 \right),$$

with variable $X \in \mathbf{R}^{n \times n}$, where $Y \in \mathbf{R}^{n \times n}$ and $\lambda > 0$ are given.

9.26 *Newton equations with linear structure.* Consider the problem of minimizing a function of the form

$$f(x) = \sum_{i=1}^{N} \psi_i(A_i x + b_i) \tag{9.63}$$

where $A_i \in \mathbf{R}^{m_i \times n}$, $b_i \in \mathbf{R}^{m_i}$, and the functions $\psi_i : \mathbf{R}^{m_i} \to \mathbf{R}$ are twice differentiable and convex. The Hessian H and gradient g of f at x are given by

$$H = \sum_{i=1}^{N} A_i^T H_i A_i, \qquad g = \sum_{i=1}^{N} A_i^T g_i. \tag{9.64}$$

where $H_i = \nabla^2 \psi_i(A_i x + b_i)$ and $g_i = \nabla \psi_i(A_i x + b_i)$.

Describe how you would implement Newton's method for minimizing f. Assume that $n \gg m_i$, the matrices A_i are very sparse, but the Hessian H is dense.

9.27 *Analytic center of linear inequalities with variable bounds.* Give the most efficient method for computing the Newton step of the function

$$f(x) = -\sum_{i=1}^{n} \log(x_i + 1) - \sum_{i=1}^{n} \log(1 - x_i) - \sum_{i=1}^{m} \log(b_i - a_i^T x),$$

with $\mathbf{dom}\, f = \{x \in \mathbf{R}^n \mid -1 \prec x \prec 1, Ax \prec b\}$, where a_i^T is the ith row of A. Assume A is dense, and distinguish two cases: $m \geq n$ and $m \leq n$. (See also exercise 9.30.)

9.28 *Analytic center of quadratic inequalities.* Describe an efficient method for computing the Newton step of the function

$$f(x) = -\sum_{i=1}^{m} \log(-x^T A_i x - b_i^T x - c_i),$$

with **dom** $f = \{x \mid x^T A_i x + b_i^T x + c_i < 0,\ i = 1, \dots, m\}$. Assume that the matrices $A_i \in \mathbf{S}_{++}^n$ are large and sparse, and $m \ll n$.

Hint. The Hessian and gradient of f at x are given by

$$H = \sum_{i=1}^{m} (2\alpha_i A_i + \alpha_i^2 (2A_i x + b_i)(2A_i x + b_i)^T), \qquad g = \sum_{i=1}^{m} \alpha_i (2A_i x + b_i),$$

where $\alpha_i = 1/(-x^T A_i x - b_i^T x - c_i)$.

9.29 *Exploiting structure in two-stage optimization.* This exercise continues exercise 4.64, which describes optimization with recourse, or two-stage optimization. Using the notation and assumptions in exercise 4.64, we assume in addition that the cost function f is a twice differentiable function of (x, z), for each scenario $i = 1, \dots, S$.

Explain how to efficiently compute the Newton step for the problem of finding the optimal policy. How does the approximate flop count for your method compare to that of a generic method (which exploits no structure), as a function of S, the number of scenarios?

Numerical experiments

9.30 *Gradient and Newton methods.* Consider the unconstrained problem

$$\text{minimize} \quad f(x) = -\sum_{i=1}^{m} \log(1 - a_i^T x) - \sum_{i=1}^{n} \log(1 - x_i^2),$$

with variable $x \in \mathbf{R}^n$, and **dom** $f = \{x \mid a_i^T x < 1,\ i = 1, \dots, m,\ |x_i| < 1,\ i = 1, \dots, n\}$. This is the problem of computing the analytic center of the set of linear inequalities

$$a_i^T x \le 1, \quad i = 1, \dots, m, \qquad |x_i| \le 1, \quad i = 1, \dots, n.$$

Note that we can choose $x^{(0)} = 0$ as our initial point. You can generate instances of this problem by choosing a_i from some distribution on $\mathbf{R}^n$.

(a) Use the gradient method to solve the problem, using reasonable choices for the backtracking parameters, and a stopping criterion of the form $\|\nabla f(x)\|_2 \le \eta$. Plot the objective function and step length versus iteration number. (Once you have determined $p^\star$ to high accuracy, you can also plot $f - p^\star$ versus iteration.) Experiment with the backtracking parameters α and β to see their effect on the total number of iterations required. Carry these experiments out for several instances of the problem, of different sizes.

(b) Repeat using Newton's method, with stopping criterion based on the Newton decrement λ^2. Look for quadratic convergence. You do not have to use an efficient method to compute the Newton step, as in exercise 9.27; you can use a general purpose dense solver, although it is better to use one that is based on a Cholesky factorization.

Hint. Use the chain rule to find expressions for $\nabla f(x)$ and $\nabla^2 f(x)$.

9.31 *Some approximate Newton methods.* The cost of Newton's method is dominated by the cost of evaluating the Hessian $\nabla^2 f(x)$ and the cost of solving the Newton system. For large problems, it is sometimes useful to replace the Hessian by a positive definite approximation that makes it easier to form and solve for the search step. In this problem we explore some common examples of this idea.

For each of the approximate Newton methods described below, test the method on some instances of the analytic centering problem described in exercise 9.30, and compare the results to those obtained using the Newton method and gradient method.

(a) *Re-using the Hessian.* We evaluate and factor the Hessian only every N iterations, where $N > 1$, and use the search step $\Delta x = -H^{-1}\nabla f(x)$, where H is the last Hessian evaluated. (We need to evaluate and factor the Hessian once every N steps; for the other steps, we compute the search direction using back and forward substitution.)

(b) *Diagonal approximation.* We replace the Hessian by its diagonal, so we only have to evaluate the n second derivatives $\partial^2 f(x)/\partial x_i^2$, and computing the search step is very easy.

9.32 *Gauss-Newton method for convex nonlinear least-squares problems.* We consider a (nonlinear) least-squares problem, in which we minimize a function of the form

$$f(x) = \frac{1}{2}\sum_{i=1}^{m} f_i(x)^2,$$

where f_i are twice differentiable functions. The gradient and Hessian of f at x are given by

$$\nabla f(x) = \sum_{i=1}^{m} f_i(x)\nabla f_i(x), \qquad \nabla^2 f(x) = \sum_{i=1}^{m} \left(\nabla f_i(x)\nabla f_i(x)^T + f_i(x)\nabla^2 f_i(x)\right).$$

We consider the case when f is convex. This occurs, for example, if each f_i is either nonnegative and convex, or nonpositive and concave, or affine.

The *Gauss-Newton method* uses the search direction

$$\Delta x_{\text{gn}} = -\left(\sum_{i=1}^{m}\nabla f_i(x)\nabla f_i(x)^T\right)^{-1}\left(\sum_{i=1}^{m} f_i(x)\nabla f_i(x)\right).$$

(We assume here that the inverse exists, *i.e.*, the vectors $\nabla f_1(x), \ldots, \nabla f_m(x)$ span $\mathbf{R}^n$.) This search direction can be considered an approximate Newton direction (see exercise 9.31), obtained by dropping the second derivative terms from the Hessian of f.

We can give another simple interpretation of the Gauss-Newton search direction Δx_{gn}. Using the first-order approximation $f_i(x+v) \approx f_i(x) + \nabla f_i(x)^T v$ we obtain the approximation

$$f(x+v) \approx \frac{1}{2}\sum_{i=1}^{m}(f_i(x) + \nabla f_i(x)^T v)^2.$$

The Gauss-Newton search step Δx_{gn} is precisely the value of v that minimizes this approximation of f. (Moreover, we conclude that Δx_{gn} can be computed by solving a linear least-squares problem.)

Test the Gauss-Newton method on some problem instances of the form

$$f_i(x) = (1/2)x^T A_i x + b_i^T x + 1,$$

with $A_i \in \mathbf{S}_{++}^n$ and $b_i^T A_i^{-1} b_i \leq 2$ (which ensures that f is convex).

Chapter 10

Equality constrained minimization

10.1 Equality constrained minimization problems

In this chapter we describe methods for solving a convex optimization problem with equality constraints,

$$\begin{array}{ll} \text{minimize} & f(x) \\ \text{subject to} & Ax = b, \end{array} \tag{10.1}$$

where $f : \mathbf{R}^n \to \mathbf{R}$ is convex and twice continuously differentiable, and $A \in \mathbf{R}^{p \times n}$ with $\mathbf{rank}\, A = p < n$. The assumptions on A mean that there are fewer equality constraints than variables, and that the equality constraints are independent. We will assume that an optimal solution $x^\star$ exists, and use $p^\star$ to denote the optimal value, $p^\star = \inf\{f(x) \mid Ax = b\} = f(x^\star)$.

Recall (from §4.2.3 or §5.5.3) that a point $x^\star \in \mathbf{dom}\, f$ is optimal for (10.1) if and only if there is a $\nu^\star \in \mathbf{R}^p$ such that

$$Ax^\star = b, \qquad \nabla f(x^\star) + A^T \nu^\star = 0. \tag{10.2}$$

Solving the equality constrained optimization problem (10.1) is therefore equivalent to finding a solution of the KKT equations (10.2), which is a set of $n + p$ equations in the $n + p$ variables $x^\star$, $\nu^\star$. The first set of equations, $Ax^\star = b$, are called the *primal feasibility equations*, which are linear. The second set of equations, $\nabla f(x^\star) + A^T \nu^\star = 0$, are called the *dual feasibility equations*, and are in general nonlinear. As with unconstrained optimization, there are a few problems for which we can solve these optimality conditions analytically. The most important special case is when f is quadratic, which we examine in §10.1.1.

Any equality constrained minimization problem can be reduced to an equivalent unconstrained problem by eliminating the equality constraints, after which the methods of chapter 9 can be used to solve the problem. Another approach is to solve the dual problem (assuming the dual function is twice differentiable) using an unconstrained minimization method, and then recover the solution of the equality constrained problem (10.1) from the dual solution. The elimination and dual methods are briefly discussed in §10.1.2 and §10.1.3, respectively.

The bulk of this chapter is devoted to extensions of Newton's method that directly handle equality constraints. In many cases these methods are preferable to methods that reduce an equality constrained problem to an unconstrained one. One reason is that problem structure, such as sparsity, is often destroyed by elimination (or forming the dual); in contrast, a method that directly handles equality constraints can exploit the problem structure. Another reason is conceptual: methods that directly handle equality constraints can be thought of as methods for directly solving the optimality conditions (10.2).

10.1.1   Equality constrained convex quadratic minimization

Consider the equality constrained convex quadratic minimization problem

$$\begin{array}{ll} \text{minimize} & f(x) = (1/2)x^T P x + q^T x + r \\ \text{subject to} & Ax = b, \end{array} \tag{10.3}$$

where $P \in \mathbf{S}^n_+$ and $A \in \mathbf{R}^{p \times n}$. This problem is important on its own, and also because it forms the basis for an extension of Newton's method to equality constrained problems.

Here the optimality conditions (10.2) are

$$Ax^\star = b, \qquad Px^\star + q + A^T \nu^\star = 0,$$

which we can write as

$$\begin{bmatrix} P & A^T \\ A & 0 \end{bmatrix} \begin{bmatrix} x^\star \\ \nu^\star \end{bmatrix} = \begin{bmatrix} -q \\ b \end{bmatrix}. \tag{10.4}$$

This set of $n + p$ linear equations in the $n + p$ variables $x^\star$, $\nu^\star$ is called the *KKT system* for the equality constrained quadratic optimization problem (10.3). The coefficient matrix is called the *KKT matrix*.

When the KKT matrix is nonsingular, there is a unique optimal primal-dual pair $(x^\star, \nu^\star)$. If the KKT matrix is singular, but the KKT system is solvable, any solution yields an optimal pair $(x^\star, \nu^\star)$. If the KKT system is not solvable, the quadratic optimization problem is unbounded below or infeasible. Indeed, in this case there exist $v \in \mathbf{R}^n$ and $w \in \mathbf{R}^p$ such that

$$Pv + A^T w = 0, \qquad Av = 0, \qquad -q^T v + b^T w > 0.$$

Let $\hat{x}$ be any feasible point. The point $x = \hat{x} + tv$ is feasible for all t and

$$\begin{aligned} f(\hat{x} + tv) &= f(\hat{x}) + t(v^T P \hat{x} + q^T v) + (1/2)t^2 v^T P v \\ &= f(\hat{x}) + t(-\hat{x}^T A^T w + q^T v) - (1/2)t^2 w^T A v \\ &= f(\hat{x}) + t(-b^T w + q^T v), \end{aligned}$$

which decreases without bound as $t \to \infty$.

Nonsingularity of the KKT matrix

Recall our assumption that $P \in \mathbf{S}_+^n$ and $\mathbf{rank}\, A = p < n$. There are several conditions equivalent to nonsingularity of the KKT matrix:

- $\mathcal{N}(P) \cap \mathcal{N}(A) = \{0\}$, *i.e.*, P and A have no nontrivial common nullspace.

- $Ax = 0$, $x \neq 0 \implies x^T P x > 0$, *i.e.*, P is positive definite on the nullspace of A.

- $F^T P F \succ 0$, where $F \in \mathbf{R}^{n \times (n-p)}$ is a matrix for which $\mathcal{R}(F) = \mathcal{N}(A)$.

(See exercise 10.1.) As an important special case, we note that if $P \succ 0$, the KKT matrix must be nonsingular.

10.1.2   Eliminating equality constraints

One general approach to solving the equality constrained problem (10.1) is to eliminate the equality constraints, as described in §4.2.4, and then solve the resulting unconstrained problem using methods for unconstrained minimization. We first find a matrix $F \in \mathbf{R}^{n \times (n-p)}$ and vector $\hat{x} \in \mathbf{R}^n$ that parametrize the (affine) feasible set:

$$\{x \mid Ax = b\} = \{Fz + \hat{x} \mid z \in \mathbf{R}^{n-p}\}.$$

Here $\hat{x}$ can be chosen as any particular solution of $Ax = b$, and $F \in \mathbf{R}^{n \times (n-p)}$ is any matrix whose range is the nullspace of A. We then form the reduced or eliminated optimization problem

$$\text{minimize} \quad \tilde{f}(z) = f(Fz + \hat{x}), \tag{10.5}$$

which is an unconstrained problem with variable $z \in \mathbf{R}^{n-p}$. From its solution $z^\star$, we can find the solution of the equality constrained problem as $x^\star = Fz^\star + \hat{x}$.

We can also construct an optimal dual variable $\nu^\star$ for the equality constrained problem, as

$$\nu^\star = -(AA^T)^{-1} A \nabla f(x^\star).$$

To show that this expression is correct, we must verify that the dual feasibility condition

$$\nabla f(x^\star) + A^T (-(AA^T)^{-1} A \nabla f(x^\star)) = 0 \tag{10.6}$$

holds. To show this, we note that

$$\begin{bmatrix} F^T \\ A \end{bmatrix} (\nabla f(x^\star) - A^T (AA^T)^{-1} A \nabla f(x^\star)) = 0,$$

where in the top block we use $F^T \nabla f(x^\star) = \nabla \tilde{f}(z^\star) = 0$ and $AF = 0$. Since the matrix on the left is nonsingular, this implies (10.6).

Example 10.1 *Optimal allocation with resource constraint.* We consider the problem

$$\begin{array}{ll} \text{minimize} & \sum_{i=1}^n f_i(x_i) \\ \text{subject to} & \sum_{i=1}^n x_i = b, \end{array}$$

where the functions $f_i : \mathbf{R} \to \mathbf{R}$ are convex and twice differentiable, and $b \in \mathbf{R}$ is a problem parameter. We interpret this as the problem of optimally allocating a single resource, with a fixed total amount b (the *budget*) to n otherwise independent activities.

We can eliminate x_n (for example) using the parametrization

$$x_n = b - x_1 - \cdots - x_{n-1},$$

which corresponds to the choices

$$\hat{x} = be_n, \qquad F = \begin{bmatrix} I \\ -\mathbf{1}^T \end{bmatrix} \in \mathbf{R}^{n \times (n-1)}.$$

The reduced problem is then

$$\text{minimize} \quad f_n(b - x_1 - \cdots - x_{n-1}) + \textstyle\sum_{i=1}^{n-1} f_i(x_i),$$

with variables $x_1, \ldots, x_{n-1}$.

Choice of elimination matrix

There are, of course, many possible choices for the elimination matrix F, which can be chosen as any matrix in $\mathbf{R}^{n \times (n-p)}$ with $\mathcal{R}(F) = \mathcal{N}(A)$. If F is one such matrix, and $T \in \mathbf{R}^{(n-p) \times (n-p)}$ is nonsingular, then $\tilde{F} = FT$ is also a suitable elimination matrix, since

$$\mathcal{R}(\tilde{F}) = \mathcal{R}(F) = \mathcal{N}(A).$$

Conversely, if F and $\tilde{F}$ are any two suitable elimination matrices, then there is some nonsingular T such that $\tilde{F} = FT$.

If we eliminate the equality constraints using F, we solve the unconstrained problem

$$\text{minimize} \quad f(Fz + \hat{x}),$$

while if $\tilde{F}$ is used, we solve the unconstrained problem

$$\text{minimize} \quad f(\tilde{F}\tilde{z} + \hat{x}) = f(F(T\tilde{z}) + \hat{x}).$$

This problem is equivalent to the one above, and is simply obtained by the change of coordinates $z = T\tilde{z}$. In other words, changing the elimination matrix can be thought of as changing variables in the reduced problem.

10.1.3   Solving equality constrained problems via the dual

Another approach to solving (10.1) is to solve the dual, and then recover the optimal primal variable $x^\star$, as described in §5.5.5. The dual function of (10.1) is

$$
\begin{aligned}
g(\nu) &= -b^T \nu + \inf_x (f(x) + \nu^T A x) \\
&= -b^T \nu - \sup_x \left((-A^T \nu)^T x - f(x) \right) \\
&= -b^T \nu - f^*(-A^T \nu),
\end{aligned}
$$

where f^* is the conjugate of f, so the dual problem is

$$\text{maximize} \quad -b^T \nu - f^*(-A^T \nu).$$

Since by assumption there is an optimal point, the problem is strictly feasible, so Slater's condition holds. Therefore strong duality holds, and the dual optimum is attained, *i.e.*, there exists a $\nu^\star$ with $g(\nu^\star) = p^\star$.

If the dual function g is twice differentiable, then the methods for unconstrained minimization described in chapter 9 can be used to maximize g. (In general, the dual function g need not be twice differentiable, even if f is.) Once we find an optimal dual variable $\nu^\star$, we reconstruct an optimal primal solution $x^\star$ from it. (This is not always straightforward; see §5.5.5.)

Example 10.2 *Equality constrained analytic center.* We consider the problem

$$\begin{array}{ll} \text{minimize} & f(x) = -\sum_{i=1}^n \log x_i \\ \text{subject to} & Ax = b, \end{array} \qquad (10.7)$$

where $A \in \mathbf{R}^{p \times n}$, with implicit constraint $x \succ 0$. Using

$$f^*(y) = \sum_{i=1}^n (-1 - \log(-y_i)) = -n - \sum_{i=1}^n \log(-y_i)$$

(with $\mathbf{dom}\, f^* = -\mathbf{R}_{++}^n$), the dual problem is

$$\text{maximize} \quad g(\nu) = -b^T \nu + n + \sum_{i=1}^n \log(A^T \nu)_i, \qquad (10.8)$$

with implicit constraint $A^T \nu \succ 0$. Here we can easily solve the dual feasibility equation, *i.e.*, find the x that minimizes $L(x, \nu)$:

$$\nabla f(x) + A^T \nu = -(1/x_1, \dots, 1/x_n) + A^T \nu = 0,$$

and so

$$x_i(\nu) = 1/(A^T \nu)_i. \qquad (10.9)$$

To solve the equality constrained analytic centering problem (10.7), we solve the (unconstrained) dual problem (10.8), and then recover the optimal solution of (10.7) via (10.9).

10.2   Newton's method with equality constraints

In this section we describe an extension of Newton's method to include equality constraints. The method is almost the same as Newton's method without constraints, except for two differences: The initial point must be feasible (*i.e.*, satisfy $x \in \mathbf{dom}\, f$ and $Ax = b$), and the definition of Newton step is modified to take the equality constraints into account. In particular, we make sure that the Newton step Δx_{nt} is a feasible direction, *i.e.*, $A\Delta x_{\mathrm{nt}} = 0$.

10.2.1   The Newton step

Definition via second-order approximation

To derive the Newton step Δx_{nt} for the equality constrained problem

$$
\begin{array}{ll}
\text{minimize} & f(x) \\
\text{subject to} & Ax = b,
\end{array}
$$

at the feasible point x, we replace the objective with its second-order Taylor approximation near x, to form the problem

$$
\begin{array}{ll}
\text{minimize} & \widehat{f}(x + v) = f(x) + \nabla f(x)^T v + (1/2)v^T \nabla^2 f(x)v \\
\text{subject to} & A(x + v) = b,
\end{array}
\tag{10.10}
$$

with variable v. This is a (convex) quadratic minimization problem with equality constraints, and can be solved analytically. We define Δx_{nt}, the Newton step at x, as the solution of the convex quadratic problem (10.10), assuming the associated KKT matrix is nonsingular. In other words, the Newton step Δx_{nt} is what must be added to x to solve the problem when the quadratic approximation is used in place of f.

From our analysis in §10.1.1 of the equality constrained quadratic problem, the Newton step Δx_{nt} is characterized by

$$
\begin{bmatrix} \nabla^2 f(x) & A^T \\ A & 0 \end{bmatrix} \begin{bmatrix} \Delta x_{\mathrm{nt}} \\ w \end{bmatrix} = \begin{bmatrix} -\nabla f(x) \\ 0 \end{bmatrix},
\tag{10.11}
$$

where w is the associated optimal dual variable for the quadratic problem. The Newton step is defined only at points for which the KKT matrix is nonsingular.

As in Newton's method for unconstrained problems, we observe that when the objective f is exactly quadratic, the Newton update $x + \Delta x_{\mathrm{nt}}$ exactly solves the equality constrained minimization problem, and in this case the vector w is the optimal dual variable for the original problem. This suggests, as in the unconstrained case, that when f is nearly quadratic, $x + \Delta x_{\mathrm{nt}}$ should be a very good estimate of the solution $x^\star$, and w should be a good estimate of the optimal dual variable $\nu^\star$.

Solution of linearized optimality conditions

We can interpret the Newton step Δx_{nt}, and the associated vector w, as the solutions of a linearized approximation of the optimality conditions

$$
Ax^\star = b, \qquad \nabla f(x^\star) + A^T \nu^\star = 0.
$$

We substitute $x + \Delta x_{\mathrm{nt}}$ for $x^\star$ and w for $\nu^\star$, and replace the gradient term in the second equation by its linearized approximation near x, to obtain the equations

$$
A(x + \Delta x_{\mathrm{nt}}) = b, \qquad \nabla f(x + \Delta x_{\mathrm{nt}}) + A^T w \approx \nabla f(x) + \nabla^2 f(x)\Delta x_{\mathrm{nt}} + A^T w = 0.
$$

Using $Ax = b$, these become

$$
A\Delta x_{\mathrm{nt}} = 0, \qquad \nabla^2 f(x)\Delta x_{\mathrm{nt}} + A^T w = -\nabla f(x),
$$

which are precisely the equations (10.11) that define the Newton step.

The Newton decrement

We define the Newton decrement for the equality constrained problem as

$$\lambda(x) = (\Delta x_{\mathrm{nt}}^T \nabla^2 f(x) \Delta x_{\mathrm{nt}})^{1/2}. \qquad (10.12)$$

This is exactly the same expression as (9.29), used in the unconstrained case, and the same interpretations hold. For example, $\lambda(x)$ is the norm of the Newton step, in the norm determined by the Hessian.

Let

$$\widehat{f}(x+v) = f(x) + \nabla f(x)^T v + (1/2)v^T \nabla^2 f(x)v$$

be the second-order Taylor approximation of f at x. The difference between $f(x)$ and the minimum of the second-order model satisfies

$$f(x) - \inf\{\widehat{f}(x+v) \mid A(x+v) = b\} = \lambda(x)^2/2, \qquad (10.13)$$

exactly as in the unconstrained case (see exercise 10.6). This means that, as in the unconstrained case, $\lambda(x)^2/2$ gives an estimate of $f(x) - p^\star$, based on the quadratic model at x, and also that $\lambda(x)$ (or a multiple of $\lambda(x)^2$) serves as the basis of a good stopping criterion.

The Newton decrement comes up in the line search as well, since the directional derivative of f in the direction Δx_{nt} is

$$\left.\frac{d}{dt}f(x+t\Delta x_{\mathrm{nt}})\right|_{t=0} = \nabla f(x)^T \Delta x_{\mathrm{nt}} = -\lambda(x)^2, \qquad (10.14)$$

as in the unconstrained case.

Feasible descent direction

Suppose that $Ax = b$. We say that $v \in \mathbf{R}^n$ is a *feasible direction* if $Av = 0$. In this case, every point of the form $x+tv$ is also feasible, i.e., $A(x+tv) = b$. We say that v is a *descent direction* for f at x, if for small $t > 0$, $f(x+tv) < f(x)$.

The Newton step is always a feasible descent direction (except when x is optimal, in which case $\Delta x_{\mathrm{nt}} = 0$). Indeed, the second set of equations that define Δx_{nt} are $A\Delta x_{\mathrm{nt}} = 0$, which shows it is a feasible direction; that it is a descent direction follows from (10.14).

Affine invariance

Like the Newton step and decrement for unconstrained optimization, the Newton step and decrement for equality constrained optimization are affine invariant. Suppose $T \in \mathbf{R}^{n\times n}$ is nonsingular, and define $\bar{f}(y) = f(Ty)$. We have

$$\nabla \bar{f}(y) = T^T \nabla f(Ty), \qquad \nabla^2 \bar{f}(y) = T^T \nabla^2 f(Ty)T,$$

and the equality constraint $Ax = b$ becomes $ATy = b$.

Now consider the problem of minimizing $\bar{f}(y)$, subject to $ATy = b$. The Newton step Δy_{nt} at y is given by the solution of

$$\begin{bmatrix} T^T\nabla^2 f(Ty)T & T^TA^T \\ AT & 0 \end{bmatrix}\begin{bmatrix} \Delta y_{\mathrm{nt}} \\ \bar{w} \end{bmatrix} = \begin{bmatrix} -T^T\nabla f(Ty) \\ 0 \end{bmatrix}.$$

Comparing with the Newton step Δx_{nt} for f at $x = Ty$, given in (10.11), we see that

$$T \Delta y_{\mathrm{nt}} = \Delta x_{\mathrm{nt}}$$

(and $w = \bar{w}$), i.e., the Newton steps at y and x are related by the same change of coordinates as $Ty = x$.

10.2.2   Newton's method with equality constraints

The outline of Newton's method with equality constraints is exactly the same as for unconstrained problems.

Algorithm 10.1 *Newton's method for equality constrained minimization.*

given starting point $x \in \mathbf{dom}\, f$ with $Ax = b$, tolerance $\epsilon > 0$.

repeat
    1. Compute the Newton step and decrement Δx_{nt}, $\lambda(x)$.
    2. *Stopping criterion.* **quit** if $\lambda^2/2 \le \epsilon$.
    3. *Line search.* Choose step size t by backtracking line search.
    4. *Update.* $x := x + t\Delta x_{\mathrm{nt}}$.

The method is called a *feasible descent method*, since all the iterates are feasible, with $f(x^{(k+1)}) < f(x^{(k)})$ (unless $x^{(k)}$ is optimal). Newton's method requires that the KKT matrix be invertible at each x; we will be more precise about the assumptions required for convergence in §10.2.4.

10.2.3   Newton's method and elimination

We now show that the iterates in Newton's method for the equality constrained problem (10.1) coincide with the iterates in Newton's method applied to the reduced problem (10.5). Suppose F satisfies $\mathcal{R}(F) = \mathcal{N}(A)$ and $\mathbf{rank}\, F = n - p$, and $\hat{x}$ satisfies $A\hat{x} = b$. The gradient and Hessian of the reduced objective function $\tilde{f}(z) = f(Fz + \hat{x})$ are

$$\nabla \tilde{f}(z) = F^T \nabla f(Fz + \hat{x}), \qquad \nabla^2 \tilde{f}(z) = F^T \nabla^2 f(Fz + \hat{x}) F.$$

From the Hessian expression, we see that the Newton step for the equality constrained problem is defined, *i.e.*, the KKT matrix

$$\begin{bmatrix} \nabla^2 f(x) & A^T \\ A & 0 \end{bmatrix}$$

is invertible, if and only if the Newton step for the reduced problem is defined, *i.e.*, $\nabla^2 \tilde{f}(z)$ is invertible.

The Newton step for the reduced problem is

$$\Delta z_{\mathrm{nt}} = -\nabla^2 \tilde{f}(z)^{-1} \nabla \tilde{f}(z) = -(F^T \nabla^2 f(x) F)^{-1} F^T \nabla f(x), \qquad (10.15)$$

where $x = Fz + \hat{x}$. This search direction for the reduced problem corresponds to the direction

$$F\Delta z_{\text{nt}} = -F(F^T \nabla^2 f(x) F)^{-1} F^T \nabla f(x)$$

for the original, equality constrained problem. We claim this is precisely the same as the Newton direction Δx_{nt} for the original problem, defined in (10.11).

To show this, we take $\Delta x_{\text{nt}} = F\Delta z_{\text{nt}}$, choose

$$w = -(AA^T)^{-1} A(\nabla f(x) + \nabla^2 f(x) \Delta x_{\text{nt}}),$$

and verify that the equations defining the Newton step,

$$\nabla^2 f(x) \Delta x_{\text{nt}} + A^T w + \nabla f(x) = 0, \qquad A\Delta x_{\text{nt}} = 0, \qquad (10.16)$$

hold. The second equation, $A\Delta x_{\text{nt}} = 0$, is satisfied because $AF = 0$. To verify the first equation, we observe that

$$\begin{bmatrix} F^T \\ A \end{bmatrix} (\nabla^2 f(x) \Delta x_{\text{nt}} + A^T w + \nabla f(x))$$
$$= \begin{bmatrix} F^T \nabla^2 f(x) \Delta x_{\text{nt}} + F^T A^T w + F^T \nabla f(x) \\ A \nabla^2 f(x) \Delta x_{\text{nt}} + AA^T w + A\nabla f(x) \end{bmatrix}$$
$$= 0.$$

Since the matrix on the left of the first line is nonsingular, we conclude that (10.16) holds.

In a similar way, the Newton decrement $\tilde{\lambda}(z)$ of $\tilde{f}$ at z and the Newton decrement of f at x turn out to be equal:

$$\begin{aligned} \tilde{\lambda}(z)^2 &= \Delta z_{\text{nt}}^T \nabla^2 \tilde{f}(z) \Delta z_{\text{nt}} \\ &= \Delta z_{\text{nt}}^T F^T \nabla^2 f(x) F \Delta z_{\text{nt}} \\ &= \Delta x_{\text{nt}}^T \nabla^2 f(x) \Delta x_{\text{nt}} \\ &= \lambda(x)^2. \end{aligned}$$

10.2.4   Convergence analysis

We saw above that applying Newton's method with equality constraints is exactly the same as applying Newton's method to the reduced problem obtained by eliminating the equality constraints. Everything we know about the convergence of Newton's method for unconstrained problems therefore transfers to Newton's method for equality constrained problems. In particular, the practical performance of Newton's method with equality constraints is exactly like the performance of Newton's method for unconstrained problems. Once $x^{(k)}$ is near $x^\star$, convergence is extremely rapid, with a very high accuracy obtained in only a few iterations.

Assumptions

We make the following assumptions.

- The sublevel set $S = \{x \mid x \in \operatorname{dom} f, \ f(x) \leq f(x^{(0)}), \ Ax = b\}$ is closed, where $x^{(0)} \in \operatorname{dom} f$ satisfies $Ax^{(0)} = b$. This is the case if f is closed (see §A.3.3).

- On the set S, we have $\nabla^2 f(x) \preceq MI$, and

$$\left\| \begin{bmatrix} \nabla^2 f(x) & A^T \\ A & 0 \end{bmatrix}^{-1} \right\|_2 \leq K, \tag{10.17}$$

  i.e., the inverse of the KKT matrix is bounded on S. (Of course the inverse must exist in order for the Newton step to be defined at each point in S.)

- For $x, \tilde{x} \in S$, $\nabla^2 f$ satisfies the Lipschitz condition $\|\nabla^2 f(x) - \nabla^2 f(\tilde{x})\|_2 \leq L\|x - \tilde{x}\|_2$.

Bounded inverse KKT matrix assumption

The condition (10.17) plays the role of the strong convexity assumption in the standard Newton method (§9.5.3, page 488). When there are no equality constraints, (10.17) reduces to the condition $\|\nabla^2 f(x)^{-1}\|_2 \leq K$ on S, so we can take $K = 1/m$, if $\nabla^2 f(x) \succeq mI$ on S, where $m > 0$. With equality constraints, the condition is not as simple as a positive lower bound on the minimum eigenvalue. Since the KKT matrix is symmetric, the condition (10.17) is that its eigenvalues, n of which are positive, and p of which are negative, are bounded away from zero.

Analysis via the eliminated problem

The assumptions above imply that the eliminated objective function $\tilde{f}$, together with the associated initial point $z^{(0)}$, where $x^{(0)} = \hat{x} + Fz^{(0)}$, satisfy the assumptions required in the convergence analysis of Newton's method for unconstrained problems, given in §9.5.3 (with different constants $\tilde{m}$, $\tilde{M}$, and $\tilde{L}$). It follows that Newton's method with equality constraints converges to $x^\star$ (and $\nu^\star$ as well).

To show that the assumptions above imply that the eliminated problem satisfies the assumptions for the unconstrained Newton method is mostly straightforward (see exercise 10.4). Here we show the one implication that is tricky: that the bounded inverse KKT condition, together with the upper bound $\nabla^2 f(x) \preceq MI$, implies that $\nabla^2 \tilde{f}(z) \succeq mI$ for some positive constant m. More specifically we will show that this inequality holds for

$$m = \frac{\sigma_{\min}(F)^2}{K^2 M}, \tag{10.18}$$

which is positive, since F is full rank.

We show this by contradiction. Suppose that $F^T H F \not\succeq mI$, where $H = \nabla^2 f(x)$. Then we can find u, with $\|u\|_2 = 1$, such that $u^T F^T H F u < m$, *i.e.*, $\|H^{1/2} F u\|_2 < m^{1/2}$. Using $AF = 0$, we have

$$\begin{bmatrix} H & A^T \\ A & 0 \end{bmatrix} \begin{bmatrix} Fu \\ 0 \end{bmatrix} = \begin{bmatrix} HFu \\ 0 \end{bmatrix},$$

and so

$$
\left\| \begin{bmatrix} H & A^T \\ A & 0 \end{bmatrix}^{-1} \right\|_2 \geq \frac{\left\| \begin{bmatrix} Fu \\ 0 \end{bmatrix} \right\|_2}{\left\| \begin{bmatrix} HFu \\ 0 \end{bmatrix} \right\|_2} = \frac{\|Fu\|_2}{\|HFu\|_2}.
$$

Using $\|Fu\|_2 \geq \sigma_{\min}(F)$ and

$$
\|HFu\|_2 \leq \|H^{1/2}\|_2 \|H^{1/2}Fu\|_2 < M^{1/2}m^{1/2},
$$

we conclude

$$
\left\| \begin{bmatrix} H & A^T \\ A & 0 \end{bmatrix}^{-1} \right\|_2 \geq \frac{\|Fu\|_2}{\|HFu\|_2} > \frac{\sigma_{\min}(F)}{M^{1/2}m^{1/2}} = K,
$$

using our expression for m given in (10.18).

Convergence analysis for self-concordant functions

If f is self-concordant, then so is $\tilde{f}(z) = f(Fz + \hat{x})$. It follows that if f is self-concordant, we have the exact same complexity estimate as for unconstrained problems: the number of iterations required to produce a solution within an accuracy ϵ is no more than

$$
\frac{20 - 8\alpha}{\alpha\beta(1 - 2\alpha)^2}(f(x^{(0)}) - p^\star) + \log_2 \log_2(1/\epsilon),
$$

where α and β are the backtracking parameters (see (9.56)).

10.3   Infeasible start Newton method

Newton's method, as described in §10.2, is a feasible descent method. In this section we describe a generalization of Newton's method that works with initial points, and iterates, that are not feasible.

10.3.1   Newton step at infeasible points

As in Newton's method, we start with the optimality conditions for the equality constrained minimization problem:

$$
Ax^\star = b, \qquad \nabla f(x^\star) + A^T \nu^\star = 0.
$$

Let x denote the current point, which we do *not* assume to be feasible, but we do assume satisfies $x \in \mathbf{dom}\, f$. Our goal is to find a step Δx so that $x + \Delta x$ satisfies (at least approximately) the optimality conditions, *i.e.*, $x + \Delta x \approx x^\star$. To do this we substitute $x + \Delta x$ for $x^\star$ and w for $\nu^\star$ in the optimality conditions, and use the first-order approximation

$$\nabla f(x + \Delta x) \approx \nabla f(x) + \nabla^2 f(x)\Delta x$$

for the gradient to obtain

$$A(x + \Delta x) = b, \qquad \nabla f(x) + \nabla^2 f(x)\Delta x + A^T w = 0.$$

This is a set of linear equations for Δx and w,

$$\left[\begin{array}{cc} \nabla^2 f(x) & A^T \\ A & 0 \end{array} \right] \left[\begin{array}{c} \Delta x \\ w \end{array} \right] = - \left[\begin{array}{c} \nabla f(x) \\ Ax - b \end{array} \right]. \tag{10.19}$$

The equations are the same as the equations (10.11) that define the Newton step at a feasible point x, with one difference: the second block component of the righthand side contains $Ax - b$, which is the residual vector for the linear equality constraints. When x is feasible, the residual vanishes, and the equations (10.19) reduce to the equations (10.11) that define the standard Newton step at a feasible point x. Thus, if x is feasible, the step Δx defined by (10.19) coincides with the Newton step described above (but defined only when x is feasible). For this reason we use the notation Δx_{nt} for the step Δx defined by (10.19), and refer to it as the Newton step at x, with no confusion.

Interpretation as primal-dual Newton step

We can give an interpretation of the equations (10.19) in terms of a *primal-dual method* for the equality constrained problem. By a primal-dual method, we mean one in which we update both the primal variable x, and the dual variable ν, in order to (approximately) satisfy the optimality conditions.

We express the optimality conditions as $r(x^\star, \nu^\star) = 0$, where $r : \mathbf{R}^n \times \mathbf{R}^p \to \mathbf{R}^n \times \mathbf{R}^p$ is defined as

$$r(x, \nu) = (r_{\text{dual}}(x, \nu), r_{\text{pri}}(x, \nu)).$$

Here

$$r_{\text{dual}}(x, \nu) = \nabla f(x) + A^T \nu, \qquad r_{\text{pri}}(x, \nu) = Ax - b$$

are the *dual residual* and *primal residual*, respectively. The first-order Taylor approximation of r, near our current estimate y, is

$$r(y + z) \approx \hat{r}(y + z) = r(y) + Dr(y)z,$$

where $Dr(y) \in \mathbf{R}^{(n+p) \times (n+p)}$ is the derivative of r, evaluated at y (see §A.4.1). We define the primal-dual Newton step Δy_{pd} as the step z for which the Taylor approximation $\hat{r}(y + z)$ vanishes, i.e.,

$$Dr(y)\Delta y_{\text{pd}} = -r(y). \tag{10.20}$$

Note that here we consider both x and ν as variables; $\Delta y_{\text{pd}} = (\Delta x_{\text{pd}}, \Delta \nu_{\text{pd}})$ gives both a primal and a dual step.

Evaluating the derivative of r, we can express (10.20) as

$$\begin{bmatrix} \nabla^2 f(x) & A^T \\ A & 0 \end{bmatrix} \begin{bmatrix} \Delta x_{\text{pd}} \\ \Delta \nu_{\text{pd}} \end{bmatrix} = - \begin{bmatrix} r_{\text{dual}} \\ r_{\text{pri}} \end{bmatrix} = - \begin{bmatrix} \nabla f(x) + A^T \nu \\ Ax - b \end{bmatrix}. \qquad (10.21)$$

Writing $\nu + \Delta \nu_{\text{pd}}$ as ν^+, we can express this as

$$\begin{bmatrix} \nabla^2 f(x) & A^T \\ A & 0 \end{bmatrix} \begin{bmatrix} \Delta x_{\text{pd}} \\ \nu^+ \end{bmatrix} = - \begin{bmatrix} \nabla f(x) \\ Ax - b \end{bmatrix}, \qquad (10.22)$$

which is exactly the same set of equations as (10.19). The solutions of (10.19), (10.21), and (10.22) are therefore related as

$$\Delta x_{\text{nt}} = \Delta x_{\text{pd}}, \qquad w = \nu^+ = \nu + \Delta \nu_{\text{pd}}.$$

This shows that the (infeasible) Newton step is the same as the primal part of the primal-dual step, and the associated dual vector w is the updated primal-dual variable $\nu^+ = \nu + \Delta \nu_{\text{pd}}$.

The two expressions for the Newton step and dual variable (or dual step), given by (10.21) and (10.22), are of course equivalent, but each reveals a different feature of the Newton step. The equation (10.21) shows that the Newton step and the associated dual step are obtained by solving a set of equations, with the primal and dual residuals as the righthand side. The equation (10.22), which is how we originally defined the Newton step, gives the Newton step and the updated dual variable, and shows that the current value of the dual variable is not needed to compute the primal step, or the updated value of the dual variable.

Residual norm reduction property

The Newton direction, at an infeasible point, is not necessarily a descent direction for f. From (10.19), we note that

$$\begin{aligned}
\left. \frac{d}{dt} f(x + t\Delta x) \right|_{t=0} &= \nabla f(x)^T \Delta x \\
&= -\Delta x^T \left(\nabla^2 f(x) \Delta x + A^T w \right) \\
&= -\Delta x^T \nabla^2 f(x) \Delta x + (Ax - b)^T w,
\end{aligned}$$

which is not necessarily negative (unless, of course, x is feasible, i.e., $Ax = b$). The primal-dual interpretation, however, shows that the norm of the residual decreases in the Newton direction, i.e.,

$$\left. \frac{d}{dt} \|r(y + t\Delta y_{\text{pd}})\|_2^2 \right|_{t=0} = 2r(y)^T Dr(y) \Delta y_{\text{pd}} = -2r(y)^T r(y).$$

Taking the derivative of the square, we obtain

$$\left. \frac{d}{dt} \|r(y + t\Delta y_{\text{pd}})\|_2 \right|_{t=0} = -\|r(y)\|_2. \qquad (10.23)$$

This allows us to use $\|r\|_2$ to measure the progress of the infeasible start Newton method, for example, in the line search. (For the standard Newton method, we use the function value f to measure progress of the algorithm, at least until quadratic convergence is attained.)

Full step feasibility property

The Newton step $\Delta x_{\rm nt}$ defined by (10.19) has the property (by construction) that

$$A(x + \Delta x_{\rm nt}) = b. \tag{10.24}$$

It follows that, if a step length of one is taken using the Newton step $\Delta x_{\rm nt}$, the following iterate will be feasible. Once x is feasible, the Newton step becomes a feasible direction, so all future iterates will be feasible, regardless of the step sizes taken.

More generally, we can analyze the effect of a damped step on the equality constraint residual $r_{\rm pri}$. With a step length $t \in [0,1]$, the next iterate is $x^+ = x + t\Delta x_{\rm nt}$, so the equality constraint residual at the next iterate is

$$r_{\rm pri}^+ = A(x + \Delta x_{\rm nt} t) - b = (1-t)(Ax - b) = (1-t)r_{\rm pri},$$

using (10.24). Thus, a damped step, with length t, causes the residual to be scaled down by a factor $1 - t$. Now suppose that we have $x^{(i+1)} = x^{(i)} + t^{(i)}\Delta x_{\rm nt}^{(i)}$, for $i = 0, \ldots, k - 1$, where $\Delta x_{\rm nt}^{(i)}$ is the Newton step at the point $x^{(i)} \in \mathbf{dom}\, f$, and $t^{(i)} \in [0,1]$. Then we have

$$r^{(k)} = \left(\prod_{i=0}^{k-1}(1 - t^{(i)})\right) r^{(0)},$$

where $r^{(i)} = Ax^{(i)} - b$ is the residual of $x^{(i)}$. This formula shows that the primal residual at each step is in the direction of the initial primal residual, and is scaled down at each step. It also shows that once a full step is taken, all future iterates are primal feasible.

10.3.2   Infeasible start Newton method

We can develop an extension of Newton's method, using the Newton step $\Delta x_{\rm nt}$ defined by (10.19), with $x^{(0)} \in \mathbf{dom}\, f$, but not necessarily satisfying $Ax^{(0)} = b$. We also use the dual part of the Newton step: $\Delta \nu_{\rm nt} - w - \nu$ in the notation of (10.19), or equivalently, $\Delta \nu_{\rm nt} = \Delta \nu_{\rm pd}$ in the notation of (10.21).

Algorithm 10.2 *Infeasible start Newton method.*

given starting point $x \in \mathbf{dom}\, f$, ν, tolerance $\epsilon > 0$, $\alpha \in (0, 1/2)$, $\beta \in (0, 1)$.

repeat
    1. Compute primal and dual Newton steps $\Delta x_{\rm nt}$, $\Delta \nu_{\rm nt}$.
    2. *Backtracking line search on* $\|r\|_2$.
        $t := 1$.
        while $\|r(x + t\Delta x_{\rm nt}, \nu + t\Delta \nu_{\rm nt})\|_2 > (1 - \alpha t)\|r(x, \nu)\|_2,$ $t := \beta t$.
    3. *Update.* $x := x + t\Delta x_{\rm nt}$, $\nu := \nu + t\Delta \nu_{\rm nt}$.
until $Ax = b$ and $\|r(x, \nu)\|_2 \leq \epsilon$.

This algorithm is very similar to the standard Newton method with feasible starting point, with a few exceptions. First, the search directions include the extra correction terms that depend on the primal residual. Second, the line search is carried out using the norm of the residual, instead of the function value f. Finally, the algorithm terminates when primal feasibility has been achieved, and the norm of the (dual) residual is small.

The line search in step 2 deserves some comment. Using the norm of the residual in the line search can increase the cost, compared to a line search based on the function value, but the increase is usually negligible. Also, we note that the line search must terminate in a finite number of steps, since (10.23) shows that the line search exit condition is satisfied for small t.

The equation (10.24) shows that if at some iteration the step length is chosen to be one, the next iterate will be feasible. Thereafter, all iterates will be feasible, and therefore the search direction for the infeasible start Newton method coincides, once a feasible iterate is obtained, with the search direction for the (feasible) Newton method described in §10.2.

There are many variations on the infeasible start Newton method. For example, we can switch to the (feasible) Newton method described in §10.2 once feasibility is achieved. (In other words, we change the line search to one based on f, and terminate when $\lambda(x)^2/2 \le \epsilon$.) Once feasibility is achieved, the infeasible start and the standard (feasible) Newton method differ only in the backtracking and exit conditions, and have very similar performance.

Using infeasible start Newton method to simplify initialization

The main advantage of the infeasible start Newton method is in the initialization required. If $\mathbf{dom}\, f = \mathbf{R}^n$, then initializing the (feasible) Newton method simply requires computing a solution to $Ax = b$, and there is no particular advantage, other than convenience, in using the infeasible start Newton method.

When $\mathbf{dom}\, f$ is not all of $\mathbf{R}^n$, finding a point in $\mathbf{dom}\, f$ that satisfies $Ax = b$ can itself be a challenge. One general approach, probably the best when $\mathbf{dom}\, f$ is complex and not known to intersect $\{z \mid Az = b\}$, is to use a phase I method (described in §11.4) to compute such a point (or verify that $\mathbf{dom}\, f$ does not intersect $\{z \mid Az = b\}$). But when $\mathbf{dom}\, f$ is relatively simple, and known to contain a point satisfying $Ax = b$, the infeasible start Newton method gives a simple alternative.

One common example occurs when $\mathbf{dom}\, f = \mathbf{R}^n_{++}$, as in the equality constrained analytic centering problem described in example 10.2. To initialize Newton's method for the problem

$$\begin{array}{ll} \text{minimize} & -\sum_{i=1}^n \log x_i \\ \text{subject to} & Ax = b, \end{array} \tag{10.25}$$

requires finding a point $x^{(0)} \succ 0$ with $Ax = b$, which is equivalent to solving a standard form LP feasibility problem. This can be carried out using a phase I method, or alternatively, using the infeasible start Newton method, with any positive initial point, *e.g.*, $x^{(0)} = \mathbf{1}$.

The same trick can be used to initialize unconstrained problems where a starting point in $\mathbf{dom}\, f$ is not known. As an example, we consider the dual of the equality constrained analytic centering problem (10.25),

$$\text{maximize} \quad g(\nu) = -b^T\nu + n + \sum_{i=1}^{n} \log(A^T\nu)_i.$$

To initialize this problem for the (feasible start) Newton method, we must find a point $\nu^{(0)}$ that satisfies $A^T\nu^{(0)} \succ 0$, *i.e.*, we must solve a set of linear inequalities. This can be done using a phase I method, or using an infeasible start Newton method, after reformulating the problem. We first express it as an equality constrained problem,

$$\begin{array}{ll} \text{maximize} & -b^T\nu + n + \sum_{i=1}^{n} \log y_i \\ \text{subject to} & y = A^T\nu, \end{array}$$

with new variable $y \in \mathbf{R}^n$. We can now use the infeasible start Newton method, starting with any positive $y^{(0)}$ (and any $\nu^{(0)}$).

The disadvantage of using the infeasible start Newton method to initialize problems for which a strictly feasible starting point is not known is that there is no clear way to detect that there exists no strictly feasible point; the norm of the residual will simply converge, slowly, to some positive value. (Phase I methods, in contrast, can determine this fact unambiguously.) In addition, the convergence of the infeasible start Newton method, before feasibility is achieved, can be slow; see §11.4.2.

10.3.3   Convergence analysis

In this section we show that the infeasible start Newton method converges to the optimal point, provided certain assumptions hold. The convergence proof is very similar to those for the standard Newton method, or the standard Newton method with equality constraints. We show that once the norm of the residual is small enough, the algorithm takes full steps (which implies that feasibility is achieved), and convergence is subsequently quadratic. We also show that the norm of the residual is reduced by at least a fixed amount in each iteration before the region of quadratic convergence is reached. Since the norm of the residual cannot be negative, this shows that within a finite number of steps, the residual will be small enough to guarantee full steps, and quadratic convergence.

Assumptions

We make the following assumptions.

- The sublevel set

$$S = \{(x, \nu) \mid x \in \mathbf{dom}\, f,\ \|r(x, \nu)\|_2 \le \|r(x^{(0)}, \nu^{(0)})\|_2\} \qquad (10.26)$$

  is closed. If f is closed, then $\|r\|_2$ is a closed function, and therefore this condition is satisfied for any $x^{(0)} \in \mathbf{dom}\, f$ and any $\nu^{(0)} \in \mathbf{R}^p$ (see exercise 10.7).

- On the set S, we have

$$\|Dr(x,\nu)^{-1}\|_2 = \left\| \begin{bmatrix} \nabla^2 f(x) & A^T \\ A & 0 \end{bmatrix}^{-1} \right\|_2 \le K, \qquad (10.27)$$

for some K.

- For (x, ν), $(\tilde{x}, \tilde{\nu}) \in S$, Dr satisfies the Lipschitz condition

$$\|Dr(x, \nu) - Dr(\tilde{x}, \tilde{\nu})\|_2 \le L\|(x, \nu) - (\tilde{x}, \tilde{\nu})\|_2.$$

(This is equivalent to $\nabla^2 f(x)$ satisfying a Lipschitz condition; see exercise 10.7.)

As we will see below, these assumptions imply that $\mathbf{dom}\, f$ and $\{z \mid Az = b\}$ intersect, and that there is an optimal point $(x^\star, \nu^\star)$.

Comparison with standard Newton method

The assumptions above are very similar to the ones made in §10.2.4 (page 529) for the analysis of the standard Newton method. The second and third assumptions, the bounded inverse KKT matrix and Lipschitz condition, are essentially the same. The sublevel set condition (10.26) for the infeasible start Newton method is, however, more general than the sublevel set condition made in §10.2.4.

As an example, consider the equality constrained maximum entropy problem

$$\begin{array}{ll} \text{minimize} & f(x) = \sum_{i=1}^{n} x_i \log x_i \\ \text{subject to} & Ax = b, \end{array}$$

with $\mathbf{dom}\, f = \mathbf{R}_{++}^n$. The objective f is *not* closed; it has sublevel sets that are not closed, so the assumptions made in the standard Newton method may not hold, at least for some initial points. The problem here is that the negative entropy function does not converge to ∞ as $x_i \to 0$. On the other hand the sublevel set condition (10.26) for the infeasible start Newton method *does* hold for this problem, since the norm of the gradient of the negative entropy function does converge to ∞ as $x_i \to 0$. Thus, the infeasible start Newton method is guaranteed to solve the equality constrained maximum entropy problem. (We do not know whether the standard Newton method can fail for this problem; we are only observing here that our convergence analysis does not hold.) Note that if the initial point satisfies the equality constraints, the only difference between the standard and infeasible start Newton methods is in the line searches, which differ only during the damped stage.

A basic inequality

We start by deriving a basic inequality. Let $y = (x, \nu) \in S$ with $\|r(y)\|_2 \ne 0$, and let $\Delta y_{\text{nt}} = (\Delta x_{\text{nt}}, \Delta \nu_{\text{nt}})$ be the Newton step at y. Define

$$t_{\max} = \inf\{t > 0 \mid y + t\Delta y_{\text{nt}} \notin S\}.$$

If $y + t\Delta y_{\text{nt}} \in S$ for all $t \ge 0$, we follow the usual convention and define $t_{\max} = \infty$. Otherwise, $t_{\max}$ is the smallest positive value of t such that $\|r(y + t\Delta y_{\text{nt}})\|_2 = \|r(y^{(0)})\|_2$. In particular, it follows that $y + t\Delta y_{\text{nt}} \in S$ for $0 \le t \le t_{\max}$.

We will show that

$$\|r(y + t\Delta y_{\text{nt}})\|_2 \le (1 - t)\|r(y)\|_2 + (K^2 L/2)t^2\|r(y)\|_2^2 \qquad (10.28)$$

for $0 \le t \le \min\{1, t_{\max}\}$.

We have

$$
\begin{aligned}
r(y + t\Delta y_{\mathrm{nt}}) &= r(y) + \int_0^1 Dr(y + \tau t\Delta y_{\mathrm{nt}})t\Delta y_{\mathrm{nt}}\,d\tau \\
&= r(y) + tDr(y)\Delta y_{\mathrm{nt}} + \int_0^1 (Dr(y + \tau t\Delta y_{\mathrm{nt}}) - Dr(y))t\Delta y_{\mathrm{nt}}\,d\tau \\
&= r(y) + tDr(y)\Delta y_{\mathrm{nt}} + e \\
&= (1 - t)r(y) + e,
\end{aligned}
$$

using $Dr(y)\Delta y_{\mathrm{nt}} = -r(y)$, and defining

$$
e = \int_0^1 (Dr(y + \tau t\Delta y_{\mathrm{nt}}) - Dr(y))t\Delta y_{\mathrm{nt}}\,d\tau.
$$

Now suppose $0 \le t \le t_{\max}$, so $y + \tau t\Delta y_{\mathrm{nt}} \in S$ for $0 \le \tau \le 1$. We can bound $\|e\|_2$ as follows:

$$
\begin{aligned}
\|e\|_2 &\le \|t\Delta y_{\mathrm{nt}}\|_2 \int_0^1 \|Dr(y + \tau t\Delta y_{\mathrm{nt}}) - Dr(y)\|_2\,d\tau \\
&\le \|t\Delta y_{\mathrm{nt}}\|_2 \int_0^1 L\|\tau t\Delta y_{\mathrm{nt}}\|_2\,d\tau \\
&= (L/2)t^2\|\Delta y_{\mathrm{nt}}\|_2^2 \\
&= (L/2)t^2\|Dr(y)^{-1}r(y)\|_2^2 \\
&\le (K^2L/2)t^2\|r(y)\|_2^2,
\end{aligned}
$$

using the Lipschitz condition on the second line, and the bound $\|Dr(y)^{-1}\|_2 \le K$ on the last. Now we can derive the bound (10.28): For $0 \le t \le \min\{1, t_{\max}\}$,

$$
\begin{aligned}
\|r(y + t\Delta y_{\mathrm{nt}})\|_2 &= \|(1 - t)r(y) + e\|_2 \\
&\le (1 - t)\|r(y)\|_2 + \|e\|_2 \\
&\le (1 - t)\|r(y)\|_2 + (K^2L/2)t^2\|r(y)\|_2^2.
\end{aligned}
$$

Damped Newton phase

We first show that if $\|r(y)\|_2 > 1/(K^2L)$, one iteration of the infeasible start Newton method reduces $\|r\|_2$ by at least a certain minimum amount.

The righthand side of the basic inequality (10.28) is quadratic in t, and monotonically decreasing between $t = 0$ and its minimizer

$$
\bar{t} = \frac{1}{K^2L\|r(y)\|_2} < 1.
$$

We must have $t_{\max} > \bar{t}$, because the opposite would imply $\|r(y + t_{\max}\Delta y_{\mathrm{nt}})\|_2 < \|r(y)\|_2$, which is false. The basic inequality is therefore valid at $t = \bar{t}$, and therefore

$$
\begin{aligned}
\|r(y + \bar{t}\Delta y_{\mathrm{nt}})\|_2 &\le \|r(y)\|_2 - 1/(2K^2L) \\
&\le \|r(y)\|_2 - \alpha/(K^2L) \\
&= (1 - \alpha\bar{t})\|r(y)\|_2,
\end{aligned}
$$

which shows that the step length $\bar{t}$ satisfies the line search exit condition. Therefore we have $t \geq \beta\bar{t}$, where t is the step length chosen by the backtracking algorithm. From $t \geq \beta\bar{t}$ we have (from the exit condition in the backtracking line search)

$$
\begin{aligned}
\|r(y + t\Delta y_{\mathrm{nt}})\|_2 &\leq (1 - \alpha t)\|r(y)\|_2 \\
&\leq (1 - \alpha\beta\bar{t})\|r(y)\|_2 \\
&= \left(1 - \frac{\alpha\beta}{K^2 L \|r(y)\|_2}\right)\|r(y)\|_2 \\
&= \|r(y)\|_2 - \frac{\alpha\beta}{K^2 L}.
\end{aligned}
$$

Thus, as long as we have $\|r(y)\|_2 > 1/(K^2 L)$, we obtain a minimum decrease in $\|r\|_2$, per iteration, of $\alpha\beta/(K^2 L)$. It follows that a maximum of

$$
\frac{\|r(y^{(0)})\|_2 K^2 L}{\alpha\beta}
$$

iterations can be taken before we have $\|r(y^{(k)})\|_2 \leq 1/(K^2 L)$.

Quadratically convergent phase

Now suppose $\|r(y)\|_2 \leq 1/(K^2 L)$. The basic inequality gives

$$
\|r(y + t\Delta y_{\mathrm{nt}})\|_2 \leq (1 - t + (1/2)t^2)\|r(y)\|_2 \tag{10.29}
$$

for $0 \leq t \leq \min\{1, t_{\max}\}$. We must have $t_{\max} > 1$, because otherwise it would follow from (10.29) that $\|r(y + t_{\max}\Delta y_{\mathrm{nt}})\|_2 < \|r(y)\|_2$, which contradicts the definition of $t_{\max}$. The inequality (10.29) therefore holds with $t = 1$, $i.e.$, we have

$$
\|r(y + \Delta y_{\mathrm{nt}})\|_2 \leq (1/2)\|r(y)\|_2 \leq (1 - \alpha)\|r(y)\|_2.
$$

This shows that the backtracking line search exit criterion is satisfied for $t = 1$, so a full step will be taken. Moreover, for all future iterations we have $\|r(y)\|_2 \leq 1/(K^2 L)$, so a full step will be taken for all following iterations.

We can write the inequality (10.28) (for $t = 1$) as

$$
\frac{K^2 L \|r(y^+)\|_2}{2} \leq \left(\frac{K^2 L \|r(y)\|_2}{2}\right)^2,
$$

where $y^+ = y + \Delta y_{\mathrm{nt}}$. Therefore, if $r(y^{+k})$ denotes the residual k steps after an iteration in which $\|r(y)\|_2 \leq 1/K^2 L$, we have

$$
\frac{K^2 L \|r(y^{+k})\|_2}{2} \leq \left(\frac{K^2 L \|r(y)\|_2}{2}\right)^{2^k} \leq \left(\frac{1}{2}\right)^{2^k},
$$

$i.e.$, we have quadratic convergence of $\|r(y)\|_2$ to zero.

To show that the sequence of iterates converges, we will show that it is a Cauchy sequence. Suppose y is an iterate satisfying $\|r(y)\|_2 \leq 1/(K^2 L)$, and y^{+k} denotes the kth iterate after y. Since these iterates are in the region of quadratic convergence, the step size is one, so we have

$$
\begin{aligned}
\|y^{+k} - y\|_2 &\leq \|y^{+k} - y^{+(k-1)}\|_2 + \cdots + \|y^+ - y\|_2 \\
&= \|Dr(y^{+(k-1)})^{-1} r(y^{+(k-1)})\|_2 + \cdots + \|Dr(y)^{-1} r(y)\|_2 \\
&\leq K\left(\|r(y^{+(k-1)})\|_2 + \cdots + \|r(y)\|_2\right) \\
&\leq K\|r(y)\|_2 \sum_{i=0}^{k-1} \left(\frac{K^2 L \|r(y)\|_2}{2}\right)^{2^i - 1} \\
&\leq K\|r(y)\|_2 \sum_{i=0}^{k-1} \left(\frac{1}{2}\right)^{2^i - 1} \\
&\leq 2K\|r(y)\|_2
\end{aligned}
$$

where in the third line we use the assumption that $\|Dr^{-1}\|_2 \leq K$ for all iterates. Since $\|r(y^{(k)})\|_2$ converges to zero, we conclude $y^{(k)}$ is a Cauchy sequence, and therefore converges. By continuity of r, the limit point $y^\star$ satisfies $r(y^\star) = 0$. This establishes our earlier claim that the assumptions at the beginning of this section imply that there is an optimal point $(x^\star, \nu^\star)$.

10.3.4   Convex-concave games

The proof of convergence for the infeasible start Newton method reveals that the method can be used for a larger class of problems than equality constrained convex optimization problems. Suppose $r : \mathbf{R}^n \to \mathbf{R}^n$ is differentiable, its derivative satisfies a Lipschitz condition on S, and $\|Dr(x)^{-1}\|_2$ is bounded on S, where

$$
S = \{x \in \mathbf{dom}\, r \mid \|r(x)\|_2 \leq \|r(x^{(0)})\|_2\}
$$

is a closed set. Then the infeasible start Newton method, started at $x^{(0)}$, converges to a solution of $r(x) = 0$ in S. In the infeasible start Newton method, we apply this to the specific case in which r is the residual for the equality constrained convex optimization problem. But it applies in several other interesting cases. One interesting example is solving a *convex-concave game*. (See §5.4.3 and exercise 5.25 for discussion of other, related games).

An unconstrained (zero-sum, two-player) game on $\mathbf{R}^p \times \mathbf{R}^q$ is defined by its *payoff function* $f : \mathbf{R}^{p+q} \to \mathbf{R}$. The meaning is that player 1 chooses a value (or move) $u \in \mathbf{R}^p$, and player 2 chooses a value (or move) $v \in \mathbf{R}^q$; based on these choices, player 1 makes a payment to player 2, in the amount $f(u, v)$. The goal of player 1 is to minimize this payment, while the goal of player 2 is to maximize it.

If player 1 makes his choice u first, and player 2 knows the choice, then player 2 will choose v to maximize $f(u, v)$, which results in a payoff of $\sup_v f(u, v)$ (assuming the supremum is achieved). If player 1 assumes that player 2 will make this choice, he should choose u to minimize $\sup_v f(u, v)$. The resulting payoff, from player 1 to player 2, will then be

$$
\inf_u \sup_v f(u, v) \tag{10.30}
$$

(assuming that the supremum is achieved). On the other hand if player 2 makes the first choice, the strategies are reversed, and the resulting payoff from player 1 to player 2 is

$$\sup_{v} \inf_{u} f(u, v). \tag{10.31}$$

The payoff (10.30) is always greater than or equal to the payoff (10.31); the difference between the two payoffs can be interpreted as the advantage afforded the player who makes the second move, with knowledge of the other player's move. We say that $(u^\star, v^\star)$ is a *solution* of the game, or a *saddle-point* for the game, if for all u, v,

$$f(u^\star, v) \leq f(u^\star, v^\star) \leq f(u, v^\star).$$

When a solution exists, there is no advantage to making the second move; $f(u^\star, v^\star)$ is the common value of both payoffs (10.30) and (10.31). (See exercise 3.14.)

The game is called *convex-concave* if for each v, $f(u, v)$ is a convex function of u, and for each u, $f(u, v)$ is a concave function of v. When f is differentiable (and convex-concave), a saddle-point for the game is characterized by $\nabla f(u^\star, v^\star) = 0$.

Solution via infeasible start Newton method

We can use the infeasible start Newton method to compute a solution of a convex-concave game with twice differentiable payoff function. We define the residual as

$$r(u, v) = \nabla f(u, v) = \left[\begin{array}{c} \nabla_u f(u, v) \\ \nabla_v f(u, v) \end{array} \right],$$

and apply the infeasible start Newton method. In the context of games, the infeasible start Newton method is simply called Newton's method (for convex-concave games).

We can guarantee convergence of the (infeasible start) Newton method provided $Dr = \nabla^2 f$ has bounded inverse, and satisfies a Lipschitz condition on the sublevel set

$$S = \{(u, v) \in \mathbf{dom}\, f \mid \|r(u, v)\|_2 \leq \|r(u^{(0)}, v^{(0)})\|_2\},$$

where $u^{(0)}$, $v^{(0)}$ are the starting players' choices.

There is a simple analog of the strong convexity condition in an unconstrained minimization problem. We say the game with payoff function f is strongly convex-concave if for some $m > 0$, we have $\nabla^2_{uu} f(u, v) \succeq mI$ and $\nabla^2_{vv} f(u, v) \preceq -mI$, for all $(u, v) \in S$. Not surprisingly, this strong convex-concave assumption implies the bounded inverse condition (exercise 10.10).

10.3.5   Examples

A simple example

We illustrate the infeasible start Newton method on the equality constrained analytic center problem (10.25). Our first example is an instance with dimensions $n = 100$ and $m = 50$, generated randomly, for which the problem is feasible and bounded below. The infeasible start Newton method is used, with initial primal and dual points $x^{(0)} = \mathbf{1}$, $\nu^{(0)} = 0$, and backtracking parameters $\alpha = 0.01$ and $\beta = 0.5$. The plot in figure 10.1 shows the norms of the primal and dual residuals separately, versus iteration number, and the plot in figure 10.2 shows the step lengths. A full Newton step is taken in iteration 8, so the primal residual becomes (almost) zero, and remains (almost) zero. After around iteration 9 or so, the (dual) residual converges quadratically to zero.

An infeasible example

We also consider a problem instance, of the same dimensions as the example above, for which $\mathbf{dom}\, f$ does not intersect $\{z \mid Az = b\}$, *i.e.*, the problem is infeasible. (This violates the basic assumption in the chapter that problem (10.1) is solvable, as well as the assumptions made in §10.2.4; the example is meant only to show what happens to the infeasible start Newton method when $\mathbf{dom}\, f$ does not intersect $\{z \mid Az = b\}$.) The norm of the residual for this example is shown in figure 10.3, and the step length in figure 10.4. Here, of course, the step lengths are never one, and the residual does not converge to zero.

A convex-concave game

Our final example involves a convex-concave game on $\mathbf{R}^{100} \times \mathbf{R}^{100}$, with payoff function

$$f(u, v) = u^T A v + b^T u + c^T v - \log(1 - u^T u) + \log(1 - v^T v), \qquad (10.32)$$

defined on

$$\mathbf{dom}\, f = \{(u, v) \mid u^T u < 1, \ v^T v < 1\}.$$

The problem data A, b, and c were randomly generated. The progress of the (infeasible start) Newton method, started at $u^{(0)} = v^{(0)} = 0$, with backtracking parameters $\alpha = 0.01$ and $\beta = 0.5$, is shown in figure 10.5.

10.4   Implementation

10.4.1   Elimination

To implement the elimination method, we have to calculate a full rank matrix F and an $\hat{x}$ such that

$$\{x \mid Ax = b\} = \{Fz + \hat{x} \mid z \in \mathbf{R}^{n-p}\}.$$

Several methods for this are described in §C.5.

10.4.2   Solving KKT systems

In this section we describe methods that can be used to compute the Newton step or infeasible Newton step, both of which involve solving a set of linear equations

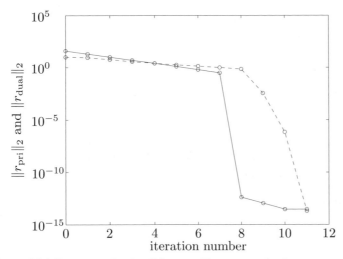

Figure 10.1 Progress of infeasible start Newton method on an equality constrained analytic centering problem with 100 variables and 50 constraints. The figure shows $\|r_{\mathrm{pri}}\|_2$ (solid line), and $\|r_{\mathrm{dual}}\|_2$ (dashed line). Note that feasibility is achieved (and maintained) after 8 iterations, and convergence is quadratic, starting from iteration 9 or so.

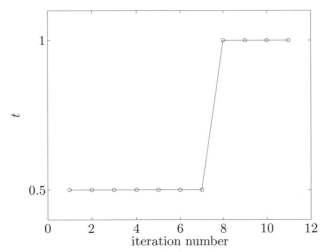

Figure 10.2 Step length versus iteration number for the same example problem. A full step is taken in iteration 8, which results in feasibility from iteration 8 on.

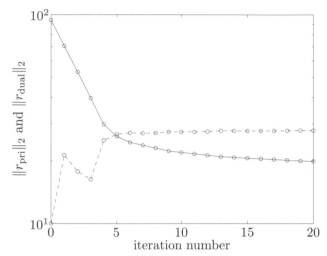

Figure 10.3 Progress of infeasible start Newton method on an equality constrained analytic centering problem with 100 variables and 50 constraints, for which $\mathbf{dom}\, f = \mathbf{R}^{100}_{++}$ does not intersect $\{z \mid Az = b\}$. The figure shows $\|r_{\mathrm{pri}}\|_2$ (solid line), and $\|r_{\mathrm{dual}}\|_2$ (dashed line). In this case, the residuals do not converge to zero.

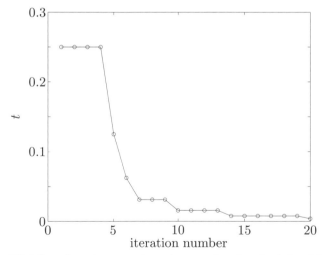

Figure 10.4 Step length versus iteration number for the infeasible example problem. No full steps are taken, and the step lengths converge to zero.

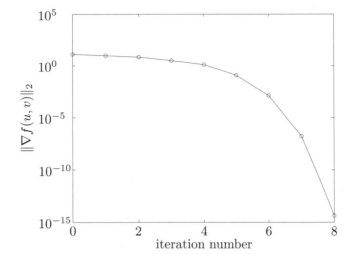

Figure 10.5 Progress of (infeasible start) Newton method on a convex-concave game. Quadratic convergence becomes apparent after about 5 iterations.

with KKT form

$$\begin{bmatrix} H & A^T \\ A & 0 \end{bmatrix} \begin{bmatrix} v \\ w \end{bmatrix} = - \begin{bmatrix} g \\ h \end{bmatrix}. \tag{10.33}$$

Here we assume $H \in \mathbf{S}_+^n$, and $A \in \mathbf{R}^{p \times n}$ with **rank** $A = p < n$. Similar methods can be used to compute the Newton step for a convex-concave game, in which the bottom right entry of the coefficient matrix is negative semidefinite (see exercise 10.13).

Solving full KKT system

One straightforward approach is to simply solve the KKT system (10.33), which is a set of $n + p$ linear equations in $n + p$ variables. The KKT matrix is symmetric, but not positive definite, so a good way to do this is to use an LDL^T factorization (see §C.3.3). If no structure of the matrix is exploited, the cost is $(1/3)(n+p)^3$ flops. This can be a reasonable approach when the problem is small (*i.e.*, n and p are not too large), or when A and H are sparse.

Solving KKT system via elimination

A method that is often better than directly solving the full KKT system is based on eliminating the variable v (see §C.4). We start by describing the simplest case, in which $H \succ 0$. Starting from the first of the KKT equations

$$Hv + A^T w = -g, \qquad Av = -h,$$

we solve for v to obtain

$$v = -H^{-1}(g + A^T w).$$

Substituting this into the second KKT equation yields $AH^{-1}(g + A^T w) = h$, so we have

$$w = (AH^{-1}A^T)^{-1}(h - AH^{-1}g).$$

These formulas give us a method for computing v and w.

The matrix appearing in the formula for w is the Schur complement S of H in the KKT matrix:

$$S = -AH^{-1}A^T.$$

Because of the special structure of the KKT matrix, and our assumption that A has rank p, the matrix S is negative definite.

Algorithm 10.3 *Solving KKT system by block elimination.*

given KKT system with $H \succ 0$.

    1. Form $H^{-1}A^T$ and $H^{-1}g$.
    2. Form Schur complement $S = -AH^{-1}A^T$.
    3. Determine w by solving $Sw = AH^{-1}g - h$.
    4. Determine v by solving $Hv = -A^T w - g$.

Step 1 can be done by a Cholesky factorization of H, followed by $p + 1$ solves, which costs $f + (p + 1)s$, where f is the cost of factoring H and s is the cost of an associated solve. Step 2 requires a $p \times n$ by $n \times p$ matrix multiplication. If we exploit no structure in this calculation, the cost is $p^2 n$ flops. (Since the result is symmetric, we only need to compute the upper triangular part of S.) In some cases special structure in A and H can be exploited to carry out step 2 more efficiently. Step 3 can be carried out by Cholesky factorization of $-S$, which costs $(1/3)p^3$ flops if no further structure of S is exploited. Step 4 can be carried out using the factorization of H already calculated in step 1, so the cost is $2np + s$ flops. The total flop count, assuming that no structure is exploited in forming or factoring the Schur complement, is

$$f + ps + p^2 n + (1/3)p^3$$

flops (keeping only dominant terms). If we exploit structure in forming or factoring S, the last two terms are even smaller.

If H can be factored efficiently, then block elimination gives us a flop count advantage over directly solving the KKT system using an LDL^T factorization. For example, if H is diagonal (which corresponds to a separable objective function), we have $f = 0$ and $s = n$, so the total cost is $p^2 n + (1/3)p^3$ flops, which grows only linearly with n. If H is banded with bandwidth $k \ll n$, then $f = nk^2$, $s = 4nk$, so the total cost is around $nk^2 + 4nkp + p^2 n + (1/3)p^3$ which still grows only linearly with n. Other structures of H that can be exploited are block diagonal (which corresponds to block separable objective function), sparse, or diagonal plus low rank; see appendix C and §9.7 for more details and examples.

Example 10.3 *Equality constrained analytic center.* We consider the problem

$$\begin{array}{ll} \text{minimize} & -\sum_{i=1}^{n} \log x_i \\ \text{subject to} & Ax = b. \end{array}$$

Here the objective is separable, so the Hessian at x is diagonal:

$$H = \mathbf{diag}(x_1^{-2}, \ldots, x_n^{-2}).$$

If we compute the Newton direction using a generic method such as an LDL^T factorization of the KKT matrix, the cost is $(1/3)(n+p)^3$ flops.

If we compute the Newton step using block elimination, the cost is $np^2 + (1/3)p^3$ flops. This is much smaller than the cost of the generic method.

In fact this cost is the same as that of computing the Newton step for the dual problem, described in example 10.2 on page 525. For the (unconstrained) dual problem, the Hessian is

$$H_{\mathrm{dual}} = -ADA^T,$$

where D is diagonal, with $D_{ii} = (A^T\nu)_i^{-2}$. Forming this matrix costs np^2 flops, and solving for the Newton step by a Cholesky factorization of $-H_{\mathrm{dual}}$ costs $(1/3)p^3$ flops.

Example 10.4 *Minimum length piecewise-linear curve subject to equality constraints.* We consider a piecewise-linear curve in $\mathbf{R}^2$ with knot points $(0,0), (1,x_1), \ldots, (n,x_n)$. To find the minimum length curve that satisfies the equality constraints $Ax = b$, we form the problem

$$\begin{array}{ll} \text{minimize} & \left(1+x_1^2\right)^{1/2} + \sum_{i=1}^{n-1}\left(1+(x_{i+1}-x_i)^2\right)^{1/2} \\ \text{subject to} & Ax = b, \end{array}$$

with variable $x \in \mathbf{R}^n$, and $A \in \mathbf{R}^{p\times n}$. In this problem, the objective is a sum of functions of pairs of adjacent variables, so the Hessian H is tridiagonal. Using block elimination, we can compute the Newton step in around $p^2 n + (1/3)p^3$ flops.

Elimination with singular H

The block elimination method described above obviously does not work when H is singular, but a simple variation on the method can be used in this more general case. The more general method is based on the following result: The KKT matrix is nonsingular if and only if $H + A^T Q A \succ 0$ for some $Q \succeq 0$, in which case, $H + A^T Q A \succ 0$ for all $Q \succ 0$. (See exercise 10.1.) We conclude, for example, that if the KKT matrix is nonsingular, then $H + A^T A \succ 0$.

Let $Q \succeq 0$ be a matrix for which $H + A^T Q A \succ 0$. Then the KKT system (10.33) is equivalent to

$$\begin{bmatrix} H + A^T Q A & A^T \\ A & 0 \end{bmatrix} \begin{bmatrix} v \\ w \end{bmatrix} = -\begin{bmatrix} g + A^T Q h \\ h \end{bmatrix},$$

which can be solved using elimination since $H + A^T Q A \succ 0$.

0.4.3   Examples

In this section we describe some longer examples, showing how structure can be exploited to efficiently compute the Newton step. We also include some numerical results.

Equality constrained analytic centering

We consider the equality constrained analytic centering problem

$$
\begin{array}{ll}
\text{minimize} & f(x) = -\sum_{i=1}^{n} \log x_i \\
\text{subject to} & Ax = b.
\end{array}
$$

(See examples 10.2 and 10.3.) We compare three methods, for a problem of size $p = 100$, $n = 500$.

The first method is Newton's method with equality constraints (§10.2). The Newton step Δx_{nt} is defined by the KKT system (10.11):

$$
\begin{bmatrix} H & A^T \\ A & 0 \end{bmatrix} \begin{bmatrix} \Delta x_{\mathrm{nt}} \\ w \end{bmatrix} = \begin{bmatrix} -g \\ 0 \end{bmatrix},
$$

where $H = \mathbf{diag}(1/x_1^2, \ldots, 1/x_n^2)$, and $g = -(1/x_1, \ldots, 1/x_n)$. As explained in example 10.3, page 546, the KKT system can be efficiently solved by elimination, i.e., by solving

$$
AH^{-1}A^T w = -AH^{-1}g,
$$

and setting $\Delta x_{\mathrm{nt}} = -H^{-1}(A^T w + g)$. In other words,

$$
\Delta x_{\mathrm{nt}} = -\mathbf{diag}(x)^2 A^T w + x,
$$

where w is the solution of

$$
A\,\mathbf{diag}(x)^2 A^T w = b. \tag{10.34}
$$

Figure 10.6 shows the error versus iteration. The different curves correspond to four different starting points. We use a backtracking line search with $\alpha = 0.1$, $\beta = 0.5$.

The second method is Newton's method applied to the dual

$$
\text{maximize} \quad g(\nu) = -b^T \nu + \sum_{i=1}^{n} \log(A^T \nu)_i + n
$$

(see example 10.2, page 525). Here the Newton step is obtained from solving

$$
A\,\mathbf{diag}(y)^2 A^T \Delta \nu_{\mathrm{nt}} = -b + Ay \tag{10.35}
$$

where $y = (1/(A^T \nu)_1, \ldots, 1/(A^T \nu)_n)$. Comparing (10.35) and (10.34) we see that both methods have the same complexity. In figure 10.7 we show the error for four different starting points. We use a backtracking line search with $\alpha = 0.1$, $\beta = 0.5$.

The third method is the infeasible start Newton method of §10.3, applied to the optimality conditions

$$
\nabla f(x^\star) + A^T \nu^\star = 0, \qquad Ax^\star = b.
$$

The Newton step is obtained by solving

$$
\begin{bmatrix} H & A^T \\ A & 0 \end{bmatrix} \begin{bmatrix} \Delta x_{\mathrm{nt}} \\ \Delta \nu_{\mathrm{nt}} \end{bmatrix} = - \begin{bmatrix} g + A^T \nu \\ Ax - b \end{bmatrix},
$$

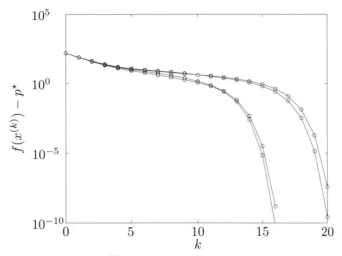

Figure 10.6 Error $f(x^{(k)}) - p^\star$ in Newton's method, applied to an equality constrained analytic centering problem of size $p = 100$, $n = 500$. The different curves correspond to four different starting points. Final quadratic convergence is clearly evident.

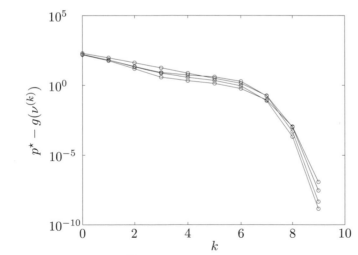

Figure 10.7 Error $|g(\nu^{(k)}) - p^\star|$ in Newton's method, applied to the dual of the equality constrained analytic centering problem.

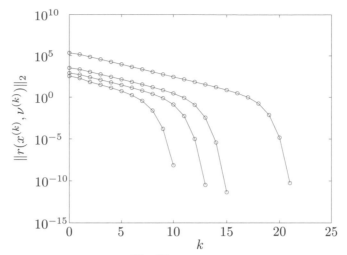

Figure 10.8 Residual $\|r(x^{(k)}, \nu^{(k)})\|_2$ in the infeasible start Newton method, applied to the equality constrained analytic centering problem.

where $H = \mathbf{diag}(1/x_1^2, \ldots, 1/x_n^2)$, and $g = -(1/x_1, \ldots, 1/x_n)$. This KKT system can be efficiently solved by elimination, at the same cost as (10.34) or (10.35). For example, if we first solve

$$A\,\mathbf{diag}(x)^2 A^T w = 2Ax - b,$$

then $\Delta\nu_{\mathrm{nt}}$ and Δx_{nt} follow from

$$\Delta\nu_{\mathrm{nt}} = w - \nu, \qquad \Delta x_{\mathrm{nt}} = x - \mathbf{diag}(x)^2 A^T w.$$

Figure 10.8 shows the norm of the residual

$$r(x, \nu) = (\nabla f(x) + A^T \nu, Ax - b)$$

versus iteration, for four different starting points. We use a backtracking line search with $\alpha = 0.1$, $\beta = 0.5$.

The figures show that for this problem, the dual method appears to be faster, but only by a factor of two or three. It takes about six iterations to reach the region of quadratic convergence, as opposed to 12–15 in the primal method and 10–20 in the infeasible start Newton method.

The methods also differ in the initialization they require. The primal method requires knowledge of a primal feasible point, *i.e.*, satisfying $Ax^{(0)} = b$, $x^{(0)} \succ 0$. The dual method requires a dual feasible point, *i.e.*, $A^T \nu^{(0)} \succ 0$. Depending on the problem, one or the other might be more readily available. The infeasible start Newton method requires no initialization; the only requirement is that $x^{(0)} \succ 0$.

Optimal network flow

We consider a connected directed graph or network with n edges and $p + 1$ nodes. We let x_j denote the flow or traffic on arc j, with $x_j > 0$ meaning flow in the direction of the arc, and $x_j < 0$ meaning flow in the direction opposite the arc. There is also a given external source (or sink) flow s_i that enters (if $s_i > 0$) or leaves (if $s_i < 0$) node i. The flow must satisfy a conservation equation, which states that at each node, the total flow entering the node, including the external sources and sinks, is zero. This conservation equation can be expressed as $\tilde{A}x = s$ where $\tilde{A} \in \mathbf{R}^{(p+1) \times n}$ is the *node incidence matrix* of the graph,

$$\tilde{A}_{ij} = \begin{cases} 1 & \text{arc } j \text{ leaves node } i \\ -1 & \text{arc } j \text{ enters node } i \\ 0 & \text{otherwise.} \end{cases}$$

The flow conservation equation $\tilde{A}x = s$ is inconsistent unless $\mathbf{1}^T s = 0$, which we assume is the case. (In other words, the total of the source flows must equal the total of the sink flows.) The flow conservation equations $\tilde{A}x = s$ are also redundant, since $\mathbf{1}^T \tilde{A} = 0$. To obtain an independent set of equations we can delete any one equation, to obtain $Ax = b$, where $A \in \mathbf{R}^{p \times n}$ is the *reduced node incidence matrix* of the graph (*i.e.*, the node incidence matrix with one row removed) and $b \in \mathbf{R}^p$ is reduced source vector (*i.e.*, s with the associated entry removed).

In summary, flow conservation is given by $Ax = b$, where A is the reduced node incidence matrix of the graph and b is the reduced source vector. The matrix A is very sparse, since each column has at most two nonzero entries (which can only be $+1$ or -1).

We will take traffic flows x as the variables, and the sources as given. We introduce the objective function

$$f(x) = \sum_{i=1}^{n} \phi_i(x_i),$$

where $\phi_i : \mathbf{R} \to \mathbf{R}$ is the flow cost function for arc i. We assume that the flow cost functions are strictly convex and twice differentiable.

The problem of choosing the best flow, that satisfies the flow conservation requirement, is

$$\begin{array}{ll} \text{minimize} & \sum_{i=1}^{n} \phi_i(x_i) \\ \text{subject to} & Ax = b. \end{array} \tag{10.36}$$

Here the Hessian H is diagonal, since the objective is separable.

We have several choices for computing the Newton step for the optimal network flow problem (10.36). The most straightforward is to solve the full KKT system, using a sparse LDL^T factorization.

For this problem it is probably better to compute the Newton step using block elimination. We can characterize the sparsity pattern of the Schur complement $S = -AH^{-1}A^T$ in terms of the graph: We have $S_{ij} \neq 0$ if and only if node i and node j are connected by an arc. It follows that if the network is sparse, *i.e.*, if each node is connected by an arc to only a few other nodes, then the Schur complement S is sparse. In this case, we can exploit sparsity in forming S, and in the associated factorization and solve steps, as well. We can expect the computational complexity of computing the Newton step to grow approximately linearly with the number of arcs (which is the number of variables).

Optimal control

We consider the problem

$$
\begin{array}{ll}
\text{minimize} & \sum_{t=1}^{N} \phi_t(z(t)) + \sum_{t=0}^{N-1} \psi_t(u(t)) \\
\text{subject to} & z(t+1) = A_t z(t) + B_t u(t), \quad t = 0, \ldots, N-1.
\end{array}
$$

Here

- $z(t) \in \mathbf{R}^k$ is the system state at time t

- $u(t) \in \mathbf{R}^l$ is the input or control action at time t

- $\phi_t : \mathbf{R}^k \to \mathbf{R}$ is the state cost function

- $\psi_t : \mathbf{R}^l \to \mathbf{R}$ is the input cost function

- N is called the *time horizon* for the problem.

We assume that the input and state cost functions are strictly convex and twice differentiable. The variables in the problem are $u(0), \ldots, u(N-1)$, and $z(1), \ldots, z(N)$. The initial state $z(0)$ is given. The linear equality constraints are called the *state equations* or *dynamic evolution equations*. We define the overall optimization variable x as

$$
x = (u(0), z(1), u(1), \ldots, u(N-1), z(N)) \in \mathbf{R}^{N(k+l)}.
$$

Since the objective is block separable (*i.e.*, a sum of functions of $z(t)$ and $u(t)$), the Hessian is block diagonal:

$$
H = \mathbf{diag}(R_0, Q_1, \ldots, R_{N-1}, Q_N),
$$

where

$$
R_t = \nabla^2 \psi_t(u(t)), \quad t = 0, \ldots, N-1, \qquad Q_t = \nabla^2 \phi_t(z(t)), \quad t = 1, \ldots, N.
$$

We can collect all the equality constraints (*i.e.*, the state equations) and express them as $Ax = b$ where

$$
A = \begin{bmatrix}
-B_0 & I & 0 & 0 & 0 & \cdots & 0 & 0 & 0 \\
0 & -A_1 & -B_1 & I & 0 & \cdots & 0 & 0 & 0 \\
0 & 0 & 0 & -A_2 & -B_2 & \cdots & 0 & 0 & 0 \\
\vdots & \vdots & \vdots & \vdots & \vdots & & \vdots & \vdots & \vdots \\
0 & 0 & 0 & 0 & 0 & \cdots & I & 0 & 0 \\
0 & 0 & 0 & 0 & 0 & \cdots & -A_{N-1} & -B_{N-1} & I
\end{bmatrix}
$$

$$
b = \begin{bmatrix}
A_0 z(0) \\
0 \\
0 \\
\vdots \\
0 \\
0
\end{bmatrix}.
$$

The number of rows of A (*i.e.*, equality constraints) is Nk.

Directly solving the KKT system for the Newton step, using a dense LDL^T factorization, would cost

$$(1/3)(2Nk + Nl)^3 = (1/3)N^3(2k + l)^3$$

flops. Using a sparse LDL^T factorization would give a large improvement, since the method would exploit the many zero entries in A and H.

In fact we can do better by exploiting the special block structure of H and A, using block elimination to compute the Newton step. The Schur complement $S = -AH^{-1}A^T$ turns out to be block tridiagonal, with $k \times k$ blocks:

$$
S = -AH^{-1}A^T
$$
$$
= \begin{bmatrix}
S_{11} & Q_1^{-1}A_1^T & 0 & \cdots & 0 & 0 \\
A_1Q_1^{-1} & S_{22} & Q_2^{-1}A_2^T & \cdots & 0 & 0 \\
0 & A_2Q_2^{-1} & S_{33} & \cdots & 0 & 0 \\
\vdots & \vdots & \vdots & \ddots & \vdots & \vdots \\
0 & 0 & 0 & \cdots & S_{N-1,N-1} & Q_{N-1}^{-1}A_{N-1}^T \\
0 & 0 & 0 & \cdots & A_{N-1}Q_{N-1}^{-1} & S_{NN}
\end{bmatrix}
$$

where

$$
\begin{aligned}
S_{11} &= -B_0R_0^{-1}B_0^T - Q_1^{-1}, \\
S_{ii} &= -A_{i-1}Q_{i-1}^{-1}A_{i-1}^T - B_{i-1}R_{i-1}^{-1}B_{i-1}^T - Q_i^{-1}, \quad i = 2, \ldots, N.
\end{aligned}
$$

In particular, S is banded, with bandwidth $2k - 1$, so we can factor it in order k^3N flops. Therefore we can compute the Newton step in order k^3N flops, assuming $k \ll N$. Note that this grows linearly with the time horizon N, whereas for a generic method, the flop count grows like N^3.

For this problem we could go one step further and exploit the block tridiagonal structure of S. Applying a standard block tridiagonal factorization method would result in the classic Riccati recursion for solving a quadratic optimal control problem. Still, using only the banded nature of S yields an algorithm that is the same order.

Analytic center of a linear matrix inequality

We consider the problem

$$
\begin{aligned}
\text{minimize} \quad & f(X) = -\log\det X \\
\text{subject to} \quad & \mathbf{tr}(A_iX) = b_i, \quad i = 1, \ldots, p,
\end{aligned}
\tag{10.37}
$$

where $X \in \mathbf{S}^n$ is the variable, $A_i \in \mathbf{S}^n$, $b_i \in \mathbf{R}$, and $\mathbf{dom}\, f = \mathbf{S}_{++}^n$. The KKT conditions for this problem are

$$
-X^{\star -1} + \sum_{i=1}^{m} \nu_i^\star A_i = 0, \qquad \mathbf{tr}(A_iX^\star) = b_i, \quad i = 1, \ldots, p.
\tag{10.38}
$$

The dimension of the variable X is $n(n + 1)/2$. We could simply ignore the special matrix structure of X, and consider it as (vector) variable $x \in \mathbf{R}^{n(n+1)/2}$, and solve the problem (10.37) using a generic method for a problem with $n(n+1)/2$ variables and p equality constraints. The cost for computing a Newton step would then be at least

$$(1/3)(n(n+1)/2 + p)^3$$

flops, which is order n^6 in n. We will see that there are a number of far more attractive alternatives.

A first option is to solve the dual problem. The conjugate of f is

$$f^*(Y) = \log \det(-Y)^{-1} - n$$

with $\mathbf{dom}\, f^* = -\mathbf{S}^n_{++}$ (see example 3.23, page 92), so the dual problem is

$$\text{maximize} \quad -b^T \nu + \log \det(\textstyle\sum_{i=1}^p \nu_i A_i) + n, \qquad (10.39)$$

with domain $\{\nu \mid \sum_{i=1}^p \nu_i A_i \succ 0\}$. This is an unconstrained problem with variable $\nu \in \mathbf{R}^p$. The optimal $X^\star$ can be recovered from the optimal $\nu^\star$ by solving the first (dual feasibility) equation in (10.38), i.e., $X^\star = (\sum_{i=1}^p \nu_i^\star A_i)^{-1}$.

Let us work out the cost of computing the Newton step for the dual problem (10.39). We have to form the gradient and Hessian of g, and then solve for the Newton step. The gradient and Hessian are given by

$$\begin{aligned}
\nabla^2 g(\nu)_{ij} &= -\mathbf{tr}(A^{-1} A_i A^{-1} A_j), \quad i,\, j = 1, \dots, p, \\
\nabla g(\nu)_i &= \mathbf{tr}(A^{-1} A_i) - b_i, \quad i = 1 \dots, p,
\end{aligned}$$

where $A = \sum_{i=1}^p \nu_i A_i$. To form $\nabla^2 g(\nu)$ and $\nabla g(\nu)$ we proceed as follows. We first form A (pn^2 flops), and $A^{-1} A_j$ for each j ($2pn^3$ flops). Then we form the matrix $\nabla^2 g(\nu)$. Each of the $p(p+1)/2$ entries of $\nabla^2 g(\nu)$ is the inner product of two matrices in $\mathbf{S}^n$, each of which costs $n(n+1)$ flops, so the total is (dropping dominated terms) $(1/2)p^2 n^2$ flops. Forming $\nabla g(\nu)$ is cheap since we already have the matrices $A^{-1} A_i$. Finally, we solve for the Newton step $-\nabla^2 g(\nu)^{-1} \nabla g(\nu)$, which costs $(1/3)p^3$ flops. All together, and keeping only the leading terms, the total cost of computing the Newton step is $2pn^3 + (1/2)p^2 n^2 + (1/3)p^3$. Note that this is order n^3 in n, which is far better than the simple primal method described above, which is order n^6.

We can also solve the primal problem more efficiently, by exploiting its special matrix structure. To derive the KKT system for the Newton step ΔX_{nt} at a feasible X, we replace $X^\star$ in the KKT conditions by $X + \Delta X_{\mathrm{nt}}$ and $\nu^\star$ by w, and linearize the first equation using the first-order approximation

$$(X + \Delta X_{\mathrm{nt}})^{-1} \approx X^{-1} - X^{-1} \Delta X_{\mathrm{nt}} X^{-1}.$$

This gives the KKT system

$$-X^{-1} + X^{-1} \Delta X_{\mathrm{nt}} X^{-1} + \sum_{i=1}^p w_i A_i = 0, \qquad \mathbf{tr}(A_i \Delta X_{\mathrm{nt}}) = 0, \quad i = 1, \dots, p.$$

$$(10.40)$$

This is a set of $n(n+1)/2 + p$ linear equations in the variables $\Delta X_{\mathrm{nt}} \in \mathbf{S}^n$ and $w \in \mathbf{R}^p$. If we solved these equations using a generic method, the cost would be order n^6.

We can use block elimination to solve the KKT system (10.40) far more efficiently. We eliminate the variable ΔX_{nt}, by solving the first equation to get

$$\Delta X_{\mathrm{nt}} = X - X \left(\sum_{i=1}^{p} w_i A_i \right) X = X - \sum_{i=1}^{p} w_i X A_i X. \qquad (10.41)$$

Substituting this expression for ΔX_{nt} into the other equation gives

$$\mathbf{tr}(A_j \Delta X_{\mathrm{nt}}) = \mathbf{tr}(A_j X) - \sum_{i=1}^{p} w_i \, \mathbf{tr}(A_j X A_i X) = 0, \quad j = 1, \ldots, p.$$

This is a set of p linear equations in w:

$$Cw = d$$

where $C_{ij} = \mathbf{tr}(A_i X A_j X)$, $d_i = \mathbf{tr}(A_i X)$. The coefficient matrix C is symmetric and positive definite, so a Cholesky factorization can be used to find w. Once we have w, we can compute ΔX_{nt} from (10.41).

The cost of this method is as follows. We form the products $A_i X$ ($2pn^3$ flops), and then form the matrix C. Each of the $p(p+1)/2$ entries of C is the inner product of two matrices in $\mathbf{R}^{n \times n}$, so forming C costs $p^2 n^2$ flops. Then we solve for $w = C^{-1} d$, which costs $(1/3)p^3$. Finally we compute ΔX_{nt}. If we use the first expression in (10.41), i.e., first compute the sum and then pre- and post-multiply with X, the cost is approximately $pn^2 + 3n^3$. All together, the total cost is $2pn^3 + p^2 n^2 + (1/3)p^3$ flops to form the Newton step for the primal problem, using block elimination. This is far better than the simple method, which is order n^6. Note also that the cost is the same as that of computing the Newton step for the dual problem.

Bibliography

The two key assumptions in our analysis of the infeasible start Newton method (the derivative Dr has a bounded inverse and satisfies a Lipschitz condition) are central to most convergence proofs of Newton's method; see Ortega and Rheinboldt [OR00] and Dennis and Schnabel [DS96].

The relative merits of solving KKT systems via direct factorization of the full system, or via elimination, have been extensively studied in the context of interior-point methods for linear and quadratic programming; see, for example, Wright [Wri97, chapter 11] and Nocedal and Wright [NW99, §16.1-2]. The Riccati recursion from optimal control can be interpreted as a method for exploiting the block tridiagonal structure in the Schur complement S of the example on page 552. This observation was made by Rao, Wright, and Rawlings [RWR98, §3.3].

Exercises

Equality constrained minimization

10.1 *Nonsingularity of the KKT matrix.* Consider the KKT matrix

$$\begin{bmatrix} P & A^T \\ A & 0 \end{bmatrix},$$

where $P \in \mathbf{S}_+^n$, $A \in \mathbf{R}^{p \times n}$, and $\mathbf{rank}\, A = p < n$.

(a) Show that each of the following statements is equivalent to nonsingularity of the KKT matrix.

- $\mathcal{N}(P) \cap \mathcal{N}(A) = \{0\}$.
- $Ax = 0$, $x \neq 0 \Longrightarrow x^T P x > 0$.
- $F^T P F \succ 0$, where $F \in \mathbf{R}^{n \times (n-p)}$ is a matrix for which $\mathcal{R}(F) = \mathcal{N}(A)$.
- $P + A^T Q A \succ 0$ for some $Q \succeq 0$.

(b) Show that if the KKT matrix is nonsingular, then it has exactly n positive and p negative eigenvalues.

10.2 *Projected gradient method.* In this problem we explore an extension of the gradient method to equality constrained minimization problems. Suppose f is convex and differentiable, and $x \in \mathbf{dom}\, f$ satisfies $Ax = b$, where $A \in \mathbf{R}^{p \times n}$ with $\mathbf{rank}\, A = p < n$. The Euclidean projection of the negative gradient $-\nabla f(x)$ on $\mathcal{N}(A)$ is given by

$$\Delta x_{\mathrm{pg}} = \underset{Au=0}{\mathrm{argmin}} \, \|-\nabla f(x) - u\|_2.$$

(a) Let (v, w) be the unique solution of

$$\begin{bmatrix} I & A^T \\ A & 0 \end{bmatrix} \begin{bmatrix} v \\ w \end{bmatrix} = \begin{bmatrix} -\nabla f(x) \\ 0 \end{bmatrix}.$$

Show that $v = \Delta x_{\mathrm{pg}}$ and $w = \mathrm{argmin}_y \, \|\nabla f(x) + A^T y\|_2$.

(b) What is the relation between the projected negative gradient Δx_{pg} and the negative gradient of the reduced problem (10.5), assuming $F^T F = I$?

(c) The *projected gradient method* for solving an equality constrained minimization problem uses the step Δx_{pg}, and a backtracking line search on f. Use the results of part (b) to give some conditions under which the projected gradient method converges to the optimal solution, when started from a point $x^{(0)} \in \mathbf{dom}\, f$ with $Ax^{(0)} = b$.

Newton's method with equality constraints

10.3 *Dual Newton method.* In this problem we explore Newton's method for solving the dual of the equality constrained minimization problem (10.1). We assume that f is twice differentiable, $\nabla^2 f(x) \succ 0$ for all $x \in \mathbf{dom}\, f$, and that for each $\nu \in \mathbf{R}^p$, the Lagrangian $L(x, \nu) = f(x) + \nu^T(Ax - b)$ has a unique minimizer, which we denote $x(\nu)$.

(a) Show that the dual function g is twice differentiable. Find an expression for the Newton step for the dual function g, evaluated at ν, in terms of f, ∇f, and $\nabla^2 f$, evaluated at $x = x(\nu)$. You can use the results of exercise 3.40.

(b) Suppose there exists a K such that

$$\left\| \begin{bmatrix} \nabla^2 f(x) & A^T \\ A & 0 \end{bmatrix}^{-1} \right\|_2 \leq K$$

for all $x \in \mathbf{dom}\, f$. Show that g is strongly concave, with $\nabla^2 g(\nu) \preceq -(1/K)I$.

10.4 *Strong convexity and Lipschitz constant of the reduced problem.* Suppose f satisfies the assumptions given on page 529. Show that the reduced objective function $\tilde{f}(z) = f(Fz+\hat{x})$ is strongly convex, and that its Hessian is Lipschitz continuous (on the associated sublevel set $\tilde{S}$). Express the strong convexity and Lipschitz constants of $\tilde{f}$ in terms of K, M, L, and the maximum and minimum singular values of F.

10.5 *Adding a quadratic term to the objective.* Suppose $Q \succeq 0$. The problem

$$\begin{array}{ll} \text{minimize} & f(x) + (Ax - b)^T Q(Ax - b) \\ \text{subject to} & Ax = b \end{array}$$

is equivalent to the original equality constrained optimization problem (10.1). Is the Newton step for this problem the same as the Newton step for the original problem?

10.6 *The Newton decrement.* Show that (10.13) holds, *i.e.*,

$$f(x) - \inf\{\widehat{f}(x + v) \mid A(x + v) = b\} = \lambda(x)^2/2.$$

Infeasible start Newton method

10.7 *Assumptions for infeasible start Newton method.* Consider the set of assumptions given on page 536.

(a) Suppose that the function f is closed. Show that this implies that the norm of the residual, $\|r(x,\nu)\|_2$, is closed.

(b) Show that Dr satisfies a Lipschitz condition if and only if $\nabla^2 f$ does.

10.8 *Infeasible start Newton method and initially satisfied equality constraints.* Suppose we use the infeasible start Newton method to minimize $f(x)$ subject to $a_i^T x = b_i$, $i = 1, \ldots, p$.

(a) Suppose the initial point $x^{(0)}$ satisfies the linear equality $a_i^T x = b_i$. Show that the linear equality will remain satisfied for future iterates, *i.e.*, if $a_i^T x^{(k)} = b_i$ for all k.

(b) Suppose that one of the equality constraints becomes satisfied at iteration k, *i.e.*, we have $a_i^T x^{(k-1)} \neq b_i$, $a_i^T x^{(k)} = b_i$. Show that at iteration k, *all* the equality constraints are satisfied.

10.9 *Equality constrained entropy maximization.* Consider the equality constrained entropy maximization problem

$$\begin{array}{ll} \text{minimize} & f(x) = \sum_{i=1}^{n} x_i \log x_i \\ \text{subject to} & Ax = b, \end{array} \tag{10.42}$$

with $\mathbf{dom}\, f = \mathbf{R}^n_{++}$ and $A \in \mathbf{R}^{p \times n}$. We assume the problem is feasible and that $\mathbf{rank}\, A = p < n$.

(a) Show that the problem has a unique optimal solution $x^\star$.

(b) Find A, b, and feasible $x^{(0)}$ for which the sublevel set

$$\{x \in \mathbf{R}^n_{++} \mid Ax = b,\ f(x) \leq f(x^{(0)})\}$$

is *not* closed. Thus, the assumptions listed in §10.2.4, page 529, are not satisfied for some feasible initial points.

(c) Show that the problem (10.42) satisfies the assumptions for the infeasible start Newton method listed in §10.3.3, page 536, for any feasible starting point.

(d) Derive the Lagrange dual of (10.42), and explain how to find the optimal solution of (10.42) from the optimal solution of the dual problem. Show that the dual problem satisfies the assumptions listed in §10.2.4, page 529, for *any* starting point.

The results of part (b), (c), and (d) do not mean the standard Newton method will fail, or that the infeasible start Newton method or dual method will work better in practice. It only means our convergence analysis for the standard Newton method does not apply, while our convergence analysis does apply to the infeasible start and dual methods. (See exercise 10.15.)

10.10 *Bounded inverse derivative condition for strongly convex-concave game.* Consider a convex-concave game with payoff function f (see page 541). Suppose $\nabla^2_{uu} f(u, v) \succeq mI$ and $\nabla^2_{vv} f(u, v) \preceq -mI$, for all $(u, v) \in \mathbf{dom}\, f$. Show that

$$\| Dr(u, v)^{-1} \|_2 = \| \nabla^2 f(u, v)^{-1} \|_2 \le 1/m.$$

Implementation

10.11 Consider the resource allocation problem described in example 10.1. You can assume the f_i are strongly convex, i.e., $f_i''(z) \ge m > 0$ for all z.

(a) Find the computational effort required to compute a Newton step for the reduced problem. Be sure to exploit the special structure of the Newton equations.

(b) Explain how to solve the problem via the dual. You can assume that the conjugate functions f_i^*, and their derivatives, are readily computable, and that the equation $f_i'(x) = \nu$ is readily solved for x, given ν. What is the computational complexity of finding a Newton step for the dual problem?

(c) What is the computational complexity of computing a Newton step for the resource allocation problem? Be sure to exploit the special structure of the KKT equations.

10.12 Describe an efficient way to compute the Newton step for the problem

$$\begin{aligned}
\text{minimize} \quad & \mathbf{tr}(X^{-1}) \\
\text{subject to} \quad & \mathbf{tr}(A_i X) = b_i, \quad i = 1, \ldots, p
\end{aligned}$$

with domain $\mathbf{S}^n_{++}$, assuming p and n have the same order of magnitude. Also derive the Lagrange dual problem and give the complexity of finding the Newton step for the dual problem.

10.13 *Elimination method for computing Newton step for convex-concave game.* Consider a convex-concave game with payoff function $f : \mathbf{R}^p \times \mathbf{R}^q \to \mathbf{R}$ (see page 541). We assume that f is *strongly convex-concave*, i.e., for all $(u, v) \in \mathbf{dom}\, f$ and some $m > 0$, we have $\nabla^2_{uu} f(u, v) \succeq mI$ and $\nabla^2_{vv} f(u, v) \preceq -mI$.

(a) Show how to compute the Newton step using Cholesky factorizations of $\nabla^2_{uu} f(u, v)$ and $-\nabla^2 f_{vv}(u, v)$. Compare the cost of this method with the cost of using an LDL$^\mathrm{T}$ factorization of $\nabla f(u, v)$, assuming $\nabla^2 f(u, v)$ is dense.

(b) Show how you can exploit diagonal or block diagonal structure in $\nabla^2_{uu} f(u, v)$ and/or $\nabla^2_{vv} f(u, v)$. How much do you save, if you assume $\nabla^2_{uv} f(u, v)$ is dense?

Numerical experiments

10.14 *Log-optimal investment.* Consider the log-optimal investment problem described in exercise 4.60, without the constraint $x \succeq 0$. Use Newton's method to compute the solution, with the following problem data: there are $n = 3$ assets, and $m = 4$ scenarios, with returns

$$p_1 = \begin{bmatrix} 2 \\ 1.3 \\ 1 \end{bmatrix}, \qquad p_2 = \begin{bmatrix} 2 \\ 0.5 \\ 1 \end{bmatrix}, \qquad p_3 = \begin{bmatrix} 0.5 \\ 1.3 \\ 1 \end{bmatrix}, \qquad p_4 = \begin{bmatrix} 0.5 \\ 0.5 \\ 1 \end{bmatrix}.$$

The probabilities of the four scenarios are given by $\pi = (1/3, 1/6, 1/3, 1/6)$.

10.15 *Equality constrained entropy maximization.* Consider the equality constrained entropy maximization problem

$$\begin{array}{ll} \text{minimize} & f(x) = \sum_{i=1}^{n} x_i \log x_i \\ \text{subject to} & Ax = b, \end{array}$$

with $\mathbf{dom}\, f = \mathbf{R}^n_{++}$ and $A \in \mathbf{R}^{p \times n}$, with $p < n$. (See exercise 10.9 for some relevant analysis.)

Generate a problem instance with $n = 100$ and $p = 30$ by choosing A randomly (checking that it has full rank), choosing $\hat{x}$ as a random positive vector (*e.g.*, with entries uniformly distributed on $[0, 1]$) and then setting $b = A\hat{x}$. (Thus, $\hat{x}$ is feasible.)

Compute the solution of the problem using the following methods.

(a) *Standard Newton method.* You can use initial point $x^{(0)} = \hat{x}$.

(b) *Infeasible start Newton method.* You can use initial point $x^{(0)} = \hat{x}$ (to compare with the standard Newton method), and also the initial point $x^{(0)} = \mathbf{1}$.

(c) *Dual Newton method, i.e.,* the standard Newton method applied to the dual problem.

Verify that the three methods compute the same optimal point (and Lagrange multiplier). Compare the computational effort per step for the three methods, assuming relevant structure is exploited. (Your implementation, however, does not need to exploit structure to compute the Newton step.)

10.16 *Convex-concave game.* Use the infeasible start Newton method to solve convex-concave games of the form (10.32), with randomly generated data. Plot the norm of the residual and step length versus iteration. Experiment with the line search parameters and initial point (which must satisfy $\|u\|_2 < 1$, $\|v\|_2 < 1$, however).

Chapter 11

Interior-point methods

11.1 Inequality constrained minimization problems

In this chapter we discuss *interior-point methods* for solving convex optimization problems that include inequality constraints,

$$
\begin{array}{ll}
\text{minimize} & f_0(x) \\
\text{subject to} & f_i(x) \le 0, \quad i = 1, \dots, m \\
& Ax = b,
\end{array} \tag{11.1}
$$

where $f_0, \dots, f_m : \mathbf{R}^n \to \mathbf{R}$ are convex and twice continuously differentiable, and $A \in \mathbf{R}^{p \times n}$ with $\mathbf{rank}\, A = p < n$. We assume that the problem is solvable, *i.e.*, an optimal $x^\star$ exists. We denote the optimal value $f_0(x^\star)$ as $p^\star$.

We also assume that the problem is strictly feasible, *i.e.*, there exists $x \in \mathcal{D}$ that satisfies $Ax = b$ and $f_i(x) < 0$ for $i = 1, \dots, m$. This means that Slater's constraint qualification holds, so there exist dual optimal $\lambda^\star \in \mathbf{R}^m$, $\nu^\star \in \mathbf{R}^p$, which together with $x^\star$ satisfy the KKT conditions

$$
\begin{array}{rcll}
Ax^\star = b, \quad f_i(x^\star) & \le & 0, & i = 1, \dots, m \\
\lambda^\star & \succeq & 0 \\
\nabla f_0(x^\star) + \sum_{i=1}^m \lambda_i^\star \nabla f_i(x^\star) + A^T \nu^\star & = & 0 \\
\lambda_i^\star f_i(x^\star) & = & 0, & i = 1, \dots, m.
\end{array} \tag{11.2}
$$

Interior-point methods solve the problem (11.1) (or the KKT conditions (11.2)) by applying Newton's method to a sequence of equality constrained problems, or to a sequence of modified versions of the KKT conditions. We will concentrate on a particular interior-point algorithm, the *barrier method*, for which we give a proof of convergence and a complexity analysis. We also describe a simple *primal-dual interior-point method* (in §11.7), but do not give an analysis.

We can view interior-point methods as another level in the hierarchy of convex optimization algorithms. Linear equality constrained quadratic problems are the simplest. For these problems the KKT conditions are a set of linear equations, which can be solved analytically. Newton's method is the next level in the hierarchy. We can think of Newton's method as a technique for solving a linear equality constrained optimization problem, with twice differentiable objective, by reducing it to a sequence of linear equality constrained quadratic problems. Interior-point methods form the next level in the hierarchy: They solve an optimization problem with linear equality and inequality constraints by reducing it to a sequence of linear equality constrained problems.

Examples

Many problems are already in the form (11.1), and satisfy the assumption that the objective and constraint functions are twice differentiable. Obvious examples are LPs, QPs, QCQPs, and GPs in convex form; another example is linear inequality constrained entropy maximization,

$$
\begin{array}{ll}
\text{minimize} & \sum_{i=1}^{n} x_i \log x_i \\
\text{subject to} & Fx \preceq g \\
& Ax = b,
\end{array}
$$

with domain $\mathcal{D} = \mathbf{R}_{++}^n$.

Many other problems do not have the required form (11.1), with twice differentiable objective and constraint functions, but can be reformulated in the required form. We have already seen many examples of this, such as the transformation of an unconstrained convex piecewise-linear minimization problem

$$
\text{minimize} \quad \max_{i=1,\dots,m} (a_i^T x + b_i)
$$

(with nondifferentiable objective), to the LP

$$
\begin{array}{ll}
\text{minimize} & t \\
\text{subject to} & a_i^T x + b_i \leq t, \quad i = 1, \dots, m
\end{array}
$$

(which has twice differentiable objective and constraint functions).

Other convex optimization problems, such as SOCPs and SDPs, are not readily recast in the required form, but can be handled by extensions of interior-point methods to problems with generalized inequalities, which we describe in §11.6.

11.2   Logarithmic barrier function and central path

Our goal is to approximately formulate the inequality constrained problem (11.1) as an equality constrained problem to which Newton's method can be applied. Our first step is to rewrite the problem (11.1), making the inequality constraints implicit in the objective:

$$
\begin{array}{ll}
\text{minimize} & f_0(x) + \sum_{i=1}^{m} I_-(f_i(x)) \\
\text{subject to} & Ax = b,
\end{array}
\tag{11.3}
$$

where $I_- : \mathbf{R} \to \mathbf{R}$ is the indicator function for the nonpositive reals,

$$
I_-(u) = \begin{cases} 0 & u \leq 0 \\ \infty & u > 0. \end{cases}
$$

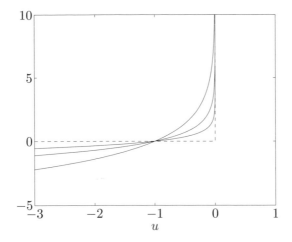

Figure 11.1 The dashed lines show the function $I_-(u)$, and the solid curves show $\widehat{I}_-(u) = -(1/t)\log(-u)$, for $t = 0.5,\ 1,\ 2$. The curve for $t = 2$ gives the best approximation.

The problem (11.3) has no inequality constraints, but its objective function is not (in general) differentiable, so Newton's method cannot be applied.

1.2.1   Logarithmic barrier

The basic idea of the barrier method is to approximate the indicator function I_- by the function

$$\widehat{I}_-(u) = -(1/t)\log(-u), \qquad \mathbf{dom}\,\widehat{I}_- = -\mathbf{R}_{++},$$

where $t > 0$ is a parameter that sets the accuracy of the approximation. Like I_-, the function $\widehat{I}_-$ is convex and nondecreasing, and (by our convention) takes on the value ∞ for $u > 0$. Unlike I_-, however, $\widehat{I}_-$ is differentiable and closed: it increases to ∞ as u increases to 0. Figure 11.1 shows the function I_-, and the approximation $\widehat{I}_-$, for several values of t. As t increases, the approximation becomes more accurate.

Substituting $\widehat{I}_-$ for I_- in (11.3) gives the approximation

$$
\begin{array}{ll}
\text{minimize} & f_0(x) + \sum_{i=1}^m -(1/t)\log(-f_i(x)) \\
\text{subject to} & Ax = b.
\end{array}
\tag{11.4}
$$

The objective here is convex, since $-(1/t)\log(-u)$ is convex and increasing in u, and differentiable. Assuming an appropriate closedness condition holds, Newton's method can be used to solve it.

The function

$$\phi(x) = -\sum_{i=1}^m \log(-f_i(x)),\tag{11.5}$$

with $\mathbf{dom}\,\phi = \{x \in \mathbf{R}^n \mid f_i(x) < 0,\ i = 1,\ldots,m\}$, is called the *logarithmic barrier* or *log barrier* for the problem (11.1). Its domain is the set of points that satisfy the inequality constraints of (11.1) strictly. No matter what value the positive parameter t has, the logarithmic barrier grows without bound if $f_i(x) \to 0$, for any i.

Of course, the problem (11.4) is only an approximation of the original problem (11.3), so one question that arises immediately is how well a solution of (11.4) approximates a solution of the original problem (11.3). Intuition suggests, and we will soon confirm, that the quality of the approximation improves as the parameter t grows.

On the other hand, when the parameter t is large, the function $f_0 + (1/t)\phi$ is difficult to minimize by Newton's method, since its Hessian varies rapidly near the boundary of the feasible set. We will see that this problem can be circumvented by solving a *sequence* of problems of the form (11.4), increasing the parameter t (and therefore the accuracy of the approximation) at each step, and starting each Newton minimization at the solution of the problem for the previous value of t.

For future reference, we note that the gradient and Hessian of the logarithmic barrier function ϕ are given by

$$\nabla\phi(x) \;=\; \sum_{i=1}^m \frac{1}{-f_i(x)}\nabla f_i(x),$$

$$\nabla^2\phi(x) \;=\; \sum_{i=1}^m \frac{1}{f_i(x)^2}\nabla f_i(x)\nabla f_i(x)^T + \sum_{i=1}^m \frac{1}{-f_i(x)}\nabla^2 f_i(x)$$

(see §A.4.2 and §A.4.4).

11.2.2 Central path

We now consider in more detail the minimization problem (11.4). It will simplify notation later on if we multiply the objective by t, and consider the equivalent problem

$$\begin{array}{ll}\text{minimize} & tf_0(x) + \phi(x) \\ \text{subject to} & Ax = b,\end{array} \tag{11.6}$$

which has the same minimizers. We assume for now that the problem (11.6) can be solved via Newton's method, and, in particular, that it has a unique solution for each $t > 0$. (We will discuss this assumption in more detail in §11.3.3.)

For $t > 0$ we define $x^\star(t)$ as the solution of (11.6). The *central path* associated with problem (11.1) is defined as the set of points $x^\star(t)$, $t > 0$, which we call the *central points*. Points on the central path are characterized by the following necessary and sufficient conditions: $x^\star(t)$ is strictly feasible, *i.e.*, satisfies

$$Ax^\star(t) = b, \qquad f_i(x^\star(t)) < 0, \quad i = 1,\ldots,m,$$

and there exists a $\hat{\nu} \in \mathbf{R}^p$ such that

$$0 \;=\; t\nabla f_0(x^\star(t)) + \nabla\phi(x^\star(t)) + A^T\hat{\nu}$$

$$= t\nabla f_0(x^\star(t)) + \sum_{i=1}^{m} \frac{1}{-f_i(x^\star(t))} \nabla f_i(x^\star(t)) + A^T \hat{\nu} \qquad (11.7)$$

holds.

Example 11.1 *Inequality form linear programming.* The logarithmic barrier function for an LP in inequality form,

$$\begin{array}{ll} \text{minimize} & c^T x \\ \text{subject to} & Ax \preceq b, \end{array} \qquad (11.8)$$

is given by

$$\phi(x) = -\sum_{i=1}^{m} \log(b_i - a_i^T x), \qquad \mathbf{dom}\, \phi = \{x \mid Ax \prec b\},$$

where $a_1^T, \ldots, a_m^T$ are the rows of A. The gradient and Hessian of the barrier function are

$$\nabla\phi(x) = \sum_{i=1}^{m} \frac{1}{b_i - a_i^T x} a_i, \qquad \nabla^2\phi(x) = \sum_{i=1}^{m} \frac{1}{(b_i - a_i^T x)^2} a_i a_i^T,$$

or, more compactly,

$$\nabla\phi(x) = A^T d, \qquad \nabla^2\phi(x) = A^T \mathbf{diag}(d)^2 A,$$

where the elements of $d \in \mathbf{R}^m$ are given by $d_i = 1/(b_i - a_i^T x)$. Since x is strictly feasible, we have $d \succ 0$, so the Hessian of ϕ is nonsingular if and only if A has rank n.

The centrality condition (11.7) is

$$tc + \sum_{i=1}^{m} \frac{1}{b_i - a_i^T x} a_i = tc + A^T d = 0. \qquad (11.9)$$

We can give a simple geometric interpretation of the centrality condition. At a point $x^\star(t)$ on the central path the gradient $\nabla\phi(x^\star(t))$, which is normal to the level set of ϕ through $x^\star(t)$, must be parallel to $-c$. In other words, the hyperplane $c^T x = c^T x^\star(t)$ is tangent to the level set of ϕ through $x^\star(t)$. Figure 11.2 shows an example with $m = 6$ and $n = 2$.

Dual points from central path

From (11.7) we can derive an important property of the central path: Every central point yields a dual feasible point, and hence a lower bound on the optimal value $p^\star$. More specifically, define

$$\lambda_i^\star(t) = -\frac{1}{t f_i(x^\star(t))}, \quad i = 1, \ldots, m, \qquad \nu^\star(t) = \hat{\nu}/t. \qquad (11.10)$$

We claim that the pair $\lambda^\star(t)$, $\nu^\star(t)$ is dual feasible.

First, it is clear that $\lambda^\star(t) \succ 0$ because $f_i(x^\star(t)) < 0$, $i = 1, \ldots, m$. By expressing the optimality conditions (11.7) as

$$\nabla f_0(x^\star(t)) + \sum_{i=1}^{m} \lambda_i^\star(t) \nabla f_i(x^\star(t)) + A^T \nu^\star(t) = 0,$$

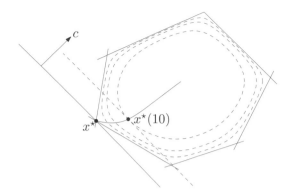

Figure 11.2 Central path for an LP with $n = 2$ and $m = 6$. The dashed curves show three contour lines of the logarithmic barrier function ϕ. The central path converges to the optimal point $x^\star$ as $t \to \infty$. Also shown is the point on the central path with $t = 10$. The optimality condition (11.9) at this point can be verified geometrically: The line $c^T x = c^T x^\star(10)$ is tangent to the contour line of ϕ through $x^\star(10)$.

we see that $x^\star(t)$ minimizes the Lagrangian

$$L(x, \lambda, \nu) = f_0(x) + \sum_{i=1}^{m} \lambda_i f_i(x) + \nu^T(Ax - b),$$

for $\lambda = \lambda^\star(t)$ and $\nu = \nu^\star(t)$, which means that $\lambda^\star(t)$, $\nu^\star(t)$ is a dual feasible pair. Therefore the dual function $g(\lambda^\star(t), \nu^\star(t))$ is finite, and

$$
\begin{aligned}
g(\lambda^\star(t), \nu^\star(t)) &= f_0(x^\star(t)) + \sum_{i=1}^{m} \lambda_i^\star(t) f_i(x^\star(t)) + \nu^\star(t)^T(Ax^\star(t) - b) \\
&= f_0(x^\star(t)) - m/t.
\end{aligned}
$$

In particular, the duality gap associated with $x^\star(t)$ and the dual feasible pair $\lambda^\star(t)$, $\nu^\star(t)$ is simply m/t. As an important consequence, we have

$$f_0(x^\star(t)) - p^\star \leq m/t,$$

i.e., $x^\star(t)$ is no more than m/t-suboptimal. This confirms the intuitive idea that $x^\star(t)$ converges to an optimal point as $t \to \infty$.

Example 11.2 *Inequality form linear programming.* The dual of the inequality form LP (11.8) is

$$
\begin{array}{ll}
\text{maximize} & -b^T \lambda \\
\text{subject to} & A^T \lambda + c = 0 \\
& \lambda \succeq 0.
\end{array}
$$

From the optimality conditions (11.9), it is clear that

$$\lambda_i^\star(t) = \frac{1}{t(b_i - a_i^T x^\star(t))}, \quad i = 1, \ldots, m,$$

is dual feasible, with dual objective value

$$-b^T \lambda^\star(t) = c^T x^\star(t) + (Ax^\star(t) - b)^T \lambda^\star(t) = c^T x^\star(t) - m/t.$$

Interpretation via KKT conditions

We can also interpret the central path conditions (11.7) as a continuous deformation of the KKT optimality conditions (11.2). A point x is equal to $x^\star(t)$ if and only if there exists λ, ν such that

$$
\begin{aligned}
Ax = b, \quad f_i(x) &\leq 0, \quad i = 1, \ldots, m \\
\lambda &\succeq 0 \\
\nabla f_0(x) + \sum_{i=1}^m \lambda_i \nabla f_i(x) + A^T \nu &= 0 \\
-\lambda_i f_i(x) &= 1/t, \quad i = 1, \ldots, m.
\end{aligned}
\tag{11.11}
$$

The only difference between the KKT conditions (11.2) and the centrality conditions (11.11) is that the complementarity condition $-\lambda_i f_i(x) = 0$ is replaced by the condition $-\lambda_i f_i(x) = 1/t$. In particular, for large t, $x^\star(t)$ and the associated dual point $\lambda^\star(t)$, $\nu^\star(t)$ 'almost' satisfy the KKT optimality conditions for (11.1).

Force field interpretation

We can give a simple mechanics interpretation of the central path in terms of potential forces acting on a particle in the strictly feasible set C. For simplicity we assume that there are no equality constraints.

We associate with each constraint the force

$$F_i(x) = -\nabla \left(-\log(-f_i(x)) \right) = \frac{1}{f_i(x)} \nabla f_i(x)$$

acting on the particle when it is at position x. The potential associated with the total force field generated by the constraints is the logarithmic barrier ϕ. As the particle moves toward the boundary of the feasible set, it is strongly repelled by the forces generated by the constraints.

Now we imagine another force acting on the particle, given by

$$F_0(x) = -t \nabla f_0(x),$$

when the particle is at position x. This objective force field acts to pull the particle in the negative gradient direction, *i.e.*, toward smaller f_0. The parameter t scales the objective force, relative to the constraint forces.

The central point $x^\star(t)$ is the point where the constraint forces exactly balance the objective force felt by the particle. As the parameter t increases, the particle is more strongly pulled toward the optimal point, but it is always trapped in C by the barrier potential, which becomes infinite as the particle approaches the boundary.

Example 11.3 *Force field interpretation for inequality form LP.* The force field associated with the ith constraint of the LP (11.8) is

$$F_i(x) = \frac{-a_i}{b_i - a_i^T x}.$$

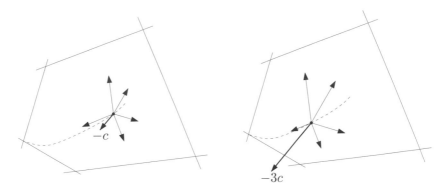

Figure 11.3 *Force field interpretation of central path.* The central path is shown as the dashed curve. The two points $x^\star(1)$ and $x^\star(3)$ are shown as dots in the left and right plots, respectively. The objective force, which is equal to $-c$ and $-3c$, respectively, is shown as a heavy arrow. The other arrows represent the constraint forces, which are given by an inverse-distance law. As the strength of the objective force varies, the equilibrium position of the particle traces out the central path.

This force is in the direction of the inward pointing normal to the constraint plane $\mathcal{H}_i = \{x \mid a_i^T x = b_i\}$, and has magnitude inversely proportional to the distance to $\mathcal{H}_i$, *i.e.*,

$$\|F_i(x)\|_2 = \frac{\|a_i\|_2}{b_i - a_i^T x} = \frac{1}{\mathbf{dist}(x, \mathcal{H}_i)}.$$

In other words, each constraint hyperplane has an associated repulsive force, given by the inverse distance to the hyperplane.

The term $tc^T x$ is the potential associated with a constant force $-tc$ on the particle. This 'objective force' pushes the particle in the direction of low cost. Thus, $x^\star(t)$ is the equilibrium position of the particle when it is subject to the inverse-distance constraint forces, and the objective force $-tc$. When t is very large, the particle is pushed almost to the optimal point. The strong objective force is balanced by the opposing constraint forces, which are large because we are near the feasible boundary.

Figure 11.3 illustrates this interpretation for a small LP with $n = 2$ and $m = 5$. The lefthand plot shows $x^\star(t)$ for $t = 1$, as well as the constraint forces acting on it, which balance the objective force. The righthand plot shows $x^\star(t)$ and the associated forces for $t = 3$. The larger value of objective force moves the particle closer to the optimal point.

11.3   The barrier method

We have seen that the point $x^\star(t)$ is m/t-suboptimal, and that a certificate of this accuracy is provided by the dual feasible pair $\lambda^\star(t)$, $\nu^\star(t)$. This suggests a very straightforward method for solving the original problem (11.1) with a guaranteed specified accuracy ϵ: We simply take $t = m/\epsilon$ and solve the equality constrained problem

$$\begin{array}{ll}\text{minimize} & (m/\epsilon)f_0(x) + \phi(x) \\ \text{subject to} & Ax = b\end{array}$$

using Newton's method. This method could be called the *unconstrained minimization method,* since it allows us to solve the inequality constrained problem (11.1) to a guaranteed accuracy by solving an unconstrained, or linearly constrained, problem. Although this method can work well for small problems, good starting points, and moderate accuracy (*i.e.,* ϵ not too small), it does not work well in other cases. As a result it is rarely, if ever, used.

1.3.1 The barrier method

A simple extension of the unconstrained minimization method does work well. It is based on solving a sequence of unconstrained (or linearly constrained) minimization problems, using the last point found as the starting point for the next unconstrained minimization problem. In other words, we compute $x^\star(t)$ for a sequence of increasing values of t, until $t \geq m/\epsilon$, which guarantees that we have an ϵ-suboptimal solution of the original problem. When the method was first proposed by Fiacco and McCormick in the 1960s, it was called the *sequential unconstrained minimization technique* (SUMT). Today the method is usually called the *barrier method* or *path-following method.* A simple version of the method is as follows.

Algorithm 11.1 *Barrier method.*

given strictly feasible x, $t := t^{(0)} > 0$, $\mu > 1$, tolerance $\epsilon > 0$.

repeat
    1. *Centering step.*
          Compute $x^\star(t)$ by minimizing $tf_0 + \phi$, subject to $Ax = b$, starting at x.
    2. *Update.* $x := x^\star(t)$.
    3. *Stopping criterion.* **quit** if $m/t < \epsilon$.
    4. *Increase t.* $t := \mu t$.

At each iteration (except the first one) we compute the central point $x^\star(t)$ starting from the previously computed central point, and then increase t by a factor $\mu > 1$. The algorithm can also return $\lambda = \lambda^\star(t)$, and $\nu = \nu^\star(t)$, a dual ϵ-suboptimal point, or certificate for x.

    We refer to each execution of step 1 as a *centering step* (since a central point is being computed) or an *outer iteration,* and to the first centering step (the computation of $x^\star(t^{(0)})$) as the *initial centering step.* (Thus the simple algorithm with $t^{(0)} = m/\epsilon$ consists of only the initial centering step.) Although any method for linearly constrained minimization can be used in step 1, we will assume that Newton's method is used. We refer to the Newton iterations or steps executed during the centering step as *inner iterations.* At each inner step, we have a primal feasible point; we have a dual feasible point, however, only at the end of each outer (centering) step.

Accuracy of centering

We should make some comments on the accuracy to which we solve the centering problems. Computing $x^\star(t)$ exactly is not necessary since the central path has no significance beyond the fact that it leads to a solution of the original problem as $t \to \infty$; inexact centering will still yield a sequence of points $x^{(k)}$ that converges to an optimal point. Inexact centering, however, means that the points $\lambda^\star(t)$, $\nu^\star(t)$, computed from (11.10), are not exactly dual feasible. This can be corrected by adding a correction term to the formula (11.10), which yields a dual feasible point provided the computed x is near the central path, i.e., $x^\star(t)$ (see exercise 11.9).

On the other hand, the cost of computing an *extremely accurate* minimizer of $tf_0 + \phi$, as compared to the cost of computing a *good* minimizer of $tf_0 + \phi$, is only marginally more, i.e., a few Newton steps at most. For this reason it is not unreasonable to assume exact centering.

Choice of μ

The choice of the parameter μ involves a trade-off in the number of inner and outer iterations required. If μ is small (i.e., near 1) then at each outer iteration t increases by a small factor. As a result the initial point for the Newton process, i.e., the previous iterate x, is a very good starting point, and the number of Newton steps needed to compute the next iterate is small. Thus for small μ we expect a small number of Newton steps per outer iteration, but of course a large number of outer iterations since each outer iteration reduces the gap by only a small amount. In this case the iterates (and indeed, the iterates of the inner iterations as well) closely follow the central path. This explains the alternate name *path-following method*.

On the other hand if μ is large we have the opposite situation. After each outer iteration t increases a large amount, so the current iterate is probably not a very good approximation of the next iterate. Thus we expect many more inner iterations. This 'aggressive' updating of t results in fewer outer iterations, since the duality gap is reduced by the large factor μ at each outer iteration, but more inner iterations. With μ large, the iterates are widely separated on the central path; the inner iterates veer way off the central path.

This trade-off in the choice of μ is confirmed both in practice and, as we will see, in theory. In practice, small values of μ (i.e., near one) result in many outer iterations, with just a few Newton steps for each outer iteration. For μ in a fairly large range, from around 3 to 100 or so, the two effects nearly cancel, so the total number of Newton steps remains approximately constant. This means that the choice of μ is not particularly critical; values from around 10 to 20 or so seem to work well. When the parameter μ is chosen to give the best worst-case bound on the total number of Newton steps required, values of μ near one are used.

Choice of $t^{(0)}$

Another important issue is the choice of initial value of t. Here the trade-off is simple: If $t^{(0)}$ is chosen too large, the first outer iteration will require too many iterations. If $t^{(0)}$ is chosen too small, the algorithm will require extra outer iterations, and possibly too many inner iterations in the first centering step.

Since $m/t^{(0)}$ is the duality gap that will result from the first centering step, one reasonable choice is to choose $t^{(0)}$ so that $m/t^{(0)}$ is approximately of the same order as $f_0(x^{(0)}) - p^\star$, or μ times this amount. For example, if a dual feasible point λ, ν is known, with duality gap $\eta = f_0(x^{(0)}) - g(\lambda, \nu)$, then we can take $t^{(0)} = m/\eta$. Thus, in the first outer iteration we simply compute a pair with the same duality gap as the initial primal and dual feasible points.

Another possibility is suggested by the central path condition (11.7). We can interpret

$$\inf_\nu \left\| t\nabla f_0(x^{(0)}) + \nabla\phi(x^{(0)}) + A^T\nu \right\|_2 \tag{11.12}$$

as a measure for the deviation of $x^{(0)}$ from the point $x^\star(t)$, and choose for $t^{(0)}$ the value that minimizes (11.12). (This value of t and ν can be found by solving a least-squares problem.)

A variation on this approach uses an affine-invariant measure of deviation between x and $x^\star(t)$ in place of the Euclidean norm. We choose t and ν that minimize

$$\alpha(t, \nu) = \left(t\nabla f_0(x^{(0)}) + \nabla\phi(x^{(0)}) + A^T\nu \right)^T H_0^{-1} \left(t\nabla f_0(x^{(0)}) + \nabla\phi(x^{(0)}) + A^T\nu \right),$$

where

$$H_0 = t\nabla^2 f_0(x^{(0)}) + \nabla^2\phi(x^{(0)}).$$

(It can be shown that $\inf_\nu \alpha(t, \nu)$ is the square of the Newton decrement of $tf_0 + \phi$ at $x^{(0)}$.) Since α is a quadratic-over-linear function of ν and t, it is convex.

Infeasible start Newton method

In one variation on the barrier method, an infeasible start Newton method (described in §10.3) is used for the centering steps. Thus, the barrier method is initialized with a point $x^{(0)}$ that satisfies $x^{(0)} \in \mathbf{dom}\, f_0$ and $f_i(x^{(0)}) < 0$, $i = 1, \ldots, m$, but not necessarily $Ax^{(0)} = b$. Assuming the problem is strictly feasible, a full Newton step is taken at some point during the first centering step, and thereafter, the iterates are all primal feasible, and the algorithm coincides with the (standard) barrier method.

1.3.2   Examples

Linear programming in inequality form

Our first example is a small LP in inequality form,

$$
\begin{array}{ll}
\text{minimize} & c^T x \\
\text{subject to} & Ax \preceq b
\end{array}
$$

with $A \in \mathbf{R}^{100 \times 50}$. The data were generated randomly, in such a way that the problem is strictly primal and dual feasible, with optimal value $p^\star = 1$.

The initial point $x^{(0)}$ is on the central path, with a duality gap of 100. The barrier method is used to solve the problem, and terminated when the duality gap is less than 10^{-6}. The centering problems are solved by Newton's method with backtracking, using parameters $\alpha = 0.01$, $\beta = 0.5$. The stopping criterion for

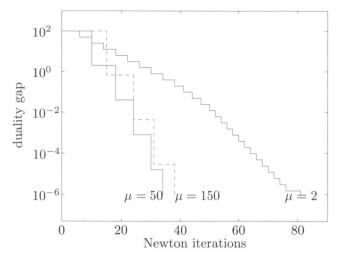

Figure 11.4 Progress of barrier method for a small LP, showing duality gap versus cumulative number of Newton steps. Three plots are shown, corresponding to three values of the parameter μ: 2, 50, and 150. In each case, we have approximately linear convergence of duality gap.

Newton's method is $\lambda(x)^2/2 \le 10^{-5}$, where $\lambda(x)$ is the Newton decrement of the function $tc^T x + \phi(x)$.

The progress of the barrier method, for three values of the parameter μ, is shown in figure 11.4. The vertical axis shows the duality gap on a log scale. The horizontal axis shows the cumulative total number of inner iterations, *i.e.*, Newton steps, which is the natural measure of computational effort. Each of the plots has a staircase shape, with each stair associated with one outer iteration. The width of each stair tread (*i.e.*, horizontal portion) is the number of Newton steps required for that outer iteration. The height of each stair riser (*i.e.*, the vertical portion) is exactly equal to (a factor of) μ, since the duality gap is reduced by the factor μ at the end of each outer iteration.

The plots illustrate several typical features of the barrier method. First of all, the method works very well, with approximately linear convergence of the duality gap. This is a consequence of the approximately constant number of Newton steps required to re-center, for each value of μ. For $\mu = 50$ and $\mu = 150$, the barrier method solves the problem with a total number of Newton steps between 35 and 40.

The plots in figure 11.4 clearly show the trade-off in the choice of μ. For $\mu = 2$, the treads are short; the number of Newton steps required to re-center is around 2 or 3. But the risers are also short, since the duality gap reduction per outer iteration is only a factor of 2. At the other extreme, when $\mu = 150$, the treads are longer, typically around 7 Newton steps, but the risers are also much larger, since the duality gap is reduced by the factor 150 in each outer iteration.

The trade-off in choice of μ is further examined in figure 11.5. We use the barrier method to solve the LP, terminating when the duality gap is smaller than 10^{-3}, for 25 values of μ between 1.2 and 200. The plot shows the total number of Newton steps required to solve the problem, as a function of the parameter μ.

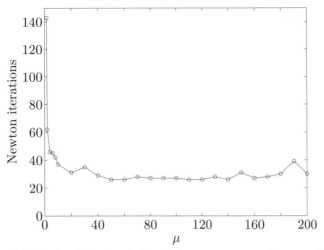

Figure 11.5 Trade-off in the choice of the parameter μ, for a small LP. The vertical axis shows the total number of Newton steps required to reduce the duality gap from 100 to 10^{-3}, and the horizontal axis shows μ. The plot shows the barrier method works well for values of μ larger than around 3, but is otherwise not sensitive to the value of μ.

This plot shows that the barrier method performs very well for a wide range of values of μ, from around 3 to 200. As our intuition suggests, the total number of Newton steps rises when μ is too small, due to the larger number of outer iterations required. One interesting observation is that the total number of Newton steps does not vary much for values of μ larger than around 3. Thus, as μ increases over this range, the decrease in the number of outer iterations is offset by an increase in the number of Newton steps per outer iteration. For even larger values of μ, the performance of the barrier method becomes less predictable (*i.e.*, more dependent on the particular problem instance). Since the performance does not improve with larger values of μ, a good choice is in the range $10 - 100$.

Geometric programming

We consider a geometric program in convex form,

$$
\begin{array}{ll}
\text{minimize} & \log\left(\sum_{k=1}^{K_0} \exp(a_{0k}^T x + b_{0k})\right) \\
\text{subject to} & \log\left(\sum_{k=1}^{K_i} \exp(a_{ik}^T x + b_{ik})\right) \leq 0, \quad i = 1, \dots, m,
\end{array}
$$

with variable $x \in \mathbf{R}^n$, and associated logarithmic barrier

$$
\phi(x) = -\sum_{i=1}^{m} \log\left(-\log \sum_{k=1}^{K_i} \exp(a_{ik}^T x + b_{ik})\right).
$$

The problem instance we consider has $n = 50$ variables and $m = 100$ inequalities (like the small LP considered above). The objective and constraint functions all

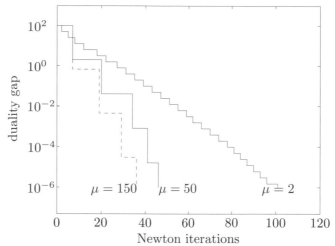

Figure 11.6 Progress of barrier method for a small GP, showing duality gap versus cumulative number of Newton steps. Again we have approximately linear convergence of duality gap.

have $K_i = 5$ terms. The problem instance was generated randomly, in such a way that it is strictly primal and dual feasible, with optimal value one.

We start with a point $x^{(0)}$ on the central path, with a duality gap of 100. The barrier method is used to solve the problem, with parameters $\mu = 2$, $\mu = 50$, and $\mu = 150$, and terminated when the duality gap is less than 10^{-6}. The centering problems are solved using Newton's method, with the same parameter values as in the LP example, *i.e.*, $\alpha = 0.01$, $\beta = 0.5$, and stopping criterion $\lambda(x)^2/2 \leq 10^{-5}$.

Figure 11.6 shows the duality gap versus cumulative number of Newton steps. This plot is very similar to the plot for LP, shown in figure 11.4. In particular, we see an approximately constant number of Newton steps required per centering step, and therefore approximately linear convergence of the duality gap.

The variation of the total number of Newton steps required to solve the problem, versus the parameter μ, is very similar to that in the LP example. For this GP, the total number of Newton steps required to reduce the duality gap below 10^{-3} is around 30 (ranging from around 20 to 40 or so) for values of μ between 10 and 200. So here, too, a good choice of μ is in the range $10 - 100$.

A family of standard form LPs

In the examples above we examined the progress of the barrier method, in terms of duality gap versus cumulative number of Newton steps, for a randomly generated instance of an LP and a GP, with similar dimensions. The results for the two examples are remarkably similar; each shows approximately linear convergence of duality gap with the number of Newton steps. We also examined the variation in performance with the parameter μ, and found essentially the same results in the two cases. For μ above around 10, the barrier method performs very well, requiring around 30 Newton steps to bring the duality gap down from 10^2 to 10^{-6}. In both cases, the choice of μ hardly affects the total number of Newton steps required (provided μ is larger than 10 or so).

In this section we examine the performance of the barrier method as a function of the problem dimensions. We consider LPs in standard form,

$$
\begin{aligned}
\text{minimize} \quad & c^T x \\
\text{subject to} \quad & Ax = b, \quad x \succeq 0
\end{aligned}
$$

with $A \in \mathbf{R}^{m \times n}$, and explore the total number of Newton steps required as a function of the number of variables n and number of equality constraints m, for a family of randomly generated problem instances. We take $n = 2m$, *i.e.*, twice as many variables as constraints.

The problems were generated as follows. The elements of A are independent and identically distributed, with zero mean, unit variance normal distribution $\mathcal{N}(0, 1)$. We take $b = Ax^{(0)}$ where the elements of $x^{(0)}$ are independent, and uniformly distributed in $[0, 1]$. This ensures that the problem is strictly primal feasible, since $x^{(0)} \succ 0$ is feasible. To construct the cost vector c, we first compute a vector $z \in \mathbf{R}^m$ with elements distributed according to $\mathcal{N}(0, 1)$ and a vector $s \in \mathbf{R}^n$ with elements from a uniform distribution on $[0, 1]$. We then take $c = A^T z + s$. This guarantees that the problem is strictly dual feasible, since $A^T z \prec c$.

The algorithm parameters we use are $\mu = 100$, and the same parameters for the centering steps in the examples above: backtracking parameters $\alpha = 0.01$, $\beta = 0.5$, and stopping criterion $\lambda(x)^2/2 \leq 10^{-5}$. The initial point is on the central path with $t^{(0)} = 1$ (*i.e.*, gap n). The algorithm is terminated when the initial duality gap is reduced by a factor 10^4, *i.e.*, after completing two outer iterations.

Figure 11.7 shows the duality gap versus iteration number for three problem instances, with dimensions $m = 50$, $m = 500$, and $m = 1000$. The plots look very much like the others, with approximately linear convergence of the duality gap. The plots show a small increase in the number of Newton steps required as the problem size grows from 50 constraints (100 variables) to 1000 constraints (2000 variables).

To examine the effect of problem size on the number of Newton steps required, we generate 100 problem instances for each of 20 values of m, ranging from $m = 10$ to $m = 1000$. We solve each of these 2000 problems using the barrier method, noting the number of Newton steps required. The results are summarized in figure 11.8, which shows the mean and standard deviation in the number of Newton steps, for each value of m. The first comment we make is that the standard deviation is around 2 iterations, and appears to be approximately independent of problem size. Since the average number of steps required is near 25, this means that the number of Newton steps required varies only around $\pm 10\%$.

The plot shows that the number of Newton steps required grows only slightly, from around 21 to around 27, as the problem dimensions increase by a factor of 100. This behavior is typical for the barrier method in general: The number of Newton steps required grows very slowly with problem dimensions, and is almost always around a few tens. Of course, the computational effort to carry out one Newton step grows with the problem dimensions.

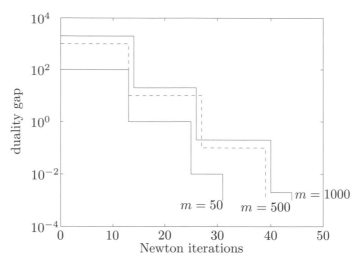

Figure 11.7 Progress of barrier method for three randomly generated standard form LPs of different dimensions, showing duality gap versus cumulative number of Newton steps. The number of variables in each problem is $n = 2m$. Here too we see approximately linear convergence of the duality gap, with a slight increase in the number of Newton steps required for the larger problems.

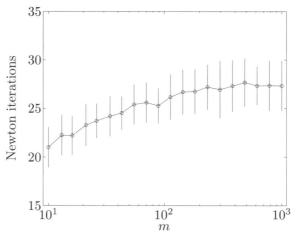

Figure 11.8 Average number of Newton steps required to solve 100 randomly generated LPs of different dimensions, with $n = 2m$. Error bars show standard deviation, around the average value, for each value of m. The growth in the number of Newton steps required, as the problem dimensions range over a 100:1 ratio, is very small.

11.3.3   Convergence analysis

Convergence analysis for the barrier method is straightforward. Assuming that $t f_0 + \phi$ can be minimized by Newton's method for $t = t^{(0)}$, $\mu t^{(0)}$, $\mu^2 t^{(0)}, \ldots$, the duality gap after the initial centering step, and k additional centering steps, is $m/(\mu^k t^{(0)})$. Therefore the desired accuracy ϵ is achieved after *exactly*

$$\left\lceil \frac{\log(m/(\epsilon t^{(0)}))}{\log \mu} \right\rceil \tag{11.13}$$

centering steps, plus the initial centering step.

It follows that the barrier method works provided the centering problem (11.6) is solvable by Newton's method, for $t \geq t^{(0)}$. For the standard Newton method, it suffices that for $t \geq t^{(0)}$, the function $t f_0 + \phi$ satisfies the conditions given in §10.2.4, page 529: its initial sublevel set is closed, the associated inverse KKT matrix is bounded, and the Hessian satisfies a Lipschitz condition. (Another set of sufficient conditions, based on self-concordance, will be discussed in detail in §11.5.) If the infeasible start Newton method is used for centering, then the conditions listed in §10.3.3, page 536, are sufficient to guarantee convergence.

Assuming that $f_0, \ldots, f_m$ are closed, a simple modification of the original problem ensures that these conditions hold. By adding a constraint of the form $\|x\|_2^2 \leq R^2$ to the problem, it follows that $t f_0 + \phi$ is strongly convex, for every $t \geq 0$; in particular convergence of Newton's method, for the centering steps, is guaranteed. (See exercise 11.4.)

While this analysis shows that the barrier method does converge, under reasonable assumptions, it does not address a basic question: As the parameter t increases, do the centering problems become more difficult (and therefore take more and more iterations)? Numerical evidence suggests that for a wide variety of problems, this is not the case; the centering problems appear to require a nearly constant number of Newton steps to solve, even as t increases. We will see (in §11.5) that this issue can be resolved, for problems that satisfy certain self-concordance conditions.

11.3.4   Newton step for modified KKT equations

In the barrier method, the Newton step Δx_{nt}, and associated dual variable are given by the linear equations

$$\begin{bmatrix} t \nabla^2 f_0(x) + \nabla^2 \phi(x) & A^T \\ A & 0 \end{bmatrix} \begin{bmatrix} \Delta x_{\mathrm{nt}} \\ \nu_{\mathrm{nt}} \end{bmatrix} = - \begin{bmatrix} t \nabla f_0(x) + \nabla \phi(x) \\ 0 \end{bmatrix}. \tag{11.14}$$

In this section we show how these Newton steps for the centering problem can be interpreted as Newton steps for directly solving the modified KKT equations

$$\begin{aligned} \nabla f_0(x) + \sum_{i=1}^m \lambda_i \nabla f_i(x) + A^T \nu &= 0 \\ -\lambda_i f_i(x) &= 1/t, \quad i = 1, \ldots, m \\ Ax &= b \end{aligned} \tag{11.15}$$

in a particular way.

To solve the modified KKT equations (11.15), which is a set of $n + p + m$ nonlinear equations in the $n + p + m$ variables x, ν, and λ, we first eliminate the variables λ_i, using $\lambda_i = -1/(t f_i(x))$. This yields

$$\nabla f_0(x) + \sum_{i=1}^{m} \frac{1}{-t f_i(x)} \nabla f_i(x) + A^T \nu = 0, \qquad Ax = b, \qquad (11.16)$$

which is a set of $n + p$ equations in the $n + p$ variables x and ν.

To find the Newton step for solving the set of nonlinear equations (11.16), we form the Taylor approximation for the nonlinear term occurring in the first equation. For v small, we have the Taylor approximation

$$\nabla f_0(x + v) + \sum_{i=1}^{m} \frac{1}{-t f_i(x + v)} \nabla f_i(x + v)$$

$$\approx \quad \nabla f_0(x) + \sum_{i=1}^{m} \frac{1}{-t f_i(x)} \nabla f_i(x) + \nabla^2 f_0(x) v$$

$$+ \sum_{i=1}^{m} \frac{1}{-t f_i(x)} \nabla^2 f_i(x) v + \sum_{i=1}^{m} \frac{1}{t f_i(x)^2} \nabla f_i(x) \nabla f_i(x)^T v.$$

The Newton step is obtained by replacing the nonlinear term in equation (11.16) by this Taylor approximation, which yields the linear equations

$$Hv + A^T \nu = -g, \qquad Av = 0, \qquad (11.17)$$

where

$$H = \nabla^2 f_0(x) + \sum_{i=1}^{m} \frac{1}{-t f_i(x)} \nabla^2 f_i(x) + \sum_{i=1}^{m} \frac{1}{t f_i(x)^2} \nabla f_i(x) \nabla f_i(x)^T$$

$$g = \nabla f_0(x) + \sum_{i=1}^{m} \frac{1}{-t f_i(x)} \nabla f_i(x).$$

Now we observe that

$$H = \nabla^2 f_0(x) + (1/t) \nabla^2 \phi(x), \qquad g = \nabla f_0(x) + (1/t) \nabla \phi(x),$$

so, from (11.14), the Newton steps Δx_{nt} and ν_{nt} in the barrier method centering step satisfy

$$t H \Delta x_{\mathrm{nt}} + A^T \nu_{\mathrm{nt}} = -t g, \qquad A \Delta x_{\mathrm{nt}} = 0.$$

Comparing this with (11.17) shows that

$$v = \Delta x_{\mathrm{nt}}, \qquad \nu = (1/t) \nu_{\mathrm{nt}}.$$

This shows that the Newton step for the centering problem (11.6) can be interpreted, after scaling the dual variable, as the Newton step for solving the modified KKT equations (11.16).

In this approach, we first eliminated the variable λ from the modified KKT equations, and then applied Newton's method to solve the resulting set of equations. Another variation on this approach is to directly apply Newton's method to the modified KKT equations, without first eliminating λ. This method yields the so-called *primal-dual search directions*, discussed in §11.7.

11.4   Feasibility and phase I methods

The barrier method requires a strictly feasible starting point $x^{(0)}$. When such a point is not known, the barrier method is preceded by a preliminary stage, called *phase I*, in which a strictly feasible point is computed (or the constraints are found to be infeasible). The strictly feasible point found during phase I is then used as the starting point for the barrier method, which is called the *phase II* stage. In this section we describe several phase I methods.

1.4.1   Basic phase I method

We consider a set of inequalities and equalities in the variables $x \in \mathbf{R}^n$,

$$f_i(x) \le 0, \quad i = 1, \ldots, m, \qquad Ax = b, \tag{11.18}$$

where $f_i : \mathbf{R}^n \to \mathbf{R}$ are convex, with continuous second derivatives. We assume that we are given a point $x^{(0)} \in \mathbf{dom}\, f_1 \cap \cdots \cap \mathbf{dom}\, f_m$, with $Ax^{(0)} = b$.

Our goal is to find a strictly feasible solution of these inequalities and equalities, or determine that none exists. To do this we form the following optimization problem:

$$
\begin{array}{ll}
\text{minimize} & s \\
\text{subject to} & f_i(x) \le s, \quad i = 1, \ldots, m \\
& Ax = b
\end{array}
\tag{11.19}
$$

in the variables $x \in \mathbf{R}^n$, $s \in \mathbf{R}$. The variable s can be interpreted as a bound on the maximum infeasibility of the inequalities; the goal is to drive the maximum infeasibility below zero.

This problem is always strictly feasible, since we can choose $x^{(0)}$ as starting point for x, and for s, we can choose any number larger than $\max_{i=1,\ldots,m} f_i(x^{(0)})$. We can therefore apply the barrier method to solve the problem (11.19), which is called the *phase I optimization problem* associated with the inequality and equality system (11.19).

We can distinguish three cases depending on the sign of the optimal value $\bar{p}^\star$ of (11.19).

1. If $\bar{p}^\star < 0$, then (11.18) has a strictly feasible solution. Moreover if (x, s) is feasible for (11.19) with $s < 0$, then x satisfies $f_i(x) < 0$. This means we do not need to solve the optimization problem (11.19) with high accuracy; we can terminate when $s < 0$.

2. If $\bar{p}^\star > 0$, then (11.18) is infeasible. As in case 1, we do not need to solve the phase I optimization problem (11.19) to high accuracy; we can terminate when a dual feasible point is found with positive dual objective (which proves that $\bar{p}^\star > 0$). In this case, we can construct the alternative that proves (11.18) is infeasible from the dual feasible point.

3. If $\bar{p}^\star = 0$ and the minimum is attained at $x^\star$ and $s^\star = 0$, then the set of inequalities is feasible, but not strictly feasible. If $\bar{p}^\star = 0$ and the minimum is not attained, then the inequalities are infeasible.

In practice it is impossible to determine exactly that $\bar{p}^\star = 0$. Instead, an optimization algorithm applied to (11.19) will terminate with the conclusion that $|\bar{p}^\star| < \epsilon$ for some small, positive ϵ. This allows us to conclude that the inequalities $f_i(x) \leq -\epsilon$ are infeasible, while the inequalities $f_i(x) \leq \epsilon$ are feasible.

Sum of infeasibilities

There are many variations on the basic phase I method just described. One method is based on minimizing the sum of the infeasibilities, instead of the maximum infeasibility. We form the problem

$$
\begin{array}{ll}
\text{minimize} & \mathbf{1}^T s \\
\text{subject to} & f_i(x) \leq s_i, \quad i = 1, \ldots, m \\
& Ax = b \\
& s \succeq 0.
\end{array}
\tag{11.20}
$$

For fixed x, the optimal value of s_i is $\max\{f_i(x), 0\}$, so in this problem we are minimizing the sum of the infeasibilities. The optimal value of (11.20) is zero and achieved if and only if the original set of equalities and inequalities is feasible.

This sum of infeasibilities phase I method has a very interesting property when the system of equalities and inequalities (11.19) is infeasible. In this case, the optimal point for the phase I problem (11.20) often violates only a small number, say r, of the inequalities. Therefore, we have computed a point that satisfies many $(m - r)$ of the inequalities, *i.e.*, we have identified a large subset of inequalities that is feasible. In this case, the dual variables associated with the strictly satisfied inequalities are zero, so we have also proved infeasibility of a subset of the inequalities. This is more informative than finding that the m inequalities, together, are mutually infeasible. (This phenomenon is closely related to ℓ_1-norm regularization, or basis pursuit, used to find sparse approximate solutions; see §6.1.2 and §6.5.4).

Example 11.4 *Comparison of phase I methods.* We apply two phase I methods to an infeasible set of inequalities $Ax \preceq b$ with dimensions $m = 100$, $n = 50$. The first method is the basic phase I method

$$
\begin{array}{ll}
\text{minimize} & s \\
\text{subject to} & Ax \preceq b + \mathbf{1}s,
\end{array}
$$

which minimizes the maximum infeasibility. The second method minimizes the sum of the infeasibilities, *i.e.*, solves the LP

$$
\begin{array}{ll}
\text{minimize} & \mathbf{1}^T s \\
\text{subject to} & Ax \preceq b + s \\
& s \succeq 0.
\end{array}
$$

Figure 11.9 shows the distributions of the infeasibilities $b_i - a_i^T x$ for these two values of x, denoted $x_{\max}$ and x_{sum}, respectively. The point $x_{\max}$ satisfies 39 of the 100 inequalities, whereas the point x_{sum} satisfies 79 of the inequalities.

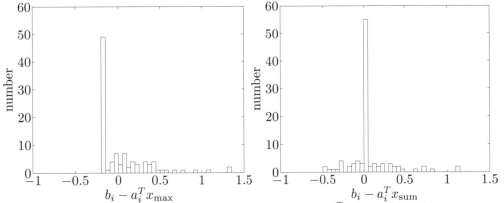

Figure 11.9 Distributions of the infeasibilities $b_i - a_i^T x$ for an infeasible set of 100 inequalities $a_i^T x \le b_i$, with 50 variables. The vector $x_{\max}$ used in the left plot was obtained by the basic phase I algorithm. It satisfies 39 of the 100 inequalities. In the right plot the vector x_{sum} was obtained by minimizing the sum of the infeasibilities. This vector satisfies 79 of the 100 inequalities.

Termination near the phase II central path

A simple variation on the basic phase I method, using the barrier method, has the property that (when the equalities and inequalities are strictly feasible) the central path for the phase I problem intersects the central path for the original optimization problem (11.1).

We assume a point $x^{(0)} \in \mathcal{D} = \mathbf{dom}\, f_0 \cap \mathbf{dom}\, f_1 \cap \cdots \cap \mathbf{dom}\, f_m$, with $Ax^{(0)} = b$ is given. We form the phase I optimization problem

$$
\begin{array}{ll}
\text{minimize} & s \\
\text{subject to} & f_i(x) \le s, \quad i = 1, \ldots, m \\
& f_0(x) \le M \\
& Ax = b,
\end{array}
\tag{11.21}
$$

where M is a constant chosen to be larger than $\max\{f_0(x^{(0)}), p^\star\}$.

We assume now that the original problem (11.1) is strictly feasible, so the optimal value $\bar{p}^\star$ of (11.21) is negative. The central path of (11.21) is characterized by

$$
\sum_{i=1}^m \frac{1}{s - f_i(x)} = \bar{t}, \qquad \frac{1}{M - f_0(x)} \nabla f_0(x) + \sum_{i=1}^m \frac{1}{s - f_i(x)} \nabla f_i(x) + A^T \nu = 0,
$$

where $\bar{t}$ is the parameter. If (x, s) is on the central path and $s = 0$, then x and ν satisfy

$$
t \nabla f_0(x) + \sum_{i=1}^m \frac{1}{-f_i(x)} \nabla f_i(x) + A^T \nu = 0
$$

for $t = 1/(M - f_0(x))$. This means that x is on the central path for the original optimization problem (11.1), with associated duality gap

$$m(M - f_0(x)) \le m(M - p^\star). \qquad (11.22)$$

11.4.2   Phase I via infeasible start Newton method

We can also carry out the phase I stage using an infeasible start Newton method, applied to a modified version of the original problem

$$
\begin{array}{ll}
\text{minimize} & f_0(x) \\
\text{subject to} & f_i(x) \le 0, \quad i = 1, \ldots, m \\
& Ax = b.
\end{array}
$$

We first express the problem in the (obviously equivalent) form

$$
\begin{array}{ll}
\text{minimize} & f_0(x) \\
\text{subject to} & f_i(x) \le s, \quad i = 1, \ldots, m \\
& Ax = b, \quad s = 0,
\end{array}
$$

with the additional variable $s \in \mathbf{R}$. To start the barrier method, we use an infeasible start Newton method to solve

$$
\begin{array}{ll}
\text{minimize} & t^{(0)} f_0(x) - \sum_{i=1}^{m} \log(s - f_i(x)) \\
\text{subject to} & Ax = b, \quad s = 0.
\end{array}
$$

This can be initialized with any $x \in \mathcal{D}$, and any $s > \max_i f_i(x)$. Provided the problem is strictly feasible, the infeasible start Newton method will eventually take an undamped step, and thereafter we will have $s = 0$, i.e., x strictly feasible.

The same trick can be applied if a point in $\mathcal{D}$, the common domain of the functions, is not known. We simply apply the infeasible start Newton method to the problem

$$
\begin{array}{ll}
\text{minimize} & t^{(0)} f_0(x + z_0) - \sum_{i=1}^{m} \log(s - f_i(x + z_i)) \\
\text{subject to} & Ax = b, \quad s = 0, \quad z_0 = 0, \quad \ldots, \quad z_m = 0
\end{array}
$$

with variables $x, z_0, \ldots, z_m$, and $s \in \mathbf{R}$. We initialize z_i so that $x + z_i \in \mathbf{dom}\, f_i$.

The main disadvantage of this approach to the phase I problem is that there is no good stopping criterion when the problem is infeasible; the residual simply fails to converge to zero.

11.4.3   Examples

We consider a family of linear feasibility problems,

$$Ax \preceq b(\gamma)$$

where $A \in \mathbf{R}^{50 \times 20}$ and $b(\gamma) = b + \gamma \Delta b$. The problem data are chosen so that the inequalities are strictly feasible for $\gamma > 0$ and infeasible for $\gamma < 0$. For $\gamma = 0$ the problem is feasible but not strictly feasible.

Figure 11.10 shows the total number of Newton steps required to find a strictly feasible point, or a certificate of infeasibility, for 40 values of γ in $[-1, 1]$. We use the basic phase I method of §11.4.1, *i.e.*, for each value of γ, we form the LP

$$
\begin{array}{ll}
\text{minimize} & s \\
\text{subject to} & Ax \preceq b(\gamma) + s\mathbf{1}.
\end{array}
$$

The barrier method is used with $\mu = 10$, and starting point $x = 0$, $s = -\min_i b_i(\gamma) + 1$. The method terminates when a point (x, s) with $s < 0$ is found, or a feasible solution z of the dual problem

$$
\begin{array}{ll}
\text{maximize} & -b(\gamma)^T z \\
\text{subject to} & A^T z = 0 \\
& \mathbf{1}^T z = 1 \\
& z \succeq 0
\end{array}
$$

is found with $-b(\gamma)^T z > 0$.

The plot shows that when the inequalities are feasible, with some margin, it takes around 25 Newton steps to produce a strictly feasible point. Conversely, when the inequalities are infeasible, again with some margin, it takes around 35 steps to produce a certificate proving infeasibility. The phase I effort increases as the set of inequalities approaches the boundary between feasible and infeasible, *i.e.*, γ near zero. When γ is very near zero, so the inequalities are very near the boundary between feasible and infeasible, the number of steps grows substantially. Figure 11.11 shows the total number of Newton steps required for values of γ near zero. The plots show an approximately logarithmic increase in the number of steps required to detect feasibility, or prove infeasibility, for problems very near the boundary between feasible and infeasible.

This example is typical: The cost of solving a set of convex inequalities and linear equalities using the barrier method is modest, and approximately constant, as long as the problem is not very close to the boundary between feasibility and infeasibility. When the problem is very close to the boundary, the number of Newton steps required to find a strictly feasible point or produce a certificate of infeasibility grows. When the problem is *exactly* on the boundary between strictly feasible and infeasible, for example, feasible but not strictly feasible, the cost becomes infinite.

Feasibility using infeasible start Newton method

We also solve the same set of feasibility problems using the infeasible start Newton method, applied to the problem

$$
\begin{array}{ll}
\text{minimize} & -\sum_{i=1}^{m} \log s_i \\
\text{subject to} & Ax + s = b(\gamma).
\end{array}
$$

We use backtracking parameters $\alpha = 0.01$, $\beta = 0.9$, and initialize with $x^{(0)} = 0$, $s^{(0)} = \mathbf{1}$, $\nu^{(0)} = 0$. We consider only feasible problems (*i.e.*, $\gamma > 0$) and terminate once a feasible point is found. (We do not consider infeasible problems, since in that case the residual simply converges to a positive number.) Figure 11.12 shows the number of Newton steps required to find a feasible point, as a function of γ.

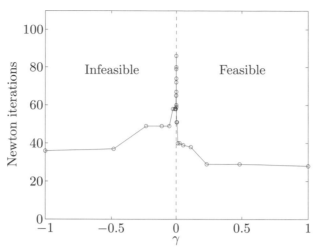

Figure 11.10 Number of Newton iterations required to detect feasibility or infeasibility of a set of linear inequalities $Ax \preceq b + \gamma \Delta b$ parametrized by $\gamma \in \mathbf{R}$. The inequalities are strictly feasible for $\gamma > 0$, and infeasible for $\gamma < 0$. For γ larger than around 0.2, about 30 steps are required to compute a strictly feasible point; for γ less than -0.5 or so, it takes around 35 steps to produce a certificate proving infeasibility. For values of γ in between, and especially near zero, more Newton steps are required to determine feasibility.

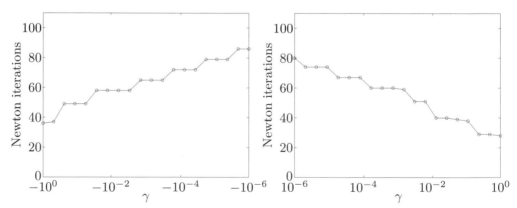

Figure 11.11 *Left.* Number of Newton iterations required to find a proof of infeasibility versus γ, for γ small and negative. *Right.* Number of Newton iterations required to find a strictly feasible point versus γ, for γ small and positive.

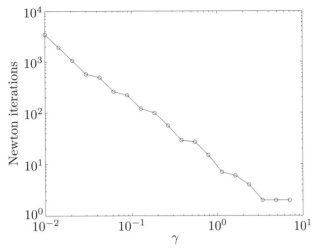

Figure 11.12 Number of iterations required to find a feasible point for a set of linear inequalities $Ax \preceq b + \gamma \Delta b$ parametrized by $\gamma \in \mathbf{R}$. The infeasible start Newton method is used, and terminated when a feasible point is found. For $\gamma = 10$, the starting point $x^{(0)} = 0$ happened to be feasible (0 iterations).

The plot shows that for γ larger than 0.3 or so, it takes fewer than 20 Newton steps to find a feasible point. In these cases the method is more efficient than a phase I method, which takes a total of around 30 Newton steps. For smaller values of γ, the number of Newton steps required grows dramatically, approximately as $1/\gamma$. For $\gamma = 0.01$, the infeasible start Newton method requires several thousand iterations to produce a feasible point. In this region the phase I approach is far more efficient, requiring only 40 iterations or so.

These results are quite typical. The infeasible start Newton method works very well provided the inequalities are feasible, and not very close to the boundary between feasible and infeasible. But when the feasible set is just barely nonempty (as is the case in this example with small γ), a phase I method is far better. Another advantage of the phase I method is that it gracefully handles the infeasible case; the infeasible start Newton method, in contrast, simply fails to converge.

11.5   Complexity analysis via self-concordance

Using the complexity analysis of Newton's method for self-concordant functions (§9.6.4, page 503, and §10.2.4, page 531), we can give a complexity analysis of the barrier method. The analysis applies to many common problems, and leads to several interesting conclusions: It gives a rigorous bound on the total number of Newton steps required to solve a problem using the barrier method, and it justifies our observation that the centering problems do not become more difficult as t increases.

11.5.1   Self-concordance assumption

We make two assumptions.

- The function $tf_0 + \phi$ is closed and self-concordant for all $t \geq t^{(0)}$.

- The sublevel sets of (11.1) are bounded.

The second assumption implies that the centering problem has bounded sublevel sets (see exercise 11.3), and, therefore, the centering problem is solvable. The bounded sublevel set assumption also implies that the Hessian of $tf_0 + \phi$ is positive definite everywhere (see exercise 11.14). While the self-concordance assumption restricts the complexity analysis to a particular class of problems, it is important to emphasize that the barrier method works well in general, whether or not the self-concordance assumption holds.

The self-concordance assumption holds for a variety of problems, including all linear and quadratic problems. If the functions f_i are linear or quadratic, then

$$tf_0 - \sum_{i=1}^{m} \log(-f_i)$$

is self-concordant for all values of $t \geq 0$ (see §9.6). The complexity analysis given below therefore applies to LPs, QPs, and QCQPs.

In other cases, it is possible to reformulate the problem so the assumption of self-concordance holds. As an example, consider the linear inequality constrained entropy maximization problem

$$\begin{array}{ll} \text{minimize} & \sum_{i=1}^{n} x_i \log x_i \\ \text{subject to} & Fx \preceq g \\ & Ax = b. \end{array}$$

The function

$$tf_0(x) + \phi(x) = t \sum_{i=1}^{n} x_i \log x_i - \sum_{i=1}^{m} \log(g_i - f_i^T x),$$

where $f_1^T, \ldots, f_m^T$ are the rows of F, is not closed (unless $Fx \preceq g$ implies $x \succeq 0$), or self-concordant. We can, however, add the redundant inequality constraints $x \succeq 0$ to obtain the equivalent problem

$$\begin{array}{ll} \text{minimize} & \sum_{i=1}^{n} x_i \log x_i \\ \text{subject to} & Fx \preceq g \\ & Ax = b \\ & x \succeq 0. \end{array} \tag{11.23}$$

For this problem we have

$$tf_0(x) + \phi(x) = t \sum_{i=1}^{n} x_i \log x_i - \sum_{i=1}^{n} \log x_i - \sum_{i=1}^{m} \log(g_i - f_i^T x),$$

which *is* self-concordant and closed, for any $t \geq 0$. (The function $ty \log y - \log y$ is self-concordant on $\mathbf{R}_{++}$, for all $t \geq 0$; see exercise 11.13.) The complexity analysis therefore applies to the reformulated linear inequality constrained entropy maximization problem (11.23).

As a more exotic example, consider the GP

$$\begin{array}{ll} \text{minimize} & f_0(x) = \log\left(\sum_{k=1}^{K_0} \exp(a_{0k}^T x + b_{0k})\right) \\ \text{subject to} & \log\left(\sum_{k=1}^{K_i} \exp(a_{ik}^T x + b_{ik})\right) \leq 0, \quad i = 1, \ldots, m. \end{array}$$

It is not clear whether or not the function

$$tf_0(x) + \phi(x) = t\log\left(\sum_{k=1}^{K_0} \exp(a_{0k}^T x + b_{0k})\right) - \sum_{i=1}^{m} \log\left(-\log\sum_{k=1}^{K_i} \exp(a_{ik}^T x + b_{ik})\right)$$

is self-concordant, so although the barrier method works, the complexity analysis of this section need not hold.

We can, however, reformulate the GP in a form that definitely satisfies the self-concordance assumption. For each (monomial) term $\exp(a_{ik}^T x + b_{ik})$ we introduce a new variable y_{ik} that serves as an upper bound,

$$\exp(a_{ik}^T x + b_{ik}) \leq y_{ik}.$$

Using these new variables we can express the GP in the form

$$\begin{array}{ll} \text{minimize} & \sum_{k=1}^{K_0} y_{0k} \\ \text{subject to} & \sum_{k=1}^{K_i} y_{ik} \leq 1, \quad i = 1, \ldots, m \\ & a_{ik}^T x + b_{ik} - \log y_{ik} \leq 0, \quad i = 0, \ldots, m, \quad k = 1, \ldots, K_i \\ & y_{ik} \geq 0, \quad i = 0, \ldots, m, \quad k = 1, \ldots, K_i. \end{array}$$

The associated logarithmic barrier is

$$\sum_{i=0}^{m}\sum_{k=1}^{K_i} \left(-\log y_{ik} - \log(\log y_{ik} - a_{ik}^T x - b_{ik})\right) - \sum_{i=1}^{m} \log\left(1 - \sum_{k=1}^{K_i} y_{ik}\right),$$

which is closed and self-concordant (example 9.8, page 500). Since the objective is linear, it follows that $tf_0 + \phi$ is closed and self-concordant for any t.

1.5.2   Newton iterations per centering step

The complexity theory of Newton's method for self-concordant functions, developed in §9.6.4 (page 503) and §10.2.4 (page 531), shows that the number of Newton iterations required to minimize a closed strictly convex self-concordant function f is bounded above by

$$\frac{f(x) - p^\star}{\gamma} + c. \tag{11.24}$$

Here x is the starting point for Newton's method, and $p^\star = \inf_x f(x)$ is the optimal value. The constant γ depends only on the backtracking parameters α and β, and is given by

$$\frac{1}{\gamma} = \frac{20 - 8\alpha}{\alpha\beta(1 - 2\alpha)^2}.$$

The constant c depends only on the tolerance ϵ_{nt},

$$c = \log_2 \log_2(1/\epsilon_{\mathrm{nt}}),$$

and can reasonably be approximated as $c = 6$. The expression (11.24) is a quite conservative bound on the number of Newton steps required, but our interest in this section is only to establish a complexity bound, concentrating on how it increases with problem size and algorithm parameters.

In this section we use this result to derive a bound on the number of Newton steps required for one outer iteration of the barrier method, *i.e.*, for computing $x^\star(\mu t)$, starting from $x^\star(t)$. To lighten the notation we use x to denote $x^\star(t)$, the current iterate, and we use x^+ to denote $x^\star(\mu t)$, the next iterate. We use λ and ν to denote $\lambda^\star(t)$ and $\nu^\star(t)$, respectively.

The self-concordance assumption implies that

$$\frac{\mu t f_0(x) + \phi(x) - \mu t f_0(x^+) - \phi(x^+)}{\gamma} + c \qquad (11.25)$$

is an upper bound on the number of Newton steps required to compute $x^+ = x^\star(\mu t)$, starting at $x = x^\star(t)$. Unfortunately we do not know x^+, and hence the upper bound (11.25), until we actually compute x^+, *i.e.*, carry out the Newton algorithm (whereupon we know the *exact* number of Newton steps required to compute $x^\star(\mu t)$, which defeats the purpose). We can, however, derive an upper bound on (11.25), as follows:

$$\mu t f_0(x) + \phi(x) - \mu t f_0(x^+) - \phi(x^+)$$

$$= \mu t f_0(x) - \mu t f_0(x^+) + \sum_{i=1}^{m} \log(-\mu t \lambda_i f_i(x^+)) - m \log \mu$$

$$\leq \mu t f_0(x) - \mu t f_0(x^+) - \mu t \sum_{i=1}^{m} \lambda_i f_i(x^+) - m - m \log \mu$$

$$= \mu t f_0(x) - \mu t \left(f_0(x^+) + \sum_{i=1}^{m} \lambda_i f_i(x^+) + \nu^T(Ax^+ - b) \right) - m - m \log \mu$$

$$\leq \mu t f_0(x) - \mu t g(\lambda, \nu) - m - m \log \mu$$

$$= m(\mu - 1 - \log \mu).$$

This chain of equalities and inequalities needs some explanation. To obtain the second line from the first, we use $\lambda_i = -1/(t f_i(x))$. In the first inequality we use the fact that $\log a \leq a - 1$ for $a > 0$. To obtain the fourth line from the third, we use $Ax^+ = b$, so the extra term $\nu^T(Ax^+ - b)$ is zero. The second inequality follows

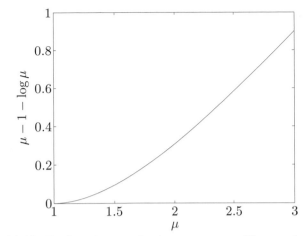

Figure 11.13 The function $\mu - 1 - \log \mu$, versus μ. The number of Newton steps required for one outer iteration of the barrier method is bounded by $(m/\gamma)(\mu - 1 - \log \mu) + c$.

from the definition of the dual function:

$$
\begin{aligned}
g(\lambda, \nu) &= \inf_z \left(f_0(z) + \sum_{i=1}^m \lambda_i f_i(z) + \nu^T(Az - b) \right) \\
&\leq f_0(x^+) + \sum_{i=1}^m \lambda_i f_i(x^+) + \nu^T(Ax^+ - b).
\end{aligned}
$$

The last line follows from $g(\lambda, \nu) = f_0(x) - m/t$.

The conclusion is that

$$
\frac{m(\mu - 1 - \log \mu)}{\gamma} + c \tag{11.26}
$$

is an upper bound on (11.25), and therefore an upper bound on the number of Newton steps required for one outer iteration of the barrier method. The function $\mu - 1 - \log \mu$ is shown in figure 11.13. For small μ it is approximately quadratic; for large μ it grows approximately linearly. This fits with our intuition that for μ near one, the number of Newton steps required to center is small, whereas for large μ, it could well grow.

The bound (11.26) shows that the number of Newton steps required in each centering step is bounded by a quantity that depends mostly on μ, the factor by which t is updated in each outer step of the barrier method, and m, the number of inequality constraints in the problem. It also depends, weakly, on the parameters α and β used in the line search for the inner iterations, and in a very weak way on the tolerance used to terminate the inner iterations. It is interesting to note that the bound does not depend on n, the dimension of the variable, or p, the number of equality constraints, or the particular values of the problem data, *i.e.*, the objective and constraint functions (provided the self-concordance assumption in §11.5.1 holds). Finally, we note that it does not depend on t; in particular, as $t \to \infty$, a uniform bound on the number of Newton steps per outer iteration holds.

11.5.3   Total number of Newton iterations

We can now give an upper bound on the total number of Newton steps in the barrier method, not counting the initial centering step (which we will analyze later, as part of phase I). We multiply (11.26), which bounds the number of Newton steps per outer iteration, by (11.13), the number of outer steps required, to obtain

$$N = \left\lceil \frac{\log(m/(t^{(0)}\epsilon))}{\log \mu} \right\rceil \left(\frac{m(\mu - 1 - \log \mu)}{\gamma} + c \right), \qquad (11.27)$$

an upper bound on the total number of Newton steps required. This formula shows that when the self-concordance assumption holds, we can bound the number of Newton steps required by the barrier method, for any value of $\mu > 1$.

If we fix μ and m, the bound N is proportional to $\log(m/(t^{(0)}\epsilon))$, which is the log of the ratio of the initial duality gap $m/t^{(0)}$ to the final duality gap ϵ, *i.e.*, the log of the required duality gap reduction. We can therefore say that the barrier method converges at least linearly, since the number of steps required to reach a given precision grows logarithmically with the inverse of the precision.

If μ, and the required duality gap reduction factor, are fixed, the bound N grows linearly with m, the number of inequalities. The bound N is independent of the other problem dimensions n and p, and the particular problem data or functions. We will see below that by a particular choice of μ, that depends on m, we can obtain a bound on the number of Newton steps that grows only as $\sqrt{m}$, instead of as m.

Finally, we analyze the bound N as a function of the algorithm parameter μ. As μ approaches one, the first term in N grows large, and therefore so does N. This is consistent with our intuition and observation that for μ near one, the number of outer iterations is very large. As μ becomes large, the bound N grows approximately as $\mu/\log \mu$, this time because the bound on the number of Newton iterations required per outer iteration grows. This, too, is consistent with our observations. As a result, the bound N has a minimum value as a function of μ.

The variation of the bound with the parameter μ is illustrated in figure 11.14, which shows the bound (11.27) versus μ for the values

$$c = 6, \qquad \gamma = 1/375, \qquad m/(t^{(0)}\epsilon) = 10^5, \qquad m = 100.$$

The bound is qualitatively consistent with intuition, and our observations: it grows very large as μ approaches one, and increases, more slowly, as μ becomes large. The bound N has a minimum at $\mu \approx 1.02$, which gives a bound on the total number of Newton iterations around 8000. The complexity analysis of Newton's method is conservative, but the basic trade-off in the choice of μ is reflected in the plot. (In practice, far larger values of μ, from around 2 to 100, work very well, and require a total number of Newton iterations on the order of a few tens.)

Choosing μ as a function of m

When μ (and the required duality gap reduction) is fixed, the bound (11.27) grows linearly with m, the number of inequalities. It turns out we can obtain a better

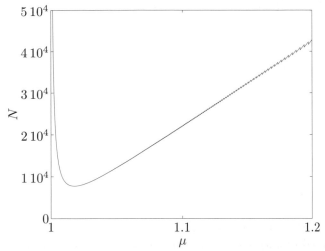

Figure 11.14 The upper bound N on the total number of Newton iterations, given by equation (11.27), for $c = 6$, $\gamma = 1/375$, $m = 100$, and a duality gap reduction factor $m/(t^{(0)}\epsilon) = 10^5$, versus the barrier algorithm parameter μ.

exponent for m by making μ a function of m. Suppose we choose

$$\mu = 1 + 1/\sqrt{m}. \qquad (11.28)$$

Then we can bound the second term in (11.27) as

$$
\begin{aligned}
\mu - 1 - \log \mu &= 1/\sqrt{m} - \log(1 + 1/\sqrt{m}) \\
&\leq 1/\sqrt{m} - 1/\sqrt{m} + 1/(2m) \\
&= 1/(2m)
\end{aligned}
$$

(using $-\log(1 + a) \leq -a + a^2/2$ for $a \geq 0$). Using concavity of the logarithm, we also have

$$\log \mu = \log(1 + 1/\sqrt{m}) \geq (\log 2)/\sqrt{m}.$$

Using these inequalities we can bound the total number of Newton steps by

$$
\begin{aligned}
N &\leq \left\lceil \frac{\log(m/(t^{(0)}\epsilon))}{\log \mu} \right\rceil \left(\frac{m(\mu - 1 - \log \mu)}{\gamma} + c \right) \\
&\leq \left\lceil \sqrt{m}\, \frac{\log(m/(t^{(0)}\epsilon))}{\log 2} \right\rceil \left(\frac{1}{2\gamma} + c \right) \\
&= \left\lceil \sqrt{m} \log_2(m/(t^{(0)}\epsilon)) \right\rceil \left(\frac{1}{2\gamma} + c \right) \\
&\leq c_1 + c_2 \sqrt{m}, \qquad (11.29)
\end{aligned}
$$

where

$$c_1 = \frac{1}{2\gamma} + c, \qquad c_2 = \log_2(m/(t^{(0)}\epsilon)) \left(\frac{1}{2\gamma} + c \right).$$

Here c_1 depends (and only weakly) on algorithm parameters for the centering Newton steps, and c_2 depends on these and the required duality gap reduction. Note that the term $\log_2(m/(t^{(0)}\epsilon))$ is exactly the number of bits of required duality gap reduction.

For fixed duality gap reduction, the bound (11.29) grows as $\sqrt{m}$, whereas the bound N in (11.27) grows like m, if the parameter μ is held constant. For this reason the barrier method, with parameter value (11.28), is said to be an *order* $\sqrt{m}$ method.

In practice, we would not use the value $\mu = 1 + 1/\sqrt{m}$, which is far too small, or even decrease μ as a function of m. Our only interest in this value of μ is that it (approximately) minimizes our (very conservative) upper bound on the number of Newton steps, and yields an overall estimate that grows as $\sqrt{m}$, instead of m.

11.5.4   Feasibility problems

In this section we analyze the complexity of a (minor) variation on the basic phase I method described in §11.4.1, used to solve a set of convex inequalities,

$$f_1(x) \leq 0, \quad \ldots, \quad f_m(x) \leq 0, \tag{11.30}$$

where $f_1, \ldots, f_m$ are convex, with continuous second derivatives. (We will consider equality constraints later.) We assume that the phase I problem

$$\begin{array}{ll} \text{minimize} & s \\ \text{subject to} & f_i(x) \leq s, \quad i = 1, \ldots, m \end{array} \tag{11.31}$$

satisfies the conditions in §11.5.1. In particular we assume that the feasible set of the inequalities (11.30) (which of course can be empty) is contained in a Euclidean ball of radius R:

$$\{x \mid f_i(x) \leq 0, \ i = 1, \ldots, m\} \subseteq \{x \mid \|x\|_2 \leq R\}.$$

We can interpret R as a prior bound on the norm of any point in the feasible set of the inequalities. This assumption implies that the sublevel sets of the phase I problem are bounded. Without loss of generality, we will start the phase I method at the point $x = 0$. We define $F = \max_i f_i(0)$, which is the maximum constraint violation, assumed to be positive (since otherwise $x = 0$ satisfies the inequalities (11.30)).

We define $\bar{p}^\star$ as the optimal value of the phase I optimization problem (11.31). The sign of $\bar{p}^\star$ determines whether or not the set of inequalities (11.30) is feasible. The magnitude of $\bar{p}^\star$ also has a meaning. If $\bar{p}^\star$ is positive and large (say, near F, the largest value it can have) it means that the set of inequalities is quite infeasible, in the sense that for each x, at least one of the inequalities is substantially violated (by at least $\bar{p}^\star$). On the other hand, if $\bar{p}^\star$ is negative and large, it means that the set of inequalities is quite feasible, in the sense that there is not only an x for which $f_i(x)$ are all nonpositive, but in fact there is an x for which $f_i(x)$ are all quite negative (no more than $\bar{p}^\star$). Thus, the magnitude $|\bar{p}^\star|$ is a measure of how clearly the set of inequalities is feasible or infeasible, and therefore related to the difficulty of determining feasibility of the inequalities (11.30). In particular, if $|\bar{p}^\star|$ is small, it means the problem is near the boundary between feasibility and infeasibility.

To determine feasibility of the inequalities, we use a variation on the basic phase I problem (11.31). We add a redundant linear inequality $a^T x \leq 1$, to obtain

$$
\begin{aligned}
\text{minimize} \quad & s \\
\text{subject to} \quad & f_i(x) \leq s, \quad i = 1, \ldots, m \\
& a^T x \leq 1.
\end{aligned}
\tag{11.32}
$$

We will specify a later. Our choice will satisfy $\|a\|_2 \leq 1/R$, so $\|x\|_2 \leq R$ implies $a^T x \leq 1$, i.e., the extra constraint is redundant.

We will choose a and s_0 so that $x = 0$, $s = s_0$ is on the central path of the problem (11.32), with a parameter value $t^{(0)}$, i.e., they minimize

$$
t^{(0)} s - \sum_{i=1}^m \log(s - f_i(x)) - \log(1 - a^T x).
$$

Setting to zero the derivative with respect to s, we get

$$
t^{(0)} = \sum_{i=1}^m \frac{1}{s_0 - f_i(0)}.
\tag{11.33}
$$

Setting to zero the gradient with respect to x yields

$$
a = -\sum_{i=1}^m \frac{1}{s_0 - f_i(0)} \nabla f_i(0).
\tag{11.34}
$$

So it remains only to pick the parameter s_0; once we have chosen s_0, the vector a is given by (11.34), and the parameter $t^{(0)}$ is given by (11.33). Since $x = 0$ and $s = s_0$ must be strictly feasible for the phase I problem (11.32), we must choose $s_0 > F$.

We must also pick s_0 to make sure that $\|a\|_2 \leq 1/R$. From (11.34), we have

$$
\|a\|_2 \leq \sum_{i=1}^m \frac{1}{s_0 - f_i(0)} \|\nabla f_i(0)\| \leq \frac{mG}{s_0 - F},
$$

where $G = \max_i \|\nabla f_i(0)\|_2$. Therefore we can take $s_0 = mGR + F$, which ensures $\|a\|_2 \leq 1/R$, so the extra linear inequality is redundant.

Using (11.33), we have

$$
t^{(0)} = \sum_{i=1}^m \frac{1}{mGR + F - f_i(0)} \geq \frac{1}{mGR},
$$

since $F = \max_i f_i(0)$. Thus $x = 0$, $s = s_0$ are on the central path for the phase I problem (11.32), with initial duality gap

$$
\frac{m+1}{t^{(0)}} \leq (m+1)mGR.
$$

To solve the original inequalities (11.30) we need to determine the sign of $\bar{p}^\star$. We can stop when either the primal objective value of (11.32) is negative, or the dual objective value is positive. One of these two cases must occur when the duality gap for (11.32) is less than $|\bar{p}^\star|$.

We use the barrier method to solve (11.32), starting from a central point with duality gap no more than $(m+1)mGR$, and terminating when (or before) the duality gap is less than $|\bar{p}^\star|$. Using the results of the previous section, this requires no more than

$$\left\lceil \sqrt{m+1} \log_2 \frac{m(m+1)GR}{|\bar{p}^\star|} \right\rceil \left(\frac{1}{2\gamma} + c \right) \tag{11.35}$$

Newton steps. (Here we take $\mu = 1 + 1/\sqrt{m+1}$, which gives a better complexity exponent for m than a fixed value of μ.)

The bound (11.35) grows only slightly faster than $\sqrt{m}$, and depends weakly on the algorithm parameters used in the centering steps. It is approximately proportional to $\log_2((GR)/|\bar{p}^\star|)$, which can be interpreted as a measure of how difficult the particular feasibility problem is, or how close it is to the boundary between feasibility and infeasibility.

Feasibility problems with equality constraints

We can apply the same analysis to feasibility problems that include equality constraints, by eliminating the equality constraints. This does not affect the self-concordance of the problem, but it does mean that G and R refer to the reduced, or eliminated, problem.

11.5.5   Combined phase I/phase II complexity

In this section we give an end-to-end complexity analysis for solving the problem

$$\begin{array}{ll} \text{minimize} & f_0(x) \\ \text{subject to} & f_i(x) \le 0, \quad i = 1, \ldots, m \\ & Ax = b \end{array}$$

using (a variation on) the barrier method. First we solve the phase I problem

$$\begin{array}{ll} \text{minimize} & s \\ \text{subject to} & f_i(x) \le s, \quad i = 1, \ldots, m \\ & f_0(x) \le M \\ & Ax = b \\ & a^T x \le 1, \end{array}$$

which we assume satisfies the self-concordance and bounded sublevel set assumptions of §11.5.1. Here we have added two redundant inequalities to the basic phase I problem. The constraint $f_0(x) \le M$ is added to guarantee that the phase I central path intersects the central path for phase II, as described in section §11.4.1 (see (11.21)). The number M is a prior bound on the optimal value of the problem. The second added constraint is the linear inequality $a^T x \le 1$, where a is chosen as described in §11.5.4. We use the barrier method to solve this problem, with $\mu = 1 + 1/\sqrt{m+2}$, and the starting points $x = 0$, $s = s_0$ given in §11.5.4.

To either find a strictly feasible point, or determine the problem is infeasible, requires no more than

$$N_{\mathrm{I}} = \left\lceil \sqrt{m+2} \log_2 \frac{(m+1)(m+2)GR}{|\bar{p}^\star|} \right\rceil \left(\frac{1}{2\gamma} + c \right) \qquad (11.36)$$

Newton steps, where G and R are as given in 11.5.4. If the problem is infeasible we are done; if it is feasible, then we find a point in phase I, associated with $s = 0$, that lies on the central path of the phase II problem

$$
\begin{array}{ll}
\text{minimize} & f_0(x) \\
\text{subject to} & f_i(x) \le 0, \quad i = 1, \ldots, m \\
& Ax = b \\
& a^T x \le 1.
\end{array}
$$

The associated initial duality gap of this initial point is no more than $(m+1)(M - p^\star)$ (see (11.22)). We assume the phase II problem also satisfies the the self-concordance and bounded sublevel set assumptions in §11.5.1.

We now proceed to phase II, again using the barrier method. We must reduce the duality gap from its initial value, which is no more than $(m+1)(M - p^\star)$, to some tolerance $\epsilon > 0$. This takes at most

$$N_{\mathrm{II}} = \left\lceil \sqrt{m+1} \log_2 \frac{(m+1)(M - p^\star)}{\epsilon} \right\rceil \left(\frac{1}{2\gamma} + c \right) \qquad (11.37)$$

Newton steps.

The total number of Newton steps is therefore no more than $N_{\mathrm{I}} + N_{\mathrm{II}}$. This bound grows with the number of inequalities m approximately as $\sqrt{m}$, and includes two terms that depend on the particular problem instance,

$$\log_2 \frac{GR}{|\bar{p}^\star|}, \qquad \log_2 \frac{M - p^\star}{\epsilon}.$$

1.5.6   Summary

The complexity analysis given in this section is mostly of theoretical interest. In particular, we remind the reader that the choice $\mu = 1 + 1/\sqrt{m}$, discussed in this section, would be a very poor one to use in practice; its only advantage is that it results in a bound that grows like $\sqrt{m}$ instead of m. Likewise, we do not recommend adding the redundant inequality $a^T x \le 1$ in practice.

The actual bounds obtained from the analysis given here are far higher than the numbers of iterations actually observed. Even the order in the bound appears to be conservative. The best bounds on the number of Newton steps grow like $\sqrt{m}$, whereas practical experience suggests that the number of Newton steps hardly grows at all with m (or any other parameter, in fact).

Still, it is comforting to know that when the self-concordance condition holds, we can give a uniform bound on the number of Newton steps required in each centering step of the barrier method. An obvious potential pitfall of the barrier method is the possibility that as t grows, the associated centering problems might become more difficult, requiring more Newton steps. While practical experience suggests that this is not the case, the uniform bound bolsters our confidence that it cannot happen.

Finally, we mention that it is not yet clear whether or not there is a practical advantage to formulating a problem so that the self-concordance condition holds. All we can say is that when the self-concordance condition holds, the barrier method will work well in practice, and we can give a worst case complexity bound.

11.6   Problems with generalized inequalities

In this section we show how the barrier method can be extended to problems with generalized inequalities. We consider the problem

$$
\begin{array}{ll}
\text{minimize} & f_0(x) \\
\text{subject to} & f_i(x) \preceq_{K_i} 0, \quad i = 1, \ldots, m \\
& Ax = b,
\end{array}
\tag{11.38}
$$

where $f_0 : \mathbf{R}^n \to \mathbf{R}$ is convex, $f_i : \mathbf{R}^n \to \mathbf{R}^{k_i}$, $i = 1, \ldots, k$, are K_i-convex, and $K_i \subseteq \mathbf{R}^{k_i}$ are proper cones. As in §11.1, we assume that the functions f_i are twice continuously differentiable, that $A \in \mathbf{R}^{p \times n}$ with $\mathbf{rank}\, A = p$, and that the problem is solvable.

The KKT conditions for problem (11.38) are

$$
\begin{array}{rcll}
Ax^\star & = & b \\
f_i(x^\star) & \preceq_{K_i} & 0, & i = 1, \ldots, m \\
\lambda_i^\star & \succeq_{K_i^*} & 0, & i = 1, \ldots, m \\
\nabla f_0(x^\star) + \sum_{i=1}^m Df_i(x^\star)^T \lambda_i^\star + A^T \nu^\star & = & 0 \\
\lambda_i^{\star T} f_i(x^\star) & = & 0, & i = 1, \ldots, m,
\end{array}
\tag{11.39}
$$

where $Df_i(x^\star) \in \mathbf{R}^{k_i \times n}$ is the derivative of f_i at $x^\star$. We will assume that problem (11.38) is strictly feasible, so the KKT conditions are necessary and sufficient conditions for optimality of $x^\star$.

The development of the method is parallel to the case with scalar constraints. Once we develop a generalization of the logarithm function that applies to general proper cones, we can define a logarithmic barrier function for the problem (11.38). From that point on, the development is essentially the same as in the scalar case. In particular, the central path, barrier method, and complexity analysis are very similar.

1.6.1 Logarithmic barrier and central path

Generalized logarithm for a proper cone

We first define the analog of the logarithm, $\log x$, for a proper cone $K \subseteq \mathbf{R}^q$. We say that $\psi : \mathbf{R}^q \to \mathbf{R}$ is a *generalized logarithm* for K if

- ψ is concave, closed, twice continuously differentiable, $\mathbf{dom}\,\psi = \mathbf{int}\,K$, and $\nabla^2 \psi(y) \prec 0$ for $y \in \mathbf{int}\,K$.

- There is a constant $\theta > 0$ such that for all $y \succ_K 0$, and all $s > 0$,

$$\psi(sy) = \psi(y) + \theta \log s.$$

In other words, ψ behaves like a logarithm along any ray in the cone K.

We call the constant θ the *degree* of ψ (since $\exp \psi$ is a homogeneous function of degree θ). Note that a generalized logarithm is only defined up to an additive constant; if ψ is a generalized logarithm for K, then so is $\psi + a$, where $a \in \mathbf{R}$. The ordinary logarithm is, of course, a generalized logarithm for $\mathbf{R}_+$.

We will use the following two properties, which are satisfied by any generalized logarithm: If $y \succ_K 0$, then

$$\nabla \psi(y) \succ_{K^*} 0, \tag{11.40}$$

which implies ψ is K-increasing (see §3.6.1), and

$$y^T \nabla \psi(y) = \theta.$$

The first property is proved in exercise 11.15. The second property follows immediately from differentiating $\psi(sy) = \psi(y) + \theta \log s$ with respect to s.

Example 11.5 *Nonnegative orthant.* The function $\psi(x) = \sum_{i=1}^{n} \log x_i$ is a generalized logarithm for $K = \mathbf{R}^n_+$, with degree n. For $x \succ 0$,

$$\nabla \psi(x) = (1/x_1, \dots, 1/x_n),$$

so $\nabla \psi(x) \succ 0$, and $x^T \nabla \psi(x) = n$.

Example 11.6 *Second-order cone.* The function

$$\psi(x) = \log \left(x_{n+1}^2 - \sum_{i=1}^{n} x_i^2 \right)$$

is a generalized logarithm for the second-order cone

$$K = \left\{ x \in \mathbf{R}^{n+1} \ \middle| \ \left(\sum_{i=1}^{n} x_i^2 \right)^{1/2} \leq x_{n+1} \right\},$$

with degree 2. The gradient of ψ at a point $x \in \mathbf{int}\, K$ is given by

$$\frac{\partial \psi(x)}{\partial x_j} = \frac{-2x_j}{\left(x_{n+1}^2 - \sum_{i=1}^n x_i^2\right)}, \quad j = 1, \dots, n$$

$$\frac{\partial \psi(x)}{\partial x_{n+1}} = \frac{2x_{n+1}}{\left(x_{n+1}^2 - \sum_{i=1}^n x_i^2\right)}.$$

The identities $\nabla \psi(x) \in \mathbf{int}\, K^* = \mathbf{int}\, K$ and $x^T \nabla \psi(x) = 2$ are easily verified.

Example 11.7 *Positive semidefinite cone.* The function $\psi(X) = \log \det X$ is a generalized logarithm for the cone $\mathbf{S}_+^p$. The degree is p, since

$$\log \det(sX) = \log \det X + p \log s$$

for $s > 0$. The gradient of ψ at a point $X \in \mathbf{S}_{++}^p$ is equal to

$$\nabla \psi(X) = X^{-1}.$$

Thus, we have $\nabla \psi(X) = X^{-1} \succ 0$, and the inner product of X and $\nabla \psi(X)$ is equal to $\mathbf{tr}(XX^{-1}) = p$.

Logarithmic barrier functions for generalized inequalities

Returning to problem (11.38), let $\psi_1, \dots, \psi_m$ be generalized logarithms for the cones $K_1, \dots, K_m$, respectively, with degrees $\theta_1, \dots, \theta_m$. We define the *logarithmic barrier function* for problem (11.38) as

$$\phi(x) = -\sum_{i=1}^m \psi_i(-f_i(x)), \qquad \mathbf{dom}\, \phi = \{x \mid f_i(x) \prec 0, \ i = 1, \dots, m\}.$$

Convexity of ϕ follows from the fact that the functions ψ_i are K_i-increasing, and the functions f_i are K_i-convex (see the composition rule of §3.6.2).

The central path

The next step is to define the central path for problem (11.38). We define the central point $x^*(t)$, for $t \geq 0$, as the minimizer of $tf_0 + \phi$, subject to $Ax = b$, *i.e.*, as the solution of

$$\begin{array}{ll} \text{minimize} & tf_0(x) - \sum_{i=1}^m \psi_i(-f_i(x)) \\ \text{subject to} & Ax = b \end{array}$$

(assuming the minimizer exists, and is unique). Central points are characterized by the optimality condition

$$t\nabla f_0(x) + \nabla \phi(x) + A^T \nu$$

$$= \; t\nabla f_0(x) + \sum_{i=1}^m Df_i(x)^T \nabla \psi_i(-f_i(x)) + A^T \nu = 0, \qquad (11.41)$$

for some $\nu \in \mathbf{R}^p$, where $Df_i(x)$ is the derivative of f_i at x.

Dual points on central path

As in the scalar case, points on the central path give dual feasible points for the problem (11.38). For $i = 1, \ldots, m$, define

$$\lambda_i^\star(t) = \frac{1}{t} \nabla \psi_i(-f_i(x^\star(t))), \tag{11.42}$$

and let $\nu^\star(t) = \nu/t$, where ν is the optimal dual variable in (11.41). We will show that $\lambda_1^\star(t), \ldots, \lambda_m^\star(t)$, together with $\nu^\star(t)$, are dual feasible for the original problem (11.38).

First, $\lambda_i^\star(t) \succ_{K_i^*} 0$, by the monotonicity property (11.40) of generalized logarithms. Second, it follows from (11.41) that the Lagrangian

$$L(x, \lambda^\star(t), \nu^\star(t)) = f_0(x) + \sum_{i=1}^m \lambda_i^\star(t)^T f_i(x) + \nu^\star(t)^T (Ax - b)$$

is minimized over x by $x = x^\star(t)$. The dual function g evaluated at $(\lambda^\star(t), \nu^\star(t))$ is therefore equal to

$$
\begin{aligned}
g(\lambda^\star(t), \nu^\star(t)) &= f_0(x^\star(t)) + \sum_{i=1}^m \lambda_i^\star(t)^T f_i(x^\star(t)) + \nu^\star(t)^T (Ax^\star(t) - b) \\
&= f_0(x^\star(t)) + (1/t) \sum_{i=1}^m \nabla \psi_i(-f_i(x^\star(t)))^T f_i(x^\star(t)) \\
&= f_0(x^\star(t)) - (1/t) \sum_{i=1}^m \theta_i,
\end{aligned}
$$

where θ_i is the degree of ψ_i. In the last line, we use the fact that $y^T \nabla \psi_i(y) = \theta_i$ for $y \succ_{K_i} 0$, and therefore

$$\lambda_i^\star(t)^T f_i(x^\star(t)) = -\theta_i/t, \quad i = 1, \ldots, m. \tag{11.43}$$

Thus, if we define

$$\bar\theta = \sum_{i=1}^m \theta_i,$$

then the primal feasible point $x^\star(t)$ and the dual feasible point $(\lambda^\star(t), \nu^\star(t))$ have duality gap $\bar\theta/t$. This is just like the scalar case, except that $\bar\theta$, the sum of the degrees of the generalized logarithms for the cones, appears in place of m, the number of inequalities.

Example 11.8 *Second-order cone programming.* We consider an SOCP with variable $x \in \mathbf{R}^n$:

$$
\begin{array}{ll}
\text{minimize} & f^T x \\
\text{subject to} & \|A_i x + b_i\|_2 \le c_i^T x + d_i, \quad i = 1, \ldots, m,
\end{array}
\tag{11.44}
$$

where $A_i \in \mathbf{R}^{n_i \times n}$. As we have seen in example 11.6, the function

$$\psi(y) = \log\left(y_{p+1}^2 - \sum_{i=1}^p y_i^2 \right)$$

is a generalized logarithm for the second-order cone in $\mathbf{R}^{p+1}$, with degree 2. The corresponding logarithmic barrier function for (11.44) is

$$\phi(x) = -\sum_{i=1}^{m} \log((c_i^T x + d_i)^2 - \|A_i x + b_i\|_2^2), \tag{11.45}$$

with $\mathbf{dom}\,\phi = \{x \mid \|A_i x + b_i\|_2 < c_i^T x + d_i, \; i = 1, \ldots, m\}$. The optimality condition on the central path is $tf + \nabla\phi(x^*(t)) = 0$, where

$$\nabla\phi(x) = -2\sum_{i=1}^{m} \frac{1}{(c_i^T x + d_i)^2 - \|A_i x + b_i\|_2^2} \left((c_i^T x + d_i)c_i - A_i^T(A_i x + b_i)\right).$$

It follows that the point

$$z_i^*(t) = -\frac{2}{t\alpha_i}(A_i x^*(t) + b_i), \qquad w_i^*(t) = \frac{2}{t\alpha_i}(c_i^T x^*(t) + d_i), \qquad i = 1, \ldots, m,$$

where $\alpha_i = (c_i^T x^*(t) + d_i)^2 - \|A_i x^*(t) + b_i\|_2^2$, is strictly feasible in the dual problem

$$\begin{array}{ll} \text{maximize} & -\sum_{i=1}^{m}(b_i^T z_i + d_i w_i) \\ \text{subject to} & \sum_{i=1}^{m}(A_i^T z_i + c_i w_i) = f \\ & \|z_i\|_2 \le w_i, \quad i = 1, \ldots, m. \end{array}$$

The duality gap associated with $x^*(t)$ and $(z^*(t), w^*(t))$ is

$$\sum_{i=1}^{m} \left((A_i x^*(t) + b_i)^T z_i^*(t) + (c_i^T x^*(t) + d_i)w_i^*(t)\right) = \frac{2m}{t},$$

which agrees with the general formula $\bar{\theta}/t$, since $\theta_i = 2$.

Example 11.9 *Semidefinite programming in inequality form.* We consider the SDP with variable $x \in \mathbf{R}^n$,

$$\begin{array}{ll} \text{minimize} & c^T x \\ \text{subject to} & F(x) = x_1 F_1 + \cdots + x_n F_n + G \preceq 0, \end{array}$$

where $G, F_1, \ldots, F_n \in \mathbf{S}^p$. The dual problem is

$$\begin{array}{ll} \text{maximize} & \mathbf{tr}(GZ) \\ \text{subject to} & \mathbf{tr}(F_i Z) + c_i = 0, \quad i = 1, \ldots, n \\ & Z \succeq 0. \end{array}$$

Using the generalized logarithm $\log \det X$ for the positive semidefinite cone $\mathbf{S}_+^p$, we have the barrier function (for the primal problem)

$$\phi(x) = \log \det(-F(x)^{-1})$$

with $\mathbf{dom}\,\phi = \{x \mid F(x) \prec 0\}$. For strictly feasible x, the gradient of ϕ is equal to

$$\frac{\partial\phi(x)}{\partial x_i} = \mathbf{tr}(-F(x)^{-1}F_i), \quad i = 1, \ldots, n,$$

which gives us the optimality conditions that characterize central points:

$$tc_i + \mathbf{tr}(-F(x^*(t))^{-1}F_i) = 0, \quad i = 1, \ldots, n.$$

Hence the matrix

$$Z^*(t) = \frac{1}{t}(-F(x^*(t)))^{-1}$$

is strictly dual feasible, and the duality gap associated with $x^*(t)$ and $Z^*(t)$ is p/t.

11.6.2 Barrier method

We have seen that the key properties of the central path generalize to problems with generalized inequalities.

- Computing a point on the central path involves minimizing a twice differentiable convex function subject to equality constraints (which can be done using Newton's method).

- With the central point $x^\star(t)$ we can associate a dual feasible point $(\lambda^\star(t), \nu^\star(t))$ with associated duality gap $\bar{\theta}/t$. In particular, $x^\star(t)$ is no more than $\bar{\theta}/t$-suboptimal.

This means we can apply the barrier method, exactly as described in §11.3, to the problem (11.38). The number of outer iterations, or centering steps, required to compute a central point with duality gap ϵ starting at $x^\star(t^{(0)})$ is equal to

$$\left\lceil \frac{\log(\bar{\theta}/(t^{(0)}\epsilon))}{\log \mu} \right\rceil,$$

plus one initial centering step. The only difference between this result and the associated one for the scalar case is that $\bar{\theta}$ takes the place of m.

Phase I and feasibility problems

The phase I methods described in §11.4 are readily extended to problems with generalized inequalities. Let $e_i \succ_{K_i} 0$ be some given, K_i-positive vectors, for $i = 1, \ldots, m$. To determine feasibility of the equalities and generalized inequalities

$$f_1(x) \preceq_{K_1} 0, \quad \ldots, \quad f_L(x) \preceq_{K_m} 0, \quad Ax = b,$$

we solve the problem

$$
\begin{array}{ll}
\text{minimize} & s \\
\text{subject to} & f_i(x) \preceq_{K_i} se_i, \quad i = 1, \ldots, m \\
& Ax = b,
\end{array}
$$

with variables x and $s \in \mathbf{R}$. The optimal value $\bar{p}^\star$ determines the feasibility of the equalities and generalized inequalities, exactly as in the case of ordinary inequalities. When $\bar{p}^\star$ is positive, any dual feasible point with positive objective gives an alternative that proves the set of equalities and generalized inequalities is infeasible (see page 270).

11.6.3 Examples

A small SOCP

We solve an SOCP

$$
\begin{array}{ll}
\text{minimize} & f^T x \\
\text{subject to} & \|A_i x + b_i\|_2 \le c_i^T x + d_i, \quad i = 1, \ldots, m,
\end{array}
$$

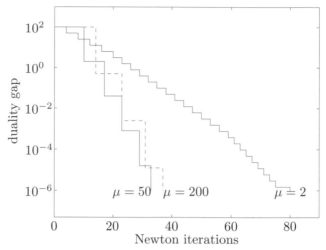

Figure 11.15 Progress of barrier method for an SOCP, showing duality gap versus cumulative number of Newton steps.

with $x \in \mathbf{R}^{50}$, $m = 50$, and $A_i \in \mathbf{R}^{5 \times 50}$. The problem instance was randomly generated, in such a way that the problem is strictly primal and dual feasible, and has optimal value $p^\star = 1$. We start with a point $x^{(0)}$ on the central path, with a duality gap of 100.

The barrier method is used to solve the problem, using the barrier function

$$\phi(x) = -\sum_{i=1}^{m} \log\left((c_i^T x + d_i)^2 - \|A_i x + b_i\|_2^2\right).$$

The centering problems are solved using Newton's method, with the same algorithm parameters as in the examples of §11.3.2: backtracking parameters $\alpha = 0.01$, $\beta = 0.5$, and a stopping criterion $\lambda(x)^2/2 \leq 10^{-5}$.

Figure 11.15 shows the duality gap versus cumulative number of Newton steps. The plot is very similar to those for linear and geometric programming, shown in figures 11.4 and 11.6, respectively. We see an approximately constant number of Newton steps required per centering step, and therefore approximately linear convergence of the duality gap. For this example, too, the choice of μ has little effect on the total number of Newton steps, provided μ is at least 10 or so. As in the examples for linear and geometric programming, a reasonable choice of μ is in the range $10 - 100$, which results in a total number of Newton steps around 30 (see figure 11.16).

A small SDP

Our next example is an SDP

$$
\begin{array}{ll}
\text{minimize} & c^T x \\
\text{subject to} & \sum_{i=1}^{n} x_i F_i + G \preceq 0
\end{array}
\tag{11.46}
$$

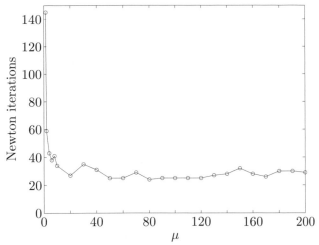

Figure 11.16 Trade-off in the choice of the parameter μ, for a small SOCP. The vertical axis shows the total number of Newton steps required to reduce the duality gap from 100 to 10^{-3}, and the horizontal axis shows μ.

with variable $x \in \mathbf{R}^{100}$, and $F_i \in \mathbf{S}^{100}$, $G \in \mathbf{S}^{100}$. The problem instance was generated randomly, in such a way that the problem is strictly primal and dual feasible, with $p^\star = 1$. The initial point is on the central path, with a duality gap of 100.

We apply the barrier method with logarithmic barrier function

$$\phi(x) = -\log\det\left(-\sum_{i=1}^{n} x_i F_i - G\right).$$

The progress of the barrier method for three values of μ is shown in figure 11.17. Note the similarity with the plots for linear, geometric, and second-order cone programming, shown in figures 11.4, 11.6, and 11.15. As in the other examples, the parameter μ has only a small effect on the efficiency, provided it is not too small. The number of Newton steps required to reduce the duality gap by a factor 10^5, versus μ, is shown in figure 11.18.

A family of SDPs

In this section we examine the performance of the barrier method as a function of the problem dimensions. We consider a family of SDPs of the form

$$\begin{array}{ll} \text{minimize} & \mathbf{1}^T x \\ \text{subject to} & A + \mathbf{diag}(x) \succeq 0, \end{array} \tag{11.47}$$

with variable $x \in \mathbf{R}^n$, and parameter $A \in \mathbf{S}^n$. The matrices A are generated as follows. For $i \geq j$, the coefficients A_{ij} are generated from independent $\mathcal{N}(0,1)$ distributions. For $i < j$, we set $A_{ij} = A_{ji}$, so $A \in \mathbf{S}^n$. We then scale A so that its (spectral) norm is one.

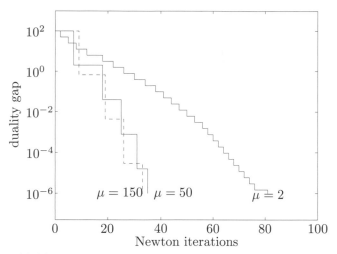

Figure 11.17 Progress of barrier method for a small SDP, showing duality gap versus cumulative number of Newton steps. Three plots are shown, corresponding to three values of the parameter μ: 2, 50, and 150.

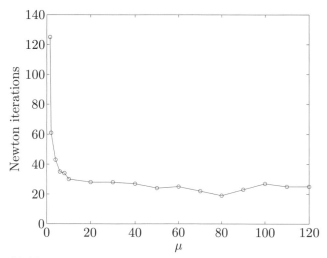

Figure 11.18 Trade-off in the choice of the parameter μ, for a small SDP. The vertical axis shows the total number of Newton steps required to reduce the duality gap from 100 to 10^{-3}, and the horizontal axis shows μ.

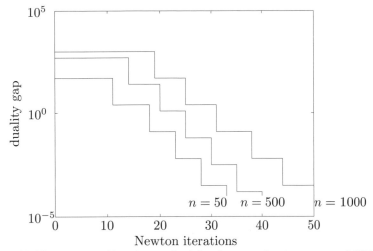

Figure 11.19 Progress of barrier method for three randomly generated SDPs of the form (11.47), with different dimensions. The plot shows duality gap versus cumulative number of Newton steps. The number of variables in each problem is n.

The algorithm parameters are $\mu = 20$, and the same parameters for the centering steps as in the examples above: backtracking parameters $\alpha = 0.01$, $\beta = 0.5$, and stopping criterion $\lambda(x)^2/2 \leq 10^{-5}$. The initial point is on the central path with $t^{(0)} = 1$ (*i.e.*, gap n). The algorithm is terminated when the initial duality gap is reduced by a factor 8000, *i.e.*, after completing three outer iterations.

Figure 11.19 shows the duality gap versus iteration number for three problem instances, with dimensions $n = 50$, $n = 500$, and $n = 1000$. The plots look very much like the others, and very much like the ones for LPs.

To examine the effect of problem size on the number of Newton steps required, we generate 100 problem instances for each of 20 values of n, ranging from $n = 10$ to $n = 1000$. We solve each of these 2000 problems using the barrier method, noting the number of Newton steps required. The results are summarized in figure 11.20, which shows the mean and standard deviation in the number of Newton steps, for each value of n. The plot looks very much like the one for LPs, shown in figure 11.8. In particular, the number of Newton steps required grows very slowly, from around 20 to 26 iterations, as the problem dimensions increase by a factor of 100.

11.6.4   Complexity analysis via self-concordance

In this section we extend the complexity analysis of the barrier method for problems with ordinary inequalities (given in §11.5), to problems with generalized inequalities. We have already seen that the number of outer iterations is given by

$$\left\lceil \frac{\log(\bar{\theta}/t^{(0)}\epsilon)}{\log\mu} \right\rceil,$$

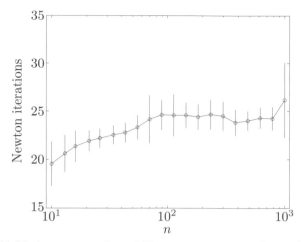

Figure 11.20 Average number of Newton steps required to solve 100 randomly generated SDPs (11.47) for each of 20 values of n, the problem size. Error bars show standard deviation, around the average value, for each value of n. The growth in the average number of Newton steps required, as the problem dimensions range over a 100:1 ratio, is very small.

plus one initial centering step. It remains to bound the number of Newton steps required in each centering step, which we will do using the complexity theory of Newton's method for self-concordant functions. For simplicity, we will exclude the cost of the initial centering.

We make the same assumptions as in §11.5: The function $tf_0 + \phi$ is closed and self-concordant for all $t \geq t^{(0)}$, and the sublevel sets of (11.38) are bounded.

Example 11.10 *Second-order cone programming.* The function

$$-\psi(x) = -\log\left(x_{p+1}^2 - \sum_{i=1}^{p} x_i^2\right),$$

is self-concordant (see example 9.8), so the logarithmic barrier function (11.45) satisfies the closedness and self-concordance assumption for the SOCP (11.44).

Example 11.11 *Semidefinite programming.* The self-concordance assumption holds for general semidefinite programs, using $\log \det X$ as generalized logarithm for the positive semidefinite cone. For example, for the standard form SDP

$$\begin{array}{ll}
\text{minimize} & \mathbf{tr}(CX) \\
\text{subject to} & \mathbf{tr}(A_i X) = b_i, \quad i = 1, \dots, p \\
& X \succeq 0,
\end{array}$$

with variable $X \in \mathbf{S}^n$, the function $t^{(0)} \mathbf{tr}(CX) - \log \det X$ is self-concordant (and closed), for any $t^{(0)} \geq 0$.

We will see that, exactly as in the scalar case, we have

$$\mu t f_0(x^\star(t)) + \phi(x^\star(t)) - \mu t f_0(x^\star(\mu t)) - \phi(x^\star(\mu t)) \leq \overline{\theta}(\mu - 1 - \log \mu). \quad (11.48)$$

Therefore when the self-concordance and bounded sublevel set conditions hold, the number of Newton steps per centering step is no more than

$$\frac{\overline{\theta}(\mu - 1 - \log \mu)}{\gamma} + c,$$

exactly as in the barrier method for problems with ordinary inequalities. Once we establish the basic bound (11.48), the complexity analysis for problems with generalized inequalities is identical to the analysis for problems with ordinary inequalities, with one exception: $\overline{\theta}$ is the sum of the degrees of the cones, instead of the number of inequalities.

Generalized logarithm for dual cone

We will use conjugates to prove the bound (11.48). Let ψ be a generalized logarithm for the proper cone K, with degree θ. The conjugate of the (convex) function $-\psi$ is

$$(-\psi)^*(v) = \sup_u \left(v^T u + \psi(u) \right).$$

This function is convex, and has domain $-K^* = \{v \mid v \prec_{K^*} 0\}$. Define $\overline{\psi}$ by

$$\overline{\psi}(v) = -(-\psi)^*(-v) = \inf_u \left(v^T u - \psi(u) \right), \qquad \mathbf{dom}\, \overline{\psi} = \mathbf{int}\, K^*. \quad (11.49)$$

The function $\overline{\psi}$ is concave, and in fact is a generalized logarithm for the dual cone K^*, with the same parameter θ (see exercise 11.17). We call $\overline{\psi}$ the *dual logarithm* associated with the generalized logarithm ψ.

From (11.49) we obtain the inequality

$$\overline{\psi}(v) + \psi(u) \leq u^T v, \quad (11.50)$$

which holds for any $u \succ_K 0$, $v \succ_{K^*} 0$, with equality holding if and only $\nabla \psi(u) = v$ (or equivalently, $\nabla \overline{\psi}(v) = u$). (This inequality is just a variation on Young's inequality, for concave functions.)

Example 11.12 *Second-order cone.* The second-order cone has generalized logarithm $\psi(x) = \log(x_{p+1}^2 - \sum_{i=1}^p x_i^2)$, with $\mathbf{dom}\, \psi = \{x \in \mathbf{R}^{p+1} \mid x_{p+1} > (\sum_{i=1}^p x_i^2)^{1/2}\}$. The associated dual logarithm is

$$\overline{\psi}(y) = \log \left(y_{p+1}^2 - \sum_{i=1}^p y_i^2 \right) + 2 - \log 4,$$

with $\mathbf{dom}\, \psi = \{y \in \mathbf{R}^{p+1} \mid y_{p+1} > (\sum_{i=1}^p y_i^2)^{1/2}\}$ (see exercise 3.36). Except for a constant, it is the same as the original generalized logarithm for the second-order cone.

Example 11.13 *Positive semidefinite cone.* The dual logarithm associated with $\psi(X) = \log \det X$, with $\mathbf{dom}\,\psi = \mathbf{S}^p_{++}$, is

$$\overline{\psi}(Y) = \log \det Y + p,$$

with domain $\mathbf{dom}\,\psi^* = \mathbf{S}^p_{++}$ (see example 3.23). Again, it is the same generalized logarithm, except for a constant.

Derivation of the basic bound

To simplify notation, we denote $x^\star(t)$ as x, $x^\star(\mu t)$ as x^+, $\lambda_i^\star(t)$ as λ_i, and $\nu^\star(t)$ as ν. From $t\lambda_i = \nabla\psi_i(-f_i(x))$ (in (11.42)) and property (11.43), we conclude that

$$\psi_i(-f_i(x)) + \overline{\psi}_i(t\lambda_i) = -t\lambda_i^T f_i(x) = \theta_i, \tag{11.51}$$

i.e., the inequality (11.50) holds with equality for the pair $u = -f_i(x)$ and $v = t\lambda_i$. The same inequality for the pair $u = -f_i(x^+)$, $v = \mu t \lambda_i$ gives

$$\psi_i(-f_i(x^+)) + \overline{\psi}_i(\mu t\lambda_i) \le -\mu t\lambda_i^T f_i(x^+),$$

which becomes, using logarithmic homogeneity of $\overline{\psi}_i$,

$$\psi_i(-f_i(x^+)) + \overline{\psi}_i(t\lambda_i) + \theta_i \log\mu \le -\mu t\lambda_i^T f_i(x^+).$$

Subtracting the equality (11.51) from this inequality, we get

$$-\psi_i(-f_i(x)) + \psi_i(-f_i(x^+)) + \theta_i \log\mu \le -\theta_i - \mu t\lambda_i^T f_i(x^+),$$

and summing over i yields

$$\phi(x) - \phi(x^+) + \overline{\theta}\log\mu \le -\overline{\theta} - \mu t \sum_{i=1}^m \lambda_i^T f_i(x^+). \tag{11.52}$$

We also have, from the definition of the dual function,

$$
\begin{aligned}
f_0(x) - \overline{\theta}/t &= g(\lambda, \nu) \\
&\le f_0(x^+) + \sum_{i=1}^m \lambda_i^T f_i(x^+) + \nu^T(Ax^+ - b) \\
&= f_0(x^+) + \sum_{i=1}^m \lambda_i^T f_i(x^+).
\end{aligned}
$$

Multiplying this inequality by μt and adding to the inequality (11.52), we get

$$\phi(x) - \phi(x^+) + \overline{\theta}\log\mu + \mu t f_0(x) - \mu\overline{\theta} \le \mu t f_0(x^+) - \overline{\theta},$$

which when re-arranged gives

$$\mu t f_0(x) + \phi(x) - \mu t f_0(x^+) - \phi(x^+) \le \overline{\theta}(\mu - 1 - \log\mu),$$

the desired inequality (11.48).

11.7 Primal-dual interior-point methods

In this section we describe a basic primal-dual interior-point method. Primal-dual interior-point methods are very similar to the barrier method, with some differences.

- There is only one loop or iteration, *i.e.*, there is no distinction between inner and outer iterations as in the barrier method. At each iteration, both the primal and dual variables are updated.

- The search directions in a primal-dual interior-point method are obtained from Newton's method, applied to modified KKT equations (*i.e.*, the optimality conditions for the logarithmic barrier centering problem). The primal-dual search directions are similar to, but not quite the same as, the search directions that arise in the barrier method.

- In a primal-dual interior-point method, the primal and dual iterates are *not* necessarily feasible.

Primal-dual interior-point methods are often more efficient than the barrier method, especially when high accuracy is required, since they can exhibit better than linear convergence. For several basic problem classes, such as linear, quadratic, second-order cone, geometric, and semidefinite programming, customized primal-dual methods outperform the barrier method. For general nonlinear convex optimization problems, primal-dual interior-point methods are still a topic of active research, but show great promise. Another advantage of primal-dual algorithms over the barrier method is that they can work when the problem is feasible, but not strictly feasible (although we will not pursue this).

In this section we present a basic primal-dual method for (11.1), without convergence analysis. We refer the reader to the references for a more thorough treatment of primal-dual methods and their convergence analysis.

1.7.1 Primal-dual search direction

As in the barrier method, we start with the modified KKT conditions (11.15), expressed as $r_t(x, \lambda, \nu) = 0$, where we define

$$r_t(x, \lambda, \nu) = \begin{bmatrix} \nabla f_0(x) + Df(x)^T \lambda + A^T \nu \\ -\operatorname{\mathbf{diag}}(\lambda) f(x) - (1/t)\mathbf{1} \\ Ax - b \end{bmatrix}, \qquad (11.53)$$

and $t > 0$. Here $f : \mathbf{R}^n \to \mathbf{R}^m$ and its derivative matrix Df are given by

$$f(x) = \begin{bmatrix} f_1(x) \\ \vdots \\ f_m(x) \end{bmatrix}, \qquad Df(x) = \begin{bmatrix} \nabla f_1(x)^T \\ \vdots \\ \nabla f_m(x)^T \end{bmatrix}.$$

If x, λ, ν satisfy $r_t(x, \lambda, \nu) = 0$ (and $f_i(x) < 0$), then $x = x^\star(t)$, $\lambda = \lambda^\star(t)$, and $\nu = \nu^\star(t)$. In particular, x is primal feasible, and λ, ν are dual feasible, with duality gap m/t. The first block component of r_t,

$$r_{\text{dual}} = \nabla f_0(x) + Df(x)^T \lambda + A^T \nu,$$

is called the *dual residual*, and the last block component, $r_{\text{pri}} = Ax - b$, is called the *primal residual*. The middle block,

$$r_{\text{cent}} = -\operatorname{\mathbf{diag}}(\lambda)f(x) - (1/t)\mathbf{1},$$

is the *centrality residual*, *i.e.*, the residual for the modified complementarity condition.

Now consider the Newton step for solving the nonlinear equations $r_t(x, \lambda, \nu) = 0$, for fixed t (without first eliminating λ, as in §11.3.4), at a point (x, λ, ν) that satisifes $f(x) \prec 0$, $\lambda \succ 0$. We will denote the current point and Newton step as

$$y = (x, \lambda, \nu), \qquad \Delta y = (\Delta x, \Delta \lambda, \Delta \nu),$$

respectively. The Newton step is characterized by the linear equations

$$r_t(y + \Delta y) \approx r_t(y) + Dr_t(y)\Delta y = 0,$$

i.e., $\Delta y = -Dr_t(y)^{-1}r_t(y)$. In terms of x, λ, and ν, we have

$$\begin{bmatrix} \nabla^2 f_0(x) + \sum_{i=1}^m \lambda_i \nabla^2 f_i(x) & Df(x)^T & A^T \\ -\operatorname{\mathbf{diag}}(\lambda)Df(x) & -\operatorname{\mathbf{diag}}(f(x)) & 0 \\ A & 0 & 0 \end{bmatrix} \begin{bmatrix} \Delta x \\ \Delta \lambda \\ \Delta \nu \end{bmatrix} = -\begin{bmatrix} r_{\text{dual}} \\ r_{\text{cent}} \\ r_{\text{pri}} \end{bmatrix}.$$
$$(11.54)$$

The *primal-dual search direction* $\Delta y_{\text{pd}} = (\Delta x_{\text{pd}}, \Delta \lambda_{\text{pd}}, \Delta \nu_{\text{pd}})$ is defined as the solution of (11.54).

The primal and dual search directions are coupled, both through the coefficient matrix and the residuals. For example, the primal search direction Δx_{pd} depends on the current value of the dual variables λ and ν, as well as x. We note also that if x satisfies $Ax = b$, *i.e.*, the primal feasibility residual r_{pri} is zero, then we have $A\Delta x_{\text{pd}} = 0$, so Δx_{pd} defines a (primal) feasible direction: for any s, $x + s\Delta x_{\text{pd}}$ will satisfy $A(x + s\Delta x_{\text{pd}}) = b$.

Comparison with barrier method search directions

The primal-dual search directions are closely related to the search directions used in the barrier method, but not quite the same. We start with the linear equations (11.54) that define the primal-dual search directions. We eliminate the variable $\Delta \lambda_{\text{pd}}$, using

$$\Delta \lambda_{\text{pd}} = -\operatorname{\mathbf{diag}}(f(x))^{-1} \operatorname{\mathbf{diag}}(\lambda)Df(x)\Delta x_{\text{pd}} + \operatorname{\mathbf{diag}}(f(x))^{-1}r_{\text{cent}},$$

which comes from the second block of equations. Substituting this into the first block of equations gives

$$\begin{bmatrix} H_{\text{pd}} & A^T \\ A & 0 \end{bmatrix} \begin{bmatrix} \Delta x_{\text{pd}} \\ \Delta \nu_{\text{pd}} \end{bmatrix}$$
$$= -\begin{bmatrix} r_{\text{dual}} + Df(x)^T \operatorname{\mathbf{diag}}(f(x))^{-1}r_{\text{cent}} \\ r_{\text{pri}} \end{bmatrix}$$
$$= -\begin{bmatrix} \nabla f_0(x) + (1/t)\sum_{i=1}^m \frac{1}{-f_i(x)}\nabla f_i(x) + A^T \nu \\ r_{\text{pri}} \end{bmatrix}, \qquad (11.55)$$

where

$$H_{\mathrm{pd}} = \nabla^2 f_0(x) + \sum_{i=1}^{m} \lambda_i \nabla^2 f_i(x) + \sum_{i=1}^{m} \frac{\lambda_i}{-f_i(x)} \nabla f_i(x) \nabla f_i(x)^T. \qquad (11.56)$$

We can compare (11.55) to the equation (11.14), which defines the Newton step for the centering problem in the barrier method with parameter t. This equation can be written as

$$
\begin{bmatrix} H_{\mathrm{bar}} & A^T \\ A & 0 \end{bmatrix} \begin{bmatrix} \Delta x_{\mathrm{bar}} \\ \nu_{\mathrm{bar}} \end{bmatrix}
$$
$$
= -\begin{bmatrix} t\nabla f_0(x) + \nabla \phi(x) \\ r_{\mathrm{pri}} \end{bmatrix}
$$
$$
= -\begin{bmatrix} t\nabla f_0(x) + \sum_{i=1}^{m} \frac{1}{-f_i(x)} \nabla f_i(x) \\ r_{\mathrm{pri}} \end{bmatrix}, \qquad (11.57)
$$

where

$$H_{\mathrm{bar}} = t\nabla^2 f_0(x) + \sum_{i=1}^{m} \frac{1}{-f_i(x)} \nabla^2 f_i(x) + \sum_{i=1}^{m} \frac{1}{f_i(x)^2} \nabla f_i(x) \nabla f_i(x)^T. \qquad (11.58)$$

(Here we give the general expression for the infeasible Newton step; if the current x is feasible, i.e., $r_{\mathrm{pri}} = 0$, then Δx_{bar} coincides with the feasible Newton step Δx_{nt} defined in (11.14).)

Our first observation is that the two systems of equations (11.55) and (11.57) are very similar. The coefficient matrices in (11.55) and (11.57) have the same structure; indeed, the matrices H_{pd} and H_{bar} are both positive linear combinations of the matrices

$$\nabla^2 f_0(x), \qquad \nabla^2 f_1(x), \ldots, \nabla^2 f_m(x), \qquad \nabla f_1(x)\nabla f_1(x)^T, \ldots, \nabla f_m(x)\nabla f_m(x)^T.$$

This means that the same method can be used to compute the primal-dual search directions and the barrier method Newton step.

We can say more about the relation between the primal-dual equations (11.55) and the barrier method equations (11.57). Suppose we divide the first block of equation (11.57) by t, and define the variable $\Delta \nu_{\mathrm{bar}} = (1/t)\nu_{\mathrm{bar}} - \nu$ (where ν is arbitrary). Then we obtain

$$
\begin{bmatrix} (1/t)H_{\mathrm{bar}} & A^T \\ A & 0 \end{bmatrix} \begin{bmatrix} \Delta x_{\mathrm{bar}} \\ \Delta \nu_{\mathrm{bar}} \end{bmatrix} = -\begin{bmatrix} \nabla f_0(x) + (1/t)\sum_{i=1}^{m} \frac{1}{-f_i(x)} \nabla f_i(x) + A^T \nu \\ r_{\mathrm{pri}} \end{bmatrix}.
$$

In this form, the righthand side is identical to the righthand side of the primal-dual equations (evaluated at the same x, λ, and ν). The coefficient matrices differ only in the $1, 1$ block:

$$H_{\mathrm{pd}} = \nabla^2 f_0(x) + \sum_{i=1}^{m} \lambda_i \nabla^2 f_i(x) + \sum_{i=1}^{m} \frac{\lambda_i}{-f_i(x)} \nabla f_i(x) \nabla f_i(x)^T,$$

$$(1/t)H_{\mathrm{bar}} = \nabla^2 f_0(x) + \sum_{i=1}^{m} \frac{1}{-tf_i(x)} \nabla^2 f_i(x) + \sum_{i=1}^{m} \frac{1}{tf_i(x)^2} \nabla f_i(x) \nabla f_i(x)^T.$$

When x and λ satisfy $-f_i(x)\lambda_i = 1/t$, the coefficient matrices, and therefore also the search directions, coincide.

11.7.2 The surrogate duality gap

In the primal-dual interior-point method the iterates $x^{(k)}$, $\lambda^{(k)}$, and $\nu^{(k)}$ are not necessarily feasible, except in the limit as the algorithm converges. This means that we cannot easily evaluate a duality gap $\eta^{(k)}$ associated with step k of the algorithm, as we do in (the outer steps of) the barrier method. Instead we define the *surrogate duality gap*, for any x that satisfies $f(x) \prec 0$ and $\lambda \succeq 0$, as

$$\hat\eta(x, \lambda) = -f(x)^T \lambda. \tag{11.59}$$

The surrogate gap $\hat\eta$ would be the duality gap, if x were primal feasible and λ, ν were dual feasible, *i.e.*, if $r_{\mathrm{pri}} = 0$ and $r_{\mathrm{dual}} = 0$. Note that the value of the parameter t that corresponds to the surrogate duality gap $\hat\eta$ is $m/\hat\eta$.

11.7.3 Primal-dual interior-point method

We can now describe the basic primal-dual interior-point algorithm.

Algorithm 11.2 *Primal-dual interior-point method.*

given x that satisfies $f_1(x) < 0, \ldots, f_m(x) < 0$, $\lambda \succ 0$, $\mu > 1$, $\epsilon_{\mathrm{feas}} > 0$, $\epsilon > 0$.

repeat

    1. *Determine t.* Set $t := \mu m/\hat\eta$.

    2. Compute primal-dual search direction Δy_{pd}.

    3. *Line search and update.*

        Determine step length $s > 0$ and set $y := y + s\Delta y_{\mathrm{pd}}$.

until $\|r_{\mathrm{pri}}\|_2 \le \epsilon_{\mathrm{feas}}$, $\|r_{\mathrm{dual}}\|_2 \le \epsilon_{\mathrm{feas}}$, and $\hat\eta \le \epsilon$.

In step 1, the parameter t is set to a factor μ times $m/\hat\eta$, which is the value of t associated with the current surrogate duality gap $\hat\eta$. If x, λ, and ν were central, with parameter t (and therefore with duality gap m/t), then in step 1 we would increase t by the factor μ, which is exactly the update used in the barrier method. Values of the parameter μ on the order of 10 appear to work well.

The primal-dual interior-point algorithm terminates when x is primal feasible and λ, ν are dual feasible (within the tolerance ϵ_{feas}) and the surrogate gap is smaller than the tolerance ϵ. Since the primal-dual interior-point method often has faster than linear convergence, it is common to choose ϵ_{feas} and ϵ small.

Line search

The line search in the primal-dual interior point method is a standard backtracking line search, based on the norm of the residual, and modified to ensure that $\lambda \succ 0$ and $f(x) \prec 0$. We denote the current iterate as x, λ, and ν, and the next iterate as x^+, λ^+, and ν^+, *i.e.*,

$$x^+ = x + s\Delta x_{\mathrm{pd}}, \qquad \lambda^+ = \lambda + s\Delta\lambda_{\mathrm{pd}}, \qquad \nu^+ = \nu + s\Delta\nu_{\mathrm{pd}}.$$

The residual, evaluated at y^+, will be denoted r^+.

We first compute the largest positive step length, not exceeding one, that gives $\lambda^+ \succeq 0$, *i.e.*,

$$\begin{aligned} s^{\max} &= \sup\{s \in [0,1] \mid \lambda + s\Delta\lambda \succeq 0\} \\ &= \min\{1, \ \min\{-\lambda_i/\Delta\lambda_i \mid \Delta\lambda_i < 0\}\}. \end{aligned}$$

We start the backtracking with $s = 0.99s^{\max}$, and multiply s by $\beta \in (0,1)$ until we have $f(x^+) \prec 0$. We continue multiplying s by β until we have

$$\|r_t(x^+, \lambda^+, \nu^+)\|_2 \le (1 - \alpha s)\|r_t(x, \lambda, \nu)\|_2.$$

Common choices for the backtracking parameters α and β are the same as those for Newton's method: α is typically chosen in the range 0.01 to 0.1, and β is typically chosen in the range 0.3 to 0.8.

One iteration of the primal-dual interior-point algorithm is the same as one step of the infeasible Newton method, applied to solving $r_t(x, \lambda, \nu) = 0$, but modified to ensure $\lambda \succ 0$ and $f(x) \prec 0$ (or, equivalently, with $\mathbf{dom}\, r_t$ restricted to $\lambda \succ 0$ and $f(x) \prec 0$). The same arguments used in the proof of convergence of the infeasible start Newton method show that the line search for the primal-dual method always terminates in a finite number of steps.

11.7.4 Examples

We illustrate the performance of the primal-dual interior-point method for the same problems considered in §11.3.2. The only difference is that instead of starting with a point on the central path, as in §11.3.2, we start the primal-dual interior-point method at a randomly generated $x^{(0)}$, that satisfies $f(x) \prec 0$, and take $\lambda_i^{(0)} = -1/f_i(x^{(0)})$, so the initial value of the surrogate gap is $\hat{\eta} = 100$. The parameter values we use for the primal-dual interior-point method are

$$\mu = 10, \qquad \beta = 0.5, \qquad \epsilon = 10^{-8}, \qquad \alpha = 0.01.$$

Small LP and GP

We first consider the small LP used in §11.3.2, with $m = 100$ inequalities and $n = 50$ variables. Figure 11.21 shows the progress of the primal-dual interior-point method. Two plots are shown: the surrogate gap $\hat{\eta}$, and the norm of the primal and dual residuals,

$$r_{\text{feas}} = \left(\|r_{\text{pri}}\|_2^2 + \|r_{\text{dual}}\|_2^2\right)^{1/2},$$

versus iteration number. (The initial point is primal feasible, so the plot shows the norm of the dual feasibility residual.) The plots show that the residual converges to zero rapidly, and becomes zero to numerical precision in 24 iterations. The surrogate gap also converges rapidly. Compared to the barrier method, the primal-dual interior-point method is faster, especially when high accuracy is required.

Figure 11.22 shows the progress of the primal-dual interior-point method on the GP considered in §11.3.2. The convergence is similar to the LP example.

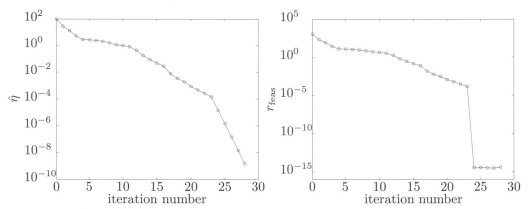

Figure 11.21 Progress of the primal-dual interior-point method for an LP, showing surrogate duality gap $\hat{\eta}$ and the norm of the primal and dual residuals, versus iteration number. The residual converges rapidly to zero within 24 iterations; the surrogate gap also converges to a very small number in about 28 iterations. The primal-dual interior-point method converges faster than the barrier method, especially if high accuracy is required.

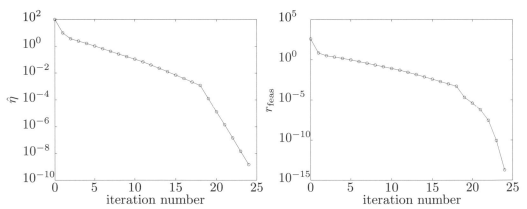

Figure 11.22 Progress of primal-dual interior-point method for a GP, showing surrogate duality gap $\hat{\eta}$ and the norm of the primal and dual residuals versus iteration number.

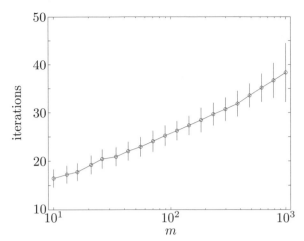

Figure 11.23 Number of iterations required to solve randomly generated standard LPs of different dimensions, with $n = 2m$. Error bars show standard deviation, around the average value, for 100 instances of each dimension. The growth in the number of iterations required, as the problem dimensions range over a 100:1 ratio, is approximately logarithmic.

A family of LPs

Here we examine the performance of the primal-dual method as a function of the problem dimensions, for the same family of standard form LPs considered in §11.3.2. We use the primal-dual interior-point method to solve the same 2000 instances, which consist of 100 instances for each value of m. The primal-dual algorithm is started at $x^{(0)} = \mathbf{1}$, $\lambda^{(0)} = \mathbf{1}$, $\nu^{(0)} = 0$, and terminated using tolerance $\epsilon = 10^{-8}$. Figure 11.23 shows the average, and standard deviation, of the number of iterations required versus m. The number of iterations ranges from 15 to 35, and grows approximately as the logarithm of m. Comparing with the results for the barrier method shown in figure 11.8, we see that the number of iterations in the primal-dual method is only slightly higher, despite the fact that we start at infeasible starting points, and solve the problem to a much higher accuracy.

1.8 Implementation

The main effort in the barrier method is computing the Newton step for the centering problem, which consists of solving sets of linear equations of the form

$$\begin{bmatrix} H & A^T \\ A & 0 \end{bmatrix} \begin{bmatrix} \Delta x_{\text{nt}} \\ \nu_{\text{nt}} \end{bmatrix} = - \begin{bmatrix} g \\ 0 \end{bmatrix}, \tag{11.60}$$

where

$$H = t\nabla^2 f_0(x) + \sum_{i=1}^{m} \frac{1}{f_i(x)^2} \nabla f_i(x) \nabla f_i(x)^T + \sum_{i=1}^{m} \frac{1}{-f_i(x)} \nabla^2 f_i(x)$$

$$g \quad = \quad t\nabla f_0(x) + \sum_{i=1}^{m} \frac{1}{-f_i(x)} \nabla f_i(x).$$

The Newton equations for the primal-dual method have exactly the same structure, so our observations in this section apply to the primal-dual method as well.

The coefficient matrix of (11.60) has KKT structure, so all of the discussion in §9.7 and §10.4 applies here. In particular, the equations can be solved by elimination, and structure such as sparsity or diagonal plus low rank can be exploited. Let us give some generic examples in which the special structure of the KKT equations can be exploited to compute the Newton step more efficiently.

Sparse problems

If the original problem is sparse, which means that the objective and every constraint function each depend on only a modest number of variables, then the gradients and Hessian matrices of the objective and constraint functions are all sparse, as is the coefficient matrix A. Provided m is not too big, the matrix H is then likely to be sparse, so a sparse matrix method can be used to compute the Newton step. The method will likely work well if there are a few relatively dense rows and columns in the KKT matrix, which would occur, for example, if there were a few equality constraints involving a large number of variables.

Separable objective and a few linear inequality constraints

Suppose the objective function is separable, and there are only a relatively small number of linear equality and inequality constraints. Then $\nabla^2 f_0(x)$ is diagonal, and the terms $\nabla^2 f_i(x)$ vanish, so the matrix H is diagonal plus low rank. Since H is easily inverted, we can solve the KKT equations efficiently. The same method can be applied whenever $\nabla^2 f_0(x)$ is easily inverted, *e.g.*, banded, sparse, or block diagonal.

11.8.1   Standard form linear programming

We first discuss the implementation of the barrier method for the standard form LP

$$\begin{array}{ll} \text{minimize} & c^T x \\ \text{subject to} & Ax = b, \quad x \succeq 0, \end{array}$$

with $A \in \mathbf{R}^{m \times n}$. The Newton equations for the centering problem

$$\begin{array}{ll} \text{minimize} & tc^T x - \sum_{i=1}^{n} \log x_i \\ \text{subject to} & Ax = b \end{array}$$

are given by

$$\begin{bmatrix} \mathbf{diag}(x)^{-2} & A^T \\ A & 0 \end{bmatrix} \begin{bmatrix} \Delta x_{\text{nt}} \\ \nu_{\text{nt}} \end{bmatrix} = \begin{bmatrix} -tc + \mathbf{diag}(x)^{-1}\mathbf{1} \\ 0 \end{bmatrix}.$$

These equations are usually solved by block elimination of Δx_{nt}. From the first equation,

$$
\begin{aligned}
\Delta x_{\mathrm{nt}} &= \mathbf{diag}(x)^2(-tc + \mathbf{diag}(x)^{-1}\mathbf{1} - A^T \nu_{\mathrm{nt}}) \\
&= -t\,\mathbf{diag}(x)^2 c + x - \mathbf{diag}(x)^2 A^T \nu_{\mathrm{nt}}.
\end{aligned}
$$

Substituting in the second equation yields

$$
A\,\mathbf{diag}(x)^2 A^T \nu_{\mathrm{nt}} = -tA\,\mathbf{diag}(x)^2 c + b.
$$

The coefficient matrix is positive definite since by assumption $\mathbf{rank}\,A = m$. Moreover if A is sparse, then usually $A\,\mathbf{diag}(x)^2 A^T$ is sparse, so a sparse Cholesky factorization can be used.

11.8.2 ℓ_1-norm approximation

Consider the ℓ_1-norm approximation problem

$$
\text{minimize} \quad \|Ax - b\|_1
$$

with $A \in \mathbf{R}^{m \times n}$. We will discuss the implementation assuming m and n are large, and A is structured, *e.g.*, sparse, and compare it with the cost of the corresponding least-squares problem

$$
\text{minimize} \quad \|Ax - b\|_2^2 \ .
$$

We start by expressing the ℓ_1-norm approximation problem as an LP by introducing auxiliary variables $y \in \mathbf{R}^m$:

$$
\begin{array}{ll}
\text{minimize} & \mathbf{1}^T y \\
\text{subject to} & \begin{bmatrix} A & -I \\ -A & -I \end{bmatrix} \begin{bmatrix} x \\ y \end{bmatrix} \preceq \begin{bmatrix} b \\ -b \end{bmatrix}.
\end{array}
$$

The Newton equation for the centering problem is

$$
\begin{bmatrix} A^T & -A^T \\ -I & -I \end{bmatrix} \begin{bmatrix} D_1 & 0 \\ 0 & D_2 \end{bmatrix} \begin{bmatrix} A & -I \\ -A & -I \end{bmatrix} \begin{bmatrix} \Delta x_{\mathrm{nt}} \\ \Delta y_{\mathrm{nt}} \end{bmatrix} = -\begin{bmatrix} A^T g_1 \\ g_2 \end{bmatrix}
$$

where

$$
D_1 = \mathbf{diag}(b - Ax + y)^{-2}, \qquad D_2 = \mathbf{diag}(-b + Ax + y)^{-2}
$$

and

$$
\begin{aligned}
g_1 &= \mathbf{diag}(b - Ax + y)^{-1}\mathbf{1} - \mathbf{diag}(-b + Ax + y)^{-1}\mathbf{1} \\
g_2 &= t\mathbf{1} - \mathbf{diag}(b - Ax + y)^{-1}\mathbf{1} - \mathbf{diag}(-b + Ax + y)^{-1}\mathbf{1}.
\end{aligned}
$$

If we multiply out the lefthand side, this can be simplified as

$$
\begin{bmatrix} A^T(D_1 + D_2)A & -A^T(D_1 - D_2) \\ -(D_1 - D_2)A & D_1 + D_2 \end{bmatrix} \begin{bmatrix} \Delta x_{\mathrm{nt}} \\ \Delta y_{\mathrm{nt}} \end{bmatrix} = -\begin{bmatrix} A^T g_1 \\ g_2 \end{bmatrix}.
$$

Applying block elimination to Δy_{nt}, we can reduce this to

$$A^T DA\Delta x_{\mathrm{nt}} = -A^T g \tag{11.61}$$

where

$$D = 4D_1 D_2 (D_1 + D_2)^{-1} = 2(\mathbf{diag}(y)^2 + \mathbf{diag}(b - Ax)^2)^{-1}$$

and

$$g = g_1 + (D_1 - D_2)(D_1 + D_2)^{-1} g_2.$$

After solving for Δx_{nt}, we obtain Δy_{nt} from

$$\Delta y_{\mathrm{nt}} = (D_1 + D_2)^{-1}(-g_2 + (D_1 - D_2)A\Delta x_{\mathrm{nt}}).$$

It is interesting to note that (11.61) are the normal equations of a weighted least-squares problem

$$\text{minimize} \quad \|D^{1/2}(A\Delta x + D^{-1}g)\|_2.$$

In other words, the cost of solving the ℓ_1-norm approximation problem is the cost of solving a relatively small number of weighted least-squares problems with the same matrix A, and weights that change at each iteration. If A has structure that allows us to solve the least-squares problem fast (for example, by exploiting sparsity), then we can solve (11.61) fast.

11.8.3   Semidefinite programming in inequality form

We consider the SDP

$$
\begin{array}{ll}
\text{minimize} & c^T x \\
\text{subject to} & \sum_{i=1}^{n} x_i F_i + G \preceq 0,
\end{array}
$$

with variable $x \in \mathbf{R}^n$, and parameters $F_1, \ldots, F_n, G \in \mathbf{S}^p$. The associated centering problem, using the log-determinant barrier function, is

$$\text{minimize} \quad tc^T x - \log\det(-\sum_{i=1}^{n} x_i F_i - G).$$

The Newton step Δx_{nt} is found from $H\Delta x_{\mathrm{nt}} = -g$, where the Hessian and gradient are given by

$$
\begin{aligned}
H_{ij} &= \mathbf{tr}(S^{-1}F_i S^{-1}F_j), \quad i,\, j = 1, \ldots, n \\
g_i &= tc_i + \mathbf{tr}(S^{-1}F_i), \quad i = 1, \ldots, n,
\end{aligned}
$$

where $S = -\sum_{i=1}^{n} x_i F_i - G$. One standard approach is to form H (and g), and then solve the Newton equation via Cholesky factorization.

We first consider the unstructured case, *i.e.*, we assume all matrices are dense. We will also just keep track of the order in the flop count, with respect to the problem dimensions n and p. We first form S, which costs order np^2 flops. We then compute the matrices $S^{-1}F_i$, for each i, via Cholesky factorization of S, and then back substitution with the columns of F_i (or forming S^{-1} and multiplying by F_i). This cost is order p^3 for each i, so the total cost is order np^3. Finally, we form H_{ij} as the inner product of the matrices $S^{-1}F_i$ and $S^{-1}F_j$, which costs order p^2 flops. Since we do this for $n(n+1)/2$ such pairs, the cost is order n^2p^2. Solving for the Newton direction costs order n^3. The dominating order is thus $\max\{np^3, n^2p^2, n^3\}$.

It is not possible, in general, to exploit sparsity in the matrices F_i and G, since H is often dense, even when F_i and G are sparse. One exception is when F_i and G have a common block diagonal structure, in which case all the operations described above can be carried out block by block.

It is often possible to exploit (common) sparsity in F_i and G to form the (dense) Hessian H more efficiently. If we can find an ordering that results in S having a reasonably sparse Cholesky factor, then we can compute the matrices $S^{-1}F_i$ efficiently, and form H_{ij} far more efficiently.

One interesting example that arises frequently is an SDP with matrix inequality

$$\mathbf{diag}(x) \preceq B.$$

This corresponds to $F_i = E_{ii}$, where E_{ii} is the matrix with i,i entry one and all others zero. In this case, the matrix H can be found very efficiently:

$$H_{ij} = (S^{-1})_{ij}^2,$$

where $S = B - \mathbf{diag}(x)$. The cost of forming H is thus the cost of forming S^{-1}, which is at most (*i.e.*, when no other structure is exploited) order n^3.

1.8.4   Network rate optimization

We consider a variation on the optimal network flow problem described in §10.4.3 (page 550), which is sometimes called the *network rate optimization problem*. The network is described as a directed graph with L arcs or links. Goods, or packets of information, travel on the network, passing through the links. The network supports n *flows*, with (nonnegative) *rates* $x_1, \ldots, x_n$, which are the optimization variables. Each flow moves along a fixed, or pre-determined, *path* (or *route*) in the network, from a source node to a destination node. Each link can support multiple flows passing through it. The *total traffic* on a link is the sum of the flow rates of the flows that travel over the link. Each link has a positive *capacity*, which is the maximum total traffic it can handle.

We can describe these link capacity limits using the *flow-link incidence matrix* $A \in \mathbf{R}^{L \times n}$, defined as

$$A_{ij} = \begin{cases} 1 & \text{flow } j \text{ passes through link } i \\ 0 & \text{otherwise.} \end{cases}$$

The total traffic on link i is then given by $(Ax)_i$, so the link capacity constraints can be expressed as $Ax \preceq c$, where c_i is the capacity of link i. Usually each path passes through only a small fraction of the total number of links, so the matrix A is sparse.

In the network rate problem the paths are fixed (and encoded in the matrix A, which is a problem parameter); the variables are the flow rates x_i. The objective is to choose the flow rates to maximize a separable utility function U, given by

$$U(x) = U_1(x_1) + \cdots + U_n(x_n).$$

We assume that each U_i (and hence, U) is concave and nondecreasing. We can think of $U_i(x_i)$ as the income derived from supporting the ith flow at rate x_i; $U(x)$ is then the total income associated with the flows. The network rate optimization problem is then

$$\begin{array}{ll} \text{maximize} & U(x) \\ \text{subject to} & Ax \preceq c, \quad x \succeq 0, \end{array} \tag{11.62}$$

which is a convex optimization problem.

Let us apply the barrier method to solve this problem. At each step we must minimize a function of the form

$$-tU(x) - \sum_{i=1}^{L} \log(c - Ax)_i - \sum_{j=1}^{n} \log x_j,$$

using Newton's method. The Newton step Δx_{nt} is found by solving the linear equations

$$(D_0 + A^T D_1 A + D_2)\Delta x_{\text{nt}} = -g,$$

where

$$\begin{array}{rcl} D_0 & = & -t\,\mathbf{diag}(U_1''(x), \ldots, U_n''(x)) \\ D_1 & = & \mathbf{diag}(1/(c - Ax)_1^2, \ldots, 1/(c - Ax)_L^2) \\ D_2 & = & \mathbf{diag}(1/x_1^2, \ldots, 1/x_n^2) \end{array}$$

are diagonal matrices, and $g \in \mathbf{R}^n$. We can describe the sparsity structure of this $n \times n$ coefficient matrix precisely:

$$(D_0 + A^T D_1 A + D_2)_{ij} \neq 0$$

if and only if flow i and flow j share a link. If the paths are relatively short, and each link has relatively few paths passing through it, then this matrix is sparse, so a sparse Cholesky factorization can be used. We can also solve the Newton system efficiently when some, but not too many, of the rows and columns are relatively dense. This occurs when a few of the flows intersect with a large number of the other flows, which might occur if a few flows are relatively long.

We can also use the matrix inversion lemma to compute the Newton step by solving a system with $L \times L$ coefficient matrix, with form

$$(D_1^{-1} + A(D_0 + D_2)^{-1} A^T)y = -A(D_0 + D_2)^{-1}g,$$

and then computing

$$\Delta x_{\text{nt}} = -(D_0 + D_2)^{-1}(g + A^T y).$$

Here too we can precisely describe the sparsity pattern:

$$(D_1^{-1} + A(D_0 + D_2)^{-1} A^T)_{ij} \neq 0$$

if and only if there is a path that passes through link i and link j. If most paths are short, this matrix is sparse. This matrix will be sparse, with a few dense rows and columns, if there are a few bottlenecks, *i.e.*, a few links over which many flows travel.

Bibliography

The early history of the barrier method is described in detail by Fiacco and McCormick [FM90, §1.2]. The method was a popular algorithm for convex optimization in the 1960s, along with closely related techniques such as the method of centers (Liêu and Huard [LH66]; see also exercise 11.11), and penalty (or exterior-point) methods [FM90, §4]. Interest declined in the 1970s amid concerns about the ill-conditioning of the Newton equations of the centering problem (11.6) for high values of t.

The barrier method regained popularity in the 1980s, after Gill, Murray, Saunders, Tomlin, and Wright [GMS$^+$86] pointed out the close connections with Karmarkar's polynomial-time projective algorithm for linear programming [Kar84]. The focus of research throughout the 1980s remained on linear (and to a lesser extent, quadratic) programming, resulting in different variations of the basic interior-point methods, and improved worst-case complexity results (see Gonzaga [Gon92]). Primal-dual methods emerged as the algorithms of choice for practical implementations (see Mehrotra [Meh92], Lustig, Marsten, and Shanno [LMS94], Wright [Wri97]).

In their 1994 book, Nesterov and Nemirovski extended the complexity theory of linear programming interior-point methods to nonlinear convex optimization problems, using the convergence theory of Newton's method for self-concordant functions. They also developed interior-point methods for problems with generalized inequalities, and discussed ways of reformulating problems to satisfy the self-concordance assumption. The geometric programming reformulation on page 587, for example, is from [NN94, §6.3.1].

As mentioned on page 585, the complexity analysis shows that, contrary to what one might expect, the centering problems in the barrier method do not become more difficult as t increases, at least not in exact arithmetic. Practical experience, supported by theoretical results (Forsgren, Gill, and Wright [FGW02, §4.3.2], Nocedal and Wright [NW99, page 525]), also indicates that the effects of ill-conditioning on the computed solution of the Newton system are more benign than thought earlier.

Recent research on interior-point methods has concentrated on extending the primal-dual methods for linear programming, which converge faster and reach higher accuracies than (primal) barrier methods, to nonlinear convex problems. One popular approach, along the lines of the simple primal-dual method of §11.7, is based on linearizing modified KKT equations for a convex optimization problem in standard form, i.e., problem (11.1). More sophisticated algorithms of this type differ from algorithm 11.2 in the strategy used to select t (which is crucial to achieve superlinear asymptotic convergence), and the line search. We refer to Wright [Wri97, chapter 8], Ralph and Wright [RW97], den Hertog [dH93], Terlaky [Ter96], and the survey by Forsgren, Gill, and Wright [FGW02, §5] for details and references.

Other authors adopt the cone programming framework as starting point for extending primal-dual interior-point methods for linear programming to convex optimization (see for example, Nesterov and Todd [NT98]). This approach has resulted in efficient and accurate primal-dual methods for semidefinite and second-order programming (see the surveys by Todd [Tod01] and Alizadeh and Goldfarb [AG03]).

As for linear programming, primal-dual methods for semidefinite programming are usually described as variations of Newton's method applied to modified KKT equations. Unlike in linear programming, however, the linearization can be carried out in many different ways, which lead to different search directions and algorithms; see Helmberg, Rendl, Vanderbei, and Wolkowicz [HRVW96], Kojima, Shindo, and Harah [KSH97], Monteiro [Mon97], Nesterov and Todd [NT98], Zhang [Zha98], Alizadeh, Haeberly, and Overton [AHO98], and Todd, Toh, and Tütüncü [TTT98].

Great progress has also been made in the area of initialization and infeasibility detection. Homogeneous self-dual formulations provide an elegant and efficient alternative to the classical two-phase approach of §11.4; see Ye, Todd, and Mizuno [YTM94], Xu, Hung, and Ye [XHY96], Andersen and Ye [AY98] and Luo, Sturm, and Zhang [LSZ00] for details.

The primal-dual interior-point methods for semidefinite and second-order cone programming have been implemented in a number of software packages, including SeDuMi [Stu99], SDPT3 [TTT02], SDPA [FKN98], CSDP [Bor02], and DSDP [BY02], A user-friendly interface to several of these codes is provided by YALMIP [Löf04].

The following books document the recent developments in this rapidly advancing field in greater detail: Vanderbei [Van96], Wright [Wri97], Roos, Terlaky, and Vial [RTV97] Ye [Ye97], Wolkowicz, Saigal, and Vandenberghe [WSV00], Ben-Tal and Nemirovski, [BTN01], Renegar [Ren01], and Peng, Roos, and Terlaky [PRT02].

Exercises

The barrier method

11.1 *Barrier method example.* Consider the simple problem

$$\begin{array}{ll}
\text{minimize} & x^2 + 1 \\
\text{subject to} & 2 \le x \le 4,
\end{array}$$

which has feasible set $[2, 4]$, and optimal point $x^\star = 2$. Plot f_0, and $tf_0 + \phi$, for several values of $t > 0$, versus x. Label $x^\star(t)$.

11.2 What happens if the barrier method is applied to the LP

$$\begin{array}{ll}
\text{minimize} & x_2 \\
\text{subject to} & x_1 \le x_2, \quad 0 \le x_2,
\end{array}$$

with variable $x \in \mathbf{R}^2$?

11.3 *Boundedness of centering problem.* Suppose the sublevel sets of (11.1),

$$\begin{array}{ll}
\text{minimize} & f_0(x) \\
\text{subject to} & f_i(x) \le 0, \quad i = 1, \dots, m \\
& Ax = b,
\end{array}$$

are bounded. Show that the sublevel sets of the associated centering problem,

$$\begin{array}{ll}
\text{minimize} & tf_0(x) + \phi(x) \\
\text{subject to} & Ax = b,
\end{array}$$

are bounded.

11.4 *Adding a norm bound to ensure strong convexity of the centering problem.* Suppose we add the constraint $x^T x \le R^2$ to the problem (11.1):

$$\begin{array}{ll}
\text{minimize} & f_0(x) \\
\text{subject to} & f_i(x) \le 0, \quad i = 1, \dots, m \\
& Ax = b \\
& x^T x \le R^2.
\end{array}$$

Let $\tilde{\phi}$ denote the logarithmic barrier function for this modified problem. Find $a > 0$ for which $\nabla^2(tf_0(x) + \tilde{\phi}(x)) \succeq aI$ holds, for all feasible x.

11.5 *Barrier method for second-order cone programming.* Consider the SOCP (without equality constraints, for simplicity)

$$\begin{array}{ll}
\text{minimize} & f^T x \\
\text{subject to} & \|A_i x + b_i\|_2 \le c_i^T x + d_i, \quad i = 1, \dots, m.
\end{array} \tag{11.63}$$

The constraint functions in this problem are not differentiable (since the Euclidean norm $\|u\|_2$ is not differentiable at $u = 0$) so the (standard) barrier method cannot be applied. In §11.6, we saw that this SOCP can be solved by an extension of the barrier method that handles generalized inequalities. (See example 11.8, page 599, and page 601.) In this exercise, we show how the standard barrier method (with scalar constraint functions) can be used to solve the SOCP.

We first reformulate the SOCP as

$$\begin{array}{ll}
\text{minimize} & f^T x \\
\text{subject to} & \|A_i x + b_i\|_2^2/(c_i^T x + d_i) \le c_i^T x + d_i, \quad i = 1, \dots, m \\
& c_i^T x + d_i \ge 0, \quad i = 1, \dots, m.
\end{array} \tag{11.64}$$

The constraint function

$$f_i(x) = \frac{\|A_i x + b_i\|_2^2}{c_i^T x + d_i} - c_i^T x - d_i$$

is the composition of a quadratic-over-linear function with an affine function, and is twice differentiable (and convex), provided we define its domain as $\mathbf{dom}\, f_i = \{x \mid c_i^T x + d_i > 0\}$. Note that the two problems (11.63) and (11.64) are not exactly equivalent. If $c_i^T x^\star + d_i = 0$ for some i, where $x^\star$ is the optimal solution of the SOCP (11.63), then the reformulated problem (11.64) is not solvable; $x^\star$ is not in its domain. Nevertheless we will see that the barrier method, applied to (11.64), produces arbitrarily accurate suboptimal solutions of (11.64), and hence also for (11.63).

(a) Form the log barrier ϕ for the problem (11.64). Compare it to the log barrier that arises when the SOCP (11.63) is solved using the barrier method for generalized inequalities (in §11.6).

(b) Show that if $tf^T x + \phi(x)$ is minimized, the minimizer $x^\star(t)$ is $2m/t$-suboptimal for the problem (11.63). It follows that the standard barrier method, applied to the reformulated problem (11.64), solves the SOCP (11.63), in the sense of producing arbitrarily accurate suboptimal solutions. This is the case even though the optimal point $x^\star$ need not be in the domain of the reformulated problem (11.64).

11.6 *General barriers.* The log barrier is based on the approximation $-(1/t) \log(-u)$ of the indicator function $\widehat{I}_-(u)$ (see §11.2.1, page 563). We can also construct barriers from other approximations, which in turn yield generalizations of the central path and barrier method. Let $h : \mathbf{R} \to \mathbf{R}$ be a twice differentiable, closed, increasing convex function, with $\mathbf{dom}\, h = -\mathbf{R}_{++}$. (This implies $h(u) \to \infty$ as $u \to 0$.) One such function is $h(u) = -\log(-u)$; another example is $h(u) = -1/u$ (for $u < 0$).

Now consider the optimization problem (without equality constraints, for simplicity)

$$\begin{array}{ll} \text{minimize} & f_0(x) \\ \text{subject to} & f_i(x) \le 0, \quad i = 1, \ldots, m, \end{array}$$

where f_i are twice differentiable. We define the *h-barrier* for this problem as

$$\phi_h(x) = \sum_{i=1}^m h(f_i(x)),$$

with domain $\{x \mid f_i(x) < 0, \ i = 1, \ldots, m\}$. When $h(u) = -\log(-u)$, this is the usual logarithmic barrier; when $h(u) = -1/u$, ϕ_h is called the *inverse barrier*. We define the *h-central path* as

$$x^\star(t) = \operatorname{argmin} t f_0(x) + \phi_h(x),$$

where $t > 0$ is a parameter. (We assume that for each t, the minimizer exists and is unique.)

(a) Explain why $t f_0(x) + \phi_h(x)$ is convex in x, for each $t > 0$.

(b) Show how to construct a dual feasible λ from $x^\star(t)$. Find the associated duality gap.

(c) For what functions h does the duality gap found in part (b) depend only on t and m (and no other problem data)?

11.7 *Tangent to central path.* This problem concerns $dx^\star(t)/dt$, which gives the tangent to the central path at the point $x^\star(t)$. For simplicity, we consider a problem without equality constraints; the results readily generalize to problems with equality constraints.

(a) Find an explicit expression for $dx^\star(t)/dt$. *Hint.* Differentiate the centrality equations (11.7) with respect to t.

(b) Show that $f_0(x^\star(t))$ decreases as t increases. Thus, the objective value in the barrier method decreases, as the parameter t is increased. (We already know that the duality gap, which is m/t, decreases as t increases.)

11.8 *Predictor-corrector method for centering problems.* In the standard barrier method, $x^\star(\mu t)$ is computed using Newton's method, starting from the initial point $x^\star(t)$. One alternative that has been proposed is to make an approximation or prediction $\widehat{x}$ of $x^\star(\mu t)$, and then start the Newton method for computing $x^\star(\mu t)$ from $\widehat{x}$. The idea is that this should reduce the number of Newton steps, since $\widehat{x}$ is (presumably) a better initial point than $x^\star(t)$. This method of centering is called a *predictor-corrector method*, since it first makes a *prediction* of what $x^\star(\mu t)$ is, then *corrects* the prediction using Newton's method.

The most widely used predictor is the first-order predictor, based on the tangent to the central path, explored in exercise 11.7. This predictor is given by

$$\widehat{x} = x^\star(t) + \frac{dx^\star(t)}{dt}(\mu t - t).$$

Derive an expression for the first-order predictor $\widehat{x}$. Compare it to the Newton update obtained, *i.e.*, $x^\star(t) + \Delta x_{\mathrm{nt}}$, where Δx_{nt} is the Newton step for $\mu t f_0(x) + \phi(x)$, at $x^\star(t)$. What can you say when the objective f_0 is linear? (For simplicity, you can consider a problem without equality constraints.)

11.9 *Dual feasible points near the central path.* Consider the problem

$$\begin{array}{ll} \text{minimize} & f_0(x) \\ \text{subject to} & f_i(x) \le 0, \quad i = 1, \dots, m, \end{array}$$

with variable $x \in \mathbf{R}^n$. We assume the functions f_i are convex and twice differentiable. (We assume for simplicity there are no equality constraints.) Recall (from §11.2.2, page 565) that $\lambda_i = -1/(t f_i(x^\star(t)))$, $i = 1, \dots, m$, is dual feasible, and in fact, $x^\star(t)$ minimizes $L(x, \lambda)$. This allows us to evaluate the dual function for λ, which turns out to be $g(\lambda) = f_0(x^\star(t)) - m/t$. In particular, we conclude that $x^\star(t)$ is m/t-suboptimal.

In this problem we consider what happens when a point x is close to $x^\star(t)$, but not quite centered. (This would occur if the centering steps were terminated early, or not carried out to full accuracy.) In this case, of course, we cannot claim that $\lambda_i = -1/(t f_i(x))$, $i = 1, \dots, m$, is dual feasible, or that x is m/t-suboptimal. However, it turns out that a slightly more complicated formula does yield a dual feasible point, provided x is close enough to centered.

Let Δx_{nt} be the Newton step at x of the centering problem

$$\text{minimize} \quad t f_0(x) - \sum_{i=1}^{m} \log(-f_i(x)).$$

A formula that often gives a dual feasible point when Δx_{nt} is small (*i.e.*, for x nearly centered) is

$$\lambda_i = \frac{1}{-t f_i(x)} \left(1 + \frac{\nabla f_i(x)^T \Delta x_{\mathrm{nt}}}{-f_i(x)} \right), \quad i = 1, \dots, m.$$

In this case, the vector x *does not* minimize $L(x, \lambda)$, so there is no general formula for the dual function value $g(\lambda)$ associated with λ. (If we have an analytical expression for the dual objective, however, we can simply evaluate $g(\lambda)$.)

Verify that for a QCQP

$$\begin{array}{ll} \text{minimize} & (1/2)x^T P_0 x + q_0^T x + r_0 \\ \text{subject to} & (1/2)x^T P_i x + q_i^T x + r_i \le 0, \quad i = 1, \dots, m, \end{array}$$

the formula for λ yields a dual feasible point (*i.e.*, $\lambda \succeq 0$ and $L(x, \lambda)$ is bounded below) when Δx_{nt} is sufficiently small.

Hint. Define

$$x_0 = x + \Delta x_{\text{nt}}, \qquad x_i = x - \frac{1}{t\lambda_i f_i(x)}\Delta x_{\text{nt}}, \quad i = 1, \ldots, m.$$

Show that

$$\nabla f_0(x_0) + \sum_{i=1}^{m} \lambda_i \nabla f_i(x_i) = 0.$$

Now use $f_i(z) \geq f_i(x_i) + \nabla f_i(x_i)^T(z - x_i)$, $i = 0, \ldots, m$, to derive a lower bound on $L(z, \lambda)$.

11.10 *Another parametrization of the central path.* We consider the problem (11.1), with central path $x^\star(t)$ for $t > 0$, defined as the solution of

$$
\begin{array}{ll}
\text{minimize} & tf_0(x) - \sum_{i=1}^{m} \log(-f_i(x)) \\
\text{subject to} & Ax = b.
\end{array}
$$

In this problem we explore another parametrization of the central path.
For $u > p^\star$, let $z^\star(u)$ denote the solution of

$$
\begin{array}{ll}
\text{minimize} & -\log(u - f_0(x)) - \sum_{i=1}^{m} \log(-f_i(x)) \\
\text{subject to} & Ax = b.
\end{array}
$$

Show that the curve defined by $z^\star(u)$, for $u > p^\star$, is the central path. (In other words, for each $u > p^\star$, there is a $t > 0$ for which $x^\star(t) = z^\star(u)$, and conversely, for each $t > 0$, there is an $u > p^\star$ for which $z^\star(u) = x^\star(t)$).

11.11 *Method of analytic centers.* In this problem we consider a variation on the barrier method, based on the parametrization of the central path described in exercise 11.10. For simplicity, we consider a problem with no equality constraints,

$$
\begin{array}{ll}
\text{minimize} & f_0(x) \\
\text{subject to} & f_i(x) \leq 0, \quad i = 1, \ldots, m.
\end{array}
$$

The method of analytic centers starts with any strictly feasible initial point $x^{(0)}$, and any $u^{(0)} > f_0(x^{(0)})$. We then set

$$u^{(1)} = \theta u^{(0)} + (1 - \theta)f_0(x^{(0)}),$$

where $\theta \in (0, 1)$ is an algorithm parameter (usually chosen small), and then compute the next iterate as

$$x^{(1)} = z^\star(u^{(1)})$$

(using Newton's method, starting from $x^{(0)}$). Here $z^\star(s)$ denotes the minimizer of

$$-\log(s - f_0(x)) - \sum_{i=1}^{m} \log(-f_i(x)),$$

which we assume exists and is unique. This process is then repeated.
The point $z^\star(s)$ is the *analytic center* of the inequalities

$$f_0(x) \leq s, \quad f_1(x) \leq 0, \ldots, f_m(x) \leq 0,$$

hence the algorithm name.
Show that the method of centers works, *i.e.*, $x^{(k)}$ converges to an optimal point. Find a stopping criterion that guarantees that x is ϵ-suboptimal, where $\epsilon > 0$.
Hint. The points $x^{(k)}$ are on the central path; see exercise 11.10. Use this to show that

$$u^+ - p^\star \leq \frac{m + \theta}{m + 1}(u - p^\star),$$

where u and u^+ are the values of u on consecutive iterations.

11.12 *Barrier method for convex-concave games.* We consider a convex-concave game with inequality constraints,

$$\begin{array}{ll} \text{minimize}_w \text{ maximize}_z & f_0(w,z) \\ \text{subject to} & f_i(w) \leq 0, \quad i = 1, \ldots, m \\ & \tilde{f}_i(z) \leq 0, \quad i = 1, \ldots, \tilde{m}. \end{array}$$

Here $w \in \mathbf{R}^n$ is the variable associated with minimizing the objective, and $z \in \mathbf{R}^{\tilde{n}}$ is the variable associated with maximizing the objective. The constraint functions f_i and $\tilde{f}_i$ are convex and differentiable, and the objective function f_0 is differentiable and convex-concave, *i.e.*, convex in w, for each z, and concave in z, for each w. We assume for simplicity that $\textbf{dom } f_0 = \mathbf{R}^n \times \mathbf{R}^{\tilde{n}}$.

A *solution* or *saddle-point* for the game is a pair $w^\star$, $z^\star$, for which

$$f_0(w^\star, z) \leq f_0(w^\star, z^\star) \leq f_0(w, z^\star)$$

holds for every feasible w and z. (For background on convex-concave games and functions, see §5.4.3, §10.3.4 and exercises 3.14, 5.24, 5.25, 10.10, and 10.13.) In this exercise we show how to solve this game using an extension of the barrier method, and the infeasible start Newton method (see §10.3).

(a) Let $t > 0$. Explain why the function

$$t f_0(w,z) - \sum_{i=1}^{m} \log(-f_i(w)) + \sum_{i=1}^{\tilde{m}} \log(-\tilde{f}_i(z))$$

is convex-concave in (w, z). We will assume that it has a unique saddle-point, $(w^\star(t), z^\star(t))$, which can be found using the infeasible start Newton method.

(b) As in the barrier method for solving a convex optimization problem, we can derive a simple bound on the suboptimality of $(w^\star(t), z^\star(t))$, which depends only on the problem dimensions, and decreases to zero as t increases. Let W and Z denote the feasible sets for w and z,

$$W = \{w \mid f_i(w) \leq 0, \ i = 1, \ldots, m\}, \qquad Z = \{z \mid \tilde{f}_i(z) \leq 0, \ i = 1, \ldots, \tilde{m}\}.$$

Show that

$$f_0(w^\star(t), z^\star(t)) \ \leq \ \inf_{w \in W} \ f_0(w, z^\star(t)) + \frac{m}{t},$$

$$f_0(w^\star(t), z^\star(t)) \ \geq \ \sup_{z \in Z} \ f_0(w^\star(t), z) - \frac{\tilde{m}}{t},$$

and therefore

$$\sup_{z \in Z} \ f_0(w^\star(t), z) - \inf_{w \in W} \ f_0(w, z^\star(t)) \leq \frac{m + \tilde{m}}{t}.$$

Self-concordance and complexity analysis

11.13 *Self-concordance and negative entropy.*

(a) Show that the negative entropy function $x \log x$ (on $\mathbf{R}_{++}$) is *not* self-concordant.

(b) Show that for any $t > 0$, $tx \log x - \log x$ is self-concordant (on $\mathbf{R}_{++}$).

11.14 *Self-concordance and the centering problem.* Let ϕ be the logarithmic barrier function of problem (11.1). Suppose that the sublevel sets of (11.1) are bounded, and that $t f_0 + \phi$ is closed and self-concordant. Show that $t \nabla^2 f_0(x) + \nabla^2 \phi(x) \succ 0$, for all $x \in \textbf{dom } \phi$. *Hint.* See exercises 9.17 and 11.3.

Barrier method for generalized inequalities

11.15 *Generalized logarithm is K-increasing.* Let ψ be a generalized logarithm for the proper cone K. Suppose $y \succ_K 0$.

(a) Show that $\nabla\psi(y) \succeq_{K^*} 0$, *i.e.*, that ψ is K-nondecreasing. *Hint.* If $\nabla\psi(y) \not\succeq_{K^*} 0$, then there is some $w \succ_K 0$ for which $w^T \nabla\psi(y) \le 0$. Use the inequality $\psi(sw) \le \psi(y) + \nabla\psi(y)^T(sw - y)$, with $s > 0$.

(b) Now show that $\nabla\psi(y) \succ_{K^*} 0$, *i.e.*, that ψ is K-increasing. *Hint.* Show that $\nabla^2\psi(y) \prec 0$, $\nabla\psi(y) \succeq_{K^*} 0$ imply $\nabla\psi(y) \succ_{K^*} 0$.

11.16 [NN94, page 41] *Properties of a generalized logarithm.* Let ψ be a generalized logarithm for the proper cone K, with degree θ. Prove that the following properties hold at any $y \succ_K 0$.

(a) $\nabla\psi(sy) = \nabla\psi(y)/s$ for all $s > 0$.

(b) $\nabla\psi(y) = -\nabla^2\psi(y)y$.

(c) $y^T \nabla\psi^2(y)y = -\theta$.

(d) $\nabla\psi(y)^T \nabla^2\psi(y)^{-1}\nabla\psi(y) = -\theta$.

11.17 *Dual generalized logarithm.* Let ψ be a generalized logarithm for the proper cone K, with degree θ. Show that the dual generalized logarithm $\overline{\psi}$, defined in (11.49), satisfies

$$\overline{\psi}(sv) = \psi(v) + \theta \log s,$$

for $v \succ_{K^*} 0$, $s > 0$.

11.18 Is the function

$$\psi(y) = \log\left(y_{n+1} - \frac{\sum_{i=1}^n y_i^2}{y_{n+1}}\right),$$

with $\mathbf{dom}\,\psi = \{y \in \mathbf{R}^{n+1} \mid y_{n+1} > \sum_{i=1}^n y_i^2\}$, a generalized logarithm for the second-order cone in $\mathbf{R}^{n+1}$?

Implementation

11.19 *Yet another method for computing the Newton step.* Show that the Newton step for the barrier method, which is given by the solution of the linear equations (11.14), can be found by solving a *larger* set of linear equations with coefficient matrix

$$\begin{bmatrix} t\nabla^2 f_0(x) + \sum_i \frac{1}{-f_i(x)}\nabla^2 f_i(x) & Df(x)^T & A^T \\ Df(x) & -\mathbf{diag}(f(x))^2 & 0 \\ A & 0 & 0 \end{bmatrix}$$

where $f(x) = (f_1(x), \ldots, f_m(x))$.

For what types of problem structure might solving this larger system be interesting?

11.20 *Network rate optimization via the dual problem.* In this problem we examine a dual method for solving the network rate optimization problem of §11.8.4. To simplify the presentation we assume that the utility functions U_i are strictly concave, with $\mathbf{dom}\,U_i = \mathbf{R}_{++}$, and that they satisfy $U_i'(x_i) \to \infty$ as $x_i \to 0$ and $U_i'(x_i) \to 0$ as $x_i \to \infty$.

(a) Express the dual problem of (11.62) in terms of the conjugate utility functions $V_i = (-U_i)^*$, defined as

$$V_i(\lambda) = \sup_{x>0}(\lambda x + U_i(x)).$$

Show that $\mathbf{dom}\,V_i = -\mathbf{R}_{++}$, and that for each $\lambda < 0$ there is a unique x with $U_i'(x) = -\lambda$.

(b) Describe a barrier method for the dual problem. Compare the complexity per iteration with the complexity of the method in §11.8.4. Distinguish the same two cases as in §11.8.4 ($A^T A$ is sparse and AA^T is sparse).

Numerical experiments

11.21 *Log-Chebyshev approximation with bounds.* We consider an approximation problem: find $x \in \mathbf{R}^n$, that satisfies the variable bounds $l \preceq x \preceq u$, and yields $Ax \approx b$, where $b \in \mathbf{R}^m$. You can assume that $l \prec u$, and $b \succ 0$ (for reasons we explain below). We let a_i^T denote the ith row of the matrix A.

We judge the approximation $Ax \approx b$ by the *maximum fractional deviation*, which is

$$\max_{i=1,\dots,n} \max\{(a_i^T x)/b_i, b_i/(a_i^T x)\} = \max_{i=1,\dots,n} \frac{\max\{a_i^T x, b_i\}}{\min\{a_i^T x, b_i\}},$$

when $Ax \succ 0$; we define the maximum fractional deviation as ∞ if $Ax \not\succ 0$.

The problem of minimizing the maximum fractional deviation is called the *fractional Chebyshev approximation problem*, or the *logarithmic Chebyshev approximation problem*, since it is equivalent to minimizing the objective

$$\max_{i=1,\dots,n} |\log a_i^T x - \log b_i|.$$

(See also exercise 6.3, part (c).)

(a) Formulate the fractional Chebyshev approximation problem (with variable bounds) as a convex optimization problem with twice differentiable objective and constraint functions.

(b) Implement a barrier method that solves the fractional Chebyshev approximation problem. You can assume an initial point $x^{(0)}$, satisfying $l \prec x^{(0)} \prec u$, $Ax^{(0)} \succ 0$, is known.

11.22 *Maximum volume rectangle inside a polyhedron.* Consider the problem described in exercise 8.16, *i.e.*, finding the maximum volume rectangle $\mathcal{R} = \{x \mid l \preceq x \preceq u\}$ that lies in a polyhedron described by a set of linear inequalities, $\mathcal{P} = \{x \mid Ax \preceq b\}$. Implement a barrier method for solving this problem. You can assume that $b \succ 0$, which means that for small $l \prec 0$ and $u \succ 0$, the rectangle $\mathcal{R}$ lies inside $\mathcal{P}$.

Test your implementation on several simple examples. Find the maximum volume rectangle that lies in the polyhedron defined by

$$A = \begin{bmatrix} 0 & -1 \\ 2 & -4 \\ 2 & 1 \\ -4 & 4 \\ -4 & 0 \end{bmatrix}, \qquad b = \mathbf{1}.$$

Plot this polyhedron, and the maximum volume rectangle that lies inside it.

11.23 *SDP bounds and heuristics for the two-way partitioning problem.* In this exercise we consider the two-way partitioning problem (5.7), described on page 219, and also in exercise 5.39:

$$\begin{array}{ll} \text{minimize} & x^T W x \\ \text{subject to} & x_i^2 = 1, \quad i = 1, \dots, n, \end{array} \tag{11.65}$$

with variable $x \in \mathbf{R}^n$. We assume, without loss of generality, that $W \in \mathbf{S}^n$ satisfies $W_{ii} = 0$. We denote the optimal value of the partitioning problem as $p^\star$, and $x^\star$ will denote an optimal partition. (Note that $-x^\star$ is also an optimal partition.)

The Lagrange dual of the two-way partitioning problem (11.65) is given by the SDP

$$\begin{array}{ll} \text{maximize} & -\mathbf{1}^T \nu \\ \text{subject to} & W + \mathbf{diag}(\nu) \succeq 0, \end{array} \tag{11.66}$$

with variable $\nu \in \mathbf{R}^n$. The dual of this SDP is

$$
\begin{array}{ll}
\text{minimize} & \mathbf{tr}(WX) \\
\text{subject to} & X \succeq 0 \\
& X_{ii} = 1, \quad i = 1, \dots, n,
\end{array}
\tag{11.67}
$$

with variable $X \in \mathbf{S}^n$. (This SDP can be interpreted as a relaxation of the two-way partitioning problem (11.65); see exercise 5.39.) The optimal values of these two SDPs are equal, and give a lower bound, which we denote $d^\star$, on the optimal value $p^\star$. Let $\nu^\star$ and $X^\star$ denote optimal points for the two SDPs.

(a) Implement a barrier method that solves the SDP (11.66) and its dual (11.67), given the weight matrix W. Explain how you obtain nearly optimal ν and X, give formulas for any Hessians and gradients that your method requires, and explain how you compute the Newton step. Test your implementation on some small problem instances, comparing the bound you find with the optimal value (which can be found by checking the objective value of all 2^n partitions). Try your implementation on a randomly chosen problem instance large enough that you cannot find the optimal partition by exhaustive search (e.g., $n = 100$).

(b) *A heuristic for partitioning.* In exercise 5.39, you found that if $X^\star$ has rank one, then it must have the form $X^\star = x^\star(x^\star)^T$, where $x^\star$ is optimal for the two-way partitioning problem. This suggests the following simple heuristic for finding a good partition (if not the best): solve the SDPs above, to find $X^\star$ (and the bound $d^\star$). Let v denote an eigenvector of $X^\star$ associated with its largest eigenvalue, and let $\hat{x} = \mathbf{sign}(v)$. The vector $\hat{x}$ is our guess for a good partition.
Try this heuristic on some small problem instances, and the large problem instance you used in part (a). Compare the objective value of your heuristic partition, $\hat{x}^T W \hat{x}$, with the lower bound $d^\star$.

(c) *A randomized method.* Another heuristic technique for finding a good partition, given the solution $X^\star$ of the SDP (11.67), is based on *randomization*. The method is simple: we generate independent samples $x^{(1)}, \dots, x^{(K)}$ from a normal distribution on $\mathbf{R}^n$, with zero mean and covariance $X^\star$. For each sample we consider the heuristic approximate solution $\hat{x}^{(k)} = \mathbf{sign}(x^{(k)})$. We then take the best among these, *i.e.*, the one with lowest cost. Try out this procedure on some small problem instances, and the large problem instance you considered in part (a).

(d) *A greedy heuristic refinement.* Suppose you are given a partition x, *i.e.*, $x_i \in \{-1, 1\}$, $i = 1, \dots, n$. How does the objective value change if we move element i from one set to the other, *i.e.*, change x_i to $-x_i$? Now consider the following simple greedy algorithm: given a starting partition x, move the element that gives the largest reduction in the objective. Repeat this procedure until no reduction in objective can be obtained by moving an element from one set to the other.
Try this heuristic on some problem instances, including the large one, starting from various initial partitions, including $x = \mathbf{1}$, the heuristic approximate solution found in part (b), and the randomly generated approximate solutions found in part (c). How much does this greedy refinement improve your approximate solutions from parts (b) and (c)?

11.24 *Barrier and primal-dual interior-point methods for quadratic programming.* Implement a barrier method, and a primal-dual method, for solving the QP (without equality constraints, for simplicity)

$$
\begin{array}{ll}
\text{minimize} & (1/2)x^T P x + q^T x \\
\text{subject to} & Ax \preceq b,
\end{array}
$$

with $A \in \mathbf{R}^{m \times n}$. You can assume a strictly feasible initial point is given. Test your codes on several examples. For the barrier method, plot the duality gap versus Newton steps. For the primal-dual interior-point method, plot the surrogate duality gap and the norm of the dual residual versus iteration number.

Appendices

Appendix A

Mathematical background

In this appendix we give a brief review of some basic concepts from analysis and linear algebra. The treatment is by no means complete, and is meant mostly to set out our notation.

A.1 Norms

A.1.1 Inner product, Euclidean norm, and angle

The *standard inner product* on $\mathbf{R}^n$, the set of real n-vectors, is given by

$$\langle x, y \rangle = x^T y = \sum_{i=1}^{n} x_i y_i,$$

for $x, y \in \mathbf{R}^n$. In this book we use the notation $x^T y$, instead of $\langle x, y \rangle$. The *Euclidean norm*, or ℓ_2-norm, of a vector $x \in \mathbf{R}^n$ is defined as

$$\|x\|_2 = (x^T x)^{1/2} = (x_1^2 + \cdots + x_n^2)^{1/2}. \tag{A.1}$$

The *Cauchy-Schwartz inequality* states that $|x^T y| \le \|x\|_2 \|y\|_2$ for any $x, y \in \mathbf{R}^n$. The (unsigned) *angle* between nonzero vectors $x, y \in \mathbf{R}^n$ is defined as

$$\angle(x, y) = \cos^{-1} \left(\frac{x^T y}{\|x\|_2 \|y\|_2} \right),$$

where we take $\cos^{-1}(u) \in [0, \pi]$. We say x and y are *orthogonal* if $x^T y = 0$.

The standard inner product on $\mathbf{R}^{m \times n}$, the set of $m \times n$ real matrices, is given by

$$\langle X, Y \rangle = \mathbf{tr}(X^T Y) = \sum_{i=1}^{m} \sum_{j=1}^{n} X_{ij} Y_{ij},$$

for $X, Y \in \mathbf{R}^{m \times n}$. (Here $\mathbf{tr}$ denotes *trace* of a matrix, *i.e.*, the sum of its diagonal elements.) We use the notation $\mathbf{tr}(X^T Y)$ instead of $\langle X, Y \rangle$. Note that the inner product of two matrices is the inner product of the associated vectors, in $\mathbf{R}^{mn}$, obtained by listing the coefficients of the matrices in some order, such as row major.

The *Frobenius norm* of a matrix $X \in \mathbf{R}^{m \times n}$ is given by

$$\|X\|_F = \left(\mathbf{tr}(X^T X)\right)^{1/2} = \left(\sum_{i=1}^{m}\sum_{j=1}^{n} X_{ij}^2\right)^{1/2}. \tag{A.2}$$

The Frobenius norm is the Euclidean norm of the vector obtained by listing the coefficients of the matrix. (The ℓ_2-norm of a matrix is a different norm; see §A.1.5.)

The standard inner product on $\mathbf{S}^n$, the set of symmetric $n \times n$ matrices, is given by

$$\langle X, Y \rangle = \mathbf{tr}(XY) = \sum_{i=1}^{n}\sum_{j=1}^{n} X_{ij}Y_{ij} = \sum_{i=1}^{n} X_{ii}Y_{ii} + 2\sum_{i<j} X_{ij}Y_{ij}.$$

A.1.2   Norms, distance, and unit ball

A function $f : \mathbf{R}^n \to \mathbf{R}$ with $\mathbf{dom}\, f = \mathbf{R}^n$ is called a *norm* if

- f is nonnegative: $f(x) \geq 0$ for all $x \in \mathbf{R}^n$

- f is definite: $f(x) = 0$ only if $x = 0$

- f is homogeneous: $f(tx) = |t| f(x)$, for all $x \in \mathbf{R}^n$ and $t \in \mathbf{R}$

- f satisfies the triangle inequality: $f(x+y) \leq f(x) + f(y)$, for all $x, y \in \mathbf{R}^n$

We use the notation $f(x) = \|x\|$, which is meant to suggest that a norm is a generalization of the absolute value on $\mathbf{R}$. When we specify a particular norm, we use the notation $\|x\|_{\mathrm{symb}}$, where the subscript is a mnemonic to indicate which norm is meant.

A norm is a measure of the *length* of a vector x; we can measure the *distance* between two vectors x and y as the length of their difference, *i.e.*,

$$\mathbf{dist}(x, y) = \|x - y\|.$$

We refer to $\mathbf{dist}(x, y)$ as the distance between x and y, in the norm $\|\cdot\|$.

The set of all vectors with norm less than or equal to one,

$$\mathcal{B} = \{x \in \mathbf{R}^n \mid \|x\| \leq 1\},$$

is called the *unit ball* of the norm $\|\cdot\|$. The unit ball satisfies the following properties:

- $\mathcal{B}$ is symmetric about the origin, *i.e.*, $x \in \mathcal{B}$ if and only if $-x \in \mathcal{B}$

- $\mathcal{B}$ is convex

- $\mathcal{B}$ is closed, bounded, and has nonempty interior

Conversely, if $C \subseteq \mathbf{R}^n$ is any set satisfying these three conditions, then it is the unit ball of a norm, which is given by

$$\|x\| = (\sup\{t \geq 0 \mid tx \in C\})^{-1}.$$

A.1.3   Examples

The simplest example of a norm is the absolute value on $\mathbf{R}$. Another simple example is the Euclidean or ℓ_2-norm on $\mathbf{R}^n$, defined above in (A.1). Two other frequently used norms on $\mathbf{R}^n$ are the *sum-absolute-value*, or ℓ_1-*norm*, given by

$$\|x\|_1 = |x_1| + \cdots + |x_n|,$$

and the *Chebyshev* or ℓ_∞-*norm*, given by

$$\|x\|_\infty = \max\{|x_1|, \ldots, |x_n|\}.$$

These three norms are part of a family parametrized by a constant traditionally denoted p, with $p \geq 1$: the ℓ_p-*norm* is defined by

$$\|x\|_p = (|x_1|^p + \cdots + |x_n|^p)^{1/p}.$$

This yields the ℓ_1-norm when $p = 1$ and the Euclidean norm when $p = 2$. It is easy to show that for any $x \in \mathbf{R}^n$,

$$\lim_{p \to \infty} \|x\|_p = \max\{|x_1|, \ldots, |x_n|\},$$

so the ℓ_∞-norm also fits in this family, as a limit.

Another important family of norms are the *quadratic norms*. For $P \in \mathbf{S}_{++}^n$, we define the P-quadratic norm as

$$\|x\|_P = (x^T P x)^{1/2} = \|P^{1/2} x\|_2.$$

The unit ball of a quadratic norm is an ellipsoid (and conversely, if the unit ball of a norm is an ellipsoid, the norm is a quadratic norm).

Some common norms on $\mathbf{R}^{m \times n}$ are the Frobenius norm, defined above in (A.2), the sum-absolute-value norm,

$$\|X\|_{\mathrm{sav}} = \sum_{i=1}^{m} \sum_{j=1}^{n} |X_{ij}|,$$

and the maximum-absolute-value norm,

$$\|X\|_{\mathrm{mav}} = \max\{|X_{ij}| \mid i = 1, \ldots, m, \ j = 1, \ldots, n\}.$$

We will encounter several other important norms of matrices in §A.1.5.

A.1.4   Equivalence of norms

Suppose that $\|\cdot\|_a$ and $\|\cdot\|_b$ are norms on $\mathbf{R}^n$. A basic result of analysis is that there exist positive constants α and β such that, for all $x \in \mathbf{R}^n$,

$$\alpha\|x\|_a \le \|x\|_b \le \beta\|x\|_a.$$

This means that the norms are *equivalent, i.e.,* they define the same set of open subsets, the same set of convergent sequences, and so on (see §A.2). (We conclude that any norms on any finite-dimensional vector space are equivalent, but on infinite-dimensional vector spaces, the result need not hold.) Using convex analysis, we can give a more specific result: If $\|\cdot\|$ is any norm on $\mathbf{R}^n$, then there exists a quadratic norm $\|\cdot\|_P$ for which

$$\|x\|_P \le \|x\| \le \sqrt{n}\|x\|_P$$

holds for all x. In other words, any norm on $\mathbf{R}^n$ can be uniformly approximated, within a factor of $\sqrt{n}$, by a quadratic norm. (See §8.4.1.)

A.1.5   Operator norms

Suppose $\|\cdot\|_a$ and $\|\cdot\|_b$ are norms on $\mathbf{R}^m$ and $\mathbf{R}^n$, respectively. We define the *operator norm* of $X \in \mathbf{R}^{m \times n}$, induced by the norms $\|\cdot\|_a$ and $\|\cdot\|_b$, as

$$\|X\|_{a,b} = \sup\left\{\|Xu\|_a \mid \|u\|_b \le 1\right\}.$$

(It can be shown that this defines a norm on $\mathbf{R}^{m \times n}$.)

When $\|\cdot\|_a$ and $\|\cdot\|_b$ are both Euclidean norms, the operator norm of X is its *maximum singular value*, and is denoted $\|X\|_2$:

$$\|X\|_2 = \sigma_{\max}(X) = (\lambda_{\max}(X^T X))^{1/2}.$$

(This agrees with the Euclidean norm on $\mathbf{R}^m$, when $X \in \mathbf{R}^{m \times 1}$, so there is no clash of notation.) This norm is also called the *spectral norm* or ℓ_2-*norm* of X.

As another example, the norm induced by the ℓ_∞-norm on $\mathbf{R}^m$ and $\mathbf{R}^n$, denoted $\|X\|_\infty$, is the *max-row-sum norm*,

$$\|X\|_\infty = \sup\left\{\|Xu\|_\infty \mid \|u\|_\infty \le 1\right\} = \max_{i=1,\ldots,m} \sum_{j=1}^n |X_{ij}|.$$

The norm induced by the ℓ_1-norm on $\mathbf{R}^m$ and $\mathbf{R}^n$, denoted $\|X\|_1$, is the *max-column-sum norm*,

$$\|X\|_1 = \max_{j=1,\ldots,n} \sum_{i=1}^m |X_{ij}|.$$

A.1.6   Dual norm

Let $\| \cdot \|$ be a norm on $\mathbf{R}^n$. The associated *dual norm*, denoted $\| \cdot \|_*$, is defined as

$$\|z\|_* = \sup\{z^T x \mid \|x\| \leq 1\}.$$

(This can be shown to be a norm.) The dual norm can be interpreted as the operator norm of z^T, interpreted as a $1 \times n$ matrix, with the norm $\| \cdot \|$ on $\mathbf{R}^n$, and the absolute value on $\mathbf{R}$:

$$\|z\|_* = \sup\{|z^T x| \mid \|x\| \leq 1\}.$$

From the definition of dual norm we have the inequality

$$z^T x \leq \|x\| \, \|z\|_*,$$

which holds for all x and z. This inequality is tight, in the following sense: for any x there is a z for which the inequality holds with equality. (Similarly, for any z there is an x that gives equality.) The dual of the dual norm is the original norm: we have $\|x\|_{**} = \|x\|$ for all x. (This need not hold in infinite-dimensional vector spaces.)

The dual of the Euclidean norm is the Euclidean norm, since

$$\sup\{z^T x \mid \|x\|_2 \leq 1\} = \|z\|_2.$$

(This follows from the Cauchy-Schwarz inequality; for nonzero z, the value of x that maximizes $z^T x$ over $\|x\|_2 \leq 1$ is $z/\|z\|_2$.)

The dual of the ℓ_∞-norm is the ℓ_1-norm:

$$\sup\{z^T x \mid \|x\|_\infty \leq 1\} = \sum_{i=1}^{n} |z_i| = \|z\|_1,$$

and the dual of the ℓ_1-norm is the ℓ_∞-norm. More generally, the dual of the ℓ_p-norm is the ℓ_q-norm, where q satisfies $1/p + 1/q = 1$, *i.e.*, $q = p/(p-1)$.

As another example, consider the ℓ_2- or spectral norm on $\mathbf{R}^{m \times n}$. The associated dual norm is

$$\|Z\|_{2*} = \sup\{\mathbf{tr}(Z^T X) \mid \|X\|_2 \leq 1\},$$

which turns out to be the sum of the singular values,

$$\|Z\|_{2*} = \sigma_1(Z) + \cdots + \sigma_r(Z) = \mathbf{tr}(Z^T Z)^{1/2},$$

where $r = \mathbf{rank}\, Z$. This norm is sometimes called the *nuclear* norm.

A.2   Analysis

A.2.1   Open and closed sets

An element $x \in C \subseteq \mathbf{R}^n$ is called an *interior* point of C if there exists an $\epsilon > 0$ for which

$$\{y \mid \|y - x\|_2 \leq \epsilon\} \subseteq C,$$

i.e., there exists a ball centered at x that lies entirely in C. The set of all points interior to C is called the *interior* of C and is denoted **int** C. (Since all norms on $\mathbf{R}^n$ are equivalent to the Euclidean norm, all norms generate the same set of interior points.) A set C is *open* if **int** $C = C$, *i.e.*, every point in C is an interior point. A set $C \subseteq \mathbf{R}^n$ is *closed* if its complement $\mathbf{R}^n \setminus C = \{x \in \mathbf{R}^n \mid x \notin C\}$ is open.

The *closure* of a set C is defined as

$$\mathbf{cl}\, C = \mathbf{R}^n \setminus \mathbf{int}(\mathbf{R}^n \setminus C),$$

i.e., the complement of the interior of the complement of C. A point x is in the closure of C if for every $\epsilon > 0$, there is a $y \in C$ with $\|x - y\|_2 \le \epsilon$.

We can also describe closed sets and the closure in terms of convergent sequences and limit points. A set C is closed if and only if it contains the limit point of every convergent sequence in it. In other words, if $x_1, x_2, \ldots$ converges to x, and $x_i \in C$, then $x \in C$. The closure of C is the set of all limit points of convergent sequences in C.

The *boundary* of the set C is defined as

$$\mathbf{bd}\, C = \mathbf{cl}\, C \setminus \mathbf{int}\, C.$$

A *boundary point* x (*i.e.*, a point $x \in \mathbf{bd}\, C$) satisfies the following property: For all $\epsilon > 0$, there exists $y \in C$ and $z \notin C$ with

$$\|y - x\|_2 \le \epsilon, \qquad \|z - x\|_2 \le \epsilon,$$

i.e., there exist arbitrarily close points in C, and also arbitrarily close points not in C. We can characterize closed and open sets in terms of the boundary operation: C is *closed* if it contains its boundary, *i.e.*, $\mathbf{bd}\, C \subseteq C$. It is *open* if it contains no boundary points, *i.e.*, $C \cap \mathbf{bd}\, C = \emptyset$.

A.2.2 Supremum and infimum

Suppose $C \subseteq \mathbf{R}$. A number a is an *upper bound* on C if for each $x \in C$, $x \le a$. The set of upper bounds on a set C is either empty (in which case we say C is unbounded above), all of $\mathbf{R}$ (only when $C = \emptyset$), or a closed infinite interval $[b, \infty)$. The number b is called the *least upper bound* or *supremum* of the set C, and is denoted $\sup C$. We take $\sup \emptyset = -\infty$, and $\sup C = \infty$ if C is unbounded above. When $\sup C \in C$, we say the supremum of C is attained or achieved.

When the set C is finite, $\sup C$ is the maximum of its elements. Some authors use the notation $\max C$ to denote supremum, when it is attained, but we follow standard mathematical convention, using $\max C$ only when the set C is finite.

We define lower bound, and infimum, in a similar way. A number a is a lower bound on $C \subseteq \mathbf{R}$ if for each $x \in C$, $a \le x$. The *infimum* (or *greatest lower bound*) of a set $C \subseteq \mathbf{R}$ is defined as $\inf C = -\sup(-C)$. When C is finite, the infimum is the minimum of its elements. We take $\inf \emptyset = \infty$, and $\inf C = -\infty$ if C is unbounded below, *i.e.*, has no lower bound.

A.3   Functions

A.3.1   Function notation

Our notation for functions is mostly standard, with one exception. When we write

$$f : A \to B$$

we mean that f is a function on the set $\mathbf{dom}\, f \subseteq A$ into the set B; in particular we can have $\mathbf{dom}\, f$ a proper subset of the set A. Thus the notation $f : \mathbf{R}^n \to \mathbf{R}^m$ means that f maps (some) n-vectors into m-vectors; it does not mean that $f(x)$ is defined for every $x \in \mathbf{R}^n$. This convention is similar to function declarations in computer languages. Specifying the data types of the input and output arguments of a function gives the *syntax* of that function; it does not guarantee that any input argument with the specified data type is valid.

As an example consider the function $f : \mathbf{S}^n \to \mathbf{R}$, given by

$$f(X) = \log \det X, \tag{A.3}$$

with $\mathbf{dom}\, f = \mathbf{S}^n_{++}$. The notation $f : \mathbf{S}^n \to \mathbf{R}$ specifies the *syntax* of f: it takes as argument a symmetric $n \times n$ matrix, and returns a real number. The notation $\mathbf{dom}\, f = \mathbf{S}^n_{++}$ specifies which symmetric $n \times n$ matrices are valid input arguments for f (*i.e.*, only positive definite ones). The formula (A.3) specifies what $f(X)$ is, for $X \in \mathbf{dom}\, f$.

A.3.2   Continuity

A function $f : \mathbf{R}^n \to \mathbf{R}^m$ is *continuous* at $x \in \mathbf{dom}\, f$ if for all $\epsilon > 0$ there exists a δ such that

$$y \in \mathbf{dom}\, f, \quad \|y - x\|_2 \le \delta \implies \|f(y) - f(x)\|_2 \le \epsilon.$$

Continuity can be described in terms of limits: whenever the sequence $x_1, x_2, \ldots$ in $\mathbf{dom}\, f$ converges to a point $x \in \mathbf{dom}\, f$, the sequence $f(x_1), f(x_2), \ldots$ converges to $f(x)$, *i.e.*,

$$\lim_{i \to \infty} f(x_i) = f(\lim_{i \to \infty} x_i).$$

A function f is *continuous* if it is continuous at every point in its domain.

A.3.3   Closed functions

A function $f : \mathbf{R}^n \to \mathbf{R}$ is said to be *closed* if, for each $\alpha \in \mathbf{R}$, the sublevel set

$$\{x \in \mathbf{dom}\, f \mid f(x) \le \alpha\}$$

is closed. This is equivalent to the condition that the epigraph of f,

$$\mathbf{epi}\, f = \{(x, t) \in \mathbf{R}^{n+1} \mid x \in \mathbf{dom}\, f, \ f(x) \le t\},$$

is closed. (This definition is general, but is usually only applied to convex functions.)

If $f : \mathbf{R}^n \to \mathbf{R}$ is continuous, and $\mathbf{dom}\, f$ is closed, then f is closed. If $f : \mathbf{R}^n \to \mathbf{R}$ is continuous, with $\mathbf{dom}\, f$ open, then f is closed if and only if f converges to ∞ along every sequence converging to a boundary point of $\mathbf{dom}\, f$. In other words, if $\lim_{i \to \infty} x_i = x \in \mathbf{bd\, dom}\, f$, with $x_i \in \mathbf{dom}\, f$, we have $\lim_{i \to \infty} f(x_i) = \infty$.

Example A.1 *Examples on* $\mathbf{R}$.

- The function $f : \mathbf{R} \to \mathbf{R}$, with $f(x) = x \log x$, $\mathbf{dom}\, f = \mathbf{R}_{++}$, is *not* closed.

- The function $f : \mathbf{R} \to \mathbf{R}$, with

$$f(x) = \begin{cases} x \log x & x > 0 \\ 0 & x = 0, \end{cases} \qquad \mathbf{dom}\, f = \mathbf{R}_{+},$$

  is closed.

- The function $f(x) = -\log x$, $\mathbf{dom}\, f = \mathbf{R}_{++}$, is closed.

A.4   Derivatives

A.4.1   Derivative and gradient

Suppose $f : \mathbf{R}^n \to \mathbf{R}^m$ and $x \in \mathbf{int\, dom}\, f$. The function f is differentiable at x if there exists a matrix $Df(x) \in \mathbf{R}^{m \times n}$ that satisfies

$$\lim_{z \in \mathbf{dom}\, f,\ z \neq x,\ z \to x} \frac{\|f(z) - f(x) - Df(x)(z - x)\|_2}{\|z - x\|_2} = 0, \tag{A.4}$$

in which case we refer to $Df(x)$ as the *derivative* (or *Jacobian*) of f at x. (There can be at most one matrix that satisfies (A.4).) The function f is *differentiable* if $\mathbf{dom}\, f$ is open, and it is differentiable at every point in its domain.

The affine function of z given by

$$f(x) + Df(x)(z - x)$$

is called the *first-order approximation* of f at (or near) x. Evidently this function agrees with f at $z = x$; when z is *close* to x, this affine function is *very close* to f.

The derivative can be found by deriving the first-order approximation of the function f at x (*i.e.*, the matrix $Df(x)$ that satisfies (A.4)), or from partial derivatives:

$$Df(x)_{ij} = \frac{\partial f_i(x)}{\partial x_j}, \qquad i = 1, \ldots, m, \quad j = 1, \ldots, n.$$

Gradient

When f is real-valued (*i.e.*, $f : \mathbf{R}^n \to \mathbf{R}$) the derivative $Df(x)$ is a $1 \times n$ matrix, *i.e.*, it is a *row* vector. Its transpose is called the *gradient* of the function:

$$\nabla f(x) = Df(x)^T,$$

which is a (column) vector, *i.e.*, in $\mathbf{R}^n$. Its components are the partial derivatives of f:

$$\nabla f(x)_i = \frac{\partial f(x)}{\partial x_i}, \quad i = 1, \dots, n.$$

The first-order approximation of f at a point $x \in \mathbf{int\,dom}\, f$ can be expressed as (the affine function of z)

$$f(x) + \nabla f(x)^T (z - x).$$

Examples

As a simple example consider the quadratic function $f : \mathbf{R}^n \to \mathbf{R}$,

$$f(x) = (1/2)x^T P x + q^T x + r,$$

where $P \in \mathbf{S}^n$, $q \in \mathbf{R}^n$, and $r \in \mathbf{R}$. Its derivative at x is the row vector $Df(x) = x^T P + q^T$, and its gradient is

$$\nabla f(x) = Px + q.$$

As a more interesting example, we consider the function $f : \mathbf{S}^n \to \mathbf{R}$, given by

$$f(X) = \log \det X, \qquad \mathbf{dom}\, f = \mathbf{S}_{++}^n.$$

One (tedious) way to find the gradient of f is to introduce a basis for $\mathbf{S}^n$, find the gradient of the associated function, and finally translate the result back to $\mathbf{S}^n$. Instead, we will directly find the first-order approximation of f at $X \in \mathbf{S}_{++}^n$. Let $Z \in \mathbf{S}_{++}^n$ be close to X, and let $\Delta X = Z - X$ (which is assumed to be small). We have

$$
\begin{aligned}
\log \det Z &= \log \det(X + \Delta X) \\
&= \log \det \left(X^{1/2}(I + X^{-1/2}\Delta X X^{-1/2})X^{1/2} \right) \\
&= \log \det X + \log \det(I + X^{-1/2}\Delta X X^{-1/2}) \\
&= \log \det X + \sum_{i=1}^{n} \log(1 + \lambda_i),
\end{aligned}
$$

where λ_i is the ith eigenvalue of $X^{-1/2}\Delta X X^{-1/2}$. Now we use the fact that ΔX is small, which implies λ_i are small, so to first order we have $\log(1 + \lambda_i) \approx \lambda_i$. Using this first-order approximation in the expression above, we get

$$
\begin{aligned}
\log \det Z &\approx \log \det X + \sum_{i=1}^{n} \lambda_i \\
&= \log \det X + \mathbf{tr}(X^{-1/2}\Delta X X^{-1/2}) \\
&= \log \det X + \mathbf{tr}(X^{-1}\Delta X) \\
&= \log \det X + \mathbf{tr}\left(X^{-1}(Z - X) \right),
\end{aligned}
$$

where we have used the fact that the sum of the eigenvalues is the trace, and the property $\mathbf{tr}(AB) = \mathbf{tr}(BA)$.

Thus, the first-order approximation of f at X is the affine function of Z given by

$$f(Z) \approx f(X) + \mathbf{tr}\left(X^{-1}(Z - X)\right).$$

Noting that the second term on the righthand side is the standard inner product of X^{-1} and $Z - X$, we can identify X^{-1} as the gradient of f at X. Thus, we can write the simple formula

$$\nabla f(X) = X^{-1}.$$

This result should not be surprising, since the derivative of $\log x$, on $\mathbf{R}_{++}$, is $1/x$.

A.4.2 Chain rule

Suppose $f : \mathbf{R}^n \to \mathbf{R}^m$ is differentiable at $x \in \mathbf{int\,dom}\,f$ and $g : \mathbf{R}^m \to \mathbf{R}^p$ is differentiable at $f(x) \in \mathbf{int\,dom}\,g$. Define the composition $h : \mathbf{R}^n \to \mathbf{R}^p$ by $h(z) = g(f(z))$. Then h is differentiable at x, with derivative

$$Dh(x) = Dg(f(x))Df(x). \tag{A.5}$$

As an example, suppose $f : \mathbf{R}^n \to \mathbf{R}$, $g : \mathbf{R} \to \mathbf{R}$, and $h(x) = g(f(x))$. Taking the transpose of $Dh(x) = Dg(f(x))Df(x)$ yields

$$\nabla h(x) = g'(f(x))\nabla f(x). \tag{A.6}$$

Composition with affine function

Suppose $f : \mathbf{R}^n \to \mathbf{R}^m$ is differentiable, $A \in \mathbf{R}^{n \times p}$, and $b \in \mathbf{R}^n$. Define $g : \mathbf{R}^p \to \mathbf{R}^m$ as $g(x) = f(Ax + b)$, with $\mathbf{dom}\,g = \{x \mid Ax + b \in \mathbf{dom}\,f\}$. The derivative of g is, by the chain rule (A.5), $Dg(x) = Df(Ax + b)A$.

When f is real-valued (i.e., $m = 1$), we obtain the formula for the gradient of a composition of a function with an affine function,

$$\nabla g(x) = A^T \nabla f(Ax + b).$$

For example, suppose that $f : \mathbf{R}^n \to \mathbf{R}$, $x, v \in \mathbf{R}^n$, and we define the function $\tilde{f} : \mathbf{R} \to \mathbf{R}$ by $\tilde{f}(t) = f(x + tv)$. (Roughly speaking, $\tilde{f}$ is f, restricted to the line $\{x + tv \mid t \in \mathbf{R}\}$.) Then we have

$$D\tilde{f}(t) = \tilde{f}'(t) = \nabla f(x + tv)^T v.$$

(The scalar $\tilde{f}'(0)$ is the *directional derivative* of f, at x, in the direction v.)

Example A.2 Consider the function $f : \mathbf{R}^n \to \mathbf{R}$, with $\mathbf{dom}\,f = \mathbf{R}^n$ and

$$f(x) = \log \sum_{i=1}^m \exp(a_i^T x + b_i),$$

where $a_1, \ldots, a_m \subset \mathbf{R}^n$, and $b_1, \ldots, b_m \in \mathbf{R}$. We can find a simple expression for its gradient by noting that it is the composition of the affine function $Ax + b$, where $A \in \mathbf{R}^{m \times n}$ with rows $a_1^T, \ldots, a_m^T$, and the function $g : \mathbf{R}^m \to \mathbf{R}$ given by $g(y) = \log(\sum_{i=1}^m \exp y_i)$. Simple differentiation (or the formula (A.6)) shows that

$$\nabla g(y) = \frac{1}{\sum_{i=1}^m \exp y_i} \begin{bmatrix} \exp y_1 \\ \vdots \\ \exp y_m \end{bmatrix}, \tag{A.7}$$

so by the composition formula we have

$$\nabla f(x) = \frac{1}{\mathbf{1}^T z} A^T z$$

where $z_i = \exp(a_i^T x + b_i)$, $i = 1, \ldots, m$.

Example A.3 We derive an expression for $\nabla f(x)$, where

$$f(x) = \log \det(F_0 + x_1 F_1 + \cdots + x_n F_n),$$

where $F_0, \ldots, F_n \in \mathbf{S}^p$, and

$$\mathbf{dom}\, f = \{x \in \mathbf{R}^n \mid F_0 + x_1 F_1 + \cdots + x_n F_n \succ 0\}.$$

The function f is the composition of the affine mapping from $x \in \mathbf{R}^n$ to $F_0 + x_1 F_1 + \cdots + x_n F_n \in \mathbf{S}^p$, with the function $\log \det X$. We use the chain rule to evaluate

$$\frac{\partial f(x)}{\partial x_i} = \mathbf{tr}(F_i \nabla \log \det(F)) = \mathbf{tr}(F^{-1} F_i),$$

where $F = F_0 + x_1 F_1 + \cdots + x_n F_n$. Thus we have

$$\nabla f(x) = \begin{bmatrix} \mathbf{tr}(F^{-1} F_1) \\ \vdots \\ \mathbf{tr}(F^{-1} F_n) \end{bmatrix}.$$

A.4.3   Second derivative

In this section we review the second derivative of a real-valued function $f : \mathbf{R}^n \to \mathbf{R}$. The second derivative or *Hessian matrix* of f at $x \in \mathbf{int\, dom}\, f$, denoted $\nabla^2 f(x)$, is given by

$$\nabla^2 f(x)_{ij} = \frac{\partial^2 f(x)}{\partial x_i \partial x_j}, \qquad i = 1, \ldots n, \quad j = 1, \ldots, n,$$

provided f is twice differentiable at x, where the partial derivatives are evaluated at x. The *second-order approximation* of f, at or near x, is the quadratic function of z defined by

$$\widehat{f}(z) = f(x) + \nabla f(x)^T (z - x) + (1/2)(z - x)^T \nabla^2 f(x)(z - x).$$

This second-order approximation satisfies

$$\lim_{z \in \mathbf{dom}\, f,\; z \neq x,\; z \to x} \frac{|f(z) - \widehat{f}(z)|}{\|z - x\|_2^2} = 0.$$

Not surprisingly, the second derivative can be interpreted as the derivative of the first derivative. If f is differentiable, the *gradient mapping* is the function $\nabla f : \mathbf{R}^n \to \mathbf{R}^n$, with $\mathbf{dom}\, \nabla f = \mathbf{dom}\, f$, with value $\nabla f(x)$ at x. The derivative of this mapping is

$$D\nabla f(x) = \nabla^2 f(x).$$

Examples

As a simple example consider the quadratic function $f : \mathbf{R}^n \to \mathbf{R}$,

$$f(x) = (1/2)x^T P x + q^T x + r,$$

where $P \in \mathbf{S}^n$, $q \in \mathbf{R}^n$, and $r \in \mathbf{R}$. Its gradient is $\nabla f(x) = Px + q$, so its Hessian is given by $\nabla^2 f(x) = P$. The second-order approximation of a quadratic function is itself.

As a more complicated example, we consider again the function $f : \mathbf{S}^n \to \mathbf{R}$, given by $f(X) = \log \det X$, with $\mathbf{dom}\, f = \mathbf{S}_{++}^n$. To find the second-order approximation (and therefore, the Hessian), we will derive a first-order approximation of the gradient, $\nabla f(X) = X^{-1}$. For $Z \in \mathbf{S}_{++}^n$ near $X \in \mathbf{S}_{++}^n$, and $\Delta X = Z - X$, we have

$$
\begin{aligned}
Z^{-1} &= (X + \Delta X)^{-1} \\
&= \left(X^{1/2} (I + X^{-1/2} \Delta X X^{-1/2}) X^{1/2} \right)^{-1} \\
&= X^{-1/2} (I + X^{-1/2} \Delta X X^{-1/2})^{-1} X^{-1/2} \\
&\approx X^{-1/2} (I - X^{-1/2} \Delta X X^{-1/2}) X^{-1/2} \\
&= X^{-1} - X^{-1} \Delta X X^{-1},
\end{aligned}
$$

using the first-order approximation $(I + A)^{-1} \approx I - A$, valid for A small.

This approximation is enough for us to identify the Hessian of f at X. The Hessian is a quadratic form on $\mathbf{S}^n$. Such a quadratic form is cumbersome to describe in the general case, since it requires four indices. But from the first-order approximation of the gradient above, the quadratic form can be expressed as

$$- \mathbf{tr}(X^{-1} U X^{-1} V),$$

where $U, V \in \mathbf{S}^n$ are the arguments of the quadratic form. (This generalizes the expression for the scalar case: $(\log x)'' = -1/x^2$.)

Now we have the second-order approximation of f near X:

$$
\begin{aligned}
f(Z) &= f(X + \Delta X) \\
&\approx f(X) + \mathbf{tr}(X^{-1} \Delta X) - (1/2)\, \mathbf{tr}(X^{-1} \Delta X X^{-1} \Delta X) \\
&\approx f(X) + \mathbf{tr}\left(X^{-1}(Z - X)\right) - (1/2)\, \mathbf{tr}\left(X^{-1}(Z - X)X^{-1}(Z - X)\right).
\end{aligned}
$$

A.4.4   Chain rule for second derivative

A general chain rule for the second derivative is cumbersome in most cases, so we will state it only for some special cases that we will need.

Composition with scalar function

Suppose $f : \mathbf{R}^n \to \mathbf{R}$, $g : \mathbf{R} \to \mathbf{R}$, and $h(x) = g(f(x))$. Simply working out the partial derivatives yields

$$\nabla^2 h(x) = g'(f(x))\nabla^2 f(x) + g''(f(x))\nabla f(x)\nabla f(x)^T. \qquad (A.8)$$

Composition with affine function

Suppose $f : \mathbf{R}^n \to \mathbf{R}$, $A \in \mathbf{R}^{n \times m}$, and $b \in \mathbf{R}^n$. Define $g : \mathbf{R}^m \to \mathbf{R}$ by $g(x) = f(Ax + b)$. Then we have

$$\nabla^2 g(x) = A^T \nabla^2 f(Ax + b)A.$$

As an example, consider the restriction of a real-valued function f to a line, *i.e.*, the function $\tilde{f}(t) = f(x + tv)$, where x and v are fixed. Then we have

$$\nabla^2 \tilde{f}(t) = \tilde{f}''(t) = v^T \nabla^2 f(x + tv)v.$$

Example A.4 We consider the function $f : \mathbf{R}^n \to \mathbf{R}$ from example A.2,

$$f(x) = \log \sum_{i=1}^{m} \exp(a_i^T x + b_i),$$

where $a_1, \ldots, a_m \in \mathbf{R}^n$, and $b_1, \ldots, b_m \in \mathbf{R}$. By noting that $f(x) = g(Ax+b)$, where $g(y) = \log(\sum_{i=1}^{m} \exp y_i)$, we can obtain a simple formula for the Hessian of f. Taking partial derivatives, or using the formula (A.8), noting that g is the composition of $\log$ with $\sum_{i=1}^{m} \exp y_i$, yields

$$\nabla^2 g(y) = \mathbf{diag}(\nabla g(y)) - \nabla g(y)\nabla g(y)^T,$$

where $\nabla g(y)$ is given in (A.7). By the composition formula we have

$$\nabla^2 f(x) = A^T \left(\frac{1}{\mathbf{1}^T z} \mathbf{diag}(z) - \frac{1}{(\mathbf{1}^T z)^2} z z^T \right) A,$$

where $z_i = \exp(a_i^T x + b_i)$, $i = 1, \ldots, m$.

A.5   Linear algebra

A.5.1   Range and nullspace

Let $A \in \mathbf{R}^{m \times n}$ (*i.e.*, A is a real matrix with m rows and n columns). The *range* of A, denoted $\mathcal{R}(A)$, is the set of all vectors in $\mathbf{R}^m$ that can be written as linear combinations of the columns of A, *i.e.*,

$$\mathcal{R}(A) = \{Ax \mid x \in \mathbf{R}^n\}.$$

The range $\mathcal{R}(A)$ is a subspace of $\mathbf{R}^m$, *i.e.*, it is itself a vector space. Its dimension is the *rank* of A, denoted **rank** A. The rank of A can never be greater than the minimum of m and n. We say A has *full rank* if **rank** $A = \min\{m,n\}$.

The *nullspace* (or *kernel*) of A, denoted $\mathcal{N}(A)$, is the set of all vectors x mapped into zero by A:

$$\mathcal{N}(A) = \{x \mid Ax = 0\}.$$

The nullspace is a subspace of $\mathbf{R}^n$.

Orthogonal decomposition induced by A

If $\mathcal{V}$ is a subspace of $\mathbf{R}^n$, its *orthogonal complement*, denoted $\mathcal{V}^\perp$, is defined as

$$\mathcal{V}^\perp = \{x \mid z^T x = 0 \text{ for all } z \in \mathcal{V}\}.$$

(As one would expect of a complement, we have $\mathcal{V}^{\perp\perp} = \mathcal{V}$.)

A basic result of linear algebra is that, for any $A \in \mathbf{R}^{m \times n}$, we have

$$\mathcal{N}(A) = \mathcal{R}(A^T)^\perp.$$

(Applying the result to A^T we also have $\mathcal{R}(A) = \mathcal{N}(A^T)^\perp$.) This result is often stated as

$$\mathcal{N}(A) \overset{\perp}{\oplus} \mathcal{R}(A^T) = \mathbf{R}^n. \tag{A.9}$$

Here the symbol $\overset{\perp}{\oplus}$ refers to *orthogonal direct sum*, *i.e.*, the sum of two subspaces that are orthogonal. The decomposition (A.9) of $\mathbf{R}^n$ is called the *orthogonal decomposition induced by A*.

A.5.2   Symmetric eigenvalue decomposition

Suppose $A \in \mathbf{S}^n$, *i.e.*, A is a real symmetric $n \times n$ matrix. Then A can be factored as

$$A = Q\Lambda Q^T, \tag{A.10}$$

where $Q \in \mathbf{R}^{n \times n}$ is *orthogonal*, *i.e.*, satisfies $Q^T Q = I$, and $\Lambda = \mathbf{diag}(\lambda_1, \ldots, \lambda_n)$. The (real) numbers λ_i are the *eigenvalues* of A, and are the roots of the *characteristic polynomial* $\det(sI - A)$. The columns of Q form an orthonormal set of *eigenvectors* of A. The factorization (A.10) is called the *spectral decomposition* or (symmetric) *eigenvalue decomposition* of A.

We order the eigenvalues as $\lambda_1 \geq \lambda_2 \geq \cdots \geq \lambda_n$. We use the notation $\lambda_i(A)$ to refer to the ith largest eigenvalue of $A \in \mathbf{S}$. We usually write the largest or maximum eigenvalue as $\lambda_1(A) = \lambda_{\max}(A)$, and the least or minimum eigenvalue as $\lambda_n(A) = \lambda_{\min}(A)$.

The determinant and trace can be expressed in terms of the eigenvalues,

$$\det A = \prod_{i=1}^{n} \lambda_i, \qquad \mathbf{tr}\, A = \sum_{i=1}^{n} \lambda_i,$$

as can the spectral and Frobenius norms,

$$\|A\|_2 = \max_{i=1,\ldots,n} |\lambda_i| = \max\{\lambda_1, -\lambda_n\}, \qquad \|A\|_F = \left(\sum_{i=1}^{n} \lambda_i^2\right)^{1/2}.$$

Definiteness and matrix inequalities

The largest and smallest eigenvalues satisfy

$$\lambda_{\max}(A) = \sup_{x \neq 0} \frac{x^T A x}{x^T x}, \qquad \lambda_{\min}(A) = \inf_{x \neq 0} \frac{x^T A x}{x^T x}.$$

In particular, for any x, we have

$$\lambda_{\min}(A) x^T x \leq x^T A x \leq \lambda_{\max}(A) x^T x,$$

with both inequalities tight for (different) choices of x.

A matrix $A \in \mathbf{S}^n$ is called *positive definite* if for all $x \neq 0$, $x^T A x > 0$. We denote this as $A \succ 0$. By the inequality above, we see that $A \succ 0$ if and only all its eigenvalues are positive, *i.e.*, $\lambda_{\min}(A) > 0$. If $-A$ is positive definite, we say A is *negative definite*, which we write as $A \prec 0$. We use $\mathbf{S}^n_{++}$ to denote the set of positive definite matrices in $\mathbf{S}^n$.

If A satisfies $x^T A x \geq 0$ for all x, we say that A is *positive semidefinite* or *nonnegative definite*. If $-A$ is in nonnegative definite, *i.e.*, if $x^T A x \leq 0$ for all x, we say that A is *negative semidefinite* or *nonpositive definite*. We use $\mathbf{S}^n_+$ to denote the set of nonnegative definite matrices in $\mathbf{S}^n$.

For $A, B \in \mathbf{S}^n$, we use $A \prec B$ to mean $B - A \succ 0$, and so on. These inequalities are called *matrix inequalities*, or generalized inequalities associated with the positive semidefinite cone.

Symmetric squareroot

Let $A \in \mathbf{S}^n_+$, with eigenvalue decomposition $A = Q \,\mathbf{diag}(\lambda_1, \ldots, \lambda_n) Q^T$. We define the (symmetric) squareroot of A as

$$A^{1/2} = Q \,\mathbf{diag}(\lambda_1^{1/2}, \ldots, \lambda_n^{1/2}) Q^T.$$

The squareroot $A^{1/2}$ is the unique symmetric positive semidefinite solution of the equation $X^2 = A$.

A.5.3   Generalized eigenvalue decomposition

The *generalized eigenvalues* of a pair of symmetric matrices $(A, B) \in \mathbf{S}^n \times \mathbf{S}^n$ are defined as the roots of the polynomial $\det(sB - A)$.

We are usually interested in matrix pairs with $B \in \mathbf{S}^n_{++}$. In this case the generalized eigenvalues are also the eigenvalues of $B^{-1/2}AB^{-1/2}$ (which are real). As with the standard eigenvalue decomposition, we order the generalized eigenvalues in nonincreasing order, as $\lambda_1 \geq \lambda_2 \geq \cdots \geq \lambda_n$, and denote the maximum generalized eigenvalue by $\lambda_{\max}(A, B)$.

When $B \in \mathbf{S}^n_{++}$, the pair of matrices can be factored as

$$A = V\Lambda V^T, \qquad B = VV^T, \tag{A.11}$$

where $V \in \mathbf{R}^{n \times n}$ is nonsingular, and $\Lambda = \mathbf{diag}(\lambda_1, \dots, \lambda_n)$, where λ_i are the generalized eigenvalues of the pair (A, B). The decomposition (A.11) is called the *generalized eigenvalue decomposition*.

The generalized eigenvalue decomposition is related to the standard eigenvalue decomposition of the matrix $B^{-1/2}AB^{-1/2}$. If $Q\Lambda Q^T$ is the eigenvalue decomposition of $B^{-1/2}AB^{-1/2}$, then (A.11) holds with $V = B^{1/2}Q$.

A.5.4   Singular value decomposition

Suppose $A \in \mathbf{R}^{m \times n}$ with $\mathbf{rank}\, A = r$. Then A can be factored as

$$A = U\Sigma V^T, \tag{A.12}$$

where $U \in \mathbf{R}^{m \times r}$ satisfies $U^T U = I$, $V \in \mathbf{R}^{n \times r}$ satisfies $V^T V = I$, and $\Sigma = \mathbf{diag}(\sigma_1, \dots, \sigma_r)$, with

$$\sigma_1 \geq \sigma_2 \geq \cdots \geq \sigma_r > 0.$$

The factorization (A.12) is called the *singular value decomposition* (SVD) of A. The columns of U are called *left singular vectors* of A, the columns of V are *right singular vectors*, and the numbers σ_i are the *singular values*. The singular value decomposition can be written

$$A = \sum_{i=1}^{r} \sigma_i u_i v_i^T,$$

where $u_i \in \mathbf{R}^m$ are the left singular vectors, and $v_i \in \mathbf{R}^n$ are the right singular vectors.

The singular value decomposition of a matrix A is closely related to the eigenvalue decomposition of the (symmetric, nonnegative definite) matrix $A^T A$. Using (A.12) we can write

$$A^T A = V\Sigma^2 V^T = \begin{bmatrix} V & \tilde{V} \end{bmatrix} \begin{bmatrix} \Sigma^2 & 0 \\ 0 & 0 \end{bmatrix} \begin{bmatrix} V & \tilde{V} \end{bmatrix}^T,$$

where $\tilde{V}$ is any matrix for which $[V \ \tilde{V}]$ is orthogonal. The righthand expression is the eigenvalue decomposition of $A^T A$, so we conclude that its nonzero eigenvalues are the singular values of A squared, and the associated eigenvectors of $A^T A$ are the right singular vectors of A. A similar analysis of AA^T shows that its nonzero eigenvalues are also the squares of the singular values of A, and the associated eigenvectors are the left singular vectors of A.

The first or largest singular value is also written as $\sigma_{\max}(A)$. It can be expressed as

$$\sigma_{\max}(A) = \sup_{x,y\neq 0} \frac{x^T A y}{\|x\|_2 \|y\|_2} = \sup_{y\neq 0} \frac{\|Ay\|_2}{\|y\|_2}.$$

The righthand expression shows that the maximum singular value is the ℓ_2 operator norm of A. The *minimum singular value* of $A \in \mathbf{R}^{m\times n}$ is given by

$$\sigma_{\min}(A) = \begin{cases} \sigma_r(A) & r = \min\{m,n\} \\ 0 & r < \min\{m,n\}, \end{cases}$$

which is positive if and only if A is full rank.

The singular values of a symmetric matrix are the absolute values of its nonzero eigenvalues, sorted into descending order. The singular values of a symmetric positive semidefinite matrix are the same as its nonzero eigenvalues.

The *condition number* of a nonsingular $A \in \mathbf{R}^{n\times n}$, denoted $\mathbf{cond}(A)$ or $\kappa(A)$, is defined as

$$\mathbf{cond}(A) = \|A\|_2 \|A^{-1}\|_2 = \sigma_{\max}(A)/\sigma_{\min}(A).$$

Pseudo-inverse

Let $A = U\Sigma V^T$ be the singular value decomposition of $A \in \mathbf{R}^{m\times n}$, with $\mathbf{rank}\, A = r$. We define the *pseudo-inverse* or *Moore-Penrose inverse* of A as

$$A^\dagger = V\Sigma^{-1}U^T \in \mathbf{R}^{n\times m}.$$

Alternative expressions are

$$A^\dagger = \lim_{\epsilon\to 0}(A^T A + \epsilon I)^{-1}A^T = \lim_{\epsilon\to 0} A^T(AA^T + \epsilon I)^{-1},$$

where the limits are taken with $\epsilon > 0$, which ensures that the inverses in the expressions exist. If $\mathbf{rank}\, A = n$, then $A^\dagger = (A^T A)^{-1}A^T$. If $\mathbf{rank}\, A = m$, then $A^\dagger = A^T(AA^T)^{-1}$. If A is square and nonsingular, then $A^\dagger = A^{-1}$.

The pseudo-inverse comes up in problems involving least-squares, minimum norm, quadratic minimization, and (Euclidean) projection. For example, $A^\dagger b$ is a solution of the least-squares problem

$$\text{minimize} \quad \|Ax - b\|_2^2$$

in general. When the solution is not unique, $A^\dagger b$ gives the solution with minimum (Euclidean) norm. As another example, the matrix $AA^\dagger = UU^T$ gives (Euclidean) projection on $\mathcal{R}(A)$. The matrix $A^\dagger A = VV^T$ gives (Euclidean) projection on $\mathcal{R}(A^T)$.

The optimal value $p^\star$ of the (general, nonconvex) quadratic optimization problem

$$\text{minimize} \quad (1/2)x^T Px + q^T x + r,$$

where $P \in \mathbf{S}^n$, can be expressed as

$$p^\star = \begin{cases} -(1/2)q^T P^\dagger q + r & P \succeq 0, \quad q \in \mathcal{R}(P) \\ -\infty & \text{otherwise.} \end{cases}$$

(This generalizes the expression $p^\star = -(1/2)q^T P^{-1}q + r$, valid for $P \succ 0$.)

A.5.5　Schur complement

Consider a matrix $X \in \mathbf{S}^n$ partitioned as

$$X = \begin{bmatrix} A & B \\ B^T & C \end{bmatrix},$$

where $A \in \mathbf{S}^k$. If $\det A \neq 0$, the matrix

$$S = C - B^T A^{-1} B$$

is called the *Schur complement* of A in X. Schur complements arise in several contexts, and appear in many important formulas and theorems. For example, we have

$$\det X = \det A \det S.$$

Inverse of block matrix

The Schur complement comes up in solving linear equations, by eliminating one block of variables. We start with

$$\begin{bmatrix} A & B \\ B^T & C \end{bmatrix} \begin{bmatrix} x \\ y \end{bmatrix} = \begin{bmatrix} u \\ v \end{bmatrix},$$

and assume that $\det A \neq 0$. If we eliminate x from the top block equation and substitute it into the bottom block equation, we obtain $v = B^T A^{-1} u + Sy$, so

$$y = S^{-1}(v - B^T A^{-1} u).$$

Substituting this into the first equation yields

$$x = \left(A^{-1} + A^{-1} B S^{-1} B^T A^{-1} \right) u - A^{-1} B S^{-1} v.$$

We can express these two equations as a formula for the inverse of a block matrix:

$$\begin{bmatrix} A & B \\ B^T & C \end{bmatrix}^{-1} = \begin{bmatrix} A^{-1} + A^{-1} B S^{-1} B^T A^{-1} & -A^{-1} B S^{-1} \\ -S^{-1} B^T A^{-1} & S^{-1} \end{bmatrix}.$$

In particular, we see that the Schur complement is the inverse of the $2,2$ block entry of the inverse of X.

Minimization and definiteness

The Schur complement arises when you minimize a quadratic form over some of the variables. Suppose $A \succ 0$, and consider the minimization problem

$$\text{minimize} \quad u^T A u + 2 v^T B^T u + v^T C v \tag{A.13}$$

with variable u. The solution is $u = -A^{-1} B v$, and the optimal value is

$$\inf_u \begin{bmatrix} u \\ v \end{bmatrix}^T \begin{bmatrix} A & B \\ B^T & C \end{bmatrix} \begin{bmatrix} u \\ v \end{bmatrix} = v^T S v. \tag{A.14}$$

From this we can derive the following characterizations of positive definiteness or semidefiniteness of the block matrix X:

- $X \succ 0$ if and only if $A \succ 0$ and $S \succ 0$.

- If $A \succ 0$, then $X \succeq 0$ if and only if $S \succeq 0$.

Schur complement with singular A

Some Schur complement results have generalizations to the case when A is singular, although the details are more complicated. As an example, if $A \succeq 0$ and $Bv \in \mathcal{R}(A)$, then the quadratic minimization problem (A.13) (with variable u) is solvable, and has optimal value

$$v^T (C - B^T A^\dagger B)v,$$

where $A^\dagger$ is the pseudo-inverse of A. The problem is unbounded if $Bv \notin \mathcal{R}(A)$ or if $A \nsucceq 0$.

The range condition $Bv \in \mathcal{R}(A)$ can also be expressed as $(I - AA^\dagger)Bv = 0$, so we have the following characterization of positive semidefiniteness of the block matrix X:

$$X \succeq 0 \quad \Longleftrightarrow \quad A \succeq 0, \quad (I - AA^\dagger)B = 0, \quad C - B^T A^\dagger B \succeq 0.$$

Here the matrix $C - B^T A^\dagger B$ serves as a generalization of the Schur complement, when A is singular.

Bibliography

Some basic references for the material in this appendix are Rudin [Rud76] for analysis, and Strang [Str80] and Meyer [Mey00] for linear algebra. More advanced linear algebra texts include Horn and Johnson [HJ85, HJ91], Parlett [Par98], Golub and Van Loan [GL89], Trefethen and Bau [TB97], and Demmel [Dem97].

The concept of closed function (§A.3.3) appears frequently in convex optimization, although the terminology varies. The term is used by Rockafellar [Roc70, page 51], Hiriart-Urruty and Lemaréchal [HUL93, volume 1, page 149], Borwein and Lewis [BL00, page 76], and Bertsekas, Nedić, and Ozdaglar [Ber03, page 28].

Appendix B

Problems involving two quadratic functions

In this appendix we consider some optimization problems that involve two quadratic, but not necessarily convex, functions. Several strong results hold for these problems, even when they are not convex.

B.1 Single constraint quadratic optimization

We consider the problem with one constraint

$$
\begin{array}{ll}
\text{minimize} & x^T A_0 x + 2b_0^T x + c_0 \\
\text{subject to} & x^T A_1 x + 2b_1^T x + c_1 \le 0,
\end{array}
\tag{B.1}
$$

with variable $x \in \mathbf{R}^n$, and problem parameters $A_i \in \mathbf{S}^n$, $b_i \in \mathbf{R}^n$, $c_i \in \mathbf{R}$. We do not assume that $A_i \succeq 0$, so problem (B.1) is not a convex optimization problem.

The Lagrangian of (B.1) is

$$
L(x, \lambda) = x^T (A_0 + \lambda A_1) x + 2(b_0 + \lambda b_1)^T x + c_0 + \lambda c_1,
$$

and the dual function is

$$
\begin{aligned}
g(\lambda) &= \inf_x L(x, \lambda) \\
&= \begin{cases}
c_0 + \lambda c_1 - (b_0 + \lambda b_1)^T (A_0 + \lambda A_1)^\dagger (b_0 + \lambda b_1) & \begin{aligned} A_0 + \lambda A_1 &\succeq 0, \\ b_0 + \lambda b_1 &\in \mathcal{R}(A_0 + \lambda A_1) \end{aligned} \\
-\infty & \text{otherwise}
\end{cases}
\end{aligned}
$$

(see §A.5.4). Using a Schur complement, we can express the dual problem as

$$
\begin{array}{ll}
\text{maximize} & \gamma \\
\text{subject to} & \lambda \ge 0 \\
& \begin{bmatrix} A_0 + \lambda A_1 & b_0 + \lambda b_1 \\ (b_0 + \lambda b_1)^T & c_0 + \lambda c_1 - \gamma \end{bmatrix} \succeq 0,
\end{array}
\tag{B.2}
$$

an SDP with two variables $\gamma, \lambda \in \mathbf{R}$.

The first result is that *strong duality holds* for problem (B.1) and its Lagrange dual (B.2), provided Slater's constraint qualification is satisfied, *i.e.*, there exists an x with $x^T A_1 x + 2b_1^T x + c_1 < 0$. In other words, if (B.1) is strictly feasible, the optimal values of (B.1) and (B.2) are equal. (A proof is given in §B.4.)

Relaxation interpretation

The dual of the SDP (B.2) is

$$
\begin{array}{ll}
\text{minimize} & \mathbf{tr}(A_0 X) + 2b_0^T x + c_0 \\
\text{subject to} & \mathbf{tr}(A_1 X) + 2b_1^T x + c_1 \leq 0 \\
& \begin{bmatrix} X & x \\ x^T & 1 \end{bmatrix} \succeq 0,
\end{array}
\tag{B.3}
$$

an SDP with variables $X \in \mathbf{S}^n$, $x \in \mathbf{R}^n$. This dual SDP has an interesting interpretation in terms of the original problem (B.1).

We first note that (B.1) is equivalent to

$$
\begin{array}{ll}
\text{minimize} & \mathbf{tr}(A_0 X) + 2b_0^T x + c_0 \\
\text{subject to} & \mathbf{tr}(A_1 X) + 2b_1^T x + c_1 \leq 0 \\
& X = xx^T.
\end{array}
\tag{B.4}
$$

In this formulation we express the quadratic terms $x^T A_i x$ as $\mathbf{tr}(A_i xx^T)$, and then introduce a new variable $X = xx^T$. Problem (B.4) has a linear objective function, one linear inequality constraint, and a nonlinear equality constraint $X = xx^T$. The next step is to replace the equality constraint by an inequality $X \succeq xx^T$:

$$
\begin{array}{ll}
\text{minimize} & \mathbf{tr}(A_0 X) + b_0^T x + c_0 \\
\text{subject to} & \mathbf{tr}(A_1 X) + b_1^T x + c_1 \leq 0 \\
& X \succeq xx^T.
\end{array}
\tag{B.5}
$$

This problem is called a *relaxation* of (B.4), since we have replaced one of the constraints with a looser constraint. Finally we note that the inequality in (B.5) can be expressed as a linear matrix inequality by using a Schur complement, which gives (B.3).

A number of interesting facts follow immediately from this interpretation of (B.3) as a relaxation of (B.1). First, it is obvious that the optimal value of (B.3) is less than or equal to the optimal value of (B.1), since we minimize the same objective function over a larger set. Second, we can conclude that if $X = xx^T$ at the optimum of (B.3), then x must be optimal in (B.1).

Combining the result above, that strong duality holds between (B.1) and (B.2) (if (B.1) is strictly feasible), with strong duality between the dual SDPs (B.2) and (B.3), we conclude that strong duality holds between the original, nonconvex quadratic problem (B.1), and the SDP relaxation (B.3), provided (B.1) is strictly feasible.

B.2   The S-procedure

The next result is a theorem of alternatives for a pair of (nonconvex) quadratic inequalities. Let $A_1, A_2 \in \mathbf{S}^n$, $b_1, b_2 \in \mathbf{R}^n$, $c_1, c_2 \in \mathbf{R}$, and suppose there exists an $\hat{x}$ with

$$\hat{x}^T A_2 \hat{x} + 2b_2^T \hat{x} + c_2 < 0.$$

Then there exists an $x \in \mathbf{R}^n$ satisfying

$$x^T A_1 x + 2b_1^T x + c_1 < 0, \qquad x^T A_2 x + 2b_2^T x + c_2 \leq 0, \tag{B.6}$$

if and only if there exists no λ such that

$$\lambda \geq 0, \qquad \begin{bmatrix} A_1 & b_1 \\ b_1^T & c_1 \end{bmatrix} + \lambda \begin{bmatrix} A_2 & b_2 \\ b_2^T & c_2 \end{bmatrix} \succeq 0. \tag{B.7}$$

In other words, (B.6) and (B.7) are strong alternatives.

   This result is readily shown to be equivalent to the result from §B.1, and a proof is given in §B.4. Here we point out that the two inequality systems are clearly weak alternatives, since (B.6) and (B.7) together lead to a contradiction:

$$\begin{aligned} 0 &\leq \begin{bmatrix} x \\ 1 \end{bmatrix}^T \left(\begin{bmatrix} A_1 & b_1 \\ b_1^T & c_1 \end{bmatrix} + \lambda \begin{bmatrix} A_2 & b_2 \\ b_2^T & c_2 \end{bmatrix} \right) \begin{bmatrix} x \\ 1 \end{bmatrix} \\ &= x^T A_1 x + 2b_1^T x + c_1 + \lambda(x^T A_2 x + 2b_2^T x + c_2) \\ &< 0. \end{aligned}$$

   This theorem of alternatives is sometimes called the *S-procedure*, and is usually stated in the following form: the implication

$$x^T F_1 x + 2g_1^T x + h_1 \leq 0 \quad \Longrightarrow \quad x^T F_2 x + 2g_2^T x + h_2 \leq 0,$$

where $F_i \in \mathbf{S}^n$, $g_i \in \mathbf{R}^n$, $h_i \in \mathbf{R}$, holds if and only if there exists a λ such that

$$\lambda \geq 0, \qquad \begin{bmatrix} F_2 & g_2 \\ g_2^T & h_2 \end{bmatrix} \preceq \lambda \begin{bmatrix} F_1 & g_1 \\ g_1^T & h_1 \end{bmatrix},$$

provided there exists a point $\hat{x}$ with $\hat{x}^T F_1 \hat{x} + 2g_1^T \hat{x} + h_1 < 0$. (Note that sufficiency is clear.)

Example B.1 *Ellipsoid containment.* An ellipsoid $\mathcal{E} \subseteq \mathbf{R}^n$ with nonempty interior can be represented as the sublevel set of a quadratic function,

$$\mathcal{E} = \{x \mid x^T F x + 2g^T x + h \leq 0\},$$

where $F \in \mathbf{S}_{++}$ and $h - g^T F^{-1} g < 0$. Suppose $\tilde{\mathcal{E}}$ is another ellipsoid with similar representation,

$$\tilde{\mathcal{E}} = \{x \mid x^T \tilde{F} x + 2\tilde{g}^T x + \tilde{h} \leq 0\},$$

with $\tilde{F} \in \mathbf{S}_{++}$, $\tilde{h} - \tilde{g}^T \tilde{F}^{-1} \tilde{g} < 0$. By the S-procedure, we see that $\mathcal{E} \subseteq \tilde{\mathcal{E}}$ if and only if there is a $\lambda > 0$ such that

$$\begin{bmatrix} \tilde{F} & \tilde{g} \\ \tilde{g}^T & \tilde{h} \end{bmatrix} \preceq \lambda \begin{bmatrix} F & g \\ g^T & h \end{bmatrix}.$$

B.3　The field of values of two symmetric matrices

The following result is the basis for the proof of the strong duality result in §B.1 and the S-procedure in §B.2. If $A, B \in \mathbf{S}^n$, then for all $X \in \mathbf{S}^n_+$, there exists an $x \in \mathbf{R}^n$ such that

$$x^T A x = \mathbf{tr}(AX), \qquad x^T B x = \mathbf{tr}(BX). \tag{B.8}$$

Remark B.1 *Geometric interpretation.* This result has an interesting interpretation in terms of the set

$$W(A, B) = \{(x^T A x, x^T B x) \mid x \in \mathbf{R}^n\},$$

which is a cone in $\mathbf{R}^2$. It is the cone generated by the set

$$F(A, B) = \{(x^T A x, x^T B x) \mid \|x\|_2 = 1\},$$

which is called the *2-dimensional field of values* of the pair (A, B). Geometrically, $W(A, B)$ is the image of the set of rank-one positive semidefinite matrices under the linear transformation $f : \mathbf{S}^n \to \mathbf{R}^2$ defined by

$$f(X) = (\mathbf{tr}(AX), \mathbf{tr}(BX)).$$

The result that for every $X \in \mathbf{S}^n_+$ there exists an x satisfying (B.8) means that

$$W(A, B) = f(\mathbf{S}^n_+).$$

In other words, $W(A, B)$ is a *convex* cone.

The proof is constructive and uses induction on the rank of X. Suppose it is true for all $X \in \mathbf{S}^n_+$ with $1 \leq \mathbf{rank}\, X \leq k$, where $k \geq 2$, that there exists an x such that (B.8) holds. Then the result also holds if $\mathbf{rank}\, X = k + 1$, as can be seen as follows. A matrix $X \in \mathbf{S}^n_+$ with $\mathbf{rank}\, X = k + 1$ can be expressed as $X = yy^T + Z$ where $y \neq 0$ and $Z \in \mathbf{S}^n_+$ with $\mathbf{rank}\, Z = k$. By assumption, there exists a z such that $\mathbf{tr}(AZ) = z^T A z$, $\mathbf{tr}(AZ) = z^T B z$. Therefore

$$\mathbf{tr}(AX) = \mathbf{tr}(A(yy^T + zz^T)), \qquad \mathbf{tr}(BX) = \mathbf{tr}(B(yy^T + zz^T)).$$

The rank of $yy^T + zz^T$ is one or two, so by assumption there exists an x such that (B.8) holds.

It is therefore sufficient to prove the result if $\mathbf{rank}\, X \leq 2$. If $\mathbf{rank}\, X = 0$ and $\mathbf{rank}\, X = 1$ there is nothing to prove. If $\mathbf{rank}\, X = 2$, we can factor X as $X = VV^T$ where $V \in \mathbf{R}^{n \times 2}$, with linearly independent columns v_1 and v_2. Without loss of generality we can assume that $V^T AV$ is diagonal. (If $V^T AV$ is not diagonal we replace V with VP where $V^T AV = P \mathbf{diag}(\lambda)P^T$ is the eigenvalue decomposition of $V^T AV$.) We will write $V^T AV$ and $V^T BV$ as

$$V^T AV = \begin{bmatrix} \lambda_1 & 0 \\ 0 & \lambda_2 \end{bmatrix}, \qquad V^T BV = \begin{bmatrix} \sigma_1 & \gamma \\ \gamma & \sigma_2 \end{bmatrix},$$

and define

$$w = \begin{bmatrix} \mathbf{tr}(AX) \\ \mathbf{tr}(BX) \end{bmatrix} = \begin{bmatrix} \lambda_1 + \lambda_2 \\ \sigma_1 + \sigma_2 \end{bmatrix}.$$

We need to show that $w = (x^T A x, x^T B x)$ for some x.

We distinguish two cases. First, assume $(0, \gamma)$ is a linear combination of the vectors (λ_1, σ_1) and (λ_2, σ_2):

$$0 = z_1 \lambda_1 + z_2 \lambda_2, \qquad \gamma = z_1 \sigma_1 + z_2 \sigma_2,$$

for some z_1, z_2. In this case we choose $x = \alpha v_1 + \beta v_2$, where α and β are determined by solving two quadratic equations in two variables

$$\alpha^2 + 2\alpha\beta z_1 = 1, \qquad \beta^2 + 2\alpha\beta z_2 = 1. \tag{B.9}$$

This will give the desired result, since

$$
\begin{aligned}
&\begin{bmatrix} (\alpha v_1 + \beta v_2)^T A (\alpha v_1 + \beta v_2) \\ (\alpha v_1 + \beta v_2)^T B (\alpha v_1 + \beta v_2) \end{bmatrix} \\
&= \alpha^2 \begin{bmatrix} \lambda_1 \\ \sigma_1 \end{bmatrix} + 2\alpha\beta \begin{bmatrix} 0 \\ \gamma \end{bmatrix} + \beta^2 \begin{bmatrix} \lambda_2 \\ \sigma_2 \end{bmatrix} \\
&= (\alpha^2 + 2\alpha\beta z_1) \begin{bmatrix} \lambda_1 \\ \sigma_1 \end{bmatrix} + (\beta^2 + 2\alpha\beta z_2) \begin{bmatrix} \lambda_2 \\ \sigma_2 \end{bmatrix} \\
&= \begin{bmatrix} \lambda_1 + \lambda_2 \\ \sigma_1 + \sigma_2 \end{bmatrix}.
\end{aligned}
$$

It remains to show that the equations (B.9) are solvable. To see this, we first note that α and β must be nonzero, so we can write the equations equivalently as

$$\alpha^2(1 + 2(\beta/\alpha)z_1) = 1, \qquad (\beta/\alpha)^2 + 2(\beta/\alpha)(z_2 - z_1) = 1.$$

The equation $t^2 + 2t(z_2 - z_1) = 1$ has a positive and a negative root. At least one of these roots (the root with the same sign as z_1) satisfies $1 + 2tz_1 > 0$, so we can choose

$$\alpha = \pm 1/\sqrt{1 + 2tz_1}, \qquad \beta = t\alpha.$$

This yields two solutions (α, β) that satisfy (B.9). (If both roots of $t^2 + 2t(z_2 - z_1) = 1$ satisfy $1 + 2tz_1 > 0$, we obtain four solutions.)

Next, assume that $(0, \gamma)$ is not a linear combination of (λ_1, σ_1) and (λ_2, σ_2). In particular, this means that (λ_1, σ_1) and (λ_2, σ_2) are linearly dependent. Therefore their sum $w = (\lambda_1 + \lambda_2, \sigma_1 + \sigma_2)$ is a nonnegative multiple of (λ_1, σ_1), or (λ_2, σ_2), or both. If $w = \alpha^2(\lambda_1, \sigma_1)$ for some α, we can choose $x = \alpha v_1$. If $w = \beta^2(\lambda_2, \sigma_2)$ for some β, we can choose $x = \beta v_2$.

B.4   Proofs of the strong duality results

We first prove the S-procedure result given in §B.2. The assumption of strict feasibility of $\hat{x}$ implies that the matrix

$$\begin{bmatrix} A_2 & b_2 \\ b_2^T & c_2 \end{bmatrix}$$

has at least one negative eigenvalue. Therefore

$$\tau \geq 0, \quad \tau \begin{bmatrix} A_2 & b_2 \\ b_2^T & c_2 \end{bmatrix} \succeq 0 \quad \Longrightarrow \quad \tau = 0.$$

We can apply the theorem of alternatives for nonstrict linear matrix inequalities, given in example 5.14, which states that (B.7) is infeasible if and only if

$$X \succeq 0, \quad \mathbf{tr}\left(X \begin{bmatrix} A_1 & b_1 \\ b_1^T & c_1 \end{bmatrix} \right) < 0, \quad \mathbf{tr}\left(X \begin{bmatrix} A_2 & b_2 \\ b_2^T & c_2 \end{bmatrix} \right) \leq 0$$

is feasible. From §B.3 this is equivalent to feasibility of

$$\begin{bmatrix} v \\ w \end{bmatrix}^T \begin{bmatrix} A_1 & b_1 \\ b_1^T & c_1 \end{bmatrix} \begin{bmatrix} v \\ w \end{bmatrix} < 0, \quad \begin{bmatrix} v \\ w \end{bmatrix}^T \begin{bmatrix} A_2 & b_2 \\ b_2^T & c_2 \end{bmatrix} \begin{bmatrix} v \\ w \end{bmatrix} \leq 0.$$

If $w \neq 0$, then $x = v/w$ is feasible in (B.6). If $w = 0$, we have $v^T A_1 v < 0$, $v^T A_2 v \leq 0$, so $x = \hat{x} + tv$ satisfies

$$
\begin{aligned}
x^T A_1 x + 2b_1^T x + c_1 &= \hat{x}^T A_1 \hat{x} + 2b_1^T \hat{x} + c_1 + t^2 v^T A_1 v + 2t(A_1 \hat{x} + b_1)^T v \\
x^T A_2 x + 2b_2^T x + c_2 &= \hat{x}^T A_2 \hat{x} + 2b_2^T \hat{x} + c_2 + t^2 v^T A_2 v + 2t(A_2 \hat{x} + b_2)^T v \\
&< 2t(A_2 \hat{x} + b_2)^T v,
\end{aligned}
$$

i.e., x becomes feasible as $t \to \pm\infty$, depending on the sign of $(A_2 \hat{x} + b_2)^T v$.

Finally, we prove the result in §B.1, *i.e.*, that the optimal values of (B.1) and (B.2) are equal if (B.1) is strictly feasible. To do this we note that γ is a lower bound for the optimal value of (B.1) if

$$x^T A_1 x + b_1^T x + c_1 \leq 0 \quad \Longrightarrow \quad x^T A_0 x + b_0^T x + c_0 \geq \gamma.$$

By the S-procedure this is true if and only if there exists a $\lambda \geq 0$ such that

$$\begin{bmatrix} A_0 & b_0 \\ b_0^T & c_0 - \gamma \end{bmatrix} + \lambda \begin{bmatrix} A_1 & b_1 \\ b_1^T & c_1 \end{bmatrix} \succeq 0,$$

i.e., γ, λ are feasible in (B.2).

Bibliography

The results in this appendix are known under different names in different disciplines. The term S-procedure is from control; see Boyd, El Ghaoui, Feron, and Balakrishnan [BEFB94, pages 23, 33] for a survey and references. Variations of the S-procedure are known in linear algebra in the context of joint diagonalization of a pair of symmetric matrices; see, for example, Calabi [Cal64] and Uhlig [Uhl79]. Special cases of the strong duality result are studied in the nonlinear programming literature on trust-region methods (Stern and Wolkowicz [SW95], Nocedal and Wright [NW99, page 78]).

Brickman [Bri61] proves that the field of values of a pair of matrices $A, B \in \mathbf{S}^n$ (*i.e.*, the set $F(A, B)$ defined in remark B.1) is a convex set if $n > 2$, and that the set $W(A, B)$ is a convex cone (for any n). Our proof in §B.3 is based on Hestenes [Hes68]. Many related results and additional references can be found in Horn and Johnson [HJ91, §1.8] and Ben-Tal and Nemirovski [BTN01, §4.10.5].

Appendix C

Numerical linear algebra background

In this appendix we give a brief overview of some basic numerical linear algebra, concentrating on methods for solving one or more sets of linear equations. We focus on direct (*i.e.*, noniterative) methods, and how problem structure can be exploited to improve efficiency. There are many important issues and methods in numerical linear algebra that we do not consider here, including numerical stability, details of matrix factorizations, methods for parallel or multiple processors, and iterative methods. For these (and other) topics, we refer the reader to the references given at the end of this appendix.

C.1 Matrix structure and algorithm complexity

We concentrate on methods for solving the set of linear equations

$$Ax = b \qquad\qquad (C.1)$$

where $A \in \mathbf{R}^{n \times n}$ and $b \in \mathbf{R}^n$. We assume A is nonsingular, so the solution is unique for all values of b, and given by $x = A^{-1}b$. This basic problem arises in many optimization algorithms, and often accounts for most of the computation. In the context of solving the linear equations (C.1), the matrix A is often called the *coefficient matrix*, and the vector b is called the *righthand side*.

The standard generic methods for solving (C.1) require a computational effort that grows approximately like n^3. These methods assume nothing more about A than nonsingularity, and so are generally applicable. For n several hundred or smaller, these generic methods are probably the best methods to use, except in the most demanding real-time applications. For n more than a thousand or so, the generic methods of solving $Ax = b$ become less practical.

Coefficient matrix structure

In many cases the coefficient matrix A has some special structure or form that can be exploited to solve the equation $Ax = b$ more efficiently, using methods tailored for the special structure. For example, in the Newton system $\nabla^2 f(x) \Delta x_{\mathrm{nt}} = -\nabla f(x)$, the coefficient matrix is symmetric and positive definite, which allows us to use a solution method that is around twice as fast as the generic method (and also has better roundoff properties). There are many other types of structure that can be exploited, with computational savings (or algorithm speedup) that is usually far more than a factor of two. In many cases, the effort is reduced to something proportional to n^2 or even n, as compared to n^3 for the generic methods. Since these methods are usually applied when n is at least a hundred, and often far larger, the savings can be dramatic.

A wide variety of coefficient matrix structures can be exploited. Simple examples related to the sparsity pattern (*i.e.*, the pattern of zero and nonzero entries in the matrix) include banded, block diagonal, or sparse matrices. A more subtle exploitable structure is diagonal plus low rank. Many common forms of convex optimization problems lead to linear equations with coefficient matrices that have these exploitable structures. (There are many other matrix structures that can be exploited, *e.g.*, Toeplitz, Hankel, and circulant, that we will not consider in this appendix.)

We refer to a generic method that does not exploit any sparsity pattern in the matrices as one for *dense matrices*. We refer to a method that does not exploit any structure at all in the matrices as one for *unstructured matrices*.

C.1.1   Complexity analysis via flop count

The cost of a numerical linear algebra algorithm is often expressed by giving the total number of *floating-point operations* or *flops* required to carry it out, as a function of various problem dimensions. We define a flop as one addition, subtraction, multiplication, or division of two floating-point numbers. (Some authors define a flop as one multiplication followed by one addition, so their flop counts are smaller by a factor up to two.) To evaluate the complexity of an algorithm, we count the total number of flops, express it as a function (usually a polynomial) of the dimensions of the matrices and vectors involved, and simplify the expression by ignoring all terms except the leading (*i.e.*, highest order or dominant) terms.

As an example, suppose that a particular algorithm requires a total of

$$m^3 + 3m^2 n + mn + 4mn^2 + 5m + 22$$

flops, where m and n are problem dimensions. We would normally simplify this flop count to

$$m^3 + 3m^2 n + 4mn^2$$

flops, since these are the leading terms in the problem dimensions m and n. If in addition we assumed that $m \ll n$, we would further simplify the flop count to $4mn^2$.

Flop counts were originally popularized when floating-point operations were relatively slow, so counting the number gave a good estimate of the total computation time. This is no longer the case: Issues such as cache boundaries and locality of reference can dramatically affect the computation time of a numerical algorithm. However, flop counts can still give us a good rough estimate of the computation time of a numerical algorithm, and how the time grows with increasing problem size. Since a flop count no longer accurately predicts the computation time of an algorithm, we usually pay most attention to its order or orders, *i.e.*, its largest exponents, and ignore differences in flop counts smaller than a factor of two or so. For example, an algorithm with flop count $5n^2$ is considered comparable to one with a flop count $4n^2$, but faster than an algorithm with flop count $(1/3)n^3$.

C.1.2   Cost of basic matrix-vector operations

Vector operations

To compute the inner product $x^T y$ of two vectors $x, y \in \mathbf{R}^n$ we form the products $x_i y_i$, and then add them, which requires n multiplies and $n-1$ additions, or $2n-1$ flops. As mentioned above, we keep only the leading term, and say that the inner product requires $2n$ flops, or even more approximately, order n flops. A scalar-vector multiplication αx, where $\alpha \in \mathbf{R}$ and $x \in \mathbf{R}^n$ costs n flops. The addition $x + y$ of two vectors $x, y \in \mathbf{R}^n$ also costs n flops.

If the vectors x and y are sparse, *i.e.*, have only a few nonzero terms, these basic operations can be carried out faster (assuming the vectors are stored using an appropriate data structure). For example, if x is a sparse vector with N nonzero entries, then the inner product $x^T y$ can be computed in $2N$ flops.

Matrix-vector multiplication

A matrix-vector multiplication $y = Ax$ where $A \in \mathbf{R}^{m \times n}$ costs $2mn$ flops: We have to calculate m components of y, each of which is the product of a row of A with x, *i.e.*, an inner product of two vectors in $\mathbf{R}^n$.

Matrix-vector products can often be accelerated by taking advantage of structure in A. For example, if A is diagonal, then Ax can be computed in n flops, instead of $2n^2$ flops for multiplication by a general $n \times n$ matrix. More generally, if A is sparse, with only N nonzero elements (out of mn), then $2N$ flops are needed to form Ax, since we can skip multiplications and additions with zero.

As a less obvious example, suppose the matrix A has rank $p \ll \min\{m, n\}$, and is represented (stored) in the factored form $A = UV$, where $U \in \mathbf{R}^{m \times p}$, $V \in \mathbf{R}^{p \times n}$. Then we can compute Ax by first computing Vx (which costs $2pn$ flops), and then computing $U(Vx)$ (which costs $2mp$ flops), so the total is $2p(m + n)$ flops. Since $p \ll \min\{m, n\}$, this is small compared to $2mn$.

Matrix-matrix multiplication

The matrix-matrix product $C = AB$, where $A \in \mathbf{R}^{m \times n}$ and $B \in \mathbf{R}^{n \times p}$, costs $2mnp$ flops. We have mp elements in C to calculate, each of which is an inner product of two vectors of length n. Again, we can often make substantial savings by taking advantage of structure in A and B. For example, if A and B are sparse, we can accelerate the multiplication by skipping additions and multiplications with zero. If $m = p$ and we know that C is symmetric, then we can calculate the matrix product in $m^2 n$ flops, since we only have to compute the $(1/2)m(m+1)$ elements in the lower triangular part.

To form the product of several matrices, we can carry out the matrix-matrix multiplications in different ways, which have different flop counts in general. The simplest example is computing the product $D = ABC$, where $A \in \mathbf{R}^{m \times n}$, $B \in \mathbf{R}^{n \times p}$, and $C \in \mathbf{R}^{p \times q}$. Here we can compute D in two ways, using matrix-matrix multiplies. One method is to first form the product AB ($2mnp$ flops), and then form $D = (AB)C$ ($2mpq$ flops), so the total is $2mp(n+q)$ flops. Alternatively, we can first form the product BC ($2npq$ flops), and then form $D = A(BC)$ ($2mnq$ flops), with a total of $2nq(m+p)$ flops. The first method is better when $2mp(n+q) < 2nq(m+p)$, *i.e.*, when

$$\frac{1}{n} + \frac{1}{q} < \frac{1}{m} + \frac{1}{p}.$$

This assumes that no structure of the matrices is exploited in carrying out matrix-matrix products.

For products of more than three matrices, there are many ways to parse the product into matrix-matrix multiplications. Although it is not hard to develop an algorithm that determines the best parsing (*i.e.*, the one with the fewest required flops) given the matrix dimensions, in most applications the best parsing is clear.

C.2   Solving linear equations with factored matrices

C.2.1   Linear equations that are easy to solve

We start by examining some cases for which $Ax = b$ is easily solved, *i.e.*, $x = A^{-1}b$ is easily computed.

Diagonal matrices

Suppose A is diagonal and nonsingular (*i.e.*, $a_{ii} \neq 0$ for all i). The set of linear equations $Ax = b$ can be written as $a_{ii}x_i = b_i$, $i = 1, \ldots, n$. The solution is given by $x_i = b_i/a_{ii}$, and can be calculated in n flops.

Lower triangular matrices

A matrix $A \in \mathbf{R}^{n \times n}$ is *lower triangular* if $a_{ij} = 0$ for $j > i$. A lower triangular matrix is called *unit lower triangular* if the diagonal elements are equal to one. A lower triangular matrix is nonsingular if and only if $a_{ii} \neq 0$ for all i.

Suppose A is lower triangular and nonsingular. The equations $Ax = b$ are

$$\begin{bmatrix} a_{11} & 0 & \cdots & 0 \\ a_{21} & a_{22} & \cdots & 0 \\ \vdots & \vdots & \ddots & \vdots \\ a_{n1} & a_{n2} & \cdots & a_{nn} \end{bmatrix} \begin{bmatrix} x_1 \\ x_2 \\ \vdots \\ x_n \end{bmatrix} = \begin{bmatrix} b_1 \\ b_2 \\ \vdots \\ b_n \end{bmatrix}.$$

From the first row, we have $a_{11}x_1 = b_1$, from which we conclude $x_1 = b_1/a_{11}$. From the second row we have $a_{21}x_1 + a_{22}x_2 = b_2$, so we can express x_2 as $x_2 = (b_2 - a_{21}x_1)/a_{22}$. (We have already computed x_1, so every number on the righthand side is known.) Continuing this way, we can express each component of x in terms of previous components, yielding the algorithm

$$
\begin{aligned}
x_1 &:= b_1/a_{11} \\
x_2 &:= (b_2 - a_{21}x_1)/a_{22} \\
x_3 &:= (b_3 - a_{31}x_1 - a_{32}x_2)/a_{33} \\
&\;\;\vdots \\
x_n &:= (b_n - a_{n1}x_1 - a_{n2}x_2 - \cdots - a_{n,n-1}x_{n-1})/a_{nn}.
\end{aligned}
$$

This procedure is called *forward substitution*, since we successively compute the components of x by substituting the known values into the next equation.

Let us give a flop count for forward substitution. We start by calculating x_1 (1 flop). We substitute x_1 in the second equation to find x_2 (3 flops), then substitute x_1 and x_2 in the third equation to find x_3 (5 flops), etc. The total number of flops is

$$1 + 3 + 5 + \cdots + (2n - 1) = n^2.$$

Thus, when A is lower triangular and nonsingular, we can compute $x = A^{-1}b$ in n^2 flops.

If the matrix A has additional structure, in addition to being lower triangular, then forward substitution can be more efficient than n^2 flops. For example, if A is sparse (or banded), with at most k nonzero entries per row, then each forward substitution step requires at most $2k+1$ flops, so the overall flop count is $2(k+1)n$, or $2kn$ after dropping the term $2n$.

Upper triangular matrices

A matrix $A \in \mathbf{R}^{n \times n}$ is *upper triangular* if A^T is lower triangular, *i.e.*, if $a_{ij} = 0$ for $j < i$. We can solve linear equations with nonsingular upper triangular coefficient matrix in a way similar to forward substitution, except that we start by calculating x_n, then x_{n-1}, and so on. The algorithm is

$$
\begin{aligned}
x_n &:= b_n/a_{nn} \\
x_{n-1} &:= (b_{n-1} - a_{n-1,n}x_n)/a_{n-1,n-1} \\
x_{n-2} &:= (b_{n-2} - a_{n-2,n-1}x_{n-1} - a_{n-2,n}x_n)/a_{n-2,n-2} \\
&\;\;\vdots \\
x_1 &:= (b_1 - a_{12}x_2 - a_{13}x_3 - \cdots - a_{1n}x_n)/a_{11}.
\end{aligned}
$$

This is called *backward substitution* or *back substitution* since we determine the coefficients in backward order. The cost to compute $x = A^{-1}b$ via backward substitution is n^2 flops. If A is upper triangular and sparse (or banded), with at most k nonzero entries per row, then back substitution costs $2kn$ flops.

Orthogonal matrices

A matrix $A \in \mathbf{R}^{n \times n}$ is *orthogonal* if $A^T A = I$, i.e., $A^{-1} = A^T$. In this case we can compute $x = A^{-1}b$ by a simple matrix-vector product $x = A^T b$, which costs $2n^2$ in general.

If the matrix A has additional structure, we can compute $x = A^{-1}b$ even more efficiently than $2n^2$ flops. For example, if A has the form $A = I - 2uu^T$, where $\|u\|_2 = 1$, we can compute

$$x = A^{-1}b = (I - 2uu^T)^T b = b - 2(u^T b)u$$

by first computing $u^T b$, then forming $b - 2(u^T b)u$, which costs $4n$ flops.

Permutation matrices

Let $\pi = (\pi_1, \dots, \pi_n)$ be a permutation of $(1, 2, \dots, n)$. The associated *permutation matrix* $A \in \mathbf{R}^{n \times n}$ is given by

$$A_{ij} = \begin{cases} 1 & j = \pi_i \\ 0 & \text{otherwise.} \end{cases}$$

In each row (or column) of a permutation matrix there is exactly one entry with value one; all other entries are zero. Multiplying a vector by a permutation matrix simply permutes its coefficients:

$$Ax = (x_{\pi_1}, \dots, x_{\pi_n}).$$

The inverse of a permutation matrix is the permutation matrix associated with the inverse permutation π^{-1}. This turns out to be A^T, which shows that permutation matrices are orthogonal.

If A is a permutation matrix, solving $Ax = b$ is very easy: x is obtained by permuting the entries of b by π^{-1}. This requires no floating point operations, according to our definition (but, depending on the implementation, might involve copying floating point numbers). We can reach the same conclusion from the equation $x = A^T b$. The matrix A^T (like A) has only one nonzero entry per row, with value one. Thus no additions are required, and the only multiplications required are by one.

C.2.2   The factor-solve method

The basic approach to solving $Ax = b$ is based on expressing A as a product of nonsingular matrices,

$$A = A_1 A_2 \cdots A_k,$$

so that

$$x = A^{-1}b = A_k^{-1}A_{k-1}^{-1} \cdots A_1^{-1}b.$$

We can compute x using this formula, working from right to left:

$$
\begin{aligned}
z_1 &:= A_1^{-1}b \\
z_2 &:= A_2^{-1}z_1 = A_2^{-1}A_1^{-1}b \\
&\;\;\vdots \\
z_{k-1} &:= A_{k-1}^{-1}z_{k-2} = A_{k-1}^{-1} \cdots A_1^{-1}b \\
x &:= A_k^{-1}z_{k-1} = A_k^{-1} \cdots A_1^{-1}b.
\end{aligned}
$$

The ith step of this process requires computing $z_i = A_i^{-1}z_{i-1}$, $i.e.$, solving the linear equations $A_i z_i = z_{i-1}$. If each of these equations is easy to solve ($e.g.$, if A_i is diagonal, lower or upper triangular, a permutation, etc.), this gives a method for computing $x = A^{-1}b$.

The step of expressing A in factored form ($i.e.$, computing the factors A_i) is called the *factorization step*, and the process of computing $x = A^{-1}b$ recursively, by solving a sequence problems of the form $A_i z_i = z_{i-1}$, is often called the *solve step*. The total flop count for solving $Ax = b$ using this factor-solve method is $f + s$, where f is the flop count for computing the factorization, and s is the total flop count for the solve step. In many cases, the cost of the factorization, f, dominates the total solve cost s. In this case, the cost of solving $Ax = b$, $i.e.$, computing $x = A^{-1}b$, is just f.

Solving equations with multiple righthand sides

Suppose we need to solve the equations

$$Ax_1 = b_1, \qquad Ax_2 = b_2, \qquad \ldots, \qquad Ax_m = b_m,$$

where $A \in \mathbf{R}^{n \times n}$ is nonsingular. In other words, we need to solve m sets of linear equations, with the same coefficient matrix, but different righthand sides. Alternatively, we can think of this as computing the matrix

$$X = A^{-1}B$$

where

$$X = \begin{bmatrix} x_1 & x_2 & \cdots & x_m \end{bmatrix} \in \mathbf{R}^{n \times m}, \qquad B = \begin{bmatrix} b_1 & b_2 & \cdots & b_m \end{bmatrix} \in \mathbf{R}^{n \times m}.$$

To do this, we first factor A, which costs f. Then for $i = 1, \ldots, m$ we compute $A^{-1}b_i$ using the solve step. Since we only factor A once, the total effort is

$$f + ms.$$

In other words, we amortize the factorization cost over the set of m solves. Had we (needlessly) repeated the factorization step for each i, the cost would be $m(f + s)$.

When the factorization cost f dominates the solve cost s, the factor-solve method allows us to solve a small number of linear systems, with the same coefficient matrix, at essentially the same cost as solving one. This is because the most expensive step, the factorization, is done only once.

We can use the factor-solve method to compute the inverse A^{-1} by solving $Ax = e_i$ for $i = 1, \ldots, n$, *i.e.*, by computing $A^{-1}I$. This requires one factorization and n solves, so the cost is $f + ns$.

C.3   LU, Cholesky, and LDL$^\mathsf{T}$ factorization

C.3.1   LU factorization

Every nonsingular matrix $A \in \mathbf{R}^{n \times n}$ can be factored as

$$A = PLU$$

where $P \in \mathbf{R}^{n \times n}$ is a permutation matrix, $L \in \mathbf{R}^{n \times n}$ is unit lower triangular, and $U \in \mathbf{R}^{n \times n}$ is upper triangular and nonsingular. This is called the *LU factorization* of A. We can also write the factorization as $P^T A = LU$, where the matrix $P^T A$ is obtained from A by re-ordering the rows. The standard algorithm for computing an LU factorization is called *Gaussian elimination with partial pivoting* or *Gaussian elimination with row pivoting*. The cost is $(2/3)n^3$ flops if no structure in A is exploited, which is the case we consider first.

Solving sets of linear equations using the LU factorization

The LU factorization, combined with the factor-solve approach, is the standard method for solving a general set of linear equations $Ax = b$.

Algorithm C.1 *Solving linear equations by LU factorization.*

given a set of linear equations $Ax = b$, with A nonsingular.
   1. *LU factorization.* Factor A as $A = PLU$ ($(2/3)n^3$ flops).
   2. *Permutation.* Solve $Pz_1 = b$ (0 flops).
   3. *Forward substitution.* Solve $Lz_2 = z_1$ (n^2 flops).
   4. *Backward substitution.* Solve $Ux = z_2$ (n^2 flops).

The total cost is $(2/3)n^3 + 2n^2$, or $(2/3)n^3$ flops if we keep only the leading term.

If we need to solve multiple sets of linear equations with different righthand sides, *i.e.*, $Ax_i = b_i$, $i = 1, \ldots, m$, the cost is

$$(2/3)n^3 + 2mn^2,$$

since we factor A once, and carry out m pairs of forward and backward substitutions. For example, we can solve two sets of linear equations, with the same coefficient matrix but different righthand sides, at essentially the same cost as solving one. We can compute the inverse A^{-1} by solving the equations $Ax_i = e_i$, where x_i is the ith column of A^{-1}, and e_i is the ith unit vector. This costs $(8/3)n^3$, *i.e.*, about $3n^3$ flops.

If the matrix A has certain structure, for example banded or sparse, the LU factorization can be computed in less than $(2/3)n^3$ flops, and the associated forward and backward substitutions can also be carried out more efficiently.

LU factorization of banded matrices

Suppose the matrix $A \in \mathbf{R}^{n \times n}$ is *banded*, *i.e.*, $a_{ij} = 0$ if $|i - j| > k$, where $k < n - 1$ is called the *bandwidth* of A. We are interested in the case where $k \ll n$, *i.e.*, the bandwidth is much smaller than the size of the matrix. In this case an LU factorization of A can be computed in roughly $4nk^2$ flops. The resulting upper triangular matrix U has bandwidth at most $2k$, and the lower triangular matrix L has at most $k + 1$ nonzeros per column, so the forward and back substitutions can be carried out in order $6nk$ flops. Therefore if A is banded, the linear equations $Ax = b$ can be solved in about $4nk^2$ flops.

LU factorization of sparse matrices

When the matrix A is sparse, the LU factorization usually includes both row and column permutations, *i.e.*, A is factored as

$$A = P_1 L U P_2,$$

where P_1 and P_2 are permutation matrices, L is lower triangular, and U is upper triangular. If the factors L and U are sparse, the forward and backward substitutions can be carried out efficiently, and we have an efficient method for solving $Ax = b$. The sparsity of the factors L and U depends on the permutations P_1 and P_2, which are chosen in part to yield relatively sparse factors.

   The cost of computing the sparse LU factorization depends in a complicated way on the size of A, the number of nonzero elements, its sparsity pattern, and the particular algorithm used, but is often dramatically smaller than the cost of a dense LU factorization. In many cases the cost grows approximately linearly with n, when n is large. This means that when A is sparse, we can solve $Ax = b$ very efficiently, often with an order approximately n.

C.3.2 Cholesky factorization

If $A \in \mathbf{R}^{n \times n}$ is symmetric and positive definite, then it can be factored as

$$A = LL^T$$

where L is lower triangular and nonsingular with positive diagonal elements. This is called the *Cholesky factorization* of A, and can be interpreted as a symmetric LU factorization (with $L = U^T$). The matrix L, which is uniquely determined by A, is called the *Cholesky factor* of A. The cost of computing the Cholesky factorization of a dense matrix, *i.e.*, without exploiting any structure, is $(1/3)n^3$ flops, half the cost of an LU factorization.

Solving positive definite sets of equations using Cholesky factorization

The Cholesky factorization can be used to solve $Ax = b$ when A is symmetric positive definite.

Algorithm C.2 *Solving linear equations by Cholesky factorization.*

given a set of linear equations $Ax = b$, with $A \in \mathbf{S}_{++}^n$.
1. *Cholesky factorization.* Factor A as $A = LL^T$ $((1/3)n^3$ flops$)$.
2. *Forward substitution.* Solve $Lz_1 = b$ $(n^2$ flops$)$.
3. *Backward substitution.* Solve $L^T x = z_1$ $(n^2$ flops$)$.

The total cost is $(1/3)n^3 + 2n^2$, or roughly $(1/3)n^3$ flops.

There are specialized algorithms, with a complexity much lower than $(1/3)n^3$, for Cholesky factorization of banded and sparse matrices.

Cholesky factorization of banded matrices

If A is symmetric positive definite and banded with bandwidth k, then its Cholesky factor L is banded with bandwidth k, and can be calculated in nk^2 flops. The cost of the associated solve step is $4nk$ flops.

Cholesky factorization of sparse matrices

When A is symmetric positive definite and sparse, it is usually factored as

$$A = PLL^T P^T,$$

where P is a permutation matrix and L is lower triangular with positive diagonal elements. We can also express this as $P^T AP = LL^T$, i.e., LL^T is the Cholesky factorization of $P^T AP$. We can interpret this as first re-ordering the variables and equations, and then forming the (standard) Cholesky factorization of the resulting permuted matrix. Since $P^T AP$ is positive definite for any permutation matrix P, we are free to choose any permutation matrix; for each choice there is a unique associated Cholesky factor L. The choice of P, however, can greatly affect the sparsity of the factor L, which in turn can greatly affect the efficiency of solving $Ax = b$. Various heuristic methods are used to select a permutation P that leads to a sparse factor L.

Example C.1 *Cholesky factorization with an arrow sparsity pattern.* Consider a sparse matrix of the form

$$A = \begin{bmatrix} 1 & u^T \\ u & D \end{bmatrix}$$

where $D \in \mathbf{R}^{n \times n}$ is positive diagonal, and $u \in \mathbf{R}^n$. It can be shown that A is positive definite if $u^T D^{-1} u < 1$. The Cholesky factorization of A is

$$\begin{bmatrix} 1 & u^T \\ u & D \end{bmatrix} = \begin{bmatrix} 1 & 0 \\ u & L \end{bmatrix} \begin{bmatrix} 1 & u^T \\ 0 & L^T \end{bmatrix} \qquad (\text{C.2})$$

where L is lower triangular with $LL^T = D - uu^T$. For general u, the matrix $D - uu^T$ is dense, so we can expect L to be dense. Although the matrix A is very sparse (most of its rows have just two nonzero elements), its Cholesky factors are almost completely dense.

On the other hand, suppose we permute the first row and column of A to the end. After this re-ordering, we obtain the Cholesky factorization

$$\begin{bmatrix} D & u \\ u^T & 1 \end{bmatrix} = \begin{bmatrix} D^{1/2} & 0 \\ u^T D^{-1/2} & \sqrt{1 - u^T D^{-1} u} \end{bmatrix} \begin{bmatrix} D^{1/2} & D^{-1/2} u \\ 0 & \sqrt{1 - u^T D^{-1} u} \end{bmatrix}.$$

Now the Cholesky factor has a diagonal 1,1 block, so it is very sparse.

This example illustrates that the re-ordering greatly affects the sparsity of the Cholesky factors. Here it was quite obvious what the best permutation is, and all good re-ordering heuristics would select this re-ordering and permute the dense row and column to the end. For more complicated sparsity patterns, it can be very difficult to find the 'best' re-ordering (*i.e.*, resulting in the greatest number of zero elements in L), but various heuristics provide good suboptimal permutations.

For the sparse Cholesky factorization, the re-ordering permutation P is often determined using only sparsity pattern of the matrix A, and not the particular numerical values of the nonzero elements of A. Once P is chosen, we can also determine the sparsity pattern of L without knowing the numerical values of the nonzero entries of A. These two steps combined are called the *symbolic factorization* of A, and form the first step in a sparse Cholesky factorization. In contrast, the permutation matrices in a sparse LU factorization do depend on the numerical values in A, in addition to its sparsity pattern.

The symbolic factorization is then followed by the *numerical factorization, i.e.*, the calculation of the nonzero elements of L. Software packages for sparse Cholesky factorization often include separate routines for the symbolic and the numerical factorization. This is useful in many applications, because the cost of the symbolic factorization is significant, and often comparable to the numerical factorization. Suppose, for example, that we need to solve m sets of linear equations

$$A_1 x = b_1, \qquad A_2 x = b_2, \qquad \ldots, \qquad A_m x = b_m$$

where the matrices A_i are symmetric positive definite, with different numerical values, but the same sparsity pattern. Suppose the cost of a symbolic factorization is f_{symb}, the cost of a numerical factorization is f_{num}, and the cost of the solve step is s. Then we can solve the m sets of linear equations in

$$f_{\text{symb}} + m(f_{\text{num}} + s)$$

flops, since we only need to carry out the symbolic factorization once, for all m sets of equations. If instead we carry out a separate symbolic factorization for each set of linear equations, the flop count is $m(f_{\text{symb}} + f_{\text{num}} + s)$.

C.3.3   LDL$^\mathsf{T}$ factorization

Every nonsingular symmetric matrix A can be factored as

$$A = PLDL^T P^T$$

where P is a permutation matrix, L is lower triangular with positive diagonal elements, and D is block diagonal, with nonsingular 1×1 and 2×2 diagonal blocks. This is called an LDL^T factorization of A. (The Cholesky factorization can be considered a special case of LDL^T factorization, with $P = I$ and $D = I$.) An LDL^T factorization can be computed in $(1/3)n^3$ flops, if no structure of A is exploited.

Algorithm C.3 *Solving linear equations by LDL^T factorization.*

given a set of linear equations $Ax = b$, with $A \in \mathbf{S}^n$ nonsingular.

1. *LDL^T factorization.* Factor A as $A = PLDL^T P$ ($(1/3)n^3$ flops).
2. *Permutation.* Solve $Pz_1 = b$ (0 flops).
3. *Forward substitution.* Solve $Lz_2 = z_1$ (n^2 flops).
4. *(Block) diagonal solve.* Solve $Dz_3 = z_2$ (order n flops).
5. *Backward substitution.* Solve $L^T z_4 = z_3$ (n^2 flops).
6. *Permutation.* Solve $P^T x = z_4$ (0 flops).

The total cost is, keeping only the dominant term, $(1/3)n^3$ flops.

LDL^T factorization of banded and sparse matrices

As with the LU and Cholesky factorizations, there are specialized methods for calculating the LDL^T factorization of a sparse or banded matrix. These are similar to the analogous methods for Cholesky factorization, with the additional factor D. In a sparse LDL^T factorization, the permutation matrix P cannot be chosen only on the basis of the sparsity pattern of A (as in a sparse Cholesky factorization); it also depends on the particular nonzero values in the matrix A.

C.4   Block elimination and Schur complements

C.4.1   Eliminating a block of variables

In this section we describe a general method that can be used to solve $Ax = b$ by first eliminating a subset of the variables, and then solving a smaller system of linear equations for the remaining variables. For a dense unstructured matrix, this approach gives no advantage. But when the submatrix of A associated with the eliminated variables is easily factored (for example, if it is block diagonal or banded) the method can be substantially more efficient than a general method.

Suppose we partition the variable $x \in \mathbf{R}^n$ into two blocks or subvectors,

$$x = \begin{bmatrix} x_1 \\ x_2 \end{bmatrix},$$

where $x_1 \in \mathbf{R}^{n_1}$, $x_2 \in \mathbf{R}^{n_2}$. We conformally partition the linear equations $Ax = b$ as

$$\begin{bmatrix} A_{11} & A_{12} \\ A_{21} & A_{22} \end{bmatrix} \begin{bmatrix} x_1 \\ x_2 \end{bmatrix} = \begin{bmatrix} b_1 \\ b_2 \end{bmatrix} \tag{C.3}$$

where $A_{11} \in \mathbf{R}^{n_1 \times n_1}$, $A_{22} \in \mathbf{R}^{n_2 \times n_2}$. Assuming that the submatrix A_{11} is invertible, we can eliminate x_1 from the equations, as follows. Using the first equation, we can express x_1 in terms of x_2:

$$x_1 = A_{11}^{-1}(b_1 - A_{12}x_2). \tag{C.4}$$

Substituting this expression into the second equation yields

$$(A_{22} - A_{21}A_{11}^{-1}A_{12})x_2 = b_2 - A_{21}A_{11}^{-1}b_1. \tag{C.5}$$

We refer to this as the *reduced equation* obtained by eliminating x_1 from the original equation. The reduced equation (C.5) and the equation (C.4) together are equivalent to the original equations (C.3). The matrix appearing in the reduced equation is called the *Schur complement* of the first block A_{11} in A:

$$S = A_{22} - A_{21}A_{11}^{-1}A_{12}$$

(see also §A.5.5). The Schur complement S is nonsingular if and only if A is nonsingular.

The two equations (C.5) and (C.4) give us an alternative approach to solving the original system of equations (C.3). We first form the Schur complement S, then find x_2 by solving (C.5), and then calculate x_1 from (C.4). We can summarize this method as follows.

Algorithm C.4 *Solving linear equations by block elimination.*

given a nonsingular set of linear equations (C.3), with A_{11} nonsingular.

1. Form $A_{11}^{-1}A_{12}$ and $A_{11}^{-1}b_1$.
2. Form $S = A_{22} - A_{21}A_{11}^{-1}A_{12}$ and $\tilde{b} = b_2 - A_{21}A_{11}^{-1}b_1$.
3. Determine x_2 by solving $Sx_2 = \tilde{b}$.
4. Determine x_1 by solving $A_{11}x_1 = b_1 - A_{12}x_2$.

Remark C.1 *Interpretation as block factor-solve.* Block elimination can be interpreted in terms of the factor-solve approach described in §C.2.2, based on the factorization

$$\begin{bmatrix} A_{11} & A_{12} \\ A_{21} & A_{22} \end{bmatrix} = \begin{bmatrix} A_{11} & 0 \\ A_{21} & S \end{bmatrix} \begin{bmatrix} I & A_{11}^{-1}A_{12} \\ 0 & I \end{bmatrix},$$

which can be considered a block LU factorization. This block LU factorization suggests the following method for solving (C.3). We first do a 'block forward substitution' to solve

$$\begin{bmatrix} A_{11} & 0 \\ A_{21} & S \end{bmatrix} \begin{bmatrix} z_1 \\ z_2 \end{bmatrix} = \begin{bmatrix} b_1 \\ b_2 \end{bmatrix},$$

and then solve

$$\begin{bmatrix} I & A_{11}^{-1}A_{12} \\ 0 & I \end{bmatrix} \begin{bmatrix} x_1 \\ x_2 \end{bmatrix} = \begin{bmatrix} z_1 \\ z_2 \end{bmatrix}$$

by 'block backward substitution'. This yields the same expressions as the block elimination method:

$$
\begin{aligned}
z_1 &= A_{11}^{-1} b_1 \\
z_2 &= S^{-1}(b_2 - A_{21} z_1) \\
x_2 &= z_2 \\
x_1 &= z_1 - A_{11}^{-1} A_{12} z_2.
\end{aligned}
$$

In fact, the modern approach to the factor-solve method is based on block factor and solve steps like these, with the block sizes optimally chosen for the processor (or processors), cache sizes, etc.

Complexity analysis of block elimination method

To analyze the (possible) advantage of solving the set of linear equations using block elimination, we carry out a flop count. We let f and s denote the cost of factoring A_{11} and carrying out the associated solve step, respectively. To keep the analysis simple we assume (for now) that A_{12}, A_{22}, and A_{21} are treated as dense, unstructured matrices. The flop counts for each of the four steps in solving $Ax = b$ using block elimination are:

1. Computing $A_{11}^{-1} A_{12}$ and $A_{11}^{-1} b_1$ requires factoring A_{11} and $n_2 + 1$ solves, so it costs $f + (n_2 + 1)s$, or just $f + n_2 s$, dropping the dominated term s.

2. Forming the Schur complement S requires the matrix multiply $A_{21}(A_{11}^{-1} A_{12})$, which costs $2n_2^2 n_1$, and an $n_2 \times n_2$ matrix subtraction, which costs n_2^2 (and can be dropped). The cost of forming $\tilde{b} = b_2 - A_{21} A_{11}^{-1} b_1$ is dominated by the cost of forming S, and so can be ignored. The total cost of step 2, ignoring dominated terms, is then $2n_2^2 n_1$.

3. To compute $x_2 = S^{-1}\tilde{b}$, we factor S and solve, which costs $(2/3)n_2^3$.

4. Forming $b_1 - A_{12} x_2$ costs $2n_1 n_2 + n_1$ flops. To compute $x_1 = A_{11}^{-1}(b_1 - A_{12} x_2)$, we can use the factorization of A_{11} already computed in step 1, so only the solve is necessary, which costs s. Both of these costs are dominated by other terms, and can be ignored.

The total cost is then
$$
f + n_2 s + 2n_2^2 n_1 + (2/3)n_2^3 \tag{C.6}
$$
flops.

Eliminating an unstructured matrix

We first consider the case when no structure in A_{11} is exploited. We factor A_{11} using a standard LU factorization, so $f = (2/3)n_1^3$, and then solve using a forward and a backward substitution, so $s = 2n_1^2$. The flop count for solving the equations via block elimination is then

$$
(2/3)n_1^3 + n_2(2n_1^2) + 2n_2^2 n_1 + (2/3)n_2^3 = (2/3)(n_1 + n_2)^3,
$$

which is the same as just solving the larger set of equations using a standard LU factorization. In other words, solving a set of equations by block elimination gives no advantage when no structure of A_{11} is exploited.

On the other hand, when the structure of A_{11} allows us to factor and solve more efficiently than the standard method, block elimination can be more efficient than applying the standard method.

Eliminating a diagonal matrix

If A_{11} is diagonal, no factorization is needed, and we can carry out a solve in n_1 flops, so we have $f = 0$ and $s = n_1$. Substituting these values into (C.6) and keeping only the leading terms yields

$$2n_2^2 n_1 + (2/3)n_2^3,$$

flops, which is far smaller than $(2/3)(n_1 + n_2)^3$, the cost using the standard method. In particular, the flop count of the standard method grows cubically in n_1, whereas for block elimination the flop count grows only linearly in n_1.

Eliminating a banded matrix

If A_{11} is banded with bandwidth k, we can carry out the factorization in about $f = 4k^2 n_1$ flops, and the solve can be done in about $s = 6kn_1$ flops. The overall complexity of solving $Ax = b$ using block elimination is

$$4k^2 n_1 + 6n_2 k n_1 + 2n_2^2 n_1 + (2/3)n_2^3$$

flops. Assuming k is small compared to n_1 and n_2, this simplifies to $2n_2^2 n_1 + (2/3)n_2^3$, the same as when A_{11} is diagonal. In particular, the complexity grows linearly in n_1, as opposed to cubically in n_1 for the standard method.

A matrix for which A_{11} is banded is sometimes called an *arrow matrix* since the sparsity pattern, when $n_1 \gg n_2$, looks like an arrow pointing down and right. Block elimination can solve linear equations with arrow structure far more efficiently than the standard method.

Eliminating a block diagonal matrix

Suppose that A_{11} is block diagonal, with (square) block sizes $m_1, \ldots, m_k$, where $n_1 = m_1 + \cdots + m_k$. In this case we can factor A_{11} by factoring each block separately, and similarly we can carry out the solve step on each block separately. Using standard methods for these we find

$$f = (2/3)m_1^3 + \cdots + (2/3)m_k^3, \qquad s = 2m_1^2 + \cdots + 2m_k^2,$$

so the overall complexity of block elimination is

$$(2/3)\sum_{i=1}^{k} m_i^3 + 2n_2 \sum_{i=1}^{k} m_i^2 + 2n_2^2 \sum_{i=1}^{k} m_i + (2/3)n_2^3.$$

If the block sizes are small compared to n_1 and $n_1 \gg n_2$, the savings obtained by block elimination is dramatic.

The linear equations $Ax = b$, where A_{11} is block diagonal, are called *partially separable* for the following reason. If the subvector x_2 is fixed, the remaining equations decouple into k sets of independent linear equations (which can be solved separately). The subvector x_2 is sometimes called the *complicating variable* since the equations are much simpler when x_2 is fixed. Using block elimination, we can solve partially separable linear equations far more efficiently than by using a standard method.

Eliminating a sparse matrix

If A_{11} is sparse, we can eliminate A_{11} using a sparse factorization and sparse solve steps, so the values of f and s in (C.6) are much less than for unstructured A_{11}. When A_{11} in (C.3) is sparse and the other blocks are dense, and $n_2 \ll n_1$, we say that A is a sparse matrix with a few dense rows and columns. Eliminating the sparse block A_{11} provides an efficient method for solving equations which are sparse except for a few dense rows and columns.

An alternative is to simply apply a sparse factorization algorithm to the entire matrix A. Most sparse solvers will handle dense rows and columns, and select a permutation that results in sparse factors, and hence fast factorization and solve times. This is more straightforward than using block elimination, but often slower, especially in applications where we can exploit structure in the other blocks (see, *e.g.*, example C.4).

Remark C.2 As already suggested in remark C.1, these two methods for solving systems with a few dense rows and columns are closely related. Applying the elimination method by factoring A_{11} and S as

$$A_{11} = P_1 L_1 U_1 P_2, \qquad S = P_3 L_2 U_2,$$

can be interpreted as factoring A as

$$\begin{bmatrix} A_{11} & A_{12} \\ A_{21} & A_{22} \end{bmatrix} =$$

$$\begin{bmatrix} P_1 & 0 \\ 0 & P_3 \end{bmatrix} \begin{bmatrix} L_1 & 0 \\ P_3^T A_{21} P_2^T U_1^{-1} & L_2 \end{bmatrix} \begin{bmatrix} U_1 & L_1^{-1} P_1^T A_{12} \\ 0 & U_2 \end{bmatrix} \begin{bmatrix} P_2 & 0 \\ 0 & I \end{bmatrix},$$

followed by forward and backward substitutions.

C.4.2 Block elimination and structure

Symmetry and positive definiteness

There are variants of the block elimination method that can be used when A is symmetric, or symmetric and positive definite. When A is symmetric, so are A_{11} and the Schur complement S, so a symmetric factorization can be used for A_{11} and S. Symmetry can also be exploited in the other operations, such as the matrix multiplies. Overall the savings over the nonsymmetric case is around a factor of two.

Positive definiteness can also be exploited in block elimination. When A is symmetric and positive definite, so are A_{11} and the Schur complement S, so Cholesky factorizations can be used.

Exploiting structure in other blocks

Our complexity analysis above assumes that we exploit no structure in the matrices A_{12}, A_{21}, A_{22}, and the Schur complement S, *i.e.*, they are treated as dense. But in many cases there is structure in these blocks that can be exploited in forming the Schur complement, factoring it, and carrying out the solve steps. In such cases the computational savings of the block elimination method over a standard method can be even higher.

Example C.2 *Block triangular equations.* Suppose that $A_{12} = 0$, *i.e.*, the linear equations $Ax = b$ have block lower triangular structure:

$$\begin{bmatrix} A_{11} & 0 \\ A_{21} & A_{22} \end{bmatrix} \begin{bmatrix} x_1 \\ x_2 \end{bmatrix} = \begin{bmatrix} b_1 \\ b_2 \end{bmatrix}.$$

In this case the Schur complement is just $S = A_{22}$, and the block elimination method reduces to block forward substitution:

$$\begin{aligned} x_1 &:= A_{11}^{-1} b_1 \\ x_2 &:= A_{22}^{-1}(b_2 - A_{21}x_1). \end{aligned}$$

Example C.3 *Block diagonal and banded systems.* Suppose that A_{11} is block diagonal, with maximum block size $l \times l$, and that A_{12}, A_{21}, and A_{22} are banded, say with bandwidth k. In this case, A_{11}^{-1} is also block diagonal, with the same block sizes as A_{11}. Therefore the product $A_{11}^{-1}A_{12}$ is also banded, with bandwidth $k + l$, and the Schur complement, $S = A_{22} - A_{21}A_{11}^{-1}A_{12}$ is banded with bandwidth $2k + l$. This means that forming the Schur complement S can be done more efficiently, and that the factorization and solve steps with S can be done efficiently. In particular, for fixed maximum block size l and bandwidth k, we can solve $Ax = b$ with a number of flops that grows linearly with n.

Example C.4 *KKT structure.* Suppose that the matrix A has *KKT structure*, *i.e.*,

$$A = \begin{bmatrix} A_{11} & A_{12} \\ A_{12}^T & 0 \end{bmatrix},$$

where $A_{11} \in \mathbf{S}_{++}^p$, and $A_{12} \in \mathbf{R}^{p \times m}$ with $\mathbf{rank}\, A_{12} = m$. Since $A_{11} \succ 0$, we can use a Cholesky factorization. The Schur complement $S = -A_{12}^T A_{11}^{-1} A_{12}$ is negative definite, so we can factor $-S$ using a Cholesky factorization.

C.4.3   The matrix inversion lemma

The idea of block elimination is to remove variables, and then solve a smaller set of equations that involve the Schur complement of the original matrix with respect to the eliminated variables. The same idea can be turned around: When we recognize a matrix as a Schur complement, we can introduce new variables, and create a larger set of equations to solve. In most cases there is no advantage to doing this, since we end up with a larger set of equations. But when the larger set of equations has some special structure that can be exploited to solve it, introducing variables can lead to an efficient method. The most common case is when another block of variables can be eliminated from the larger matrix.

We start with the linear equations

$$(A + BC)x = b, \tag{C.7}$$

where $A \in \mathbf{R}^{n \times n}$ is nonsingular, and $B \in \mathbf{R}^{n \times p}$, $C \in \mathbf{R}^{p \times n}$. We introduce a new variable $y = Cx$, and rewrite the equations as

$$Ax + By = b, \qquad y = Cx,$$

or, in matrix form,

$$\begin{bmatrix} A & B \\ C & -I \end{bmatrix} \begin{bmatrix} x \\ y \end{bmatrix} = \begin{bmatrix} b \\ 0 \end{bmatrix}. \tag{C.8}$$

Note that our original coefficient matrix, $A + BC$, is the Schur complement of $-I$ in the larger matrix that appears in (C.8). If we were to eliminate the variable y from (C.8), we would get back the original equation (C.7).

In some cases, it can be more efficient to solve the larger set of equations (C.8) than the original, smaller set of equations (C.7). This would be the case, for example, if A, B, and C were relatively sparse, but the matrix $A + BC$ were far less sparse.

After introducing the new variable y, we can eliminate the original variable x from the larger set of equations (C.8), using $x = A^{-1}(b - By)$. Substituting this into the second equation $y = Cx$, we obtain

$$(I + CA^{-1}B)y = CA^{-1}b,$$

so that

$$y = (I + CA^{-1}B)^{-1}CA^{-1}b.$$

Using $x = A^{-1}(b - By)$, we get

$$x = \left(A^{-1} - A^{-1}B(I + CA^{-1}B)^{-1}CA^{-1} \right) b. \tag{C.9}$$

Since b is arbitrary, we conclude that

$$(A + BC)^{-1} = A^{-1} - A^{-1}B \left(I + CA^{-1}B \right)^{-1} CA^{-1}.$$

This is known as the *matrix inversion lemma*, or the *Sherman-Woodbury-Morrison formula*.

The matrix inversion lemma has many applications. For example if p is small (or even just not very large), it gives us a method for solving $(A + BC)x = b$, provided we have an efficient method for solving $Au = v$.

Diagonal or sparse plus low rank

Suppose that A is diagonal with nonzero diagonal elements, and we want to solve an equation of the form (C.7). The straightforward solution would consist in first forming the matrix $D = A + BC$, and then solving $Dx = b$. If the product BC is dense, then the complexity of this method is $2pn^2$ flops to form $A + BC$, plus $(2/3)n^3$ flops for the LU factorization of D, so the total cost is

$$2pn^2 + (2/3)n^3$$

flops. The matrix inversion lemma suggests a more efficient method. We can calculate x by evaluating the expression (C.9) from right to left, as follows. We first evaluate $z = A^{-1}b$ (n flops, since A is diagonal). Then we form the matrix $E = I + CA^{-1}B$ ($2p^2n$ flops). Next we solve $Ew = Cz$, which is a set of p linear equations in p variables. The cost is $(2/3)p^3$ flops, plus $2pn$ to form Cz. Finally, we evaluate $x = z - A^{-1}Bw$ ($2pn$ flops for the matrix-vector product Bw, plus lower order terms). The total cost is

$$2p^2n + (2/3)p^3$$

flops, dropping dominated terms. Comparing with the first method, we see that the second method is more efficient when $p < n$. In particular if p is small and fixed, the complexity grows linearly with n.

Another important application of the matrix inversion lemma occurs when A is sparse and nonsingular, and the matrices B and C are dense. Again we can compare two methods. The first method is to form the (dense) matrix $A + BC$, and to solve (C.7) using a dense LU factorization. The cost of this method is $2pn^2 + (2/3)n^3$ flops. The second method is based on evaluating the expression (C.9), using a sparse LU factorization of A. Specifically, suppose that f is the cost of factoring A as $A = P_1LUP_2$, and s is the cost of solving the factored system $P_1LUP_2x = d$. We can evaluate (C.9) from right to left as follows. We first factor A, and solve $p + 1$ linear systems

$$Az = b, \qquad AD = B,$$

to find $z \in \mathbf{R}^n$, and $D \in \mathbf{R}^{n \times p}$. The cost is $f + (p + 1)s$ flops. Next, we form the matrix $E = I + CD$, and solve

$$Ew = Cz,$$

which is a set of p linear equations in p variables w. The cost of this step is $2p^2n + (2/3)p^3$ plus lower order terms. Finally, we evaluate $x = z - Dw$, at a cost of $2pn$ flops. This gives us a total cost of

$$f + ps + 2p^2n + (2/3)p^3$$

flops. If $f \ll (2/3)n^3$ and $s \ll 2n^2$, this is much lower than the complexity of the first method.

Remark C.3   *The augmented system approach.* A different approach to exploiting sparse plus low rank structure is to solve (C.8) directly using a sparse LU-solver. The system (C.8) is a set of $p + n$ linear equations in $p + n$ variables, and is sometimes called the *augmented* system associated with (C.7). If A is very sparse and p is small, then solving the augmented system using a sparse solver can be much faster than solving the system (C.7) using a dense solver.

The augmented system approach is closely related to the method that we described above. Suppose

$$A = P_1 LU P_2$$

is a sparse LU factorization of A, and

$$I + CA^{-1}B = P_3 \tilde{L}\tilde{U}$$

is a dense LU factorization of $I + CA^{-1}B$. Then

$$
\begin{bmatrix} A & B \\ C & -I \end{bmatrix}
= \begin{bmatrix} P_1 & 0 \\ 0 & P_3 \end{bmatrix}
\begin{bmatrix} L & 0 \\ P_3^T CP_2^T U^{-1} & -\tilde{L} \end{bmatrix}
\begin{bmatrix} U & L^{-1}P_1^T B \\ 0 & \tilde{U} \end{bmatrix}
\begin{bmatrix} P_2 & 0 \\ 0 & I \end{bmatrix},
\tag{C.10}
$$

and this factorization can be used to solve the augmented system. It can be verified that this is equivalent to the method based on the matrix inversion lemma that we described above.

Of course, if we solve the augmented system using a sparse LU solver, we have no control over the permutations that are selected. The solver might choose a factorization different from (C.10), and more expensive to compute. In spite of this, the augmented system approach remains an attractive option. It is easier to implement than the method based on the matrix inversion lemma, and it is numerically more stable.

Low rank updates

Suppose $A \in \mathbf{R}^{n \times n}$ is nonsingular, $u, v \in \mathbf{R}^n$ with $1 + v^T A^{-1}u \neq 0$, and we want to solve two sets of linear equations

$$Ax = b, \qquad (A + uv^T)\tilde{x} = b.$$

The solution $\tilde{x}$ of the second system is called a *rank-one update* of x. The matrix inversion lemma allows us to calculate the rank-one update $\tilde{x}$ very cheaply, once we have computed x. We have

$$
\begin{aligned}
\tilde{x} &= (A + uv^T)^{-1}b \\
&= (A^{-1} - \frac{1}{1 + v^T A^{-1}u} A^{-1}uv^T A^{-1})b \\
&= x - \frac{v^T x}{1 + v^T A^{-1}u} A^{-1}u.
\end{aligned}
$$

We can therefore solve both systems by factoring A, computing $x = A^{-1}b$ and $w = A^{-1}u$, and then evaluating

$$\tilde{x} = x - \frac{v^T x}{1 + v^T w}w.$$

The overall cost is $f + 2s$, as opposed to $2(f + s)$ if we were to solve for $\tilde{x}$ from scratch.

C.5   Solving underdetermined linear equations

To conclude this appendix, we mention a few important facts about *underdetermined* linear equations

$$Ax = b, \tag{C.11}$$

where $A \in \mathbf{R}^{p \times n}$ with $p < n$. We assume that $\mathbf{rank}\, A = p$, so there is at least one solution for all b.

In many applications it is sufficient to find just one particular solution $\hat{x}$. In other situations we might need a complete parametrization of all solutions as

$$\{x \mid Ax = b\} = \{Fz + \hat{x} \mid z \in \mathbf{R}^{n-p}\} \tag{C.12}$$

where F is a matrix whose columns form a basis for the nullspace of A.

Inverting a nonsingular submatrix of A

The solution of the underdetermined system is straightforward if a $p \times p$ nonsingular submatrix of A is known. We start by assuming that the first p columns of A are independent. Then we can write the equation $Ax = b$ as

$$Ax = \begin{bmatrix} A_1 & A_2 \end{bmatrix} \begin{bmatrix} x_1 \\ x_2 \end{bmatrix} = A_1 x_1 + A_2 x_2 = b,$$

where $A_1 \in \mathbf{R}^{p \times p}$ is nonsingular. We can express x_1 as

$$x_1 = A_1^{-1}(b - A_2 x_2) = A_1^{-1}b - A_1^{-1}A_2 x_2.$$

This expression allows us to easily calculate a solution: we simply take $\hat{x}_2 = 0$, $\hat{x}_1 = A_1^{-1}b$. The cost is equal to the cost of solving one square set of p linear equations $A_1 \hat{x}_1 = b$.

We can also parametrize all solutions of $Ax = b$, using $x_2 \in \mathbf{R}^{n-p}$ as a free parameter. The general solution of $Ax = b$ can be expressed as

$$x = \begin{bmatrix} x_1 \\ x_2 \end{bmatrix} = \begin{bmatrix} -A_1^{-1}A_2 \\ I \end{bmatrix} x_2 + \begin{bmatrix} A_1^{-1}b \\ 0 \end{bmatrix}.$$

This gives a parametrization of the form (C.12) with

$$F = \begin{bmatrix} -A_1^{-1}A_2 \\ I \end{bmatrix}, \qquad \hat{x} = \begin{bmatrix} A_1^{-1}b \\ 0 \end{bmatrix}.$$

To summarize, assume that the cost of factoring A_1 is f and the cost of solving one system of the form $A_1 x = d$ is s. Then the cost of finding one solution of (C.11) is $f + s$. The cost of parametrizing all solutions (*i.e.*, calculating F and $\hat{x}$) is $f + s(n - p + 1)$.

Now we consider the general case, when the first p columns of A need not be independent. Since $\mathbf{rank}\, A = p$, we can select a set of p columns of A that is independent, permute them to the front, and then apply the method described above. In other words, we find a permutation matrix P such that the first p columns of $\tilde{A} = AP$ are independent, *i.e.*,

$$\tilde{A} = AP = \left[\begin{array}{cc} A_1 & A_2 \end{array} \right],$$

where A_1 is invertible. The general solution of $\tilde{A}\tilde{x} = b$, where $\tilde{x} = P^T x$, is then given by

$$\tilde{x} = \left[\begin{array}{c} -A_1^{-1} A_2 \\ I \end{array} \right] \tilde{x}_2 + \left[\begin{array}{c} A_1^{-1} b \\ 0 \end{array} \right].$$

The general solution of $Ax = b$ is then given by

$$x = P\tilde{x} = P \left[\begin{array}{c} -A_1^{-1} A_2 \\ I \end{array} \right] z + P \left[\begin{array}{c} A_1^{-1} b \\ 0 \end{array} \right],$$

where $z \in \mathbf{R}^{n-p}$ is a free parameter. This idea is useful when it is easy to identify a nonsingular or easily inverted submatrix of A, for example, a diagonal matrix with nonzero diagonal elements.

The QR factorization

If $C \in \mathbf{R}^{n \times p}$ with $p \leq n$ and $\mathbf{rank}\, C = p$, then it can be factored as

$$C = \left[\begin{array}{cc} Q_1 & Q_2 \end{array} \right] \left[\begin{array}{c} R \\ 0 \end{array} \right],$$

where $Q_1 \in \mathbf{R}^{n \times p}$ and $Q_2 \in \mathbf{R}^{n \times (n-p)}$ satisfy

$$Q_1^T Q_1 = I, \qquad Q_2^T Q_2 = I, \qquad Q_1^T Q_2 = 0,$$

and $R \in \mathbf{R}^{p \times p}$ is upper triangular with nonzero diagonal elements. This is called the *QR factorization* of C. The QR factorization can be calculated in $2p^2(n - p/3)$ flops. (The matrix Q is stored in a factored form that makes it possible to efficiently compute matrix-vector products Qx and $Q^T x$.)

The QR factorization can be used to solve the underdetermined set of linear equations (C.11). Suppose

$$A^T = \left[\begin{array}{cc} Q_1 & Q_2 \end{array} \right] \left[\begin{array}{c} R \\ 0 \end{array} \right]$$

is the QR factorization of A^T. Substituting in the equations it is clear that $\hat{x} = Q_1 R^{-T} b$ satisfies the equations:

$$A\hat{x} = R^T Q_1^T Q_1 R^{-T} b = b.$$

Moreover, the columns of Q_2 form a basis for the nullspace of A, so the complete solution set can be parametrized as

$$\{x = \hat{x} + Q_2 z \mid z \in \mathbf{R}^{n-p}\}.$$

The QR factorization method is the most common method for solving underdetermined equations. One drawback is that it is difficult to exploit sparsity. The factor Q is usually dense, even when C is very sparse.

LU factorization of a rectangular matrix

If $C \in \mathbf{R}^{n \times p}$ with $p \leq n$ and $\mathbf{rank}\, C = p$, then it can be factored as

$$C = PLU$$

where $P \in \mathbf{R}^{n \times n}$ is a permutation matrix, $L \in \mathbf{R}^{n \times p}$ is unit lower triangular (*i.e.*, $l_{ij} = 0$ for $i < j$ and $l_{ii} = 1$), and $U \in \mathbf{R}^{p \times p}$ is nonsingular and upper triangular. The cost is $(2/3)p^3 + p^2(n - p)$ flops if no structure in C is exploited.

If the matrix C is sparse, the LU factorization usually includes row and column permutations, *i.e.*, we factor C as

$$C = P_1 L U P_2$$

where $P_1,\ P_2 \in \mathbf{R}^{p \times p}$ are permutation matrices. The LU factorization of a sparse rectangular matrix can be calculated very efficiently, at a cost that is much lower than for dense matrices.

The LU factorization can be used to solve underdetermined sets of linear equations. Suppose $A^T = PLU$ is the LU factorization of the matrix A^T in (C.11), and we partition L as

$$L = \begin{bmatrix} L_1 \\ L_2 \end{bmatrix},$$

where $L_1 \in \mathbf{R}^{p \times p}$ and $L_2 \in \mathbf{R}^{(n-p) \times p}$. It is easily verified that the solution set can be parametrized as (C.12) with

$$\hat{x} = P \begin{bmatrix} L_1^{-T} U^{-T} b \\ 0 \end{bmatrix}, \qquad F = P \begin{bmatrix} -L_1^{-T} L_2^T \\ I \end{bmatrix}.$$

Bibliography

Standard references for dense numerical linear algebra are Golub and Van Loan [GL89], Demmel [Dem97], Trefethen and Bau [TB97], and Higham [Hig96]. The sparse Cholesky factorization is covered in George and Liu [GL81]. Duff, Erisman, and Reid [DER86] and Duff [Duf93] discuss the sparse LU and LDL^T factorizations. The books by Gill, Murray, and Wright [GMW81, §2.2], Wright [Wri97, chapter 11], and Nocedal and Wright [NW99, §A.2] include introductions to numerical linear algebra that focus on problems arising in numerical optimization.

High-quality implementations of common dense linear algebra algorithms are included in the LAPACK package [ABB$^+$99]. LAPACK is built upon the *Basic Linear Algebra Subprograms* (BLAS), a library of routines for basic vector and matrix operations that can be easily customized to take advantage of specific computer architectures. Several codes for solving sparse linear equations are also available, including SPOOLES [APWW99], SuperLU [DGL03], UMFPACK [Dav03], and WSMP [Gup00], to mention only a few.

References

[ABB+99] E. Anderson, Z. Bai, C. Bischof, S. Blackford, J. Demmel, J. Dongarra, J. Du Croz, A. Greenbaum, S. Hammarling, A. McKenney, and D. Sorensen. *LAPACK Users' Guide*. Society for Industrial and Applied Mathematics, third edition, 1999. Available from `www.netlib.org/lapack`.

[AE61] K. J. Arrow and A. C. Enthoven. Quasi-concave programming. *Econometrica*, 29(4):779–800, 1961.

[AG03] F. Alizadeh and D. Goldfarb. Second-order cone programming. *Mathematical Programming Series B*, 95:3–51, 2003.

[AHO98] F. Alizadeh, J.-P. A. Haeberly, and M. L. Overton. Primal-dual interior-point methods for semidefinite programming: Convergence rates, stability and numerical results. *SIAM Journal on Optimization*, 8(3):746–768, 1998.

[Ali91] F. Alizadeh. *Combinatorial Optimization with Interior-Point Methods and Semi-Definite Matrices*. PhD thesis, University of Minnesota, 1991.

[And70] T. W. Anderson. Estimation of covariance matrices which are linear combinations or whose inverses are linear combinations of given matrices. In R. C. Bose et al., editor, *Essays in Probability and Statistics*, pages 1–24. University of North Carolina Press, 1970.

[APWW99] C. Ashcraft, D. Pierce, D. K. Wah, and J. Wu. *The Reference Manual for SPOOLES Version 2.2: An Object Oriented Software Library for Solving Sparse Linear Systems of Equations*, 1999. Available from `www.netlib.org/linalg/spooles/spooles.2.2.html`.

[AY98] E. D. Andersen and Y. Ye. A computational study of the homogeneous algorithm for large-scale convex optimization. *Computational Optimization and Applications*, 10:243–269, 1998.

[Bar02] A. Barvinok. *A Course in Convexity*, volume 54 of *Graduate Studies in Mathematics*. American Mathematical Society, 2002.

[BB65] E. F. Beckenbach and R. Bellman. *Inequalities*. Springer, second edition, 1965.

[BB91] S. Boyd and C. Barratt. *Linear Controller Design: Limits of Performance*. Prentice-Hall, 1991.

[BBI71] A. Berman and A. Ben-Israel. More on linear inequalities with applications to matrix theory. *Journal of Mathematical Analysis and Applications*, 33:482–496, 1971.

[BD77] P. J. Bickel and K. A. Doksum. *Mathematical Statistics*. Holden-Day, 1977.

[BDX04] S. Boyd, P. Diaconis, and L. Xiao. Fastest mixing Markov chain on a graph. *SIAM Review*, 46(4):667–689, 2004.

[BE93] S. Boyd and L. El Ghaoui. Method of centers for minimizing generalized eigenvalues. *Linear Algebra and Its Applications*, 188:63–111, 1993.

[BEFB94]   S. Boyd, L. El Ghaoui, E. Feron, and V. Balakrishnan. *Linear Matrix Inequalities in System and Control Theory*. Society for Industrial and Applied Mathematics, 1994.

[Ber73]   A. Berman. *Cones, Matrices and Mathematical Programming*. Springer, 1973.

[Ber90]   M. Berger. Convexity. *The American Mathematical Monthly*, 97(8):650–678, 1990.

[Ber99]   D. P. Bertsekas. *Nonlinear Programming*. Athena Scientific, second edition, 1999.

[Ber03]   D. P. Bertsekas. *Convex Analysis and Optimization*. Athena Scientific, 2003. With A. Nedić and A. E. Ozdaglar.

[BF48]   T. Bonnesen and W. Fenchel. *Theorie der konvexen Körper*. Chelsea Publishing Company, 1948. First published in 1934.

[BF63]   R. Bellman and K. Fan. On systems of linear inequalities in Hermitian matrix variables. In V. L. Klee, editor, *Convexity*, volume VII of *Proceedings of the Symposia in Pure Mathematics*, pages 1–11. American Mathematical Society, 1963.

[BGT81]   R. G. Bland, D. Goldfarb, and M. J. Todd. The ellipsoid method: A survey. *Operations Research*, 29(6):1039–1091, 1981.

[BI69]   A. Ben-Israel. Linear equations and inequalities on finite dimensional, real or complex vector spaces: A unified theory. *Journal of Mathematical Analysis and Applications*, 27:367–389, 1969.

[Bjö96]   A. Björck. *Numerical Methods for Least Squares Problems*. Society for Industrial and Applied Mathematics, 1996.

[BKMR98]   A. Brooke, D. Kendrick, A. Meeraus, and R. Raman. *GAMS: A User's Guide*. The Scientific Press, 1998.

[BL00]   J. M. Borwein and A. S. Lewis. *Convex Analysis and Nonlinear Optimization*. Springer, 2000.

[BN78]   O. Barndorff-Nielsen. *Information and Exponential Families in Statistical Theory*. John Wiley & Sons, 1978.

[Bon94]   J. V. Bondar. Comments on and complements to *Inequalities: Theory of Majorization and Its Applications*. *Linear Algebra and Its Applications*, 199:115–129, 1994.

[Bor02]   B. Borchers. *CSDP User's Guide*, 2002. Available from www.nmt.edu/~borchers/csdp.html.

[BP94]   A. Berman and R. J. Plemmons. *Nonnegative Matrices in the Mathematical Sciences*. Society for Industrial and Applied Mathematics, 1994. First published in 1979 by Academic Press.

[Bri61]   L. Brickman. On the field of values of a matrix. *Proceedings of the American Mathematical Society*, 12:61–66, 1961.

[BS00]   D. Bertsimas and J. Sethuraman. Moment problems and semidefinite optimization. In H. Wolkowicz, R. Saigal, and L. Vandenberghe, editors, *Handbook of Semidefinite Programming*, chapter 16, pages 469–510. Kluwer Academic Publishers, 2000.

[BSS93]   M. S. Bazaraa, H. D. Sherali, and C. M. Shetty. *Nonlinear Programming. Theory and Algorithms*. John Wiley & Sons, second edition, 1993.

[BT97]   D. Bertsimas and J. N. Tsitsiklis. *Introduction to Linear Optimization*. Athena Scientific, 1997.

[BTN98]   A. Ben-Tal and A. Nemirovski. Robust convex optimization. *Mathematics of Operations Research*, 23(4):769–805, 1998.

[BTN99]   A. Ben-Tal and A. Nemirovski. Robust solutions of uncertain linear programs. *Operations Research Letters*, 25(1):1–13, 1999.

[BTN01]   A. Ben-Tal and A. Nemirovski. *Lectures on Modern Convex Optimization. Analysis, Algorithms, and Engineering Applications*. Society for Industrial and Applied Mathematics, 2001.

[BY02]   S. J. Benson and Y. Ye. *DSDP — A Software Package Implementing the Dual-Scaling Algorithm for Semidefinite Programming*, 2002. Available from `www-unix.mcs.anl.gov/~benson`.

[BYT99]   E. Bai, Y. Ye, and R. Tempo. Bounded error parameter estimation: A sequential analytic center approach. *IEEE Transactions on Automatic control*, 44(6):1107–1117, 1999.

[Cal64]   E. Calabi. Linear systems of real quadratic forms. *Proceedings of the American Mathematical Society*, 15(5):844–846, 1964.

[CDS01]   S. S. Chen, D. L. Donoho, and M. A. Saunders. Atomic decomposition by basis pursuit. *SIAM Review*, 43(1):129–159, 2001.

[CGGS98]   S. Chandrasekaran, G. H. Golub, M. Gu, and A. H. Sayed. Parameter estimation in the presence of bounded data uncertainties. *SIAM Journal of Matrix Analysis and Applications*, 19(1):235–252, 1998.

[CH53]   R. Courant and D. Hilbert. *Method of Mathematical Physics. Volume 1*. Interscience Publishers, 1953. Tranlated and revised from the 1937 German original.

[CK77]   B. D. Craven and J. J. Koliha. Generalizations of Farkas' theorem. *SIAM Journal on Numerical Analysis*, 8(6), 1977.

[CT91]   T. M. Cover and J. A. Thomas. *Elements of Information Theory*. John Wiley & Sons, 1991.

[Dan63]   G. B. Dantzig. *Linear Programming and Extensions*. Princeton University Press, 1963.

[Dav63]   C. Davis. Notions generalizing convexity for functions defined on spaces of matrices. In V. L. Klee, editor, *Convexity*, volume VII of *Proceedings of the Symposia in Pure Mathematics*, pages 187–201. American Mathematical Society, 1963.

[Dav03]   T. A. Davis. *UMFPACK User Guide*, 2003. Available from `www.cise.ufl.edu/research/sparse/umfpack`.

[DDB95]   M. A. Dahleh and I. J. Diaz-Bobillo. *Control of Uncertain Systems: A Linear Programming Approach*. Prentice-Hall, 1995.

[Deb59]   G. Debreu. *Theory of Value: An Axiomatic Analysis of Economic Equilibrium*. Yale University Press, 1959.

[Dem97]   J. W. Demmel. *Applied Numerical Linear Algebra*. Society for Industrial and Applied Mathematics, 1997.

[DER86]   I. S. Duff, A. M. Erismann, and J. K. Reid. *Direct Methods for Sparse Matrices*. Clarendon Press, 1986.

[DGL03]   J. W. Demmel, J. R. Gilbert, and X. S. Li. *SuperLU Users' Guide*, 2003. Available from `crd.lbl.gov/~xiaoye/SuperLU`.

[dH93]   D. den Hertog. *Interior Point Approach to Linear, Quadratic and Convex Programming*. Kluwer, 1993.

[DHS99]   R. O. Duda, P. E. Hart, and D. G. Stork. *Pattern Classification*. John Wiley & Sons, second edition, 1999.

[Dik67]   I. Dikin. Iterative solution of problems of linear and quadratic programming. *Soviet Mathematics Doklady*, 8(3):674–675, 1967.

[DLW00]   T. N. Davidson, Z.-Q. Luo, and K. M. Wong. Design of orthogonal pulse shapes for communications via semidefinite programming. *IEEE Transactions on Signal Processing*, 48(5):1433–1445, 2000.

[DP00]   G. E. Dullerud and F. Paganini. *A Course in Robust Control Theory: A Convex Approach*. Springer, 2000.

[DPZ67]   R. J. Duffin, E. L. Peterson, and C. Zener. *Geometric Programming. Theory and Applications*. John Wiley & Sons, 1967.

[DS96]   J. E. Dennis and R. S. Schnabel. *Numerical Methods for Unconstrained Optimization and Nonlinear Equations*. Society for Industrial and Applied Mathematics, 1996. First published in 1983 by Prentice-Hall.

[Duf93]   I. S. Duff. The solution of augmented systems. In D. F. Griffiths and G. A. Watson, editors, *Numerical Analysis 1993. Proceedings of the 15th Dundee Conference*, pages 40–55. Longman Scientific & Technical, 1993.

[Eck80]   J. G. Ecker. Geometric programming: Methods, computations and applications. *SIAM Review*, 22(3):338–362, 1980.

[Egg58]   H. G. Eggleston. *Convexity*. Cambridge University Press, 1958.

[EL97]   L. El Ghaoui and H. Lebret. Robust solutions to least-squares problems with uncertain data. *SIAM Journal of Matrix Analysis and Applications*, 18(4):1035–1064, 1997.

[EM75]   J. Elzinga and T. G. Moore. A central cutting plane algorithm for the convex programming problem. *Mathematical Programming Studies*, 8:134–145, 1975.

[EN00]   L. El Ghaoui and S. Niculescu, editors. *Advances in Linear Matrix Inequality Methods in Control*. Society for Industrial and Applied Mathematics, 2000.

[EOL98]   L. El Ghaoui, F. Oustry, and H. Lebret. Robust solutions to uncertain semidefinite programs. *SIAM Journal on Optimization*, 9(1):33–52, 1998.

[ET99]   I. Ekeland and R. Témam. *Convex Analysis and Variational Inequalities*. Classics in Applied Mathematics. Society for Industrial and Applied Mathematics, 1999. Originally published in 1976.

[Far02]   J. Farkas. Theorie der einfachen Ungleichungen. *Journal für die Reine und Angewandte Mathematik*, 124:1–27, 1902.

[FD85]   J. P. Fishburn and A. E. Dunlop. TILOS: A posynomial programming approach to transistor sizing. In *IEEE International Conference on Computer-Aided Design: ICCAD-85. Digest of Technical Papers*, pages 326–328. IEEE Computer Society Press, 1985.

[Fcn83]   W. Fenchel. Convexity through the ages. In P. M. Gruber and J. M. Wills, editors, *Convexity and Its Applications*, pages 120–130. Birkhäuser Verlag, 1983.

[FGK99]   R. Fourer, D. M. Gay, and B. W. Kernighan. *AMPL: A Modeling Language for Mathematical Programming*. Duxbury Press, 1999.

[FGW02]   A. Forsgren, P. E. Gill, and M. H. Wright. Interior methods for nonlinear optimization. *SIAM Review*, 44(4):525–597, 2002.

[FKN98]   K. Fujisawa, M. Kojima, and K. Nakata. *SDPA User's Manual*, 1998. Available from grid.r.dendai.ac.jp/sdpa.

[FL01]   M. Florenzano and C. Le Van. *Finite Dimensional Convexity and Optimization*. Number 13 in Studies in Economic Theory. Springer, 2001.

[FM90]   A. V. Fiacco and G. P. McCormick. *Nonlinear Programming. Sequential Unconstrained Minimization Techniques*. Society for Industrial and Applied Mathematics, 1990. First published in 1968 by Research Analysis Corporation.

[Fre56]   R. J. Freund. The introduction of risk into a programming model. *Econometrica*, 24(3):253–263, 1956.

[FW56]   M. Frank and P. Wolfe. An algorithm for quadratic programming. *Naval Research Logistics Quarterly*, 3:95–110, 1956.

[Gau95]   C. F. Gauss. *Theory of the Combination of Observations Least Subject to Errors*. Society for Industrial and Applied Mathematics, 1995. Translated from original 1820 manuscript by G. W. Stewart.

[GI03a]   D. Goldfarb and G. Iyengar. Robust convex quadratically constrained programs. *Mathematical Programming Series B*, 97:495–515, 2003.

[GI03b]   D. Goldfarb and G. Iyengar. Robust portfolio selection problems. *Mathematics of Operations Research*, 28(1):1–38, 2003.

[GKT51]   D. Gale, H. W. Kuhn, and A. W. Tucker. Linear programming and the theory of games. In T. C. Koopmans, editor, *Activity Analysis of Production and Allocation*, volume 13 of *Cowles Commission for Research in Economics Monographs*, pages 317–335. John Wiley & Sons, 1951.

[GL81]   A. George and J. W.-H. Liu. *Computer solution of large sparse positive definite systems*. Prentice-Hall, 1981.

[GL89]   G. Golub and C. F. Van Loan. *Matrix Computations*. Johns Hopkins University Press, second edition, 1989.

[GLS88]   M. Grötschel, L. Lovasz, and A. Schrijver. *Geometric Algorithms and Combinatorial Optimization*. Springer, 1988.

[GLY96]   J.-L. Goffin, Z.-Q. Luo, and Y. Ye. Complexity analysis of an interior cutting plane method for convex feasibility problems. *SIAM Journal on Optimization*, 6:638–652, 1996.

[GMS+86]   P. E. Gill, W. Murray, M. A. Saunders, J. A. Tomlin, and M. H. Wright. On projected newton barrier methods for linear programming and an equivalence to Karmarkar's projective method. *Mathematical Programming*, 36:183–209, 1986.

[GMW81]   P. E. Gill, W. Murray, and M. H. Wright. *Practical Optimization*. Academic Press, 1981.

[Gon92]   C. C. Gonzaga. Path-following methods for linear programming. *SIAM Review*, 34(2):167–224, 1992.

[Gow85]   J. C. Gower. Properties of Euclidean and non-Euclidean distance matrices. *Linear Algebra and Its Applications*, 67:81–97, 1985.

[Gup00]   A. Gupta. *WSMP: Watson Sparse Matrix Package. Part I — Direct Solution of Symmetric Sparse Systems. Part II — Direct Solution of General Sparse Systems*, 2000. Available from `www.cs.umn.edu/~agupta/wsmp`.

[GW95]   M. X. Goemans and D. P. Williamson. Improved approximation algorithms for maximum cut and satisfiability problems using semidefinite programming. *Journal of the Association for Computing Machinery*, 42(6):1115–1145, 1995.

[Han98]   P. C. Hansen. *Rank-Deficient and Discrete Ill-Posed Problems. Numerical Aspects of Linear Inversion*. Society for Industrial and Applied Mathematics, 1998.

[HBL01]   M. del Mar Hershenson, S. P. Boyd, and T. H. Lee. Optimal design of a CMOS op-amp via geometric programming. *IEEE Transactions on Computer-Aided Design of Integrated Circuits and Systems*, 20(1):1–21, 2001.

[Hes68]   M. R. Hestenes. Pairs of quadratic forms. *Linear Algebra and Its Applications*, 1:397–407, 1968.

[Hig96]   N. J. Higham. *Accuracy and Stability of Numerical Algorithms*. Society for Industrial and Applied Mathematics, 1996.

[Hil57]   C. Hildreth. A quadratic programming procedure. *Naval Research Logistics Quarterly*, 4:79–85, 1957.

[HJ85]   R. A. Horn and C. A. Johnson. *Matrix Analysis*. Cambridge University Press, 1985.

[HJ91]   R. A. Horn and C. A. Johnson. *Topics in Matrix Analysis*. Cambridge University Press, 1991.

[HLP52]   G. H. Hardy, J. E. Littlewood, and G. Pólya. *Inequalities*. Cambridge University Press, second edition, 1952.

[HP94]   R. Horst and P. Pardalos. *Handbook of Global Optimization*. Kluwer, 1994.

[HRVW96]   C. Helmberg, F. Rendl, R. Vanderbei, and H. Wolkowicz. An interior-point method for semidefinite programming. *SIAM Journal on Optimization*, 6:342–361, 1996.

[HTF01]   T. Hastie, R. Tibshirani, and J. Friedman. *The Elements of Statistical Learning. Data Mining, Inference, and Prediction*. Springer, 2001.

[Hub64]   P. J. Huber. Robust estimation of a location parameter. *The Annals of Mathematical Statistics*, 35(1):73–101, 1964.

[Hub81]   P. J. Huber. *Robust Statistics*. John Wiley & Sons, 1981.

[HUL93]   J.-B. Hiriart-Urruty and C. Lemaréchal. *Convex Analysis and Minimization Algorithms*. Springer, 1993. Two volumes.

[HUL01]   J.-B. Hiriart-Urruty and C. Lemaréchal. *Fundamentals of Convex Analysis*. Springer, 2001. Abridged version of *Convex Analysis and Minimization Algorithms* volumes 1 and 2.

[Isi64]   K. Isii. Inequalities of the types of Chebyshev and Cramér-Rao and mathematical programming. *Annals of The Institute of Statistical Mathematics*, 16:277–293, 1964.

[Jar94]   F. Jarre. Optimal ellipsoidal approximations around the analytic center. *Applied Mathematics and Optimization*, 30:15–19, 1994.

[Jen06]   J. L. W. V. Jensen. Sur les fonctions convexes et les inégalités entre les valeurs moyennes. *Acta Mathematica*, 30:175–193, 1906.

[Joh85]   F. John. Extremum problems with inequalities as subsidiary conditions. In J. Moser, editor, *Fritz John, Collected Papers*, pages 543–560. Birkhäuser Verlag, 1985. First published in 1948.

[Kan52]   L. V. Kantorovich. *Functional Analysis and Applied Mathematics*. National Bureau of Standards, 1952. Translated from Russian by C. D. Benster. First published in 1948.

[Kan60]   L. V. Kantorovich. Mathematical methods of organizing and planning production. *Management Science*, 6(4):366–422, 1960. Translated from Russian. First published in 1939.

[Kar84]   N. Karmarkar. A new polynomial-time algorithm for linear programming. *Combinatorica*, 4(4):373–395, 1984.

[Kel60]   J. E. Kelley. The cutting-plane method for solving convex programs. *Journal of the Society for Industrial and Applied Mathematics*, 8(4):703–712, 1960.

[Kle63]   V. L. Klee, editor. *Convexity*, volume 7 of *Proceedings of Symposia in Pure Mathematics*. American Mathematical Society, 1963.

[Kle71]   V. Klee. What is a convex set? *The American Mathematical Monthly*, 78(6):616–631, 1971.

[KN77]   M. G. Krein and A. A. Nudelman. *The Markov Moment Problem and Extremal Problems*. American Mathematical Society, 1977. Translated from Russian. First published in 1973.

[Koo51]   T. C. Koopmans, editor. *Activity Analysis of Production and Allocation*, volume 13 of *Cowles Commission for Research in Economics Monographs*. John Wiley & Sons, 1951.

[KS66]   S. Karlin and W. J. Studden. *Tchebycheff Systems: With Applications in Analysis and Statistics*. John Wiley & Sons, 1966.

[KSH97]   M. Kojima, S. Shindoh, and S. Hara. Interior-point methods for the monotone semidefinite linear complementarity problem in symmetric matrices. *SIAM Journal on Optimization*, 7(1):86–125, 1997.

[KSH00]   T. Kailath, A. H. Sayed, and B. Hassibi. *Linear Estimation*. Prentice-Hall, 2000.

[KSJA91]   J. M. Kleinhaus, G. Sigl, F. M. Johannes, and K. J. Antreich. GORDIAN: VLSI placement by quadratic programming and slicing optimization. *IEEE Transactions on Computer-Aided Design of Integrated Circuits and Systems*, 10(3):356–200, 1991.

[KT51]   H. W. Kuhn and A. W. Tucker. Nonlinear programming. In J. Neyman, editor, *Proceedings of the Second Berkeley Symposium on Mathematical Statistics and Probability*, pages 481–492. University of California Press, 1951.

[Kuh76]   H. W. Kuhn. Nonlinear programming. A historical view. In R. W. Cottle and C. E. Lemke, editors, *Nonlinear Programming*, volume 9 of *SIAM-AMS Proceedings*, pages 1–26. American Mathematical Society, 1976.

[Las95]   J. B. Lasserre. A new Farkas lemma for positive semidefinite matrices. *IEEE Transactions on Automatic Control*, 40(6):1131–1133, 1995.

[Las02]   J. B. Lasserre. Bounds on measures satisfying moment conditions. *The Annals of Applied Probability*, 12(3):1114–1137, 2002.

[Lay82]   S. R. Lay. *Convex Sets and Their Applications*. John Wiley & Sons, 1982.

[LH66]   B. Liêu and P. Huard. La méthode des centres dans un espace topologique. *Numerische Mathematik*, 8:56–67, 1966.

[LH95]   C. L. Lawson and R. J. Hanson. *Solving Least Squares Problems*. Society for Industrial and Applied Mathematics, 1995. First published in 1974 by Prentice-Hall.

[LMS94]   I. J. Lustig, R. E. Marsten, and D. F. Shanno. Interior point methods for linear programming: Computational state of the art. *ORSA Journal on Computing*, 6(1):1–14, 1994.

[LO96]   A. S. Lewis and M. L. Overton. Eigenvalue optimization. *Acta Numerica*, 5:149–190, 1996.

[Löf04]   J. Löfberg. YALMIP : A toolbox for modeling and optimization in MATLAB. In *Proceedings of the IEEE International Symposium on Computer Aided Control Systems Design*, pages 284–289, 2004. Available from `control.ee.ethz.ch/~joloef/yalmip.php`.

[Löw34]   K. Löwner. Über monotone Matrixfunktionen. *Mathematische Zeitschrift*, 38:177–216, 1934.

[LSZ00]   Z.-Q. Luo, J. F. Sturm, and S. Zhang. Conic convex programming and self-dual embedding. *Optimization Methods and Software*, 14:169–218, 2000.

[Lue68]   D. G. Luenberger. Quasi-convex programming. *SIAM Journal on Applied Mathematics*, 16(5), 1968.

[Lue69]   D. G. Luenberger. *Optimization by Vector Space Methods*. John Wiley & Sons, 1969.

[Lue84]   D. G. Luenberger. *Linear and Nonlinear Programming*. Addison-Wesley, second edition, 1984.

[Lue95]   D. G. Luenberger. *Microeconomic Theory*. McGraw-Hill, 1995.

[Lue98]   D. G. Luenberger. *Investment Science*. Oxford University Press, 1998.

[Luo03]   Z.-Q. Luo. Applications of convex optimization in signal processing and
            digital communication. *Mathematical Programming Series B*, 97:177–207,
            2003.

[LVBL98]   M. S. Lobo, L. Vandenberghe, S. Boyd, and H. Lebret. Applications of second-
            order cone programming. *Linear Algebra and Its Applications*, 284:193–228,
            1998.

[Man65]   O. Mangasarian. Linear and nonlinear separation of patterns by linear pro-
            gramming. *Operations Research*, 13(3):444–452, 1965.

[Man94]   O. Mangasarian. *Nonlinear Programming*. Society for Industrial and Applied
            Mathematics, 1994. First published in 1969 by McGraw-Hill.

[Mar52]   H. Markowitz. Portfolio selection. *The Journal of Finance*, 7(1):77–91, 1952.

[Mar56]   H. Markowitz. The optimization of a quadratic function subject to linear
            constraints. *Naval Research Logistics Quarterly*, 3:111–133, 1956.

[MDW$^+$02] W.-K. Ma, T. N. Davidson, K. M. Wong, Z.-Q. Luo, and P.-C. Ching. Quasi-
            maximum-likelihood multiuser detection using semi-definite relaxation with
            application to synchronous CDMA. *IEEE Transactions on Signal Processing*,
            50:912–922, 2002.

[Meh92]   S. Mehrotra. On the implementation of a primal-dual interior point method.
            SIAM Journal on Optimization, 2(4):575–601, 1992.

[Mey00]   C. D. Meyer. *Matrix Analysis and Applied Linear Algebra*. Society for In-
            dustrial and Applied Mathematics, 2000.

[ML57]   M. Marcus and L. Lopes. Inequalities for symmetric functions and Hermitian
            matrices. *Canadian Journal of Mathematics*, 9:305–312, 1957.

[MO60]   A. W. Marshall and I. Olkin. Multivariate Chebyshev inequalities. *Annals
            of Mathematical Statistics*, 32(4):1001–1014, 1960.

[MO79]   A. W. Marshall and I. Olkin. *Inequalities: Theory of Majorization and Its
            Applications*. Academic Press, 1979.

[Mon97]   R. D. C. Monteiro. Primal-dual path-following algorithms for semidefinite
            programming. *SIAM Journal on Optimization*, 7(3):663–678, 1997.

[MOS02]   MOSEK ApS. *The MOSEK Optimization Tools. User's Manual and Refer-
            ence*, 2002. Available from www.mosek.com.

[Mot33]   T. Motzkin. *Beiträge zur Theorie der linearen Ungleichungen*. PhD thesis,
            University of Basel, 1933.

[MP68]   R. F. Meyer and J. W. Pratt. The consistent assessment and fairing of pref-
            erence functions. *IEEE Transactions on Systems Science and Cybernetics*,
            4(3):270–278, 1968.

[MR95]   R. Motwani and P. Raghavan. *Randomized Algorithms*. Cambridge University
            Press, 1995.

[MZ89]   M. Morari and E. Zafiriou. *Robust Process Control*. Prentice-Hall, 1989.

[Nes98]   Y. Nesterov. Semidefinite relaxations and nonconvex quadratic optimization.
            Optimization Methods and Software, 9(1-3):141–160, 1998.

[Nes00]   Y. Nesterov. Squared functional systems and optimization problems. In
            J. Frenk, C. Roos, T. Terlaky, and S. Zhang, editors, *High Performance
            Optimization Techniques*, pages 405–440. Kluwer, 2000.

[Nik54]   H. Nikaidô. On von Neumann's minimax theorem. *Pacific Journal of Math-
            ematics*, 1954.

[NN94]   Y. Nesterov and A. Nemirovskii. *Interior-Point Polynomial Methods in Convex Programming*. Society for Industrial and Applied Mathematics, 1994.

[NT98]   Y. E. Nesterov and M. J. Todd. Primal-dual interior-point methods for self-scaled cones. *SIAM Journal on Optimization*, 8(2):324–364, 1998.

[NW99]   J. Nocedal and S. J. Wright. *Numerical Optimization*. Springer, 1999.

[NWY00]   Y. Nesterov, H. Wolkowicz, and Y. Ye. Semidefinite programming relaxations of nonconvex quadratic optimization. In H. Wolkowicz, R. Saigal, and L. Vandenberghe, editors, *Handbook of Semidefinite Programming*, chapter 13, pages 361–419. Kluwer Academic Publishers, 2000.

[NY83]   A. Nemirovskii and D. Yudin. *Problem Complexity and Method Efficiency in Optimization*. John Wiley & Sons, 1983.

[OR00]   J. M. Ortega and W. C. Rheinboldt. *Iterative Solution of Nonlinear Equations in Several Variables*. Society for Industrial and Applied Mathematics, 2000. First published in 1970 by Academic Press.

[Par71]   V. Pareto. *Manual of Political Economy*. A. M. Kelley Publishers, 1971. Translated from the French edition. First published in Italian in 1906.

[Par98]   B. N. Parlett. *The Symmetric Eigenvalue Problem*. Society for Industrial and Applied Mathematics, 1998. First published in 1980 by Prentice-Hall.

[Par00]   P. A. Parrilo. *Structured Semidefinite Programs and Semialgebraic Geometry Methods in Robustness and Optimization*. PhD thesis, California Institute of Technology, 2000.

[Par03]   P. A. Parrilo. Semidefinite programming relaxations for semialgebraic problems. *Mathematical Programming Series B*, 96:293–320, 2003.

[Pet76]   E. L. Peterson. Geometric programming. *SIAM Review*, 18(1):1–51, 1976.

[Pin95]   J. Pinter. *Global Optimization in Action*, volume 6 of *Nonconvex Optimization and Its Applications*. Kluwer, 1995.

[Pol87]   B. T. Polyak. *Introduction to Optimization*. Optimization Software, 1987. Translated from Russian.

[Pon67]   J. Ponstein. Seven kinds of convexity. *SIAM Review*, 9(1):115–119, 1967.

[Pré71]   A. Prékopa. Logarithmic concave measures with application to stochastic programming. *Acta Scientiarum Mathematicarum*, 32:301–315, 1971.

[Pré73]   A. Prékopa. On logarithmic concave measures and functions. *Acta Scientiarum Mathematicarum*, 34:335–343, 1973.

[Pré80]   A. Prékopa. Logarithmic concave measures and related topics. In M. A. H. Dempster, editor, *Stochastic Programming*, pages 63–82. Academic Press, 1980.

[Pro01]   J. G. Proakis. *Digital Communications*. McGraw-Hill, fourth edition, 2001.

[PRT02]   J. Peng, C. Roos, and T. Terlaky. *Self-Regularity. A New Paradigm for Primal-Dual Interior-Point Algorithms*. Princeton University Press, 2002.

[PS98]   C. H. Papadimitriou and K. Steiglitz. *Combinatorial Optimization. Algorithms and Complexity*. Dover Publications, 1998. First published in 1982 by Prentice-Hall.

[PSU88]   A. L. Peressini, F. E. Sullivan, and J. J. Uhl. *The Mathematics of Nonlinear Programming*. Undergraduate Texts in Mathematics. Springer, 1988.

[Puk93]   F. Pukelsheim. *Optimal Design of Experiments*. Wiley & Sons, 1993.

[Ren01]   J. Renegar. *A Mathematical View of Interior-Point Methods in Convex Optimization*. Society for Industrial and Applied Mathematics, 2001.

[Roc70]   R. T. Rockafellar. *Convex Analysis*. Princeton University Press, 1970.

[Roc89]   R. T. Rockafellar. *Conjugate Duality and Optimization*. Society for Industrial and Applied Mathematics, 1989. First published in 1974.

[Roc93]   R. T. Rockafellar. Lagrange multipliers and optimality. *SIAM Review*, 35:183–283, 1993.

[ROF92]   L. Rudin, S. J. Osher, and E. Fatemi. Nonlinear total variation based noise removal algorithms. *Physica D*, 60:259–268, 1992.

[Ros65]   J. B. Rosen. Pattern separation by convex programming. *Journal of Mathematical Analysis and Applications*, 10:123–134, 1965.

[Ros99]   S. M. Ross. *An Introduction to Mathematical Finance: Options and Other Topics*. Cambridge University Press, 1999.

[RTV97]   C. Roos, T. Terlaky, and J.-Ph. Vial. *Theory and Algorithms for Linear Optimization. An Interior Point Approach*. John Wiley & Sons, 1997.

[Rud76]   W. Rudin. *Principles of Mathematical Analysis*. McGraw-Hill, 1976.

[RV73]   A. W. Roberts and D. E. Varberg. *Convex Functions*. Academic Press, 1973.

[RW97]   D. Ralph and S. J. Wright. Superlinear convergence of an interior-point method for monotone variational inequalities. In M. C. Ferris and J.-S. Pang, editors, *Complementarity and Variational Problems: State of the Art*, pages 345–385. Society for Industrial and Applied Mathematics, 1997.

[RWR98]   C. V. Rao, S. J. Wright, and J. B. Rawlings. Application of interior-point methods to model predictive control. *Journal of Optimization Theory and Applications*, 99(3):723–757, 1998.

[Sch35]   I. J. Schoenberg. Remarks to Maurice Fréchet's article "Sur la définition axiomatique d'une classe d'espaces distanciés vectoriellement applicable sur l'espace de Hilbert". *Annals of Mathematics*, 38(3):724–732, 1935.

[Sch82]   S. Schaible. Bibliography in fractional programming. *Zeitschrift für Operations Research*, 26:211–241, 1982.

[Sch83]   S. Schaible. Fractional programming. *Zeitschrift für Operations Research*, 27:39–54, 1983.

[Sch86]   A. Schrijver. *Theory of Linear and Integer Programming*. John Wiley & Sons, 1986.

[Sch91]   L. L. Scharf. *Statistical Signal Processing. Detection, Estimation, and Time Series Analysis*. Addison Wesley, 1991. With Cédric Demeure.

[SDJ91]   G. Sigl, K. Doll, and F. M. Johannes. Analytical placement: A linear or quadratic objective function? In *Proceedings of the 28th ACM/IEEE Design Automation Conference*, pages 427–432, 1991.

[SGC97]   C. Scherer, P. Gahinet, and M. Chilali. Multiobjective output-feedback control via LMI optimization. *IEEE Transactions on Automatic Control*, 42(7):896–906, 1997.

[She99]   N. Sherwani. *Algorithms for VLSI Design Automation*. Kluwer Academic Publishers, third edition, 1999.

[Sho85]   N. Z. Shor. *Minimization Methods for Non-differentiable Functions*. Springer Series in Computational Mathematics. Springer, 1985.

[Sho91]   N. Z. Shor. The development of numerical methods for nonsmooth optimization in the USSR. In J. K. Lenstra, A. H. G. Rinnooy Kan, and A. Schrijver, editors, *History of Mathematical Programming. A Collection of Personal Reminiscences*, pages 135–139. Centrum voor Wiskunde en Informatica and North-Holland, Amsterdam, 1991.

[Son86]   G. Sonnevend. An 'analytical centre' for polyhedrons and new classes of global algorithms for linear (smooth, convex) programming. In *Lecture Notes in Control and Information Sciences*, volume 84, pages 866–878. Springer, 1986.

[SPV99]   A. Seifi, K. Ponnambalam, and J. Vlach. A unified approach to statistical design centering of integrated circuits with correlated parameters. *IEEE Transactions on Circuits and Systems — I. Fundamental Theory and Applications*, 46(1):190–196, 1999.

[SRVK93]   S. S. Sapatnekar, V. B. Rao, P. M. Vaidya, and S.-M. Kang. An exact solution to the transistor sizing problem for CMOS circuits using convex optimization. *IEEE Transactions on Computer-Aided Design of Integrated Circuits and Systems*, 12(11):1621–1634, 1993.

[SS01]   B. Schölkopf and A. Smola. *Learning with Kernels: Support Vector Machines, Regularization, Optimization, and Beyond*. MIT Press, 2001.

[Str80]   G. Strang. *Linear Algebra and its Applications*. Academic Press, 1980.

[Stu99]   J. F. Sturm. Using SEDUMI 1.02, a MATLAB toolbox for optimization over symmetric cones. *Optimization Methods and Software*, 11-12:625–653, 1999. Available from `sedumi.mcmaster.ca`.

[SW70]   J. Stoer and C. Witzgall. *Convexity and Optimization in Finite Dimensions I*. Springer-Verlag, 1970.

[SW95]   R. J. Stern and H. Wolkowicz. Indefinite trust region subproblems and nonsymmetric eigenvalue perturbations. *SIAM Journal on Optimization*, 15:286–313, 1995.

[TA77]   A. N. Tikhonov and V. Y. Arsenin. *Solutions of Ill-Posed Problems*. V. H. Winston & Sons, 1977. Translated from Russian.

[TB97]   L. N. Trefethen and D. Bau, III. *Numerical Linear Algebra*. Society for Industrial and Applied Mathematics, 1997.

[Ter96]   T. Terlaky, editor. *Interior Point Methods of Mathematical Programming*, volume 5 of *Applied Optimization*. Kluwer Academic Publishers, 1996.

[Tib96]   R. Tibshirani. Regression shrinkage and selection via the lasso. *Journal of the Royal Statistical Society, Series B*, 58(1):267–288, 1996.

[Tik90]   V. M. Tikhomorov. Convex analysis. In R. V. Gamkrelidze, editor, *Analysis II: Convex Analysis and Approximation Theory*, volume 14, pages 1–92. Springer, 1990.

[Tit75]   D. M. Titterington. Optimal design: Some geometrical aspects of D-optimality. *Biometrika*, 62(2):313–320, 1975.

[TKE88]   S. Tarasov, L. Khachiyan, and I. Èrlikh. The method of inscribed ellipsoids. *Soviet Mathematics Doklady*, 37(1):226–230, 1988.

[Tod01]   M. J. Todd. Semidefinite optimization. *Acta Numerica*, 10:515–560, 2001.

[Tod02]   M. J. Todd. The many facets of linear programming. *Mathematical Programming Series B*, 91:417–436, 2002.

[TTT98]   M. J. Todd, K. C. Toh, and R. H. Tütüncü. On the Nesterov-Todd direction in semidefinite programming. *SIAM Journal on Optimization*, 8(3):769–796, 1998.

[TTT02]   K. C. Toh, R. H. Tütüncü, and M. J. Todd. *SDPT3. A Matlab software for semidefinite-quadratic-linear programming*, 2002. Available from `www.math.nus.edu.sg/~mattohkc/sdpt3.html`.

[Tuy98]   H. Tuy. *Convex Analysis and Global Optimization*, volume 22 of *Nonconvex Optimization and Its Applications*. Kluwer, 1998.

[Uhl79]   F. Uhlig. A recurring theorem about pairs of quadratic forms and extensions. A survey. *Linear Algebra and Its Applications*, 25:219–237, 1979.

[Val64]   F. A. Valentine. *Convex Sets*. McGraw-Hill, 1964.

[Van84]   G. N. Vanderplaats. *Numerical Optimization Techniques for Engineering Design*. McGraw-Hill, 1984.

[Van96]   R. J. Vanderbei. *Linear Programming: Foundations and Extensions*. Kluwer, 1996.

[Van97]   R. J. Vanderbei. *LOQO User's Manual*, 1997. Available from `www.orfe.princeton.edu/~rvdb`.

[Vap00]   V. N. Vapnik. *The Nature of Statistical Learning Theory*. Springer, second edition, 2000.

[Vav91]   S. A. Vavasis. *Nonlinear Optimization: Complexity Issues*. Oxford University Press, 1991.

[VB95]   L. Vandenberghe and S. Boyd. Semidefinite programming. *SIAM Review*, pages 49–95, 1995.

[vN63]   J. von Neumann. Discussion of a maximum problem. In A. H. Taub, editor, *John von Neumann. Collected Works*, volume VI, pages 89–95. Pergamon Press, 1963. Unpublished working paper from 1947.

[vN46]   J. von Neumann. A model of general economic equilibrium. *Review of Economic Studies*, 13(1):1–9, 1945-46.

[vNM53]   J. von Neumann and O. Morgenstern. *Theory of Games and Economic Behavior*. Princeton University Press, third edition, 1953. First published in 1944.

[vT84]   J. van Tiel. *Convex Analysis. An Introductory Text*. John Wiley & Sons, 1984.

[Web71]   A. Weber. *Theory of the Location of Industries*. Russell & Russell, 1971. Translated from German by C. J. Friedrich. First published in 1929.

[Web94]   R. Webster. *Convexity*. Oxford University Press, 1994.

[Whi71]   P. Whittle. *Optimization under Constraints*. John Wiley & Sons, 1971.

[Wol81]   H. Wolkowicz. Some applications of optimization in matrix theory. *Linear Algebra and Its Applications*, 40:101–118, 1981.

[Wri97]   S. J. Wright. *Primal-Dual Interior-Point Methods*. Society for Industrial and Applied Mathematics, 1997.

[WSV00]   H. Wolkowicz, R. Saigal, and L. Vandenberghe, editors. *Handbook of Semidefinite Programming*. Kluwer Academic Publishers, 2000.

[XHY96]   X. Xu, P. Hung, and Y. Ye. A simplified homogeneous and self-dual linear programming algorithm and its implementation. *Annals of Operations Research*, 62:151–172, 1996.

[Ye97]   Y. Ye. *Interior Point Algorithms. Theory and Analysis*. John Wiley & Sons, 1997.

[Ye99]   Y. Ye. Approximating quadratic programming with bound and quadratic constraints. *Mathematical Programming*, 84:219–226, 1999.

[YTM94]   Y. Ye, M. J. Todd, and S. Mizuno. An $O(\sqrt{n}L)$-iteration homogeneous and self-dual linear programming algorithm. *Mathematics of Operations Research*, 19:53–67, 1994.

[Zen71]   C. Zener. *Engineering Design by Geometric Programming*. John Wiley & Sons, 1971.

[Zha98]   Y. Zhang. On extending some primal-dual interior-point algorithms from linear programming to semidefinite programming. *SIAM Journal on Optimization*, 8(2):365–386, 1998.

Notation

Some specific sets

$\mathbf{R}$	Real numbers.
$\mathbf{R}^n$	Real n-vectors ($n \times 1$ matrices).
$\mathbf{R}^{m \times n}$	Real $m \times n$ matrices.
$\mathbf{R}_+$, $\mathbf{R}_{++}$	Nonnegative, positive real numbers.
$\mathbf{C}$	Complex numbers.
$\mathbf{C}^n$	Complex n-vectors.
$\mathbf{C}^{m \times n}$	Complex $m \times n$ matrices.
$\mathbf{Z}$	Integers.
$\mathbf{Z}_+$	Nonnegative integers.
$\mathbf{S}^n$	Symmetric $n \times n$ matrices.
$\mathbf{S}^n_+$, $\mathbf{S}^n_{++}$	Symmetric positive semidefinite, positive definite, $n \times n$ matrices.

Vectors and matrices

$\mathbf{1}$	Vector with all components one.
e_i	ith standard basis vector.
I	Identity matrix.
X^T	Transpose of matrix X.
X^H	Hermitian (complex conjugate) transpose of matrix X.
$\mathbf{tr}\, X$	Trace of matrix X.
$\lambda_i(X)$	ith largest eigenvalue of symmetric matrix X.
$\lambda_{\max}(X)$, $\lambda_{\min}(X)$	Maximum, minimum eigenvalue of symmetric matrix X.
$\sigma_i(X)$	ith largest singular value of matrix X.
$\sigma_{\max}(X)$, $\sigma_{\min}(X)$	Maximum, minimum singular value of matrix X.
$X^\dagger$	Moore-Penrose or pseudo-inverse of matrix X.
$x \perp y$	Vectors x and y are orthogonal: $x^T y = 0$.
$V^\perp$	Orthogonal complement of subspace V.
$\mathbf{diag}(x)$	Diagonal matrix with diagonal entries $x_1, \ldots, x_n$.
$\mathbf{diag}(X, Y, \ldots)$	Block diagonal matrix with diagonal blocks $X, Y, \ldots$.
$\mathbf{rank}\, A$	Rank of matrix A.
$\mathcal{R}(A)$	Range of matrix A.
$\mathcal{N}(A)$	Nullspace of matrix A.

Norms and distances

$\|\cdot\|$	A norm.
$\|\cdot\|_*$	Dual of norm $\|\cdot\|$.
$\|x\|_2$	Euclidean (or ℓ_2-) norm of vector x.
$\|x\|_1$	ℓ_1-norm of vector x.
$\|x\|_\infty$	ℓ_∞-norm of vector x.
$\|X\|_2$	Spectral norm (maximum singular value) of matrix X.
$B(c, r)$	Ball with center c and radius r.
$\mathbf{dist}(A, B)$	Distance between sets (or points) A and B.

Generalized inequalities

$x \preceq y$	Componentwise inequality between vectors x and y.
$x \prec y$	Strict componentwise inequality between vectors x and y
$X \preceq Y$	Matrix inequality between symmetric matrices X and Y.
$X \prec Y$	Strict matrix inequality between symmetric matrices X and Y.
$x \preceq_K y$	Generalized inequality induced by proper cone K.
$x \prec_K y$	Strict generalized inequality induced by proper cone K.
$x \preceq_{K^*} y$	Dual generalized inequality.
$x \prec_{K^*} y$	Dual strict generalized inequality.

Topology and convex analysis

$\mathbf{card}\, C$	Cardinality of set C.
$\mathbf{int}\, C$	Interior of set C.
$\mathbf{relint}\, C$	Relative interior of set C.
$\mathbf{cl}\, C$	Closure of set C.
$\mathbf{bd}\, C$	Boundary of set C: $\mathbf{bd}\, C = \mathbf{cl}\, C \setminus \mathbf{int}\, C$.
$\mathbf{conv}\, C$	Convex hull of set C.
$\mathbf{aff}\, C$	Affine hull of set C.
K^*	Dual cone associated with K.
I_C	Indicator function of set C.
S_C	Support function of set C.
f^*	Conjugate function of f.

Probability

$\mathbf{E}\, X$	Expected value of random vector X.
$\mathbf{prob}\, S$	Probability of event S.
$\mathbf{var}\, X$	Variance of scalar random variable X.
$\mathcal{N}(c, \Sigma)$	Gaussian distribution with mean c, covariance (matrix) Σ.
Φ	Cumulative distribution function of $\mathcal{N}(0, 1)$ random variable.

Functions and derivatives

$f : A \to B$	f is a function on the set $\mathbf{dom}\, f \subseteq A$ into the set B.
$\mathbf{dom}\, f$	Domain of function f.
$\mathbf{epi}\, f$	Epigraph of function f.
∇f	Gradient of function f.
$\nabla^2 f$	Hessian of function f.
Df	Derivative (Jacobian) matrix of function f.

Index